2D NMR-BASED
ORGANIC SPECTROSCOPY PROBLEMS

MICHAEL T. HUGGINS
JEROME E. GURST
DAVID A. LIGHTNER

Prentice Hall
Boston Columbus Indianapolis New York San Francisco Upper Saddle River
Amsterdam Cape Town Dubai London Madrid Milan Munich Paris Montréal Toronto
Delhi Mexico City São Paulo Sydney Hong Kong Seoul Singapore Taipei Tokyo

Editor in Chief, Chemistry: Adam Jaworski
Marketing Manager: Erin Gardner
Assistant Editor: Carol DuPont
Marketing Assistant: Nicola Houston
Managing Editor, Chemistry and Geosciences: Gina M. Cheselka
Project Manager: Wendy A. Perez
Supplement Cover Designer: Seventeenth Street Studios
Operations Specialist: Maura Zaldivar

Pearson Prentice Hall
Pearson Education, Inc.
Upper Saddle River, New Jersey 07458

Printed in the United States of America
10 9 8 7 6 5 4 3 2 1

ISBN-13: 978-0-13-604206-8
ISBN-10: 0-13-604206-6

Prentice Hall
is an imprint of

PEARSON

www.pearsonhighered.com

TABLE OF CONTENTS

About the Authors

MICHAEL T. HUGGINS earned a B.S. degree in Chemistry from the University of West Florida (UWF) in 1996. In the summer 1995, he participated in a NSF Research Experience for Undergraduates program at the University of Nevada, Reno (UNR) and then completed his PhD at UNR in 2000 under the direction of Dr. David A. Lightner. During his PhD studies, Michael was a R.C. Fuson Graduate Fellow. As a graduate student, he received both the teaching assistant and research assistant of the year awards. Michael moved on to the University of Texas at Austin in 2000 where he worked with Dr. Jonathan L. Sessler as a postdoctoral fellow and visiting lecturer. In 2002, he was appointed as an assistant professor in organic chemistry at the University of West Florida and in 2004 was awarded the Distinguished Teaching Award for the College of Arts and Sciences at UWF. Michael was promoted to Associated Professor in 2007. He teaches courses in spectroscopy, supramolecular chemistry, and organic chemistry.

JEROME E. GURST first saw NMR spectra as a senior undergraduate at Dartmouth College. Spectra were recorded on a strip chart recorder, with spinning sidebands used for calibration. As a graduate student at Stanford with Carl Djerassi, he learned to measure spectra using one of the first Varian A60's. After postdoctoral stints with H.C. Brown at Purdue and Kurt Mislow at Princeton, as well as a visiting appointment at the University of Oregon, he was selected to open the University of West Florida in 1967 as the first organic chemistry professor. After more than 40 years, and four generations of NMR spectrometers, he retired at the end of 2007. During his career, he was also a Visiting Faculty member at Emory University, Dartmouth College, and the University of Nevada, Reno. Jerome emphasized the use of spectroscopy, particularly in education, in his publications. He coauthored a book on Circular Dichroism and Conformational Analysis.

DAVID A. LIGHTNER is R.C. Fuson Professor of Chemistry, Regents' Research Professor, and Adjunct Professor of Biochemistry at the University of Nevada, Reno. He has been recognized for distinguished contributions in relating chiroptical properties to stereochemistry and for clarifying the molecular mechanisms of phototherapy for neonatal jaundice, a medical treatment given to millions of newborn babies each year. His honors include University of Nevada Foundation Professor (1987), the first recipient of the Outstanding Research Award for the State of Nevada (1992), and election as Fellow of the American Association for the Advancement of Science (1996). He is the author of more than 350 research publications and coauthor of or contributor to 12 books, former associate editor of *Photochemistry & Photobiology*, and a member of the editorial advisory board of *Monatshefte für Chemie* (Chemical Monthly).

Chapter 1
Introduction

The problems are found in four sections that we have defined loosely, with some equivocation as: less-challenging (Chapter 2), challenging (Chapter 3), more-challenging (Chapter 4), and special problems (Chapter 5). In some problems, the high resolution mass spectral (HRMS) molecular ion peak exact mass is given. These values have an error range of ±0.006 amu. In others, only the integral mass is provided. We want to remind you of the "odd-even" rule: compounds with an odd number of nitrogen atoms will have an odd molecular weight while an even number (or no) nitrogen atoms must yield an even molecular mass. Of course, you must use 35 and 37 for chlorine rather than the average 35.547 found on a periodic table, or 79 and 81 for bromine rather than the average 80. It is useful to keep in mind the isotopic distribution for chlorine and bromine – 3:1 for ^{35}Cl:^{37}Cl and 1:1 for ^{79}Br:^{81}Br. When percent composition for a sample is given, the accepted error range is ±0.3%. So, 60% C means that the actual percent carbon is in the range of 59.7% to 60.3%.

In some problems, a key absorbance band peak in the infrared is supplied – as might often be available to supplement the NMR data. This is generally done to help with identification or confirmation of functional groups present in the molecule or substitution pattern on an aromatic ring.

NMR spectra were measured using a Varian Mercury Plus NMR spectrometer operating at a proton frequency of 300 MHz. Samples (~40 mg) were dissolved in approximately 0.7 mL of a deuterated solvent. When possible, deuteriochloroform ($CDCl_3$) was used, but DMSO-d_6, acetone-d_6 or benzene-d_6 was used for an occasional sample. ^{1}H NMR spectra were referenced by setting TMS at 0 δ (0 ppm) while ^{13}C spectra are referenced from the center peak of the signal arising from the deuterated solvent; *i.e.*, the peak at 77.0 δ was used as the reference peak for samples measured in deuteriochloroform. The references below, particularly the article by Gottlieb, *et al.*, are extremely useful for locating solvent peaks. Remember that while we used solvents which are >99.7% "deuterated," there is a small amount of $CHCl_3$ in the $CDCl_3$, and this will give rise to signals in the proton NMR spectra.

The samples are typical laboratory samples and, as such, may contain impurities such as residual solvents from the synthesis. Again, the Gottlieb manuscript in the *Journal of Organic Chemistry* (see below) is very useful for ferreting out these signals. We are showing you real, undoctored spectra replete with all the "warts and pimples." In some cases, integrations aren't perfect, electronic "noise" generates cross-peaks in gCOSY spectra that really are "impossible" since they are at frequencies at which there are no real signals, phasing is not always perfect, *etc.*

The problems presented in Chapter 5 have some additional features that separate them from the other problems. In some instances, the problems are connected via a synthetic sequence; *e.g.* 5.11 was prepared from 4.24. Problems 5.06-5.10 contain NOESY data which allow one to study the configuration and conformation of the molecules.

1D-NMR Data

We have utilized a variety of NMR spectra: ^{1}H, ^{13}C, ^{19}F, ^{31}P, and a variety of different pulse sequences such as APT and edited DEPT. All problems contain proton and carbon data. The presence of ^{19}F or ^{31}P spectra should be used as an indication of the presence of these elements in the molecule.

^{1}H-NMR – In proton spectra, we have tried to exclude known impurity signals by not including them in the electronic integration results provided. So a signal with no indicated integral is an impurity. Expansions are provided when warranted based on peak resolution. In many instances, the expansions include peak labels that allow for the calculation of the *J*-values associated with ^{1}H-^{1}H coupling.

^{13}C-NMR – In the carbon-13 spectra, impurity peaks are not labeled with a peak position label. Expansions are provided when warranted based on peak resolution.

APT – APT stands for Attached Proton Test. In this experiment, the carbon signals are either positive or negative. An odd number of attached protons (methines or methyls) will be in one direction while the methylenes and quaternary carbons (even number of attached protons) will be in the opposite sense. Locate the solvent peaks, which will have zero H's attached (D's are not H's!) to determine if odds are up, or down, in a particular spectrum.

Edited DEPT – While many presentations use DEPT-45 and DEPT-135 spectra, we use a software routine which compares the DEPT-45, DEPT-135, and full carbon spectra to yield the "edited DEPT" which shows only methyl carbons on the top line, methylene carbons on the next line followed by methine carbons with the full spectrum on the bottom line. A signal in the bottom line that does not reappear on one of the upper lines must be a quaternary carbon. Recall that quaternary C's generally give signals which are reduced in intensity when compared to protonated carbons. While this is a very useful presentation, the addition/subtraction routines used in the generation of the final plot are not perfect and sometimes "ghost" peaks will be seen – an intense (real) peak on the methyl line with a small residual line at the same frequency on another line. This residual may even be negative. Some judgement on the part of the user is required.

2D-NMR Data

A wide variety of 2D spectroscopic techniques are presented – gCOSY, gHSQC, gHMQC, gHMBC, and NOESY. The "g" in front of some of the acronyms indicated that we used gradient shim technology for the measurement to improve the results.

gCOSY – These spectra allow one to detect proton-proton coupling by looking for "off-diagonal correlation" signals. 3-Bond coupling is the most prevalent but longer range coupling can also be observed. And one should keep in mind that the dihedral angle between coupled hydrogens can have a dramatic effect on whether or not 3-bond coupling is measurable. The Karplus relationship is a useful way to quantify *J*-values and bond angles.

gHSQC or gHMQC – In gHSQC spectra, one can correlate the carbon spectrum with the proton spectrum. A closed circle (dot) below a carbon indicates that the carbon is either a methyl or a methine carbon, while open circles indicate that the carbon is a methylene carbon. The two H's attached to that carbon may or may not resonate at the same frequency (diastereotopic hydrogens). A proton showing no correlation with a carbon atom must be attached to oxygen, nitrogen, or sulfur. When presenting the data, we often do not show areas where there are no correlations. For example, unless we observed that a carbon resonating at >160 ppm had a correlation with a proton, we did not print the area of the spectrum beyond 160 ppm. You may assume that we have shown you all the relevant data. If you are wondering whether a carbon signal at 200 ppm has a proton correlation, and we don't show it, it is not there. gHMQC is a slight variant on gHSQC used for samples measured in DMSO-d_6.

gHMBC – gHMBC spectra show 2-bond, 3-bond and sometimes longer C-H coupling, and are often very important in the final structural elucidation.

NOESY – Non-bonding, or "through space" proton-proton interactions can often be seen by using the nuclear Overhauser effect (nOe). Like gCOSY spectra, off-diagonal cross peaks are an indication that two protons (or groups of protons) are in close physical proximity to each other.

We have refrained from providing a more detailed introduction to the spectroscopic methods employed because they can be found in most textbooks of organic spectroscopy, to mention only a few:

- *Organic Structural Spectroscopy* (Lambert, Gronert, Cooks, Lightner, Shurvell), Prentice-Hall, NY 2011
- *Organic Structure Analysis* (Crews, Rodríguez, Jaspars), Oxford Univ. Press, NY 1998
- *Spectroscopic Methods in Organic Chemistry* (Hesse, Meier and Zeeh), Georg Thieme Verlang, NY 1997

- *Organic Spectroscopic Structure Determination: A Problem-Based Learning Approach* (Taber, D.F.), Oxford Univ. Press, 2007
- *Nuclear Magnetic Resonance: An Introduction to Principles, Applications, and Experimental Methods* (Lambert, J.B.; Mazzola, E.P.), Prentice-Hall, 2004
- *Basic One- and Two-Dimensional NMR Spectroscopy*, 4th ed. (Friebolin, H.), Wiley-VCH, 2005
- *Understanding NMR Spectroscopy* (Keeler, J.), J. Wiley & Sons, 2005
- *NMR - From Spectra to Structures* (Mitchell, T.M.; Costisella, B.), Springer Verlag 2007
- *NMR Spectroscopy Explained: Simplified Theory, Applications and Examples for Organic Chemistry and Structural Biology* (Jacobsen, N.H.), J. Wiley & Sons, 2007
- *One-dimensional and Two-dimensional NMR Spectra by Modern Pulse Techniques* (ed.: Nakanishi), University Science Books, Mill Valley, CA 1990

For useful NMR Tables, see:

- *Structural Determination of Organic Compounds. Tables of Spectral Data*, 4th ed. (Pretsch, E.; Bühlmann, P.; Badertscher, M.), Springer, NY 2009

And for a useful tabulated source of NMR signals derived from common contaminants in less than analytically pure samples, see:

- Gottlieb, H.; Kotlyar, V.; Nudelman, A. *J. Org. Chem. Soc.* **1997**, *62*, 7512-7515

For tables of high resolution mass spectral data, consult:

- *Mass and Abundance Tables for Use in High Resolution Mass Spectrometry* (Beynon and Williams), Elsevier, NY 1963
- *Computation of Molecular Formulas for Mass Spectrometry* (J. Lederberg), Holden-Day, San Francisco 1964
- *Tables for Use in High Resolution Mass Spectrometry* (Binks, R.; Littler, J.S.; Cleaver, R.L.), Heyden & Son, London 1970

See also websites for high resolution mass spectral correlations:

- http://www.chm.davidson.edu/java/beynon/beynon.html
- http://www.cem.msu.edu/~reusch/Virtualtext/Spectrpy/MassSpec/masspec1.htm
- http://fiehnlab.ucdavis.edu/staff/kind/Metabolomics/MS-Adduct-Calculator/

A Final Comment

In some cases, based on the data provided, it may be possible to propose alternative structures which are in reasonable agreement with the data. And, it may not always be possible to assign the stereochemistry, *i.e., endo vs exo*.

Chapter 2
Less-Challenging Problems

Problem 2.01 Given the spectroscopic data below (Proton, Carbon, gHSQC, and gHMBC), provide the structure of the molecule. (HR-MS: 210.1409 amu; IR: 750 cm^{-1})

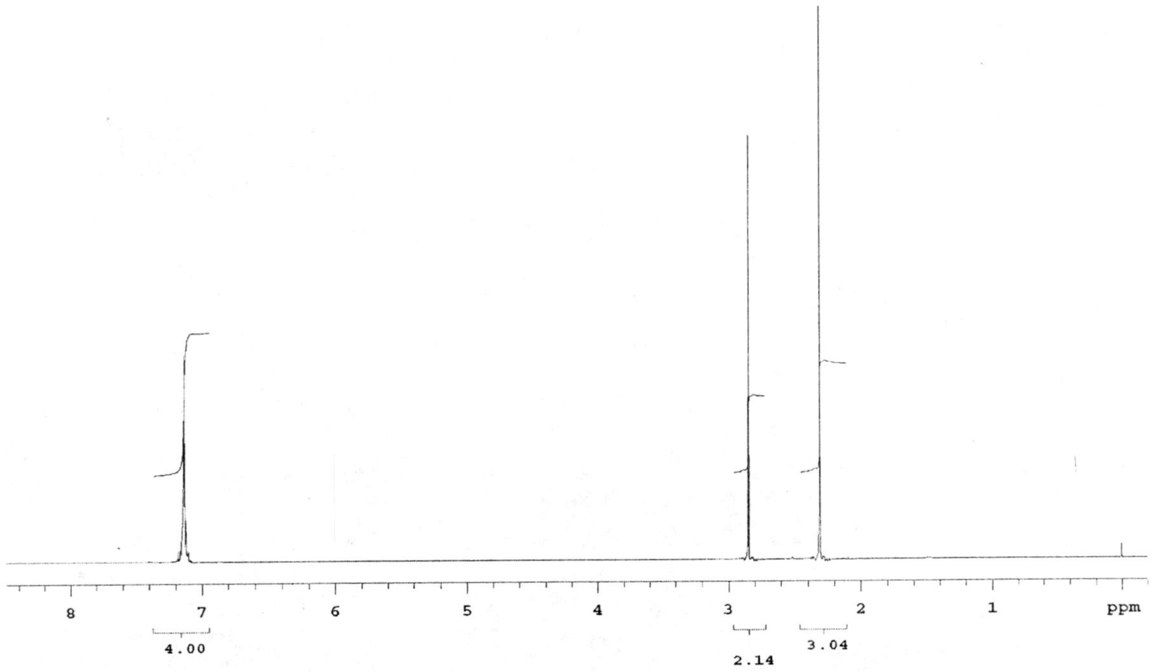

^{1}H-NMR Spectrum in $CDCl_3$

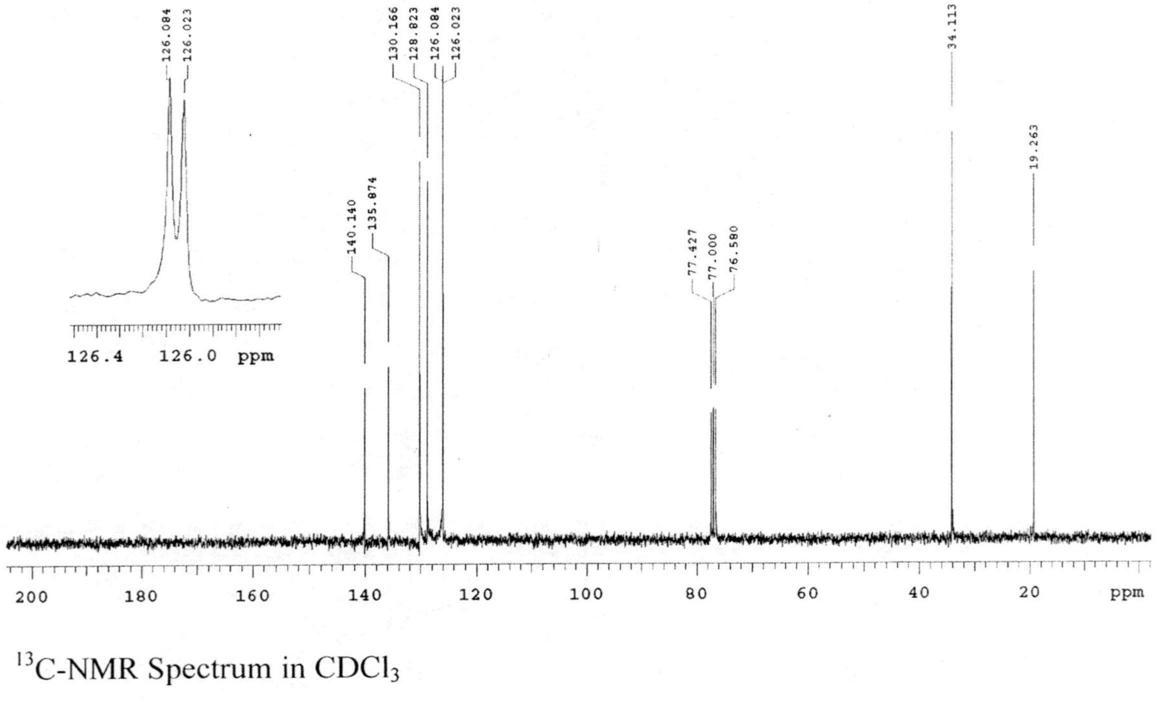

^{13}C-NMR Spectrum in $CDCl_3$

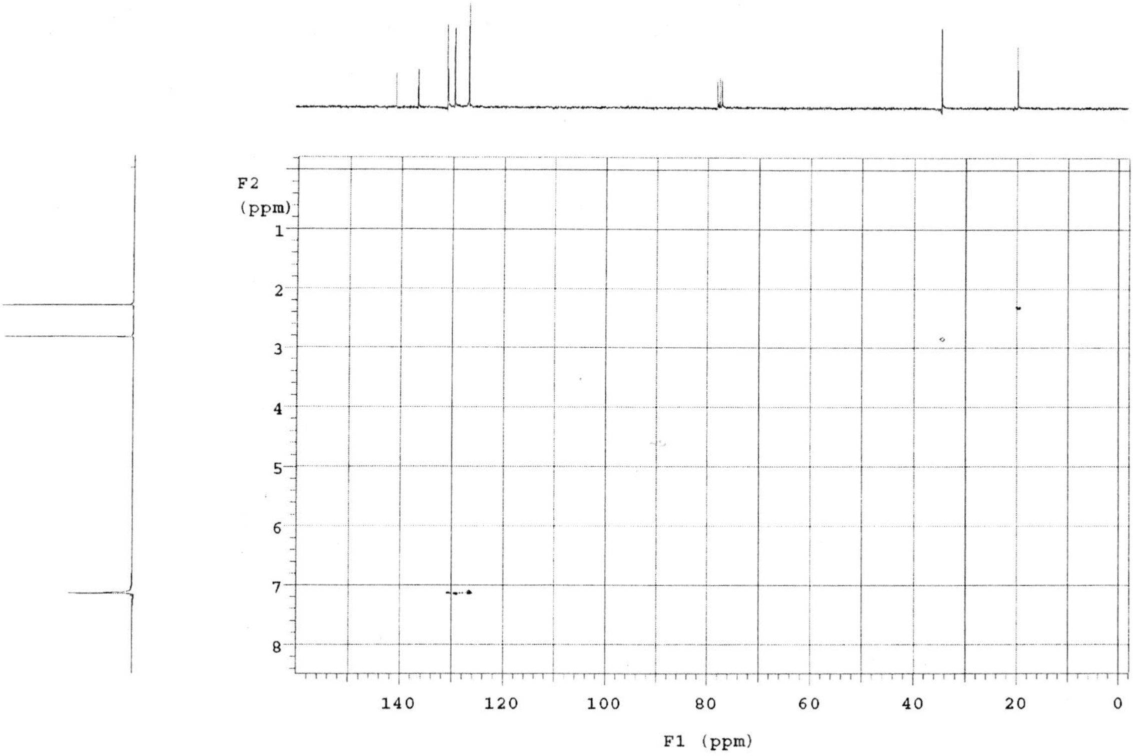

gHSQC Spectrum in $CDCl_3$

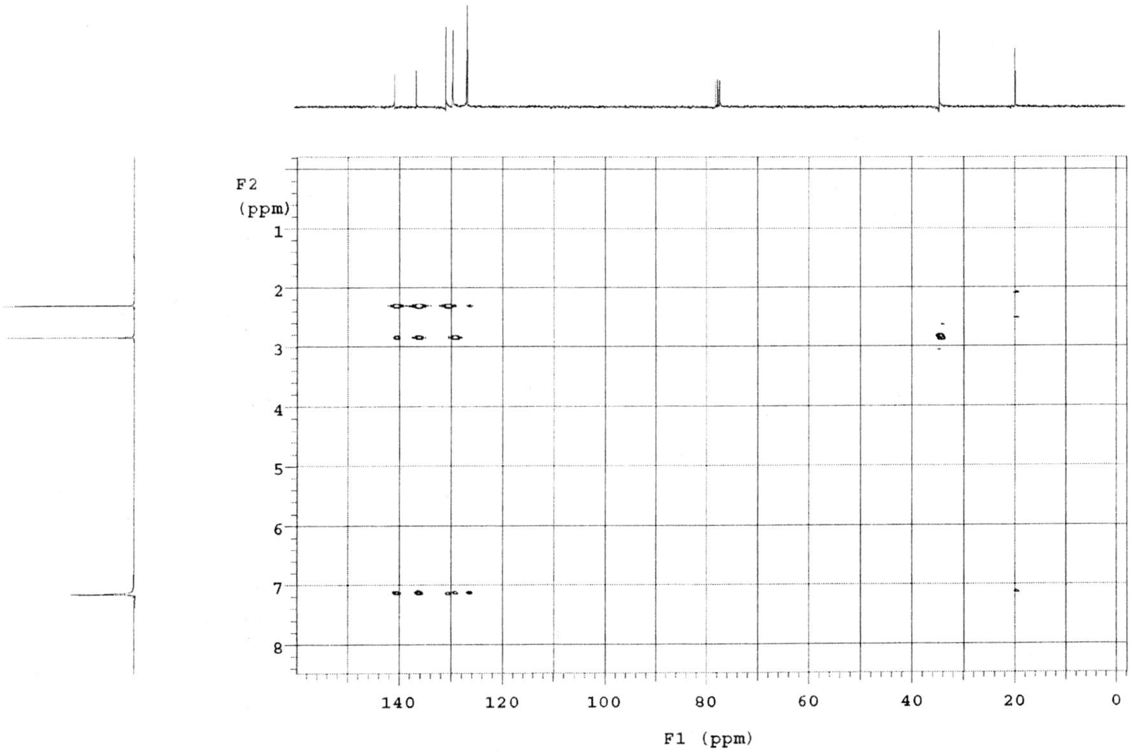

gHMBC Spectrum in $CDCl_3$

Problem 2.02 Given the spectroscopic data below (Proton, Edited DEPT, gHMQC, and gHMBC), provide the structure of the molecule. (IR: 795 cm^{-1}; peak near 5 ppm in ^{1}H-NMR spectrum disappears upon addition of D_2O)

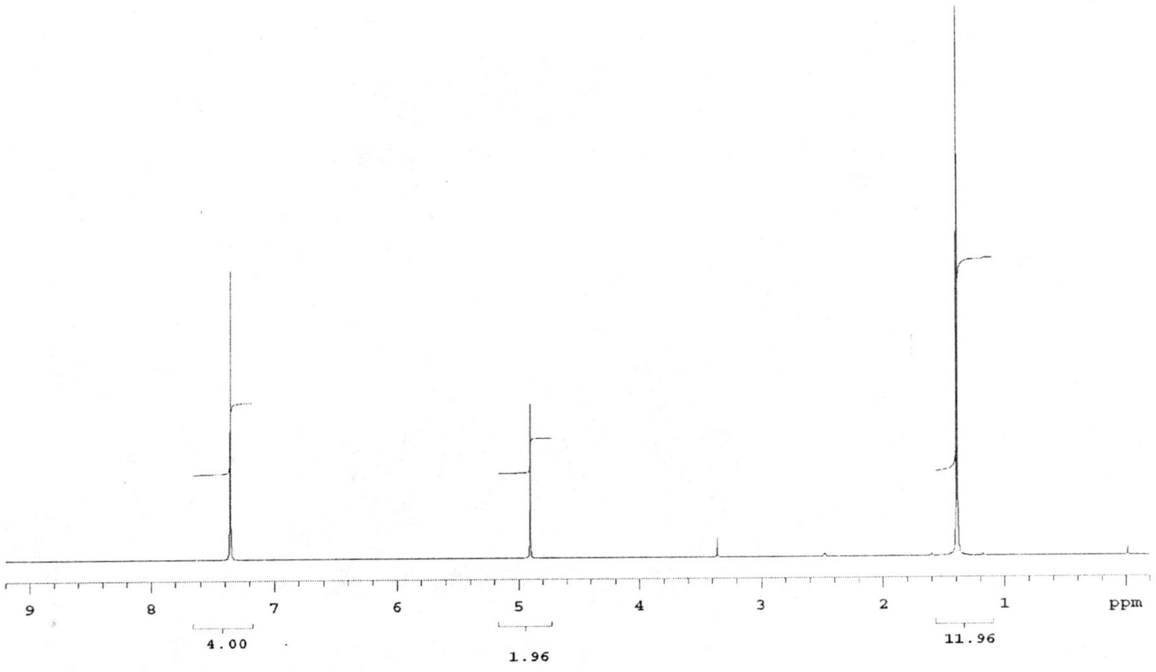

^{1}H-NMR Spectrum in DMSO-d_6

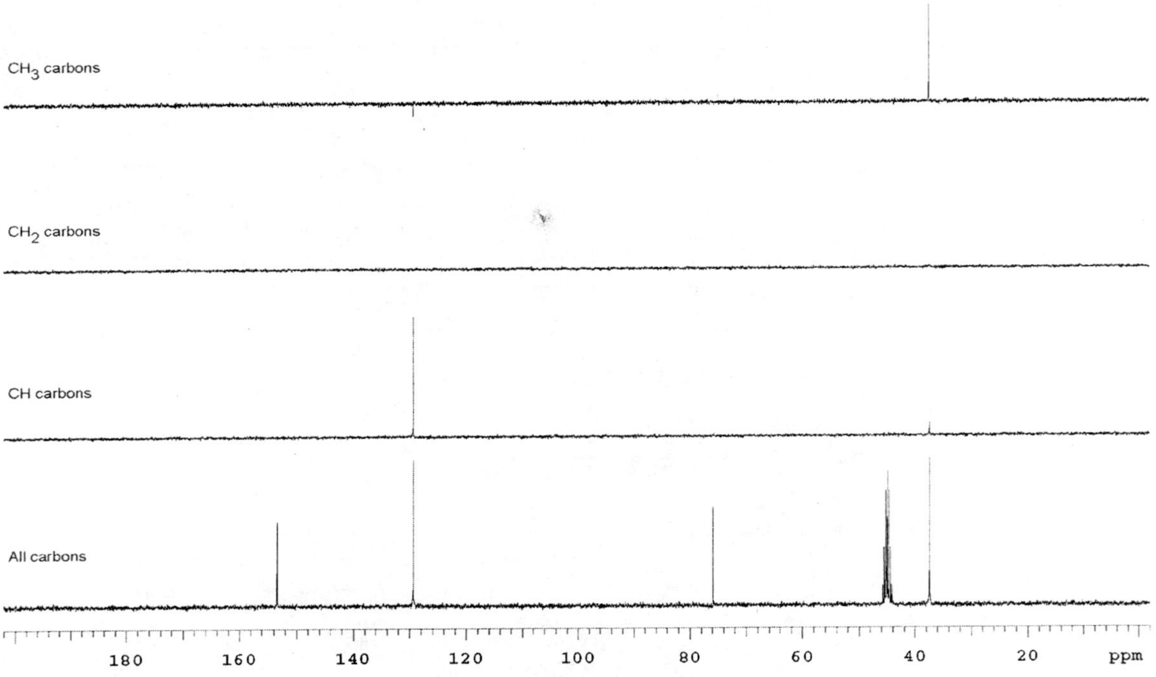

Edited DEPT ^{13}C-NMR Spectrum in DMSO-d_6

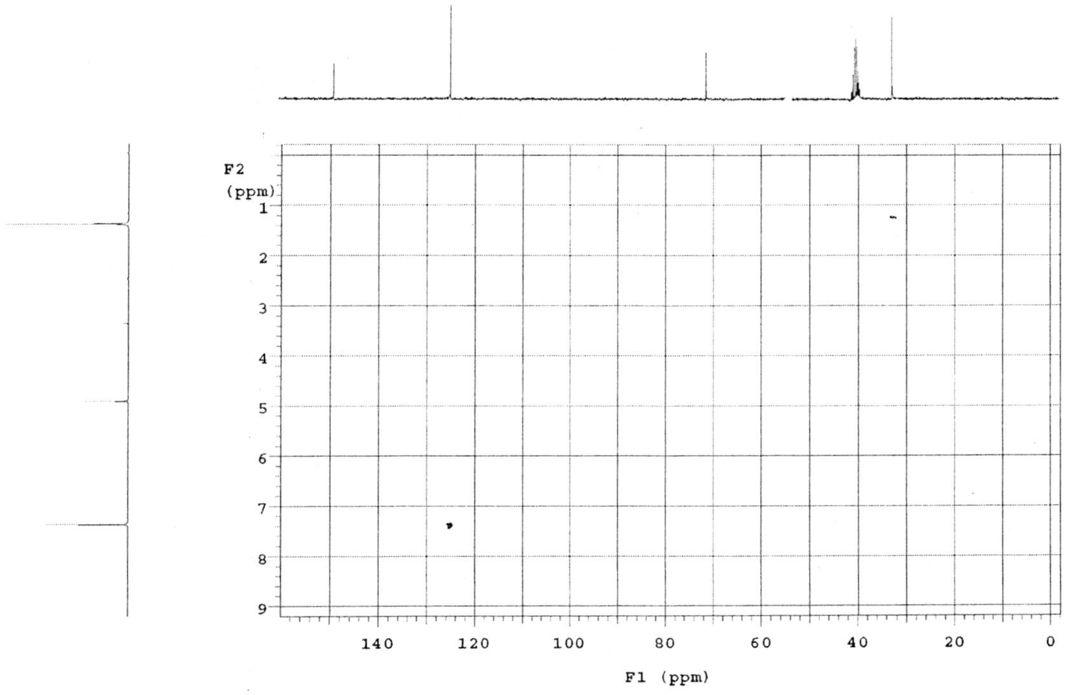

gHSQC Spectrum in DMSO-d_6

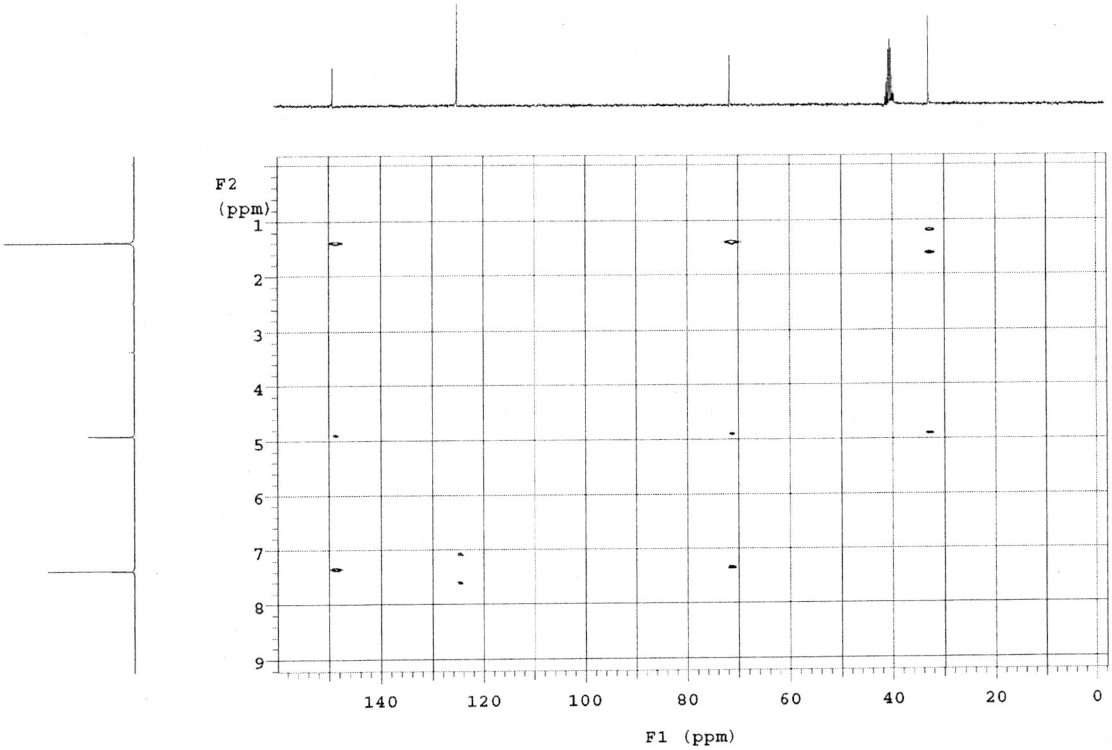

gHMBC Spectrum in DMSO-d_6

Problem 2.03 Given the spectroscopic data below (Proton, Carbon, Edited DEPT, gHSQC, and gCOSY), provide the structure of the molecule. (IR: 3600 and 1150 cm^{-1})

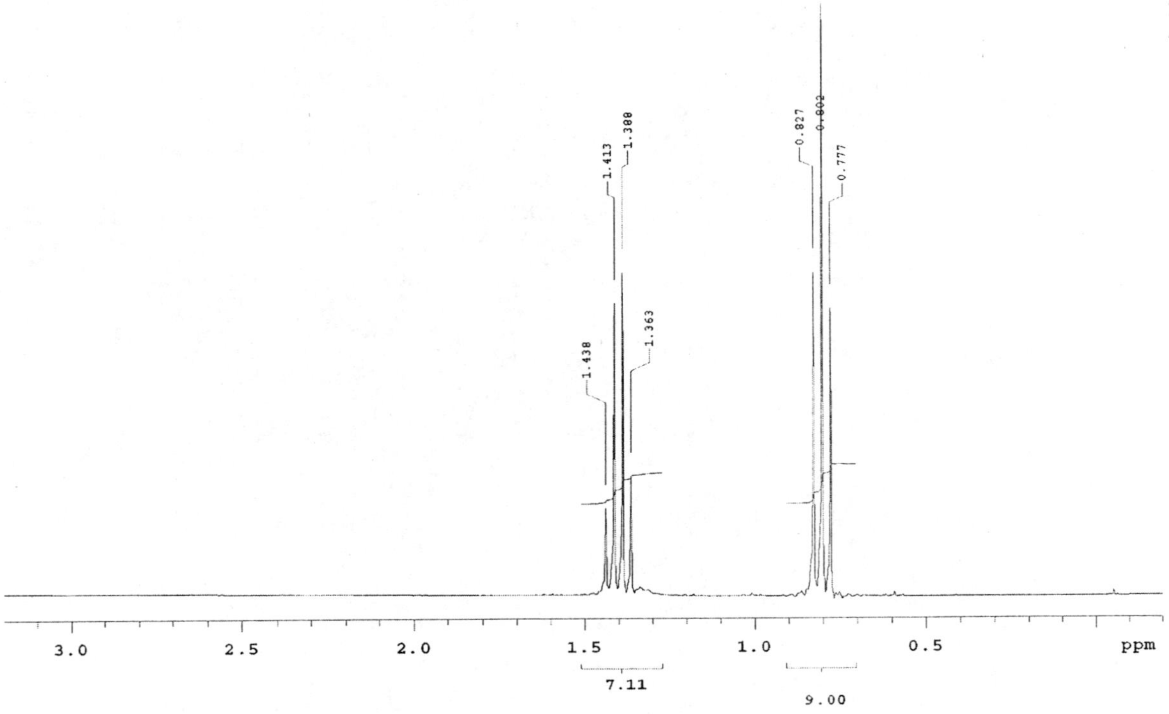

^{1}H-NMR Spectrum in $CDCl_3$

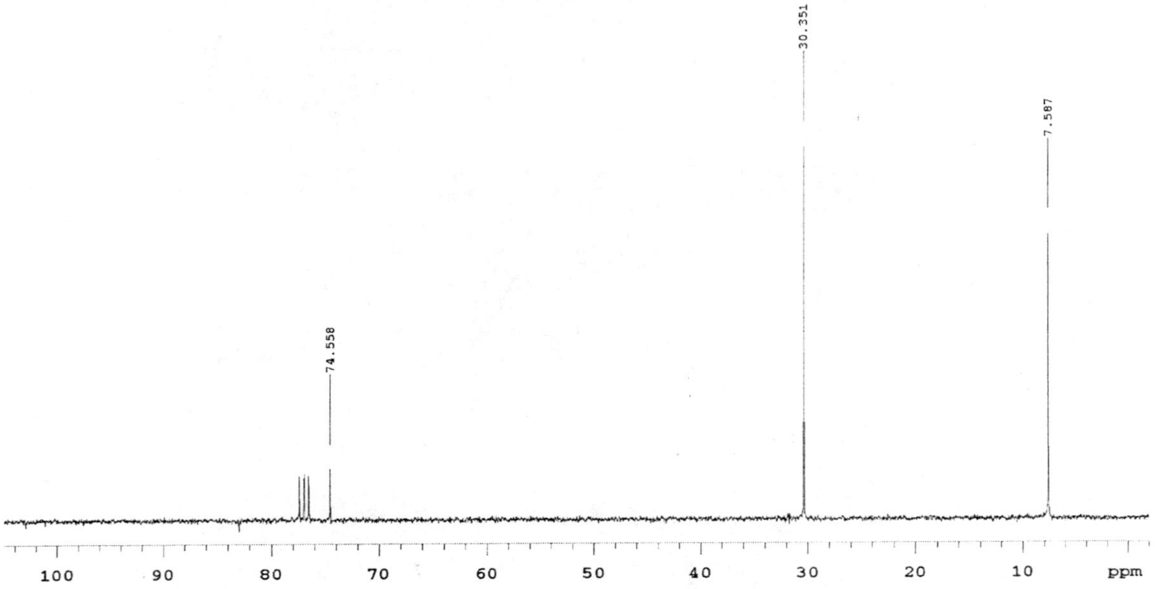

^{13}C-NMR Spectrum in $CDCl_3$

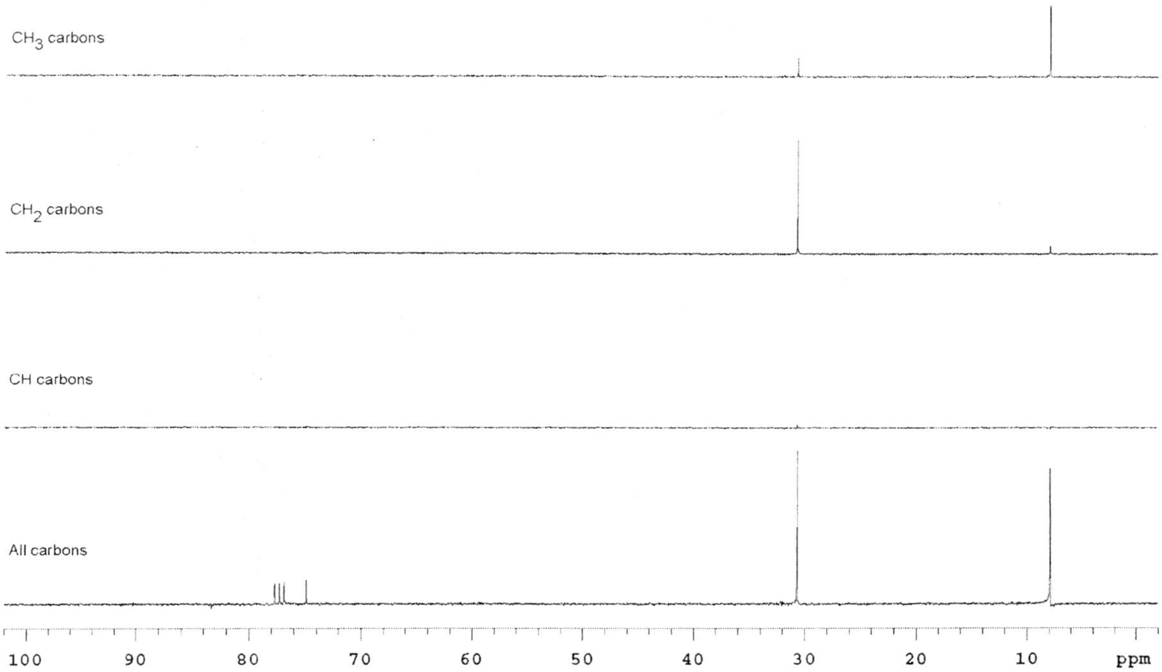

Edited DEPT ^{13}C-NMR Spectrum in $CDCl_3$

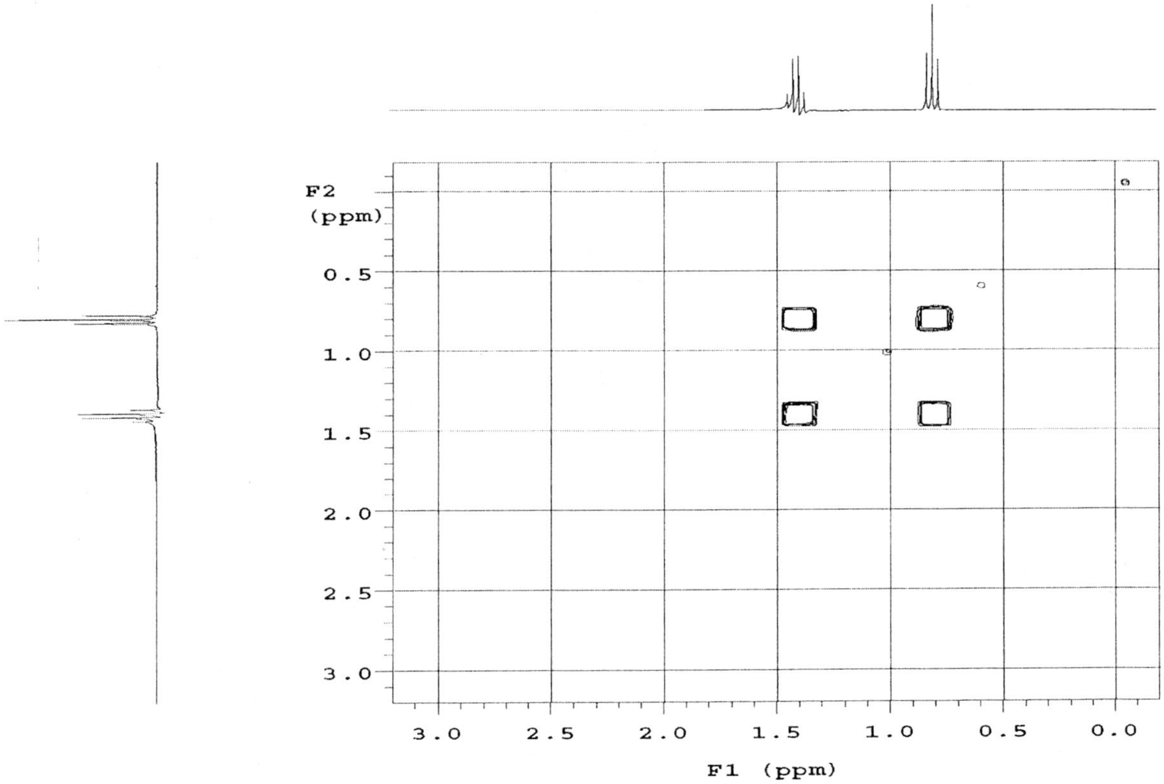

gCOSY Spectrum in $CDCl_3$

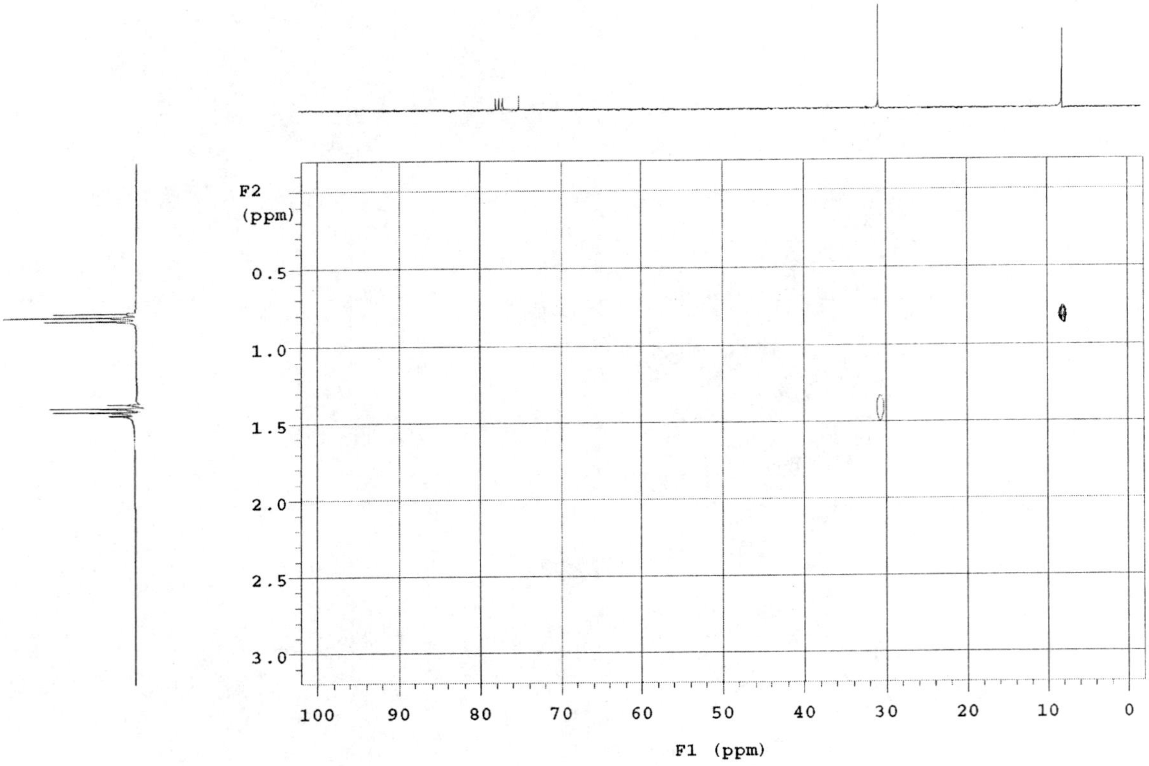

gHSQC Spectrum in $CDCl_3$

Problem 2.04 Given the spectroscopic data below (Proton, Carbon, gHMQC, gCOSY, and gHMBC), provide the structure of the molecule. (MW= 182 g/mol) Note: Sample contains small amount of residual solvents and minor impurities.

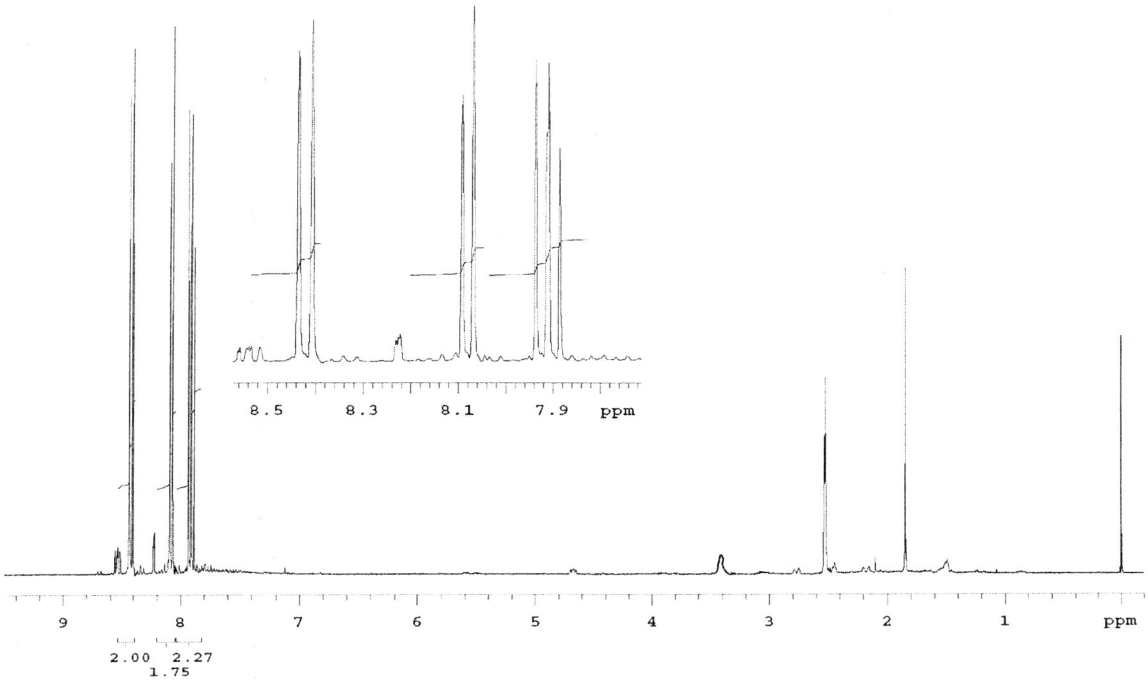

^{1}H-NMR Spectrum in DMSO-d_6

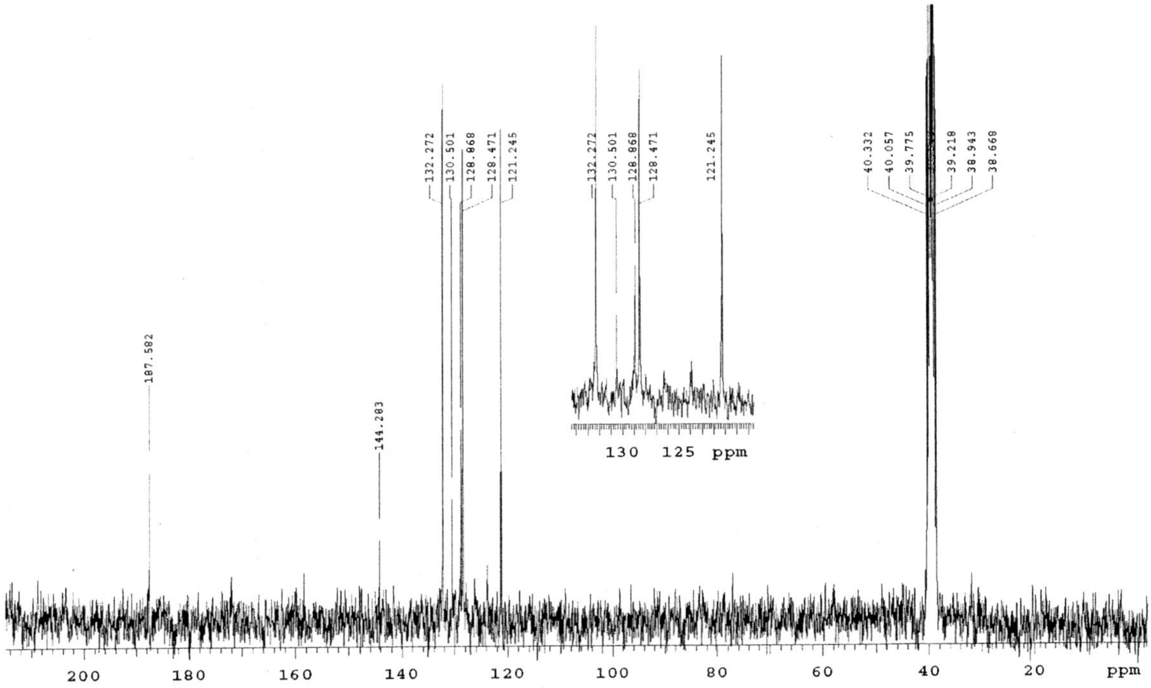

^{13}C-NMR Spectrum in DMSO-d_6

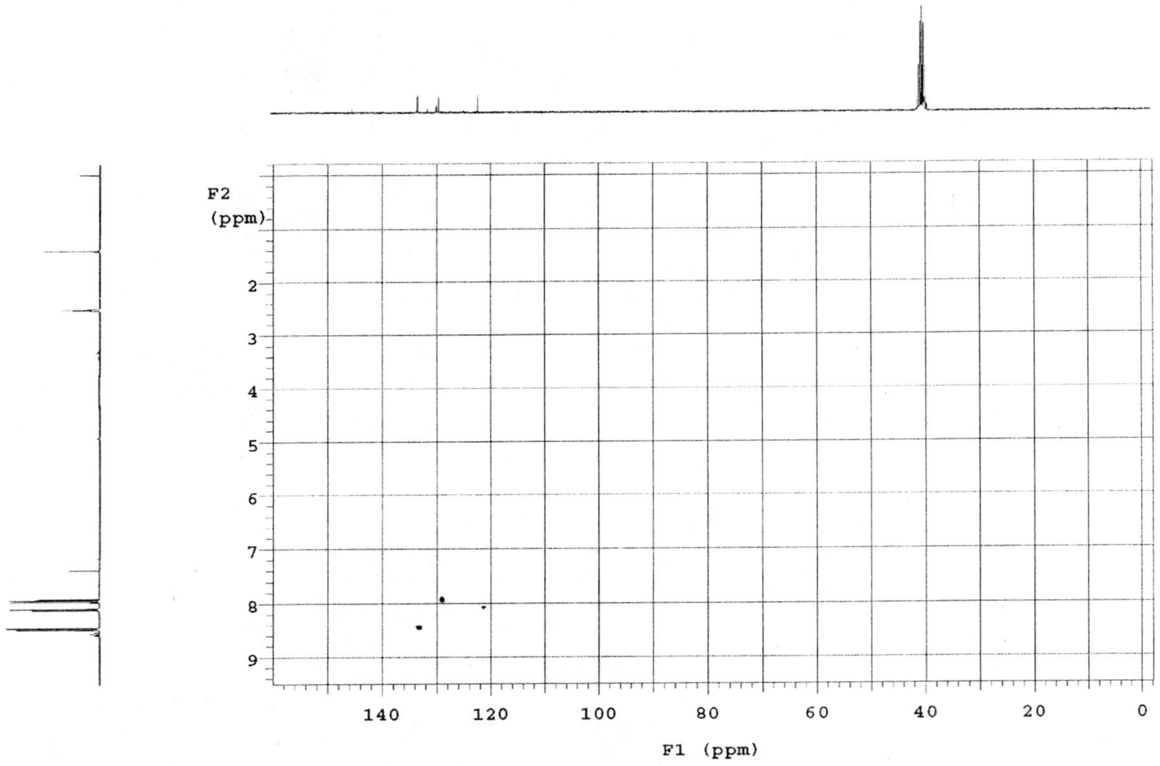

gHMQC Spectrum in DMSO-d_6

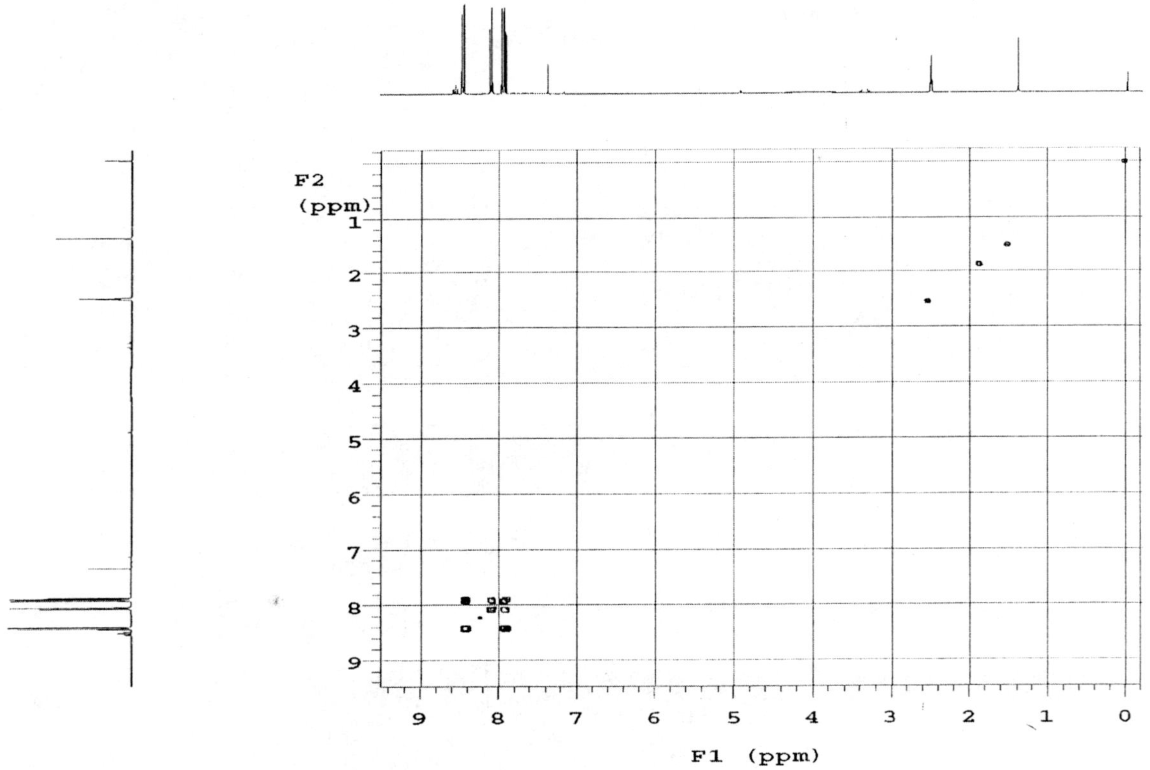

gCOSY Spectrum in DMSO-d_6

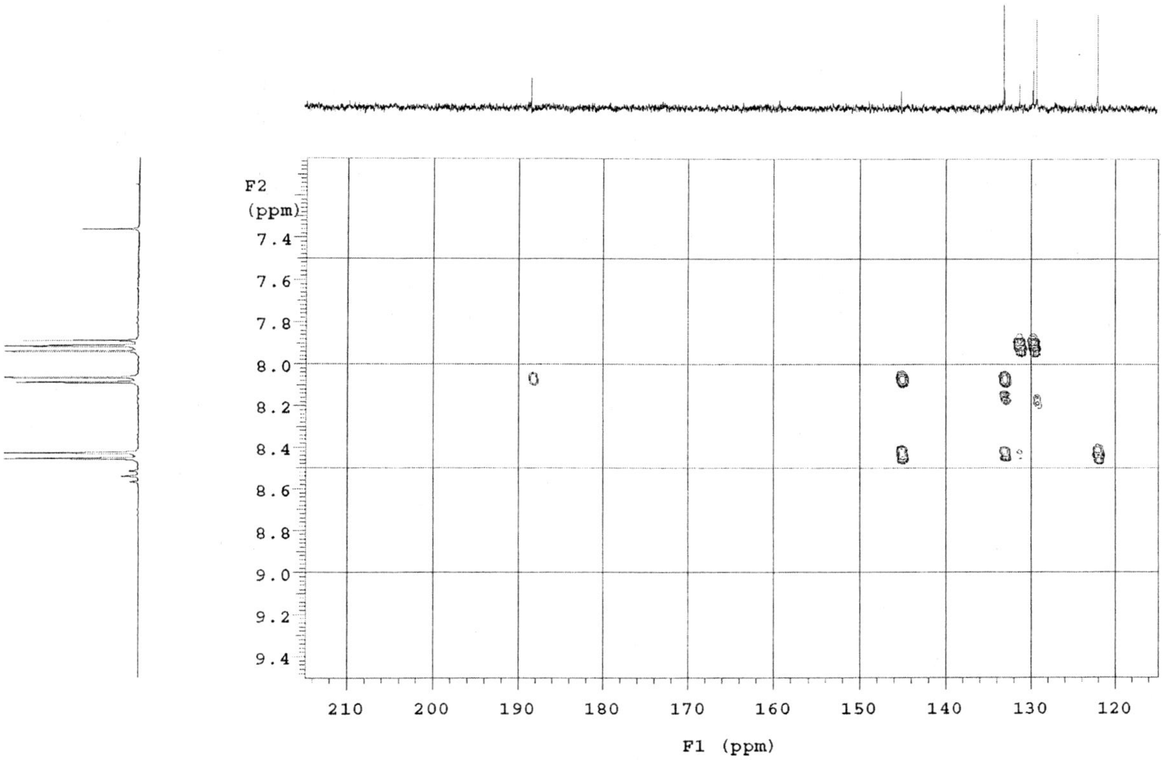

gHMBC Spectrum in DMSO-d_6

Problem 2.05 Given the spectroscopic data below (Proton, Carbon, gHMQC, and gCOSY), provide the structure of the molecule. (MW = 208 g/mol)

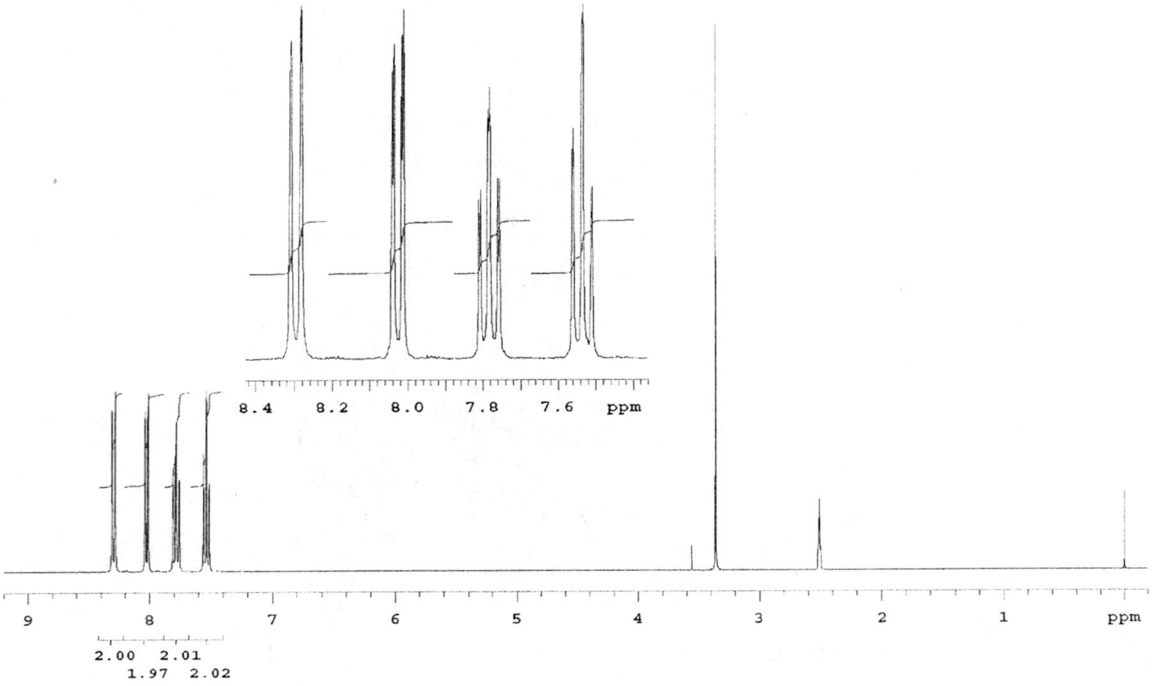

^{1}H-NMR Spectrum in DMSO-d_6

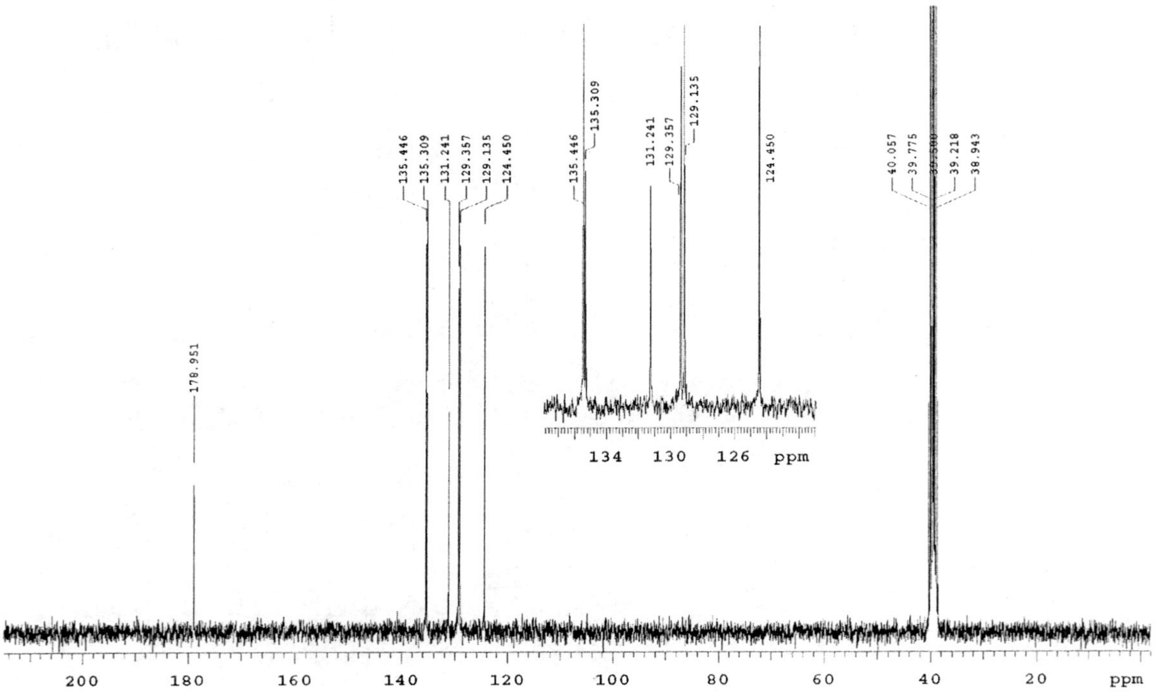

^{13}C-NMR Spectrum in DMSO-d_6

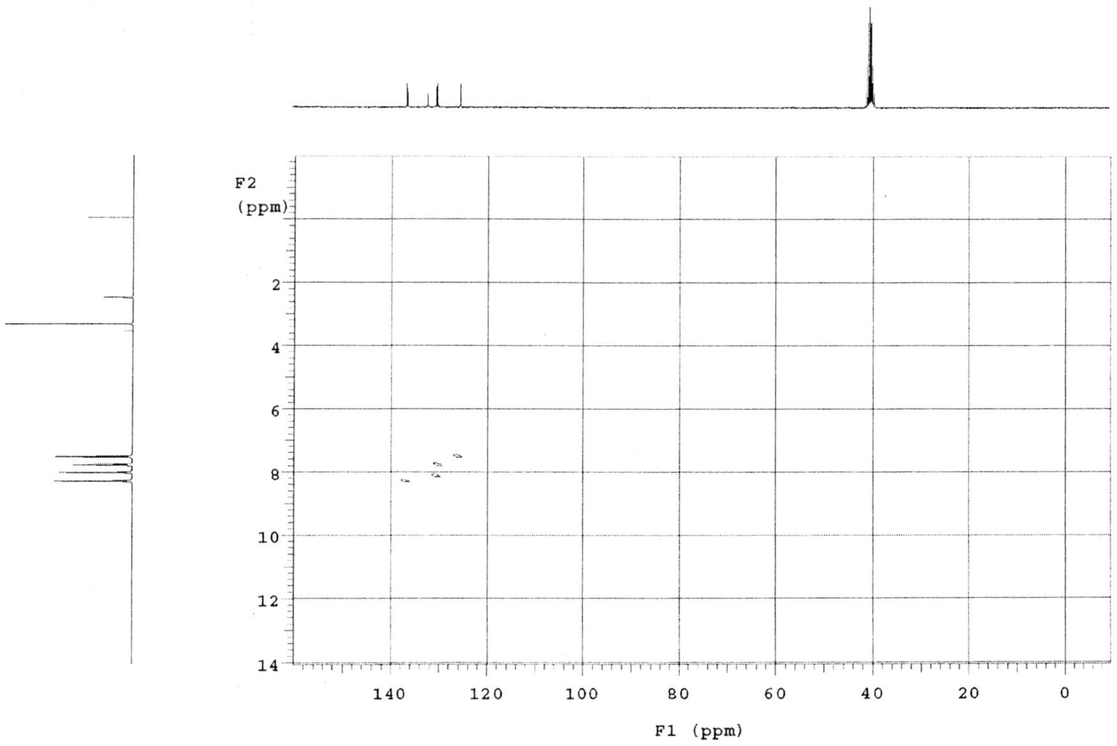

gHMQC Spectrum in DMSO-d_6

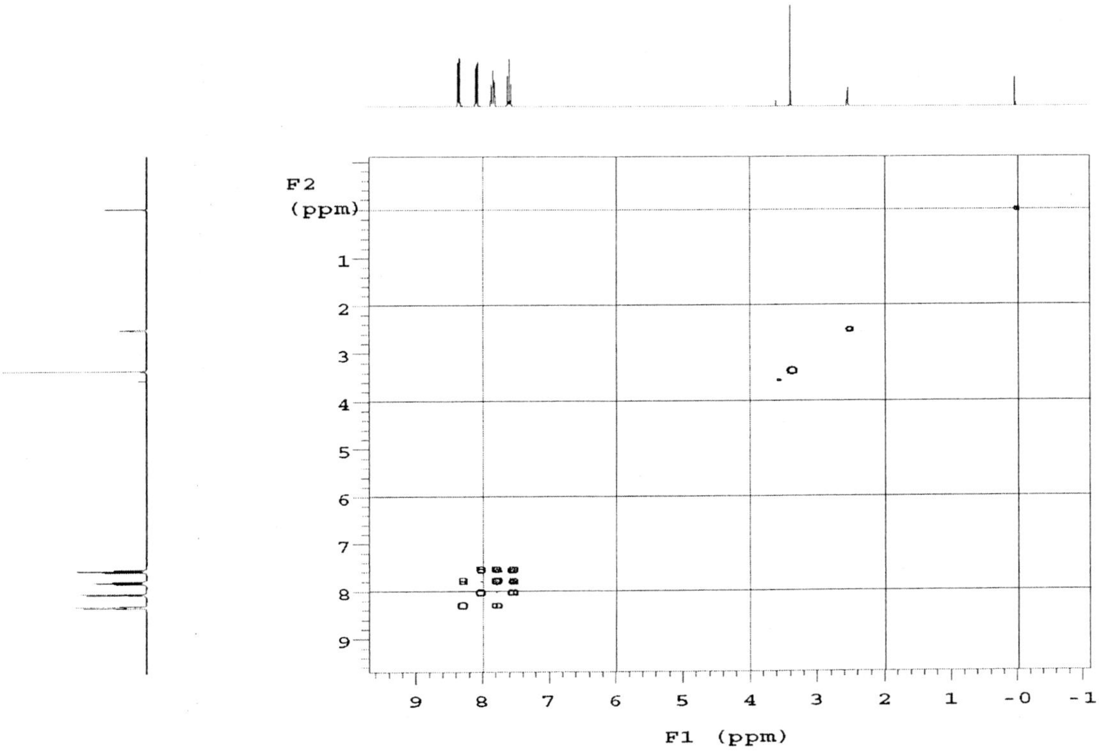

gCOSY Spectrum in DMSO-d_6

Problem 2.06 Given the spectroscopic data below (Proton, Carbon, gHSQC, gCOSY, and gHMBC), provide the structure of the molecule.

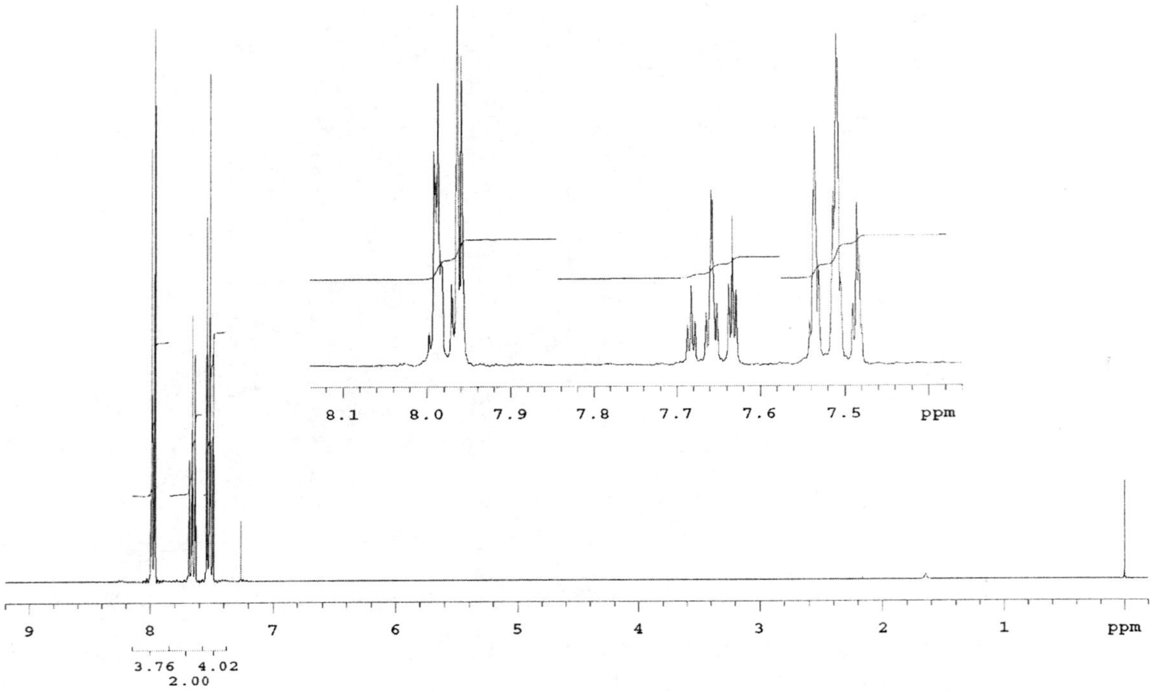

^{1}H-NMR Spectrum in $CDCl_3$

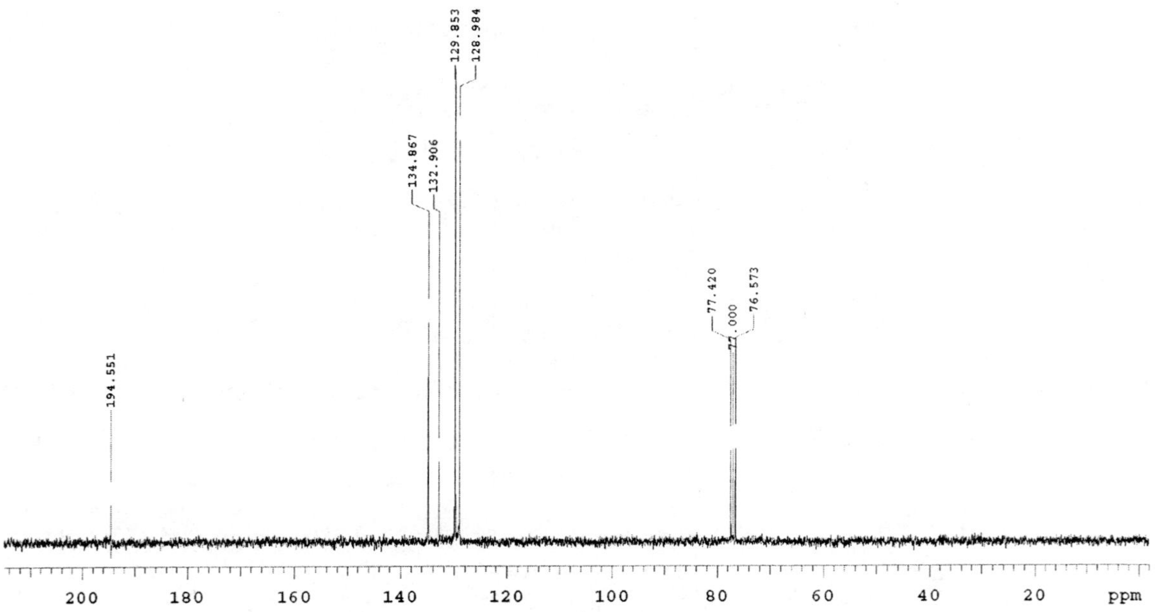

^{13}C-NMR Spectrum in $CDCl_3$

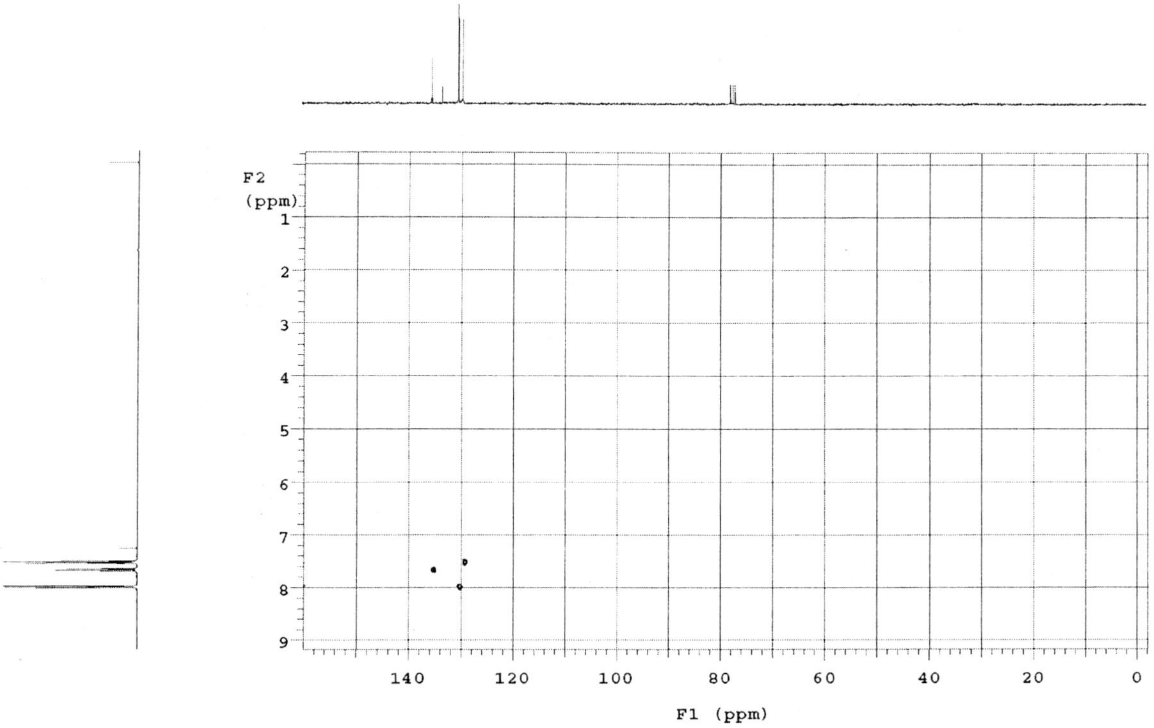

gHSQC Spectrum in $CDCl_3$

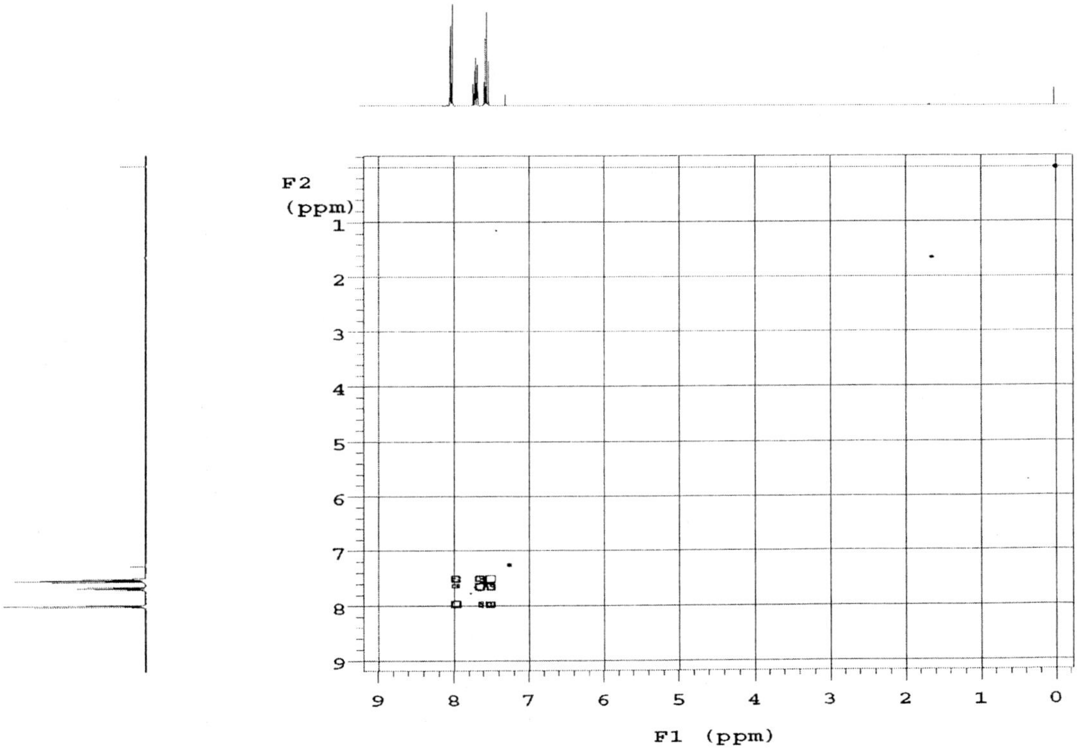

gCOSY Spectrum in $CDCl_3$

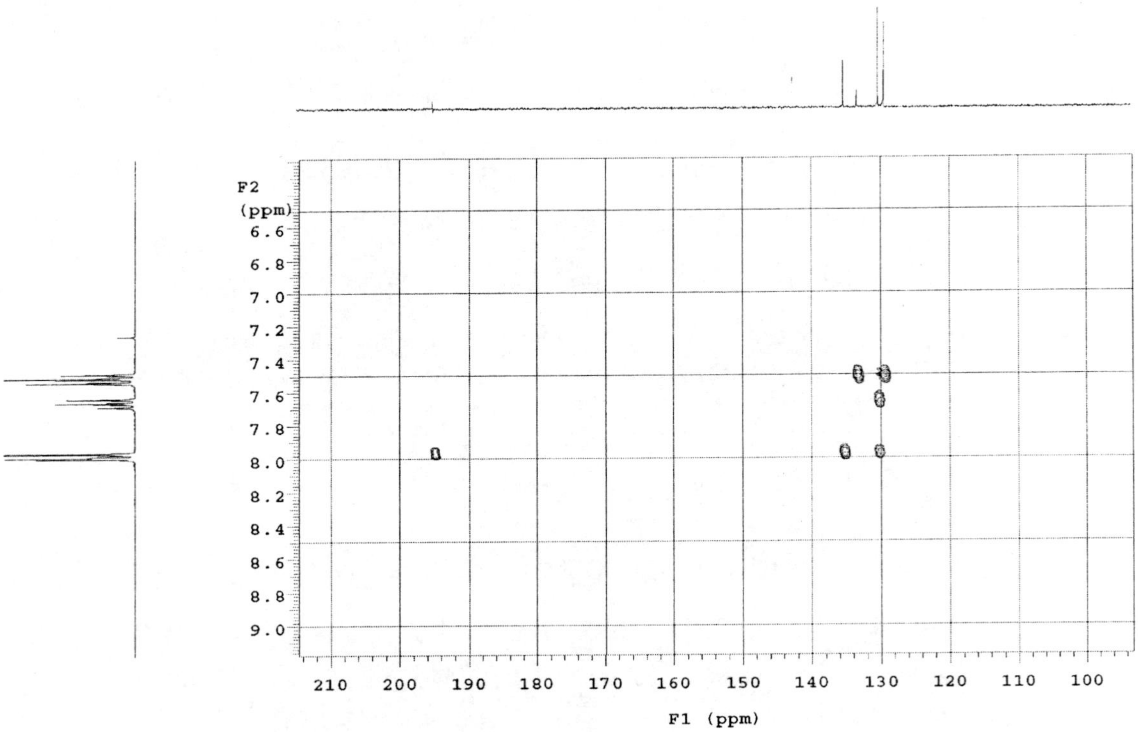

gHMBC Spectrum in $CDCl_3$

Problem 2.07 Given the spectroscopic data below (Proton, Carbon, gHMQC, and gCOSY), provide the structure of the molecule. (MW: 146 g/mol)

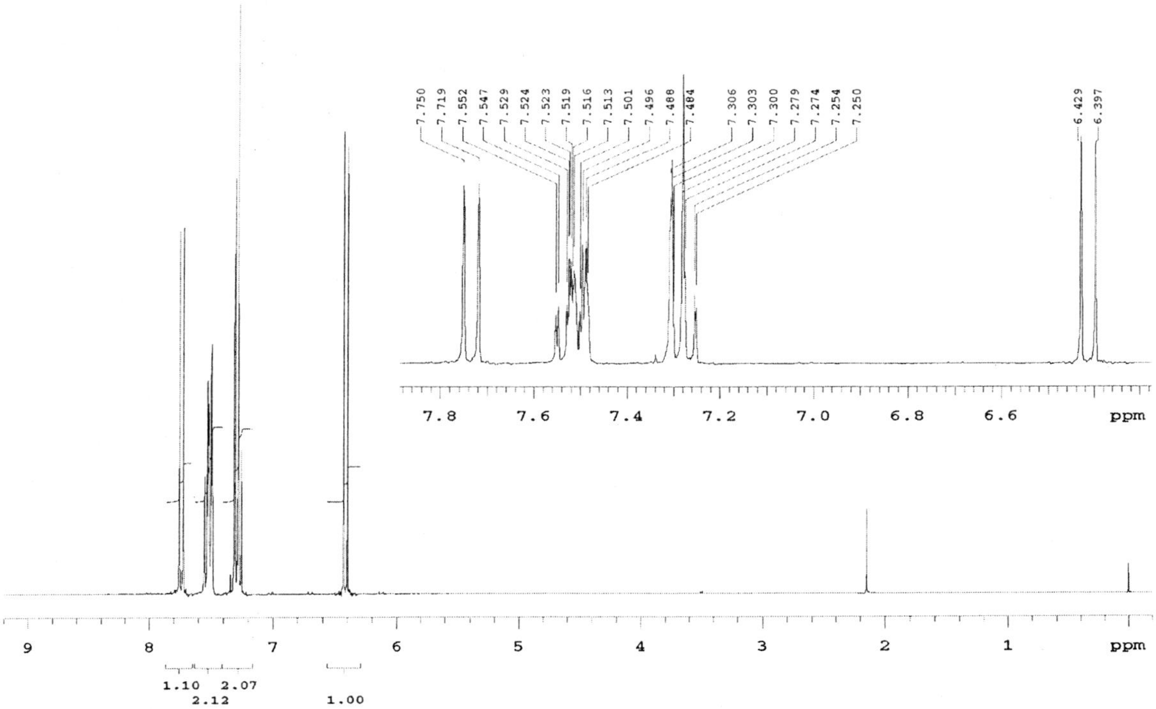

^{1}H-NMR Spectrum in $CDCl_3$

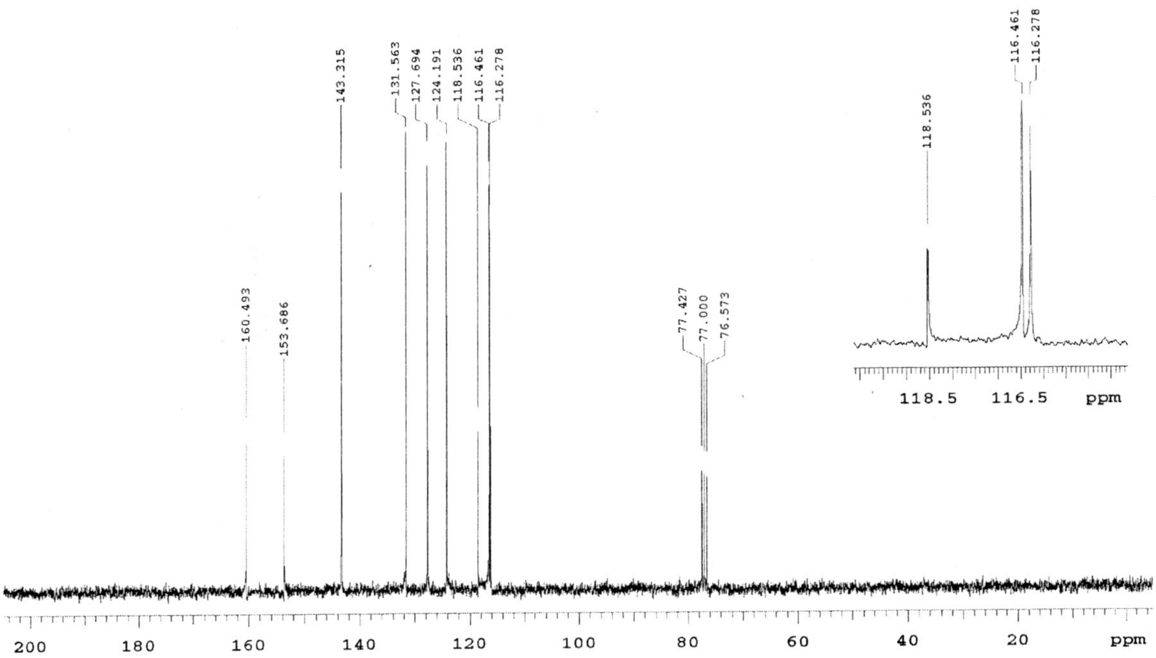

^{13}C-NMR Spectrum in $CDCl_3$

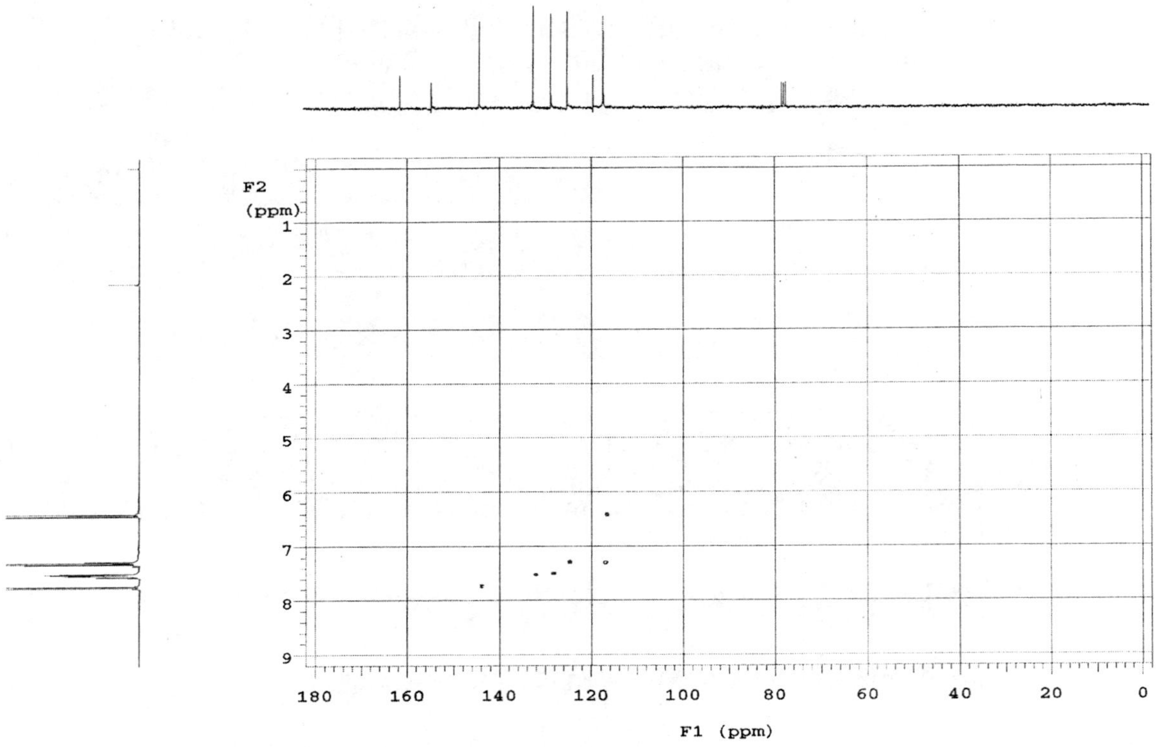

gHSQC Spectrum in $CDCl_3$

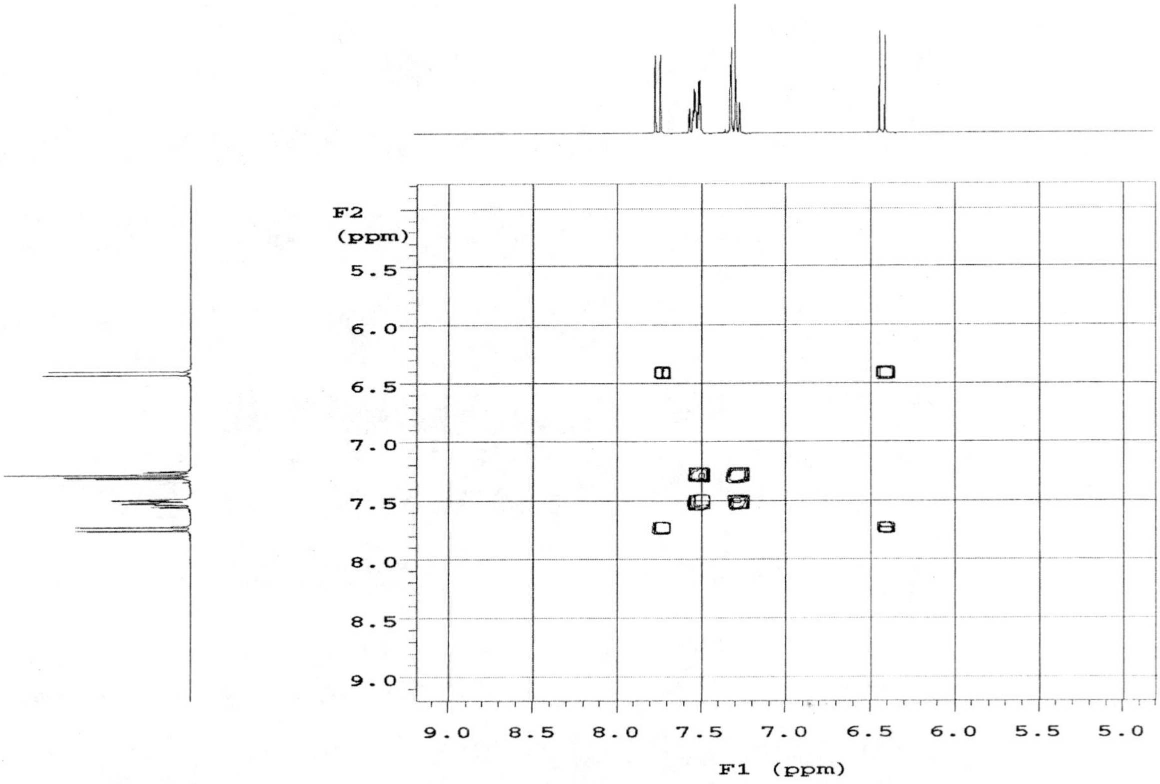

gCOSY Spectrum in $CDCl_3$

Problem 2.08 Given the spectroscopic data below (Proton, Carbon, gHSQC, and gCOSY), provide the structure of the molecule.

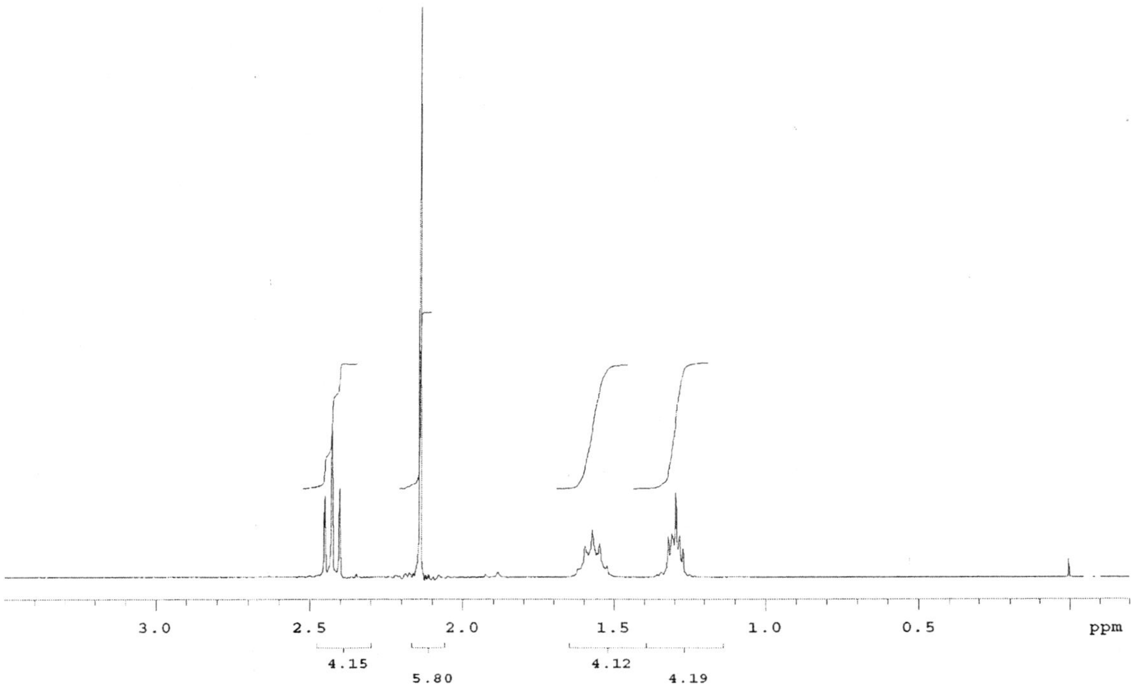

^{1}H-NMR Spectrum in $CDCl_3$

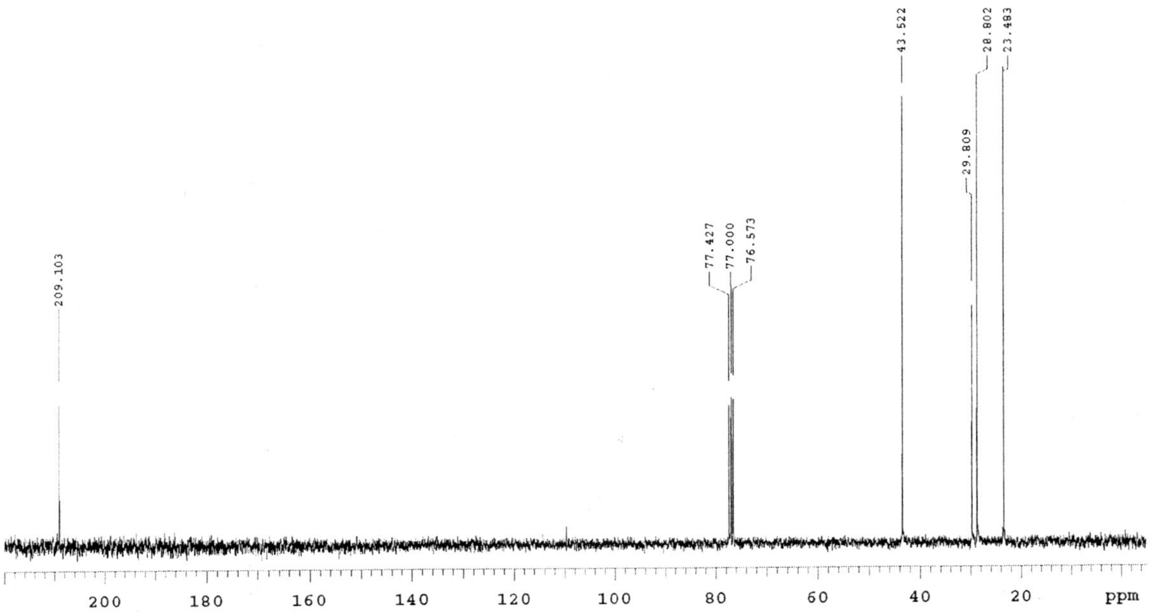

^{13}C-NMR Spectrum in $CDCl_3$

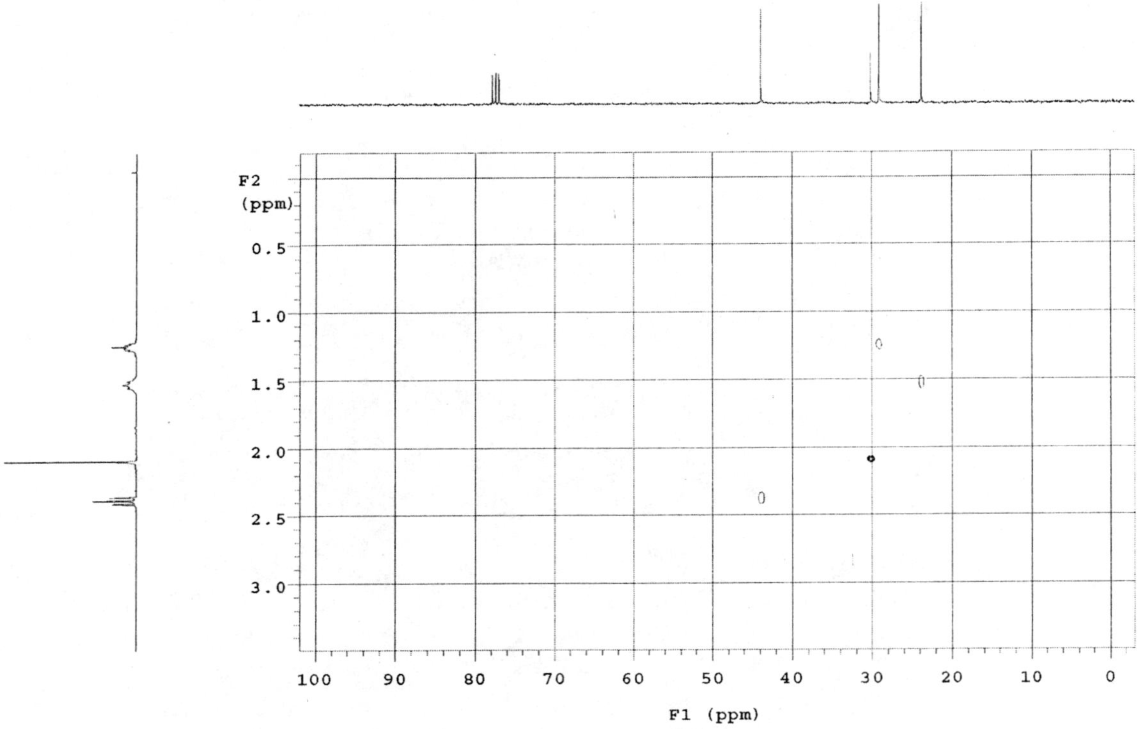

gHSQC Spectrum in $CDCl_3$

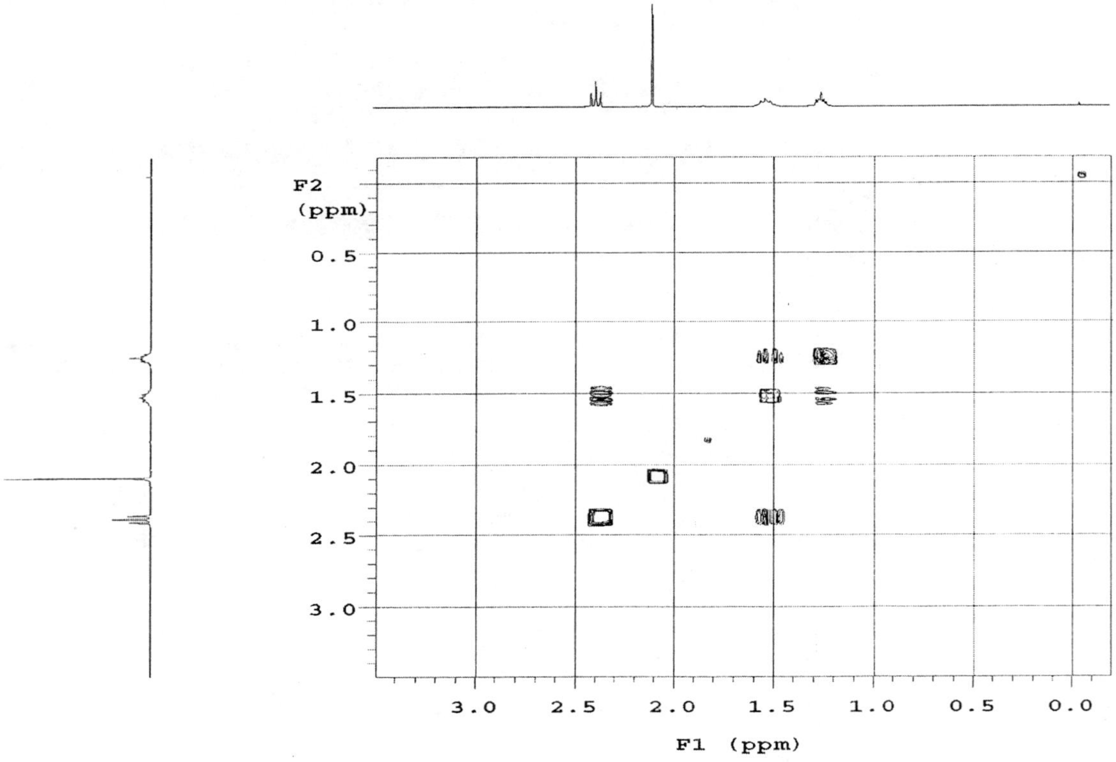

gCOSY Spectrum in $CDCl_3$

Problem 2.09 Given the spectroscopic data below (Proton, Carbon, Edited DEPT, gHSQC, and gCOSY), provide the structure of the molecule.

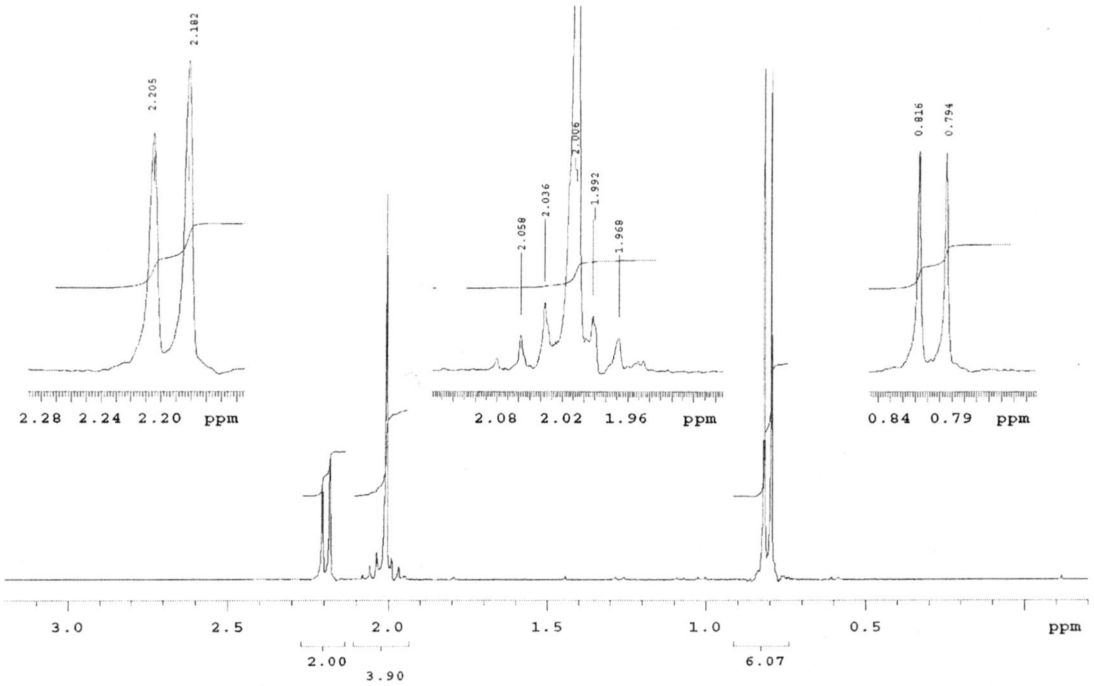

^{1}H-NMR Spectrum in $CDCl_3$

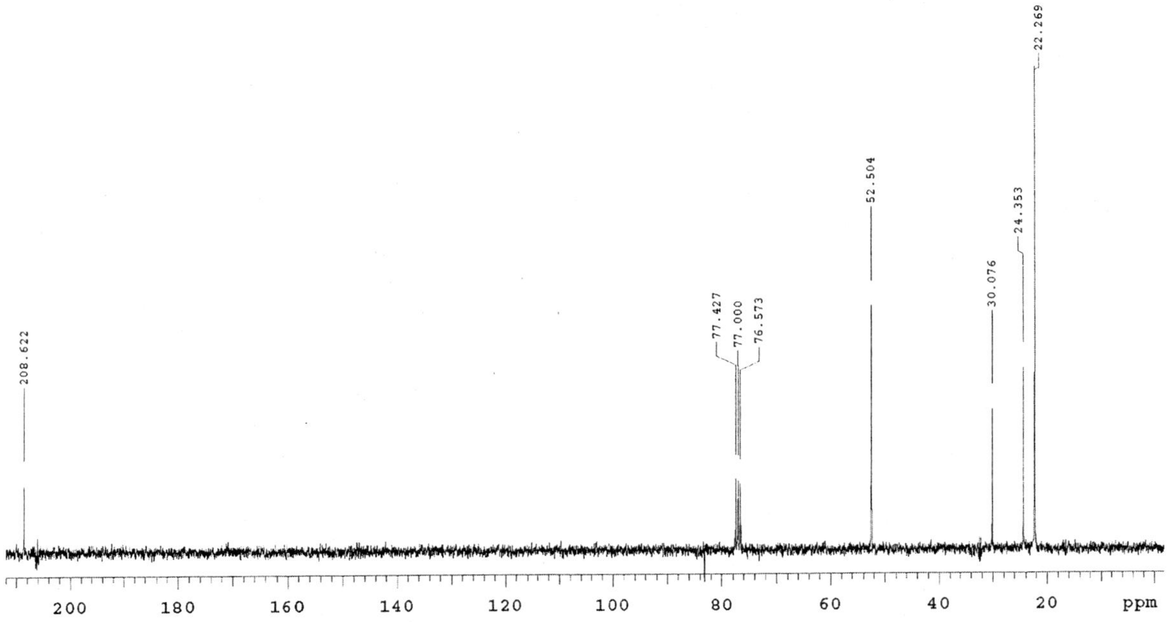

^{13}C-NMR Spectrum in $CDCl_3$

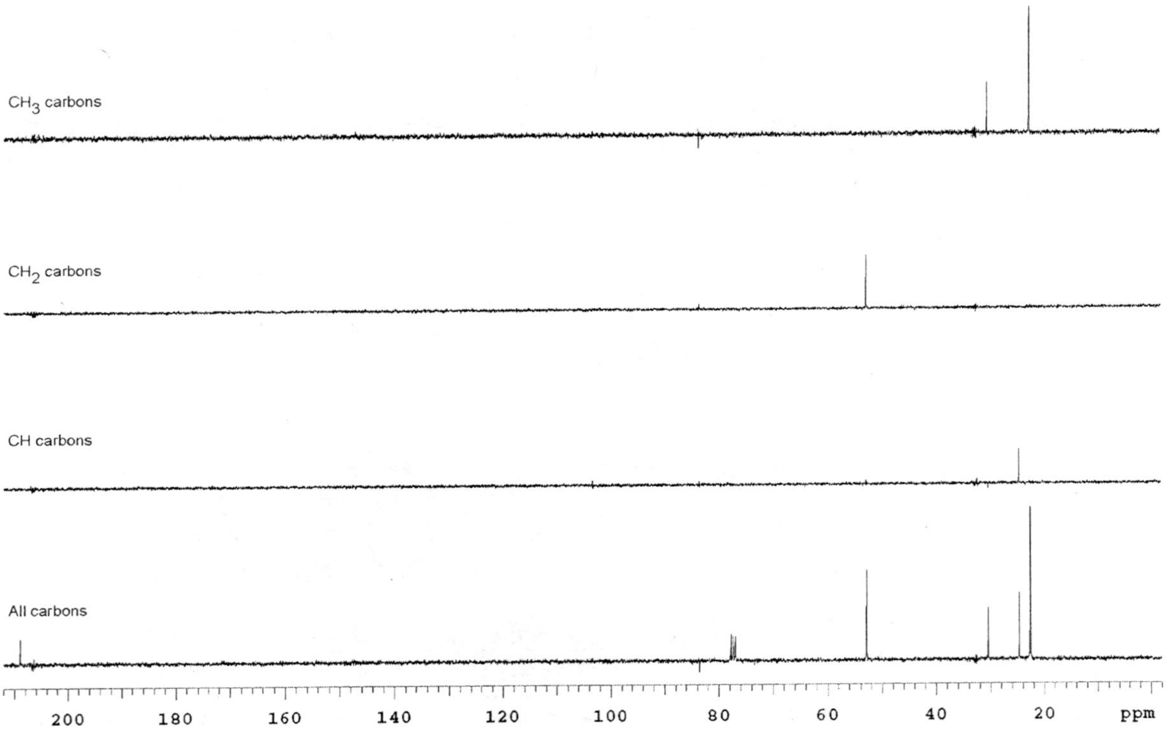

Edited DEPT ^{13}C-NMR Spectrum in $CDCl_3$

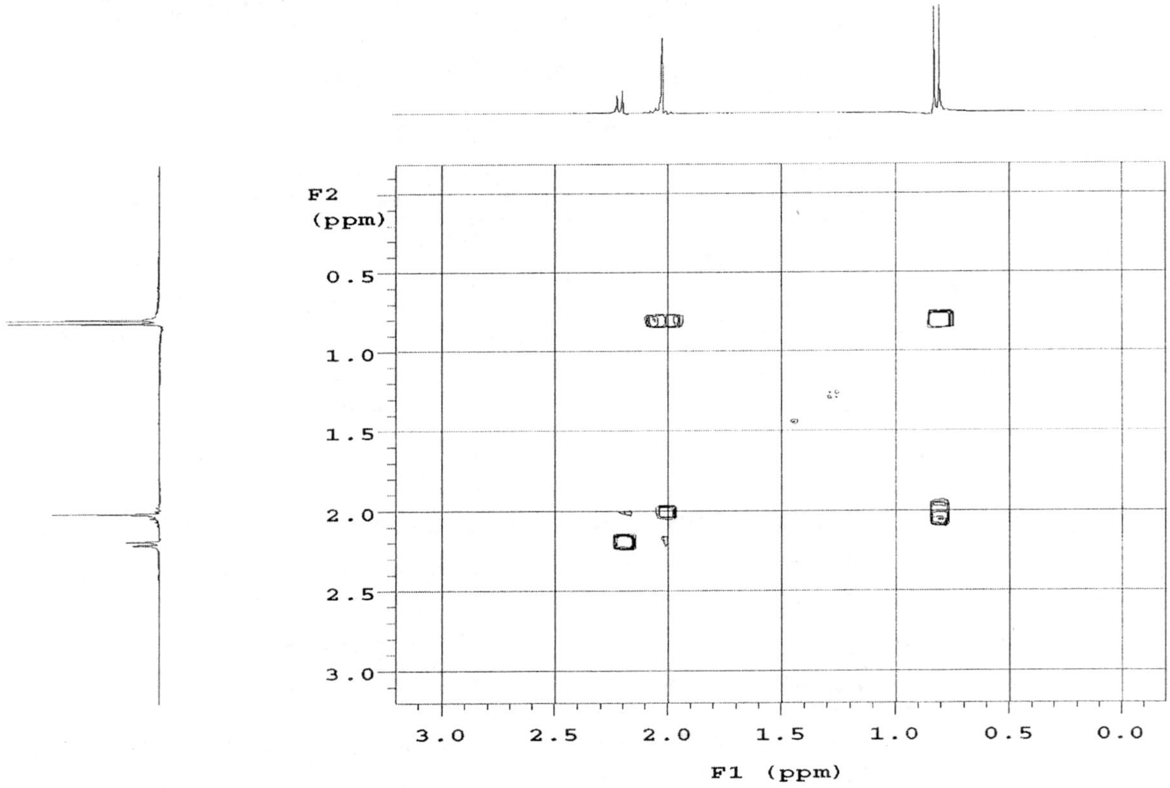

gCOSY Spectrum in $CDCl_3$

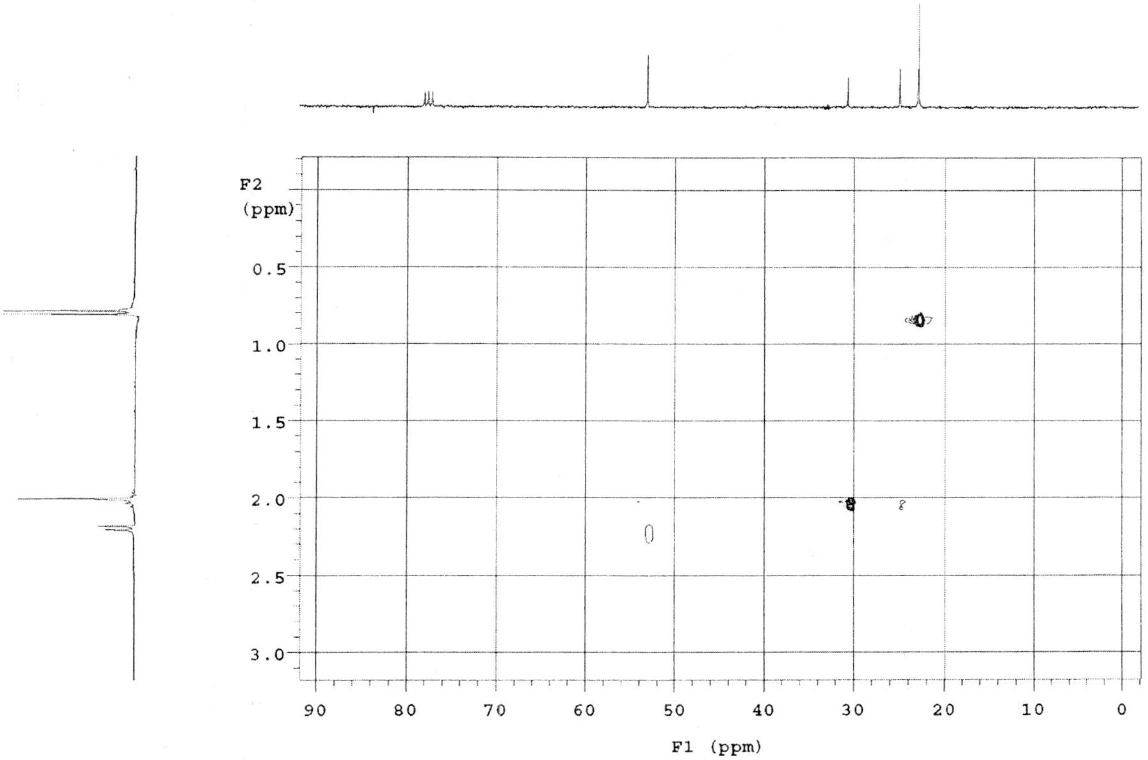

gHSQC Spectrum in $CDCl_3$

Problem 2.10 Given the spectroscopic data below (Proton, Carbon, Edited DEPT, gHSQC, and gCOSY), provide the structure of the molecule.

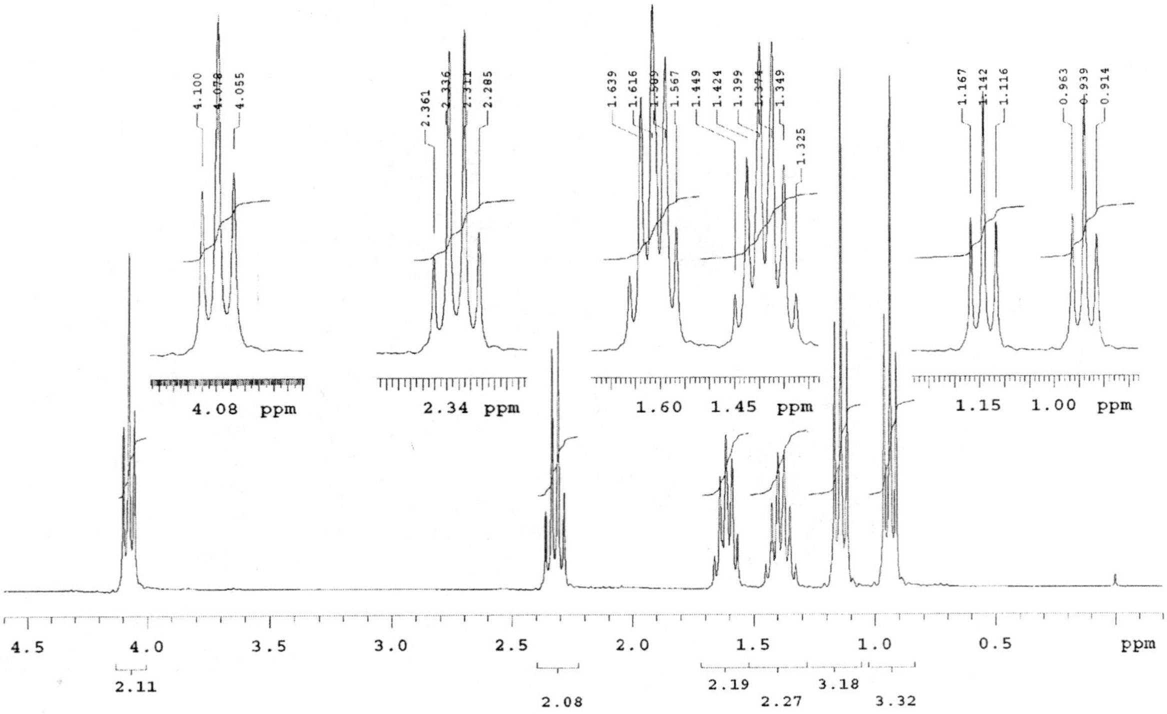

^{1}H-NMR Spectrum in $CDCl_3$

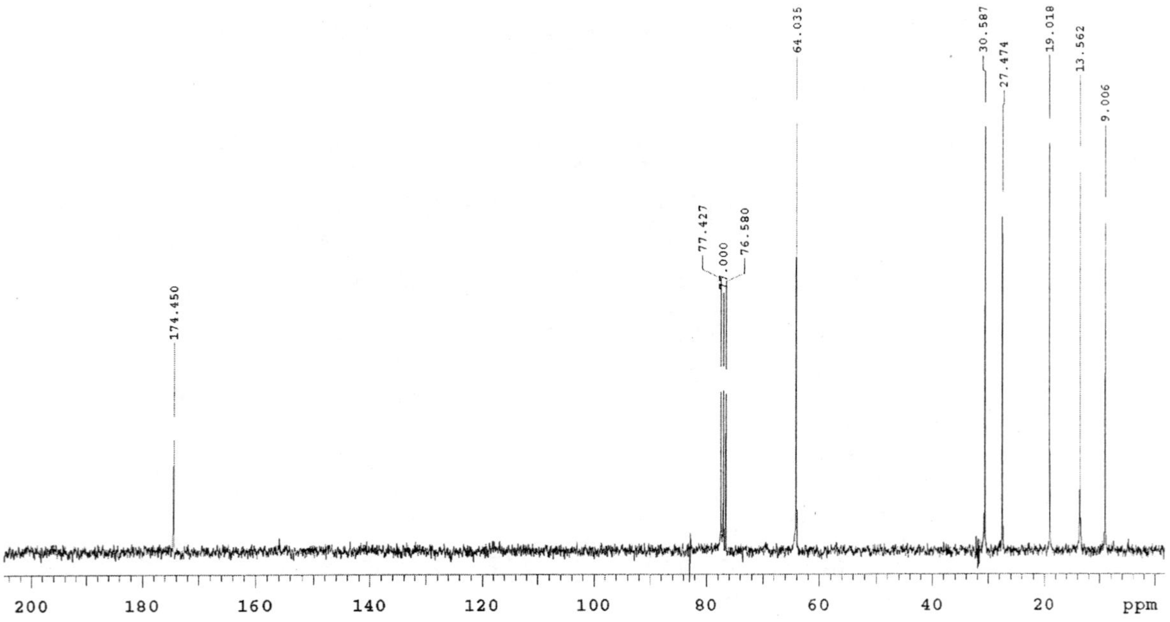

^{13}C-NMR Spectrum in $CDCl_3$

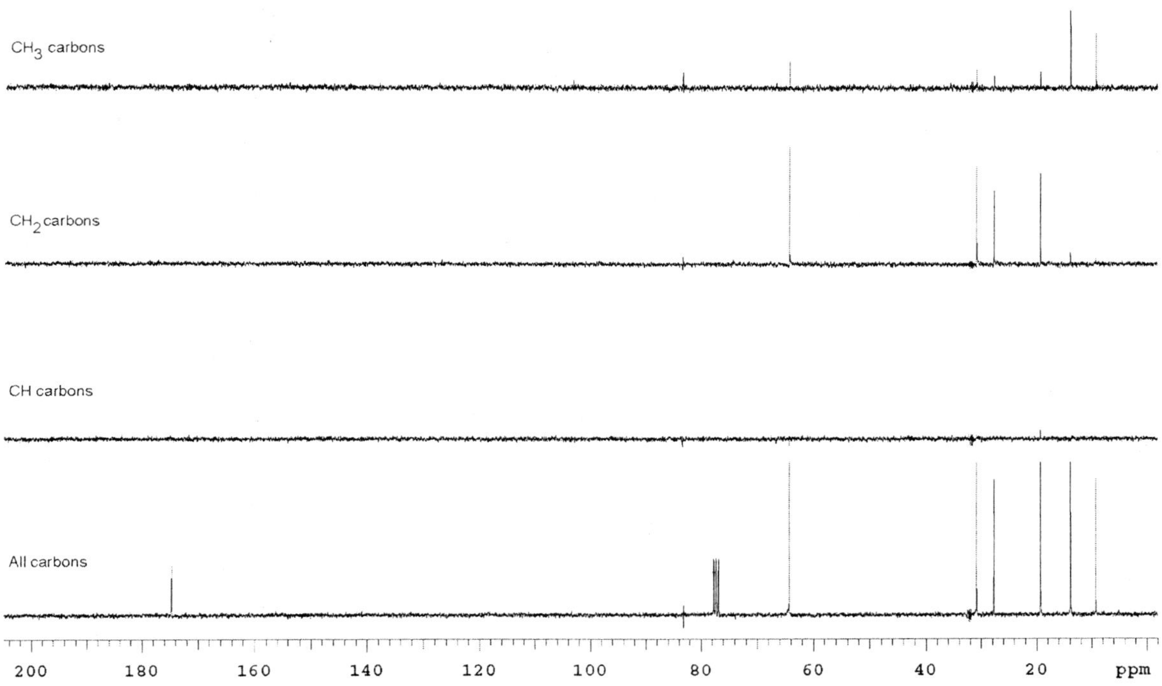

Edited DEPT ^{13}C-NMR Spectrum in $CDCl_3$

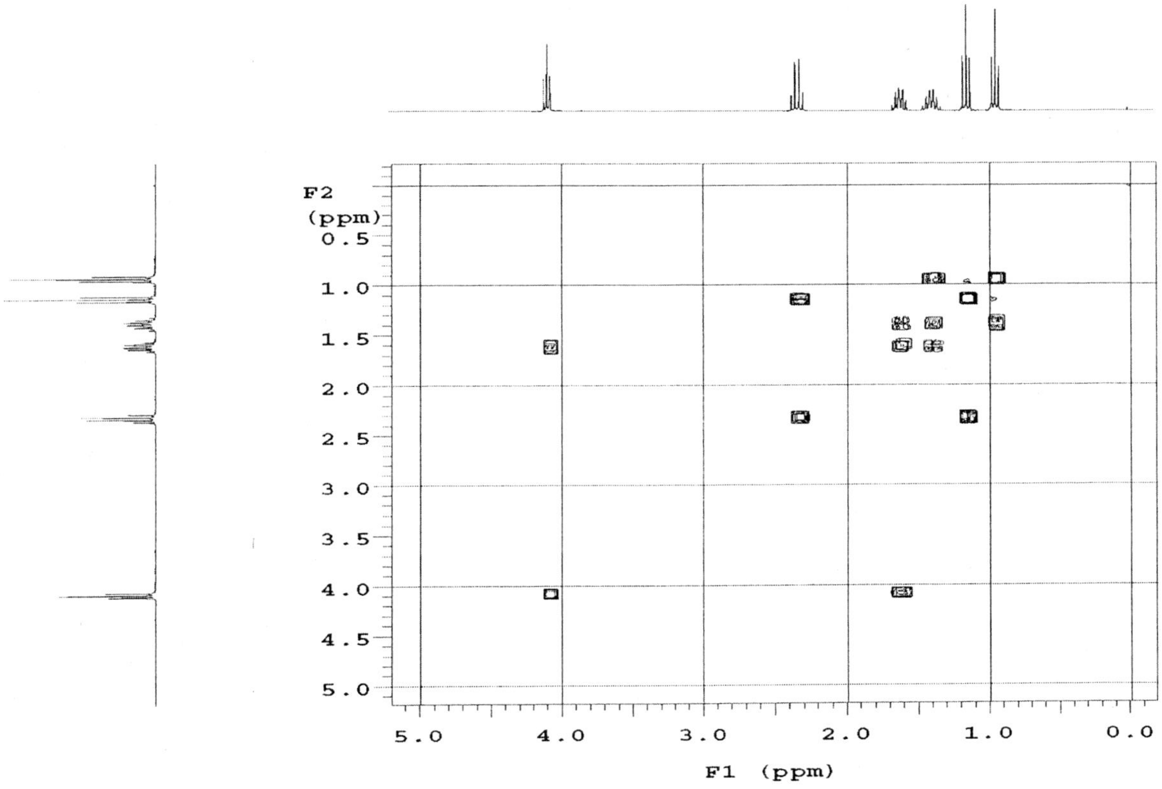

gCOSY Spectrum in $CDCl_3$

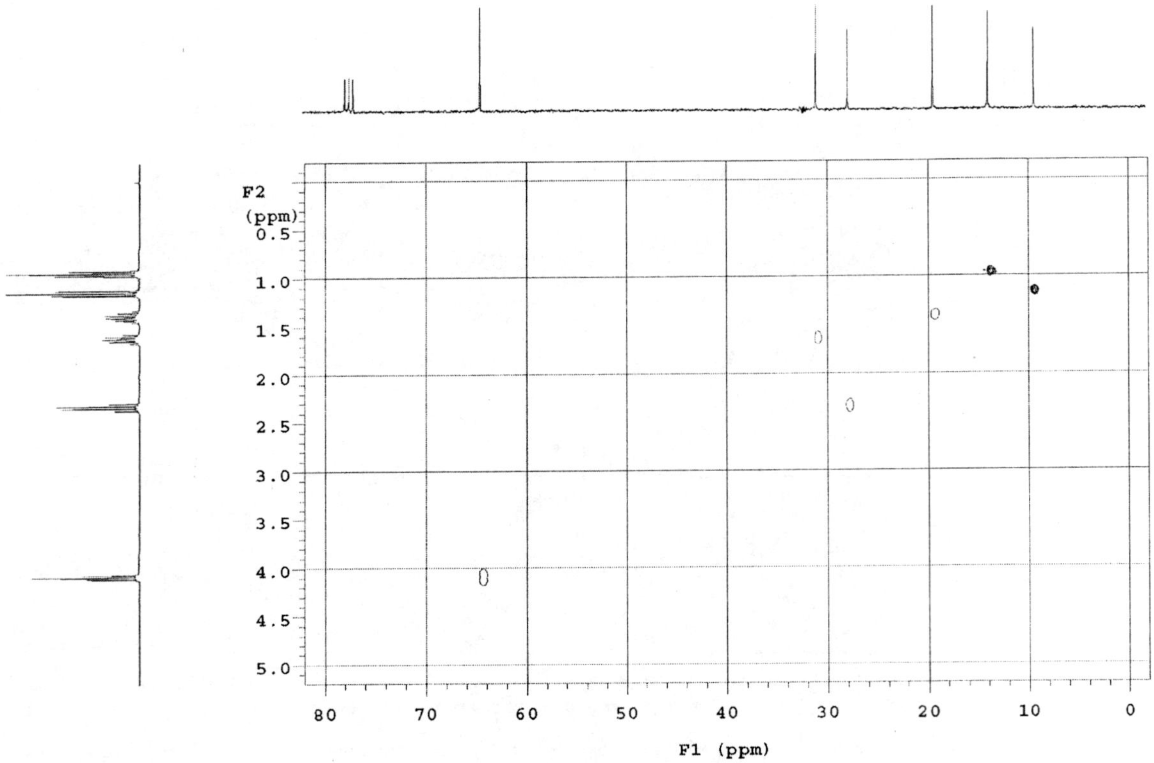

gHSQC Spectrum in $CDCl_3$

Problem 2.11 Given the spectroscopic data below (Proton, Carbon, and gHMQC), provide the structure of the molecule. (MW: 186 & 188 g/mol)

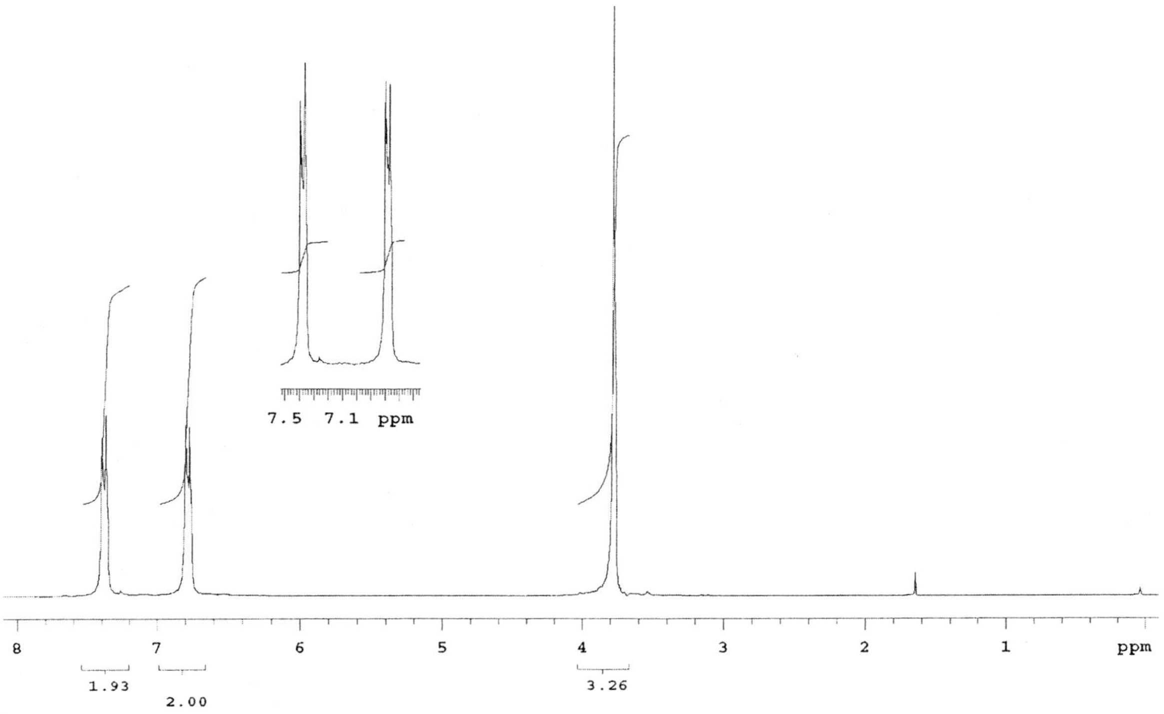

^{1}H-NMR Spectrum in $CDCl_3$

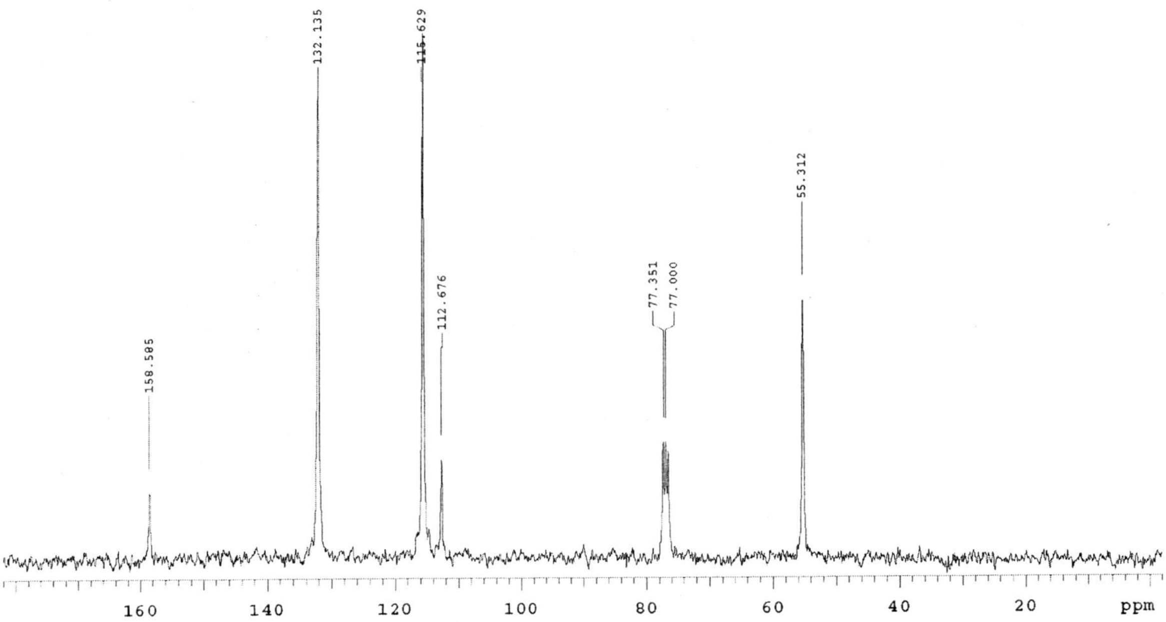

^{13}C-NMR Spectrum in $CDCl_3$

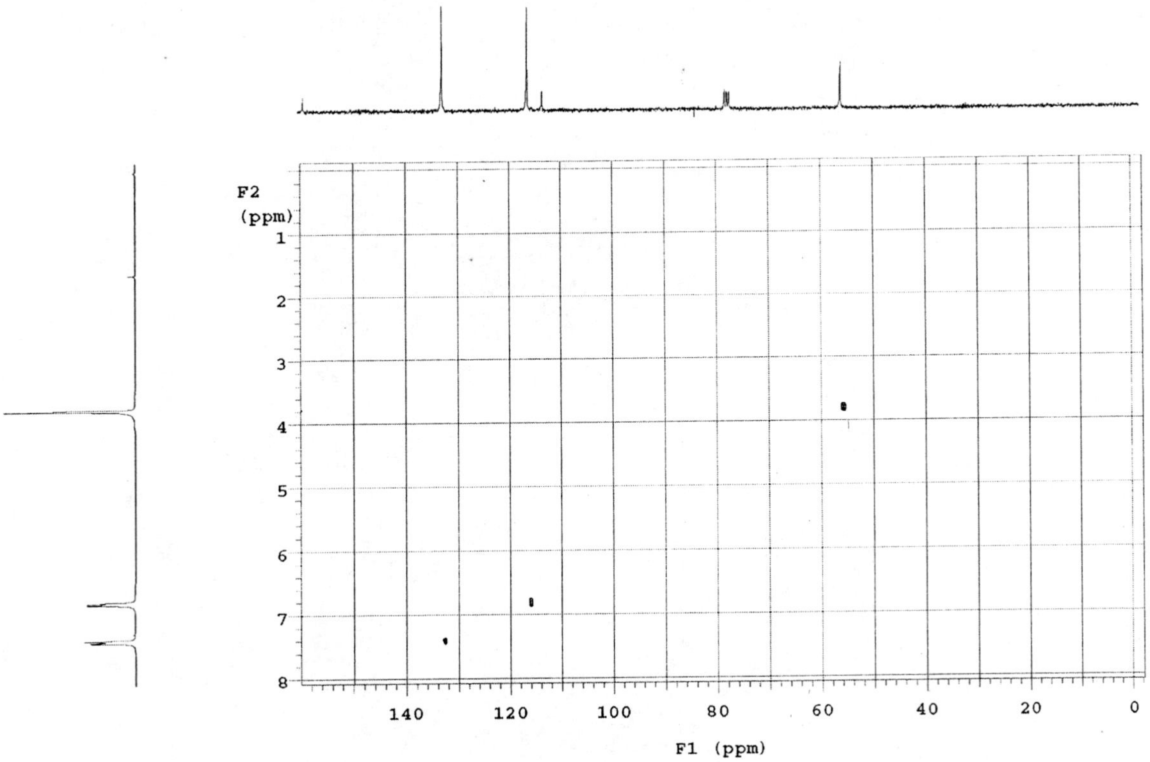

gHSQC Spectrum in $CDCl_3$

Problem 2.12 Given the spectroscopic data below (Proton, Carbon, gHSQC, and gCOSY), provide the structure of the molecule. (MW: 159 & 161 g/mol)

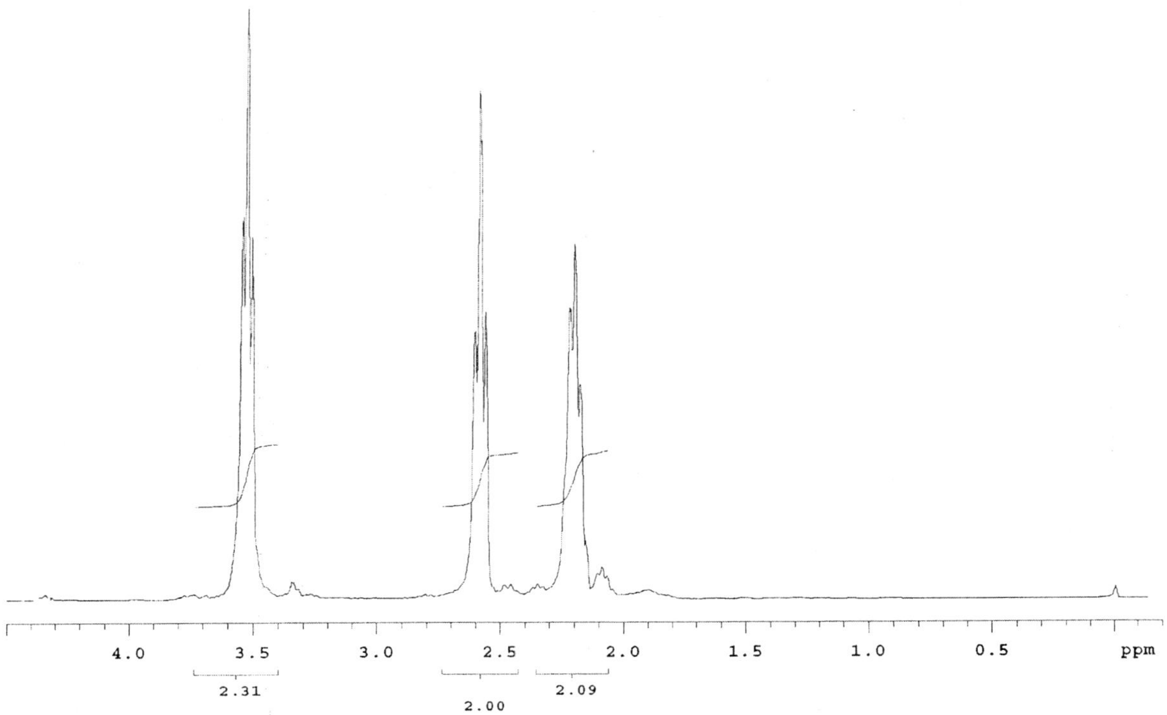

^{1}H-NMR Spectrum in $CDCl_3$

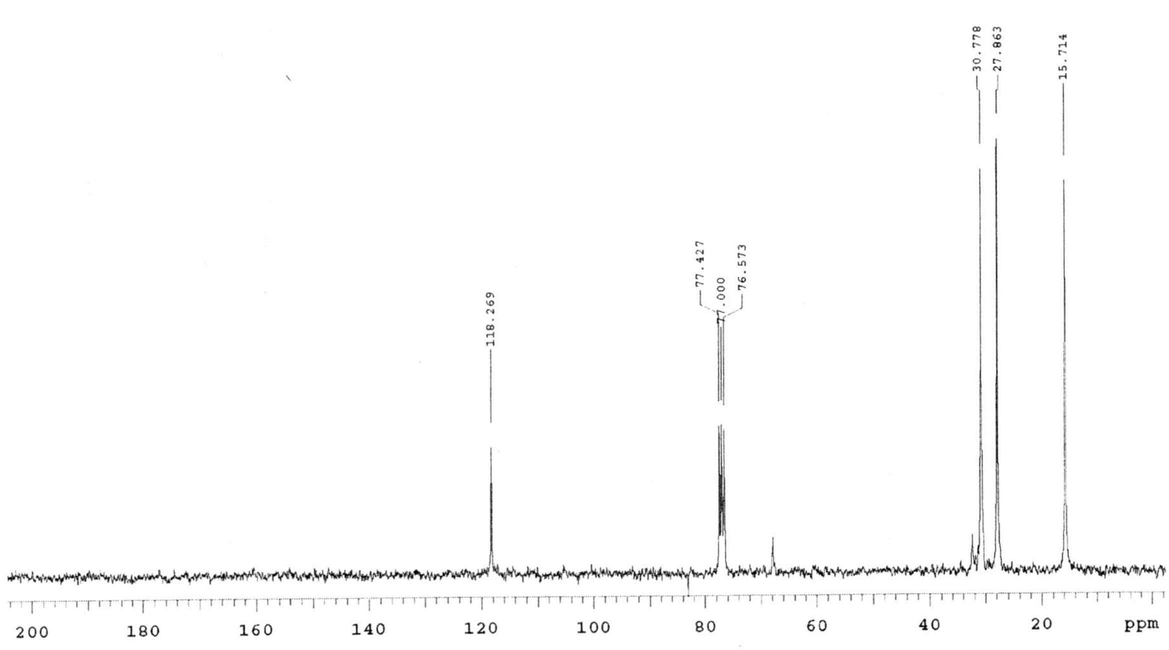

^{13}C-NMR Spectrum in $CDCl_3$

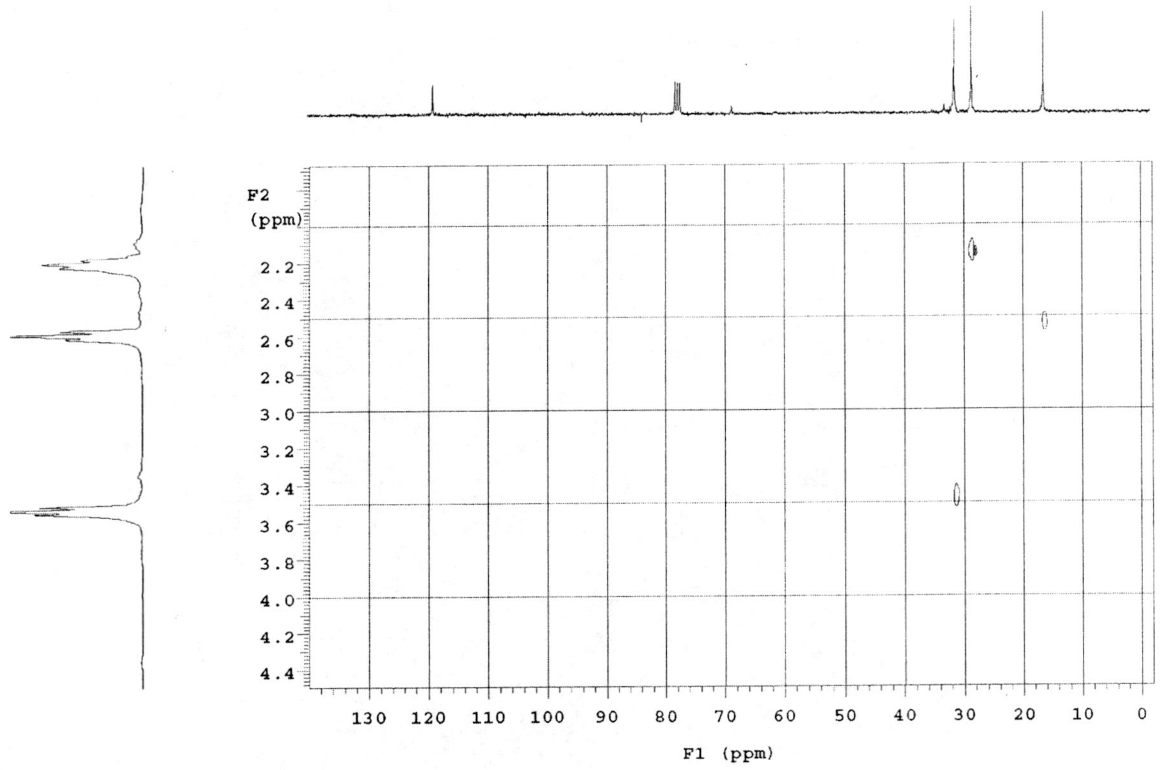

gHSQC Spectrum in $CDCl_3$

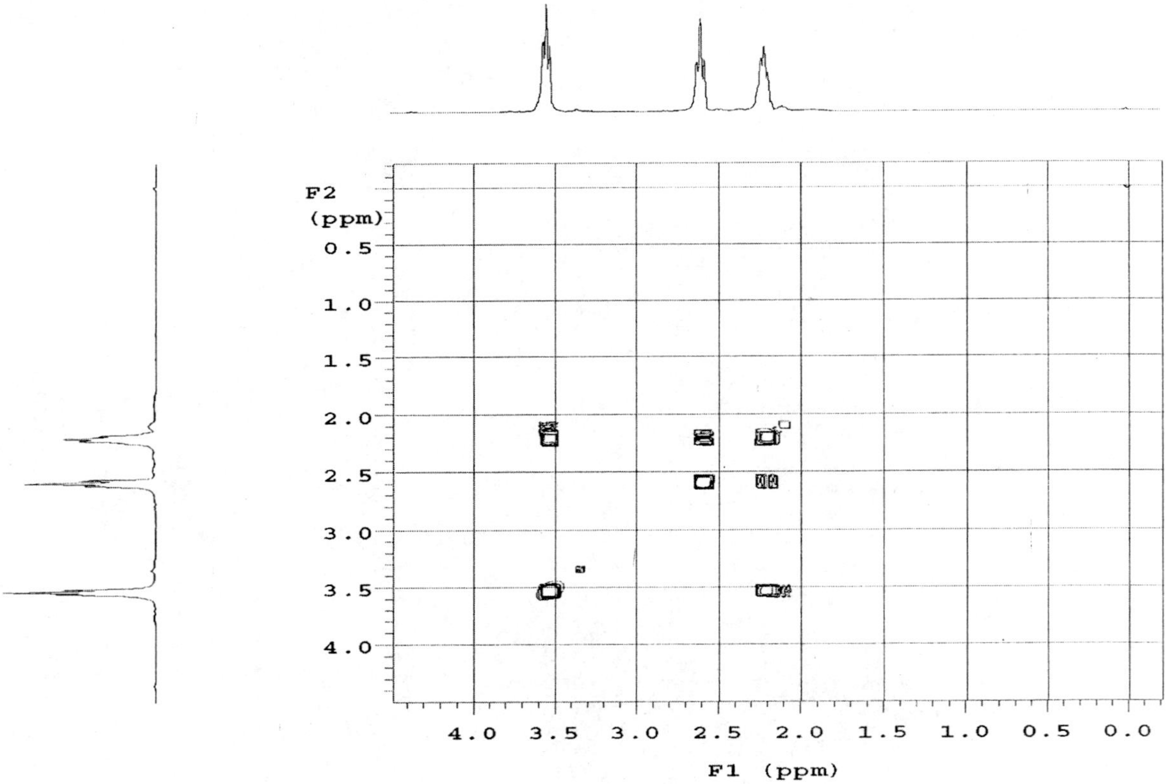

gCOSY Spectrum in $CDCl_3$

Problem 2.13 Given the spectroscopic data below (Proton and Carbon), provide the structure of the molecule. ($C_5H_4O_3$)

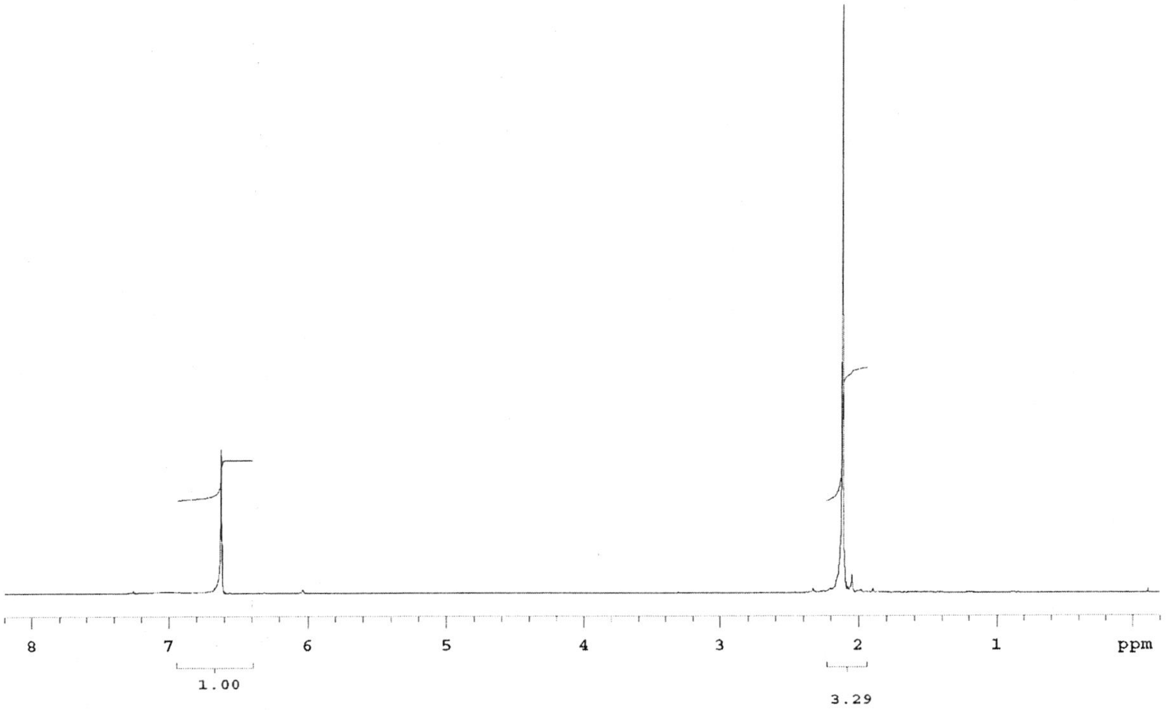

^{1}H-NMR Spectrum in $CDCl_3$

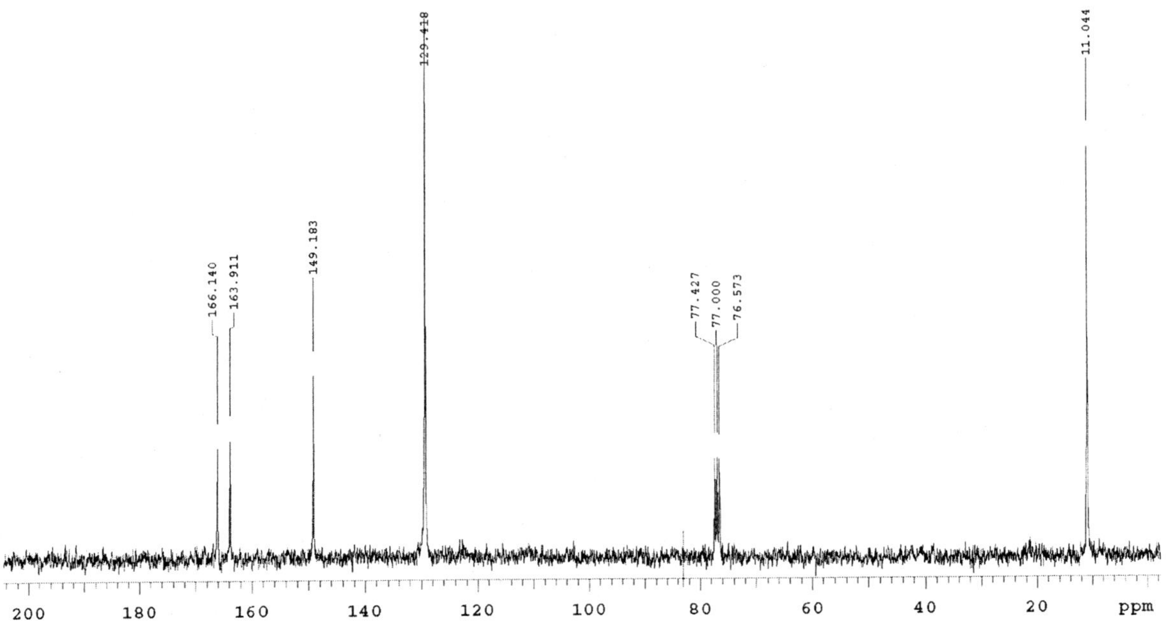

^{13}C-NMR Spectrum in $CDCl_3$

Problem 2.14 Given the spectroscopic data below (Proton, Edited DEPT, gHSQC, and gCOSY), provide the structure of the molecule.

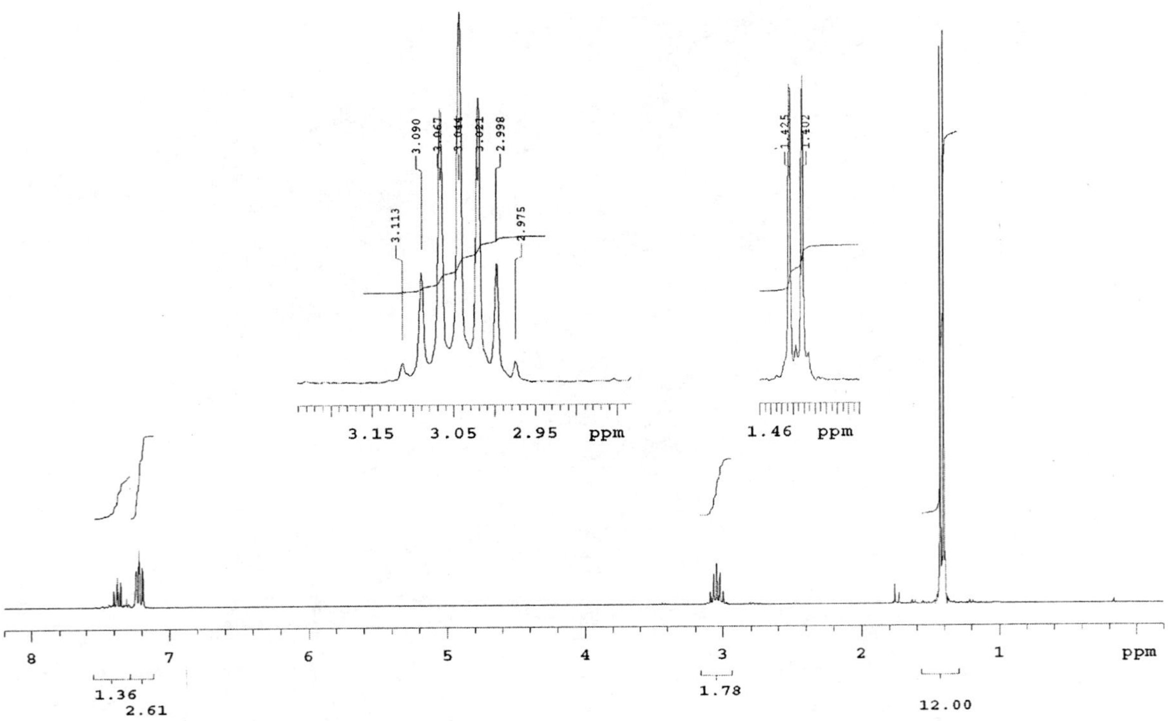

^{1}H-NMR Spectrum in $CDCl_3$

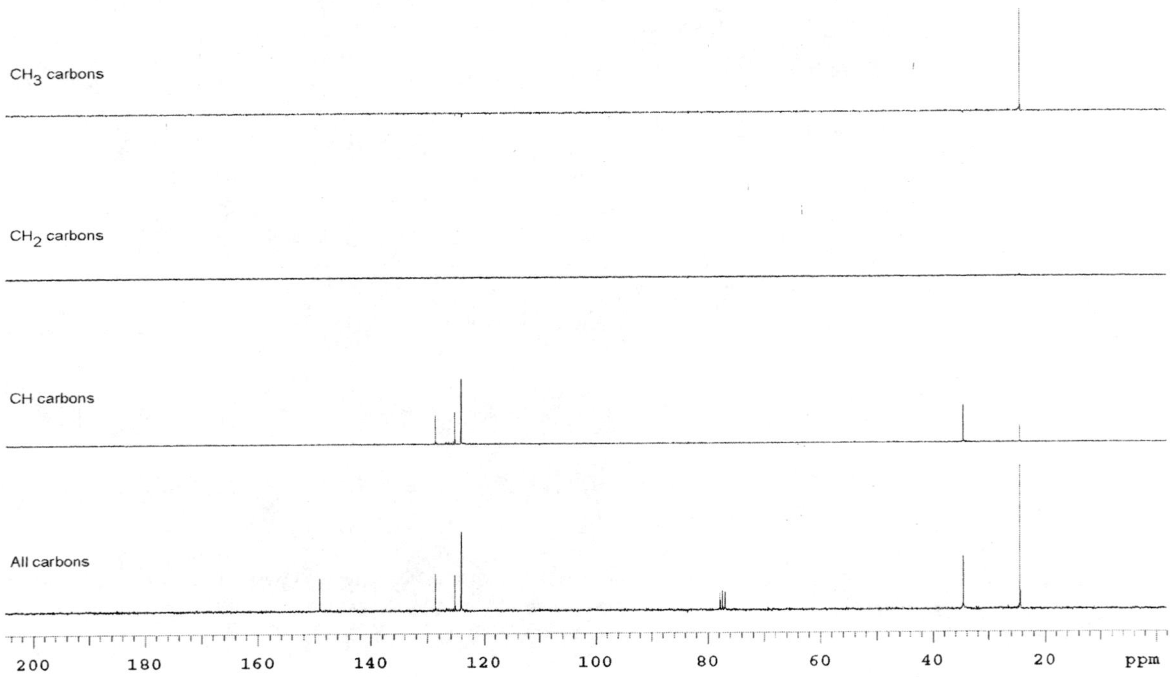

Edited DEPT ^{13}C-NMR Spectrum in $CDCl_3$

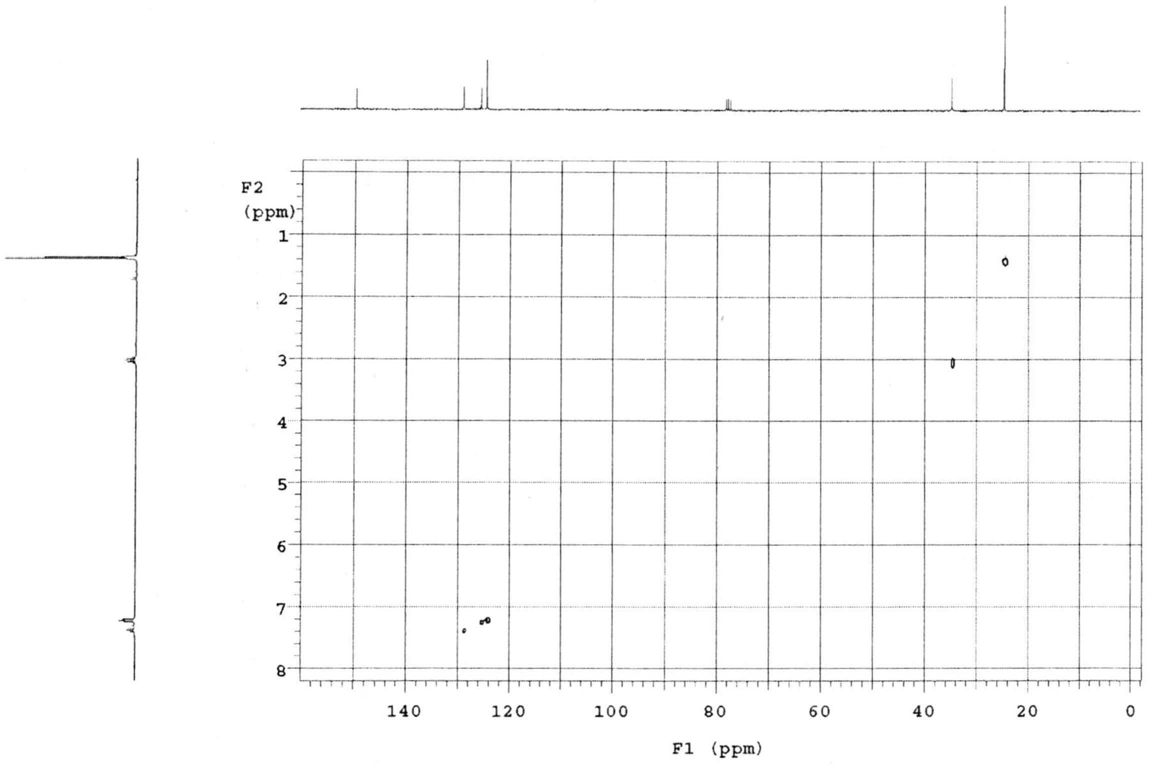

gHSQC Spectrum in $CDCl_3$

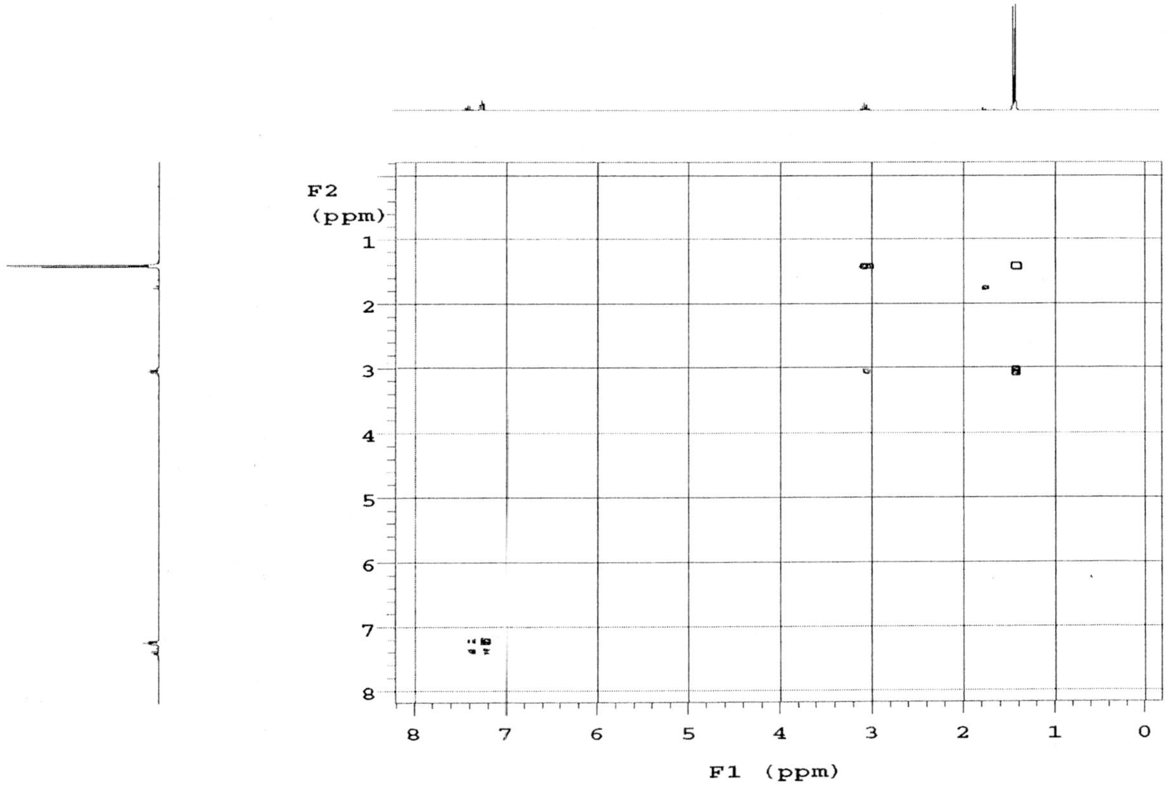

gCOSY Spectrum in $CDCl_3$

Problem 2.15 Given the spectroscopic data below (Proton, Carbon, Edited DEPT, gHSQC, and gCOSY), provide the structure of the molecule.

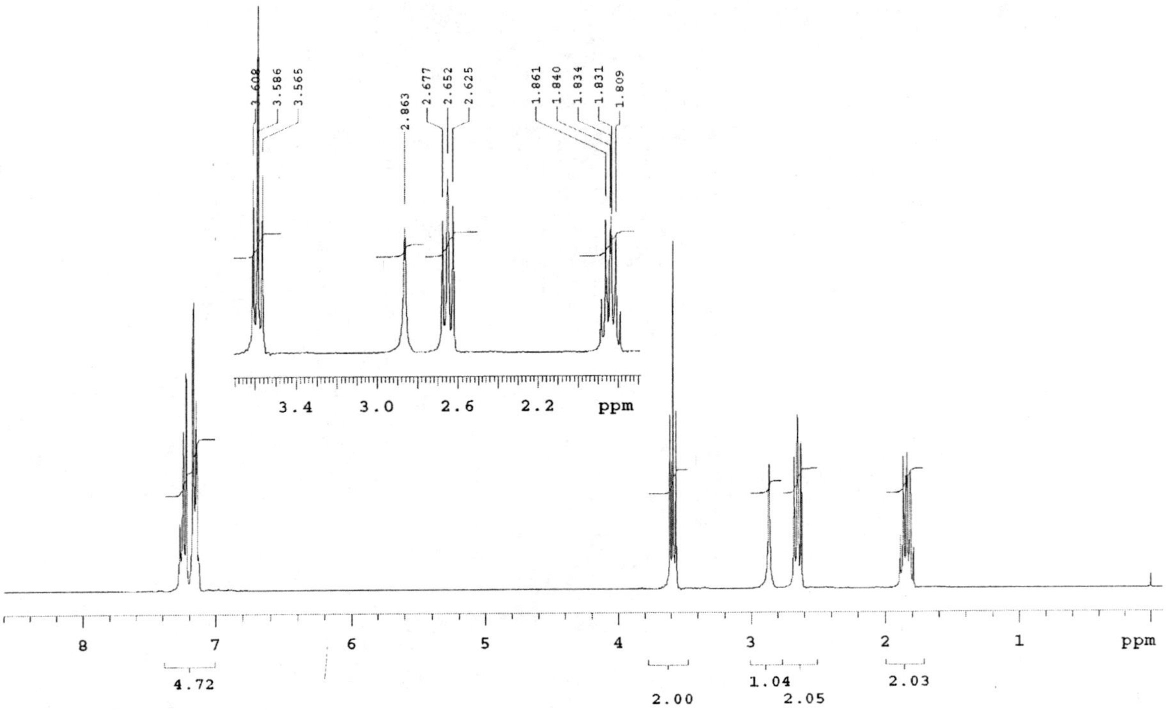

^{1}H-NMR Spectrum in $CDCl_3$

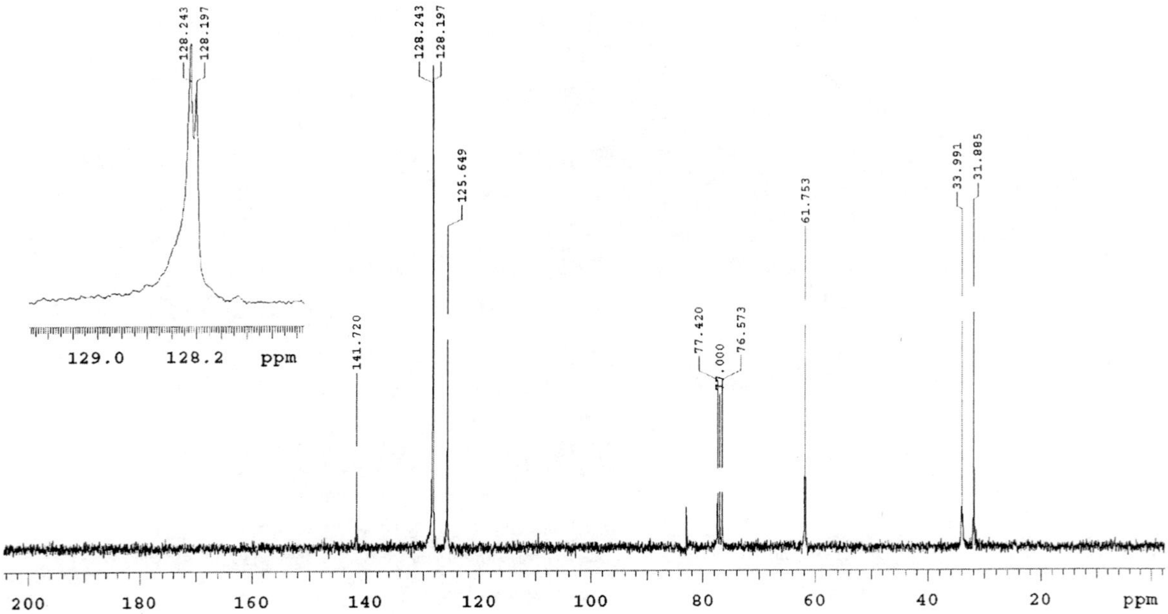

^{13}C-NMR Spectrum in $CDCl_3$

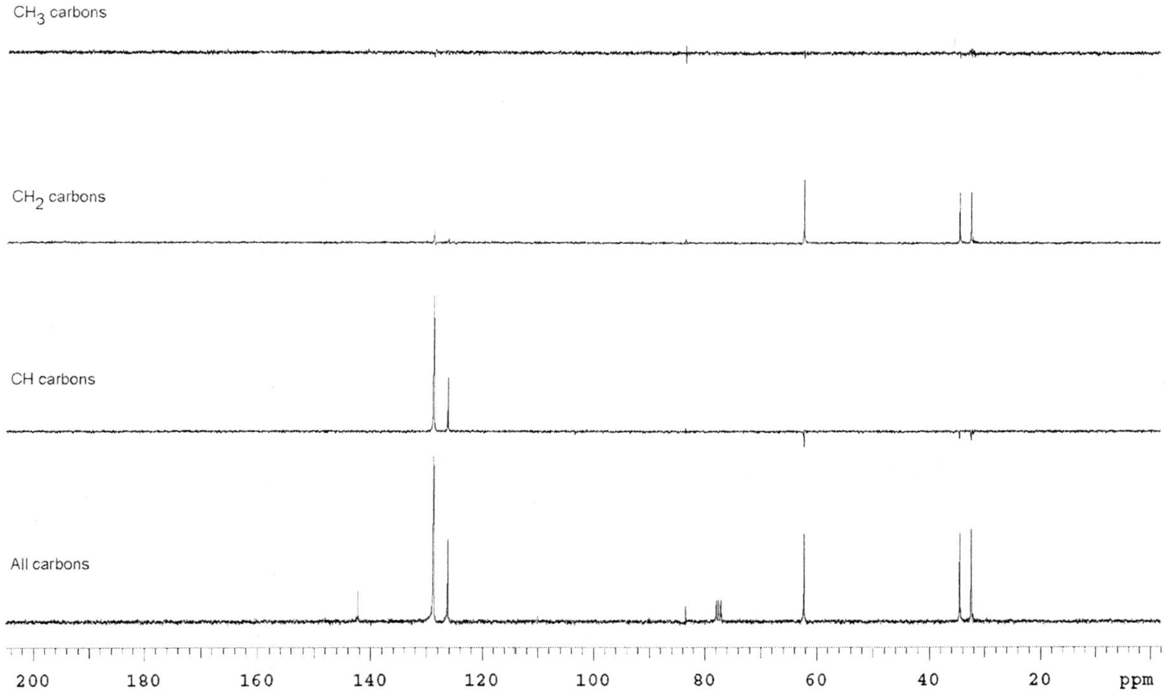

Edited DEPT ^{13}C-NMR Spectrum in $CDCl_3$

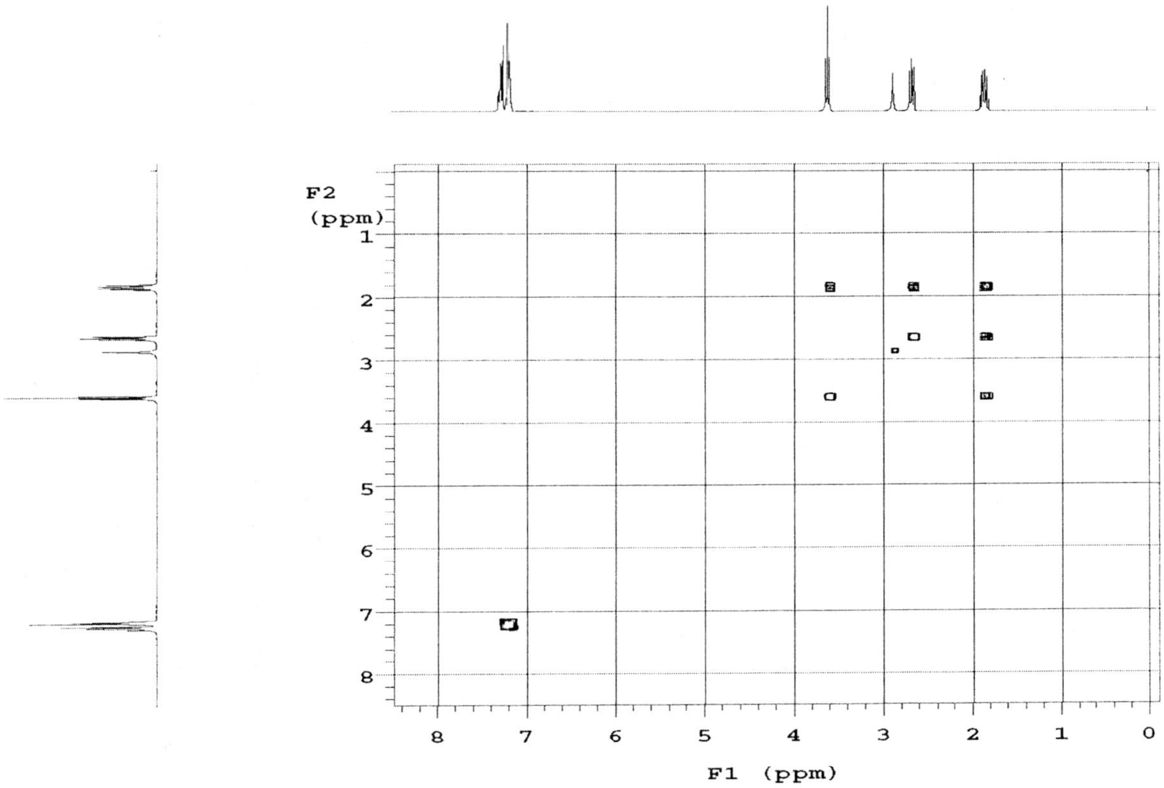

gCOSY Spectrum in $CDCl_3$

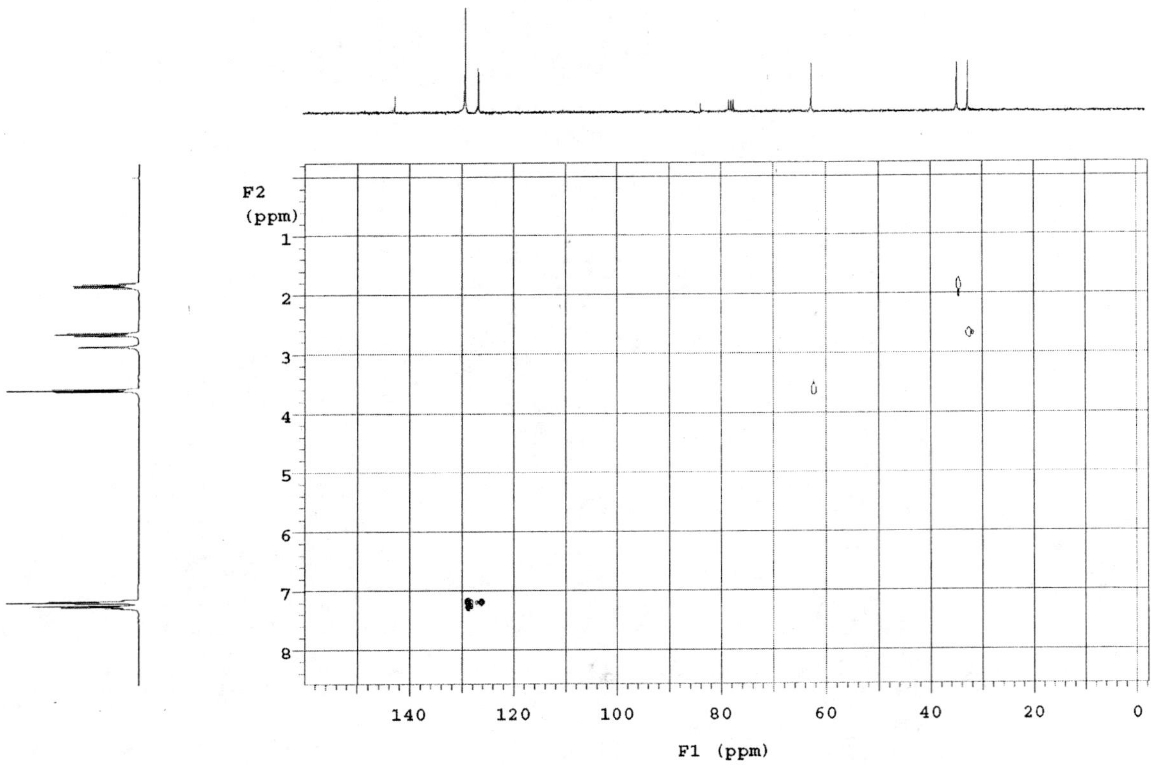

gHSQC Spectrum in $CDCl_3$

Problem 2.16 Given the spectroscopic data below (Proton, Carbon, Edited DEPT, gHSQC, and gCOSY), provide the structure of the molecule.

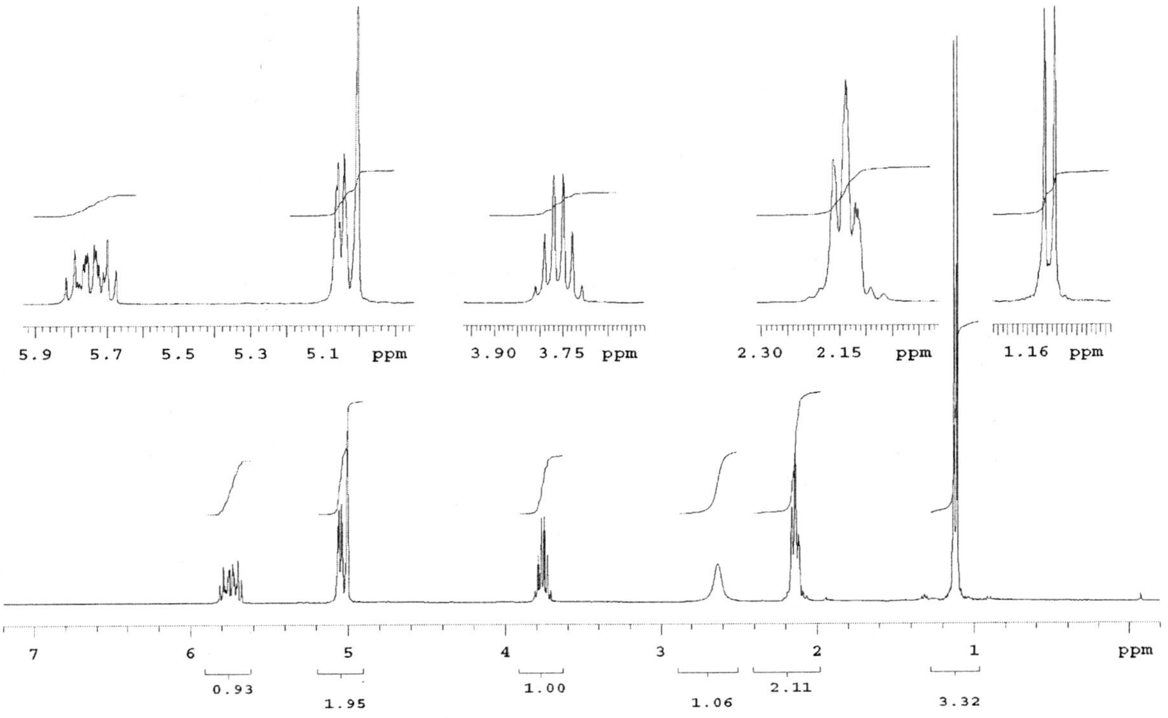

^{1}H-NMR Spectrum in $CDCl_3$

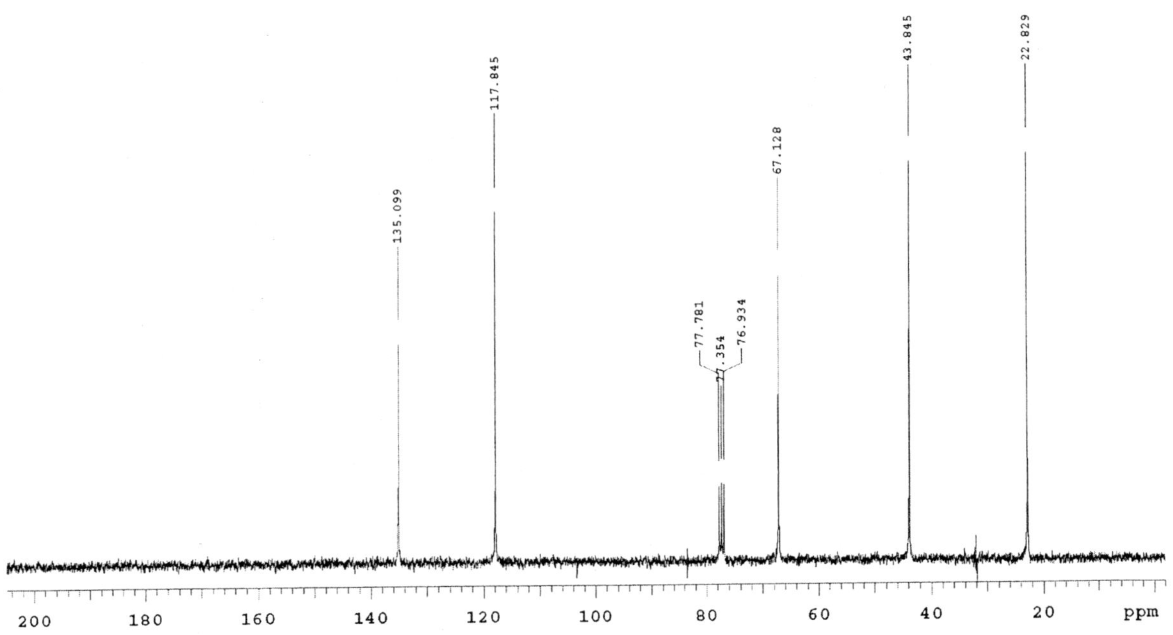

^{13}C-NMR Spectrum in $CDCl_3$

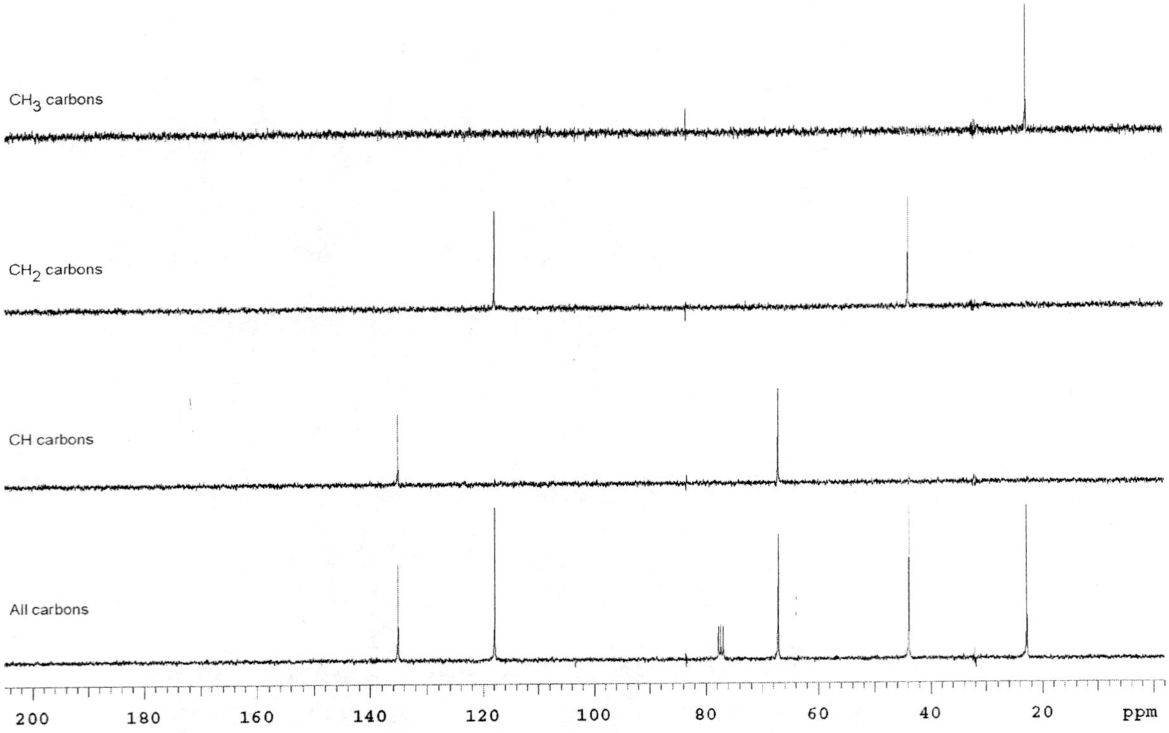

Edited DEPT ^{13}C-NMR Spectrum in $CDCl_3$

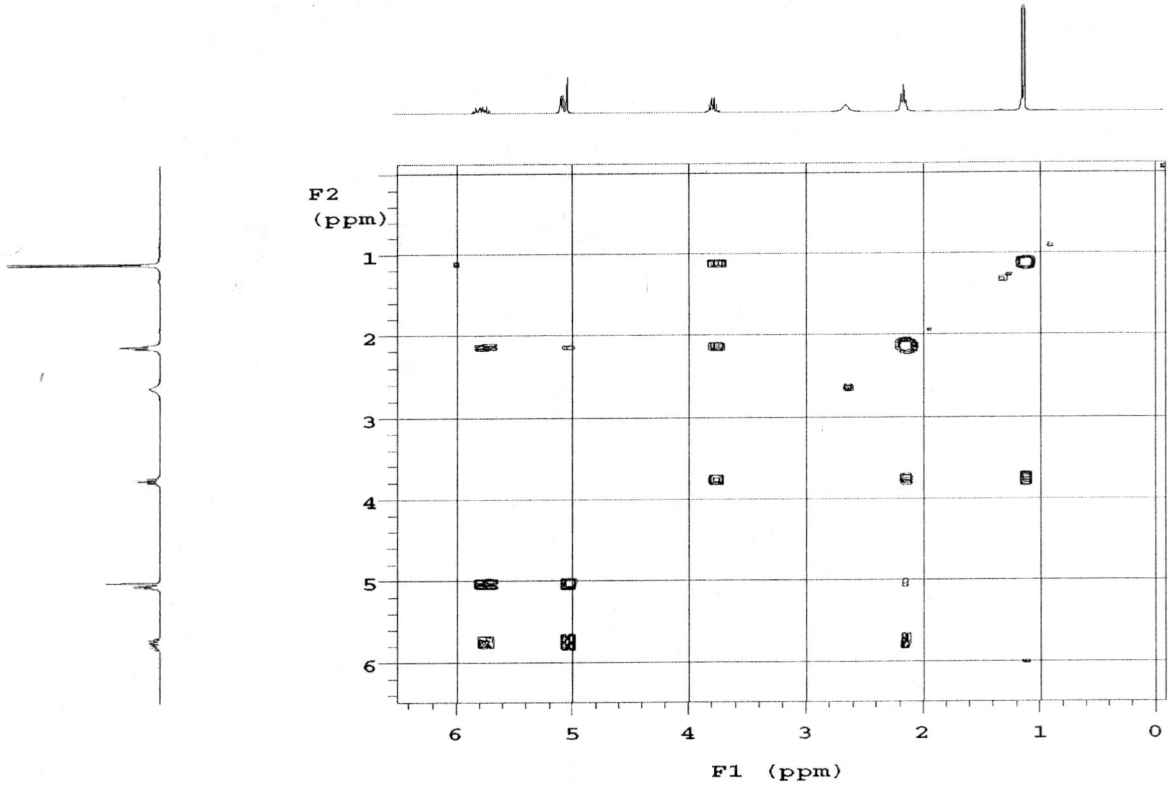

gCOSY Spectrum in $CDCl_3$

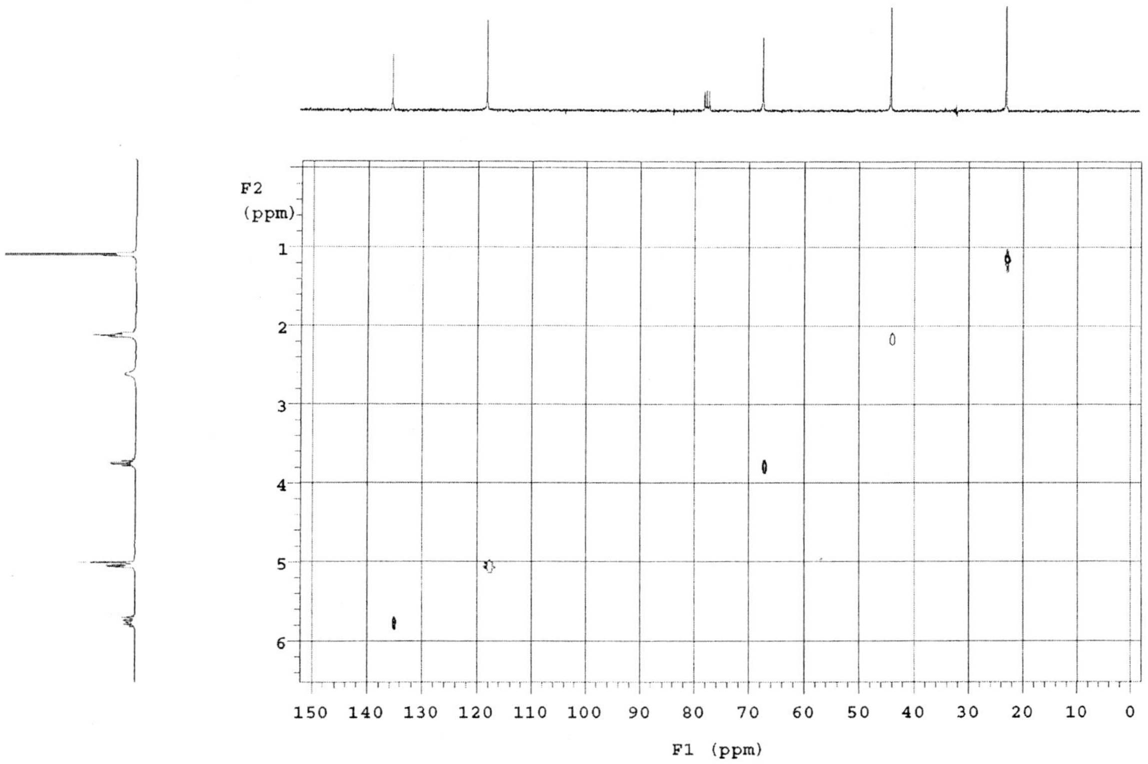

gHSQC Spectrum in $CDCl_3$

Problem 2.17 Given the spectroscopic data below (Proton, Carbon, Edited DEPT, gHSQC, and gCOSY), provide the structure of the molecule. (MW: 135 g/mol)

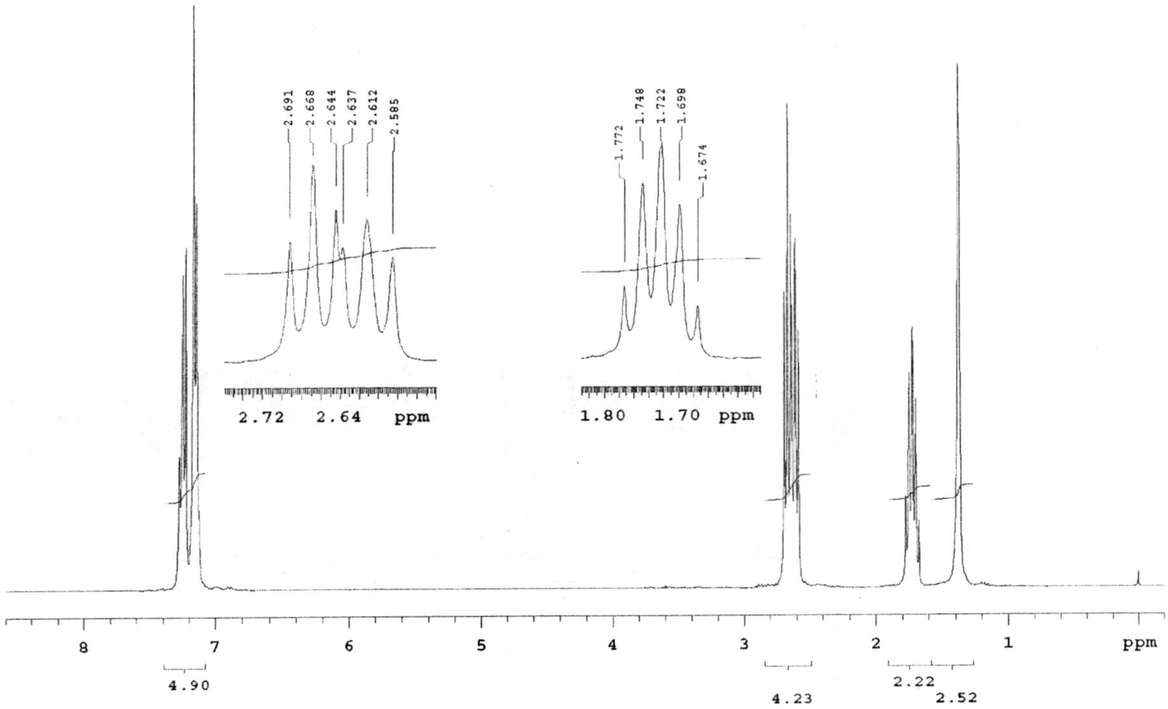

^{1}H-NMR Spectrum in $CDCl_3$

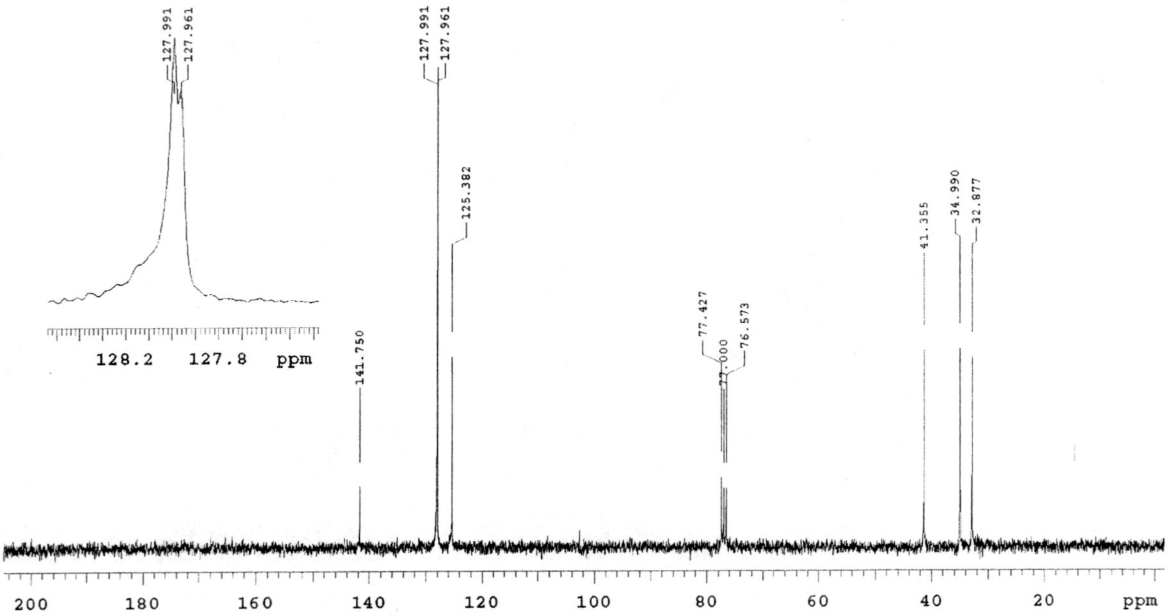

^{13}C-NMR Spectrum in $CDCl_3$

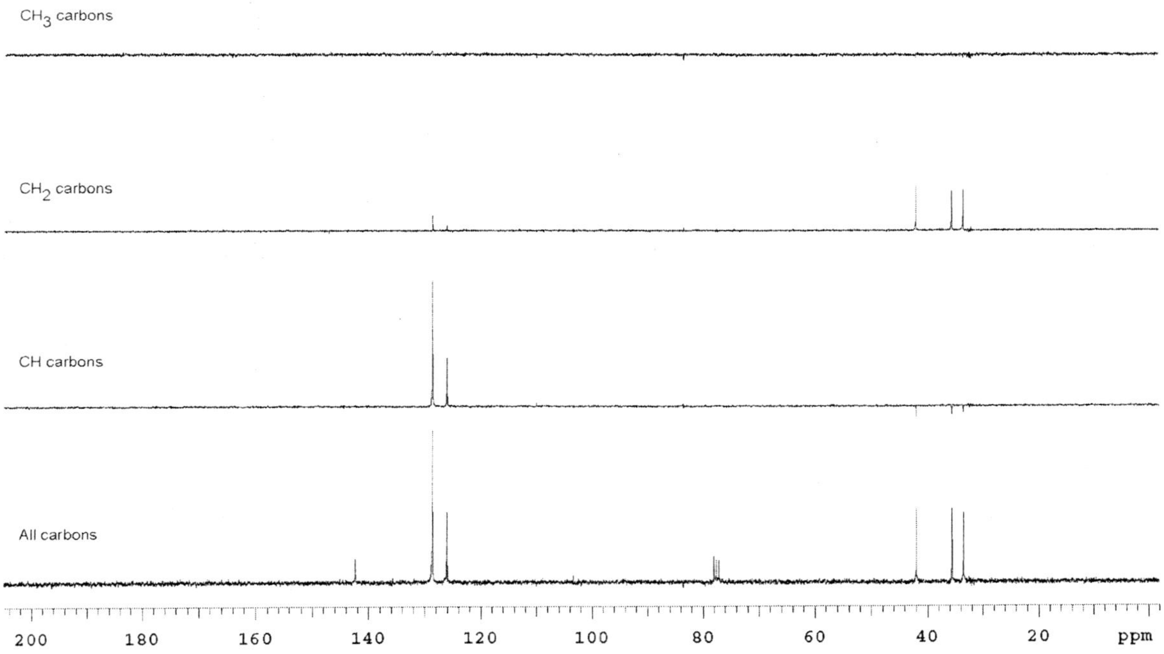

Edited DEPT ^{13}C-NMR Spectrum in $CDCl_3$

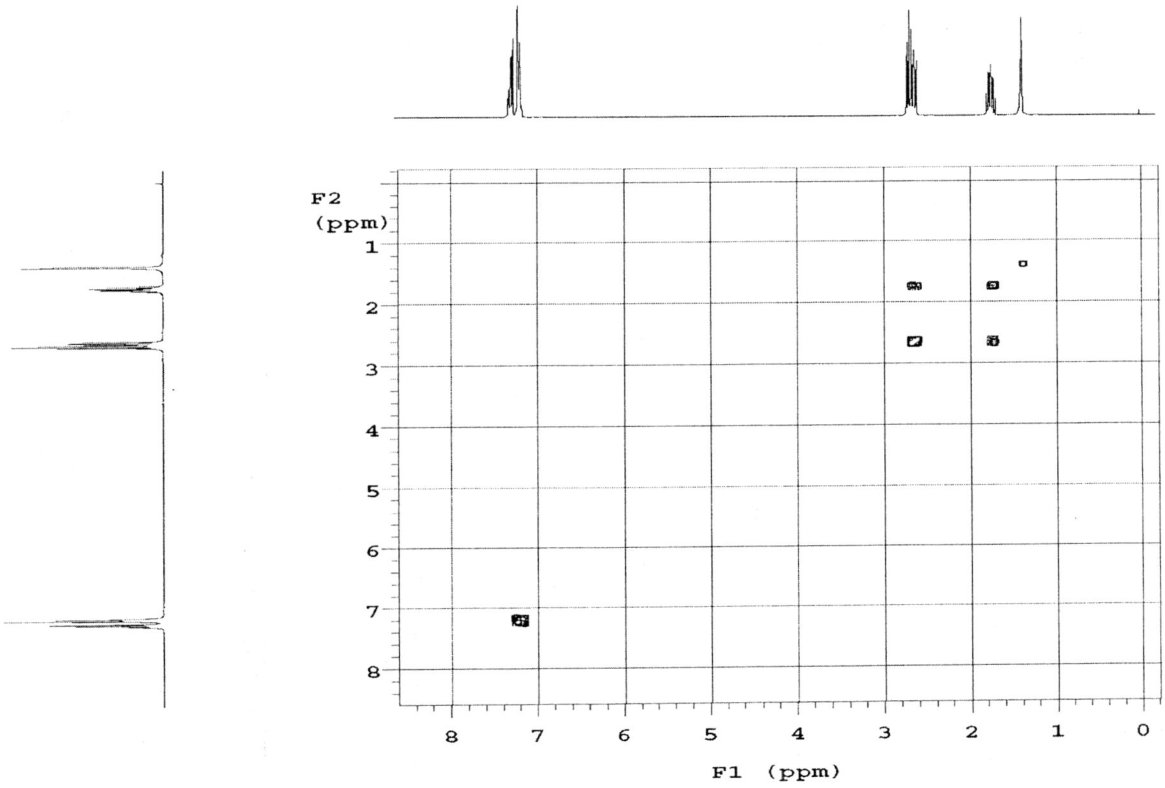

gCOSY Spectrum in $CDCl_3$

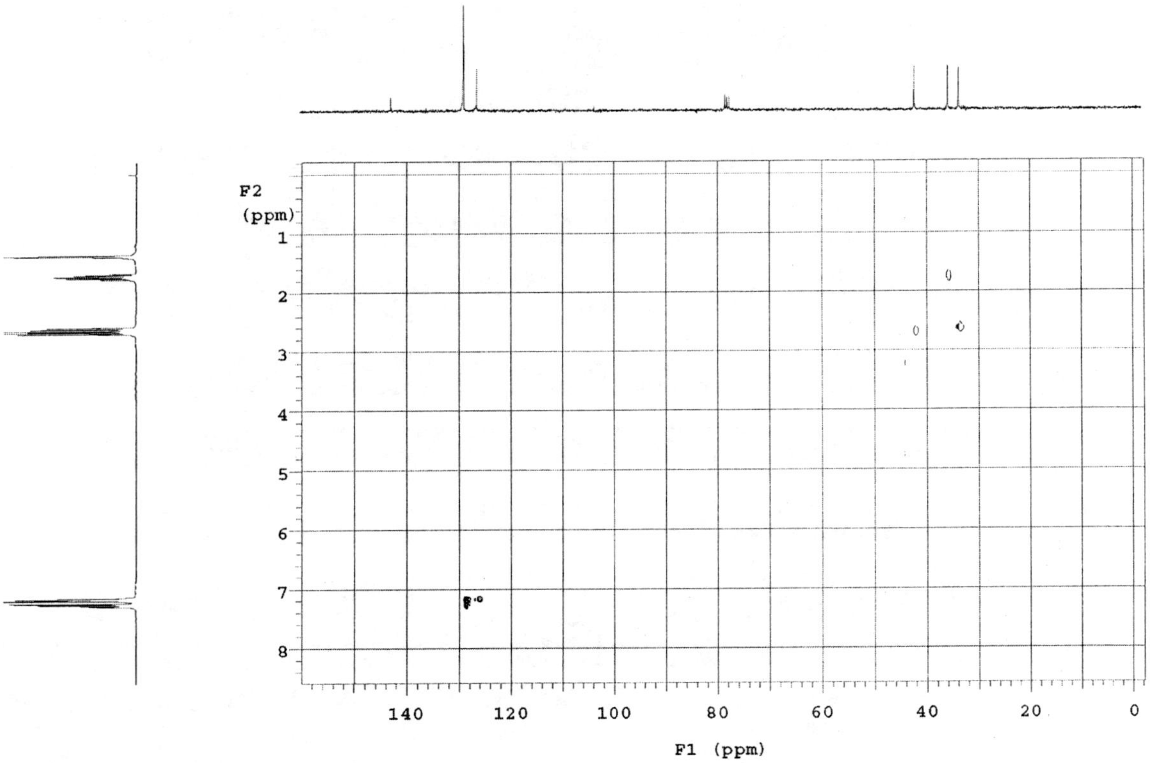

gHSQC Spectrum in $CDCl_3$

Problem 2.18 Given the spectroscopic data below (Proton, Carbon, gHMQC, and gCOSY), provide the structure of the molecule.

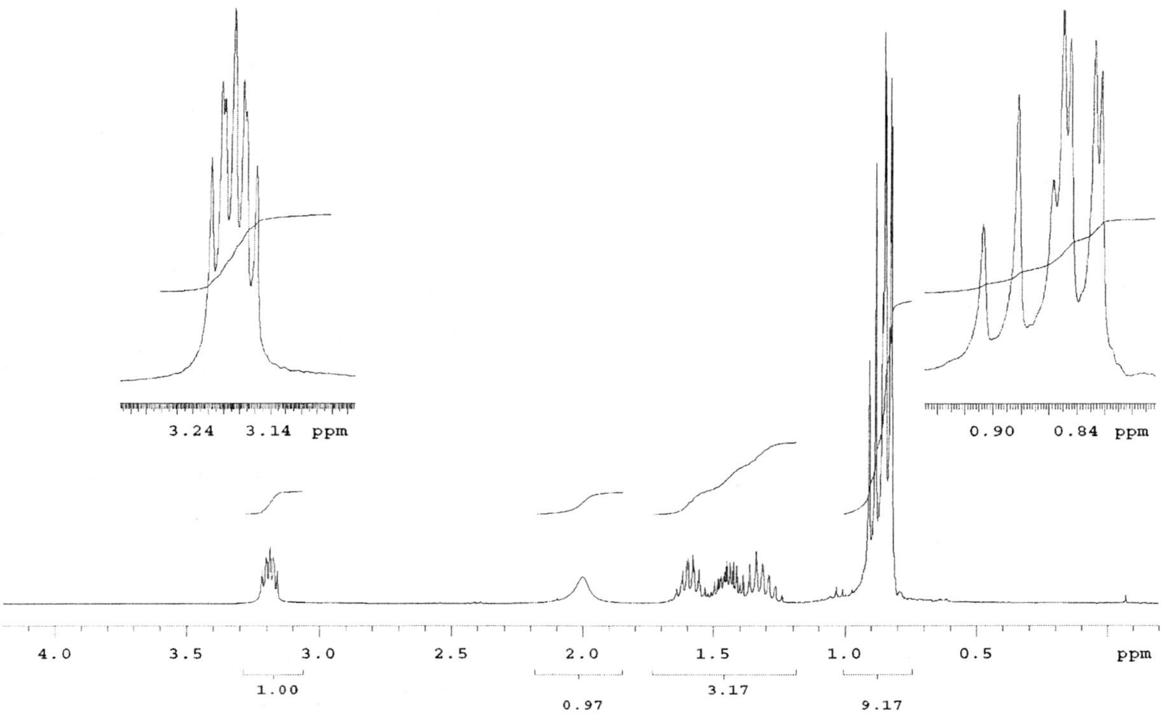

^{1}H-NMR Spectrum in $CDCl_3$

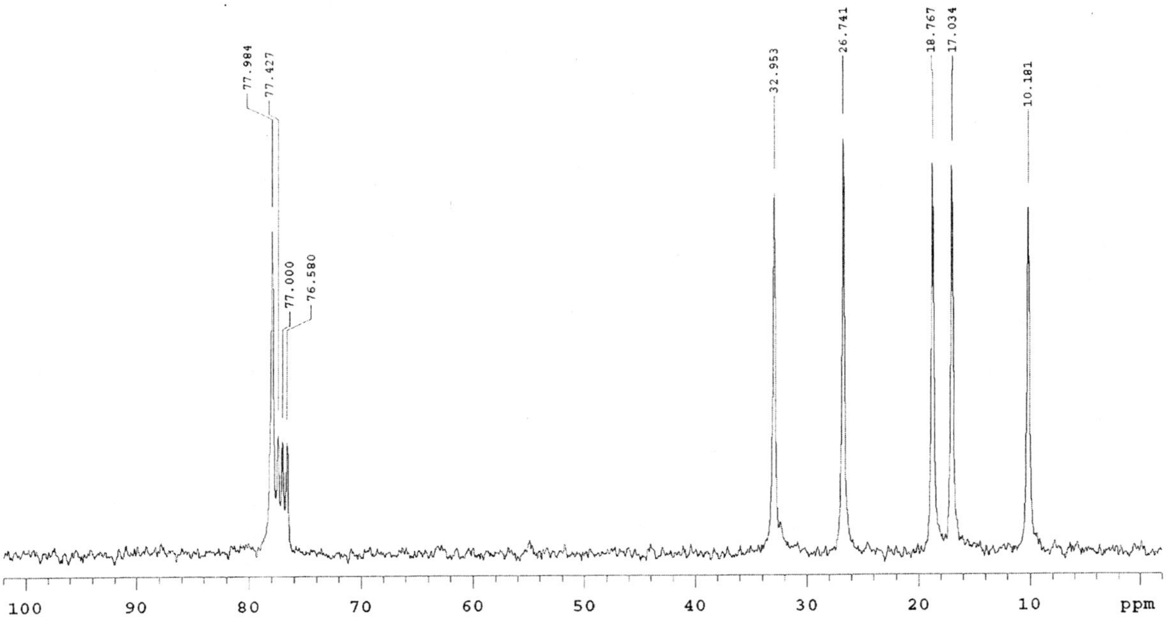

^{13}C-NMR Spectrum in $CDCl_3$

Problem 2.19 Given the spectroscopic data below (Proton, Carbon, Edited DEPT, gHSQC, and gCOSY), provide the structure of the molecule.

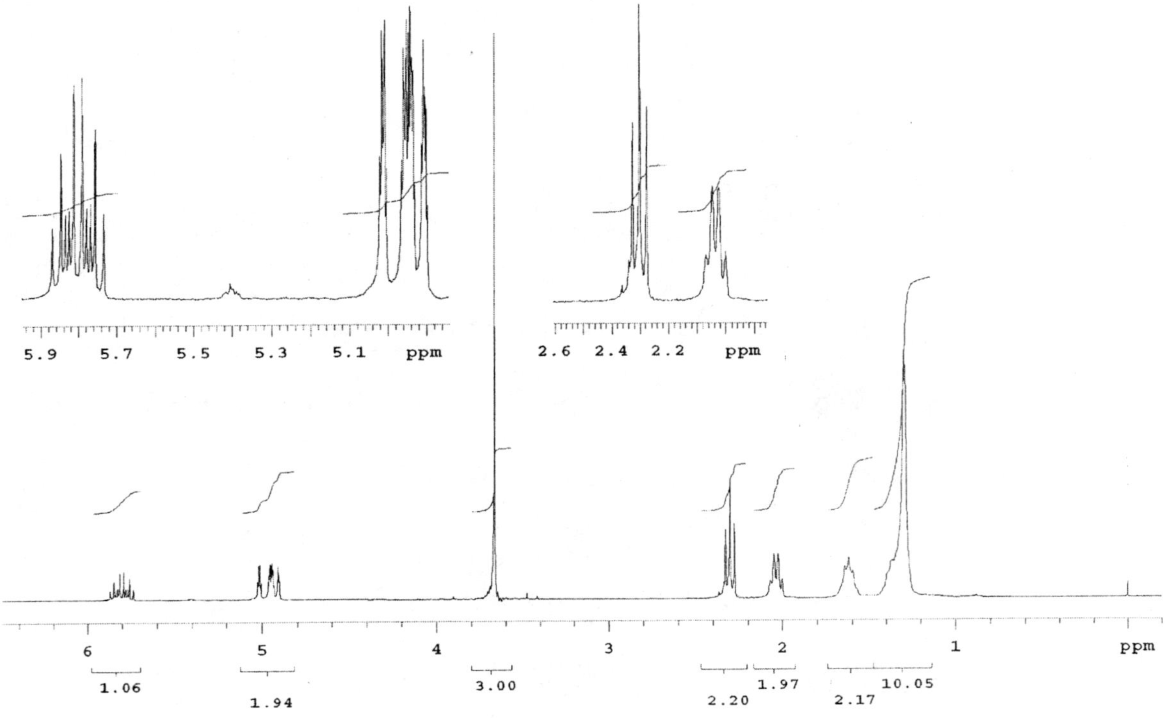

^{1}H-NMR Spectrum in $CDCl_3$

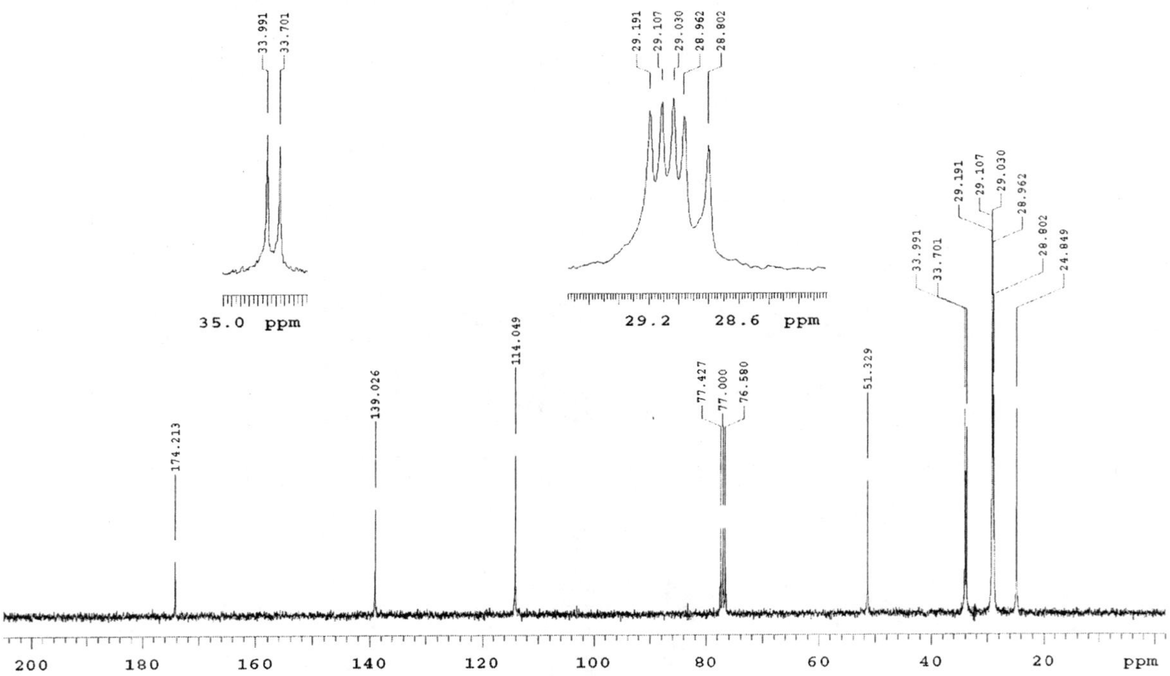

^{13}C-NMR Spectrum in $CDCl_3$

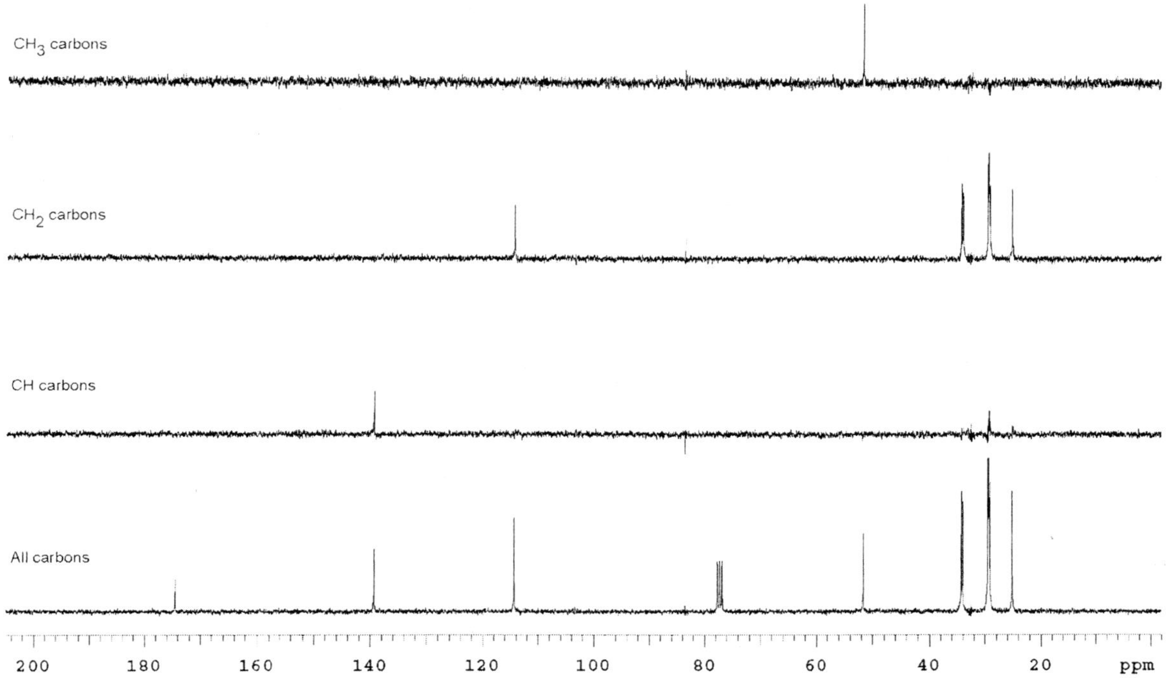

Edited DEPT ^{13}C-NMR Spectrum in $CDCl_3$

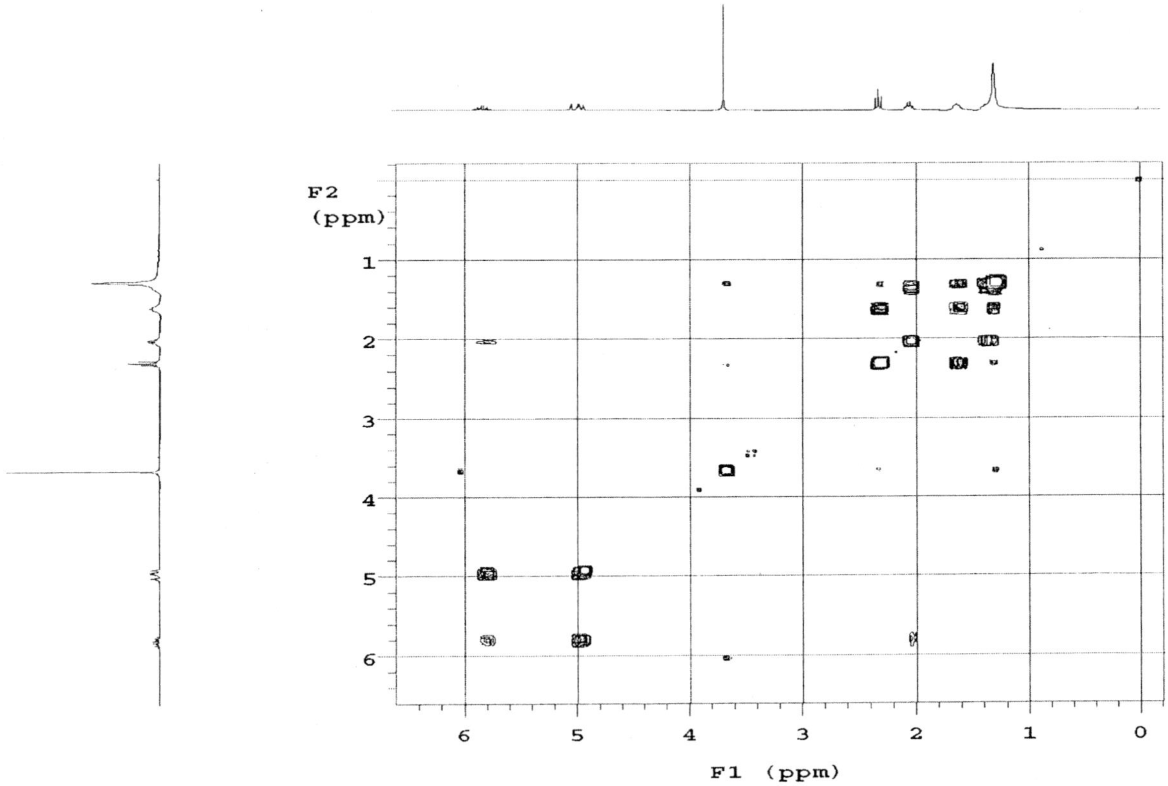

gCOSY Spectrum in $CDCl_3$

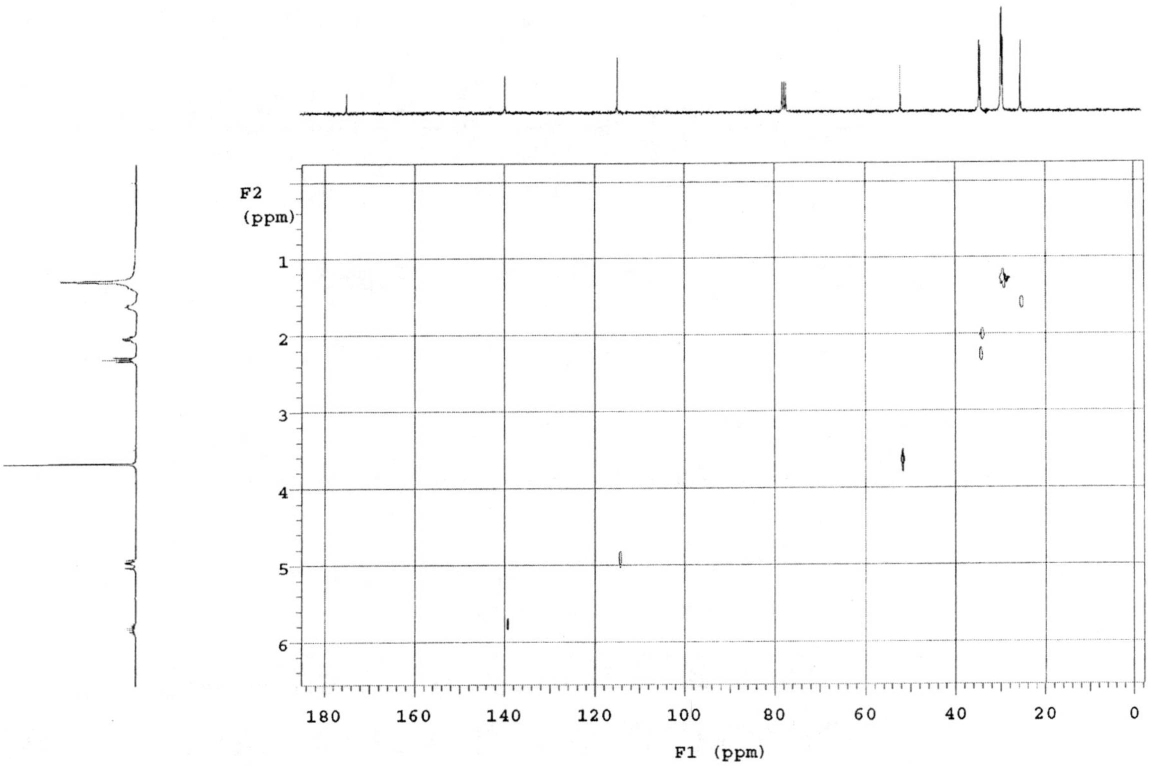

gHSQC Spectrum in $CDCl_3$

Problem 2.20 Given the spectroscopic data below (Proton, Carbon, gHSQC, and gCOSY), provide the structure of the molecule.

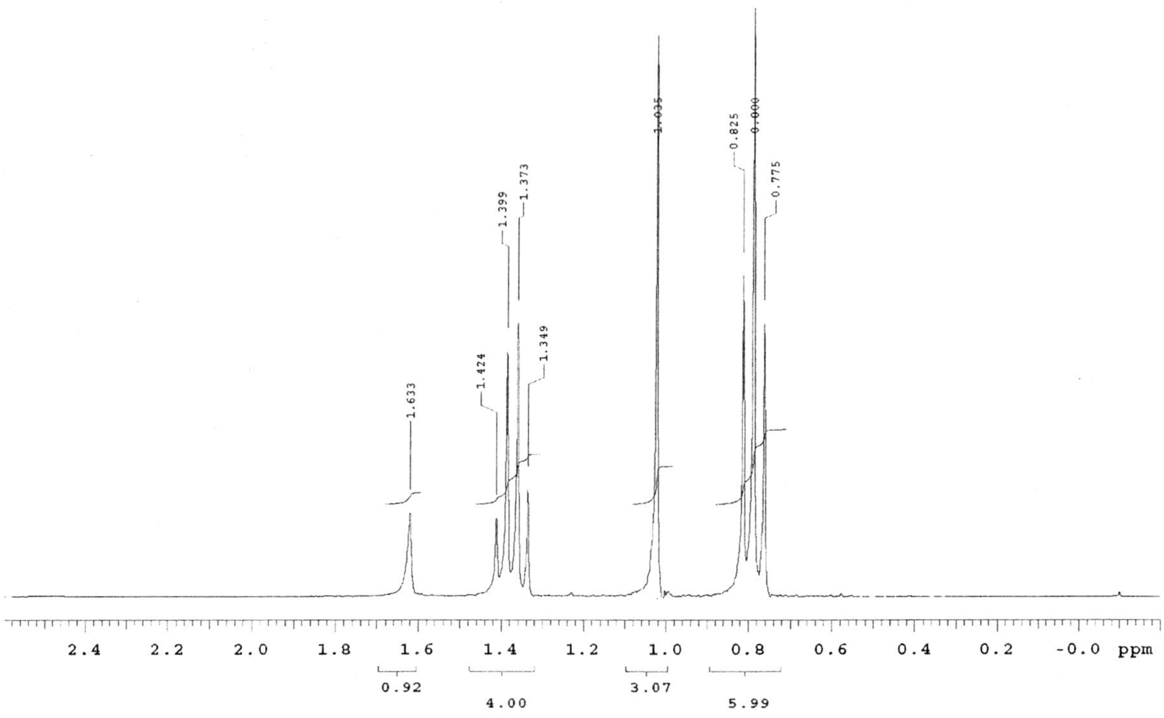

^{1}H-NMR Spectrum in $CDCl_3$

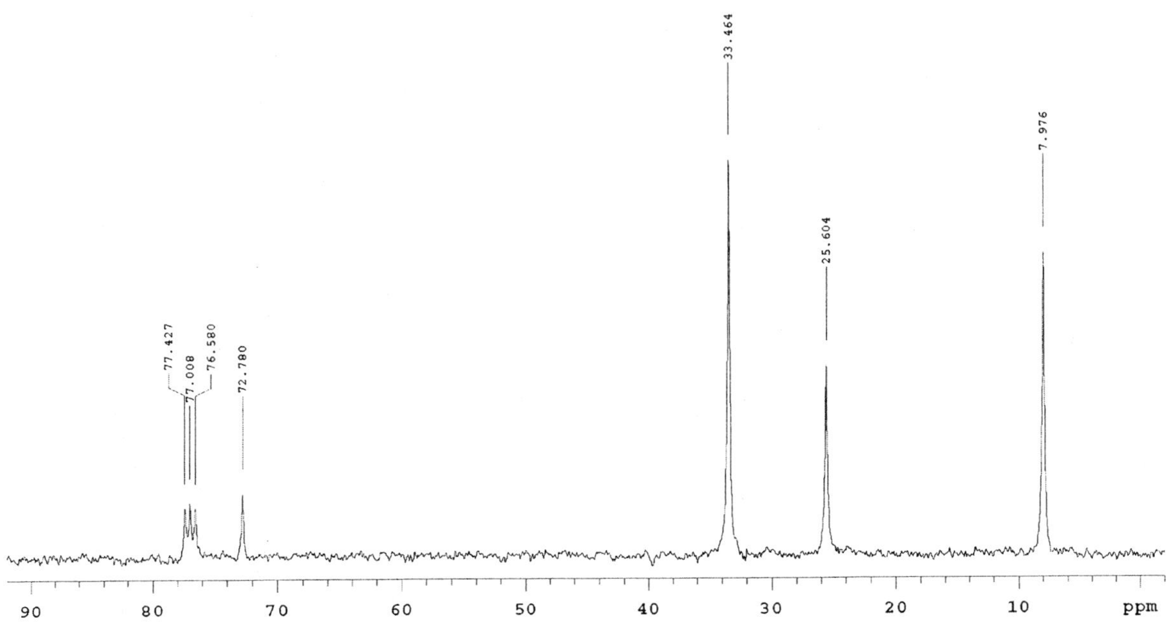

^{13}C-NMR Spectrum in $CDCl_3$

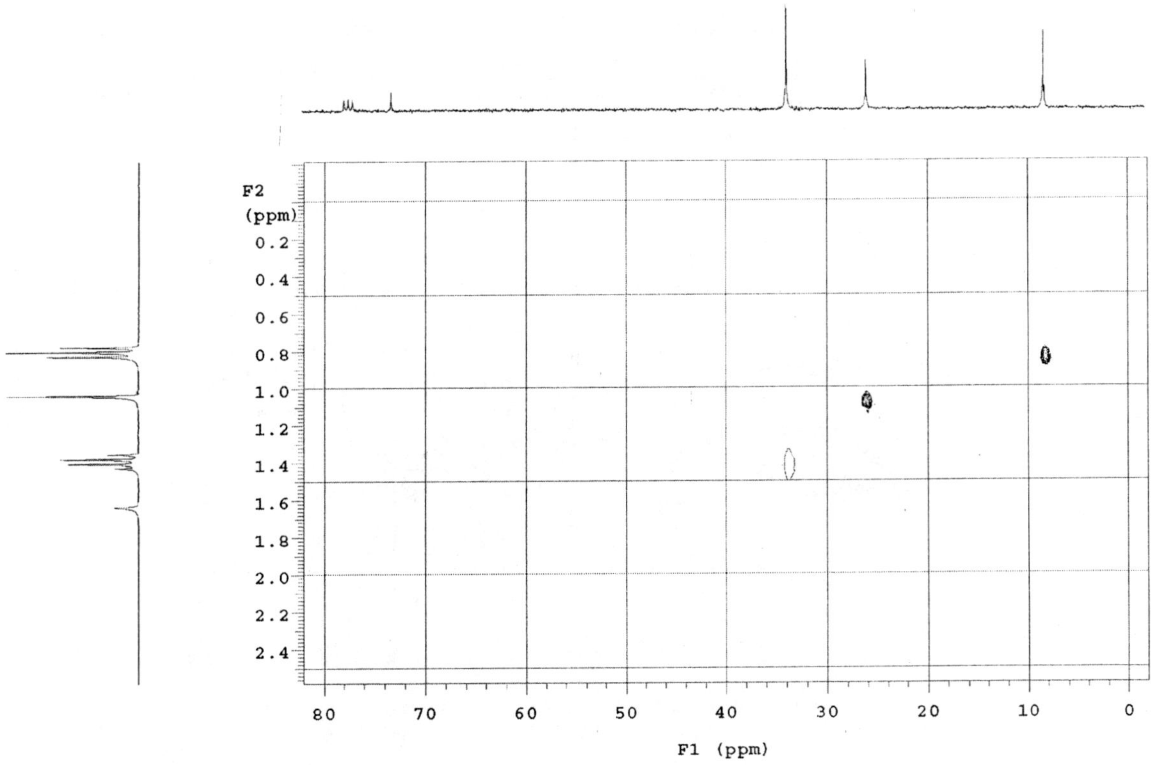

gHSQC Spectrum in $CDCl_3$

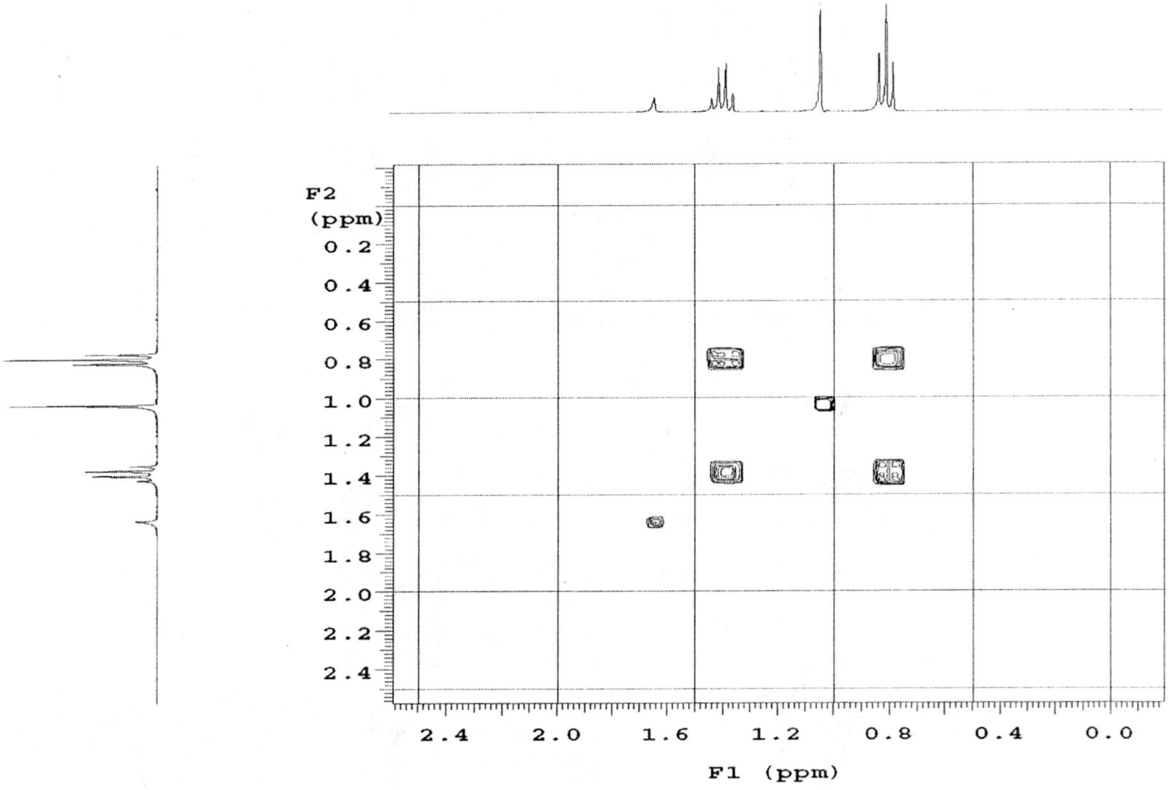

gCOSY Spectrum in $CDCl_3$

Problem 2.21 Given the spectroscopic data below (Proton, Carbon, Edited DEPT, gHSQC, and gCOSY), provide the structure of the molecule.

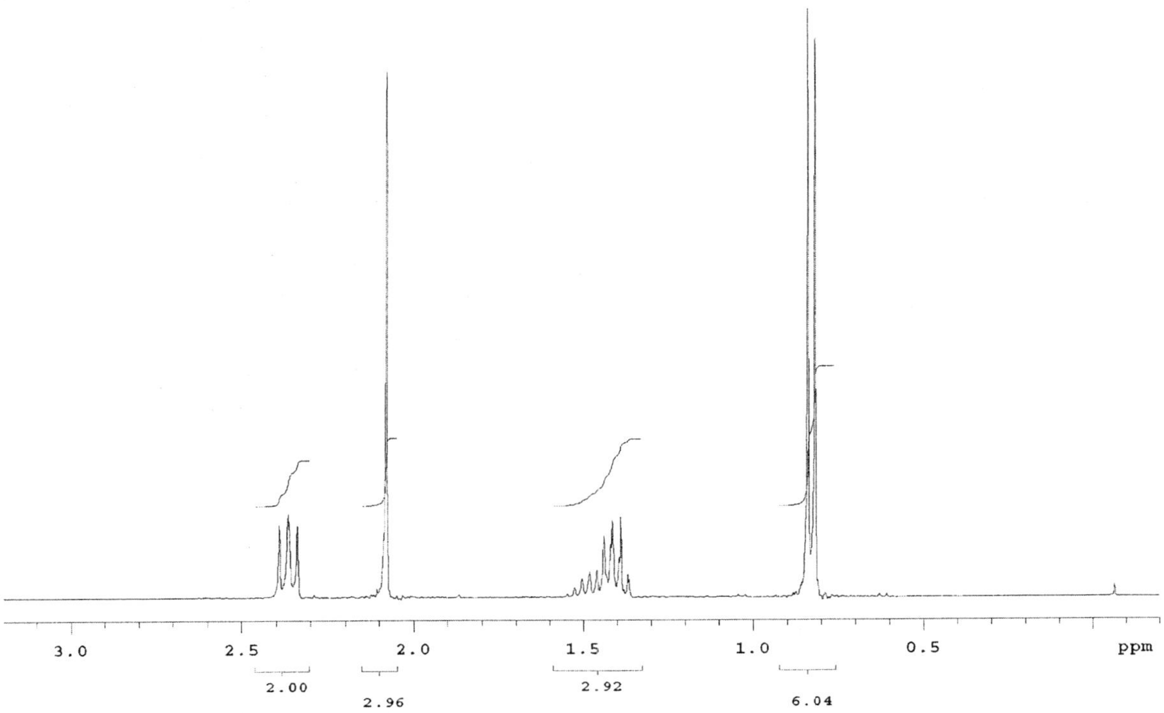

^{1}H-NMR Spectrum in $CDCl_3$

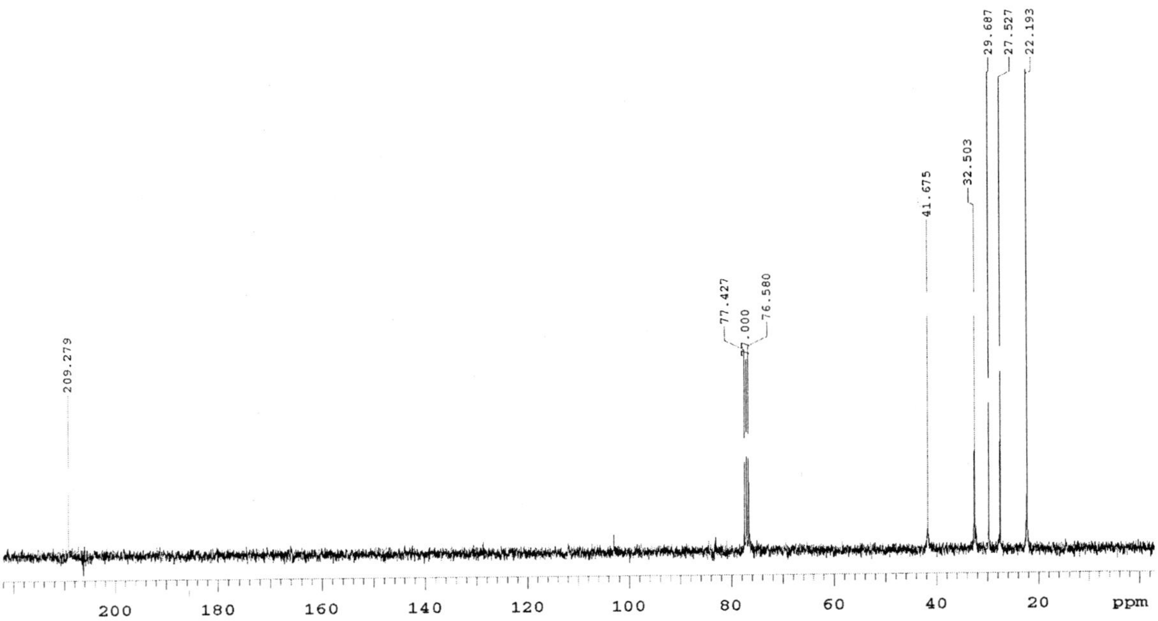

^{13}C-NMR Spectrum in $CDCl_3$

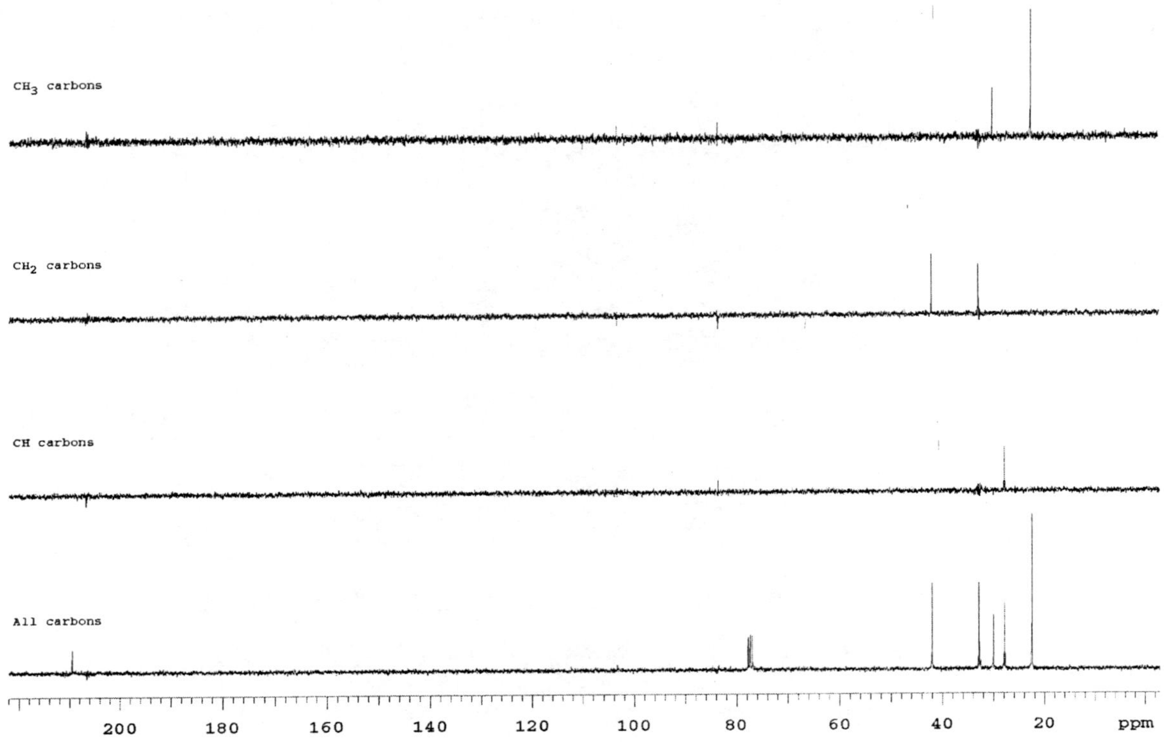

Edited DEPT ^{13}C-NMR Spectrum in $CDCl_3$

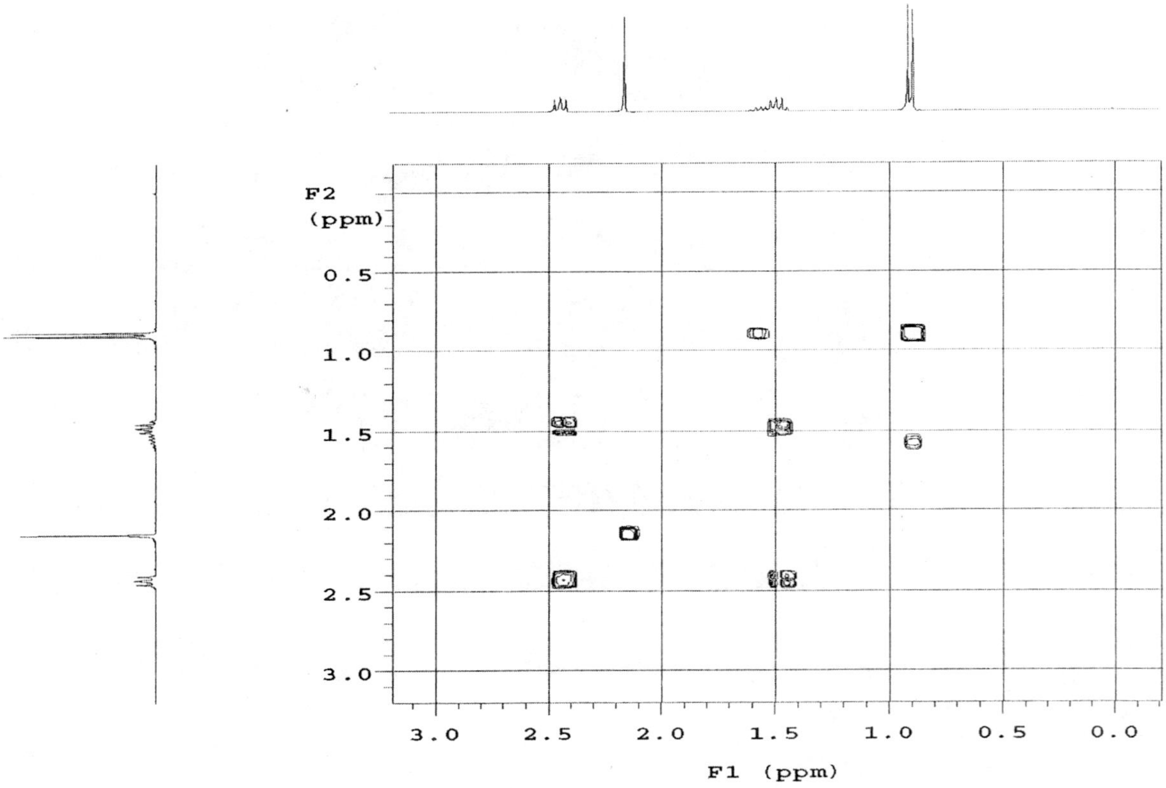

gCOSY Spectrum in $CDCl_3$

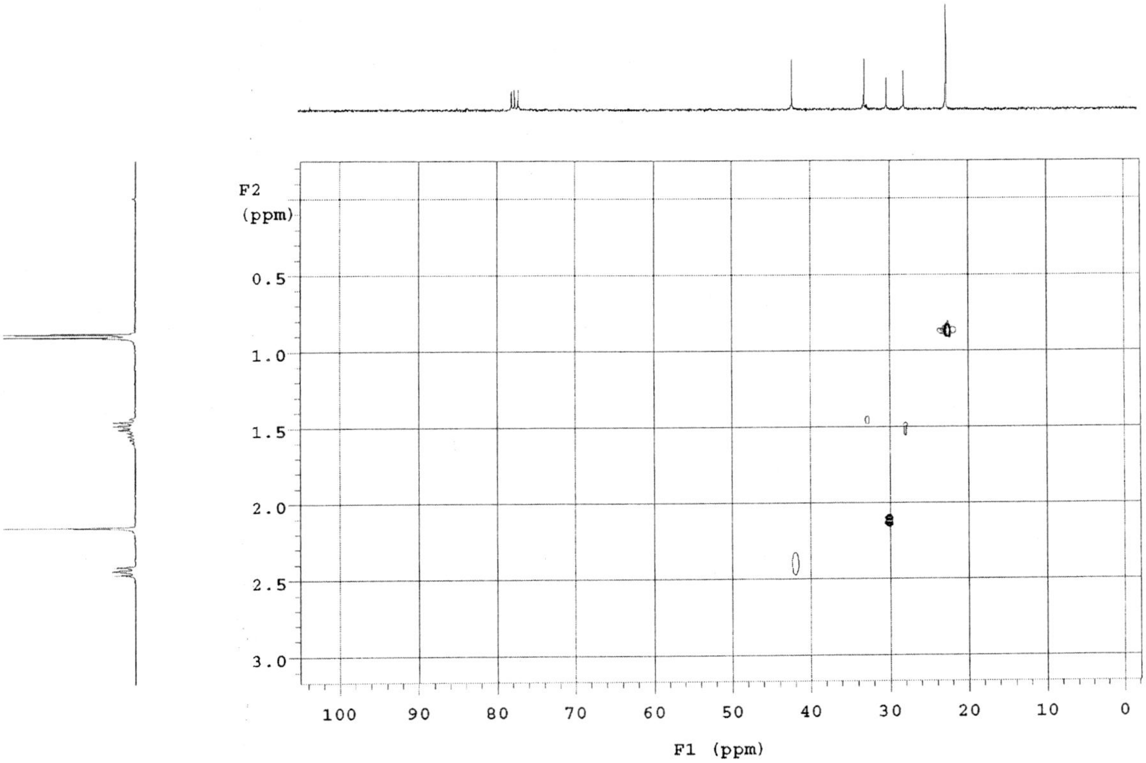

gHSQC Spectrum in $CDCl_3$

Problem 2.22 Given the spectroscopic data below (Proton, Carbon, gHSQC, and gCOSY), provide the structure of the molecule. (MW = 75 g/mol)

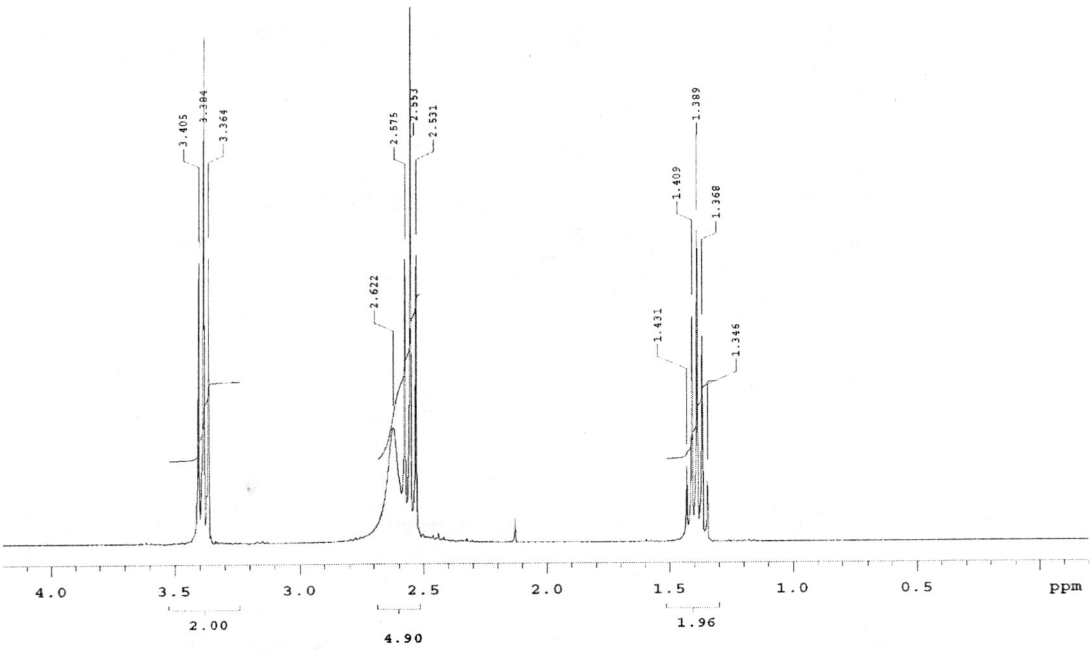

^{1}H-NMR Spectrum in $CDCl_3$

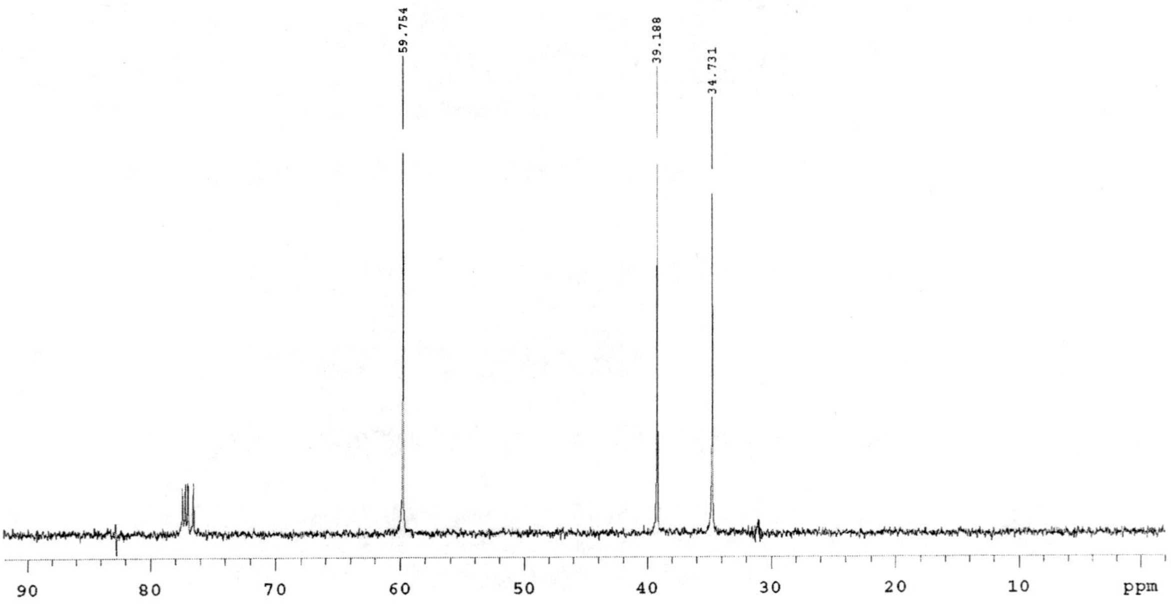

^{13}C-NMR Spectrum in $CDCl_3$

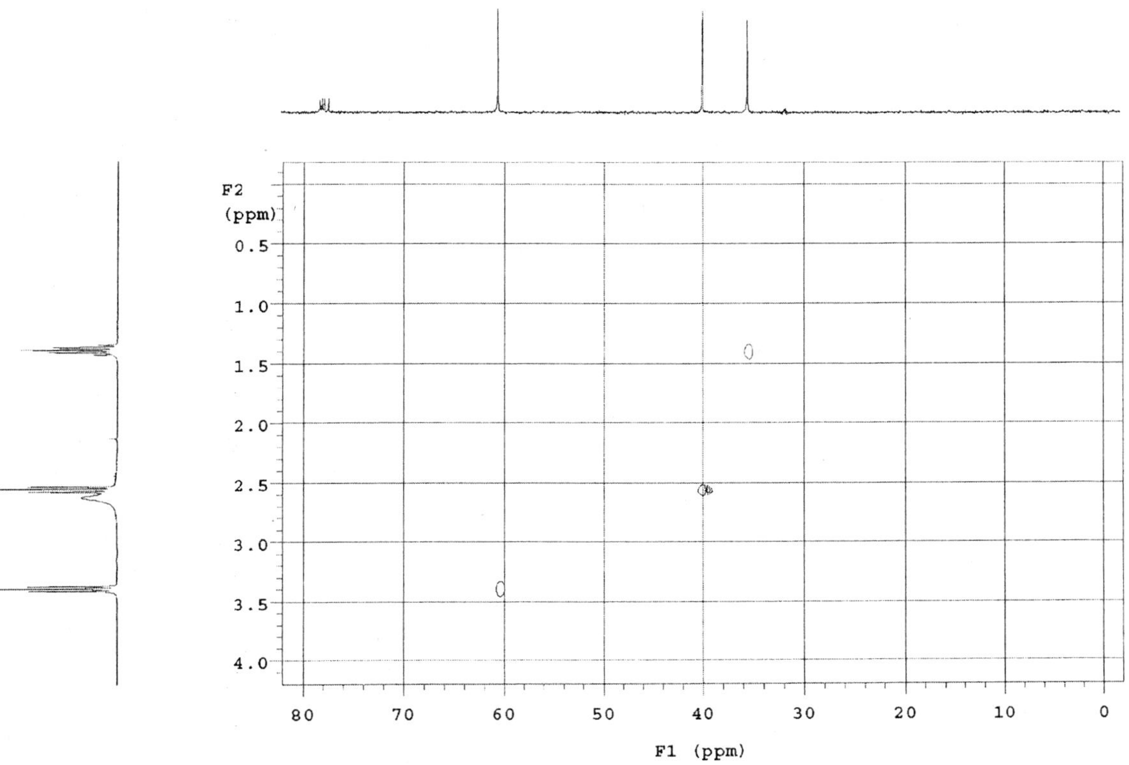

gHSQC Spectrum in $CDCl_3$

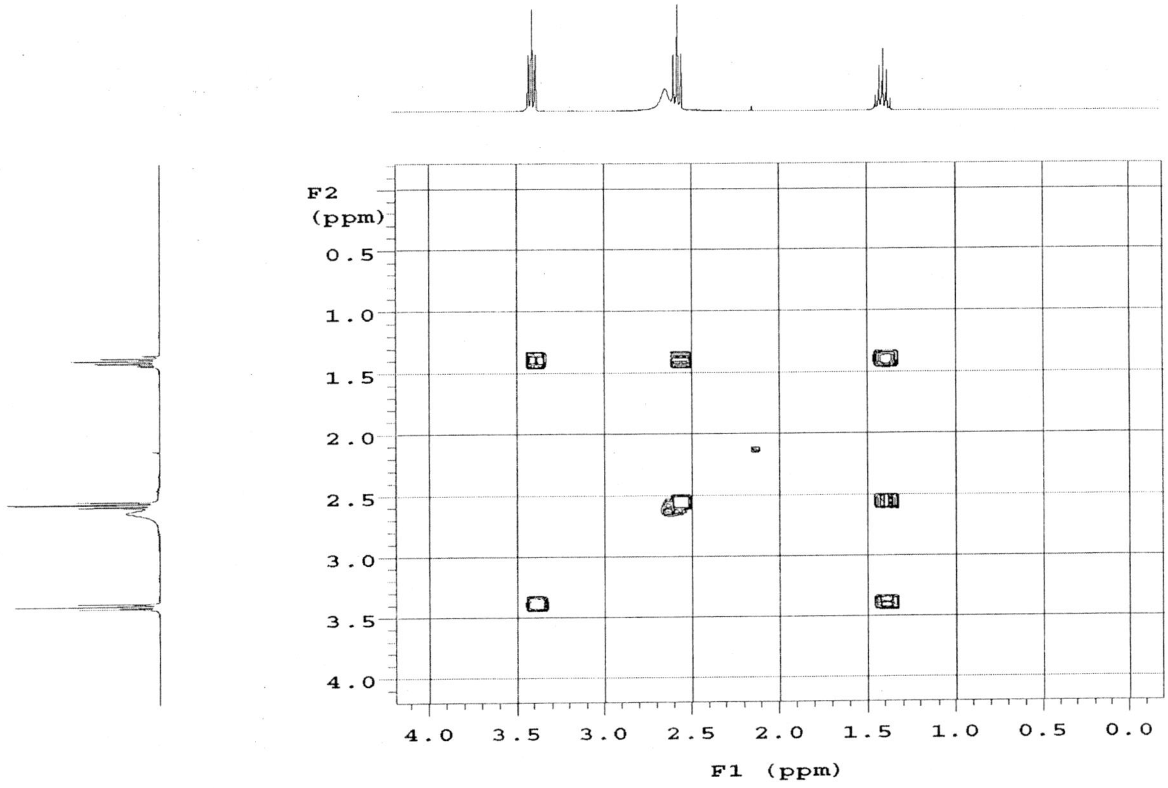

gCOSY Spectrum in $CDCl_3$

Problem 2.23 Given the spectroscopic data below (Proton, Carbon, Edited DEPT, gHSQC, and gCOSY), provide the structure of the molecule.

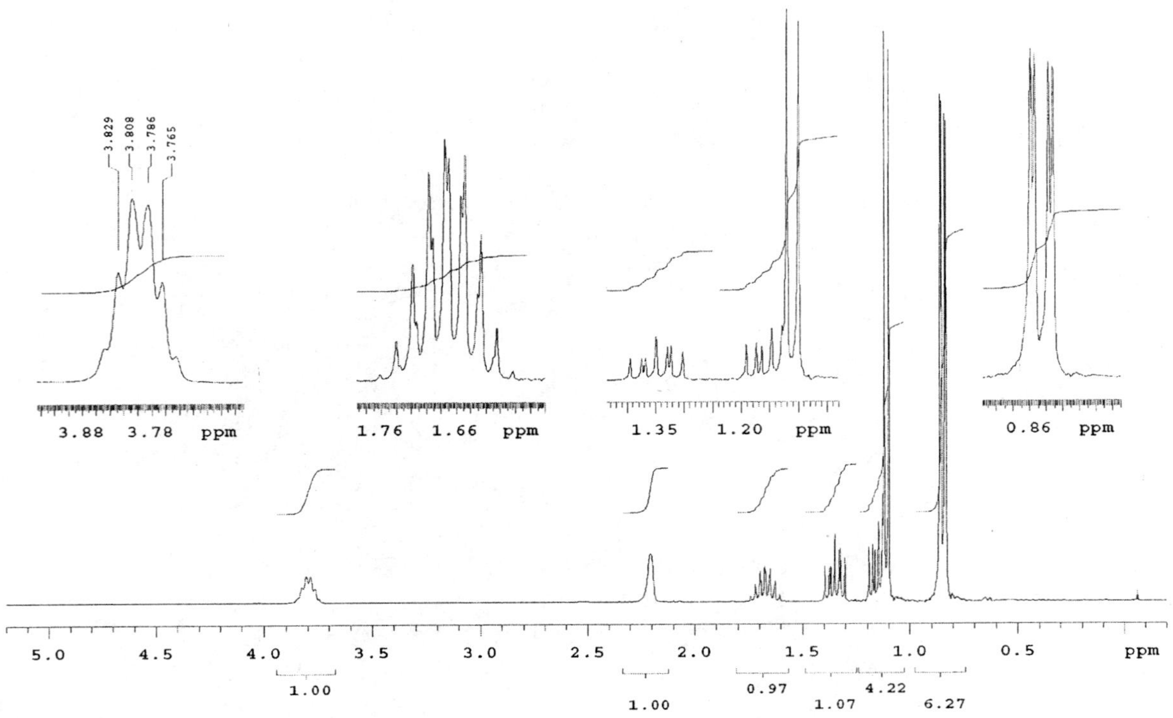

^{1}H-NMR Spectrum in $CDCl_3$

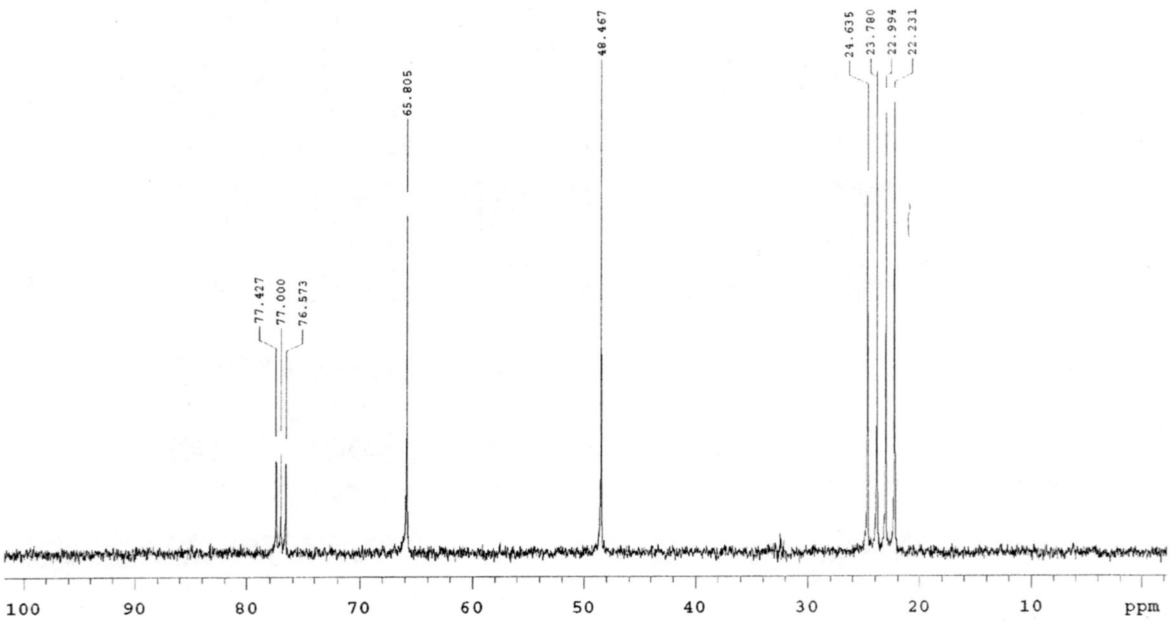

^{13}C-NMR Spectrum in $CDCl_3$

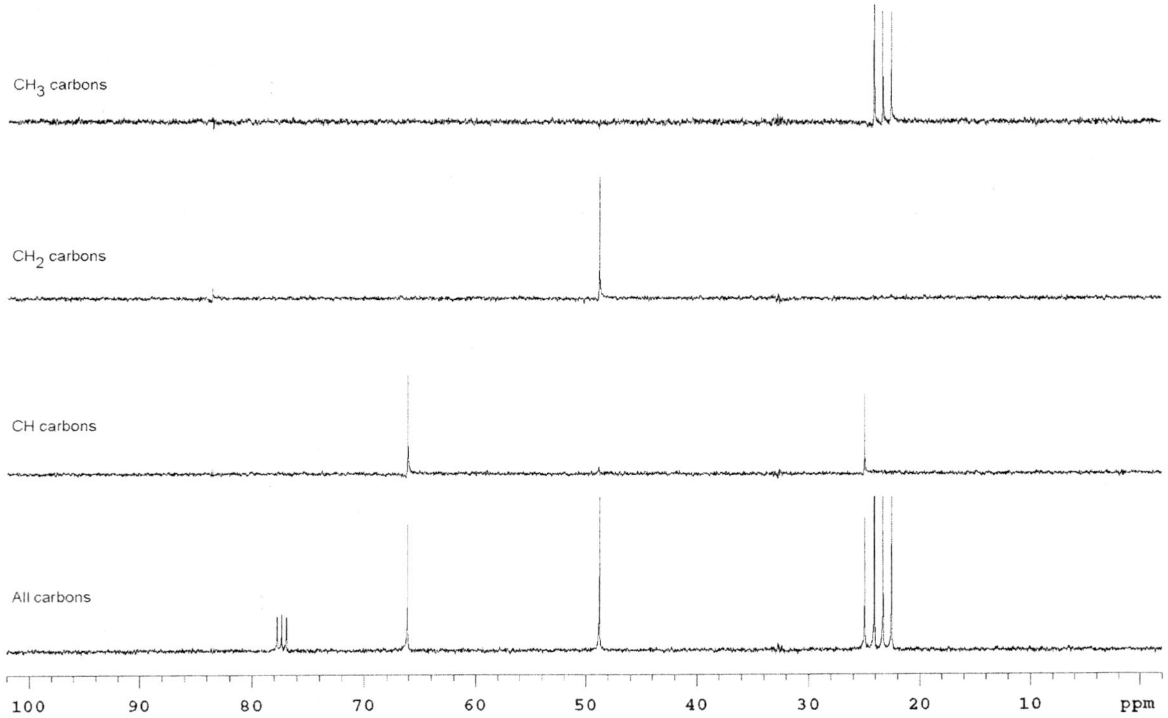

Edited DEPT ^{13}C-NMR Spectrum in $CDCl_3$

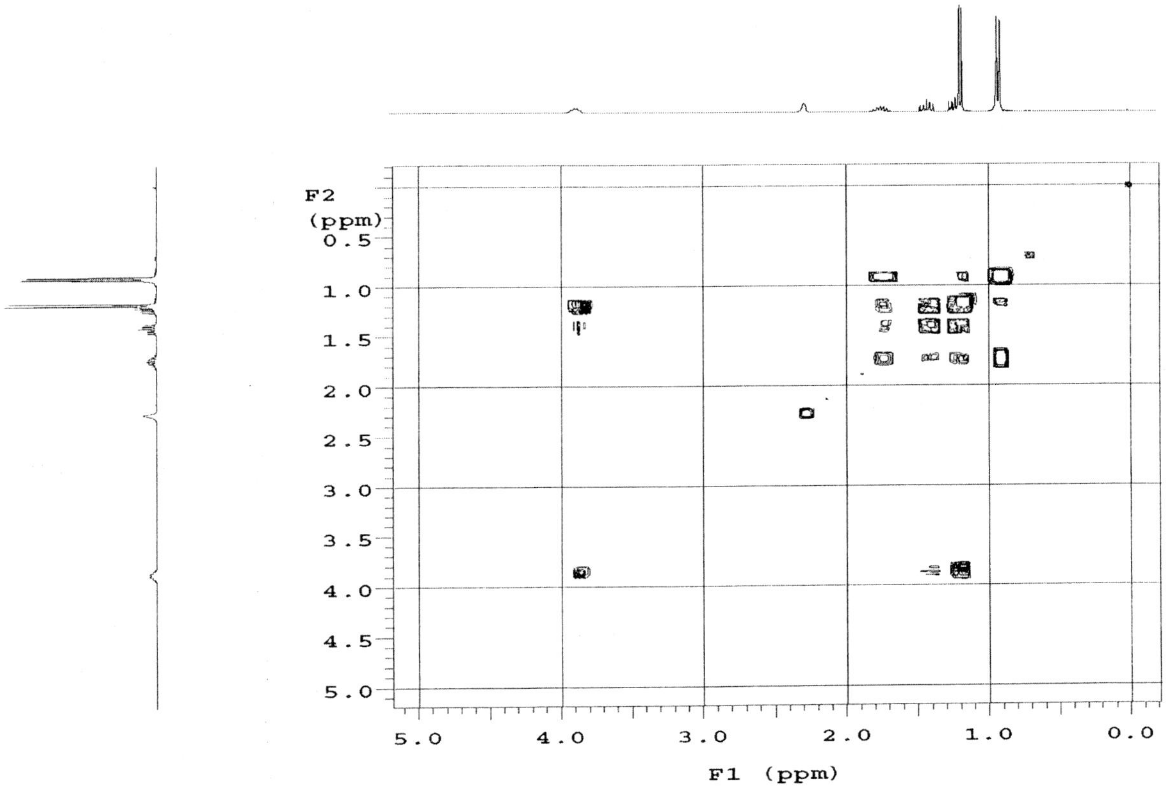

gCOSY Spectrum in $CDCl_3$

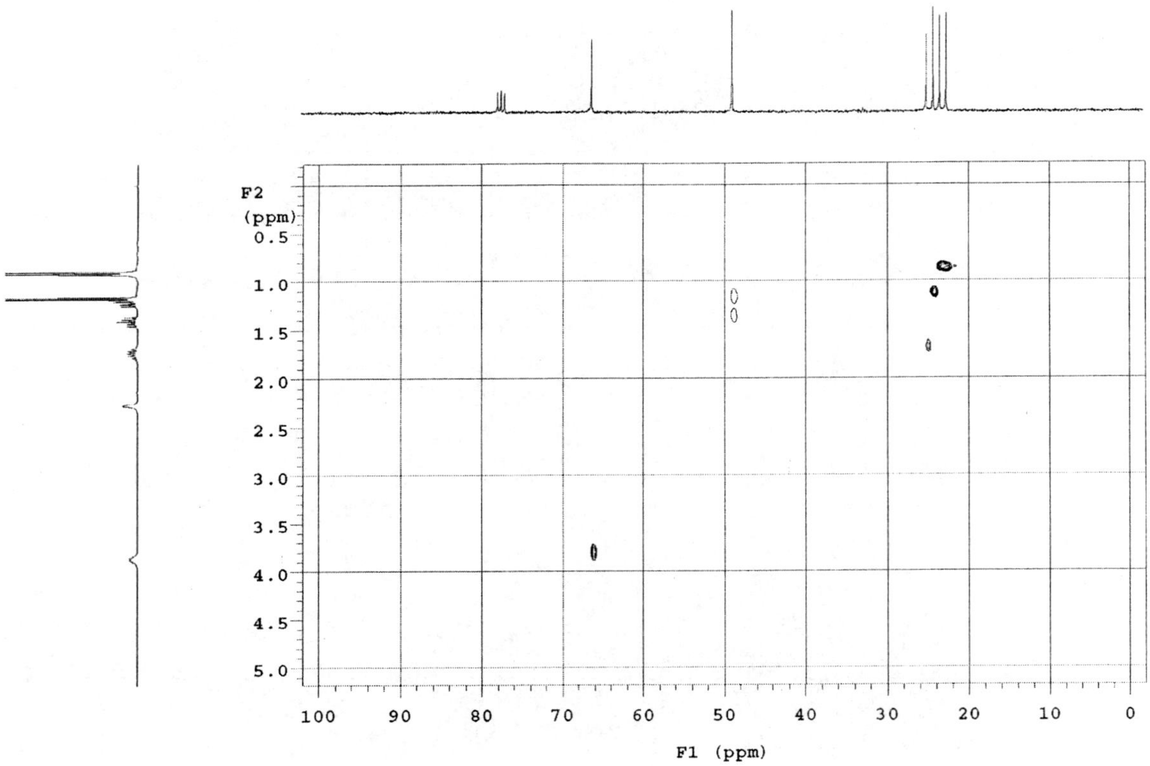

gHSQC Spectrum in $CDCl_3$

Problem 2.24 Given the spectroscopic data below (Proton, Edited DEPT, gHSQC, and gCOSY), provide the structure of the molecule.

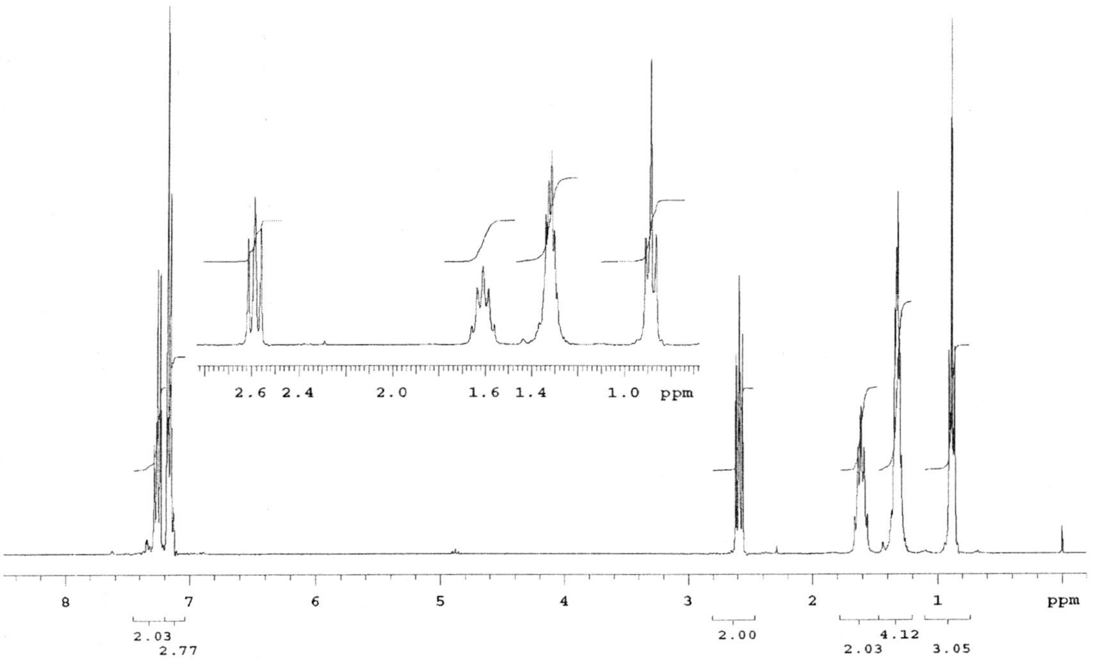

^{1}H-NMR Spectrum in $CDCl_3$

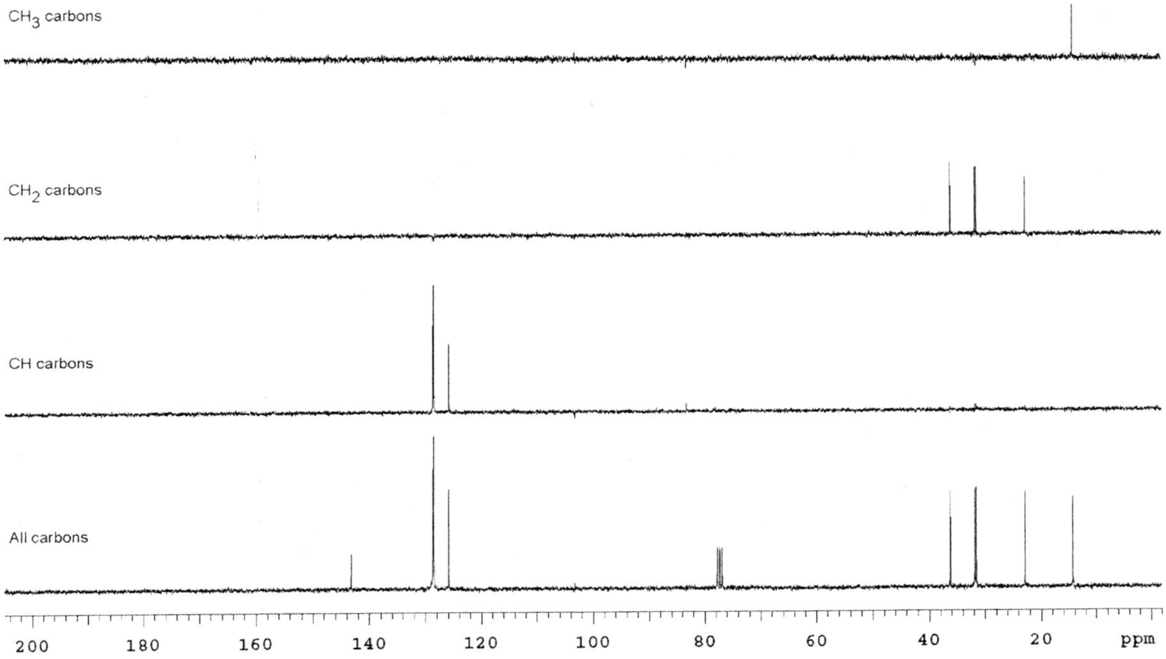

Edited DEPT ^{13}C-NMR Spectrum in $CDCl_3$

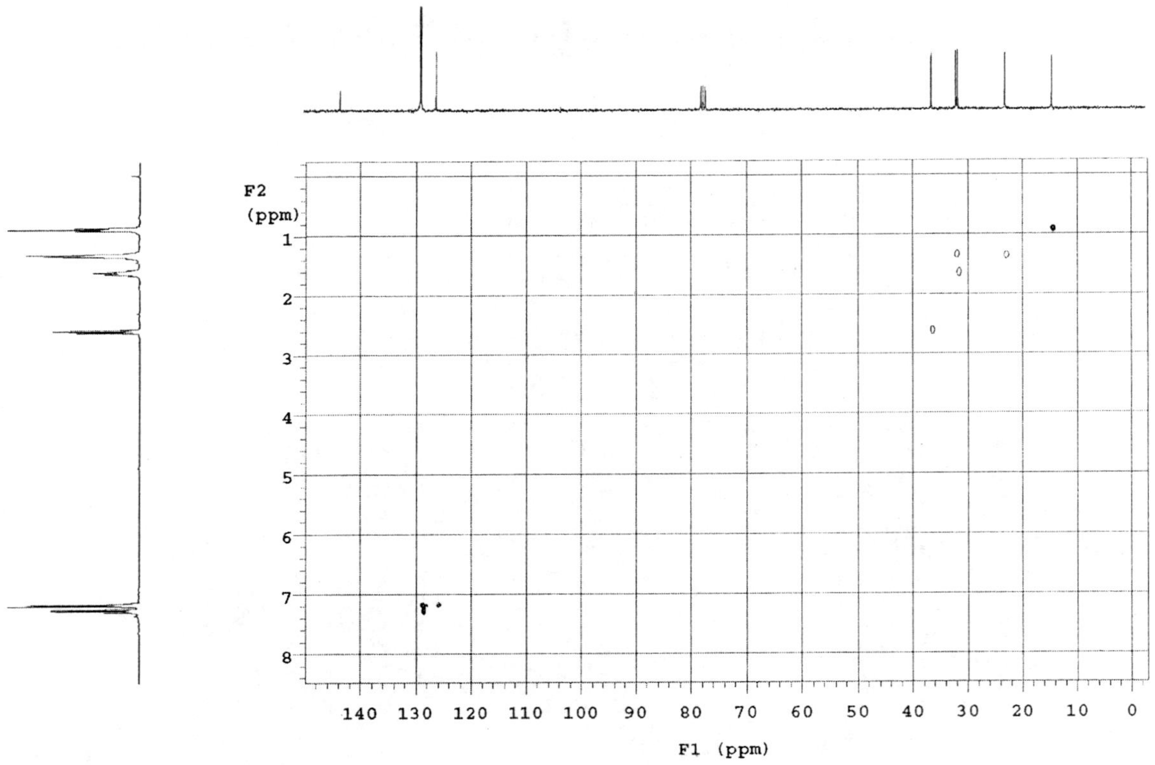

gHSQC Spectrum in $CDCl_3$

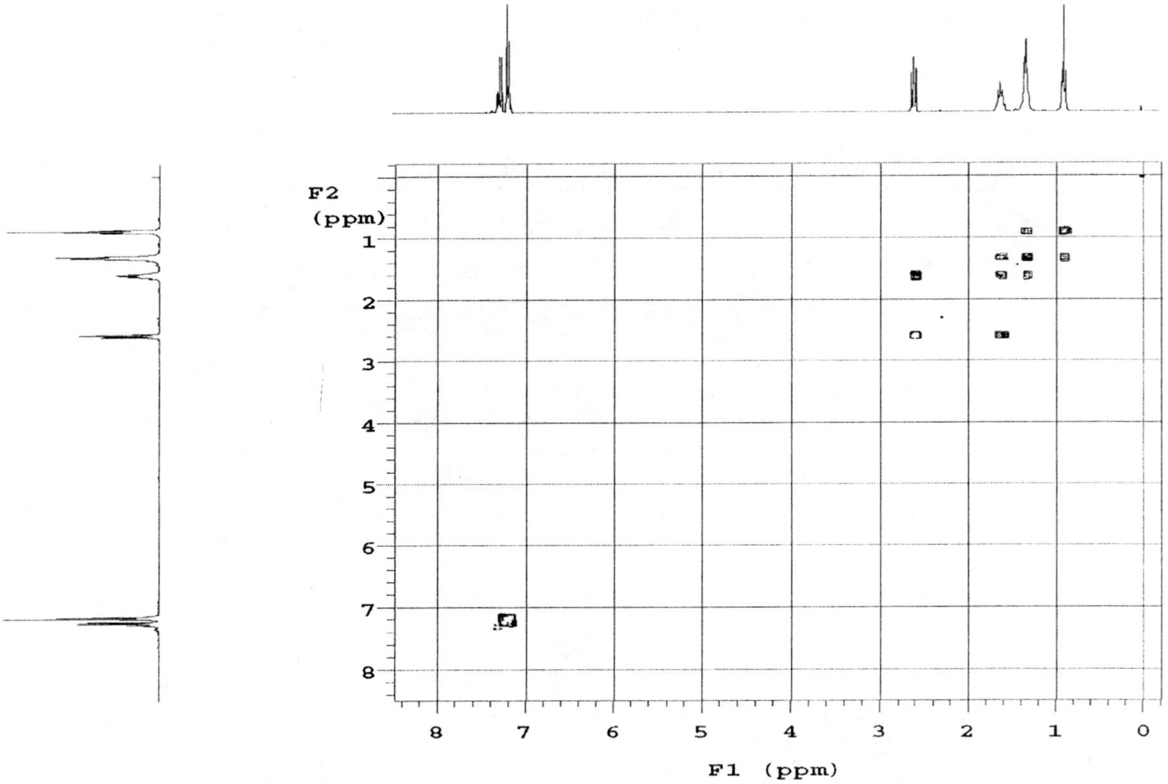

gCOSY Spectrum in $CDCl_3$

Problem 2.25 Given the spectroscopic data below (Proton, Carbon, Edited DEPT, gHSQC, and gCOSY), provide the structure of the molecule.

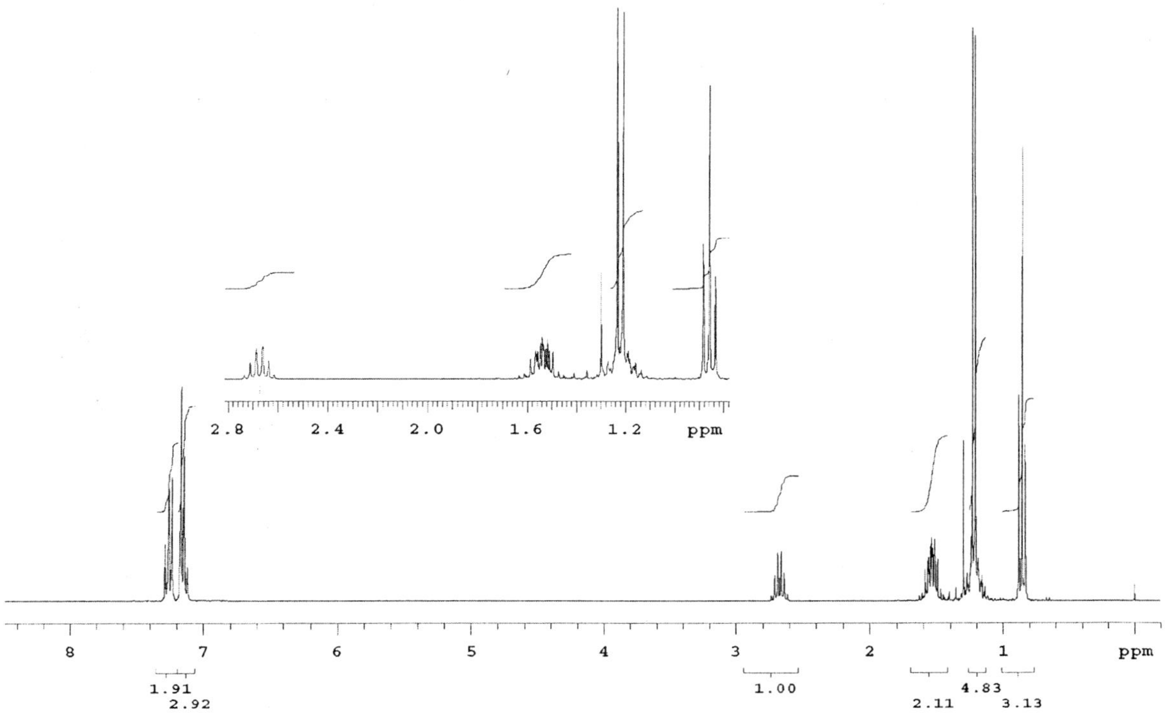

^{1}H-NMR Spectrum in $CDCl_3$

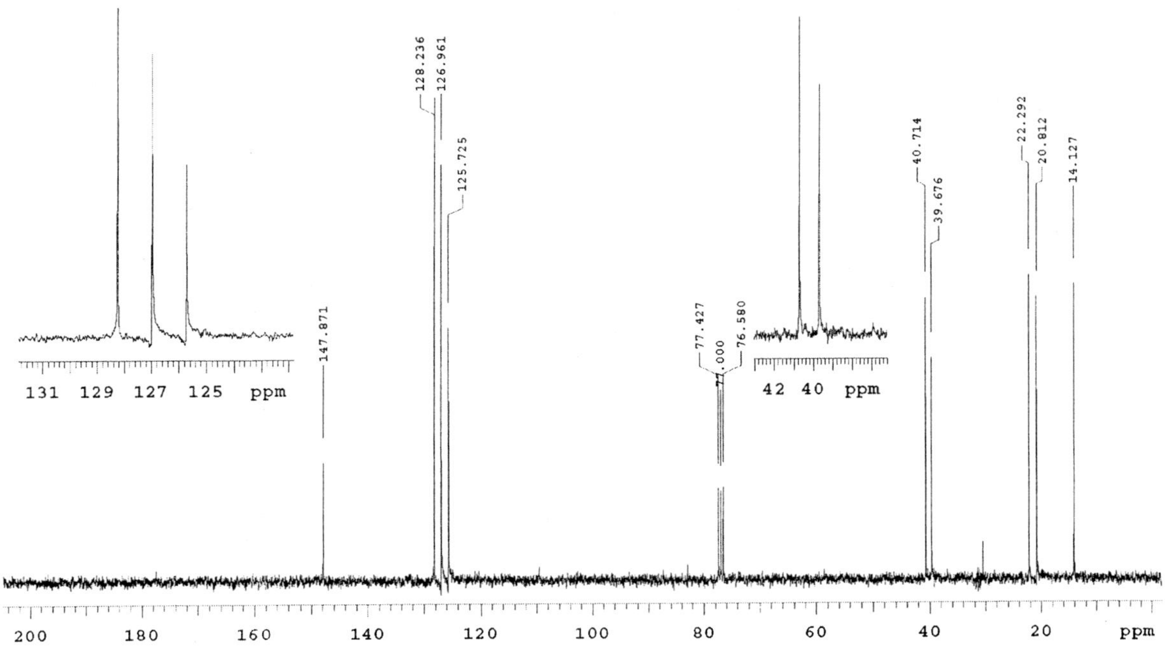

^{13}C-NMR Spectrum in $CDCl_3$

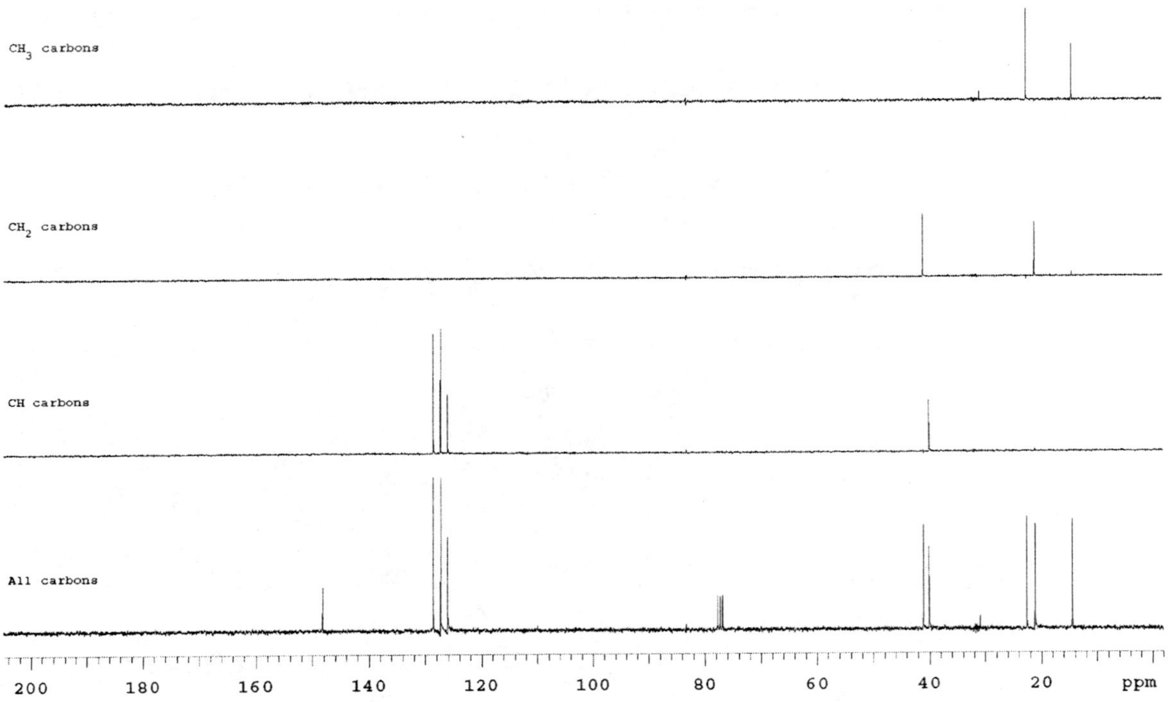

Edited DEPT ^{13}C-NMR Spectrum in $CDCl_3$

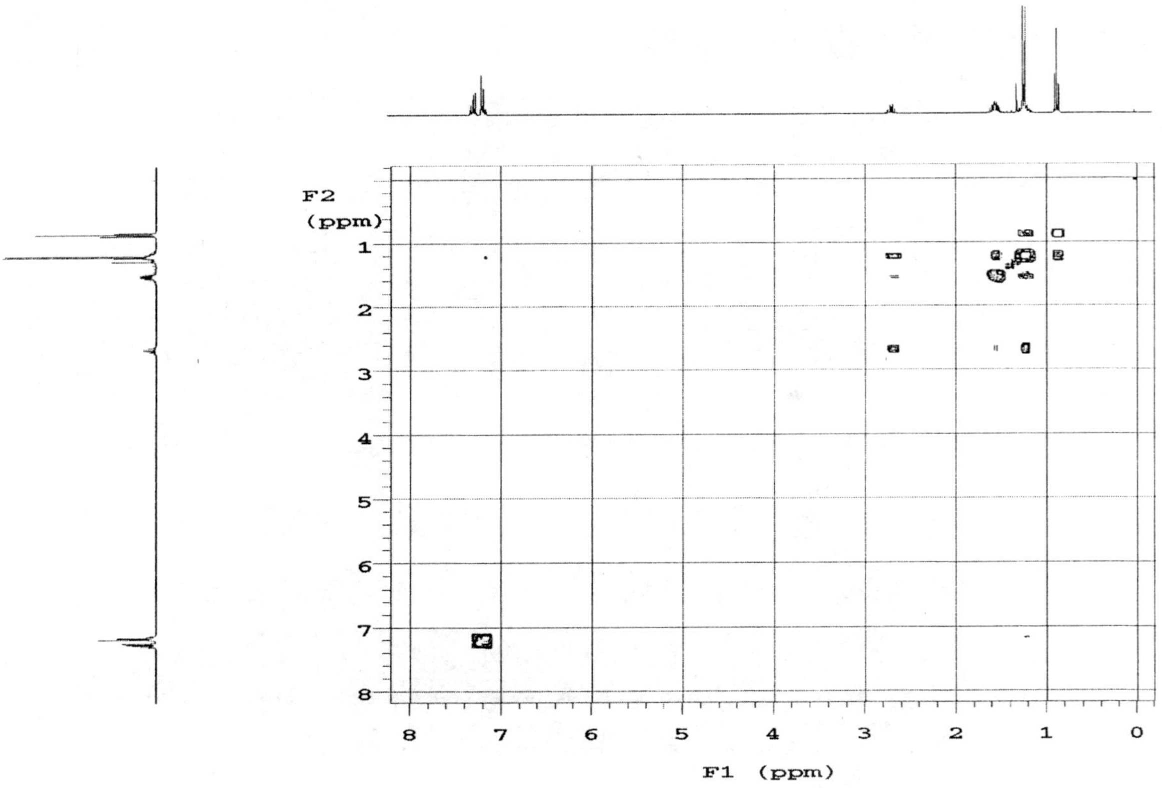

gCOSY Spectrum in $CDCl_3$

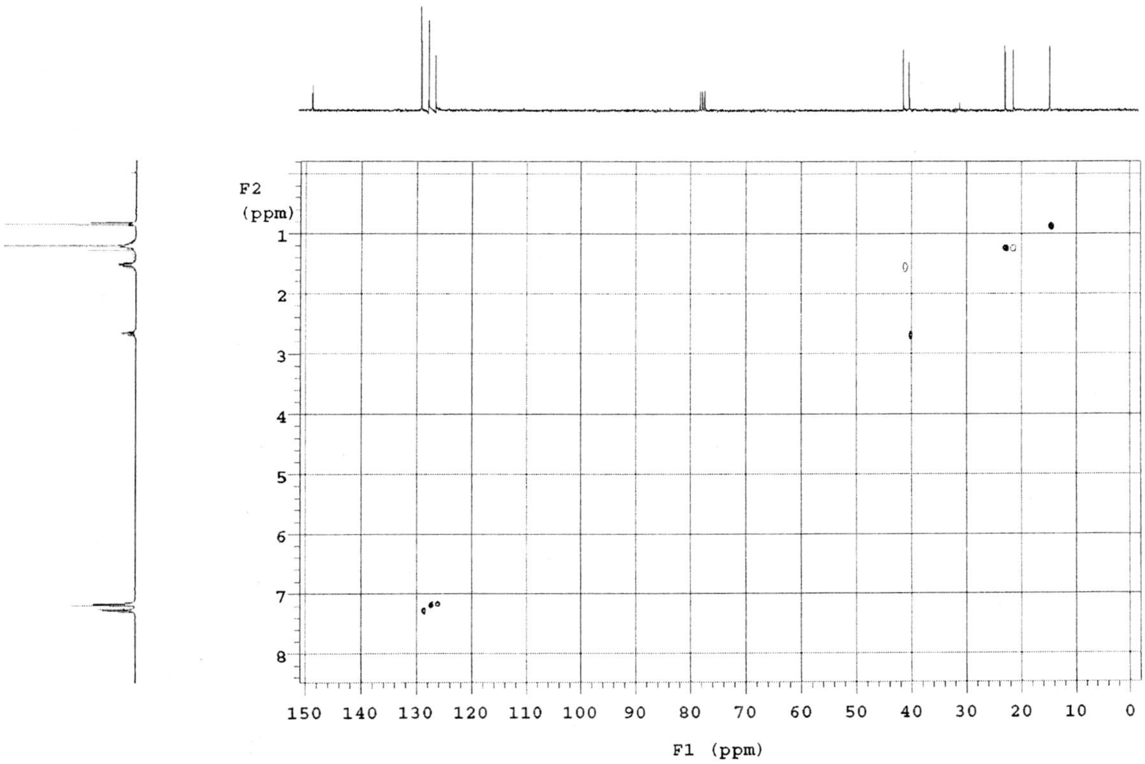

gHSQC Spectrum in $CDCl_3$

Problem 2.26 Given the spectroscopic data below (Proton, Edited DEPT, gHSQC, and gCOSY), provide the structure of the molecule.

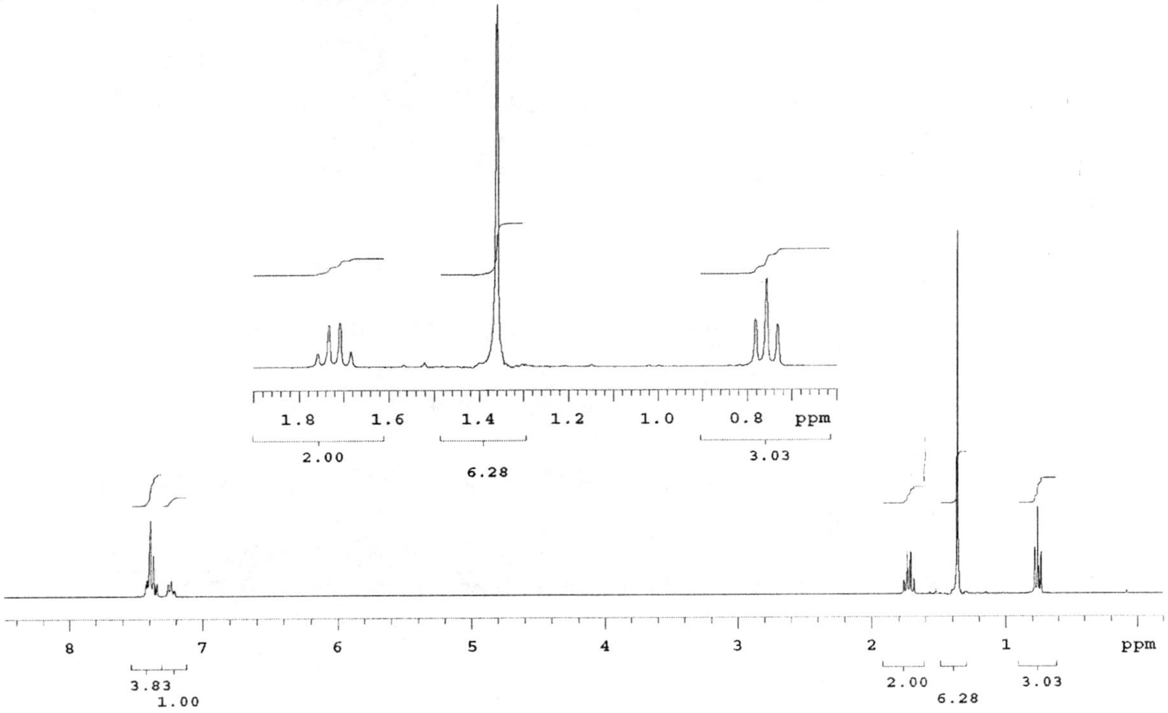

^{1}H-NMR Spectrum in $CDCl_3$

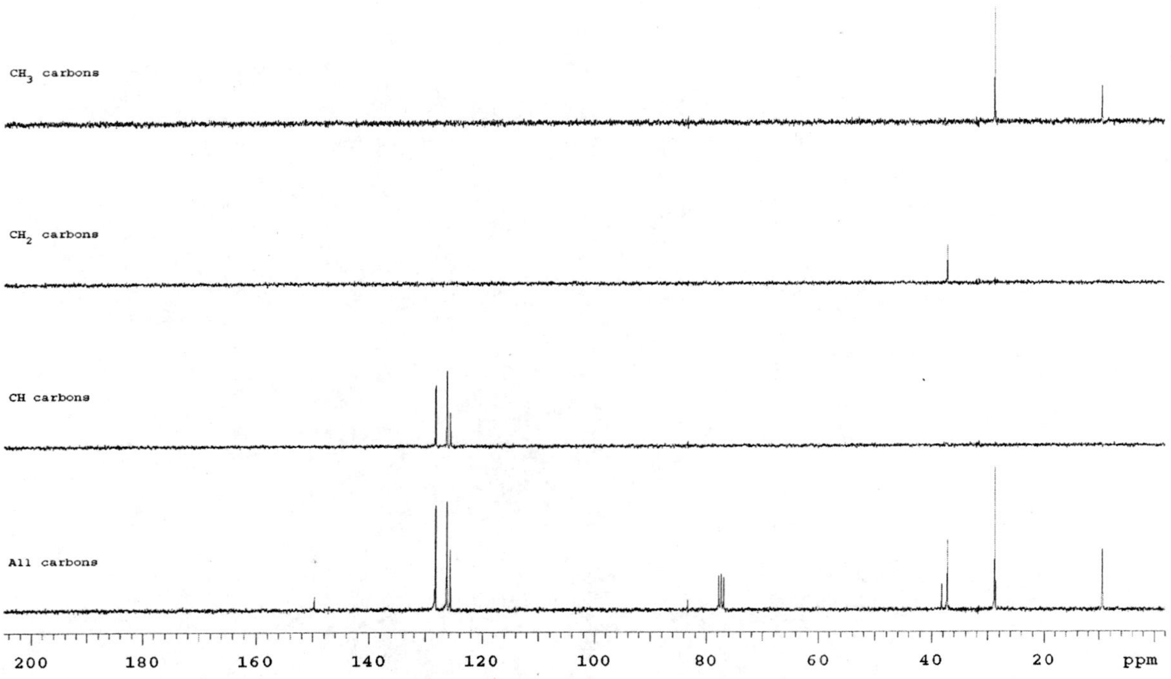

Edited DEPT ^{13}C-NMR Spectrum in $CDCl_3$

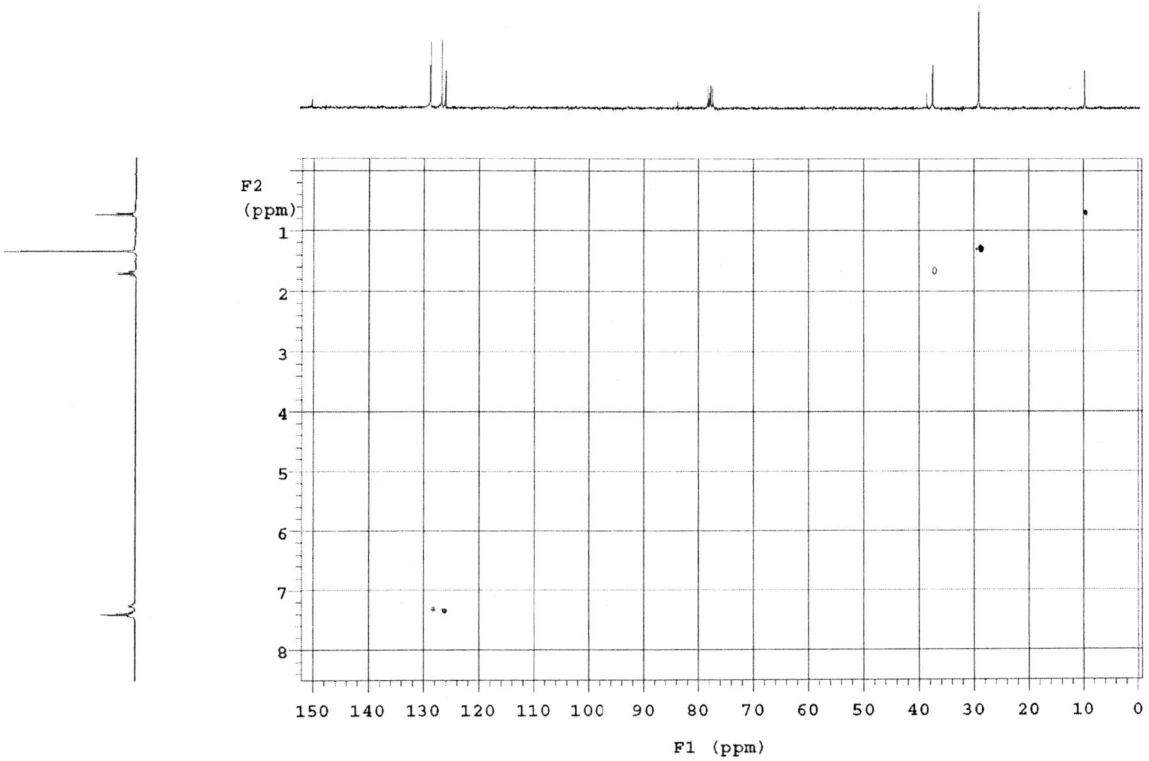

gHSQC Spectrum in $CDCl_3$

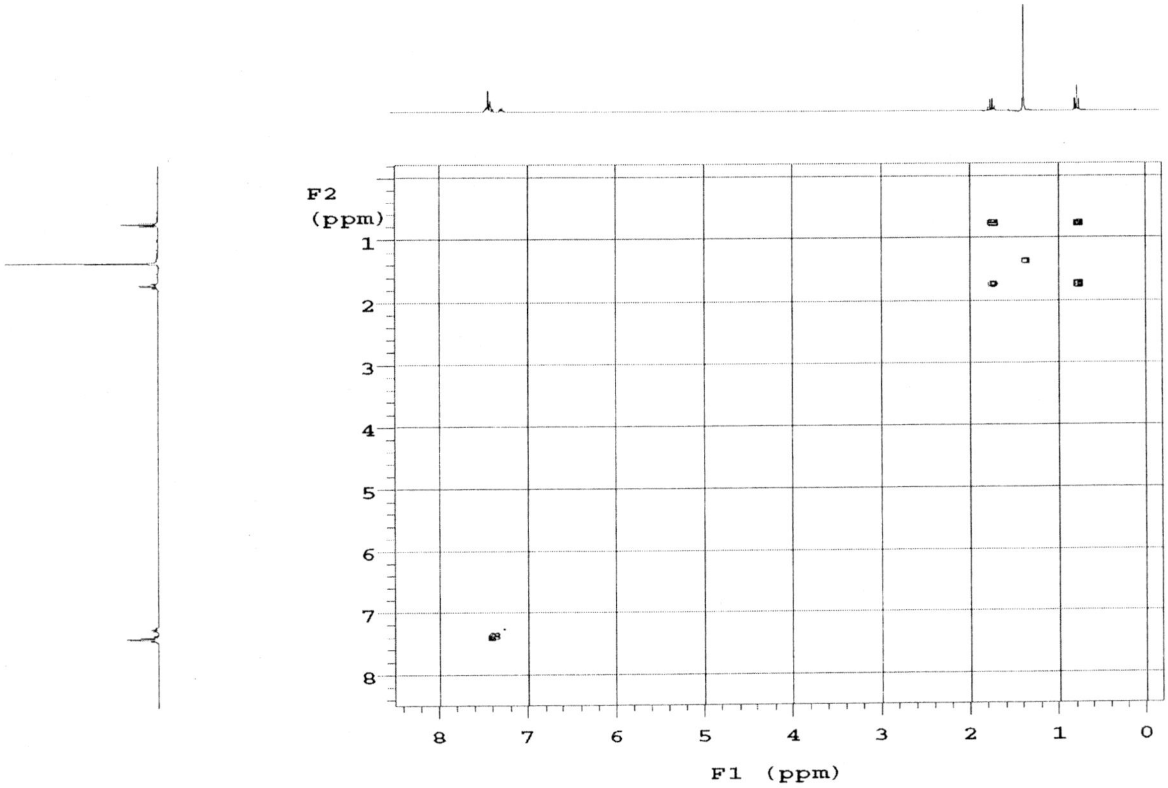

gCOSY Spectrum in $CDCl_3$

Problem 2.27 Given the spectroscopic data below (Proton, Edited DEPT, gHSQC, gHMBC, and gCOSY), provide the structure of the molecule.

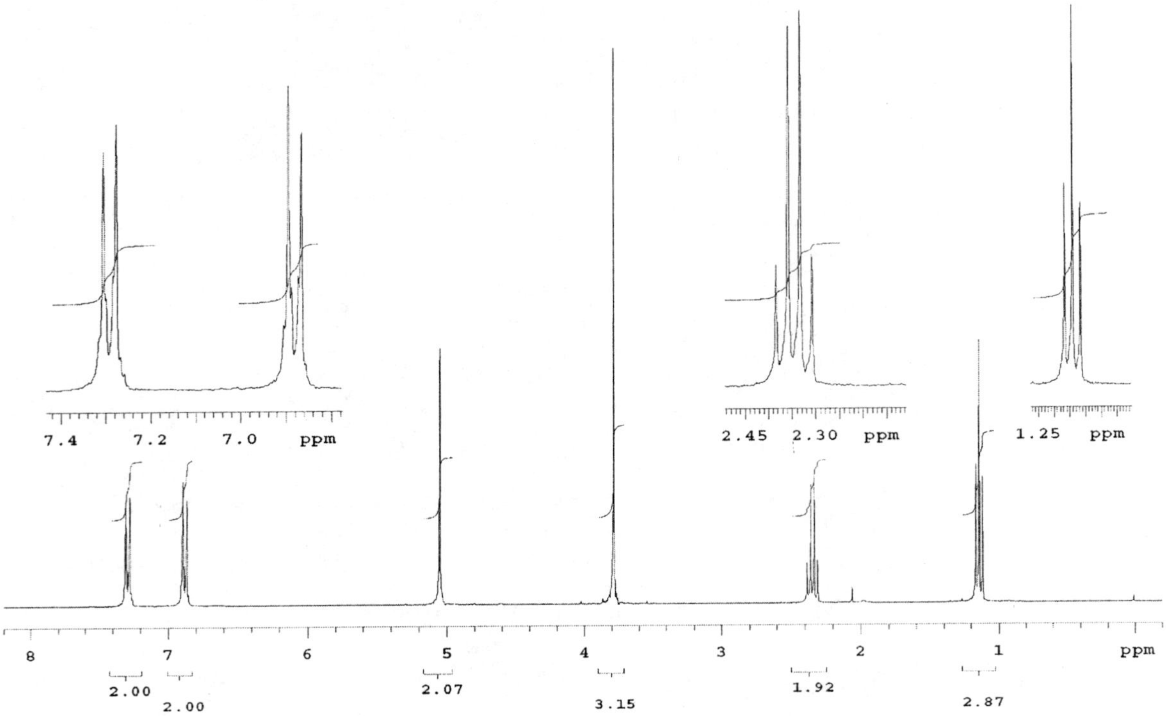

^{1}H-NMR Spectrum in $CDCl_3$

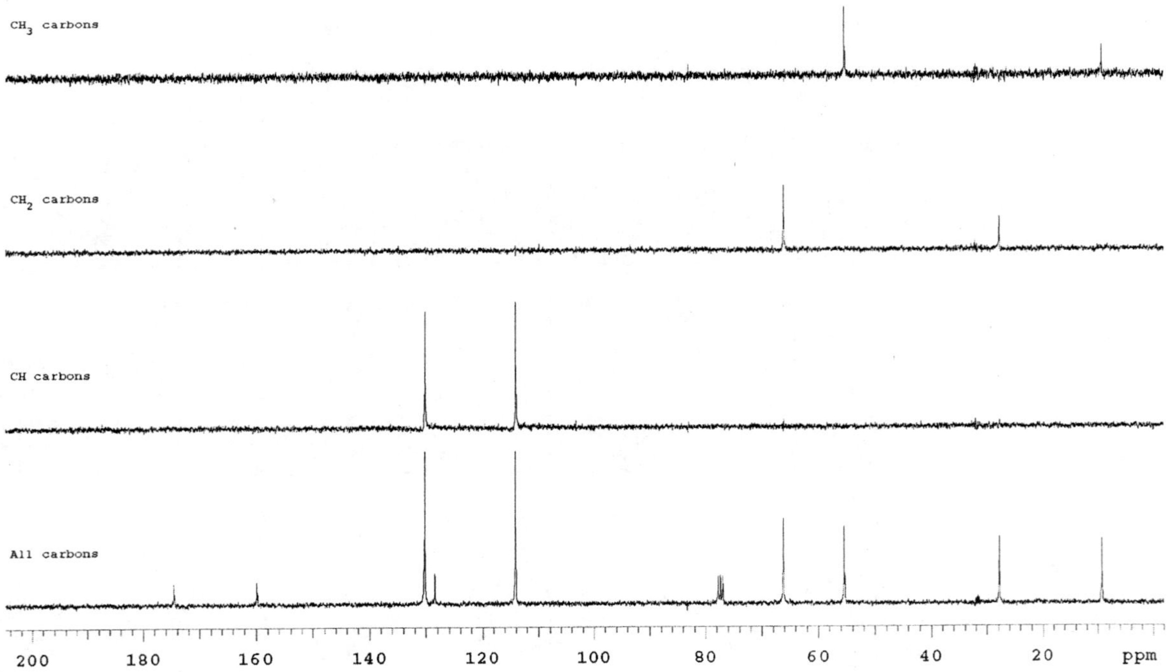

Edited DEPT ^{13}C-NMR Spectrum in $CDCl_3$

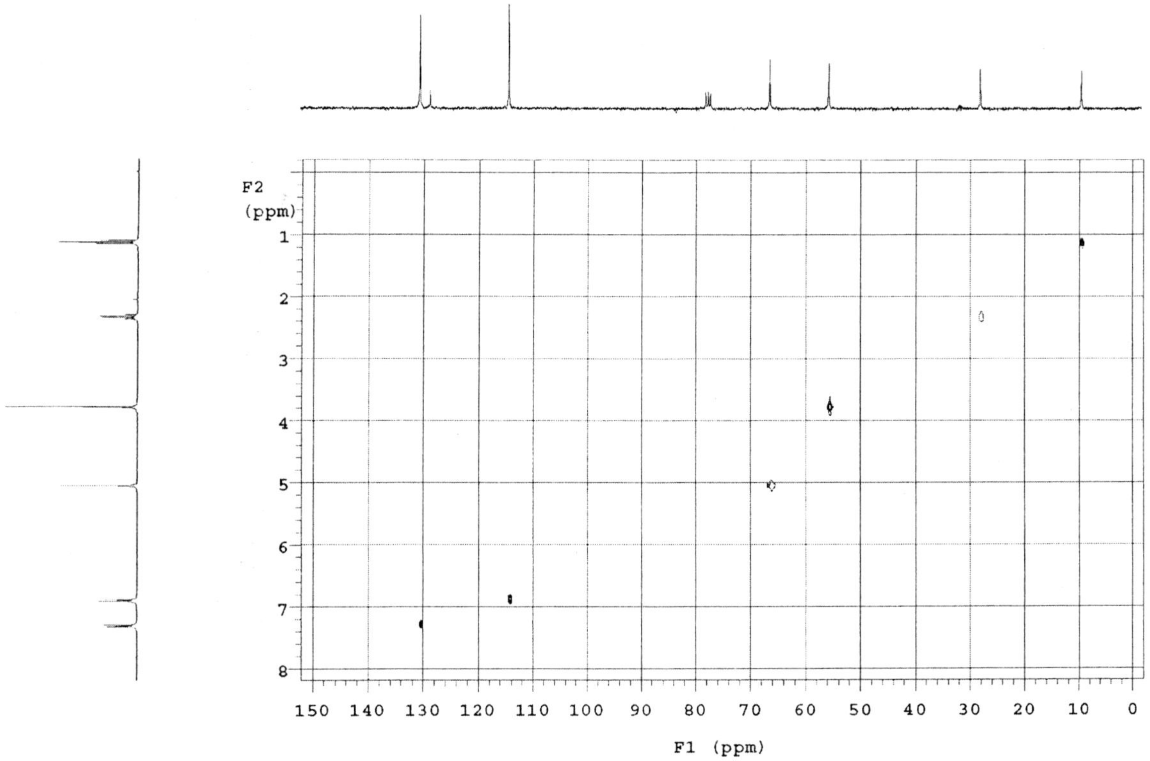

gHSQC Spectrum in $CDCl_3$

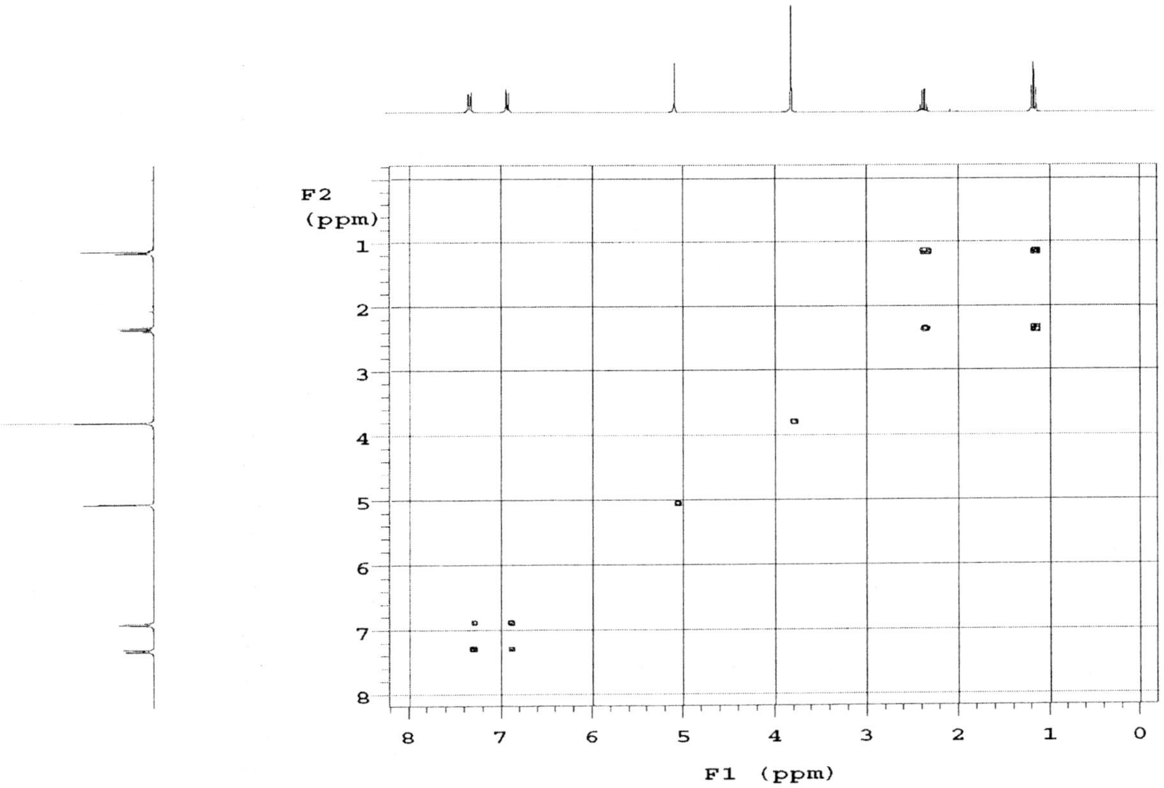

gCOSY Spectrum in $CDCl_3$

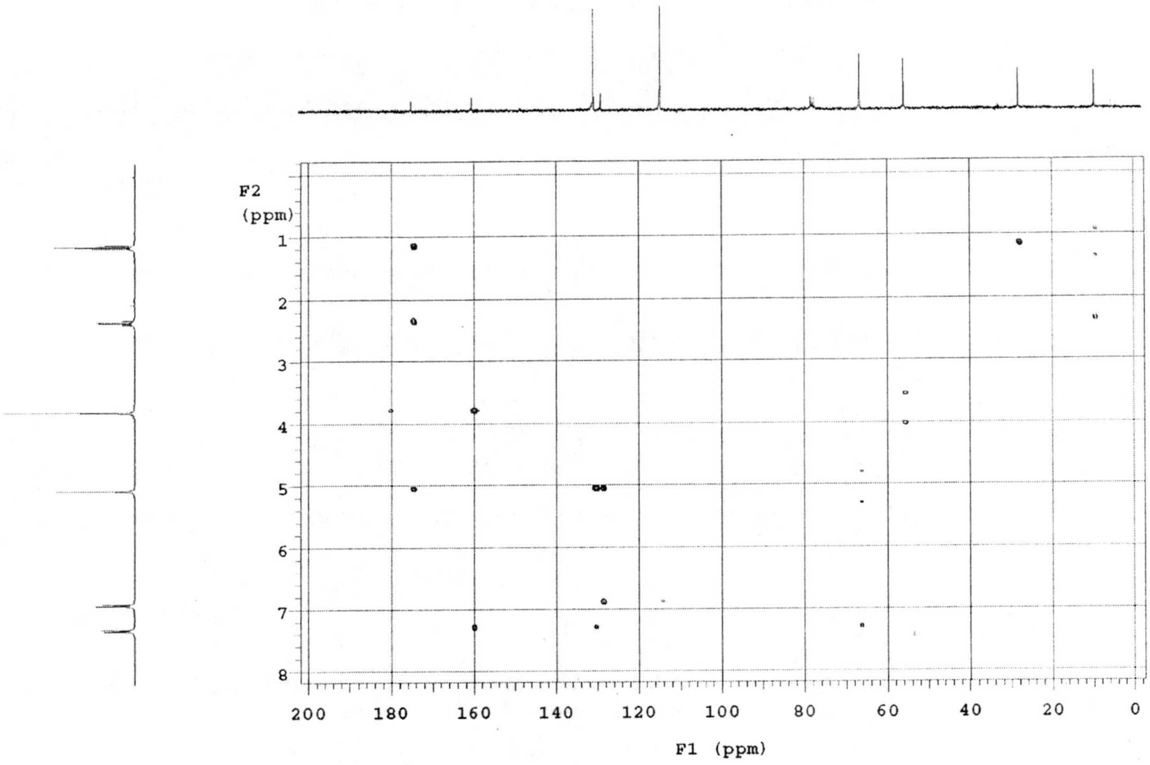

gHMBC Spectrum in $CDCl_3$

Problem 2.28 Given the spectroscopic data below (Proton, Carbon, Edited DEPT, gHMQC, and gCOSY), provide the structure of the molecule.

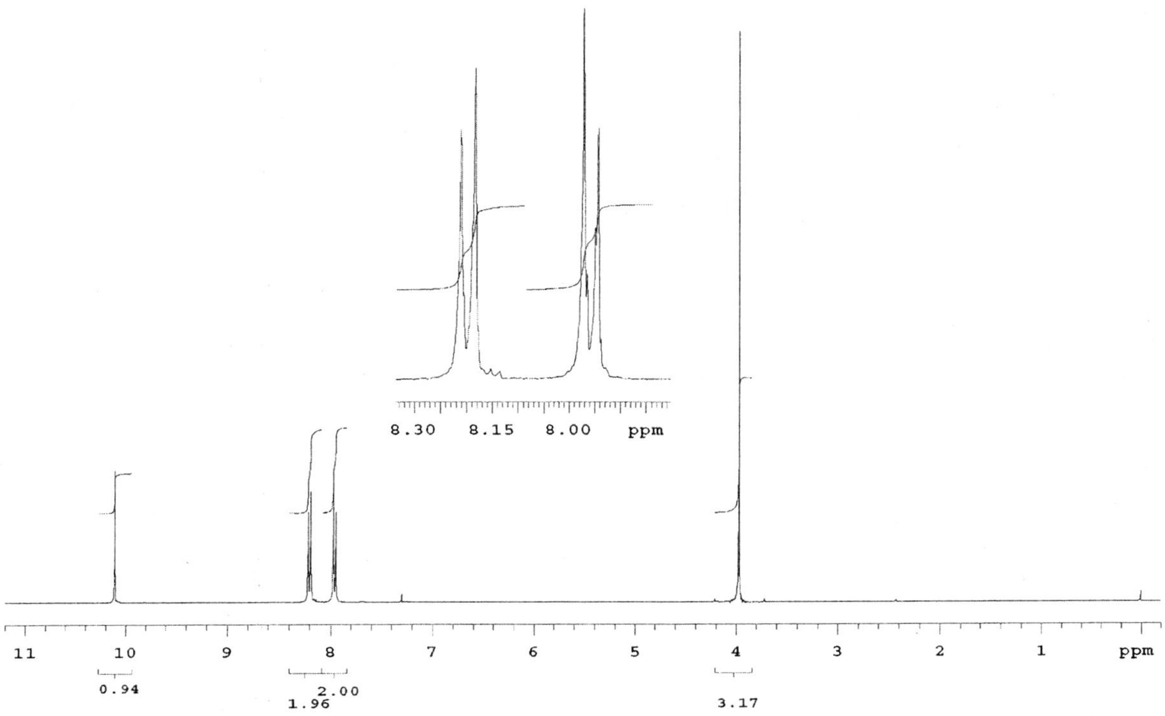

^{1}H-NMR Spectrum in $CDCl_3$

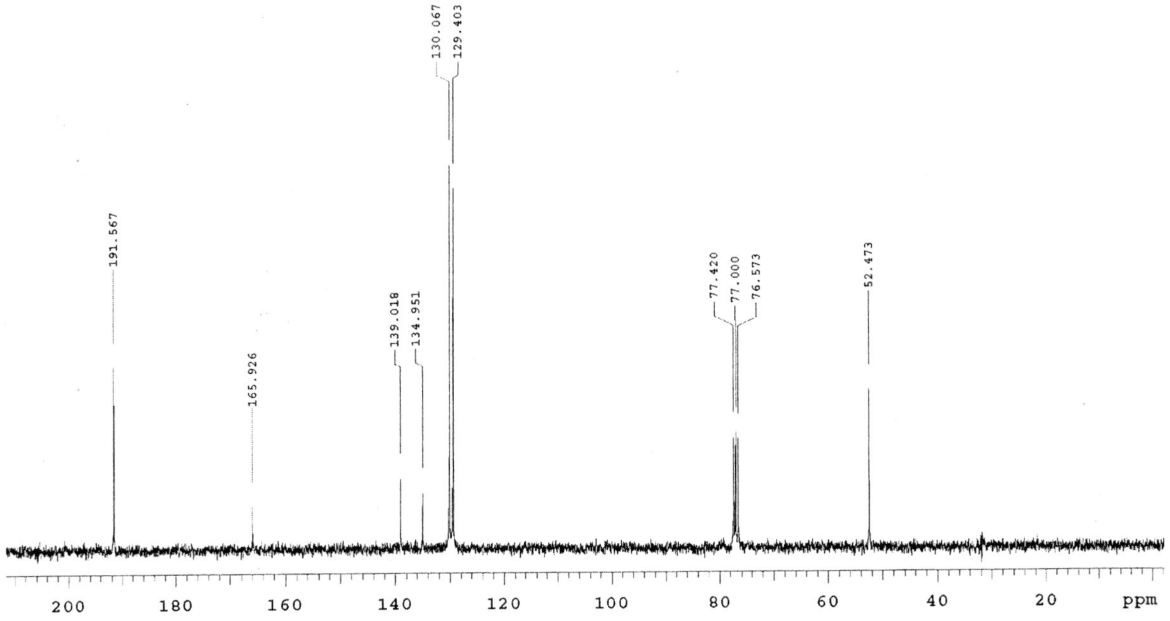

^{13}C-NMR Spectrum in $CDCl_3$

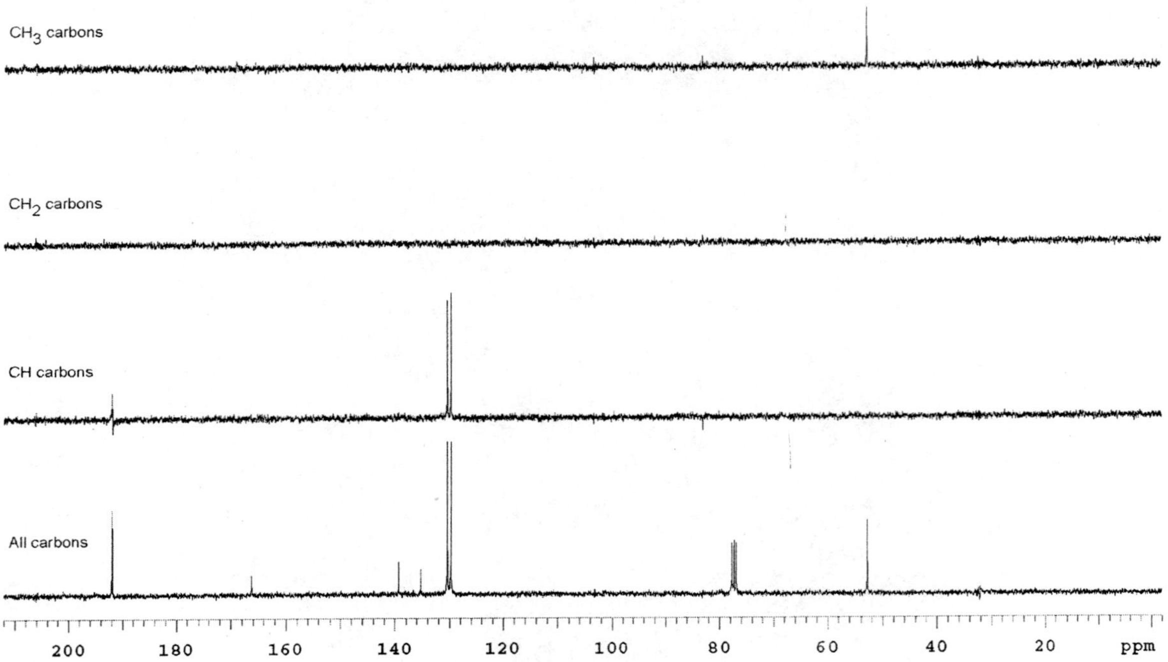

Edited DEPT ^{13}C-NMR Spectrum in $CDCl_3$

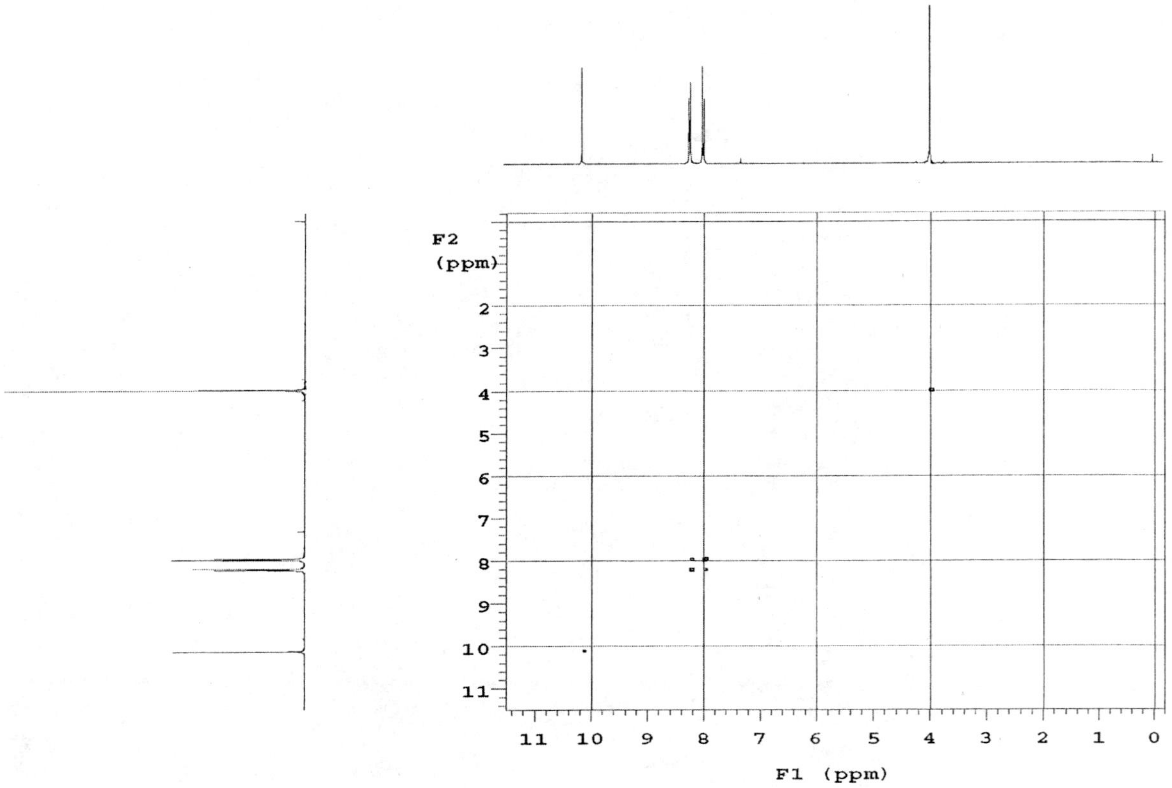

gCOSY Spectrum in $CDCl_3$

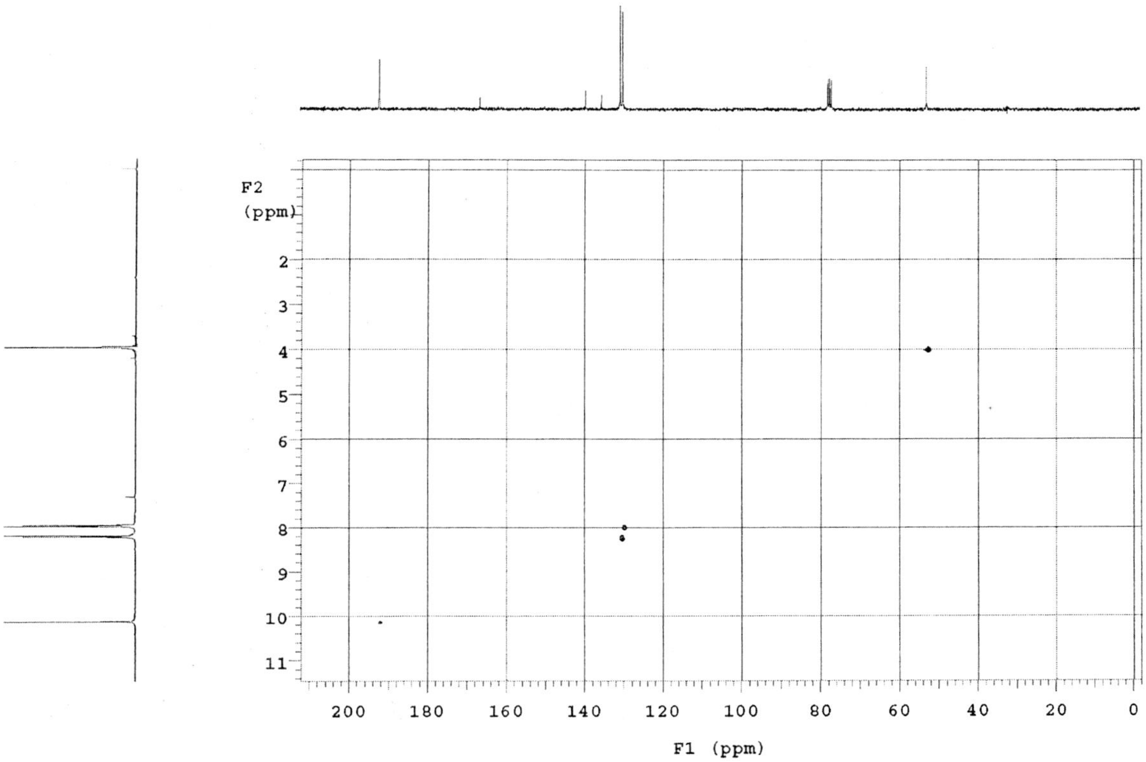

gHSQC Spectrum in $CDCl_3$

Problem 2.29 Given the spectroscopic data below (Proton, Carbon, and gHSQC), provide the structure of the molecule. (MW: 186 & 188 g/mol; IR: 705 & 825 cm^{-1})

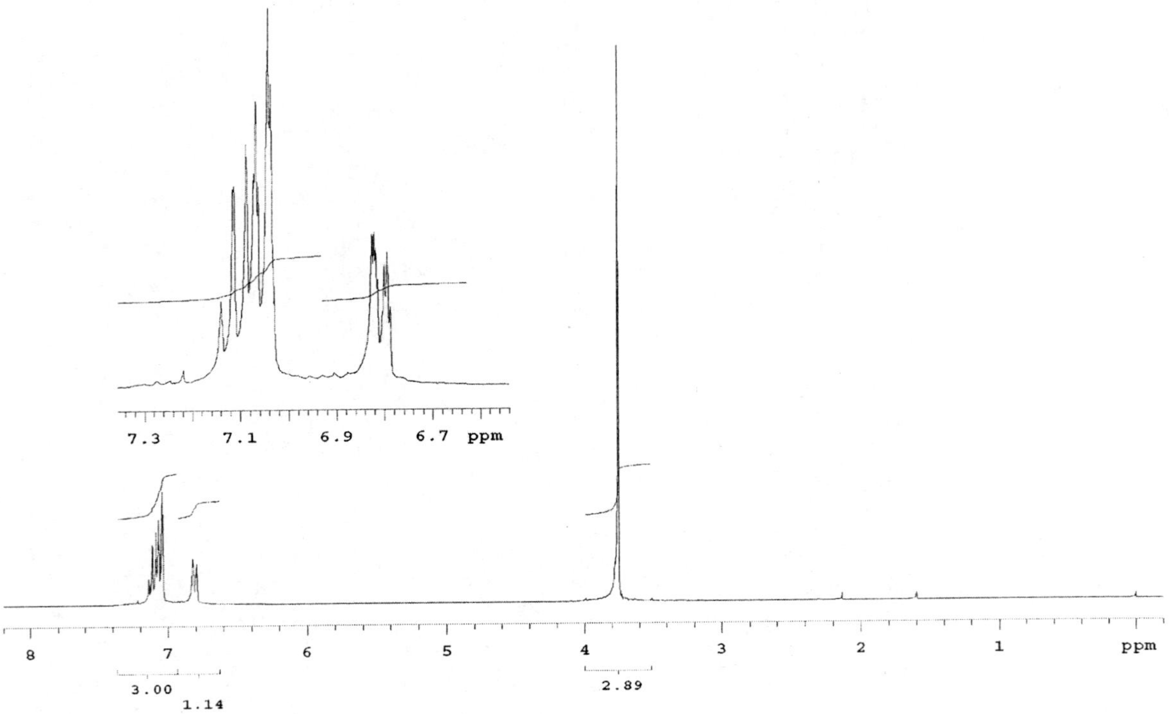

^{1}H-NMR Spectrum in $CDCl_3$

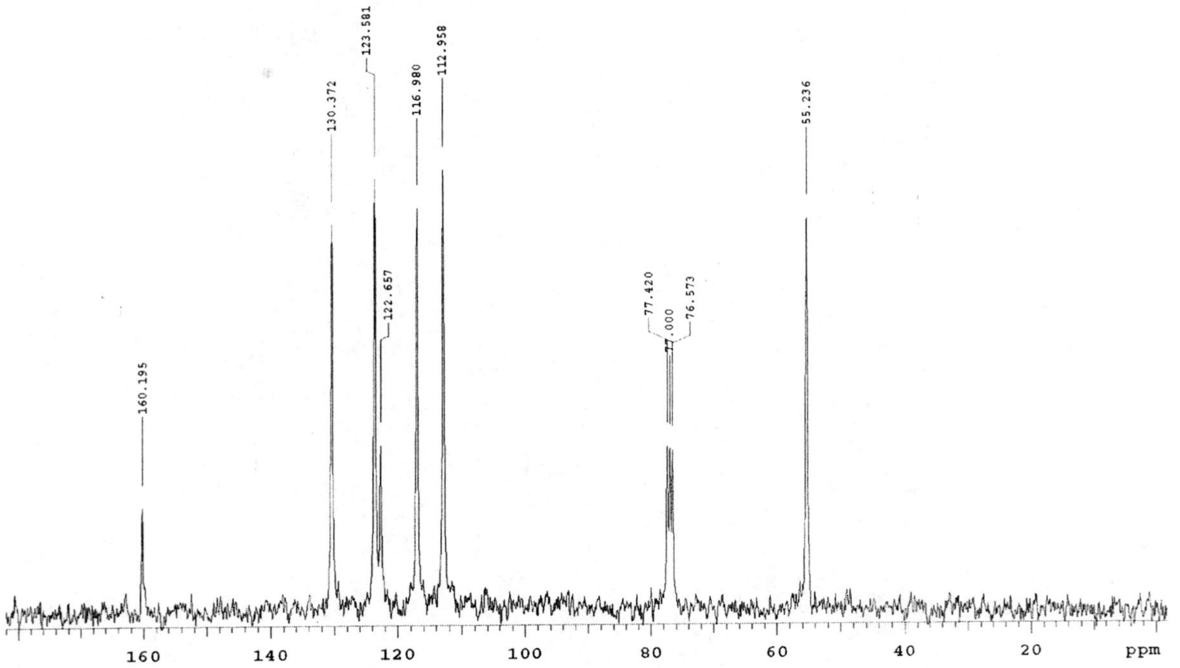

^{13}C-NMR Spectrum in $CDCl_3$

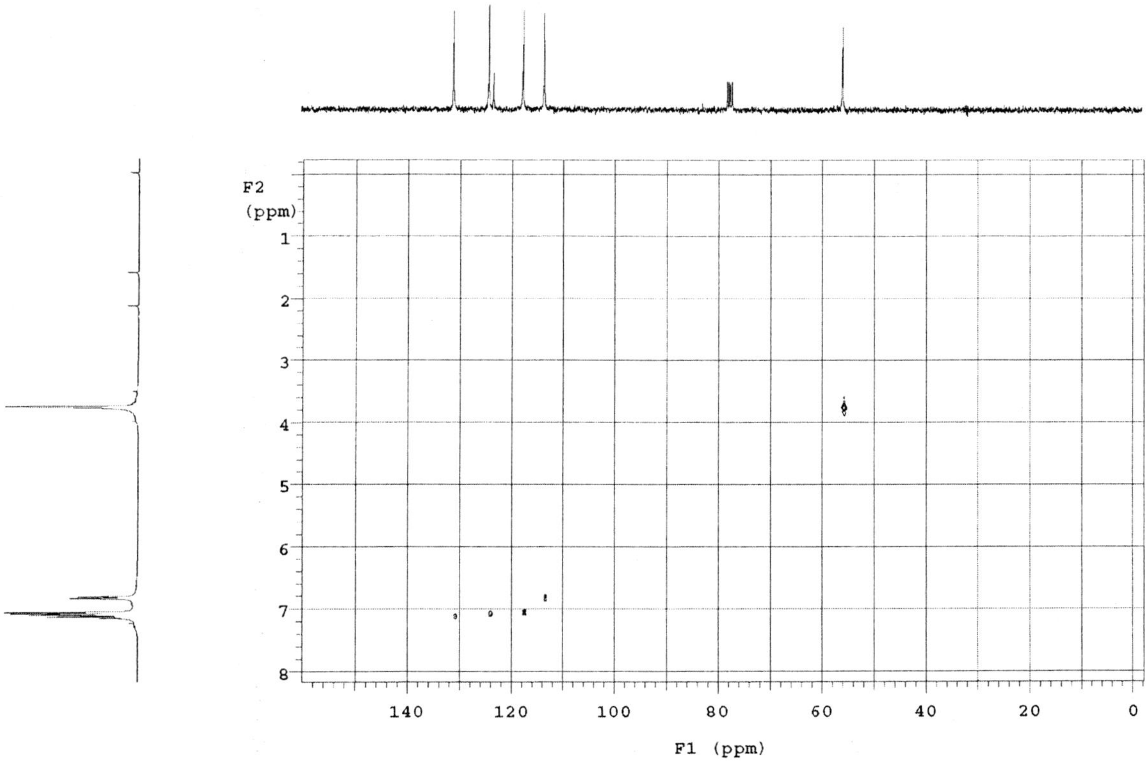

gHSQC Spectrum in $CDCl_3$

Problem 2.30 Given the spectroscopic data below (Proton, Carbon, gHSQC, and gHMBC), provide the structure of the molecule. From vapor phase osmometry, it is known that the MW is in the range of 240 to 255 g/mol.

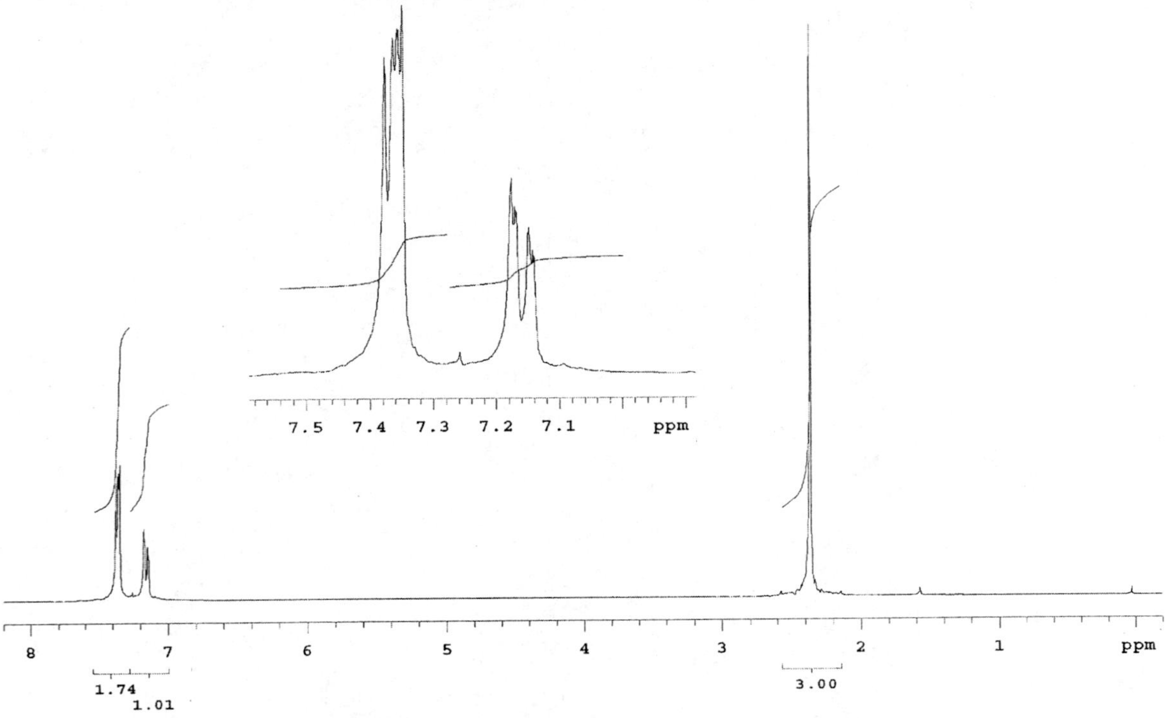

^{1}H-NMR Spectrum in $CDCl_3$

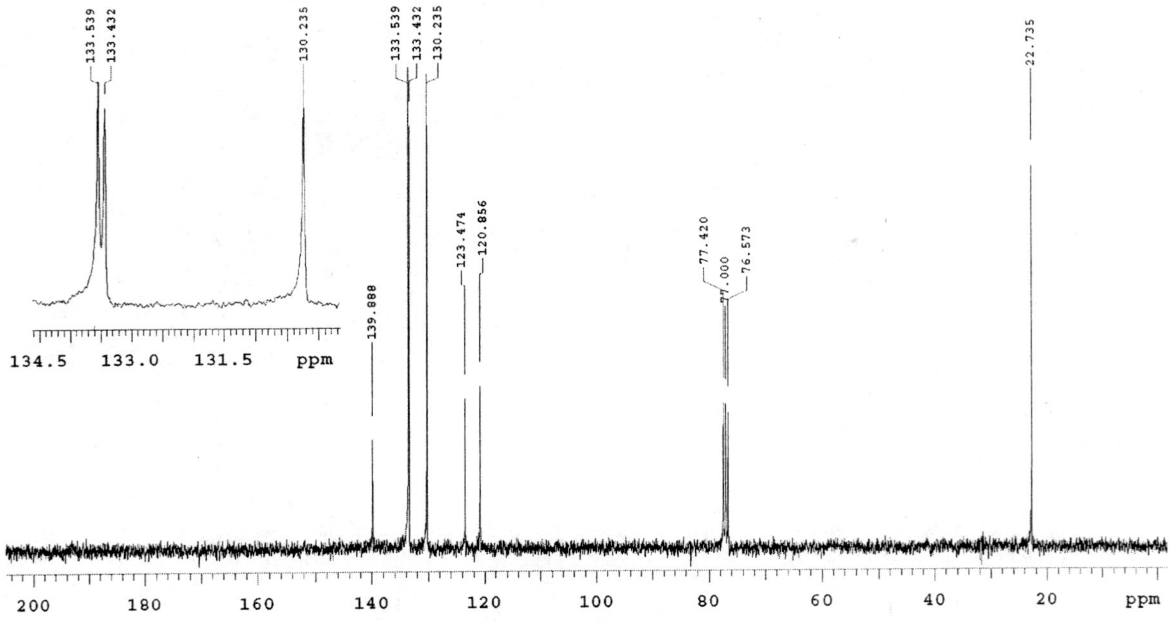

^{13}C-NMR Spectrum in $CDCl_3$

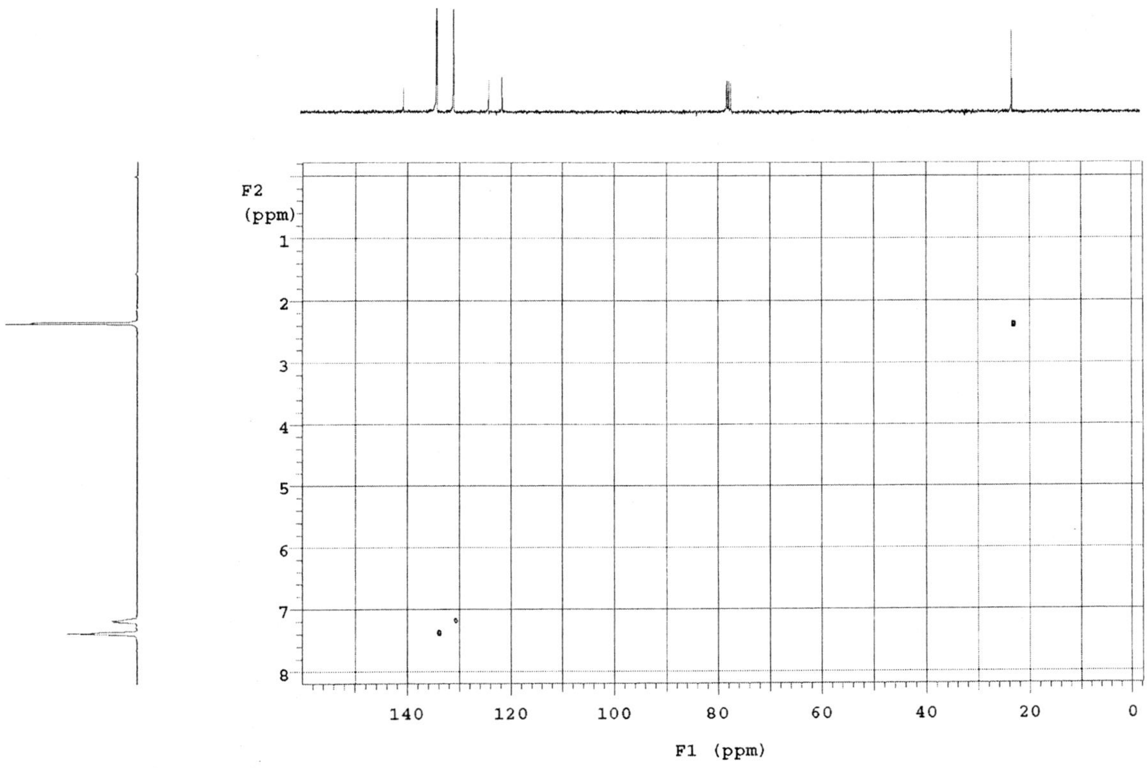

gHSQC Spectrum in $CDCl_3$

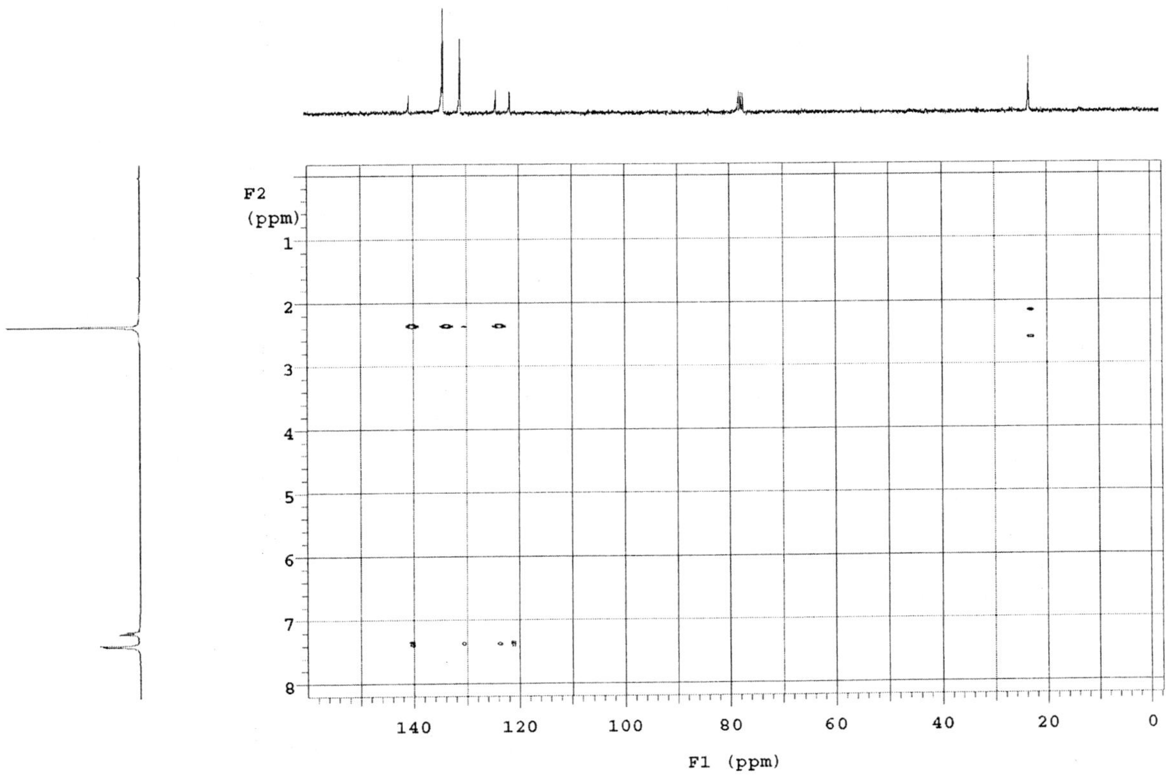

gHMBC Spectrum in $CDCl_3$

Problem 2.31 Given the spectroscopic data below (Proton, Carbon, Edited DEPT, and gHSQC), provide the structure of the molecule.

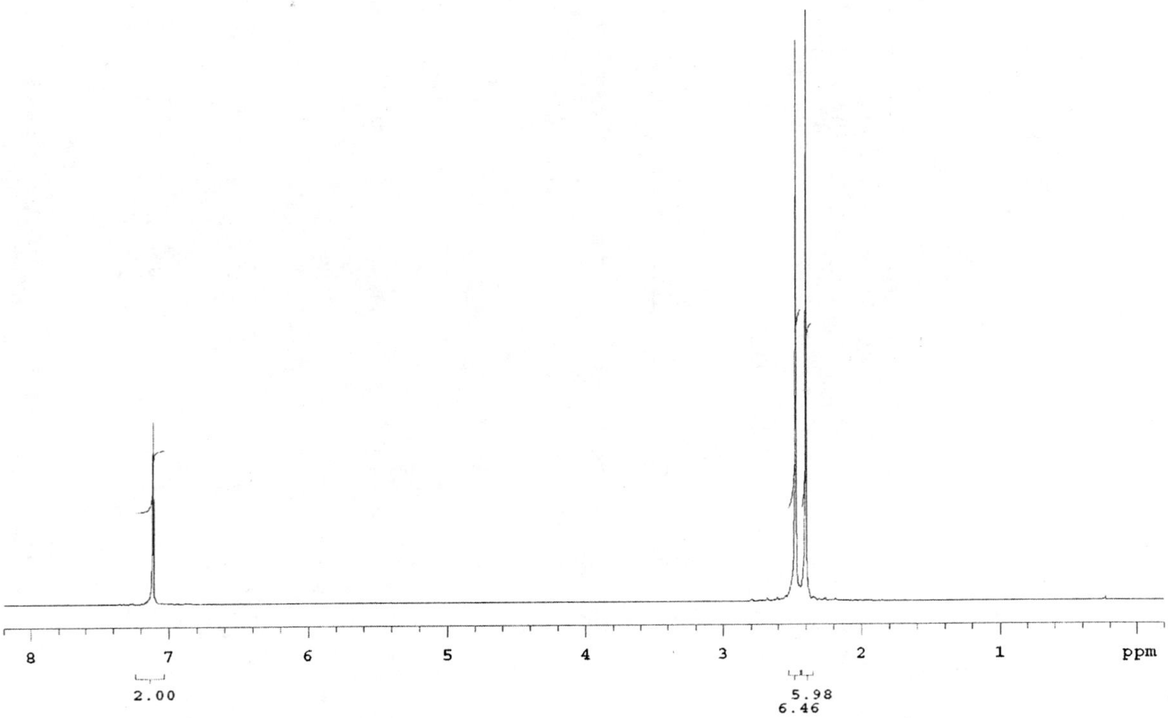

^{1}H-NMR Spectrum in $CDCl_3$

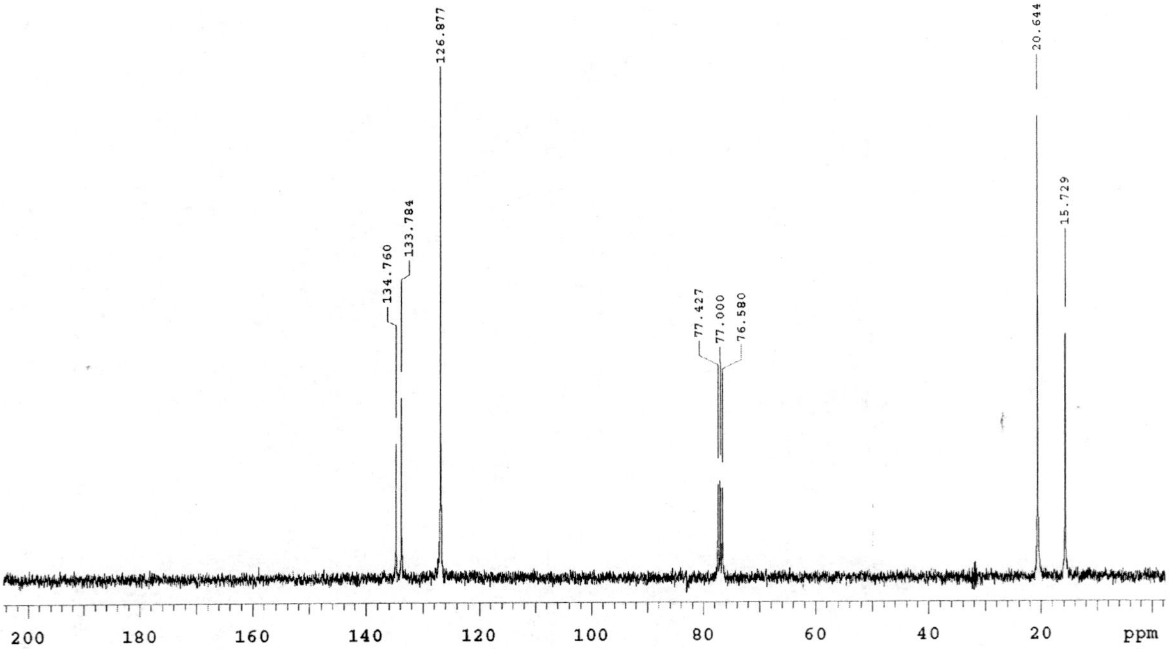

^{13}C-NMR Spectrum in $CDCl_3$

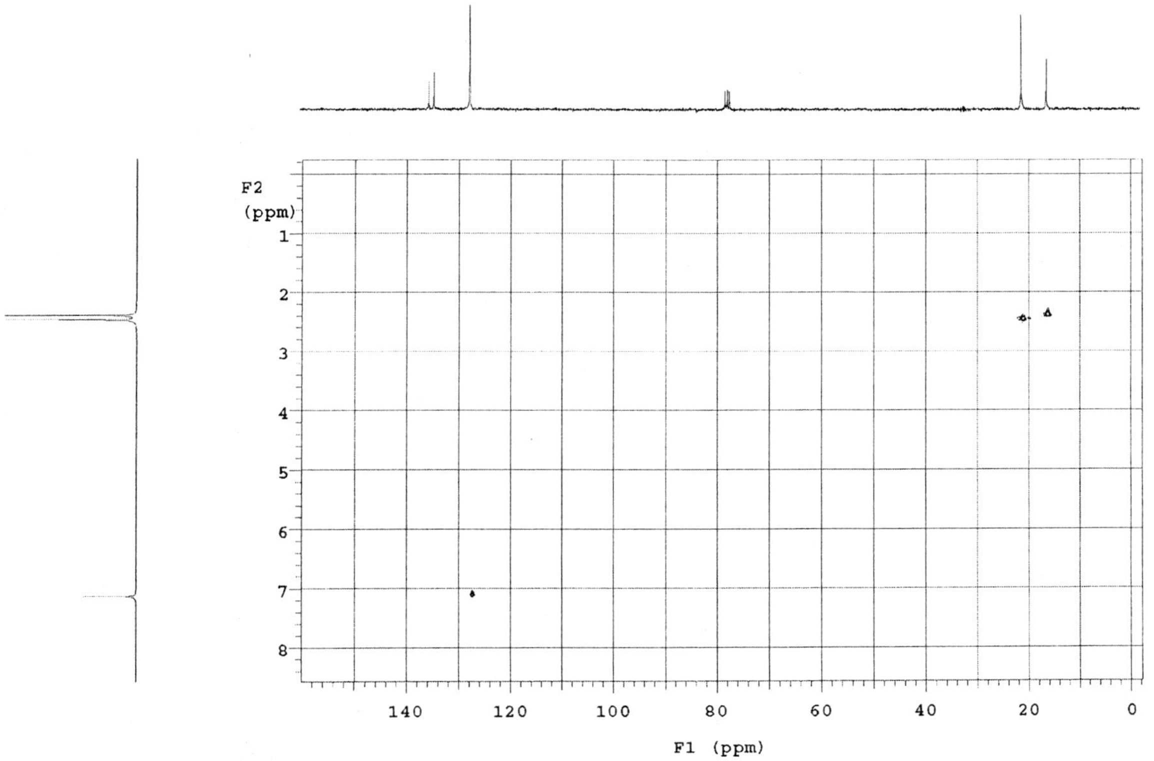

gHSQC Spectrum in $CDCl_3$

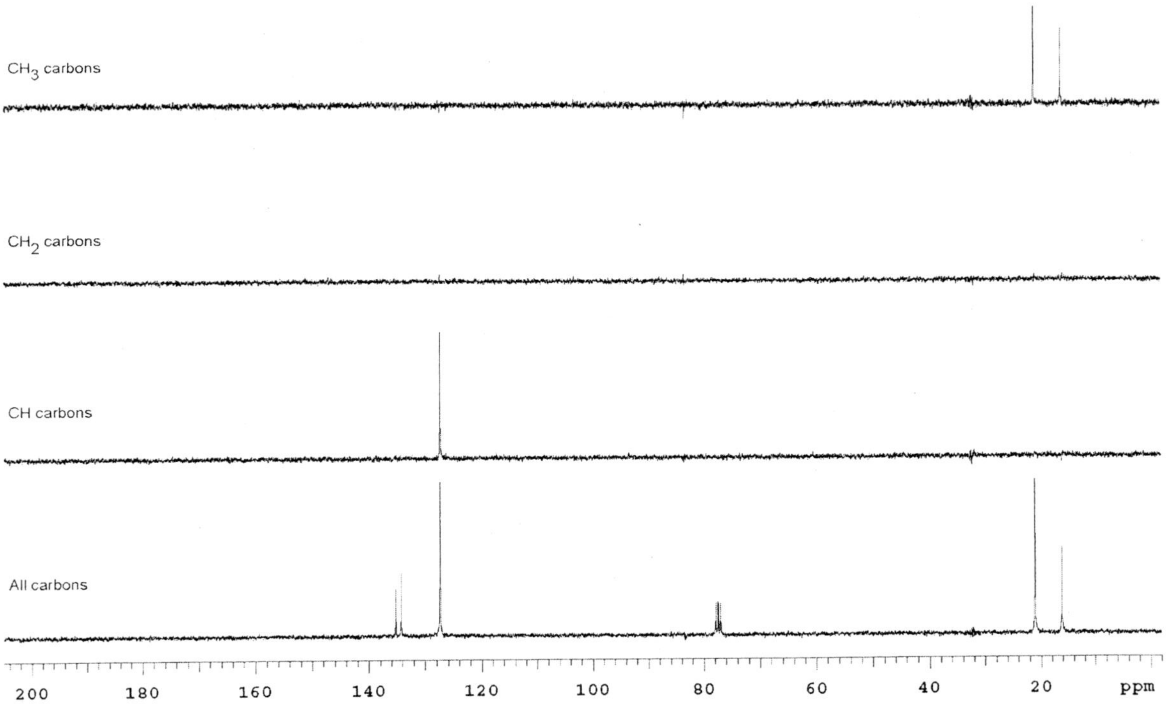

Edited DEPT ^{13}C-NMR Spectrum in $CDCl_3$

Problem 2.32 Given the spectroscopic data below (Proton, Carbon, Edited DEPT, and gHSQC), provide the structure of the molecule.

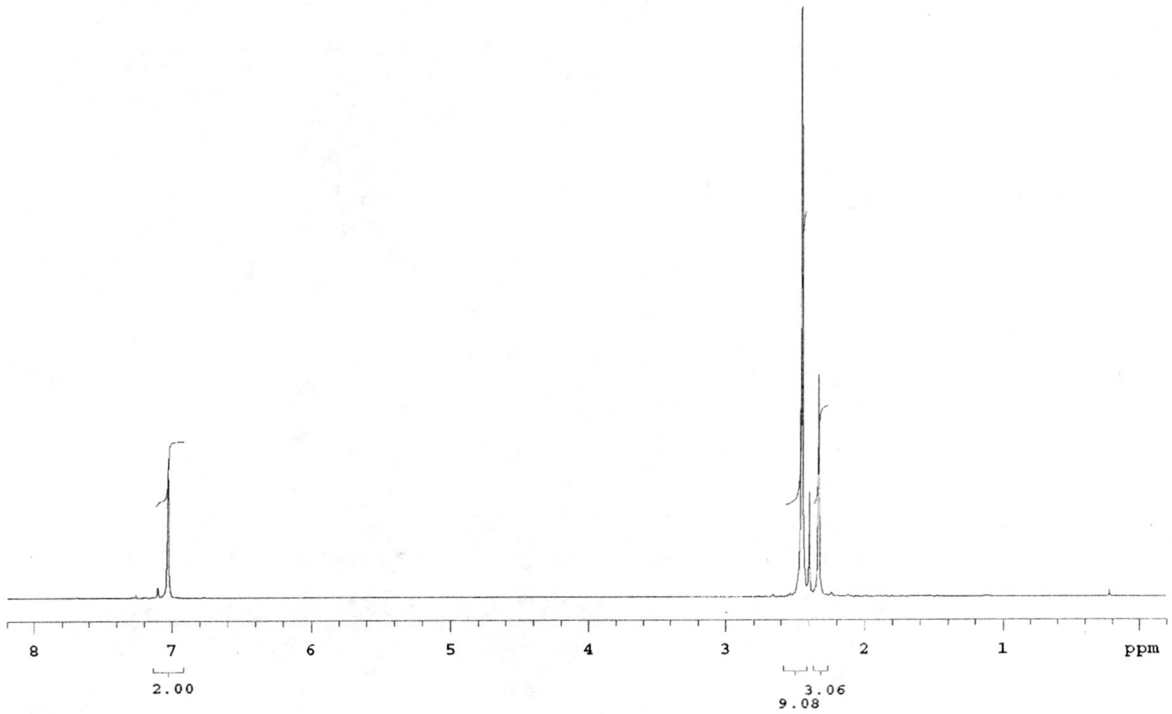

^{1}H-NMR Spectrum in $CDCl_3$

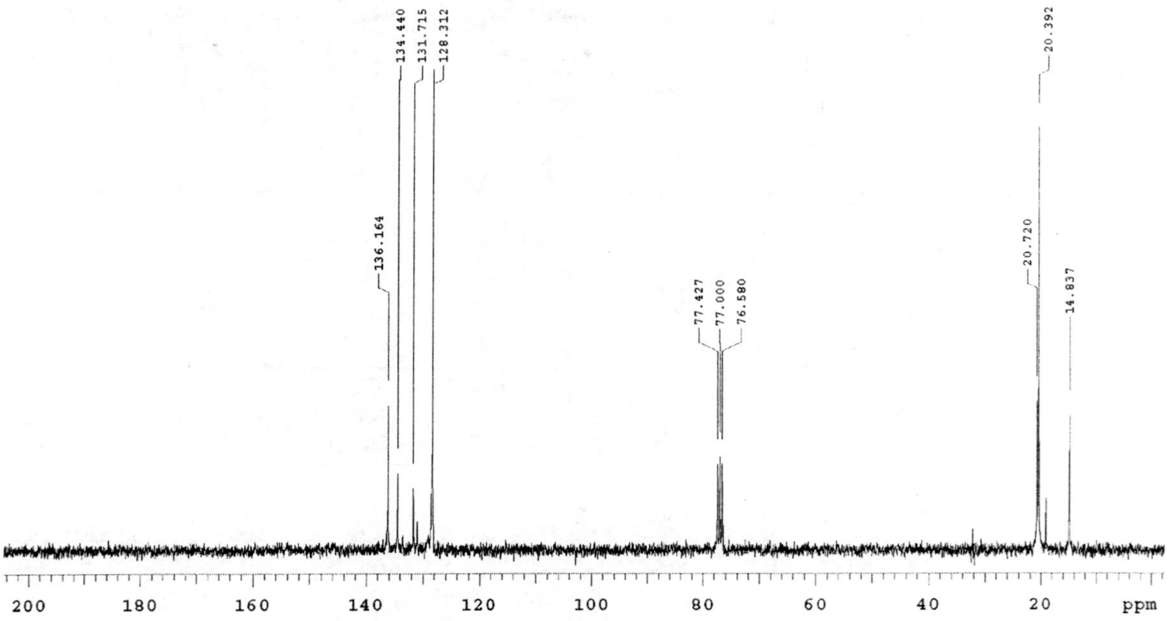

^{13}C-NMR Spectrum in $CDCl_3$

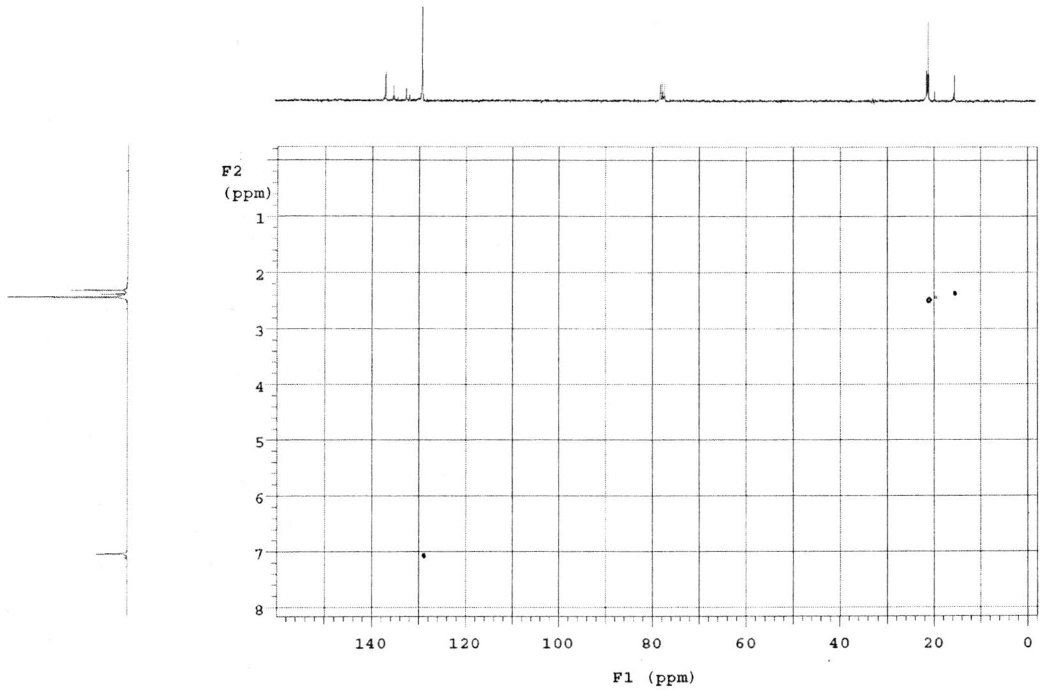

gHSQC Spectrum in $CDCl_3$

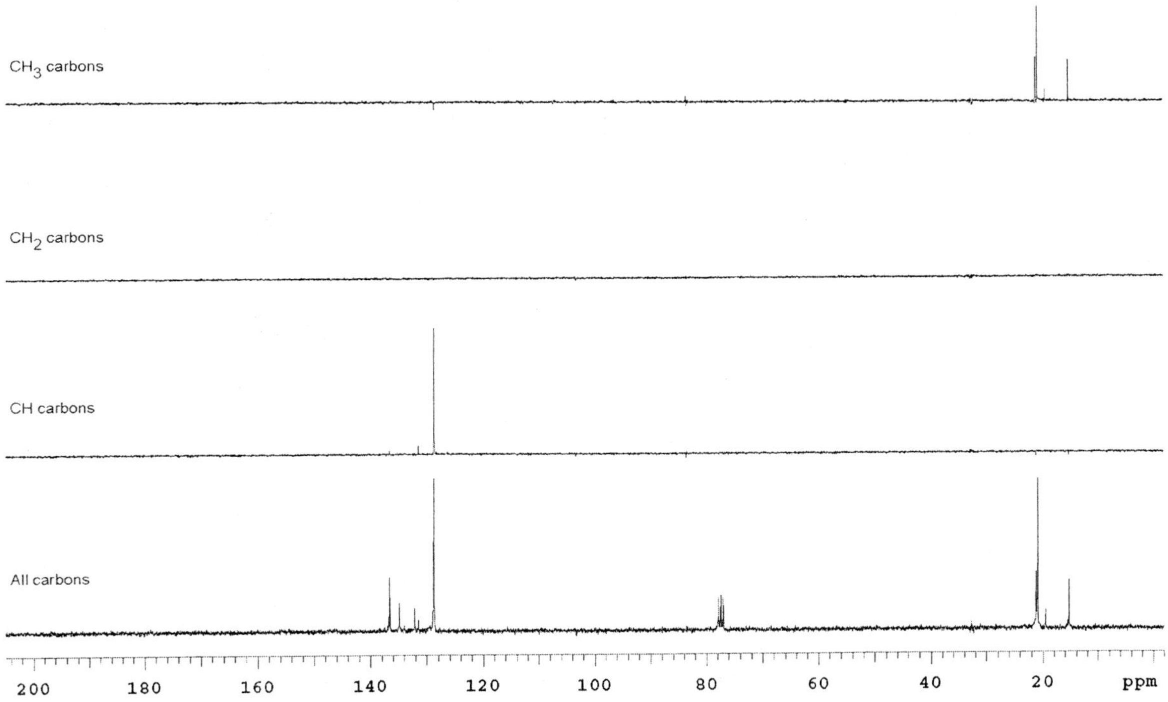

Edited DEPT ^{13}C-NMR Spectrum in $CDCl_3$

Problem 2.33 Given the spectroscopic data below (Proton, Carbon, and gHSQC), provide the structure of the molecule. (C: 61.75%; H: 12.90%; N: 11.82%)

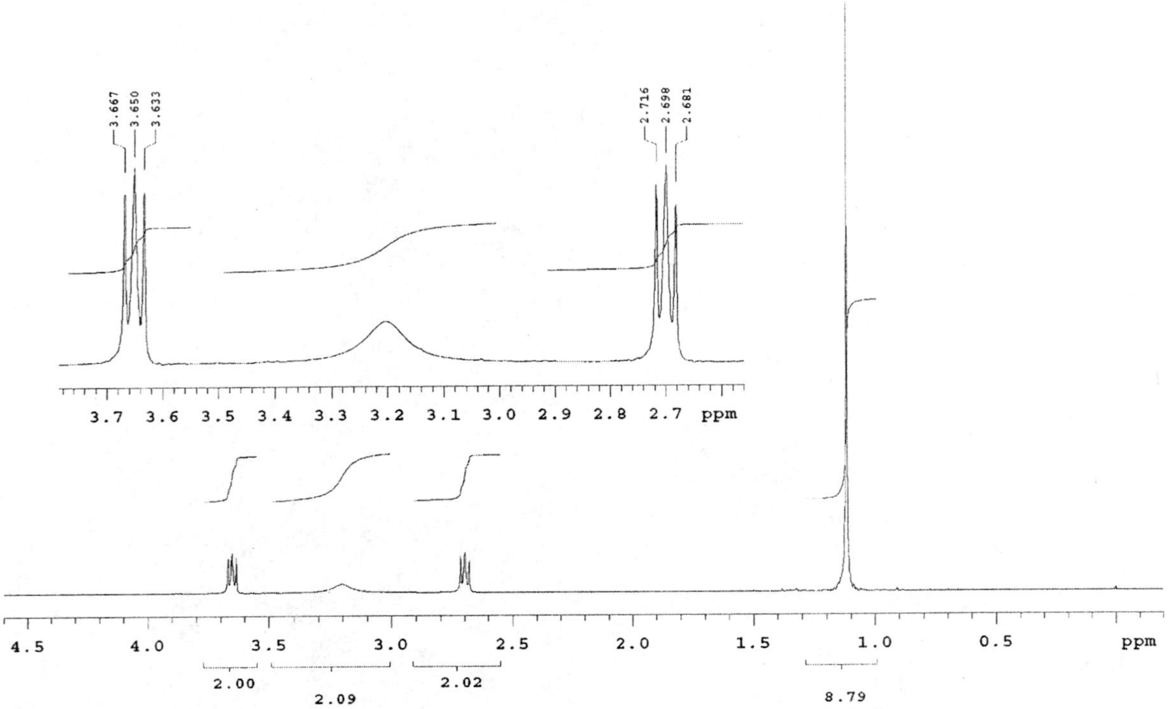

^{1}H-NMR Spectrum in $CDCl_3$

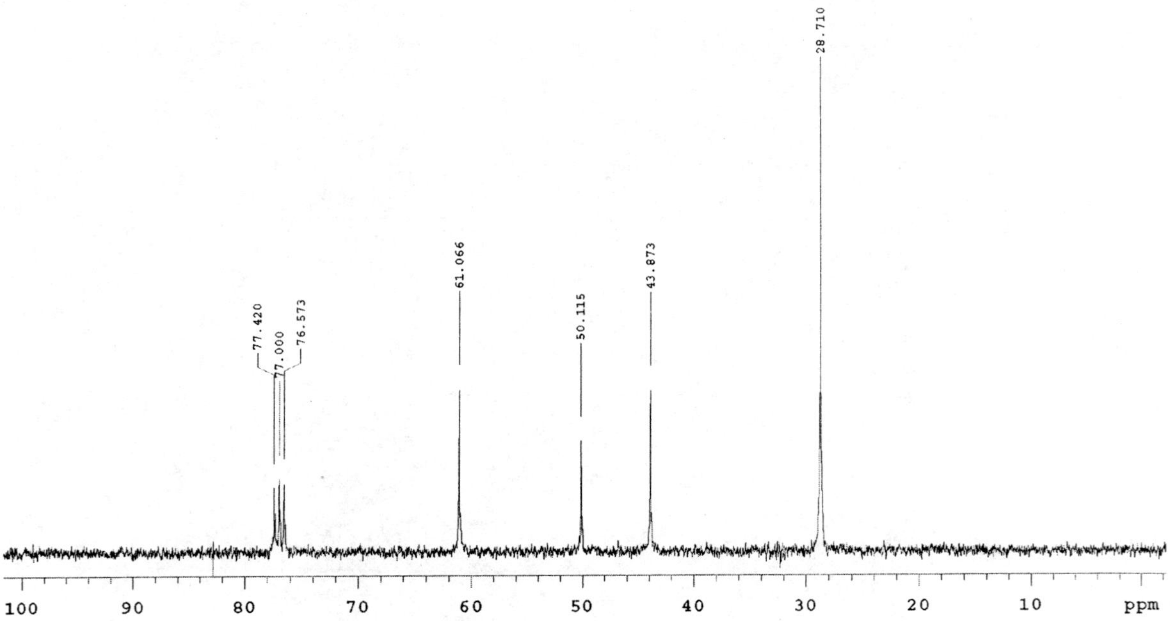

^{13}C-NMR Spectrum in $CDCl_3$

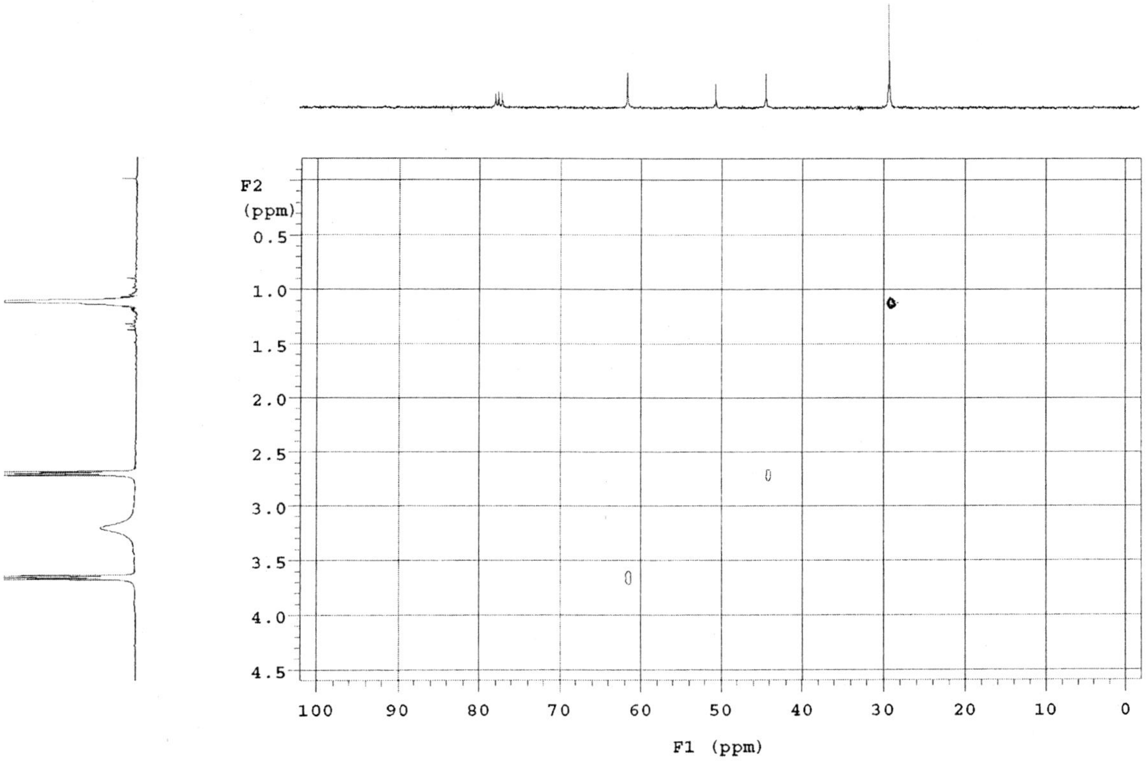

gHSQC Spectrum in $CDCl_3$

Problem 2.34 Given the spectroscopic data below (Proton, Carbon, gHSQC, and gHMBC), provide the structure of the molecule. (This red compound has MW of 182 g/mol.)

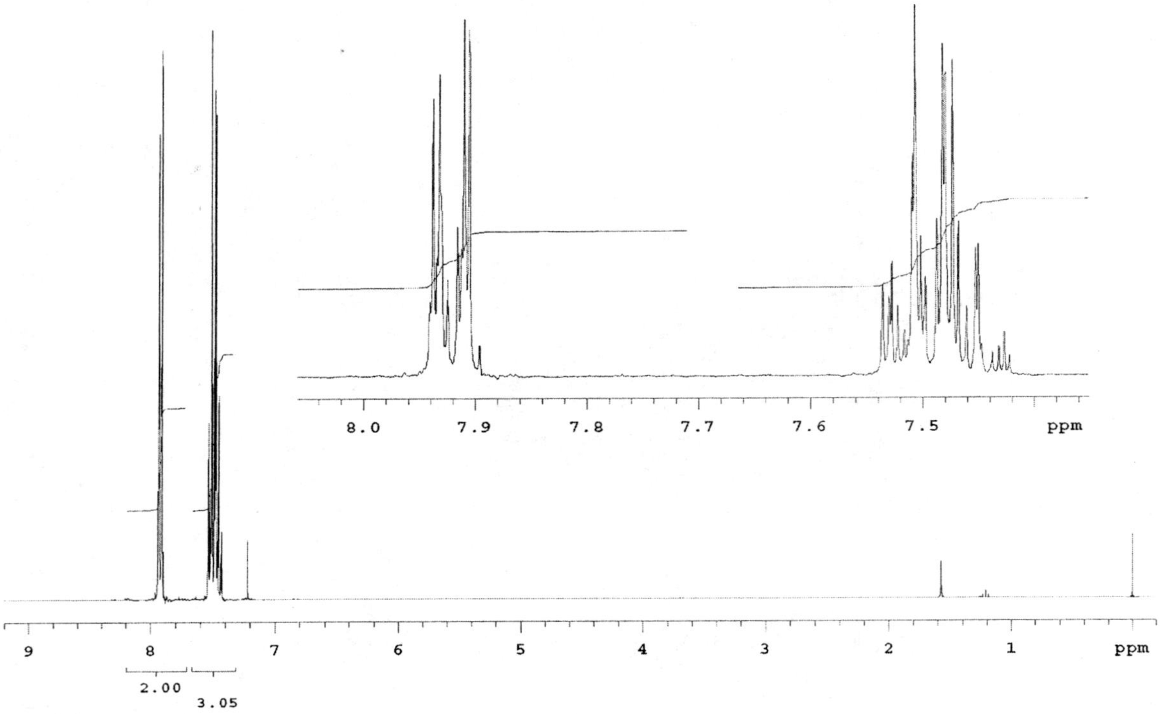

^{1}H-NMR Spectrum in $CDCl_3$

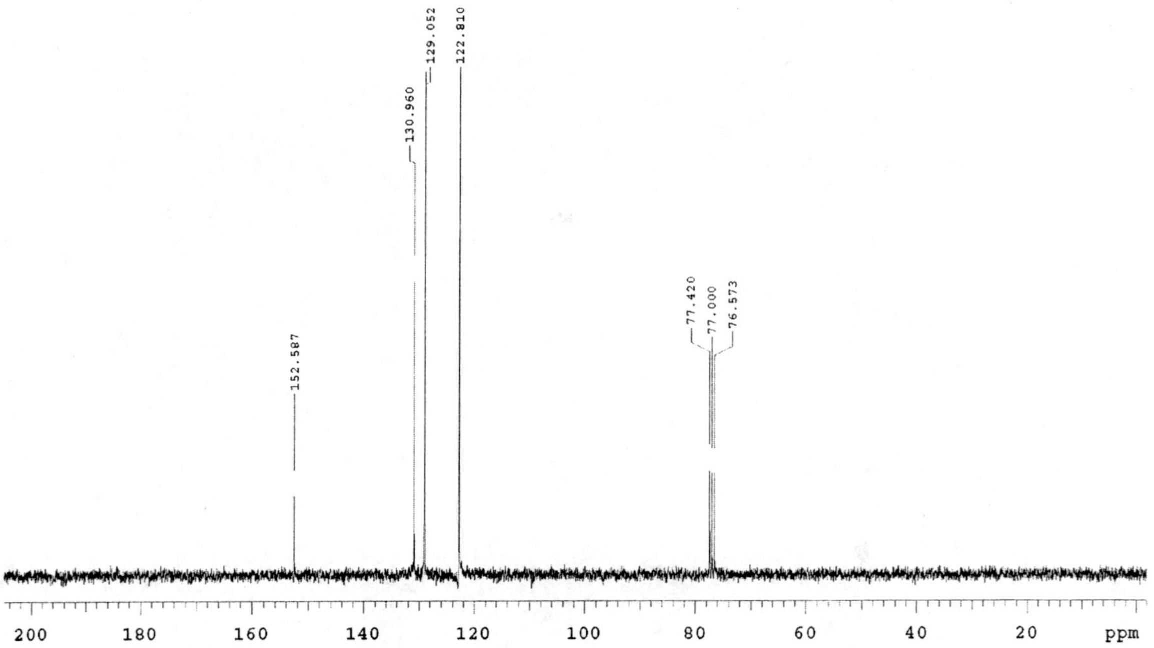

^{13}C-NMR Spectrum in $CDCl_3$

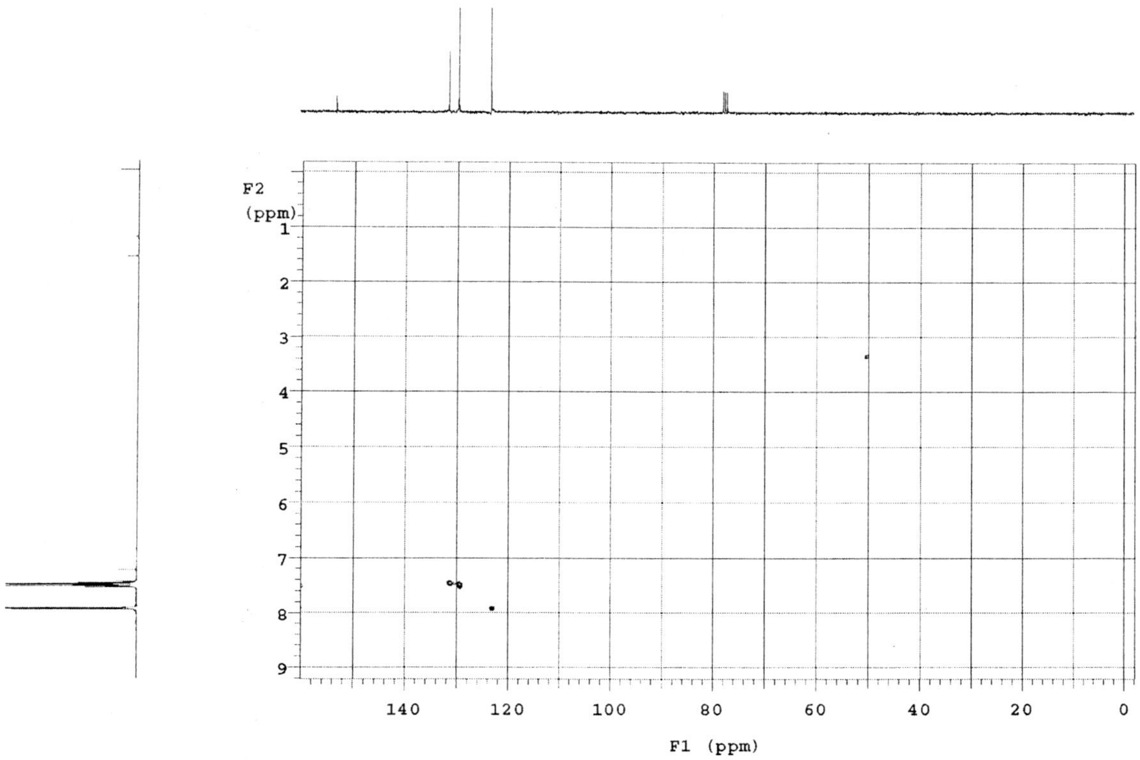

gHSQC Spectrum in $CDCl_3$

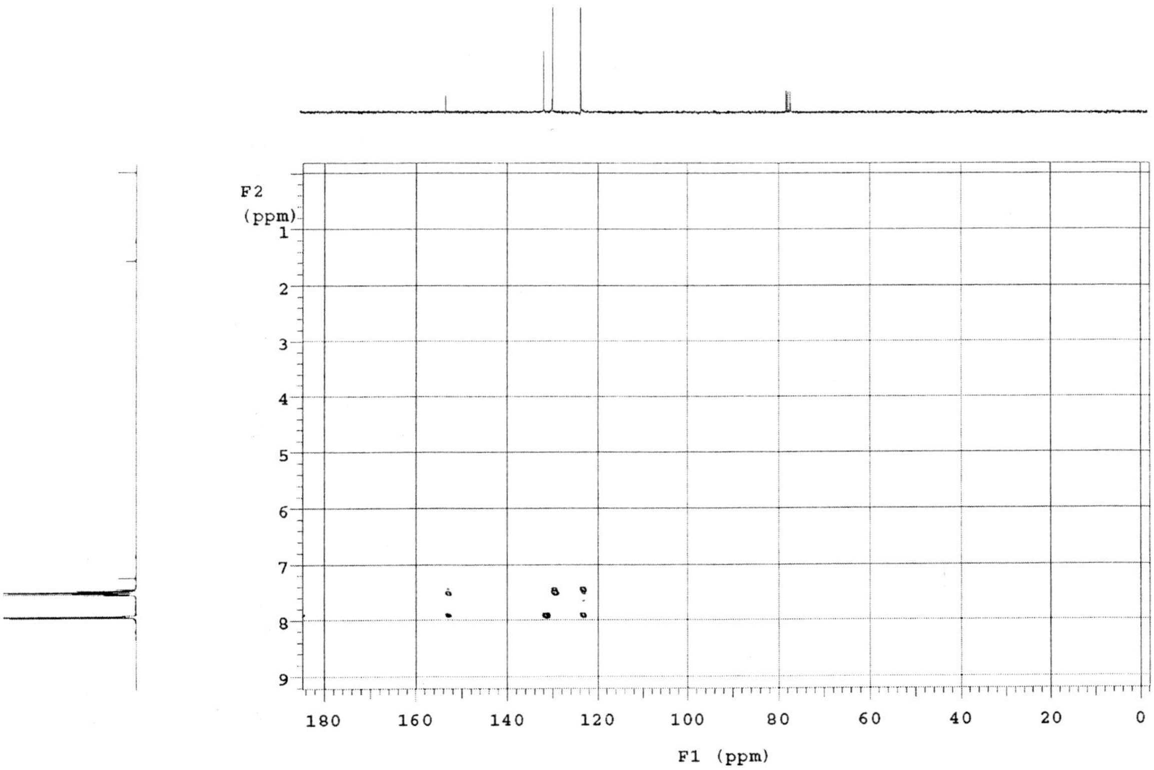

gHMBC Spectrum in $CDCl_3$

Problem 2.35 Given the spectroscopic data below (Proton, Carbon, gHSQC, and gHMBC), provide the structure of the molecule. This compound does not readily decarboxylate upon heating.

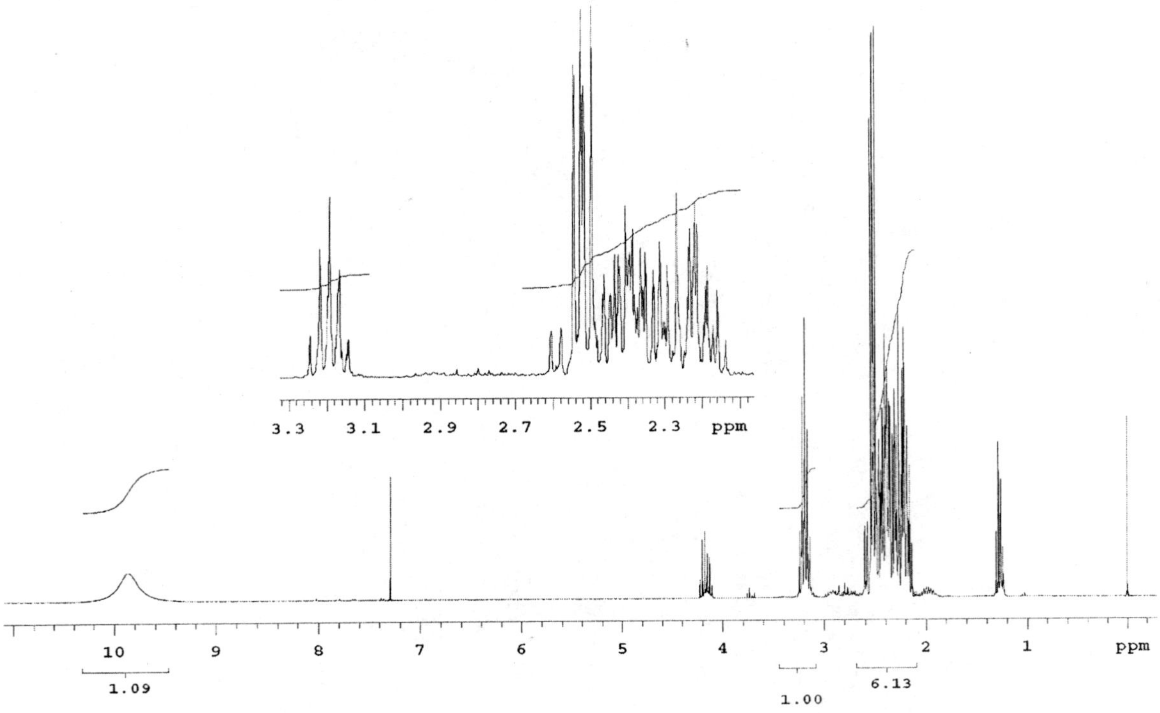

^{1}H-NMR Spectrum in $CDCl_3$

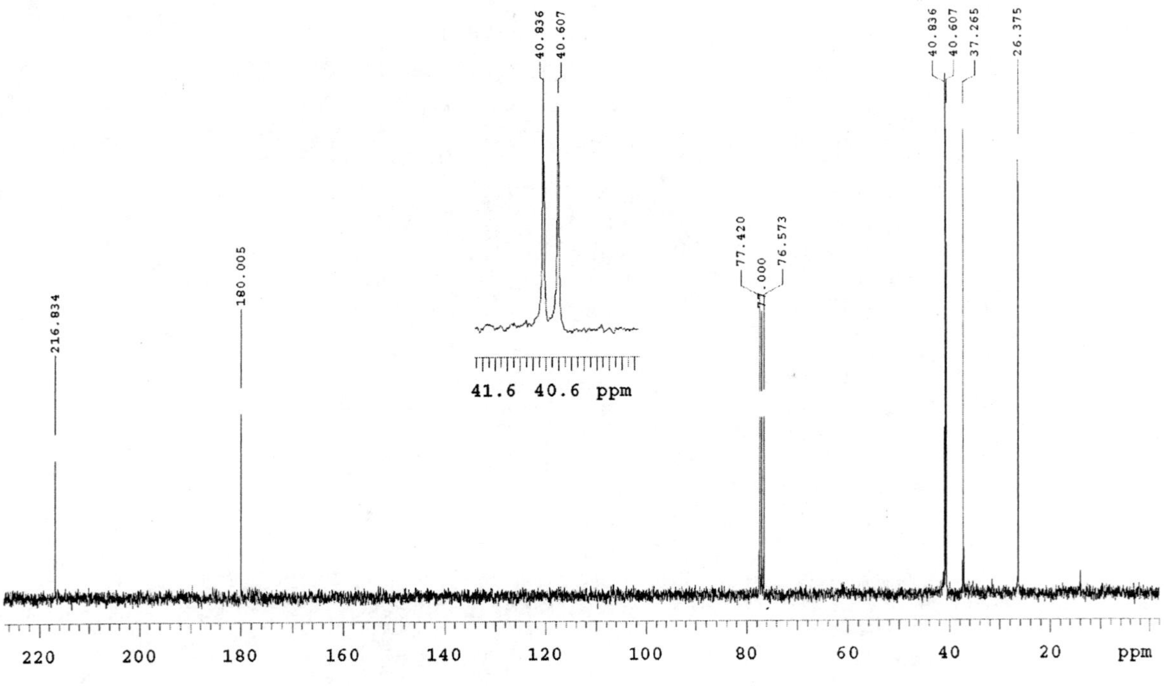

^{13}C-NMR Spectrum in $CDCl_3$

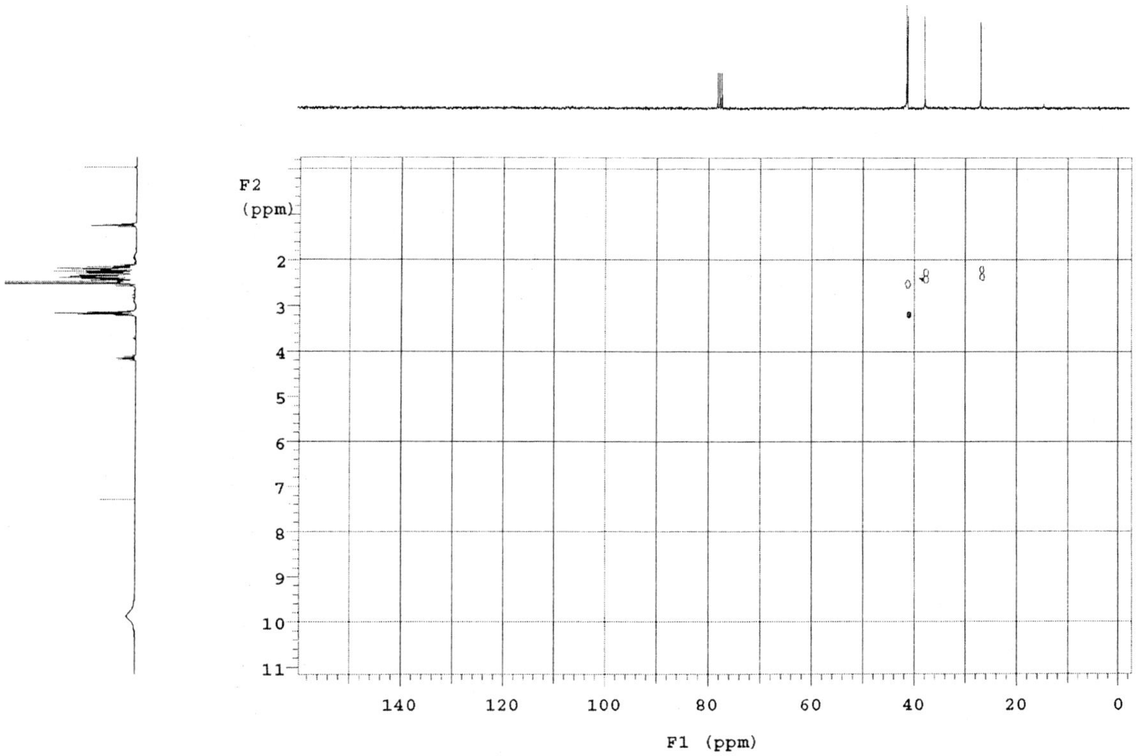

gHSQC Spectrum in $CDCl_3$

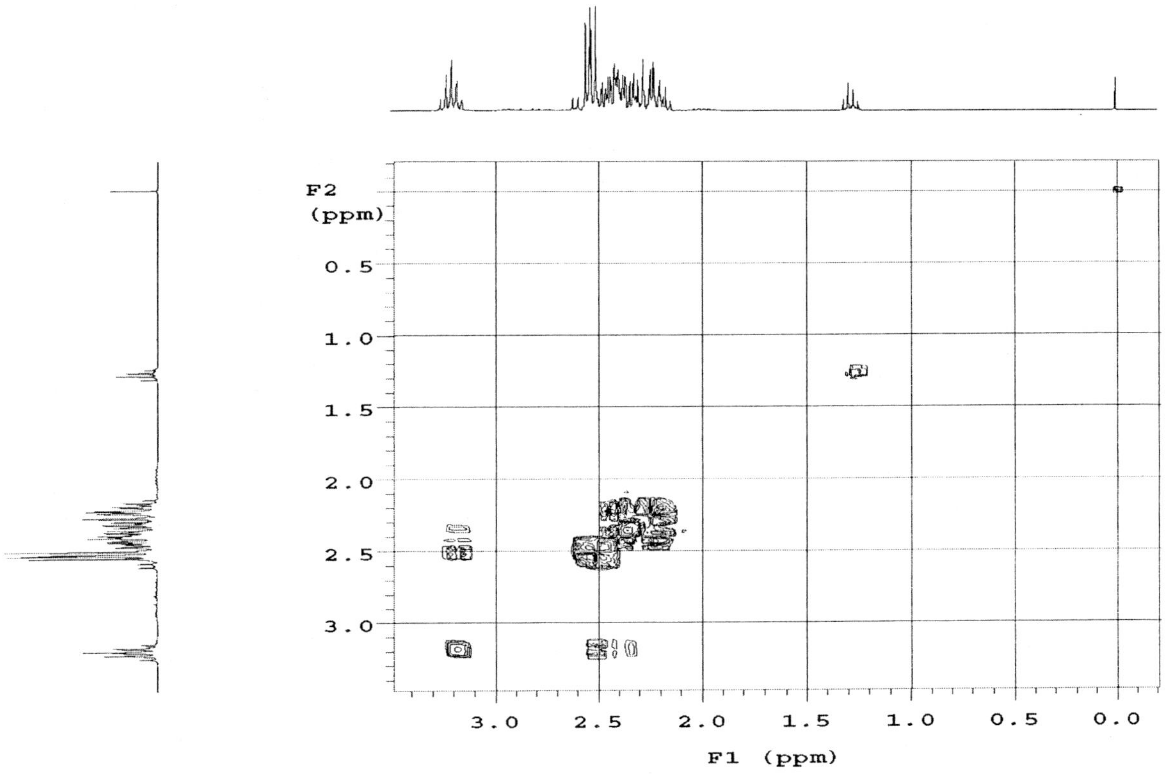

gCOSY Spectrum in $CDCl_3$

Problem 2.36 Given the spectroscopic data below (Proton, Carbon, gHSQC, and gHMBC), provide the structure of the molecule.

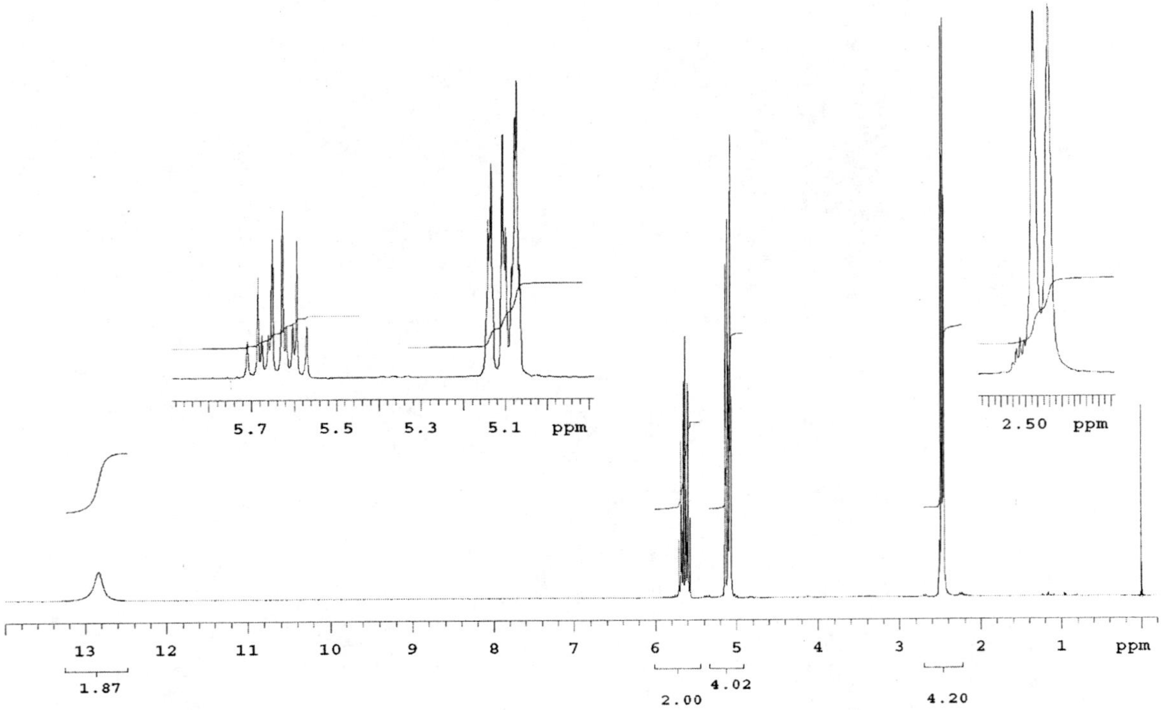

^{1}H-NMR Spectrum in DMSO-d_6

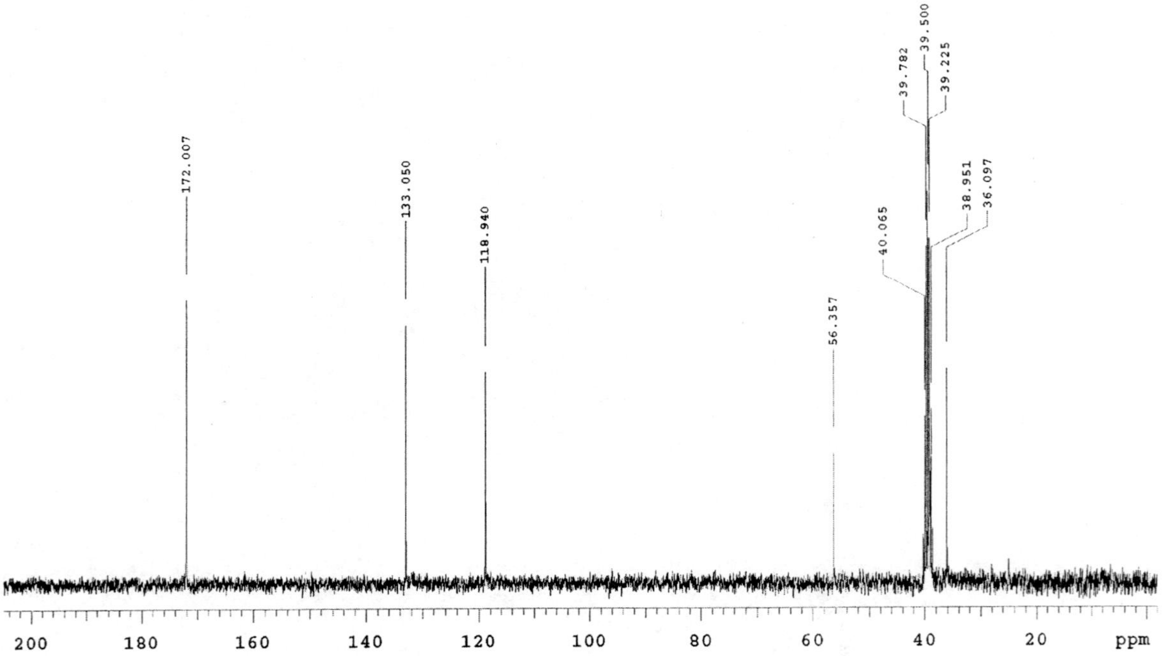

^{13}C-NMR Spectrum in DMSO-d_6

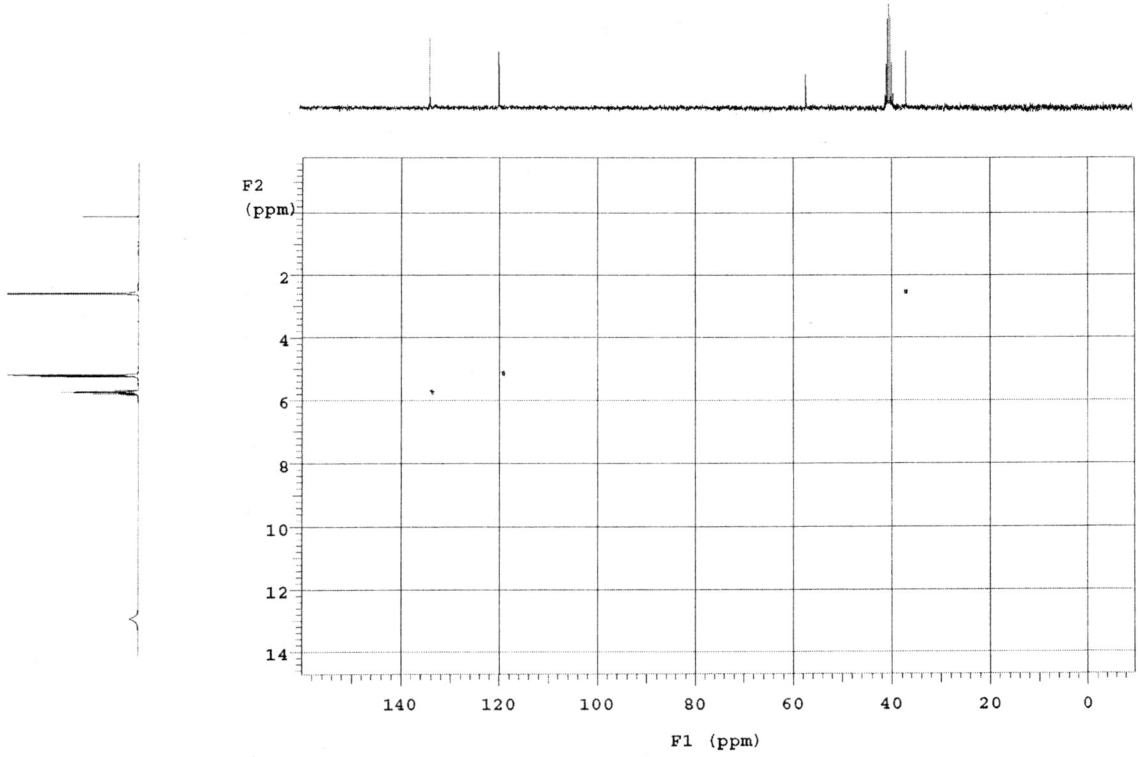

gHSQC Spectrum in DMSO-d_6

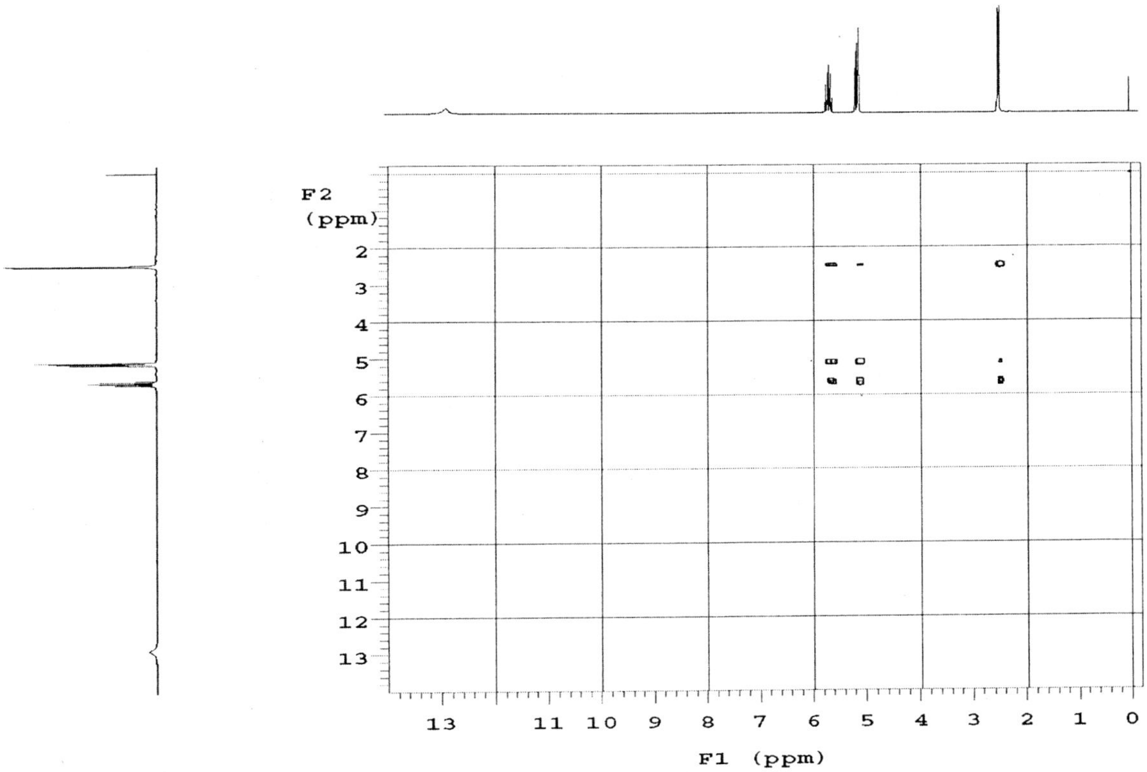

gCOSY Spectrum in DMSO-d_6

Problem 2.37 Given the spectroscopic data below (Proton, Carbon, and gHSQC), provide the structure of the molecule.

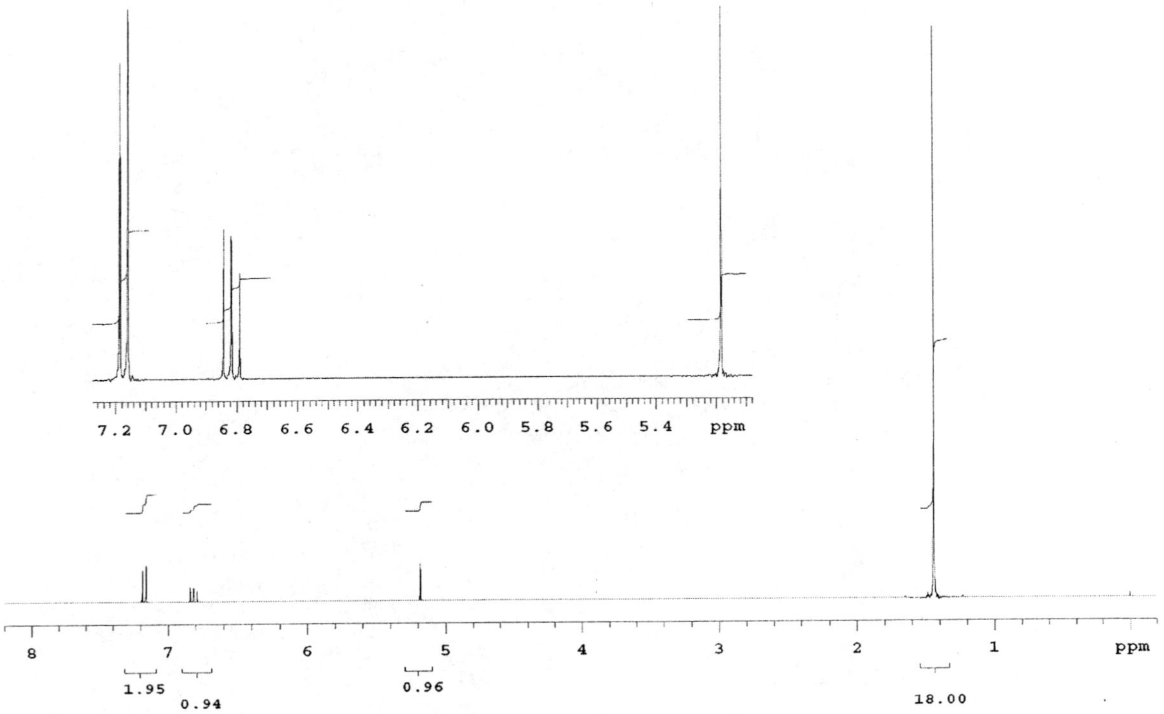

^{1}H-NMR Spectrum in $CDCl_3$

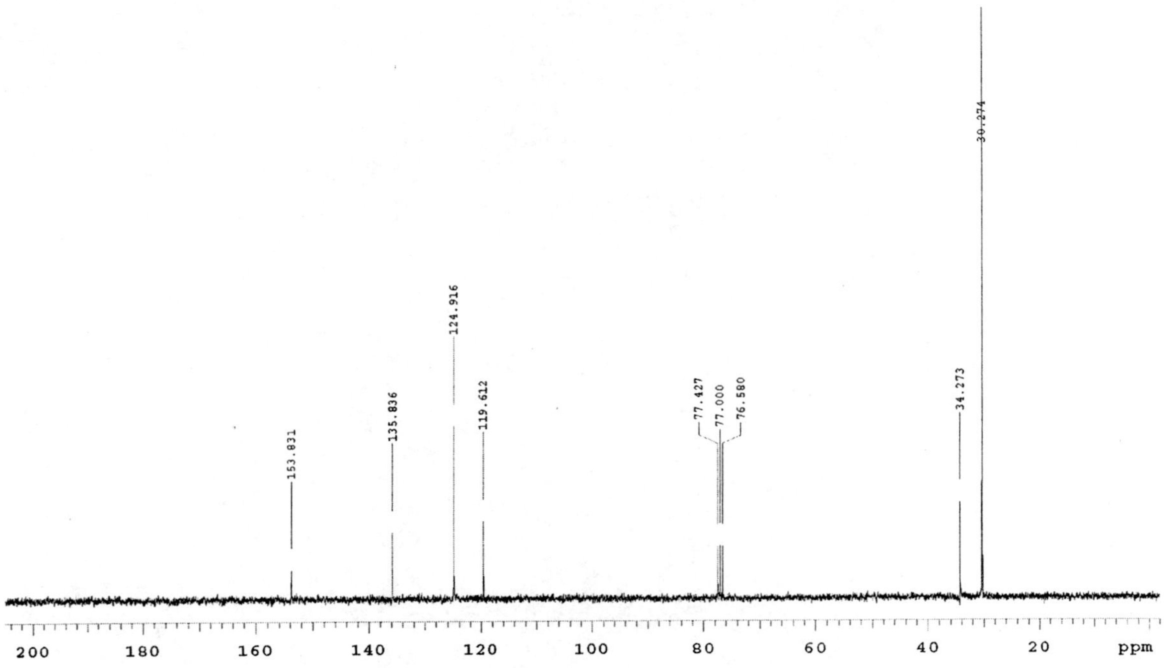

^{13}C-NMR Spectrum in $CDCl_3$

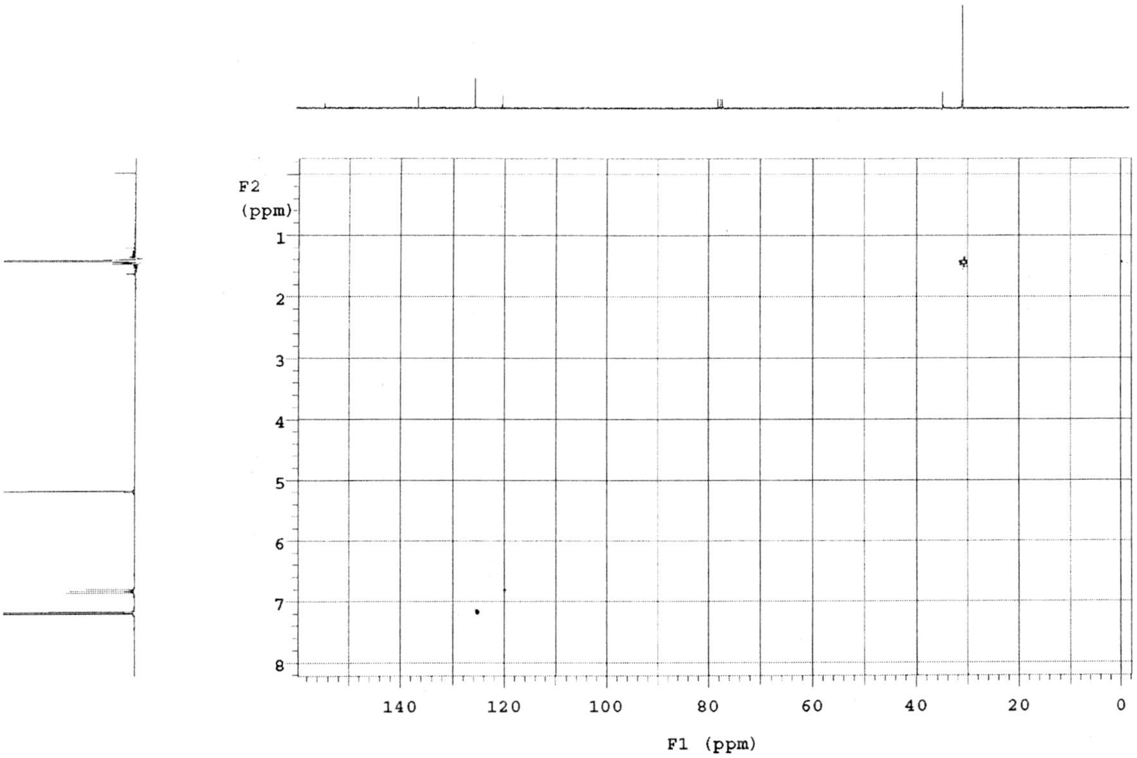

gHSQC Spectrum in $CDCl_3$

Problem 2.38 Given the spectroscopic data below (Proton, Carbon, Edited DEPT, and gHSQC), provide the structure of the molecule. (MW: 186 & 188 g/mol)

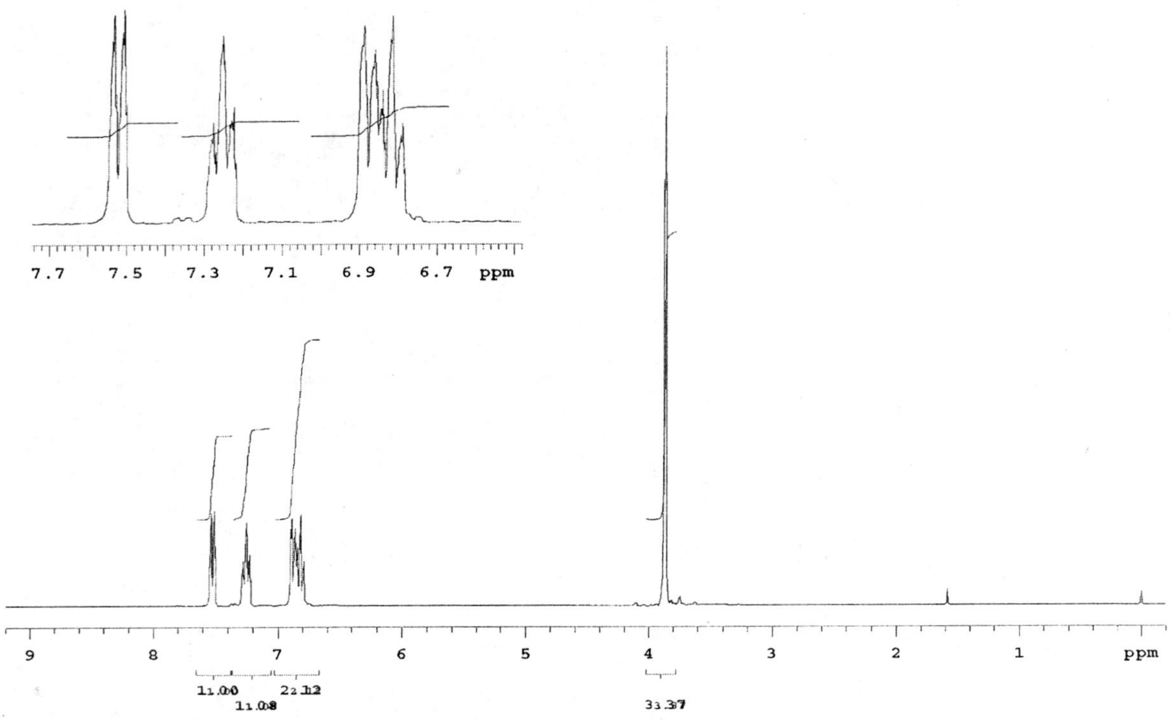

^{1}H-NMR Spectrum in $CDCl_3$

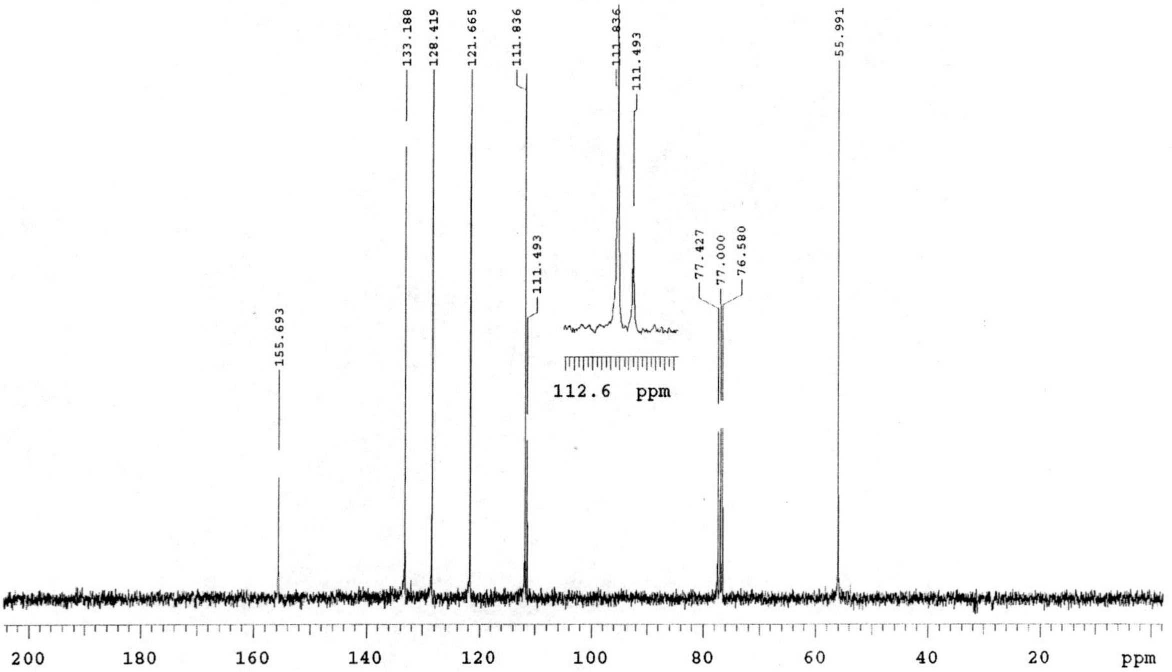

^{13}C-NMR Spectrum in $CDCl_3$

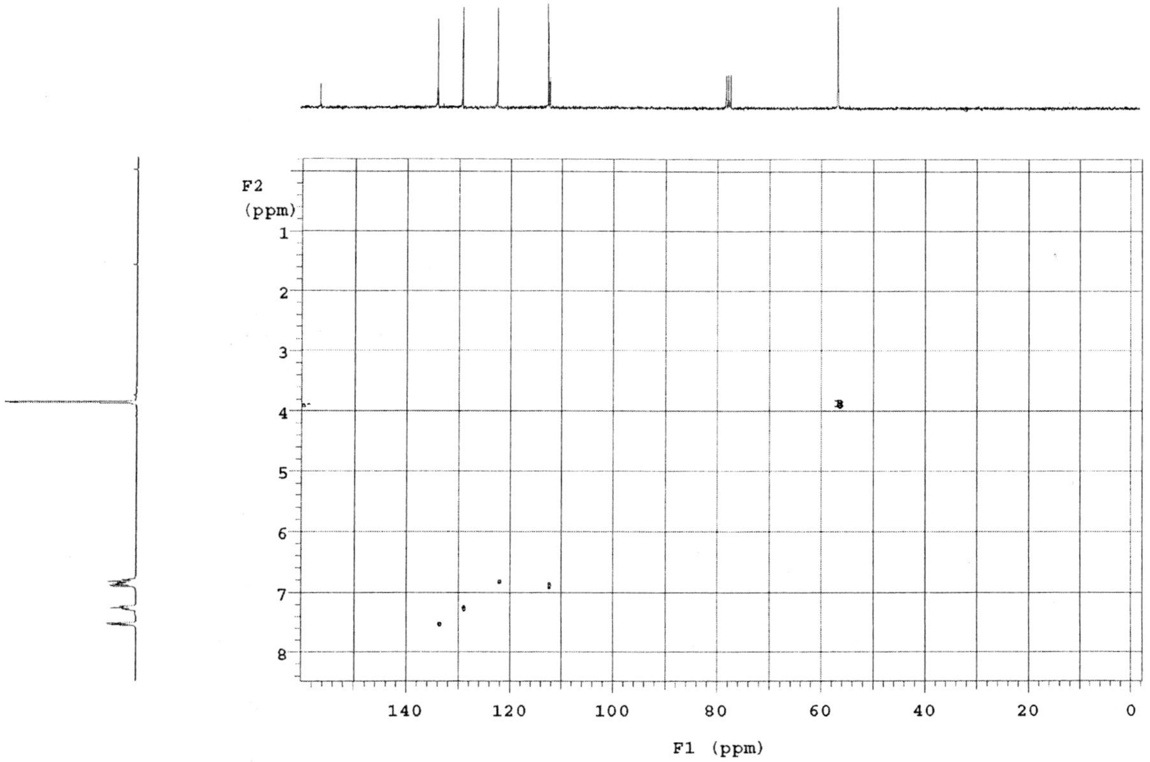

gHSQC Spectrum in $CDCl_3$

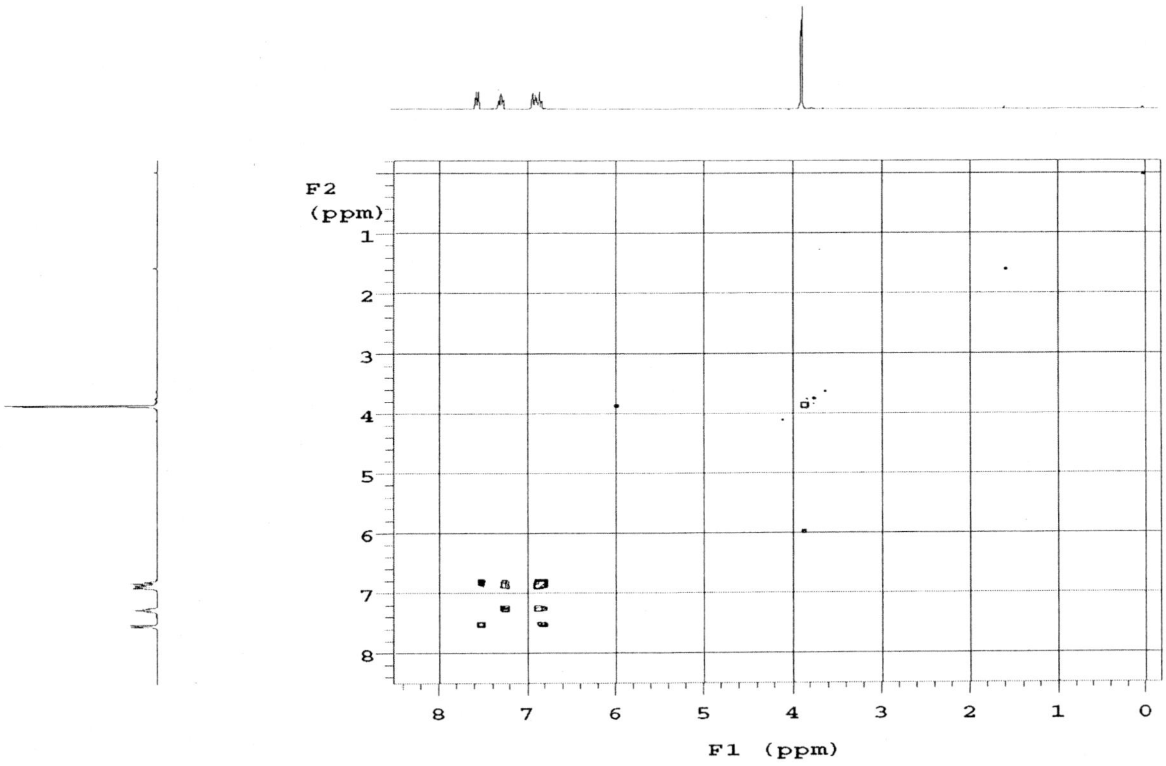

gCOSY Spectrum in $CDCl_3$

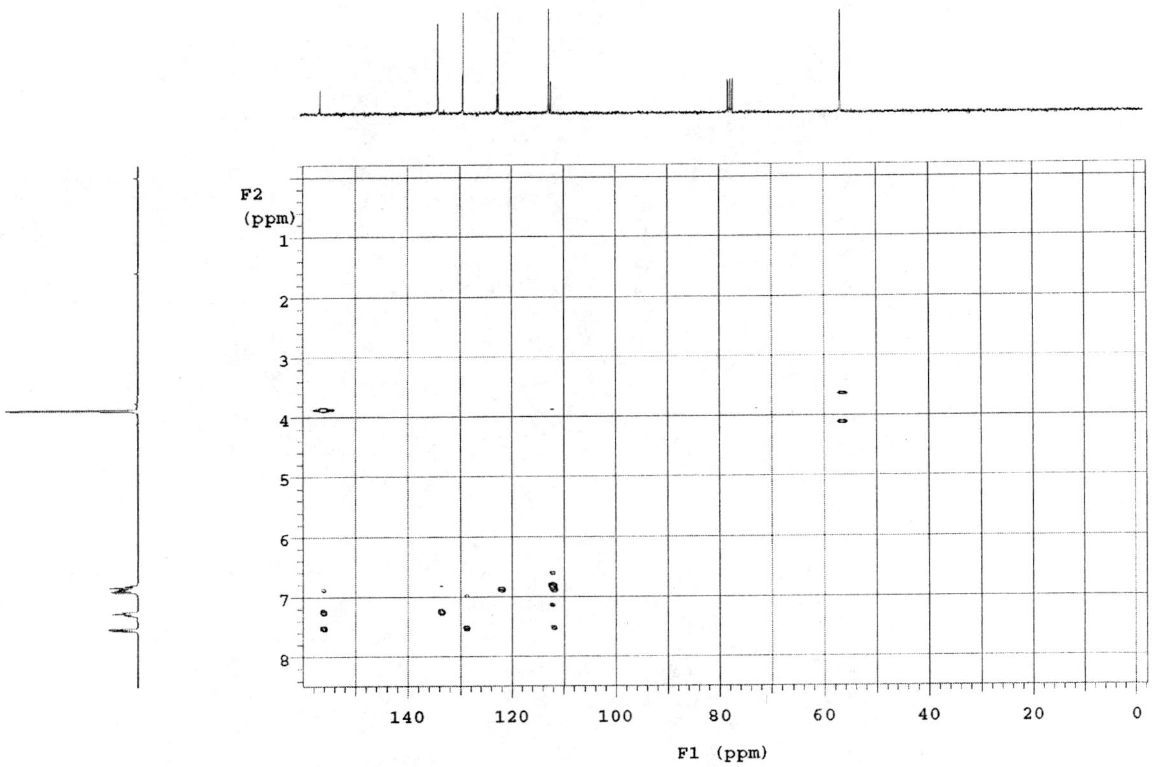

gHMBC Spectrum in $CDCl_3$

Problem 2.39 Given the spectroscopic data below (Proton, Carbon, Edited DEPT, and gHSQC), provide the structure of the molecule.

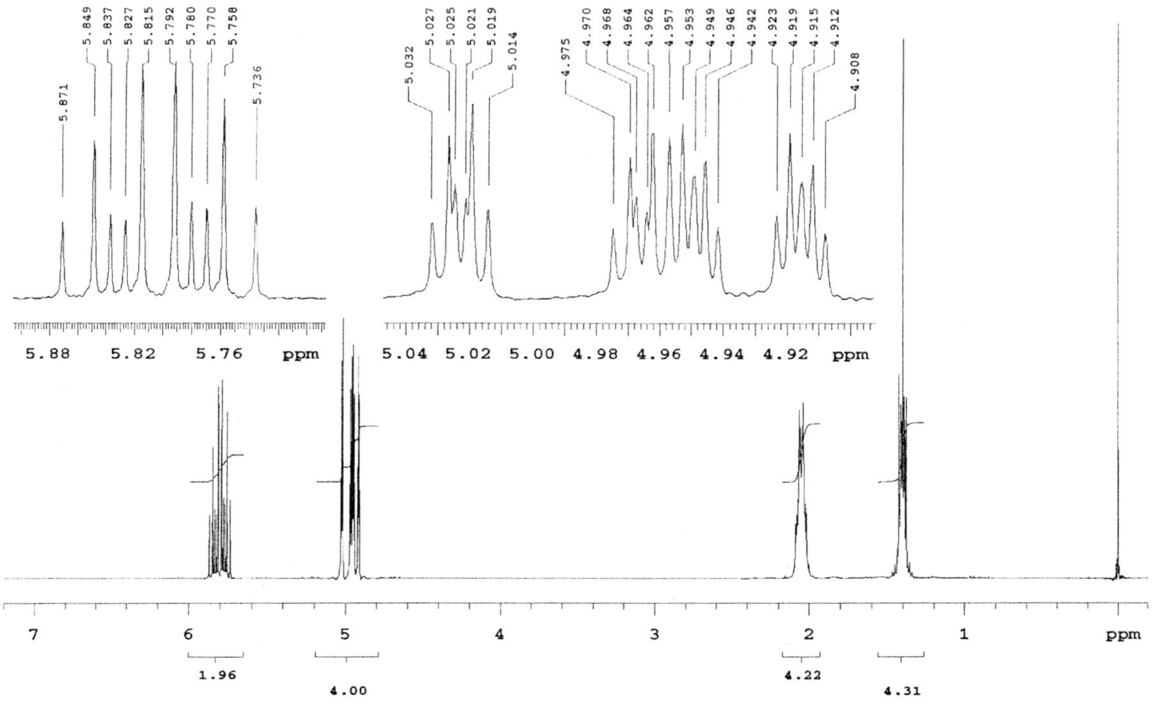

^{1}H-NMR Spectrum in $CDCl_3$

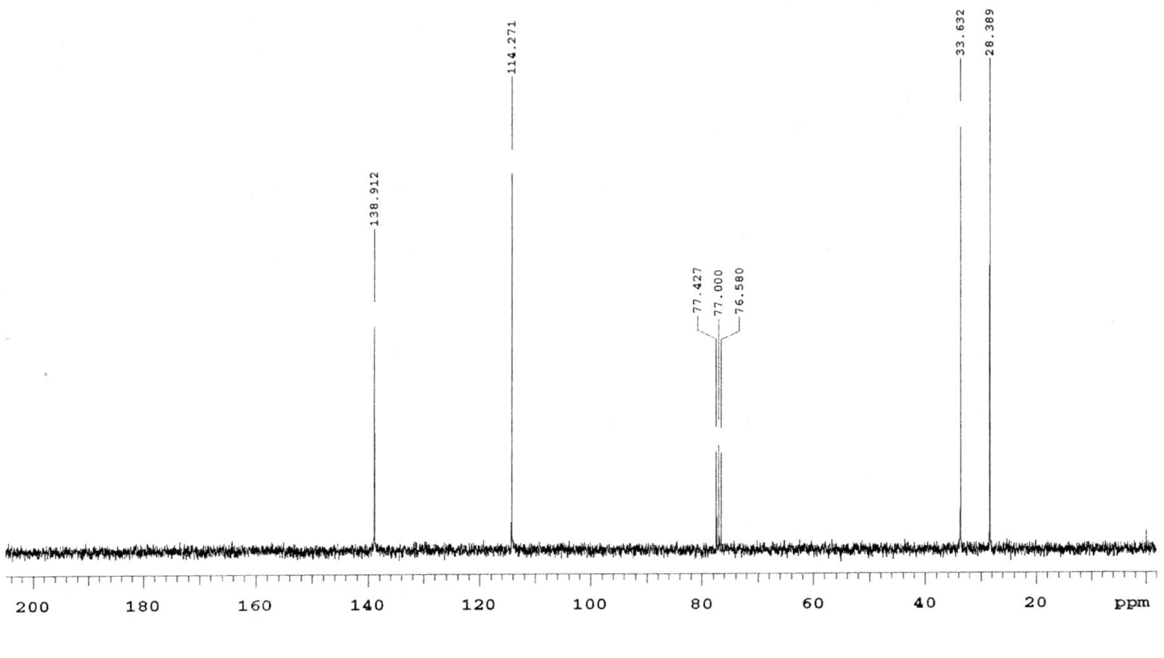

^{13}C-NMR Spectrum in $CDCl_3$

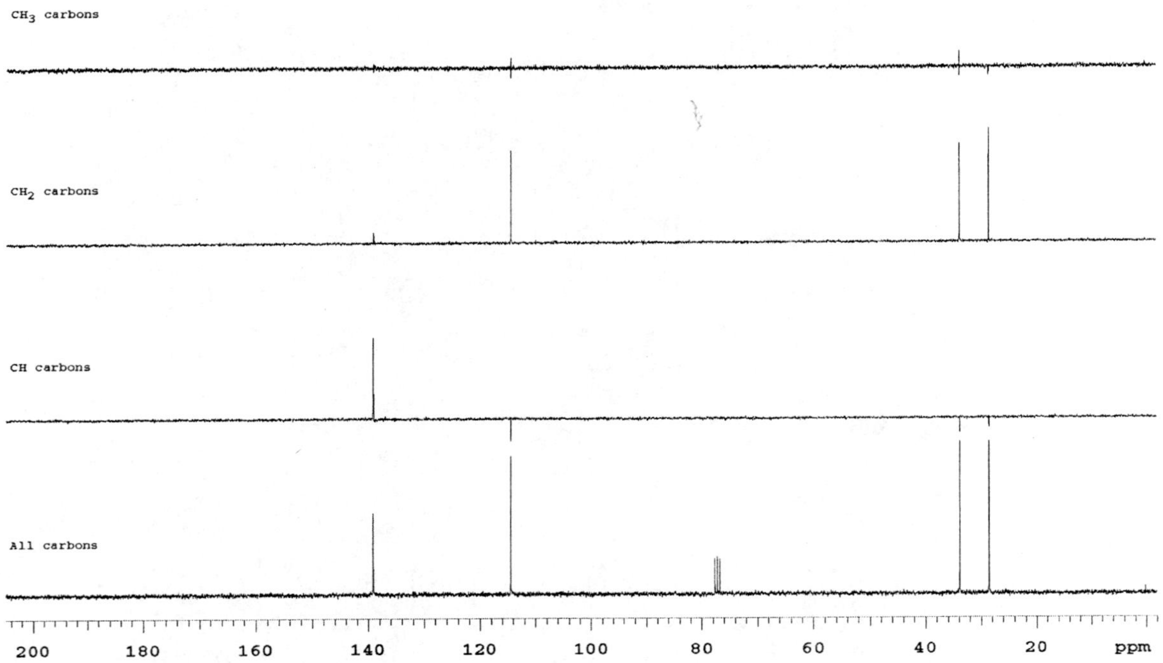

Edit DEPT ^{13}C-NMR Spectrum in $CDCl_3$

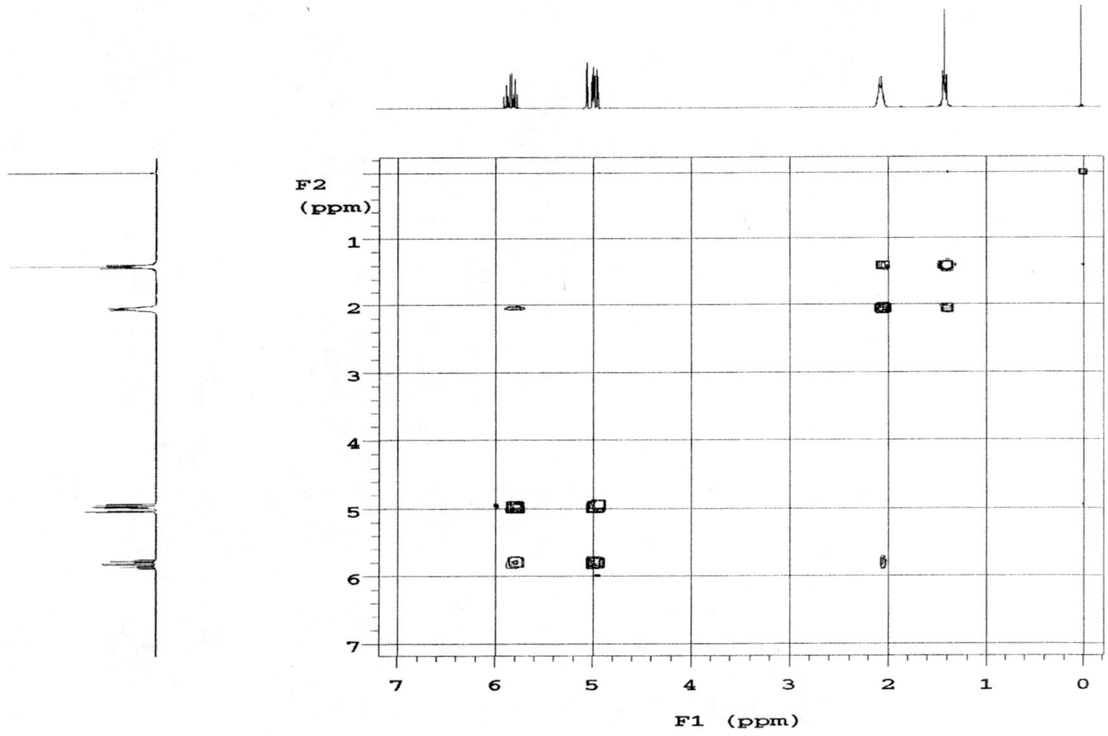

gCOSY Spectrum in $CDCl_3$

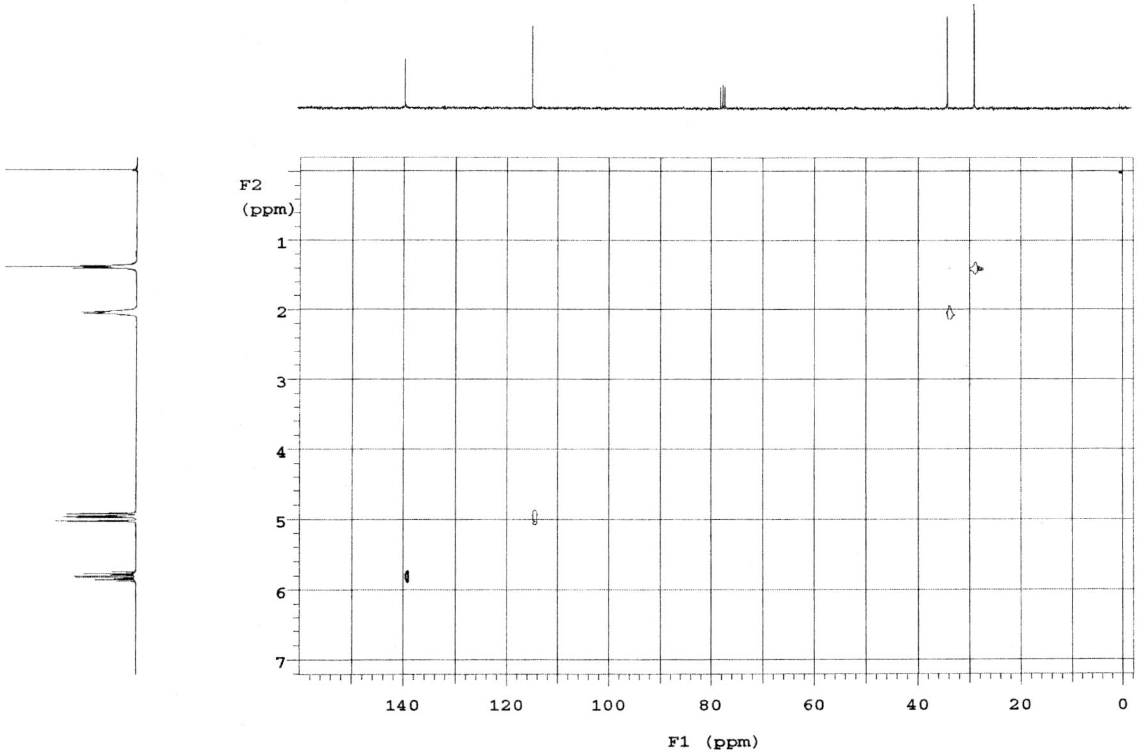

gHSQC Spectrum in $CDCl_3$

Problem 2.40 Given the spectroscopic data below (Proton, Carbon, Edited DEPT, and gHSQC), provide the structure of the molecule.

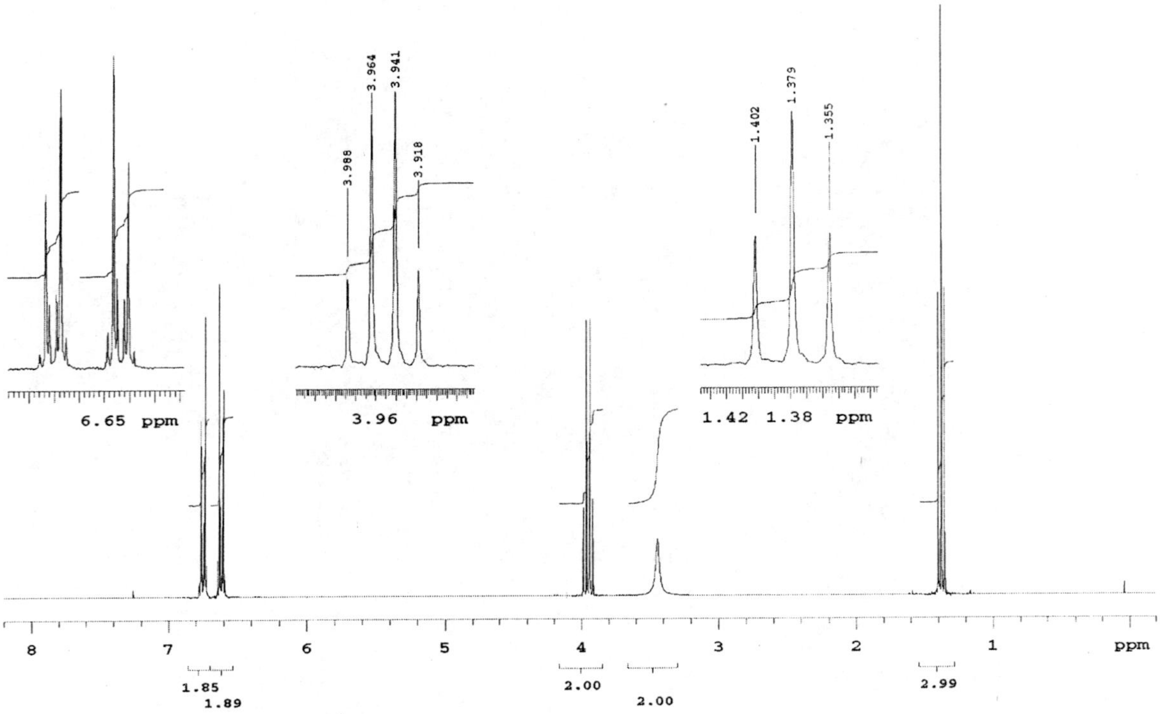

^{1}H-NMR Spectrum in $CDCl_3$

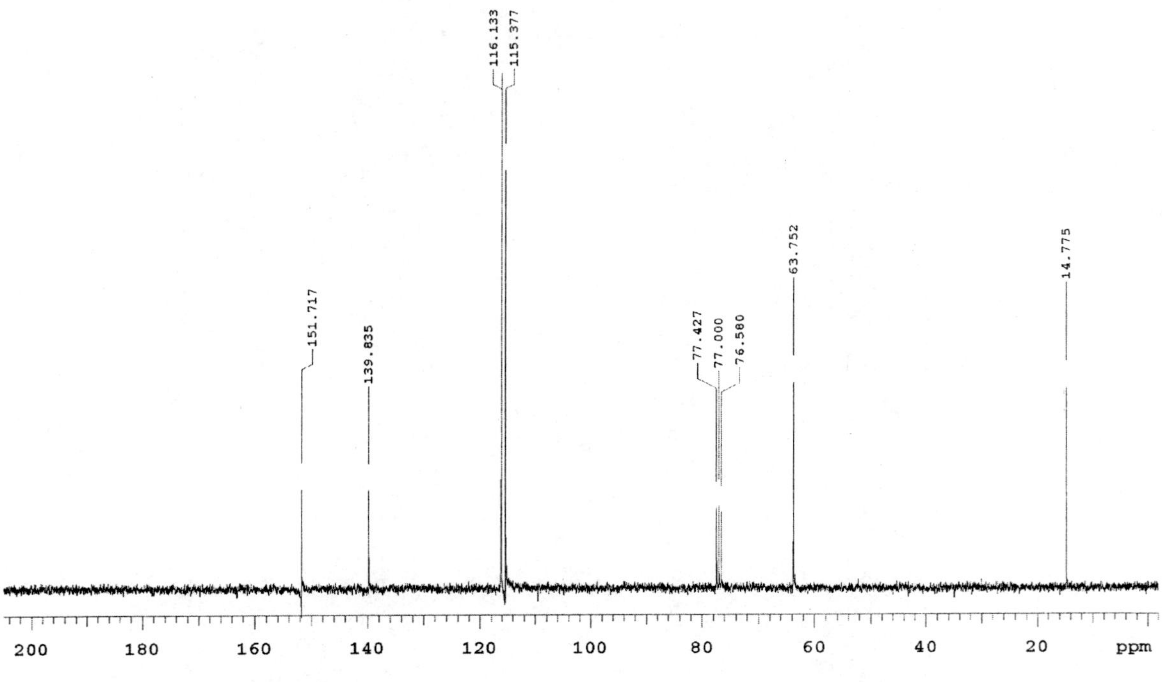

^{13}C-NMR Spectrum in $CDCl_3$

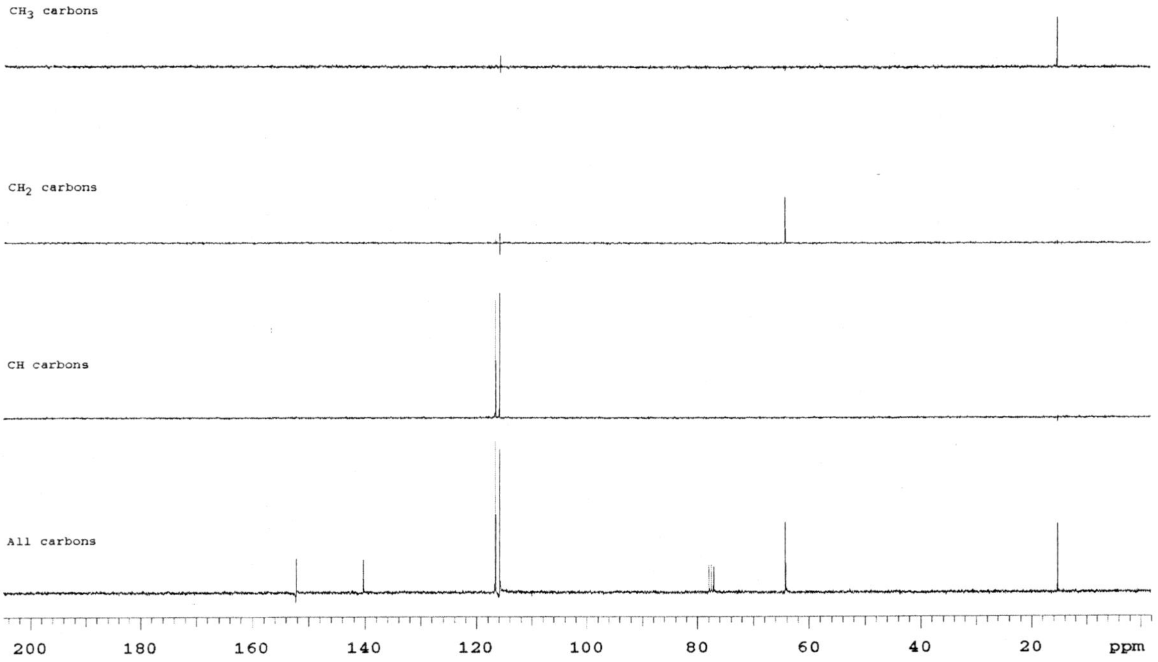

Edit DEPT ^{13}C-NMR Spectrum in $CDCl_3$

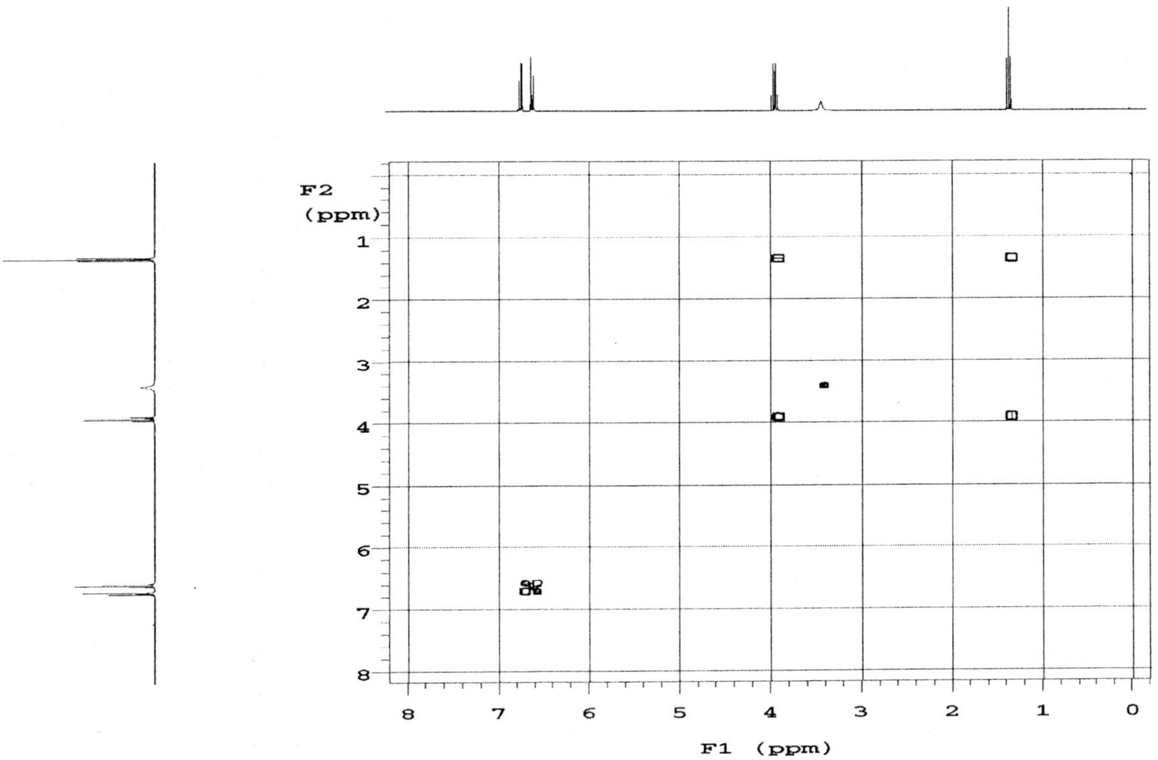

gCOSY Spectrum in $CDCl_3$

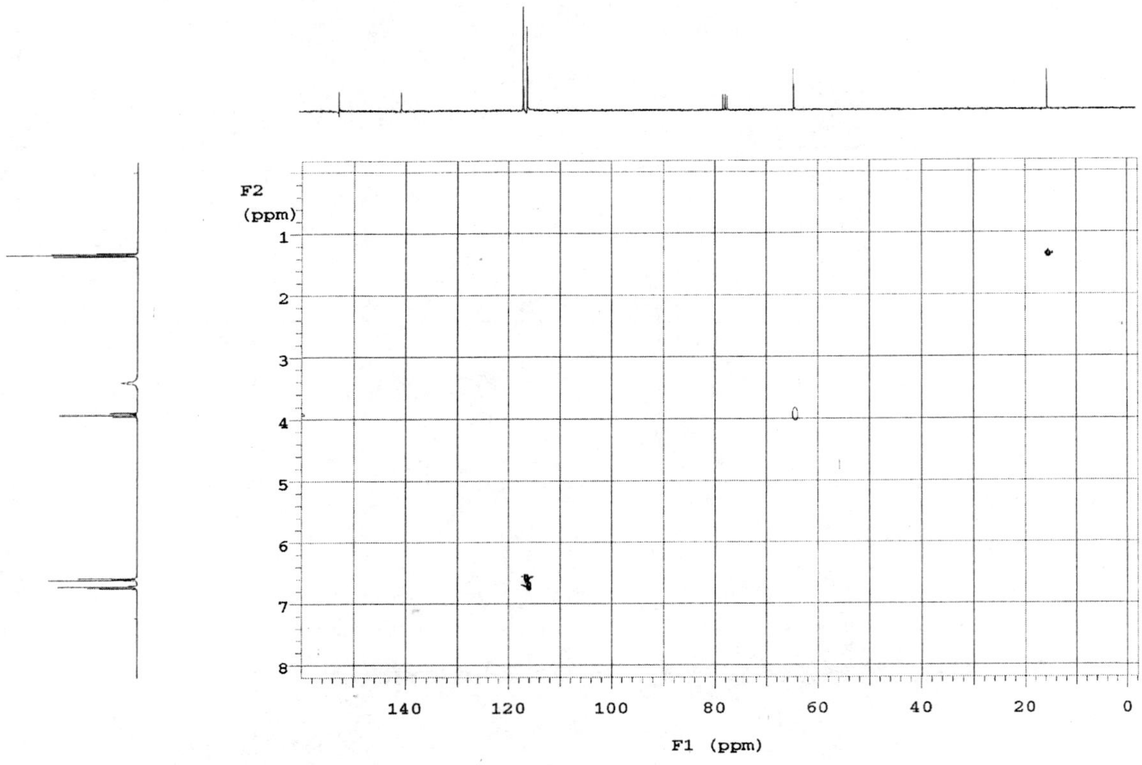

gHSQC Spectrum in $CDCl_3$

Problem 2.41 Given the spectroscopic data below (Proton, Carbon, Edited DEPT, and gHSQC), provide the structure of the molecule. From vapor pressure osmometry, the MW was determined to be 225 ± 10%.

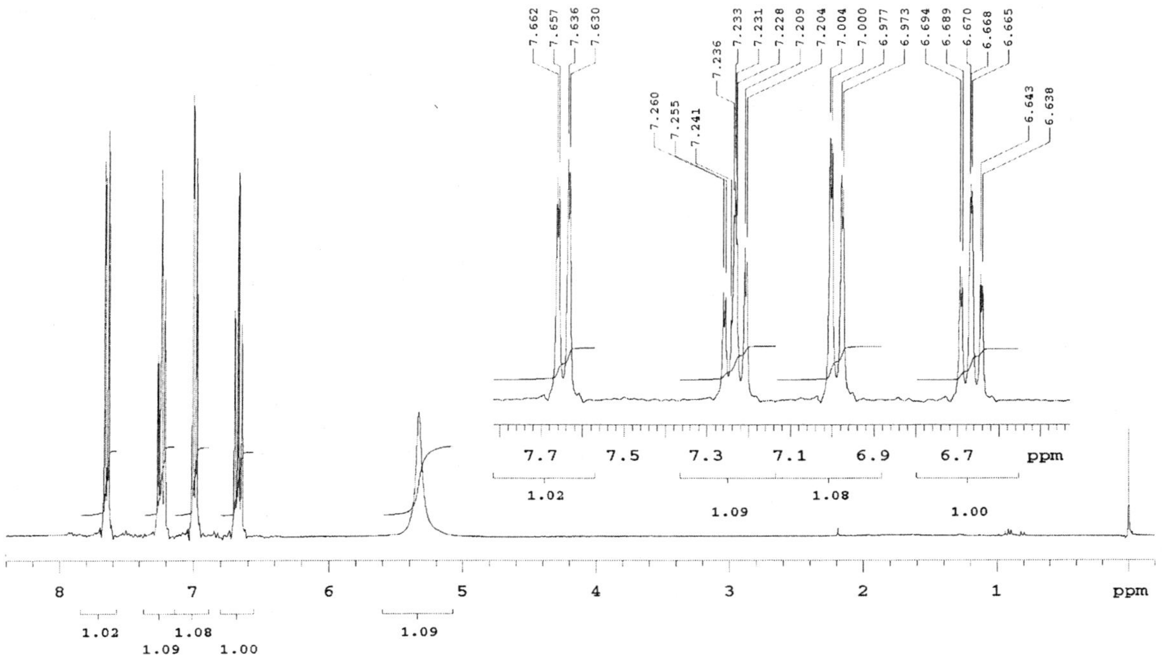

^{1}H-NMR Spectrum in $CDCl_3$

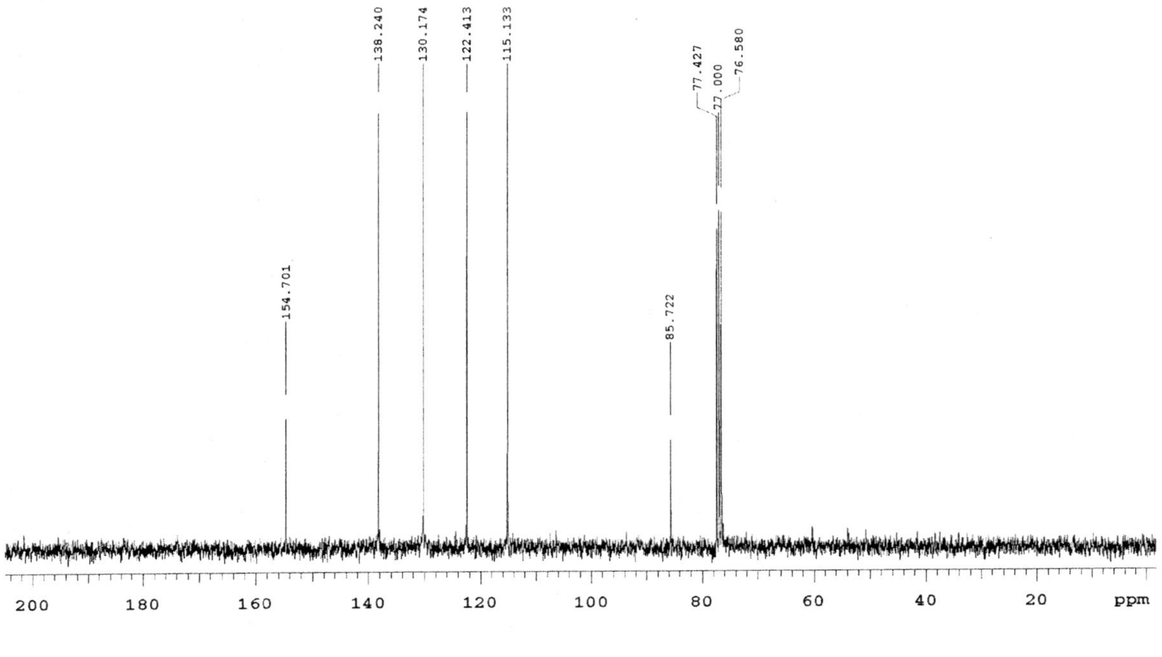

^{13}C-NMR Spectrum in $CDCl_3$

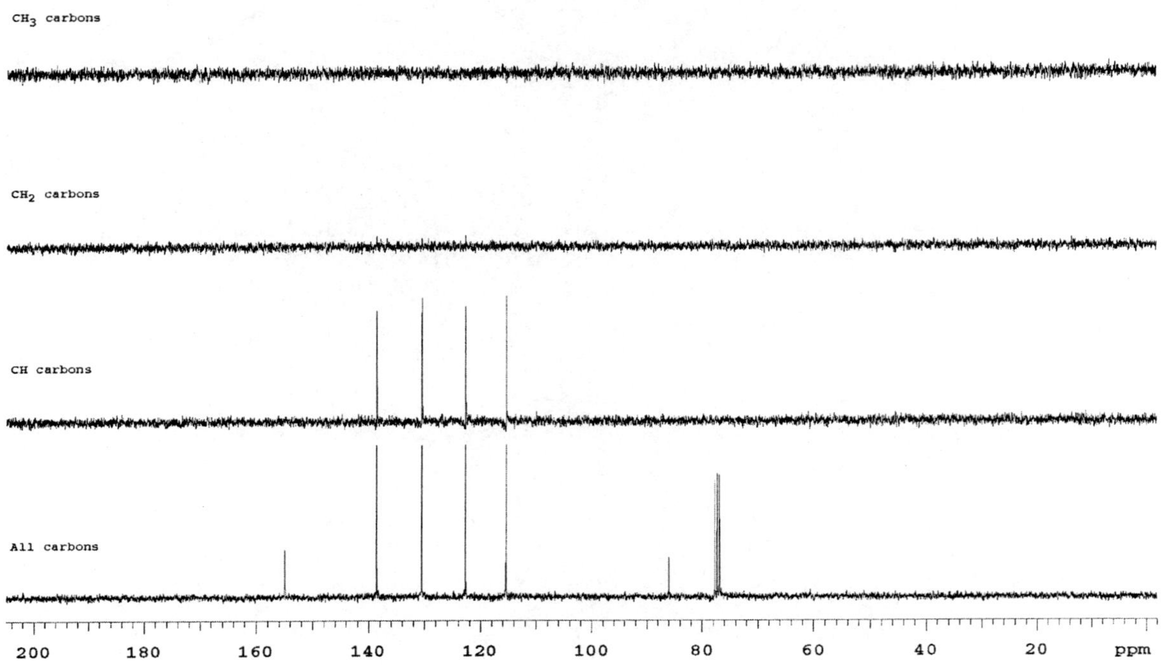

Edit DEPT ^{13}C-NMR Spectrum in $CDCl_3$

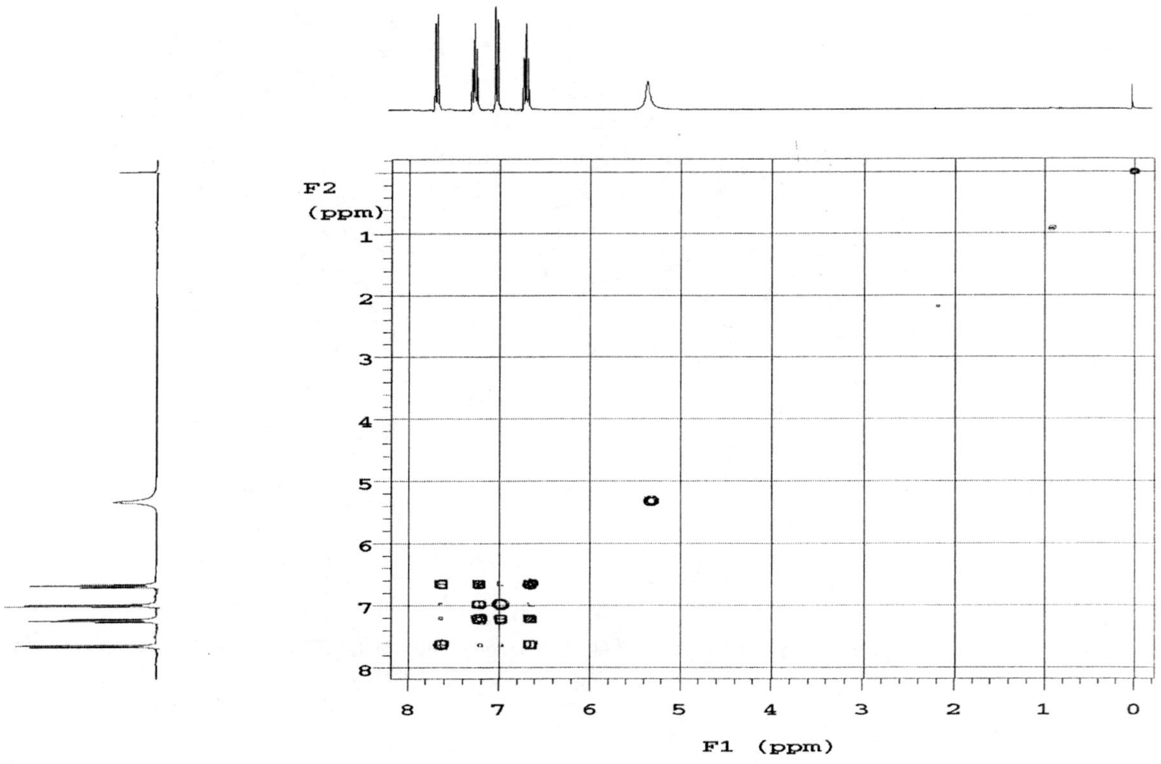

gCOSY Spectrum in $CDCl_3$

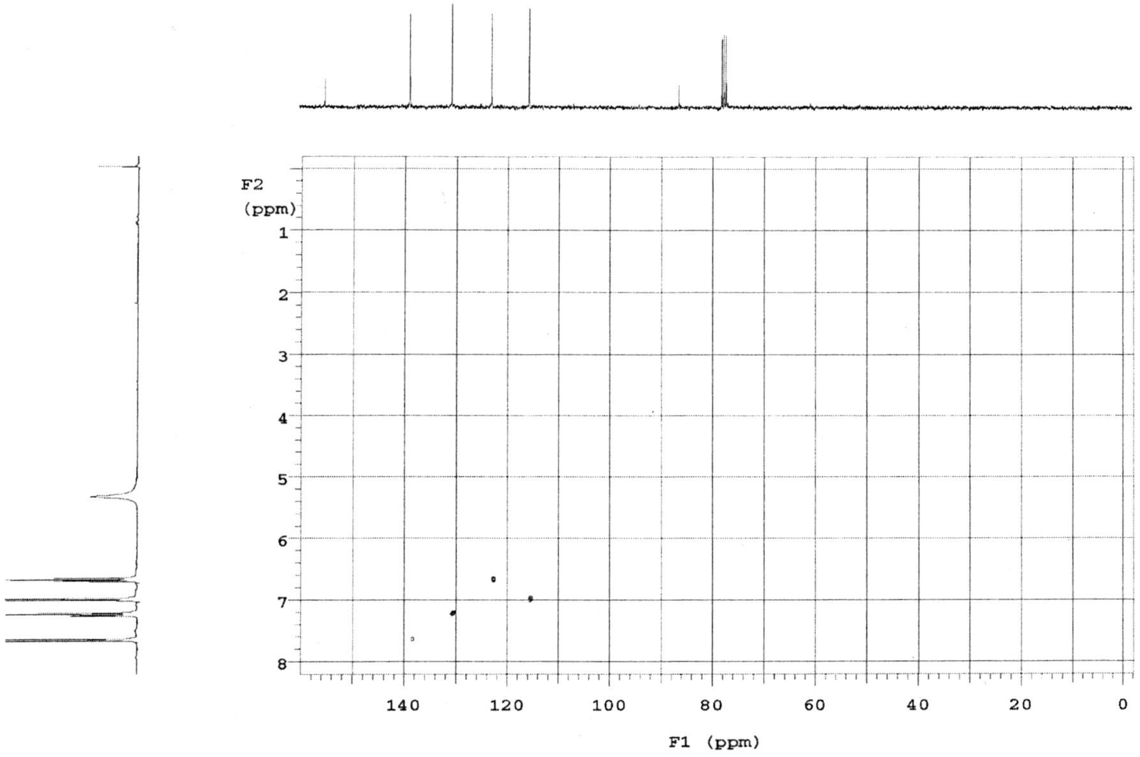

gHSQC Spectrum in $CDCl_3$

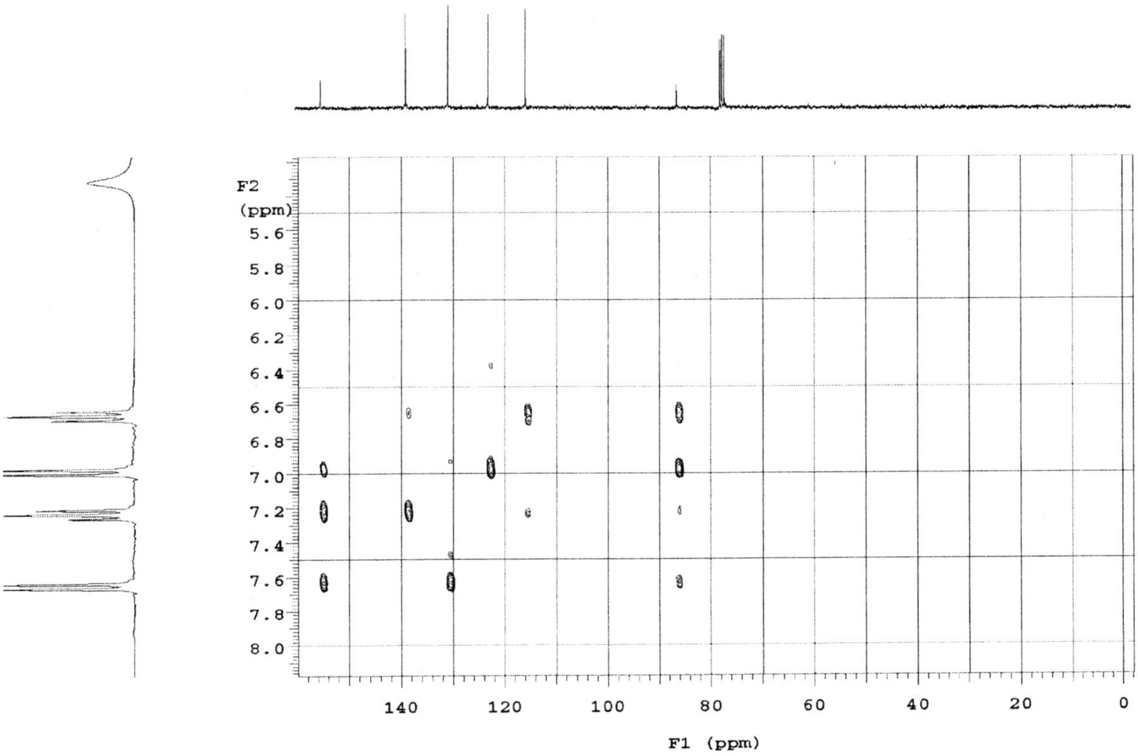

gHMBC Spectrum in $CDCl_3$

Problem 2.42 Given the spectroscopic data below (Proton, Carbon, Edited DEPT, and gHSQC), provide the structure of the molecule.

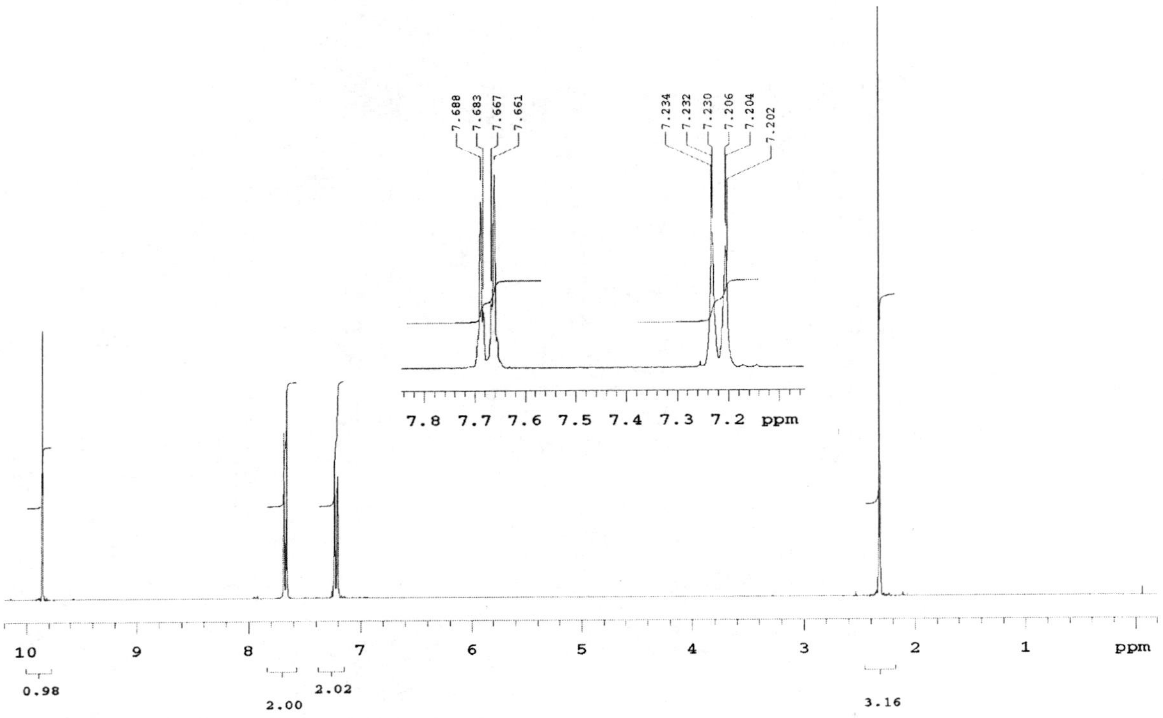

^{1}H-NMR Spectrum in $CDCl_3$

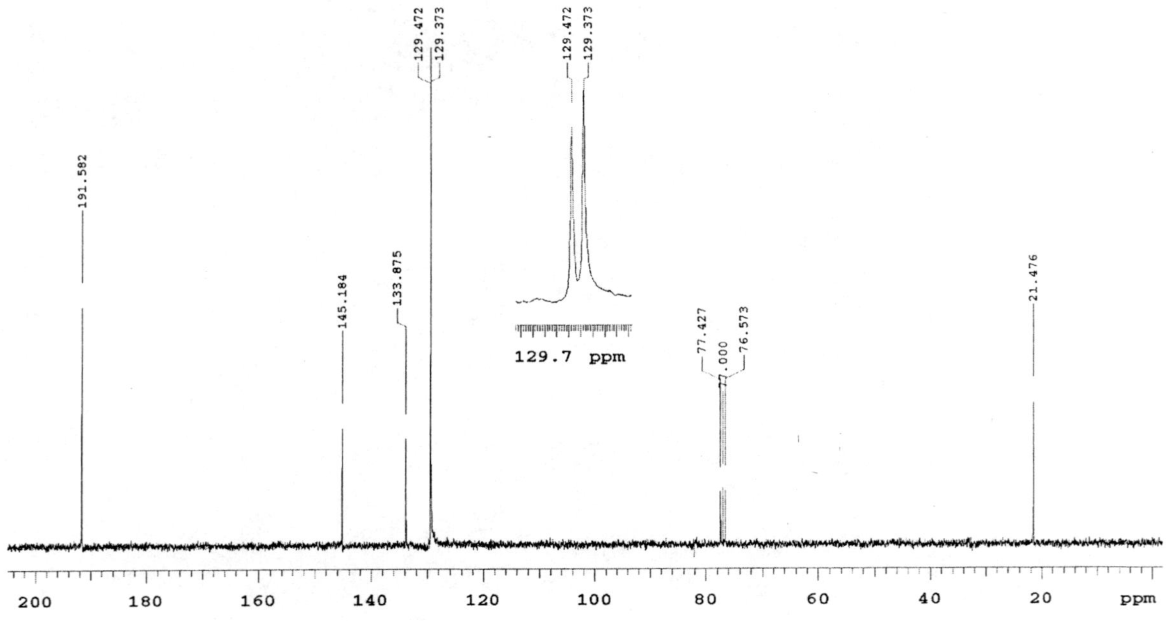

^{13}C-NMR Spectrum in $CDCl_3$

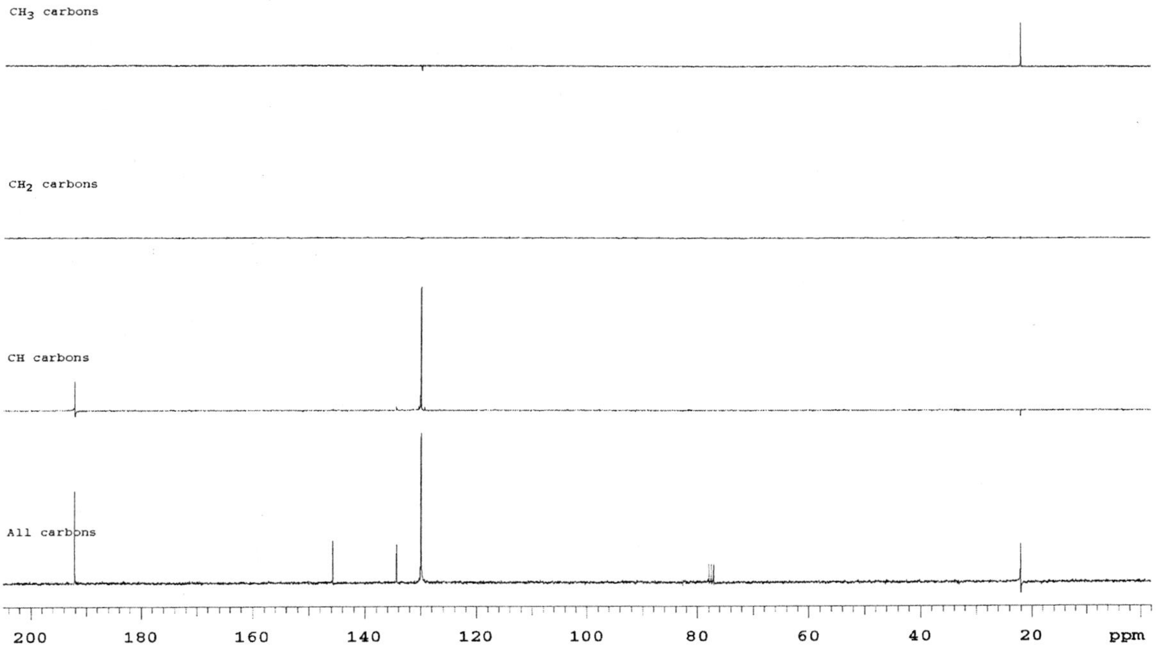

Edit DEPT ^{13}C-NMR Spectrum in $CDCl_3$

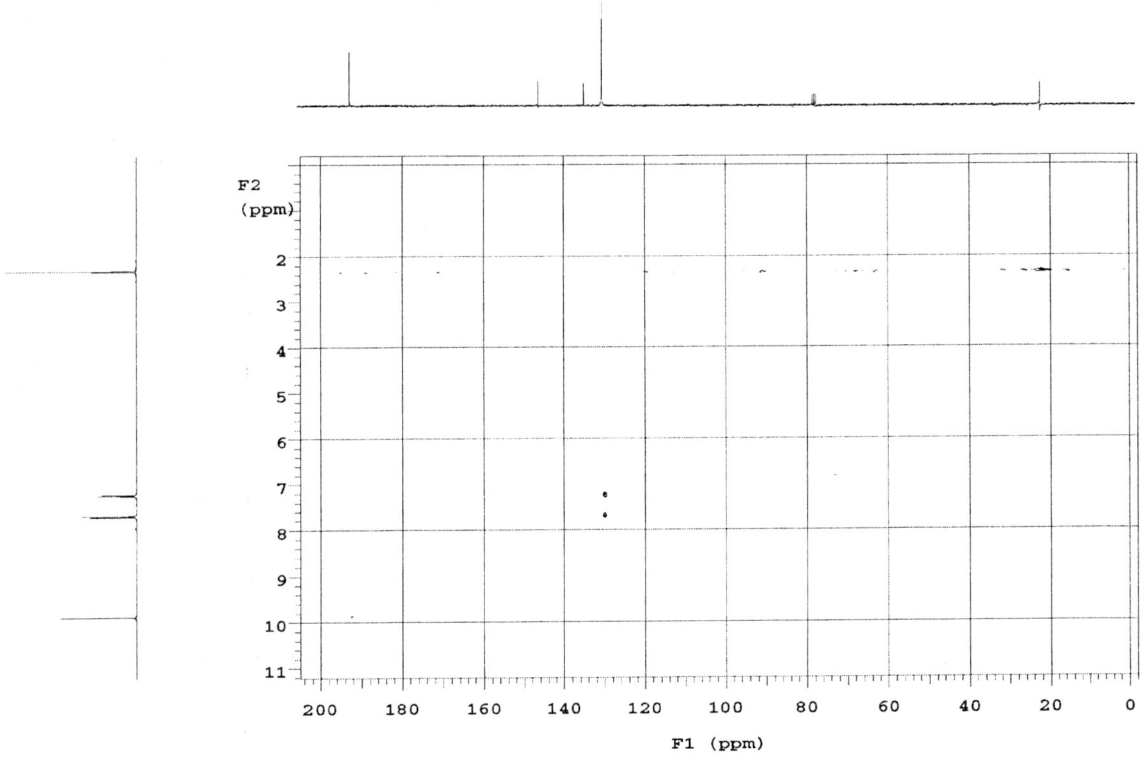

gHSQC Spectrum in $CDCl_3$

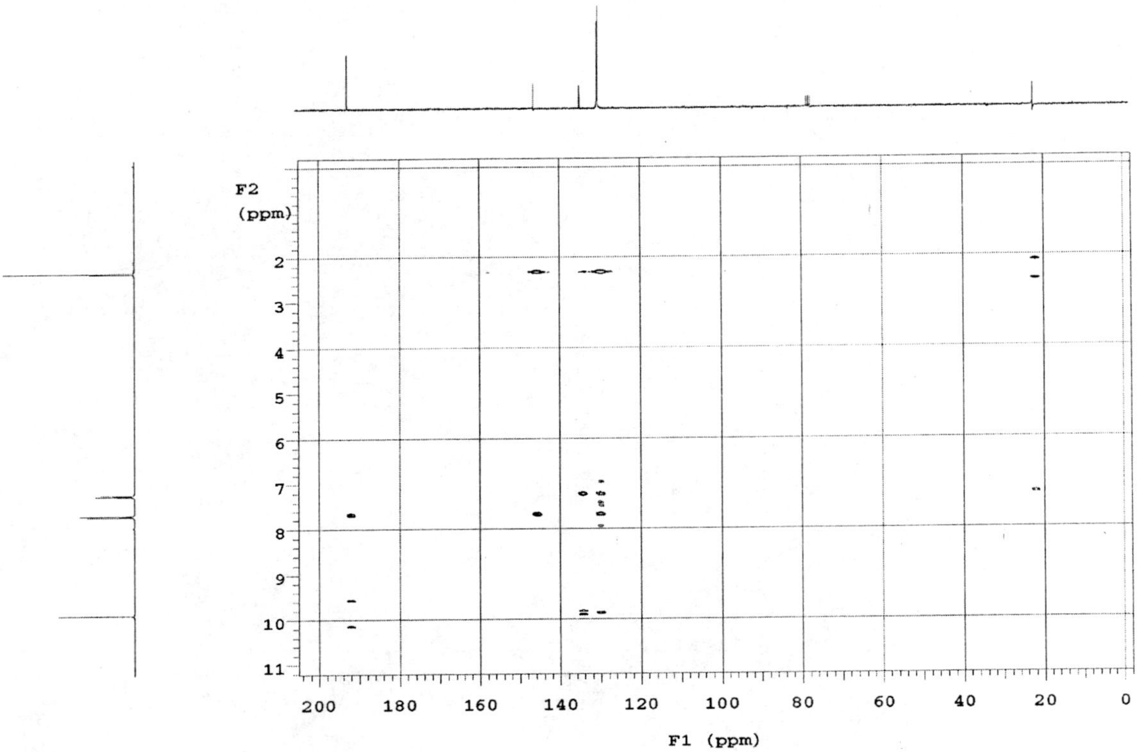

gHMBC Spectrum in $CDCl_3$

Problem 2.43 Given the spectroscopic data below (Proton, Carbon, Edited DEPT, and gHSQC), provide the structure of the molecule.

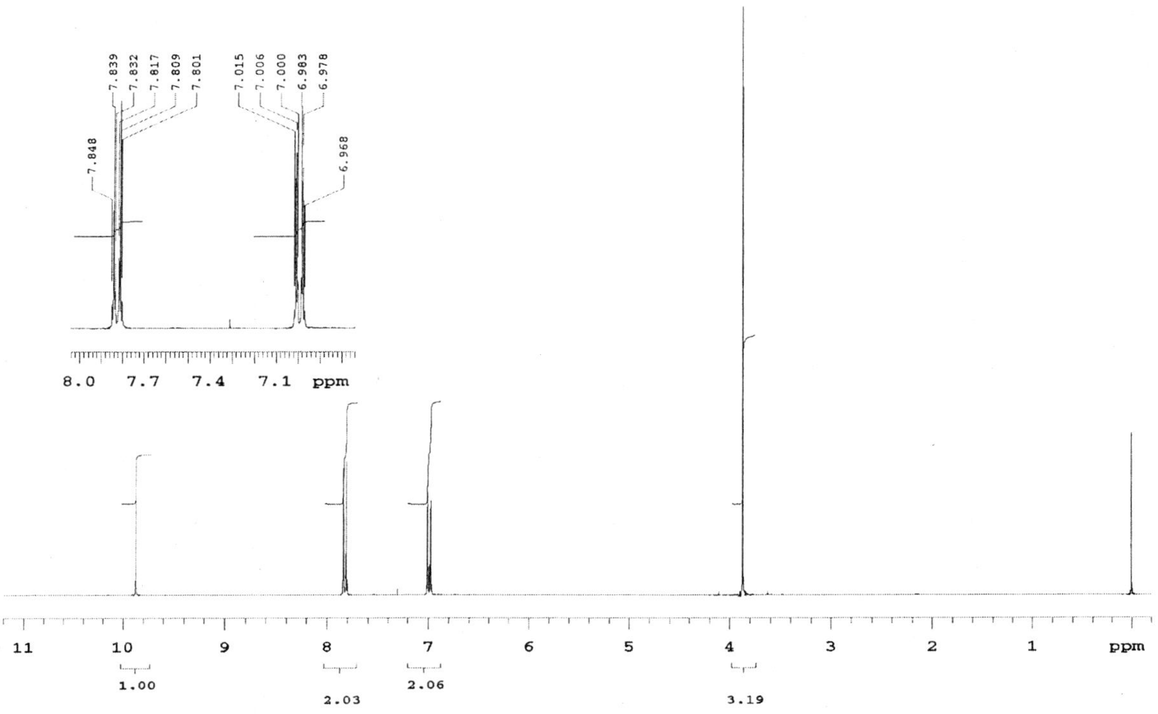

^{1}H-NMR Spectrum in $CDCl_3$

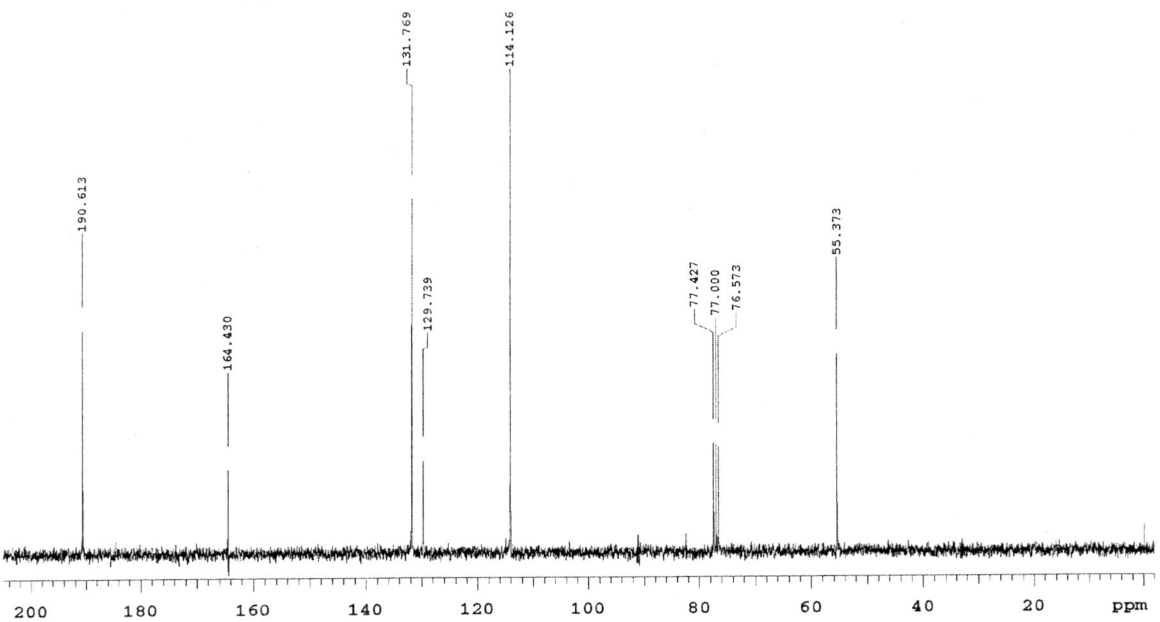

^{13}C-NMR Spectrum in $CDCl_3$

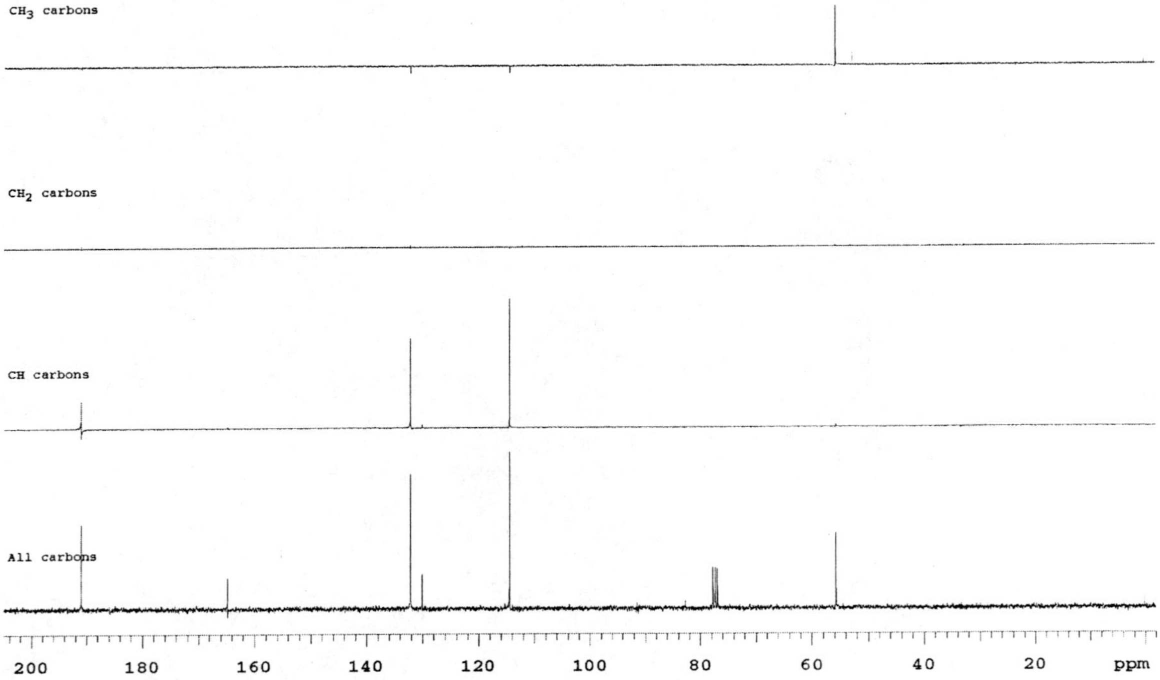

Edit DEPT ^{13}C-NMR Spectrum in $CDCl_3$

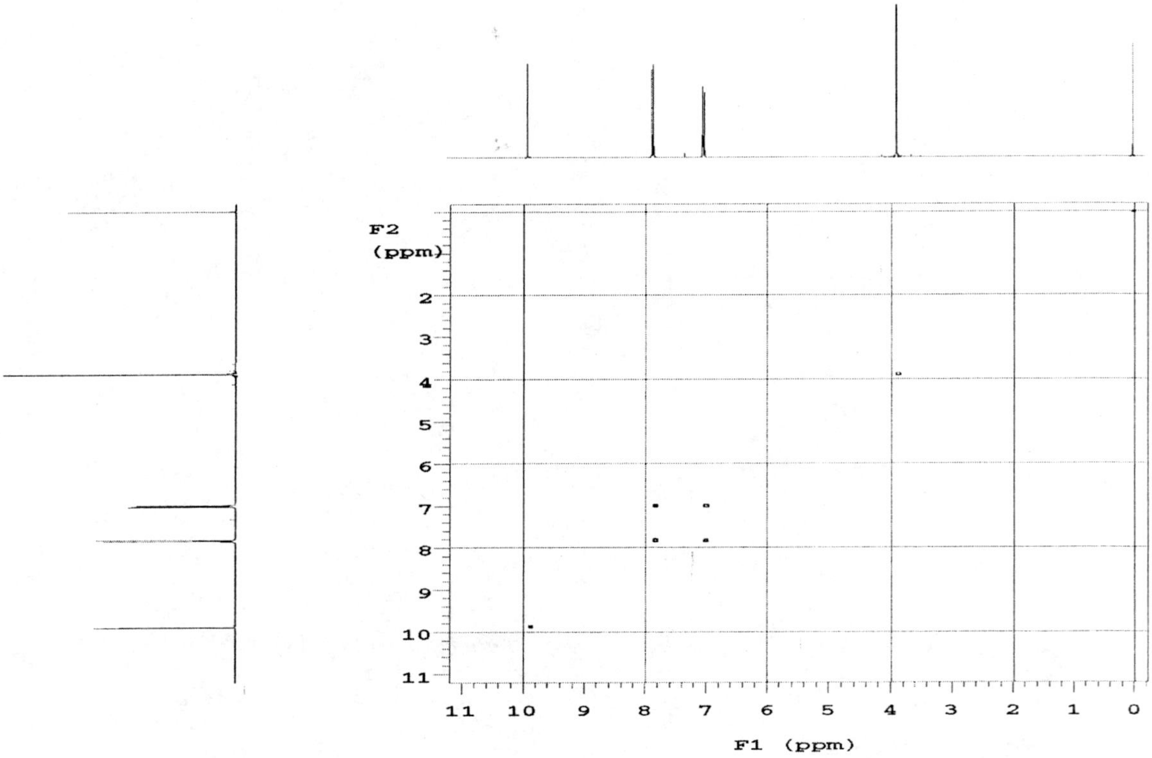

gCOSY Spectrum in $CDCl_3$

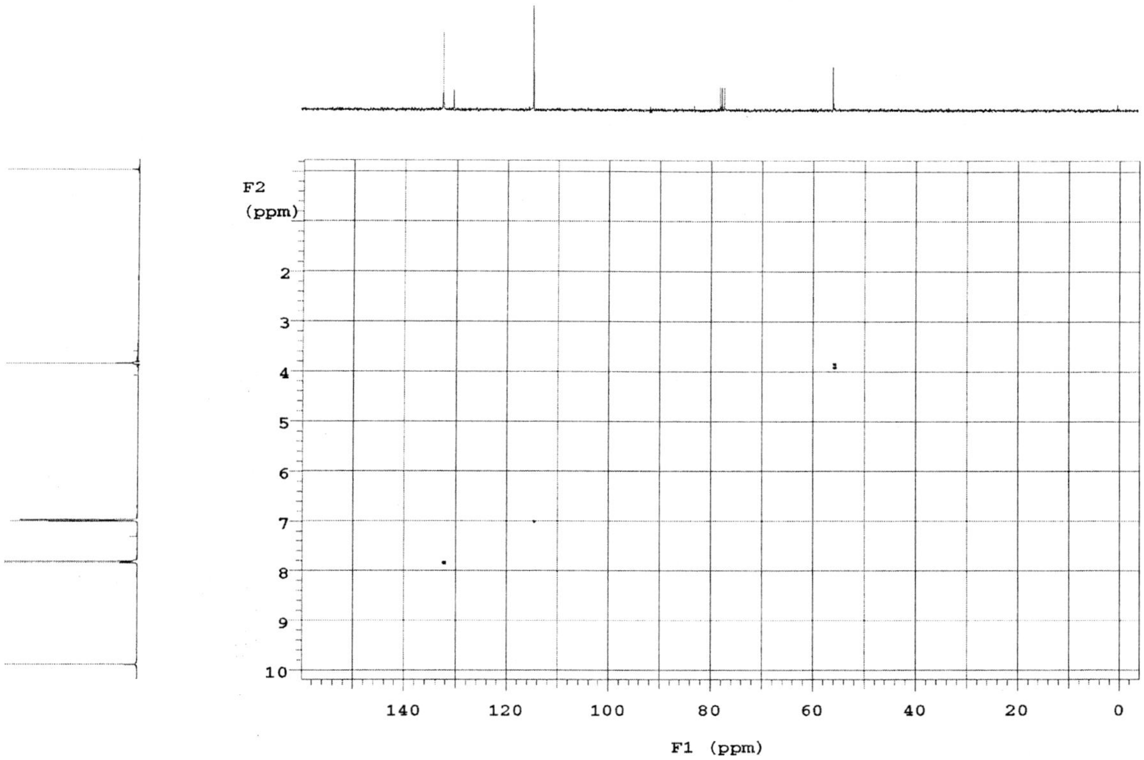

gHSQC Spectrum in $CDCl_3$

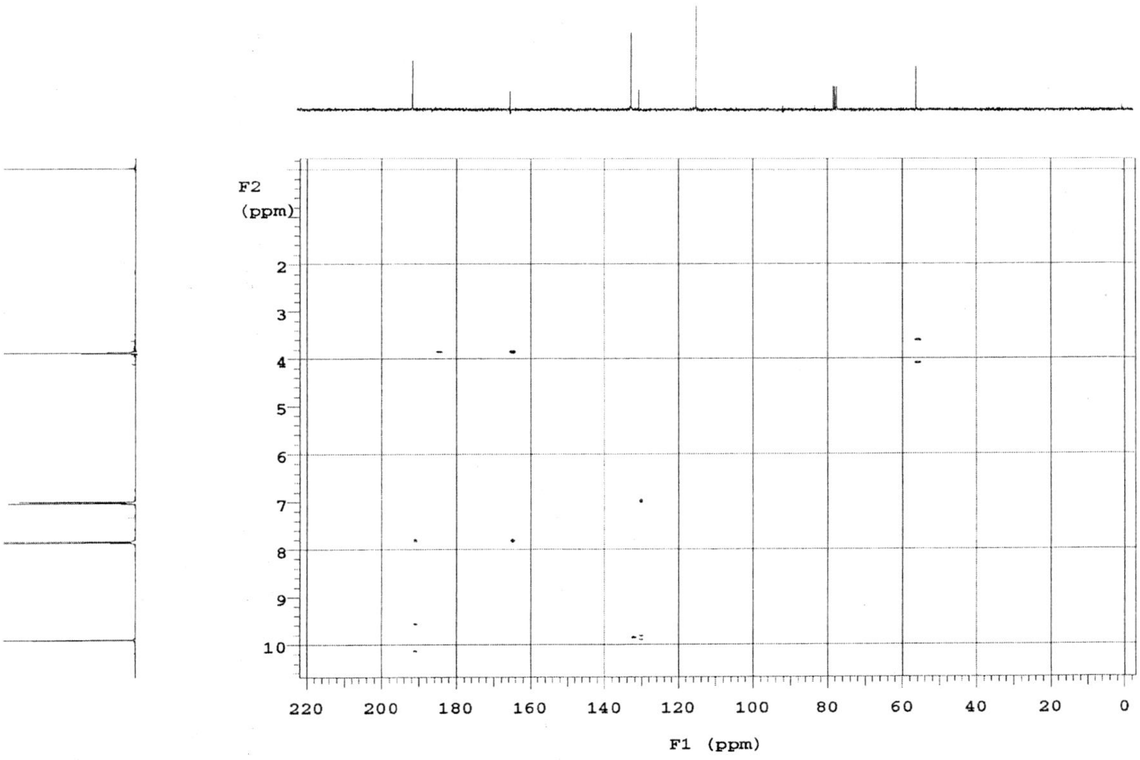

gHMBC Spectrum in $CDCl_3$

Problem 2.44 Given the spectroscopic data below (Proton, Carbon, Edited DEPT, and gHSQC), provide the structure of the molecule. Mass Spectrum shows two parent ions of equal intensity separated by two amu. (IR: 695 & 812 cm^{-1})

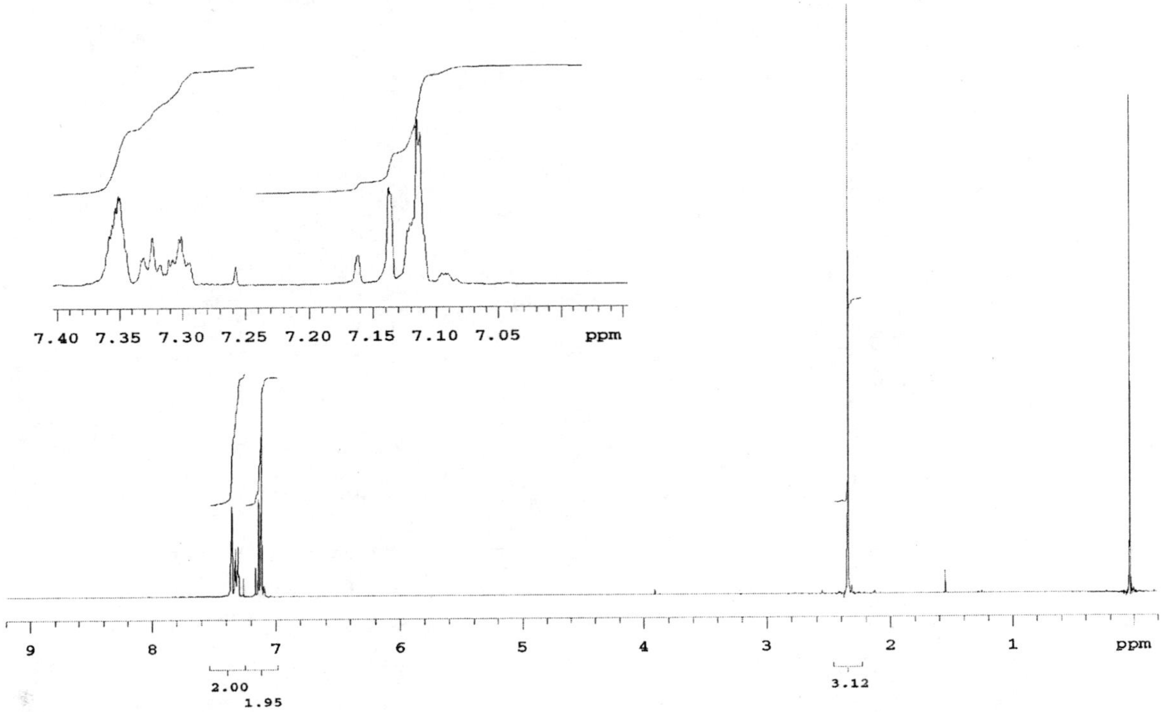

^{1}H-NMR Spectrum in $CDCl_3$

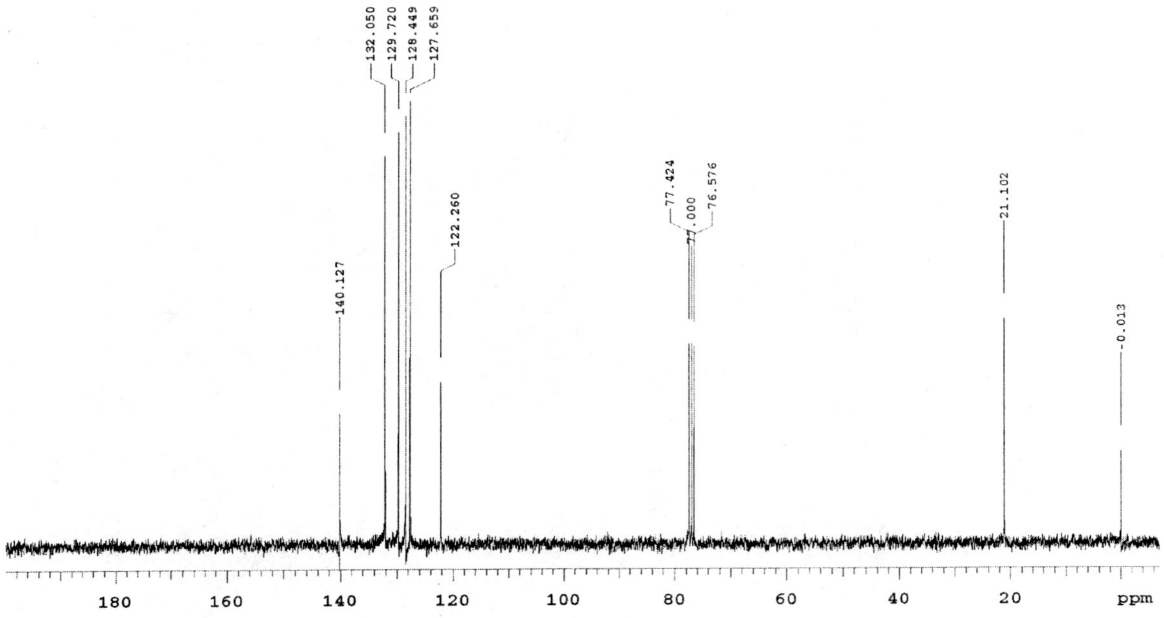

^{13}C-NMR Spectrum in $CDCl_3$

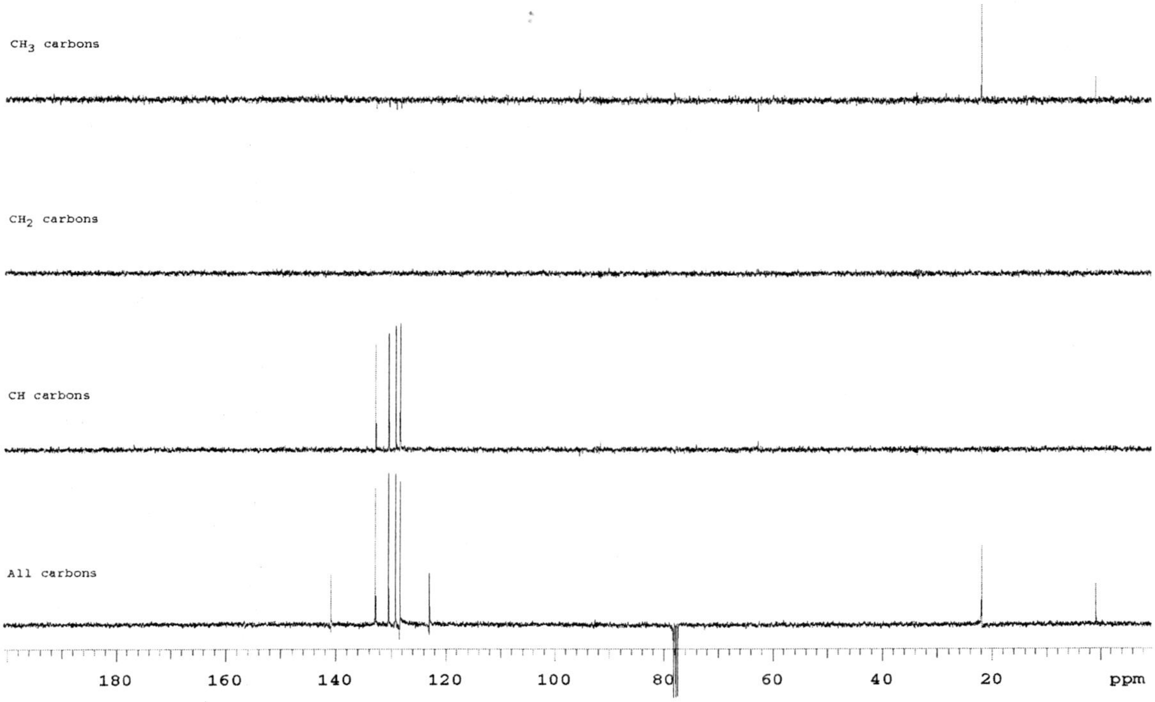

Edit DEPT ^{13}C-NMR Spectrum in $CDCl_3$

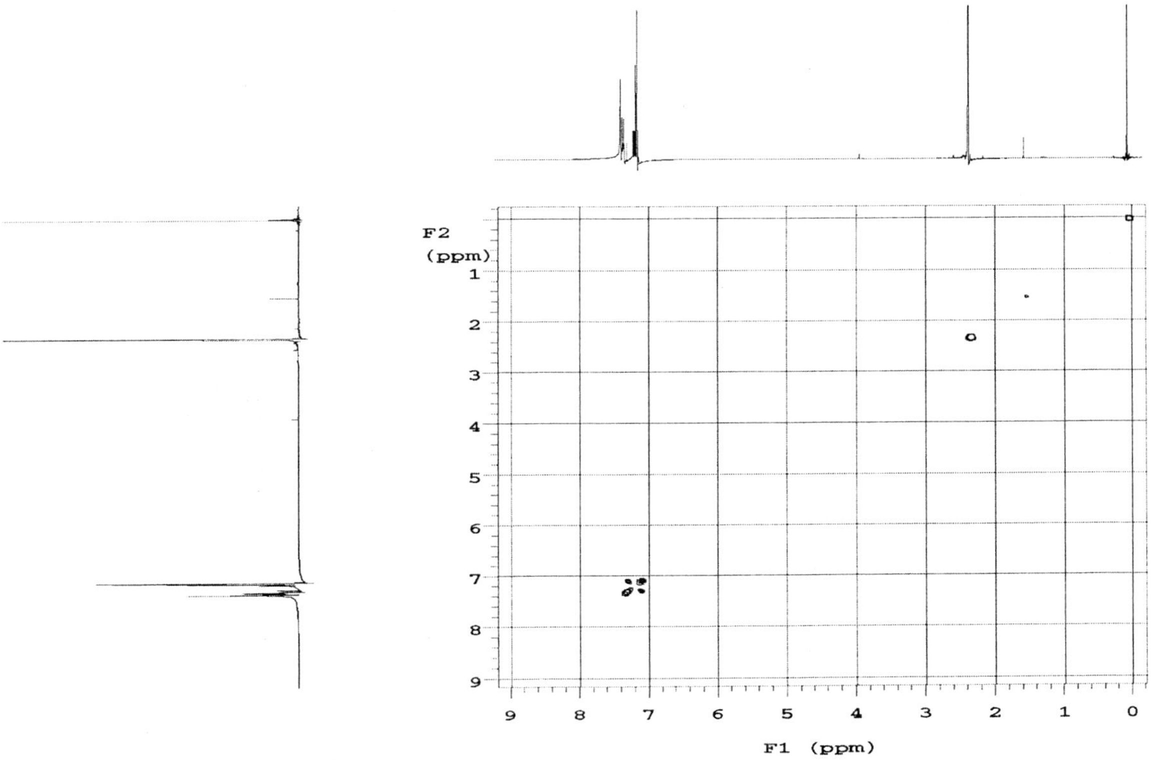

gCOSY Spectrum in $CDCl_3$

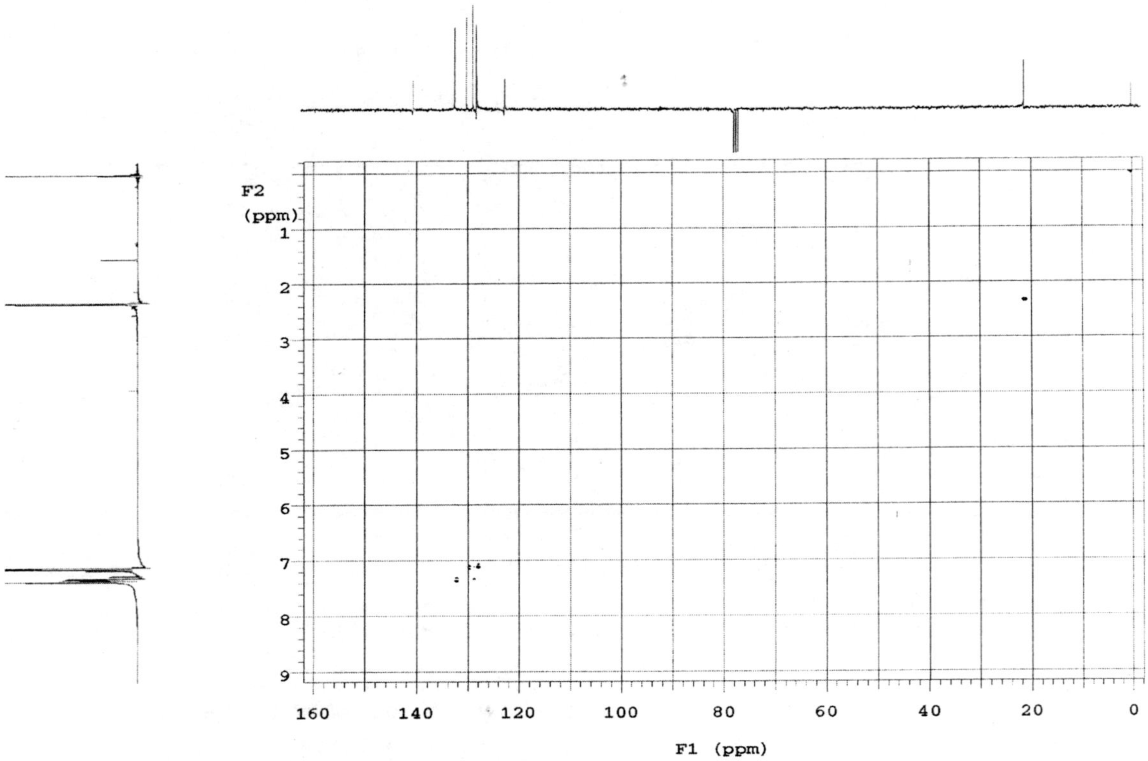

gHSQC Spectrum in $CDCl_3$

Problem 2.45 Given the spectroscopic data below (Proton, Carbon, Edited DEPT, and gHSQC), provide the structure of the molecule. (Empirical Formula: $C_5H_{10}O$)

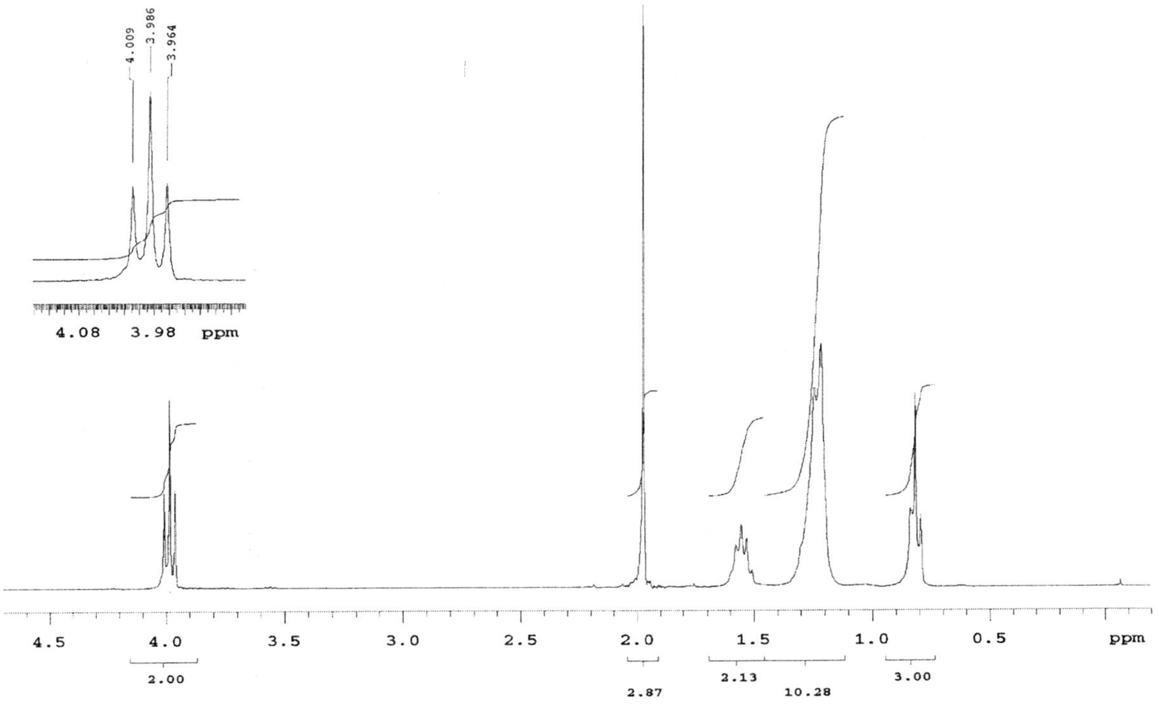

^{1}H-NMR Spectrum in $CDCl_3$

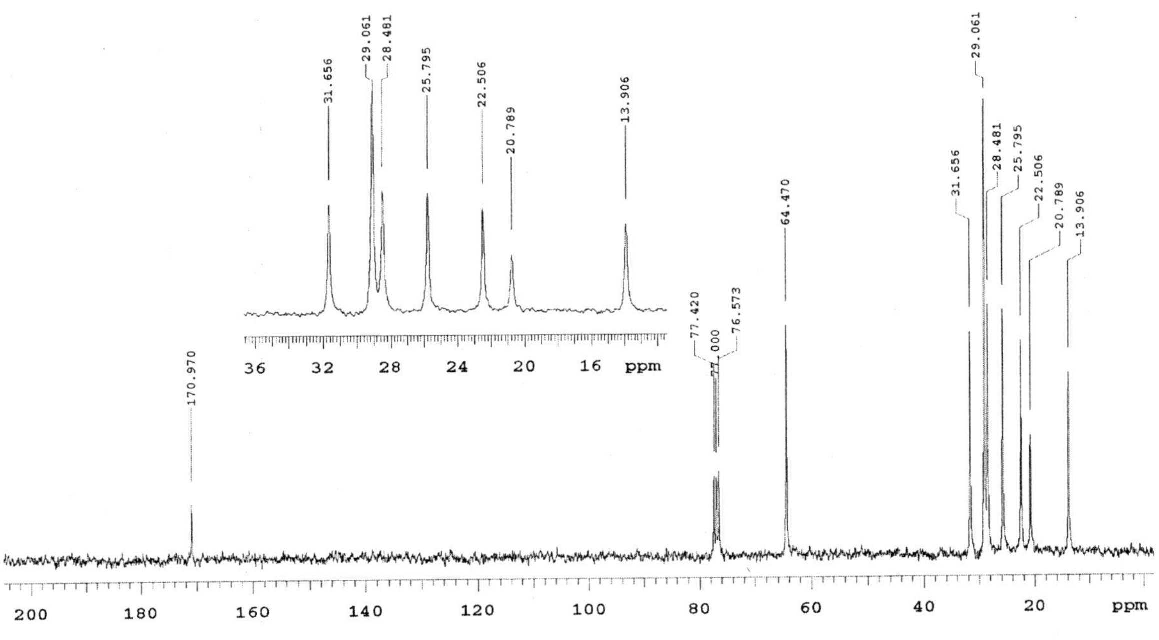

^{13}C-NMR Spectrum in $CDCl_3$

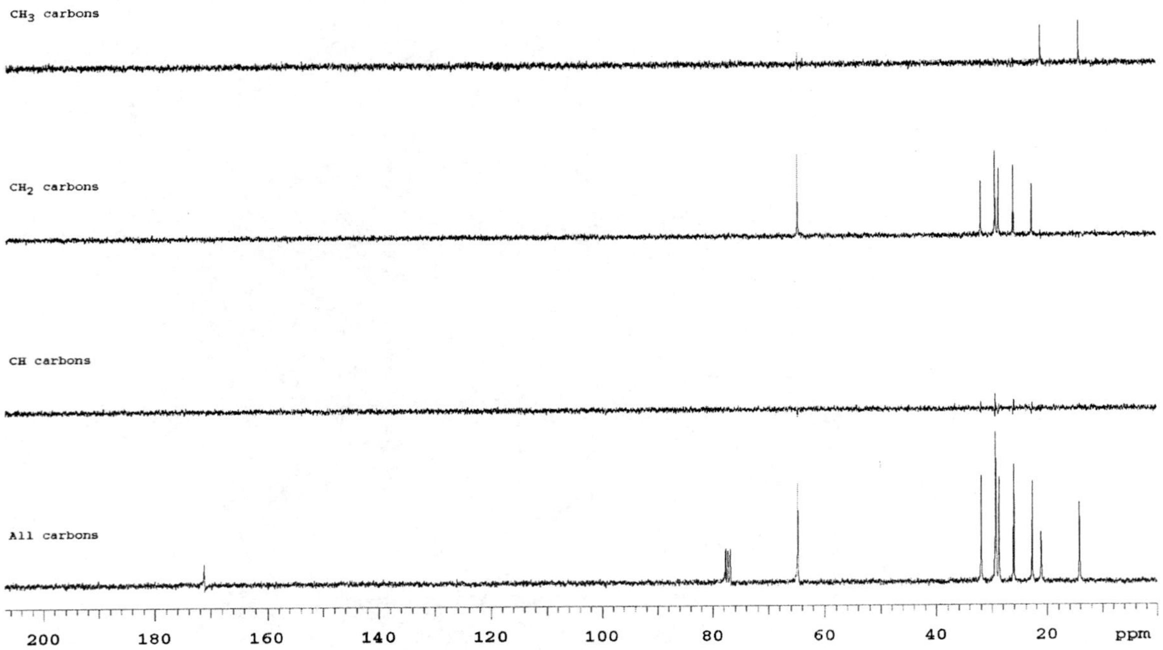

Edit DEPT ^{13}C-NMR Spectrum in $CDCl_3$

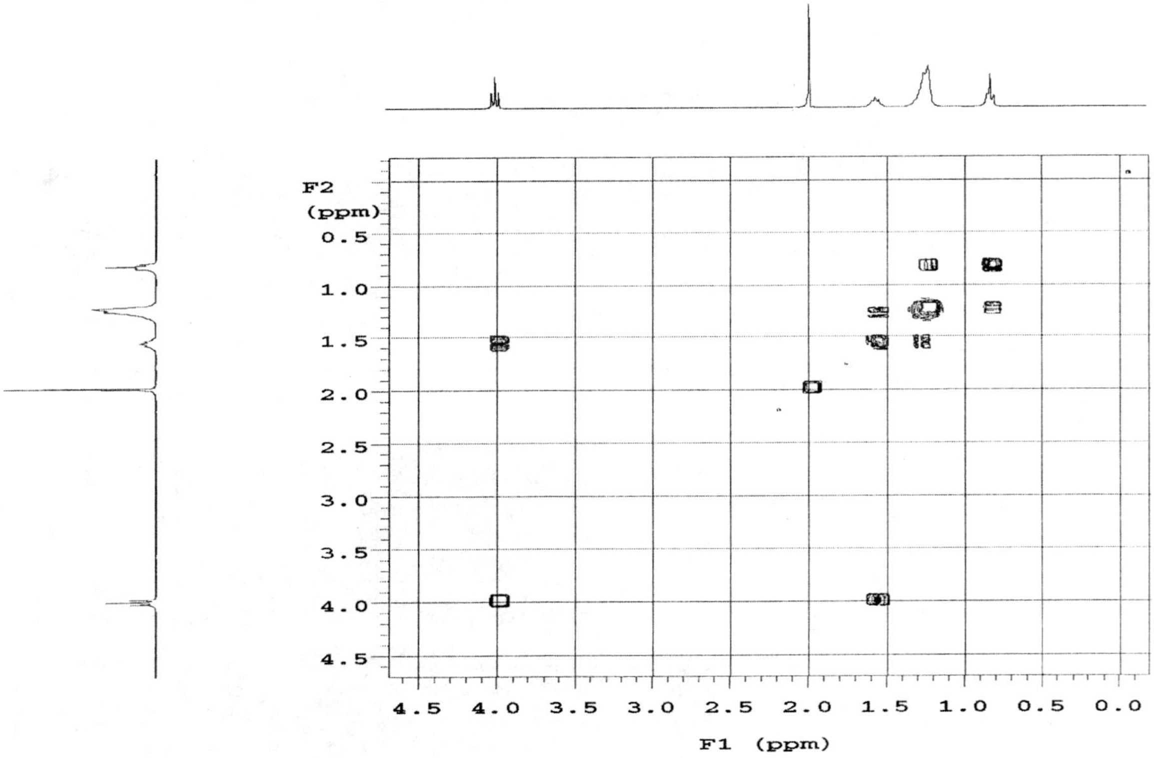

gCOSY Spectrum in $CDCl_3$

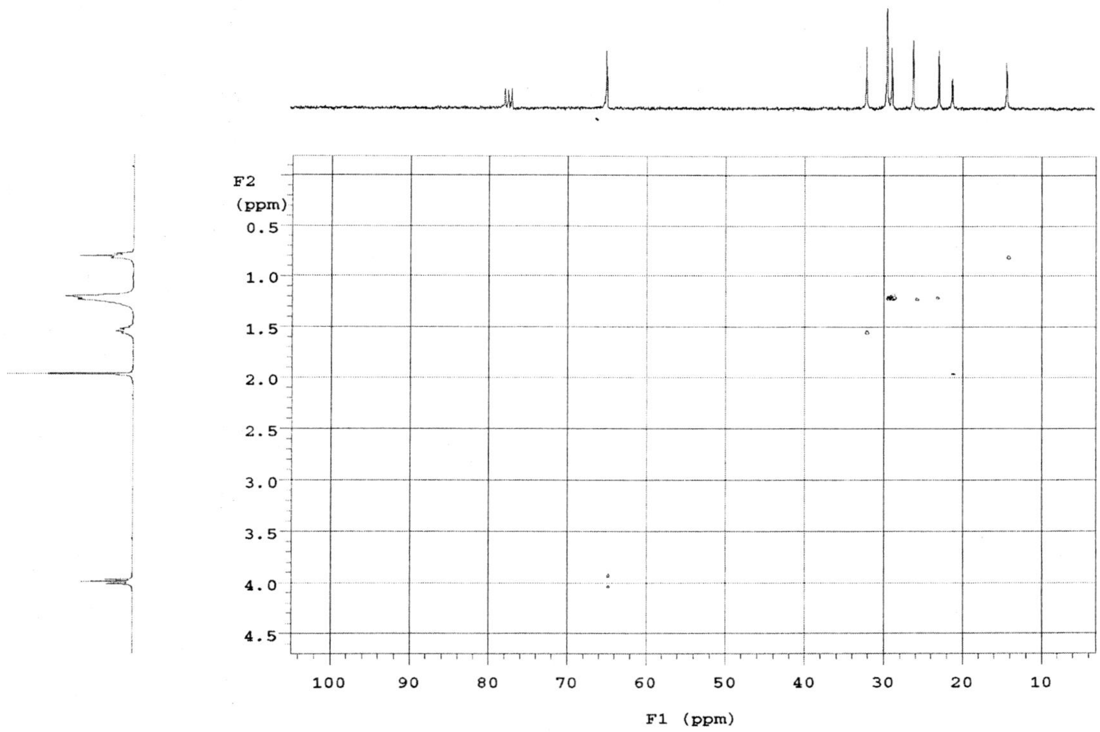

gHSQC Spectrum in $CDCl_3$

Problem 2.46 Given the spectroscopic data below (Proton, Carbon, Edited DEPT, and gHSQC), provide the structure of the molecule.

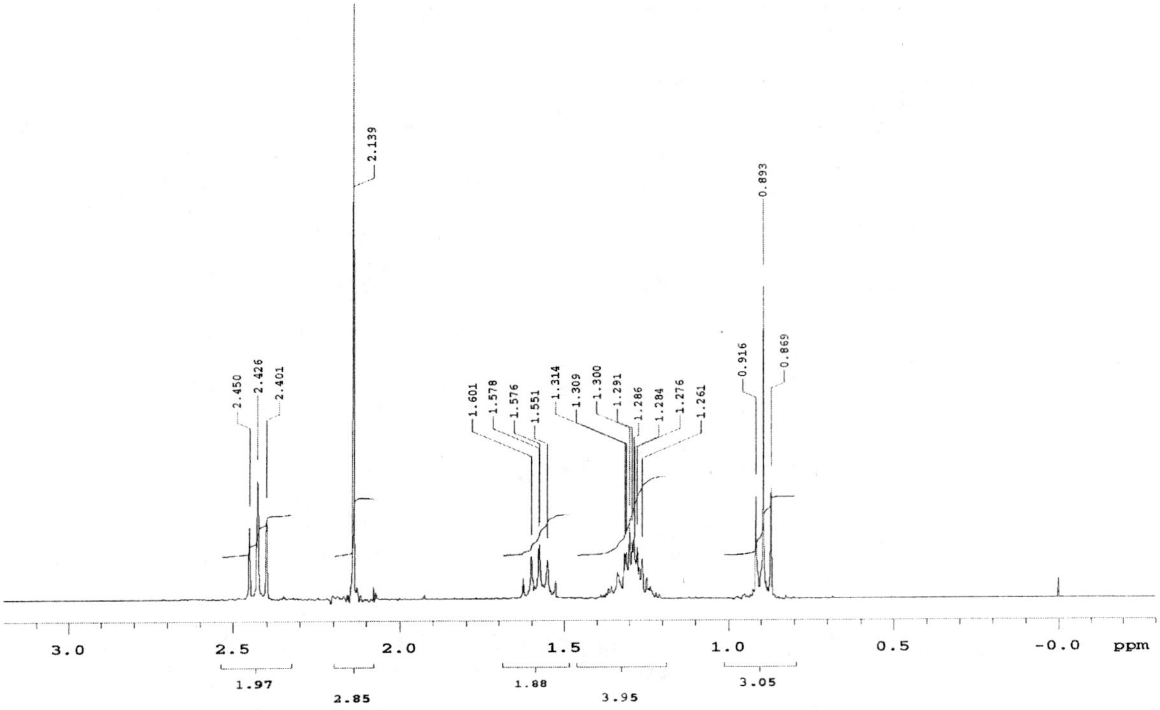

^{1}H-NMR Spectrum in $CDCl_3$

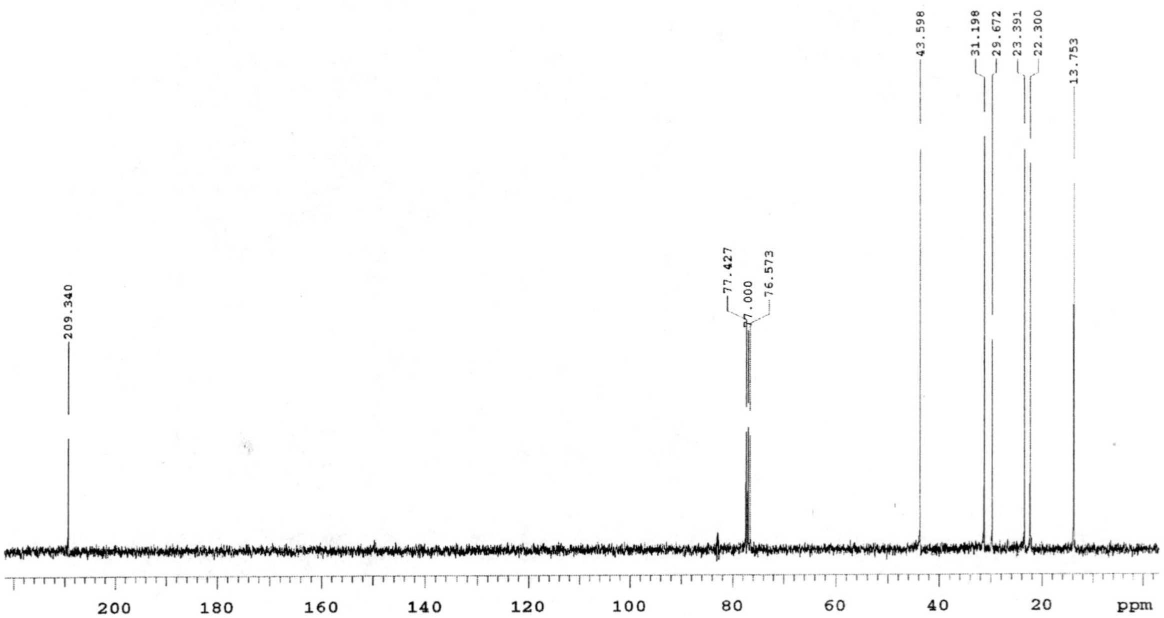

^{13}C-NMR Spectrum in $CDCl_3$

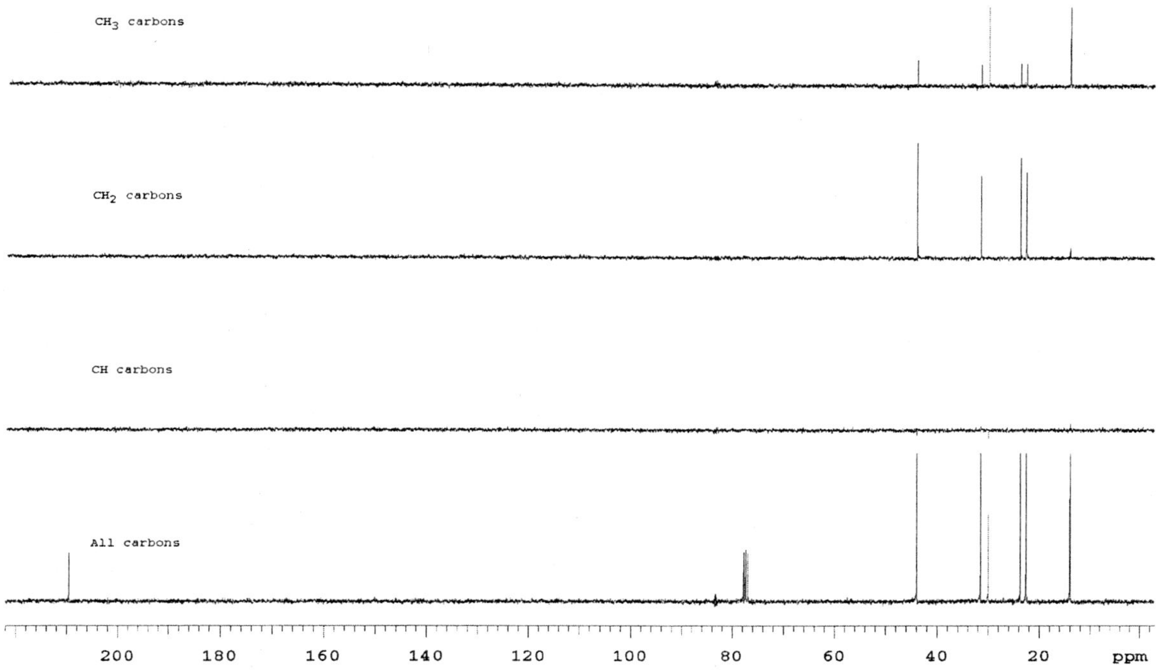

Edited DEPT ^{13}C-NMR Spectrum in $CDCl_3$ (Caution: subtraction artifacts)

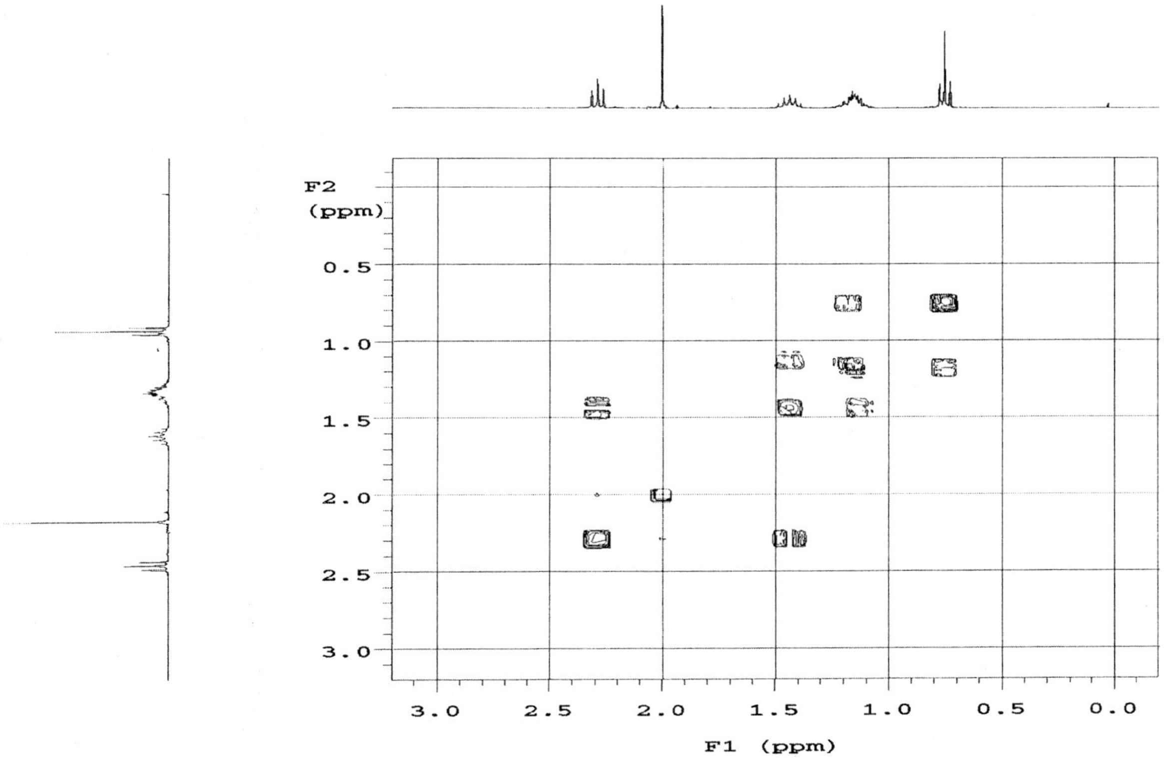

gCOSY Spectrum in $CDCl_3$

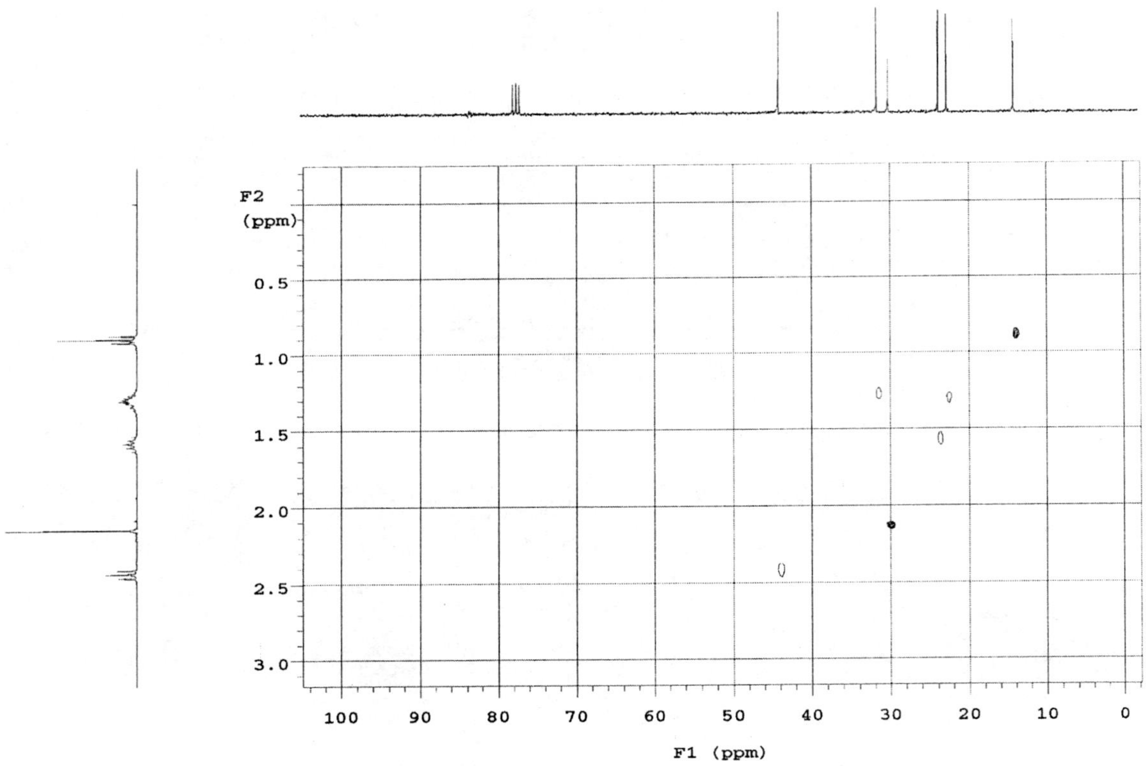

gHSQC Spectrum in $CDCl_3$

Problem 2.47 Given the spectroscopic data below (Proton, Carbon, Edited DEPT, and gHSQC), provide the structure of the molecule. (HR-MS: 181.0375 amu)

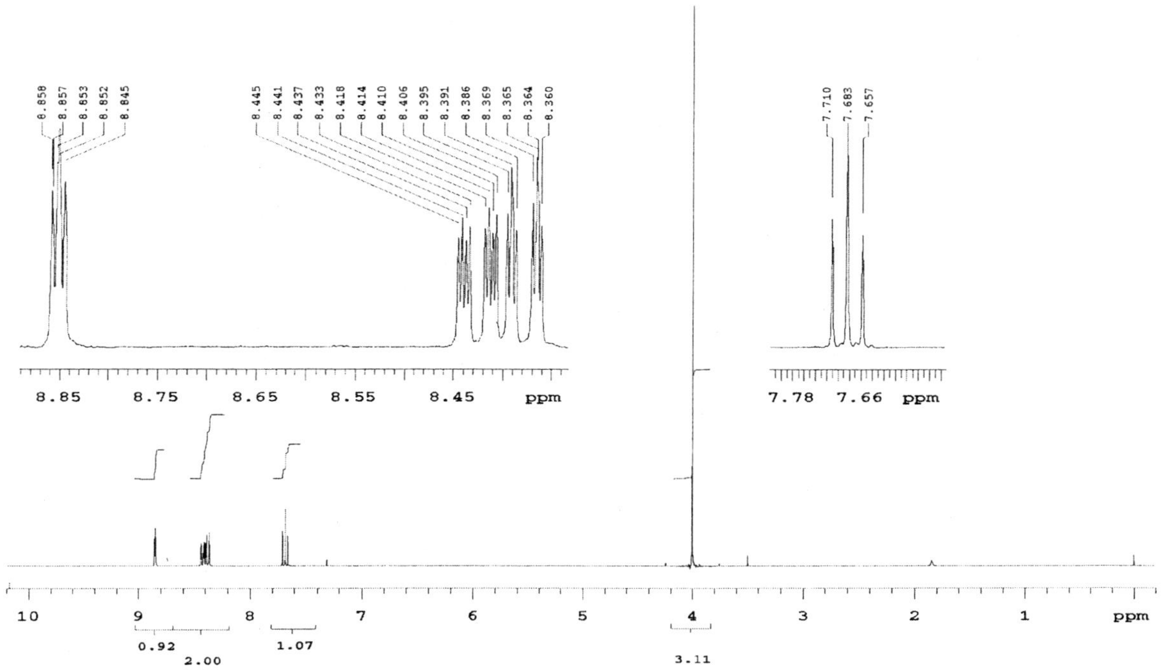

^{1}H-NMR Spectrum in $CDCl_3$

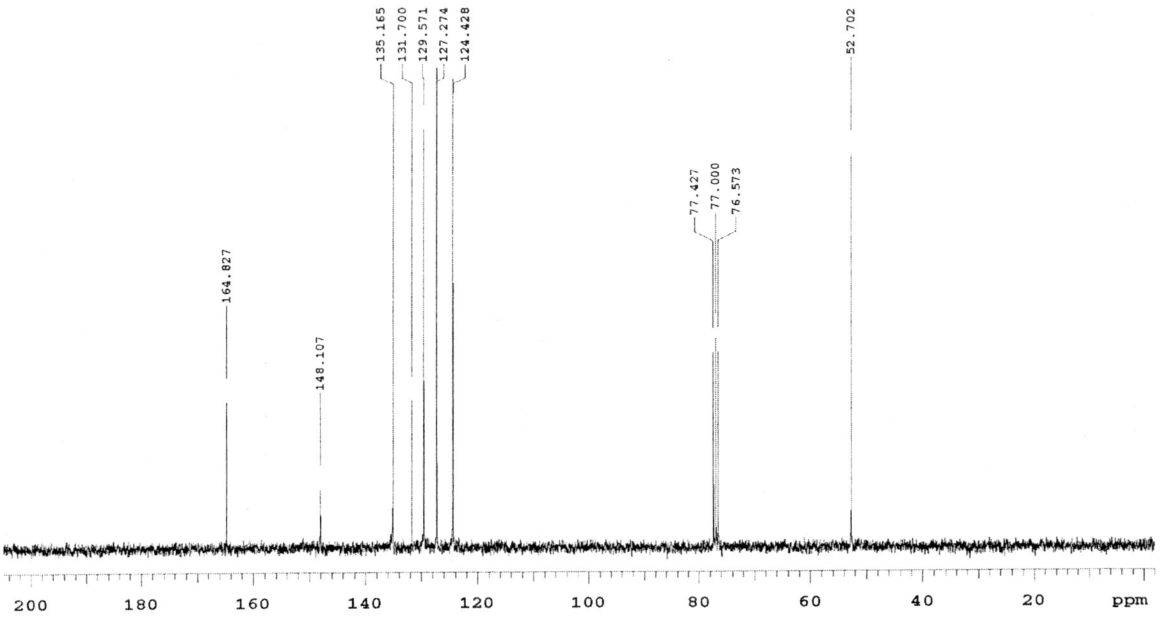

^{13}C-NMR Spectrum in $CDCl_3$

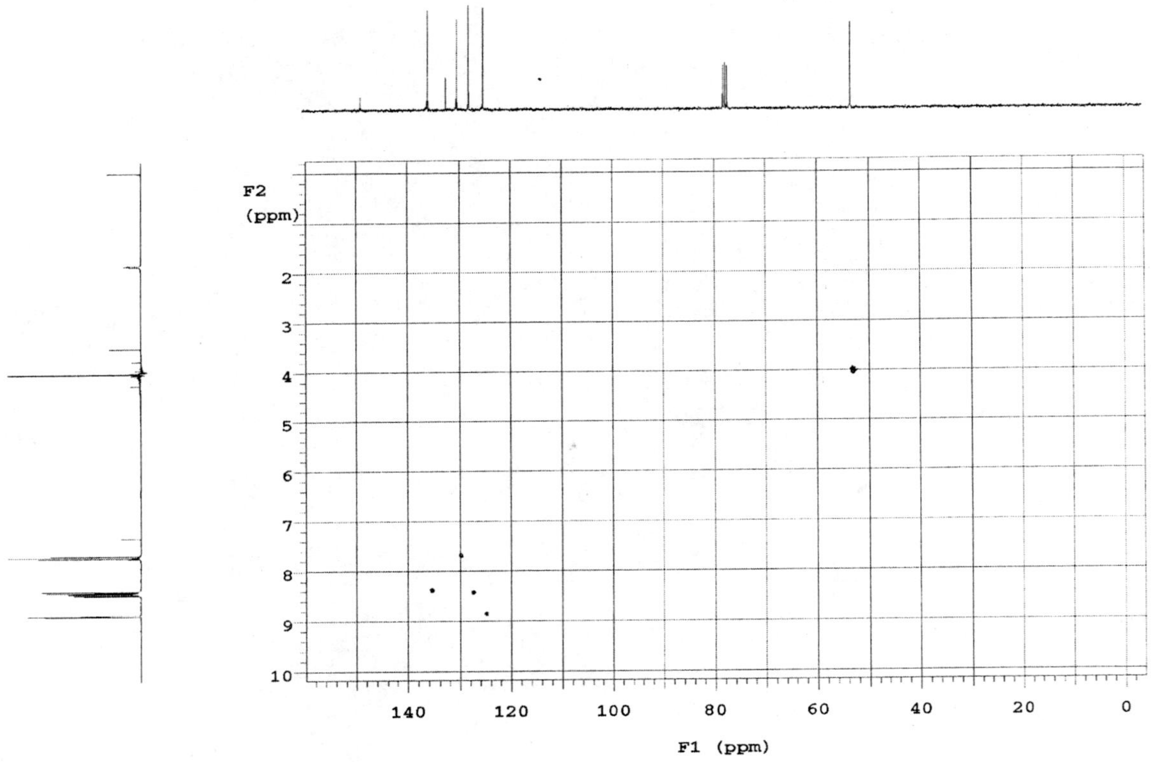

gHSQC Spectrum in $CDCl_3$

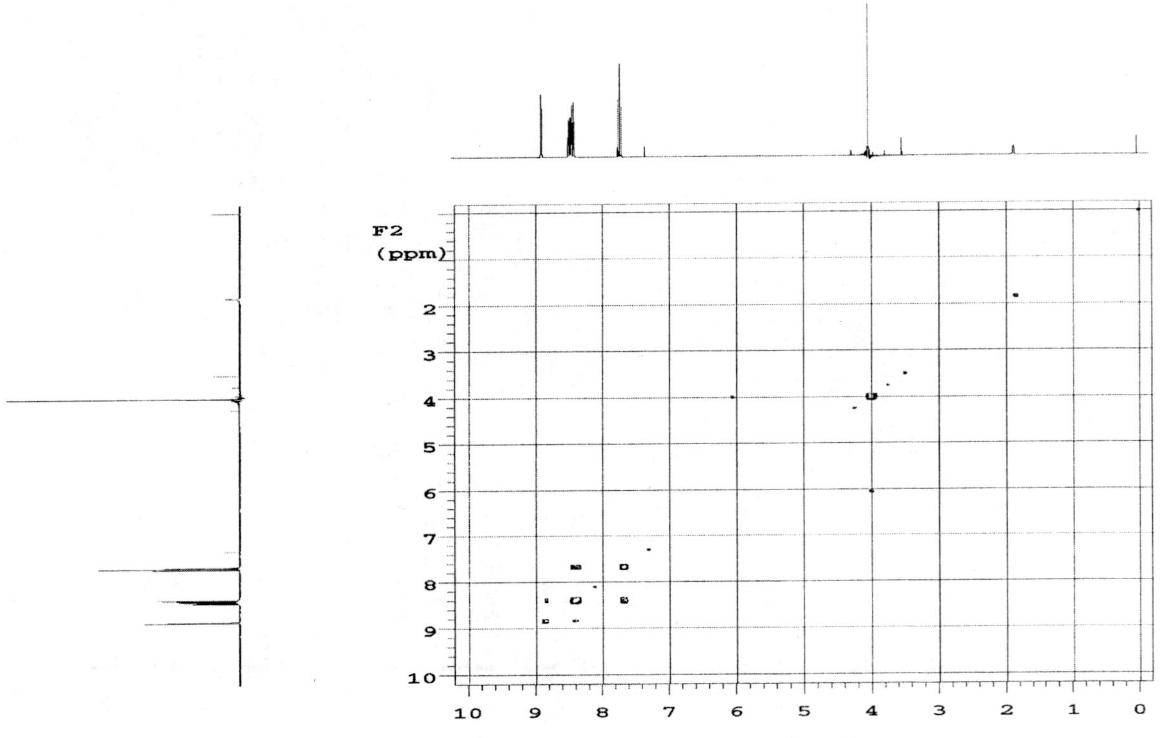

gCOSY Spectrum in $CDCl_3$

Problem 2.48 Given the spectroscopic data below (Proton, Carbon, Edited DEPT, and gHSQC), provide the structure of the molecule.

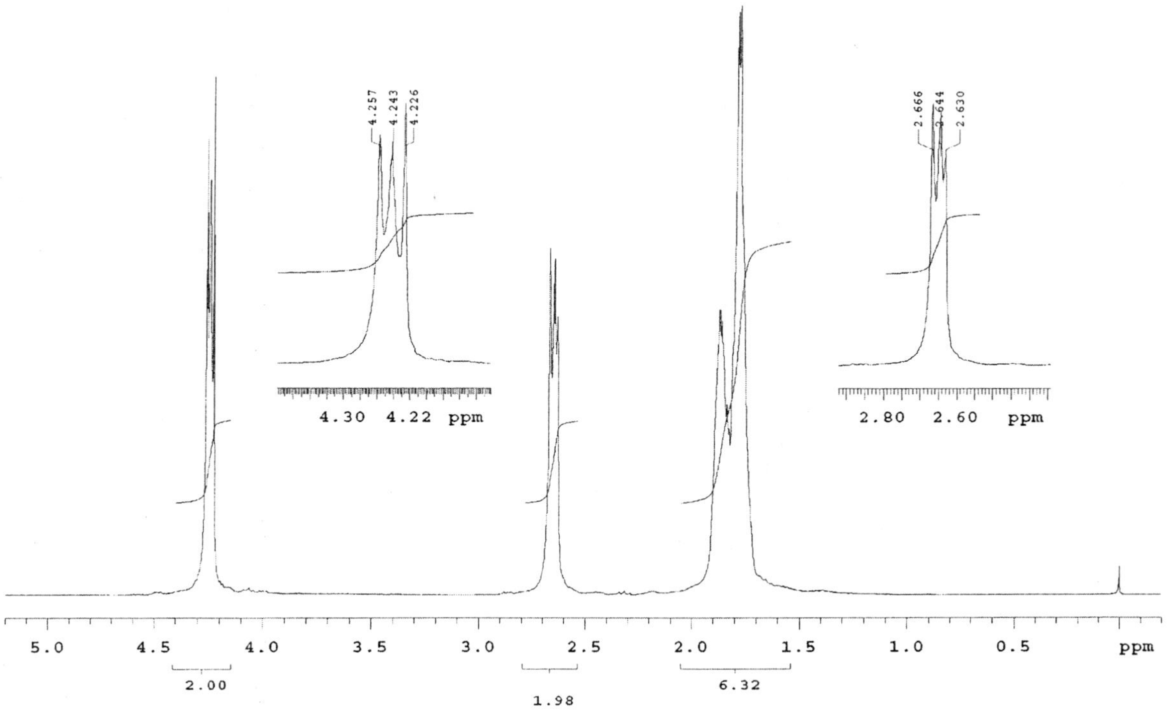

^{1}H-NMR Spectrum in $CDCl_3$

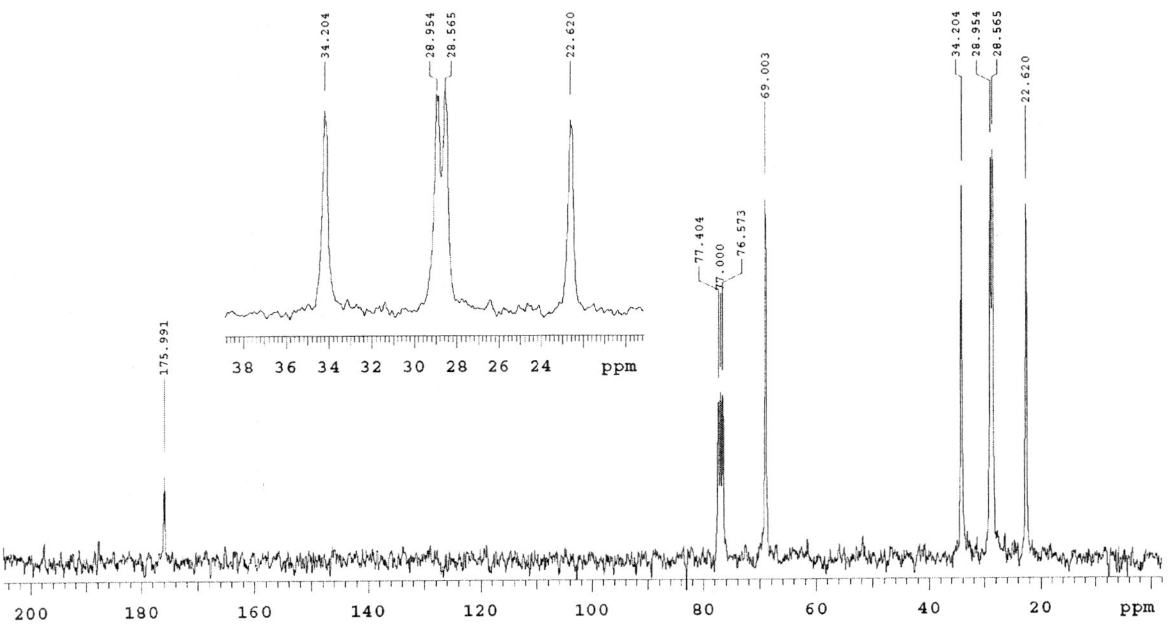

^{13}C-NMR Spectrum in $CDCl_3$

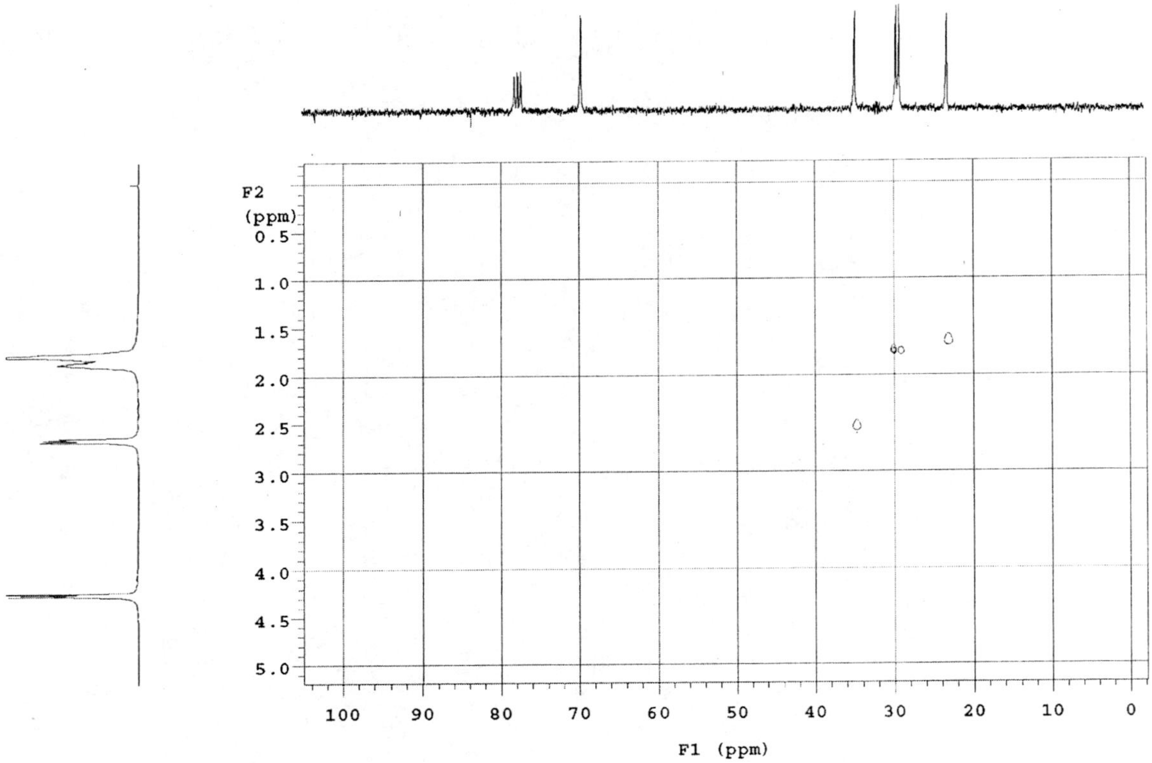

gHSQC Spectrum in $CDCl_3$

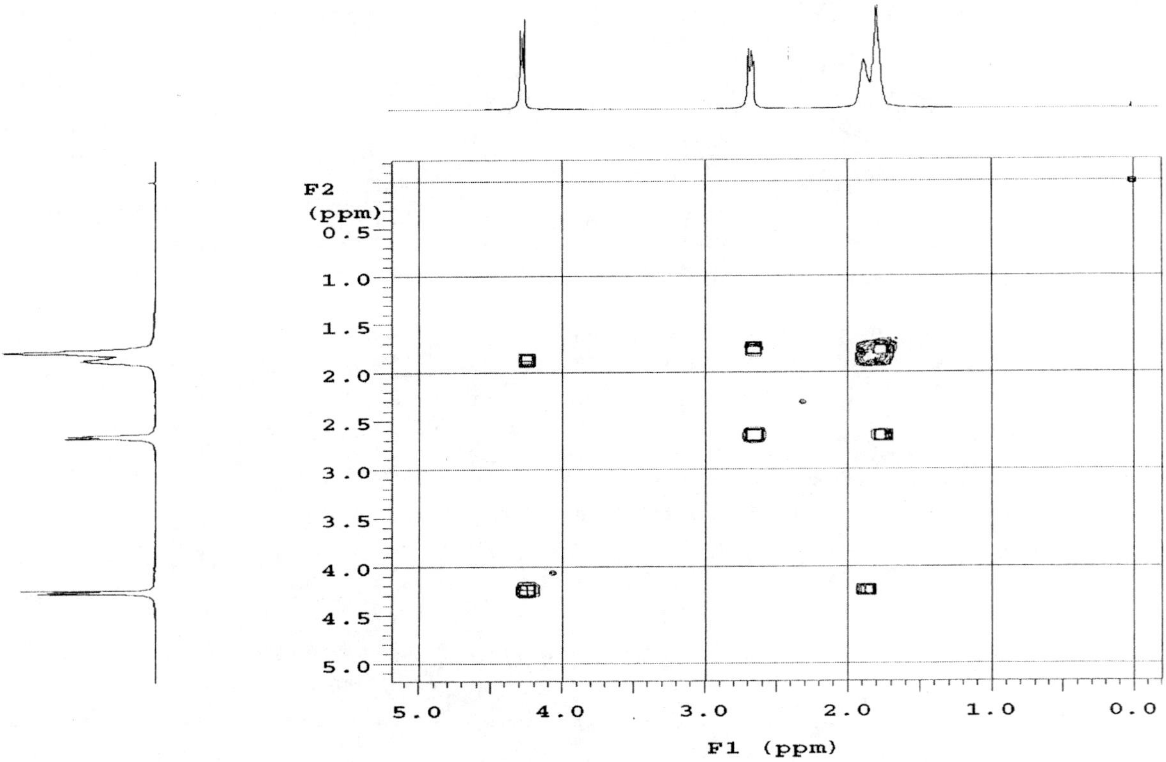

gCOSY Spectrum in $CDCl_3$

Problem 2.49 Given the spectroscopic data below (Proton, Carbon, Edited DEPT, and gHSQC), provide the structure of the molecule.

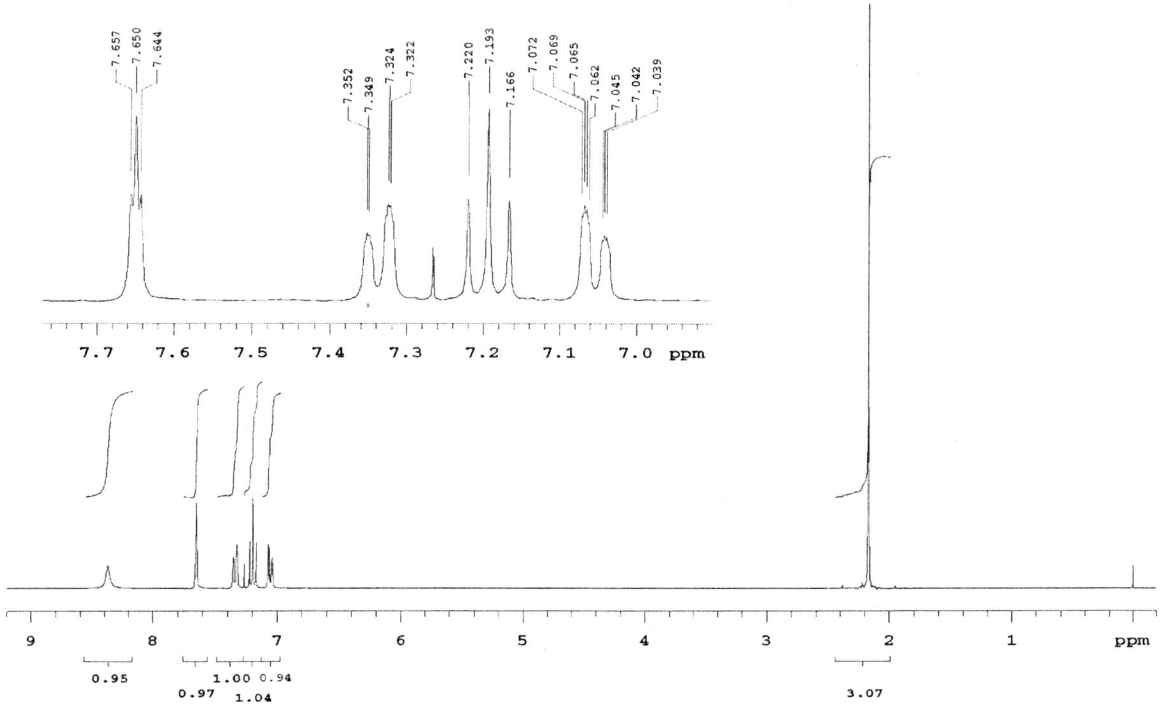

^{1}H-NMR Spectrum in $CDCl_3$

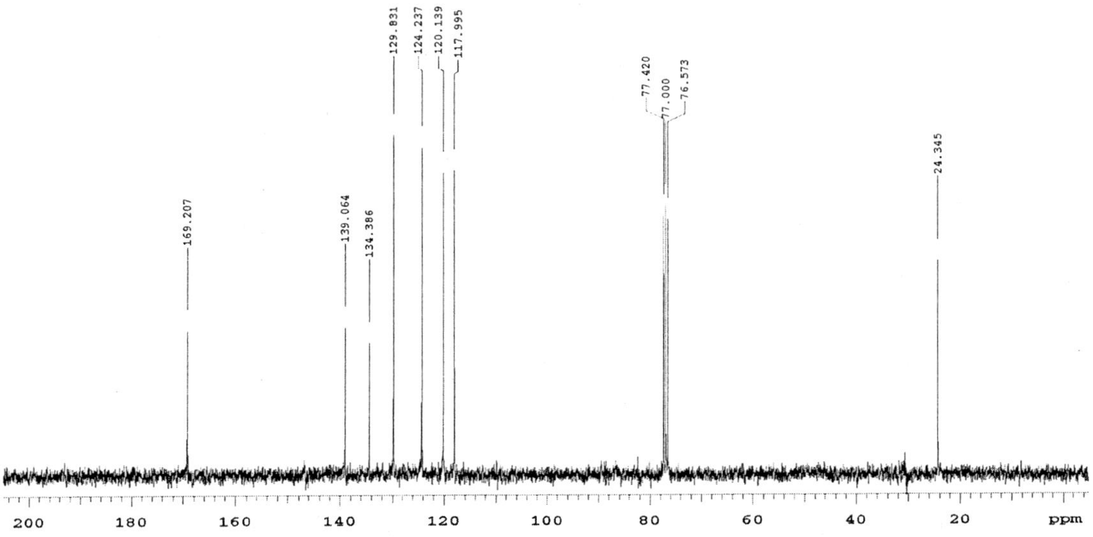

^{13}C-NMR Spectrum in $CDCl_3$

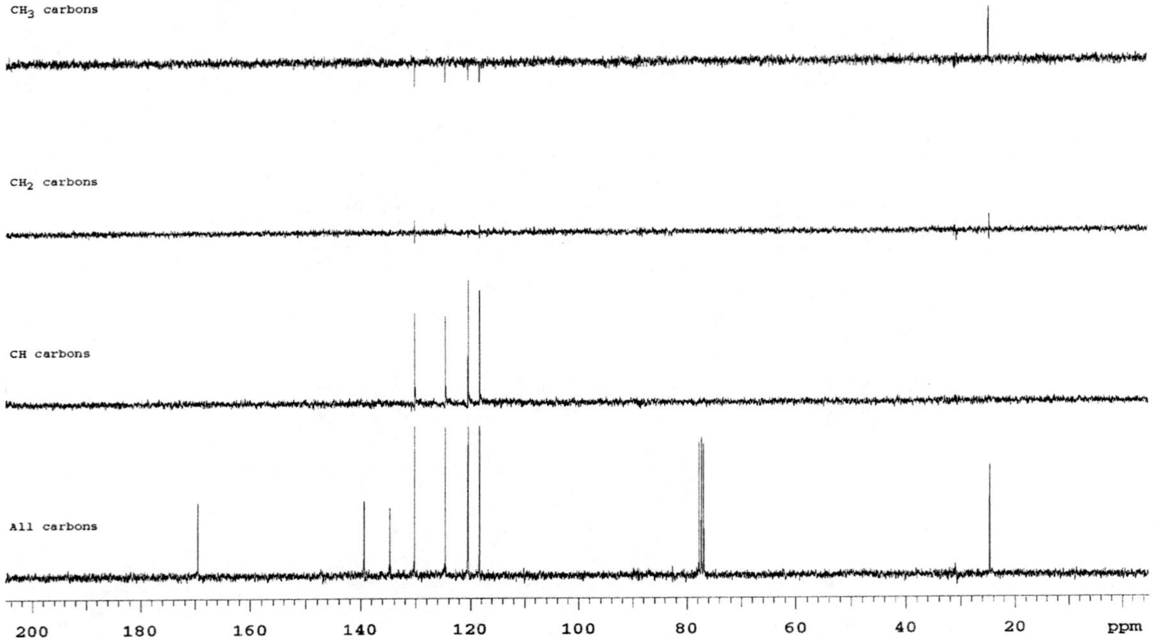

Edited DEPT ^{13}C-NMR Spectrum in $CDCl_3$

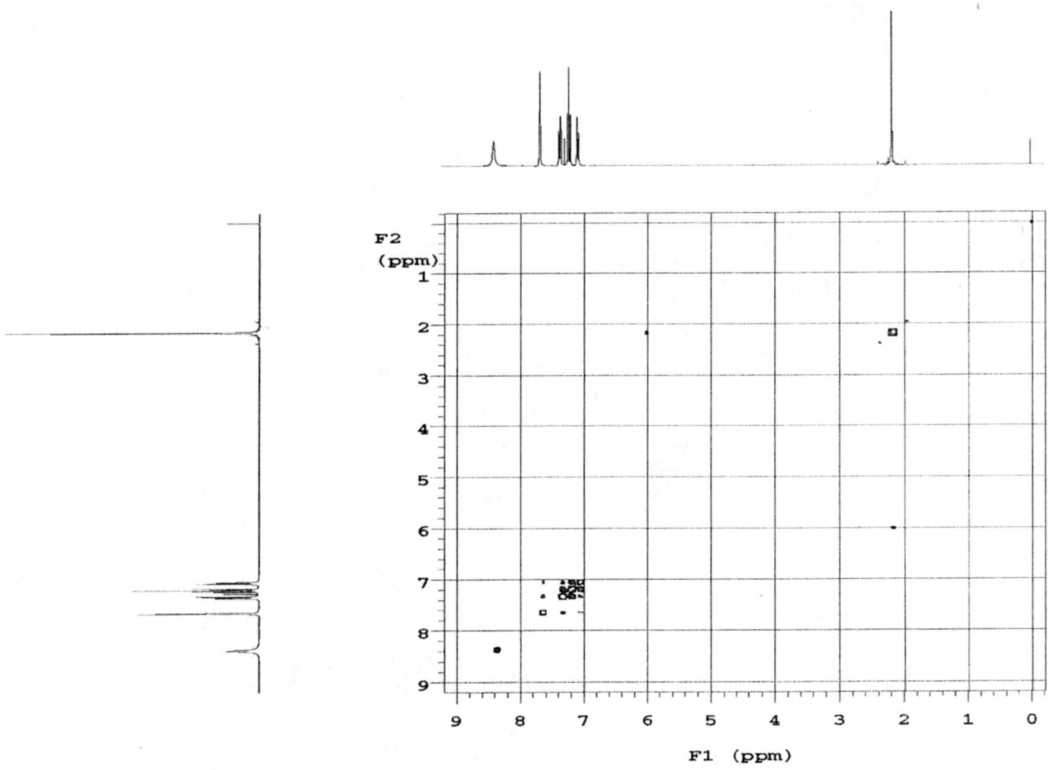

gCOSY Spectrum in $CDCl_3$

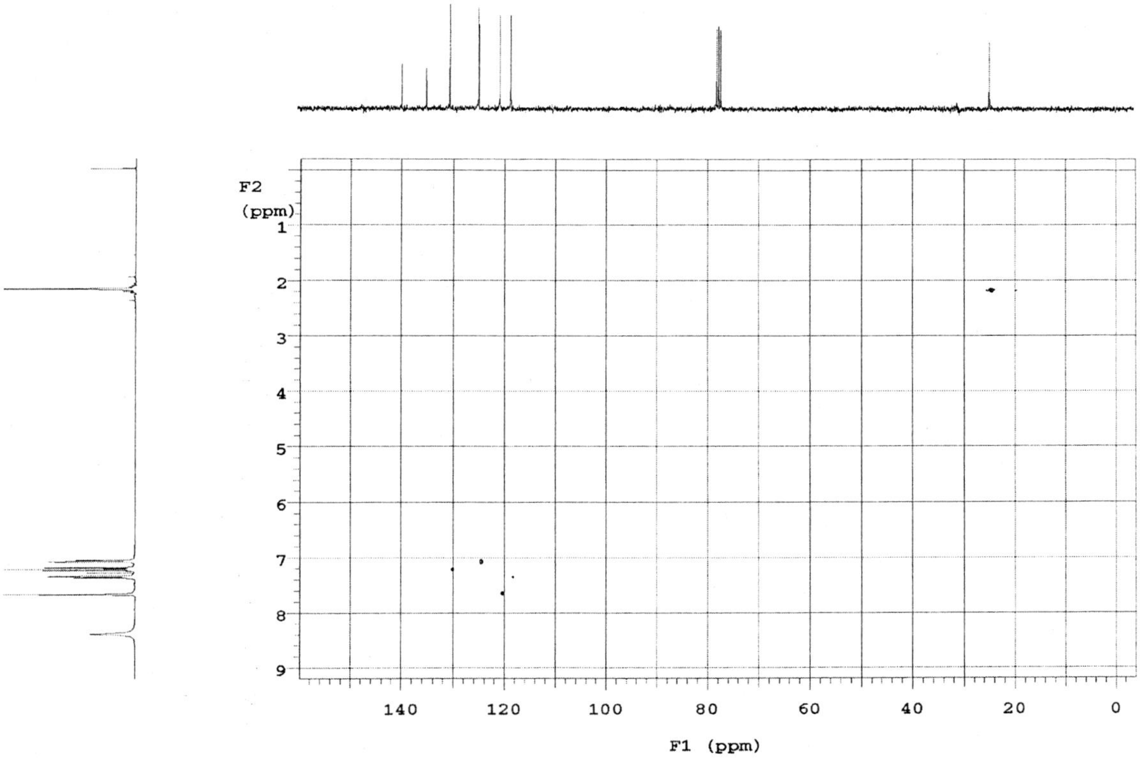

gHSQC Spectrum in $CDCl_3$

Problem 2.50 Given the spectroscopic data below (Proton, Carbon, Edited DEPT, gHSQC, and gHMBC), provide the structure of the molecule. (HRMS: 226.1934) These spectra arise from a mixture (~50:50) of the meso form and the racemate.

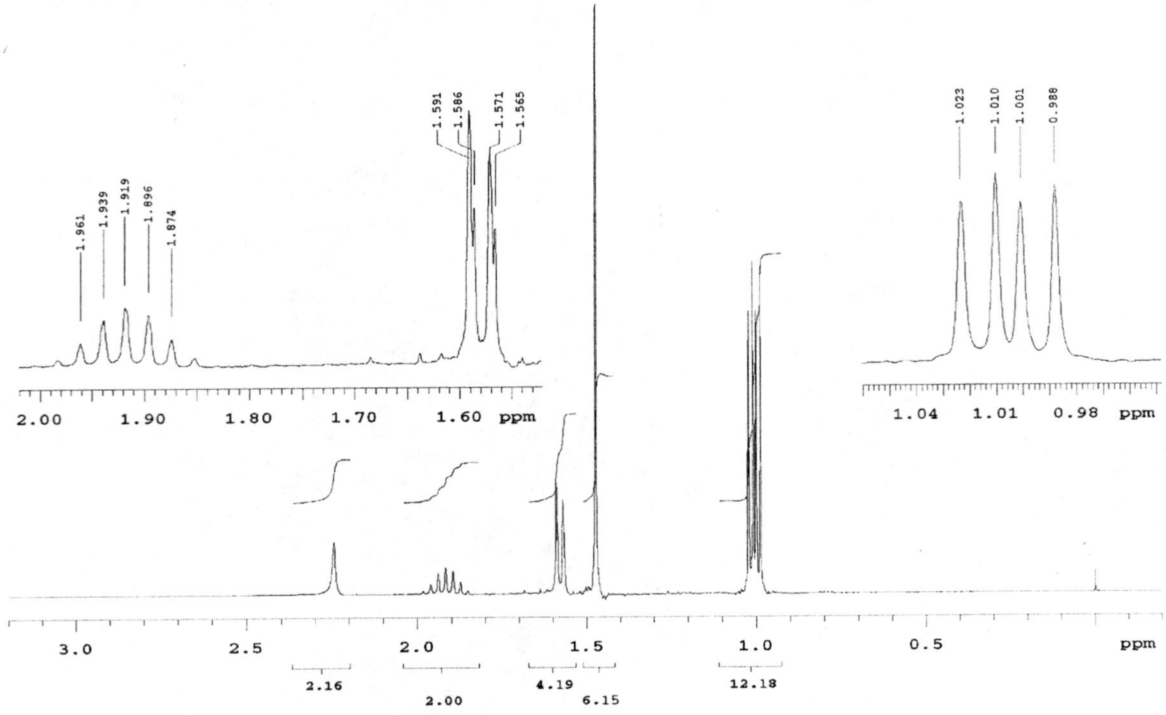

^{1}H-NMR Spectrum in $CDCl_3$

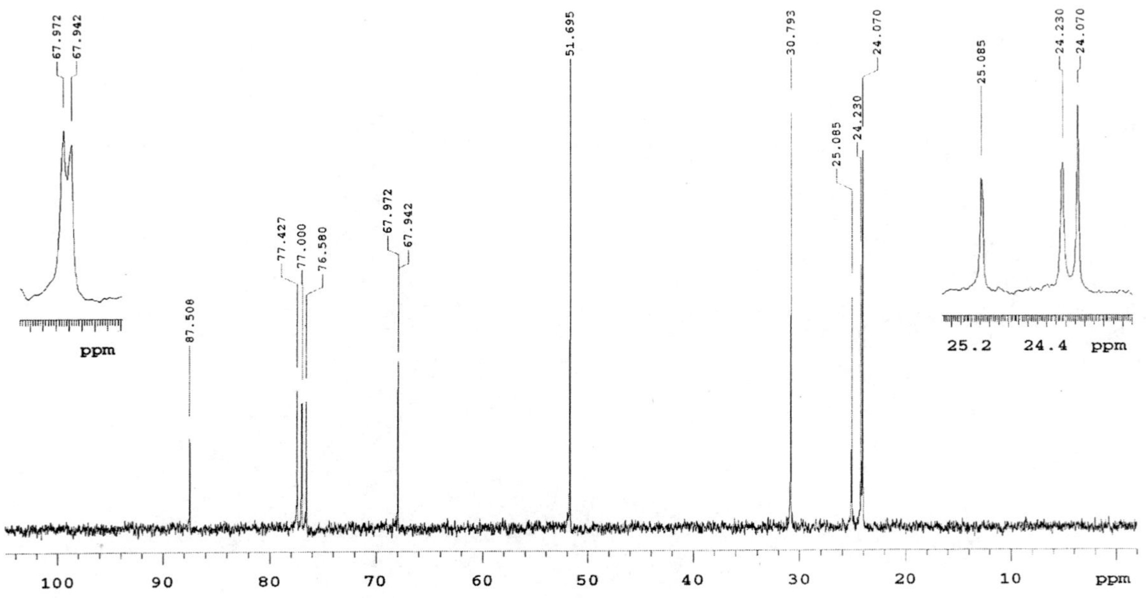

^{13}C-NMR Spectrum in $CDCl_3$

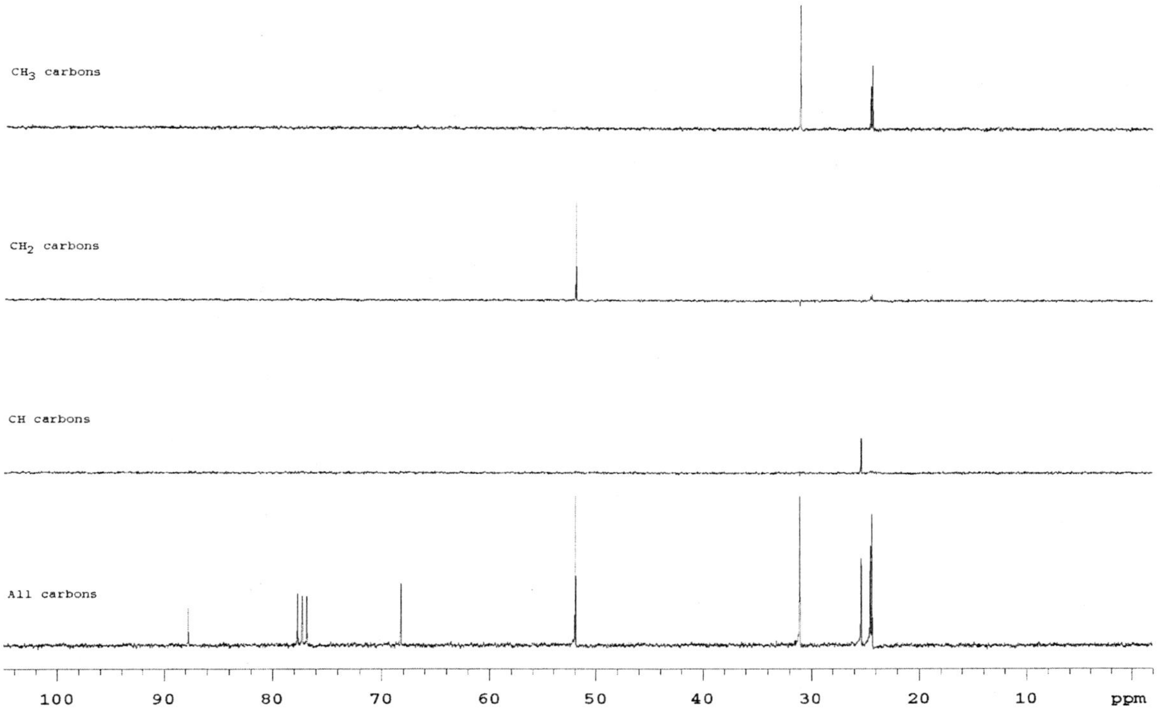

Edited DEPT ^{13}C-NMR Spectrum in $CDCl_3$

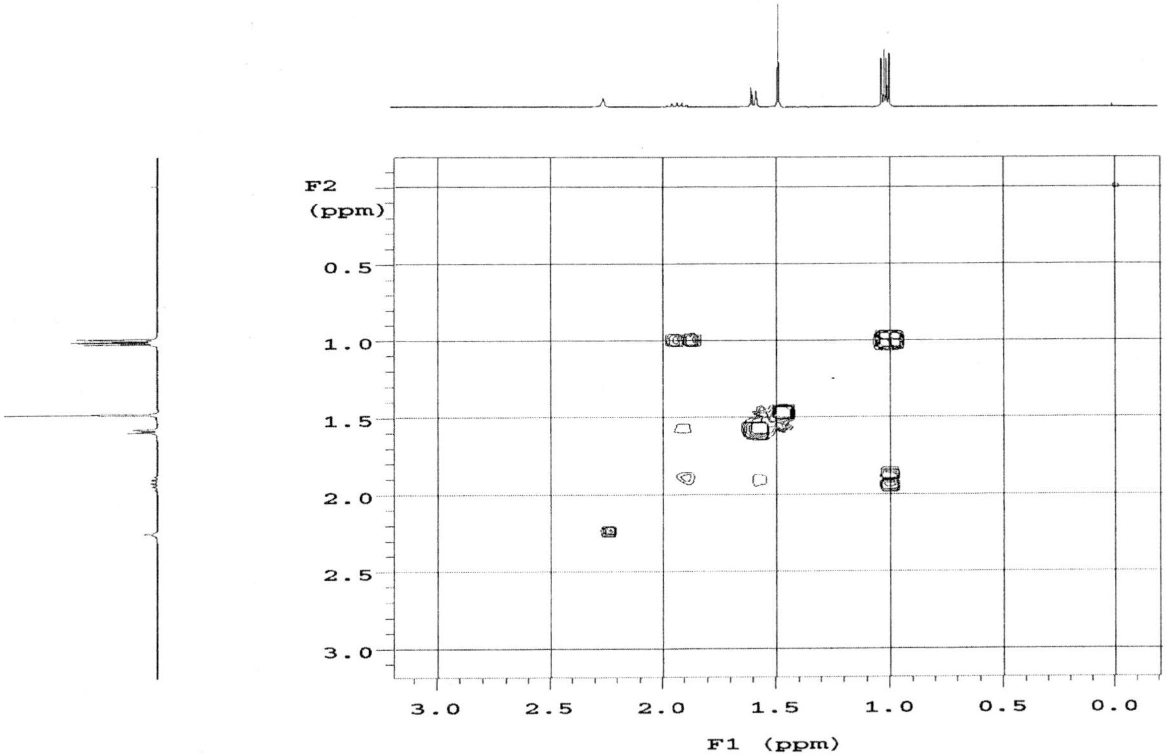

gCOSY Spectrum in $CDCl_3$

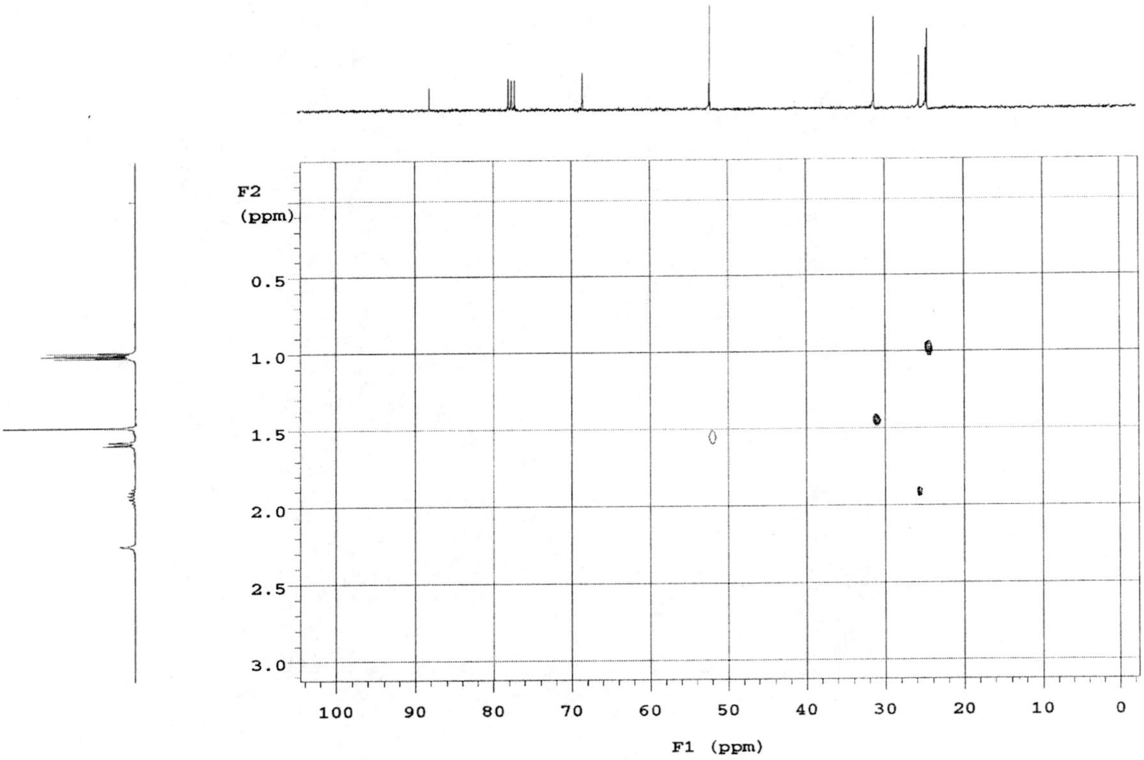

gHSQC Spectrum in $CDCl_3$

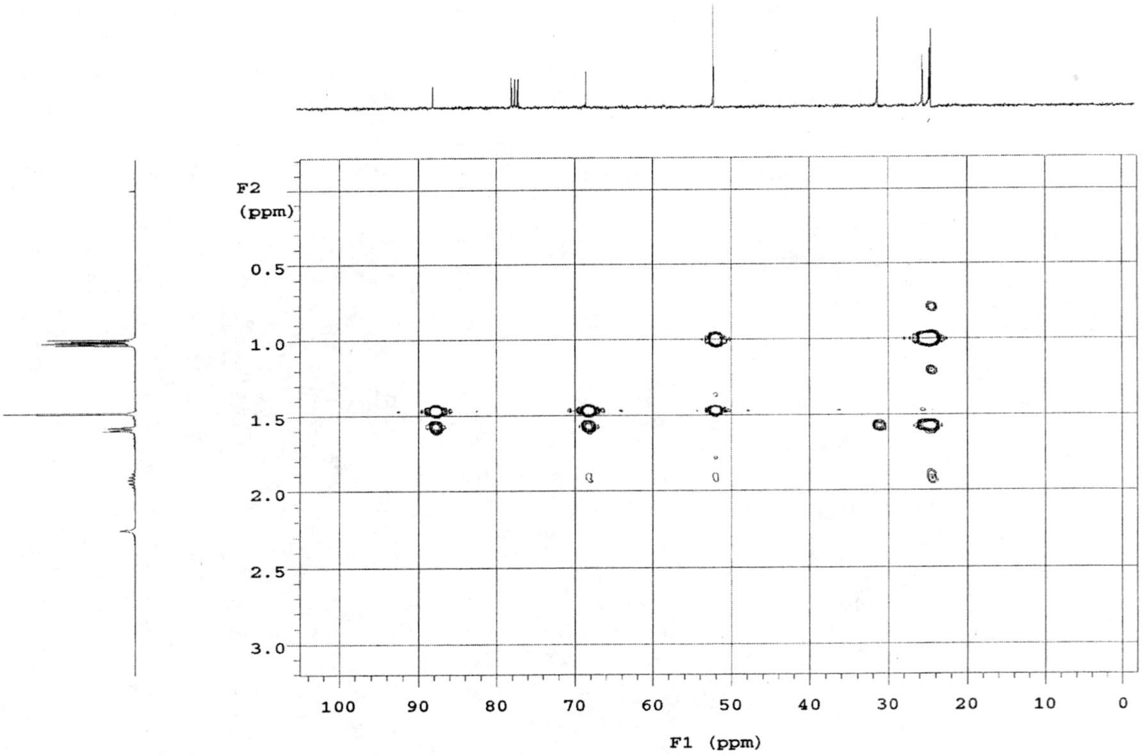

gHMBC Spectrum in $CDCl_3$

Chapter 3 Challenging Problems

Problem 3.01 Given the spectroscopic data below (Proton, Carbon, gHSQC, gCOSY, and gHMBC), provide the structure of the molecule. (HR-MS: 180.1514 amu)

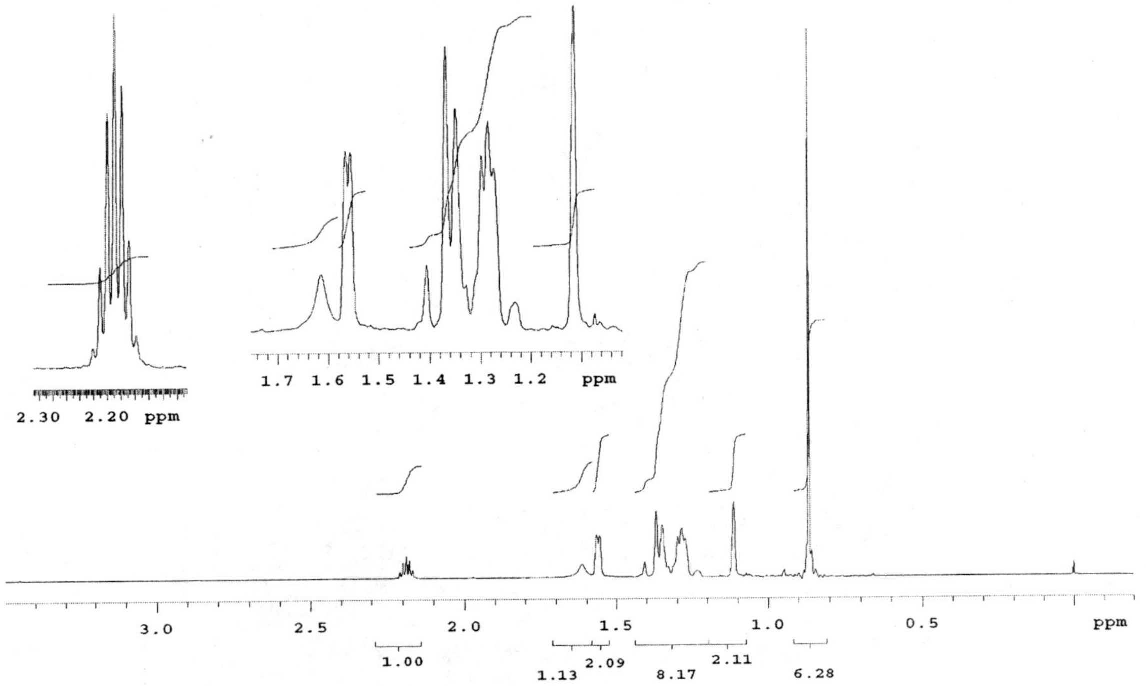

^{1}H-NMR Spectrum in $CDCl_3$

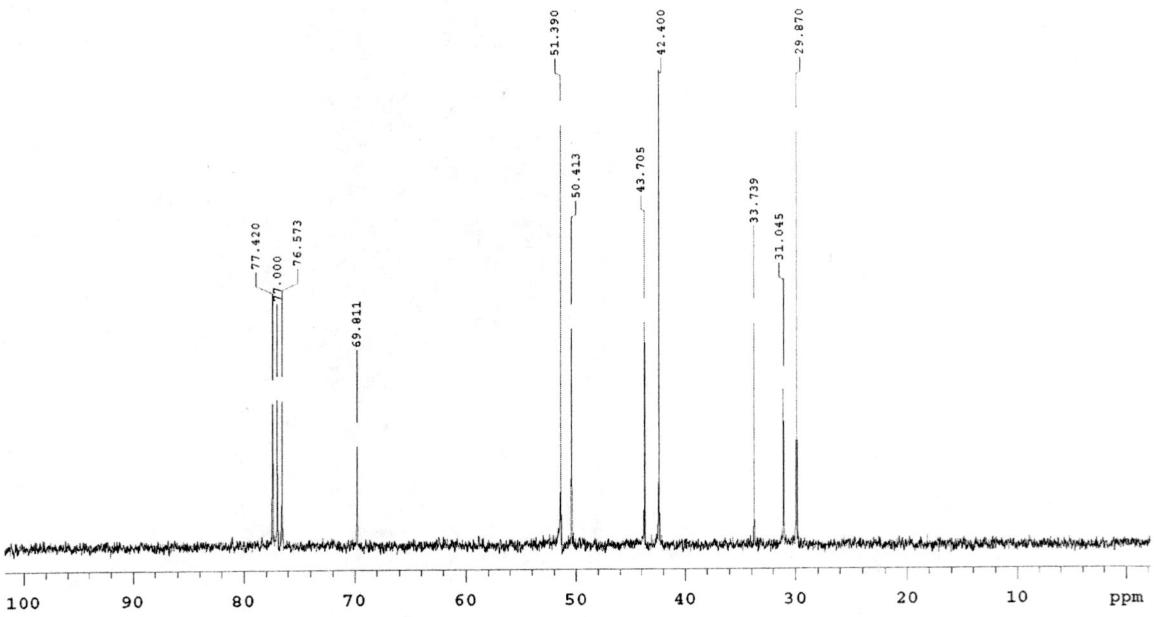

^{13}C-NMR Spectrum in $CDCl_3$

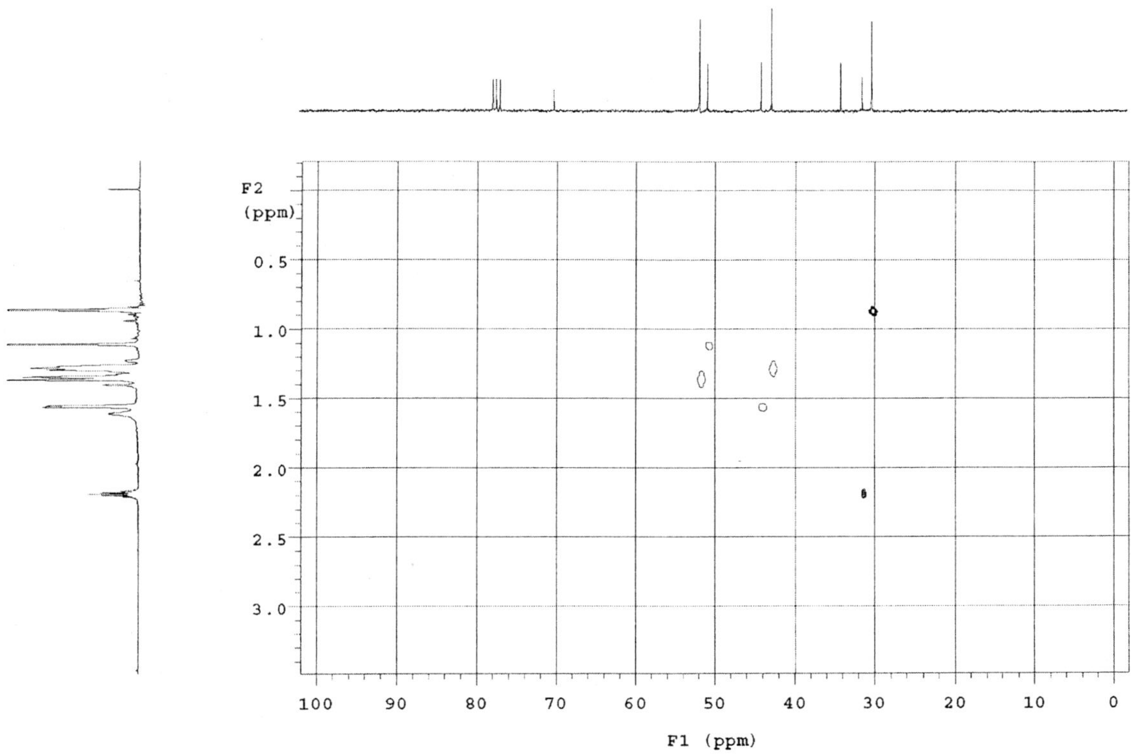

gHSQC Spectrum in $CDCl_3$

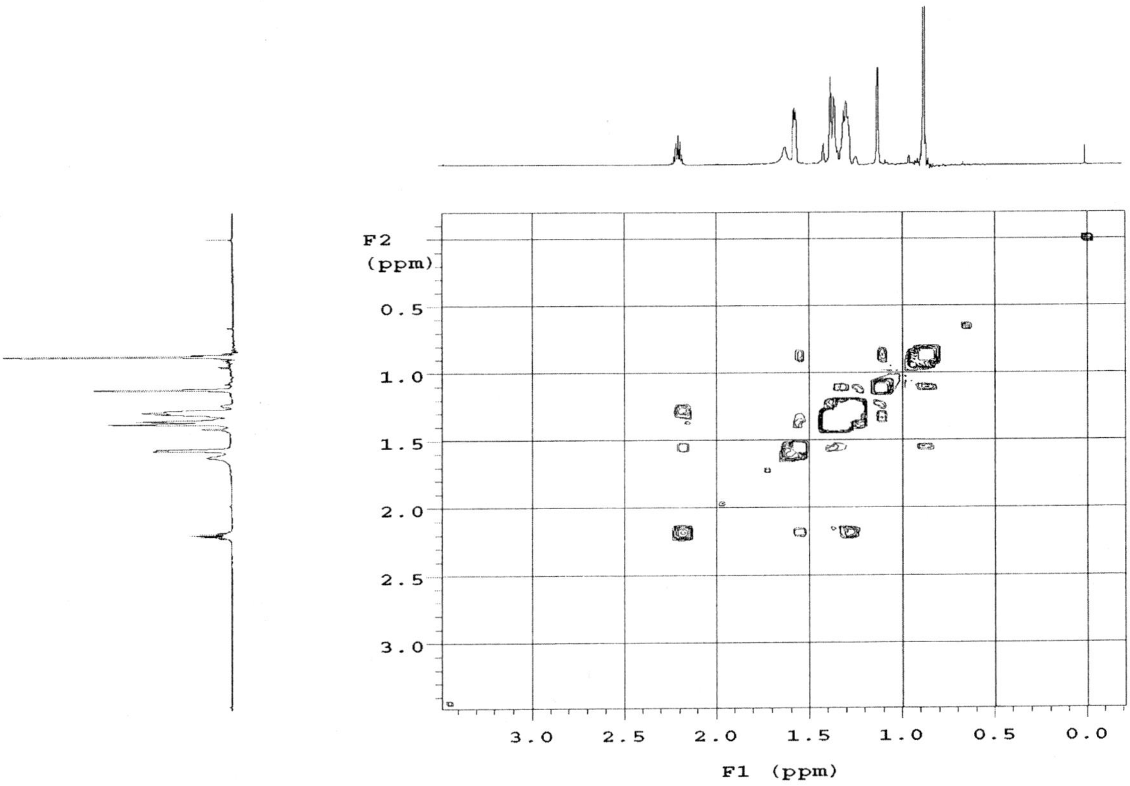

gCOSY Spectrum in $CDCl_3$

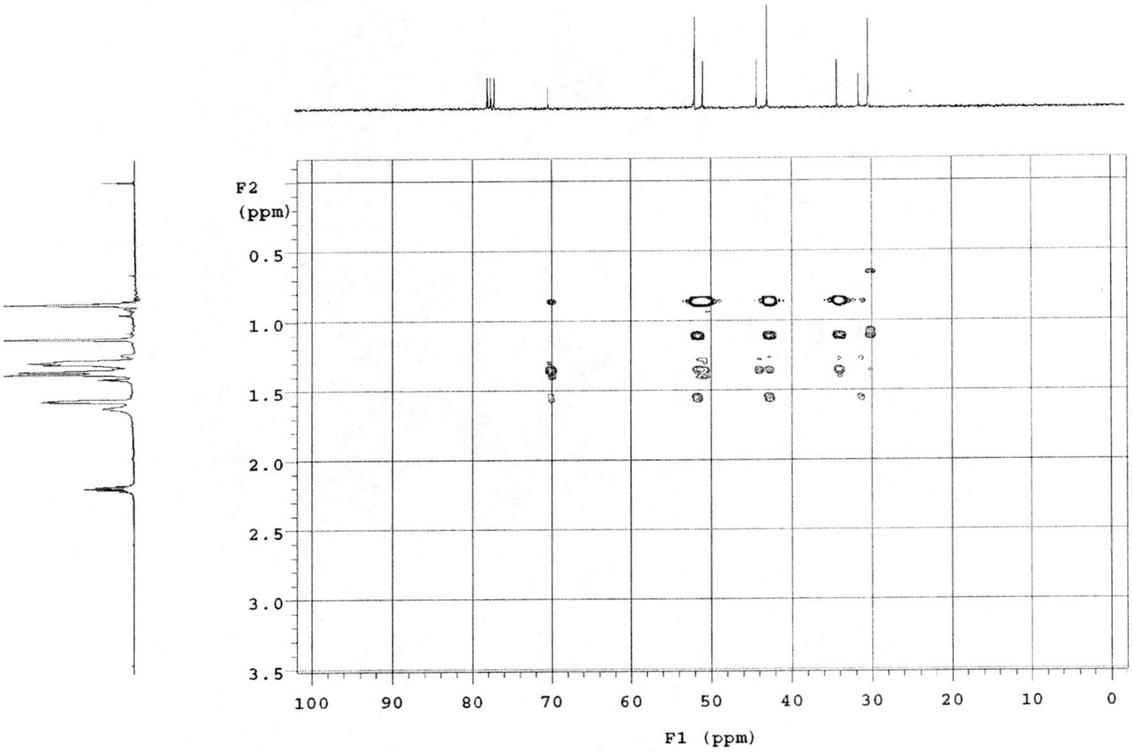

gHMBC Spectrum in $CDCl_3$

Problem 3.02 Given the spectroscopic data below (Proton, Carbon, gHSQC, gCOSY, and gHMBC), provide the structure of the molecule. (HR-MS: 252.1514 amu)

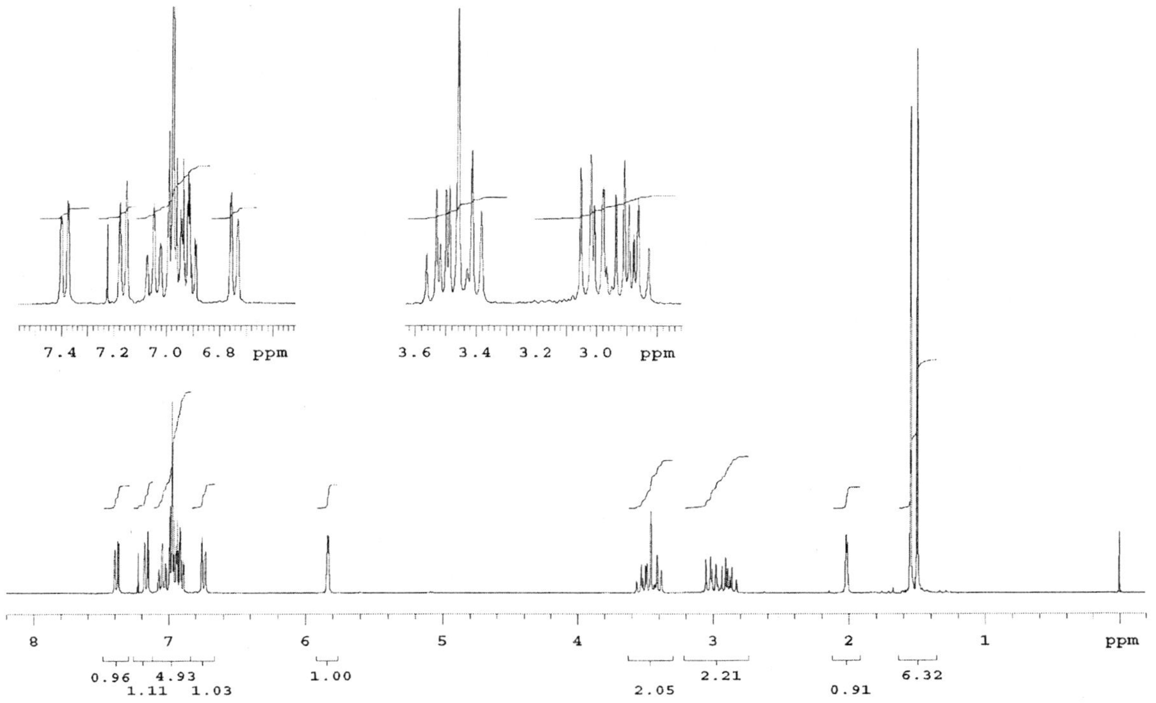

^{1}H-NMR Spectrum in $CDCl_3$

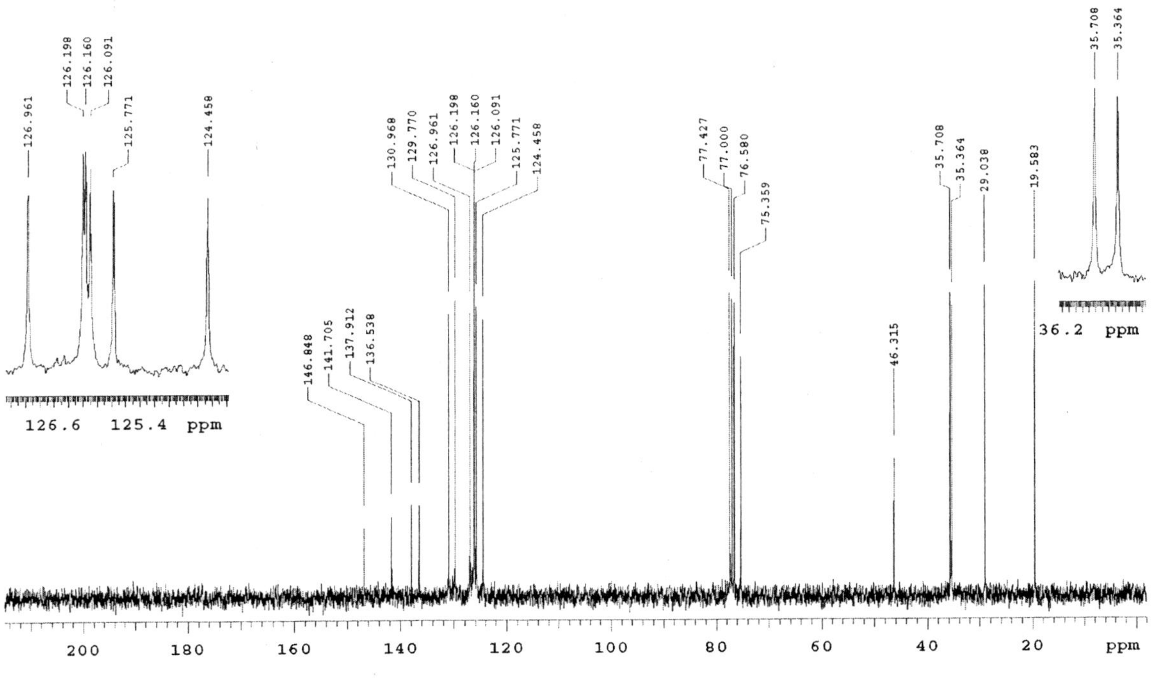

^{13}C-NMR Spectrum in $CDCl_3$

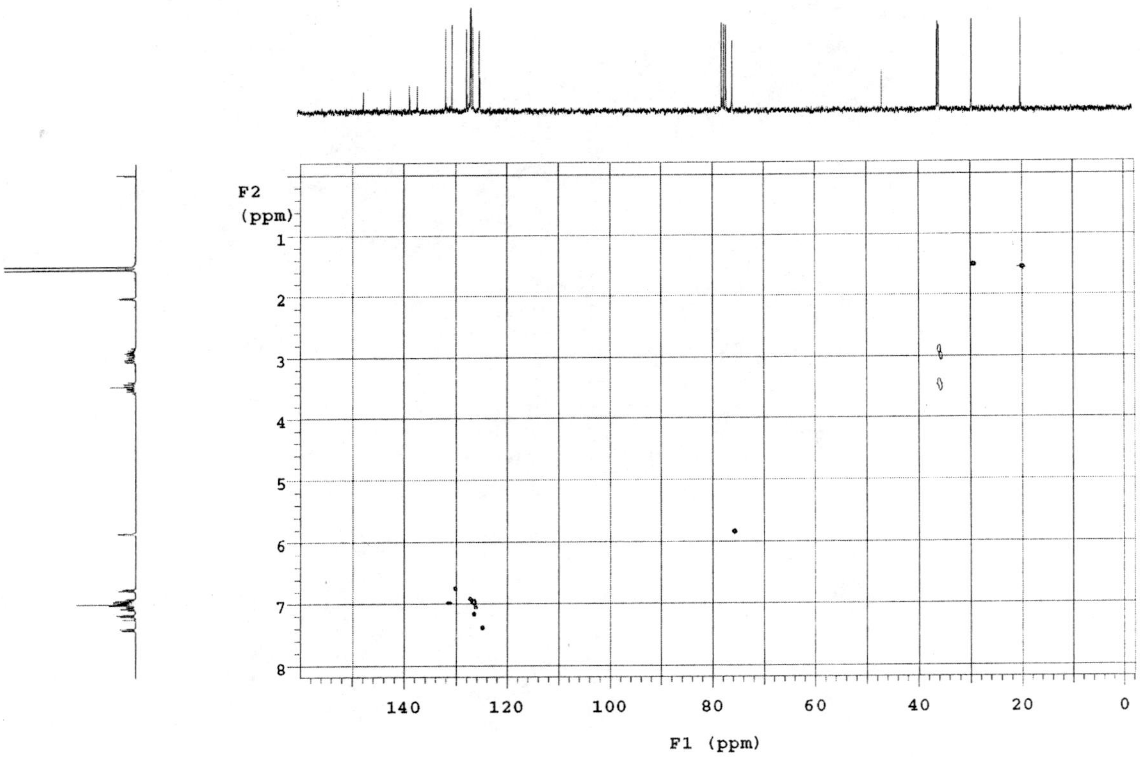

gHSQC Spectrum in $CDCl_3$

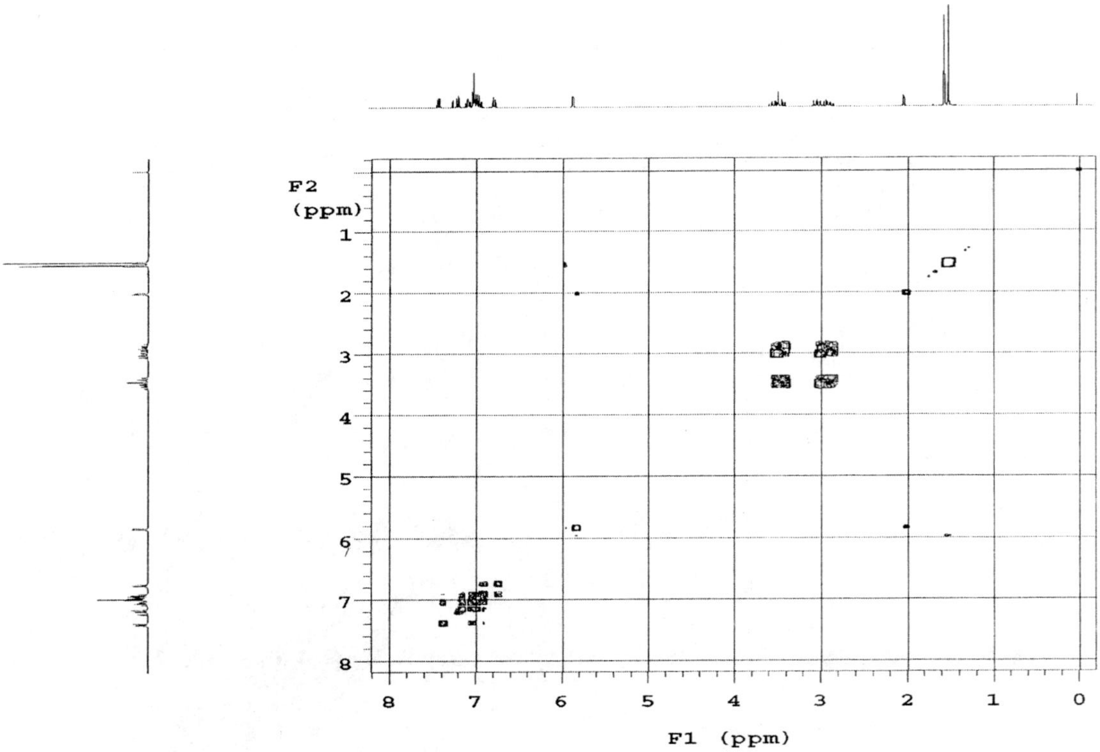

gCOSY Spectrum in $CDCl_3$

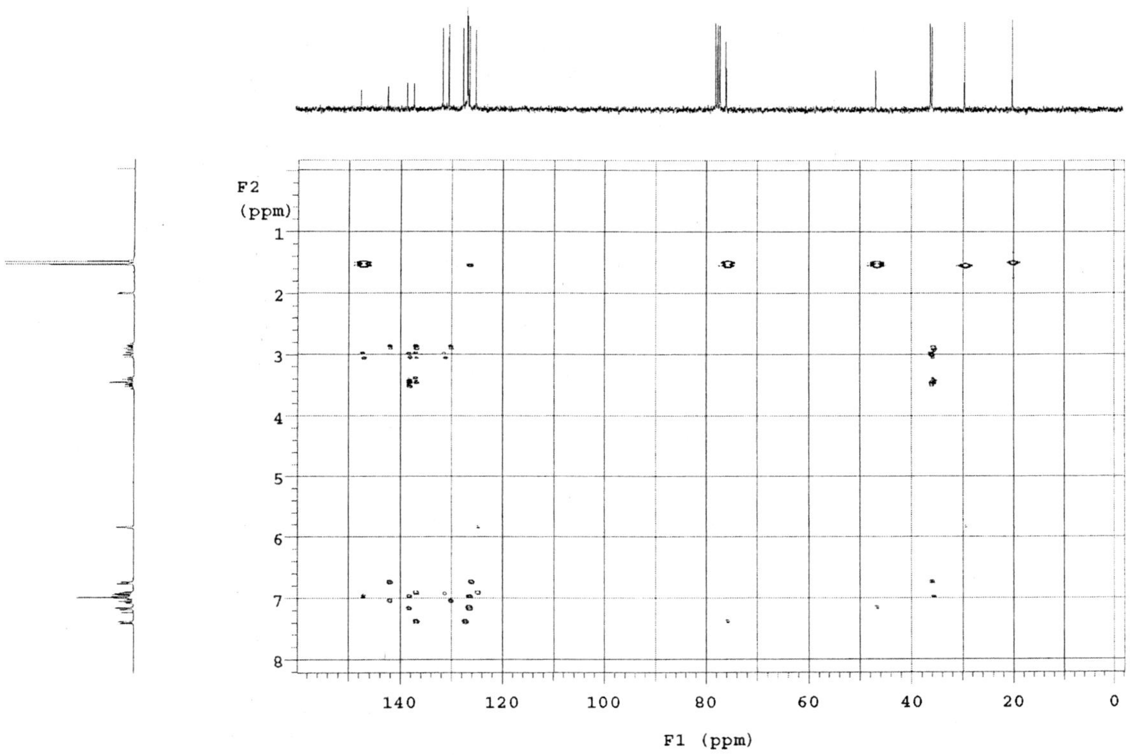

gHMBC Spectrum in $CDCl_3$

Problem 3.03 Given the spectroscopic data below (Proton, Carbon, gHSQC, and gHMBC), provide the structure of the molecule. (HR-MS: 164.0837 amu) Hint: A nuclear Overhauser effect is expected between the two signals at approximately 2.6 ppm.

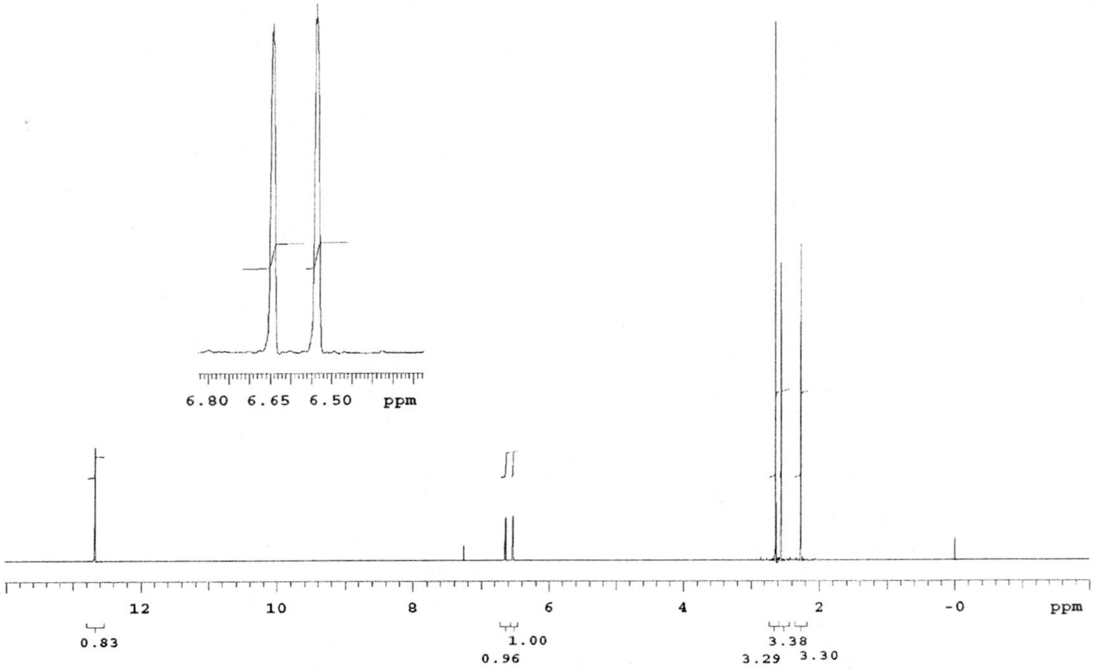

^{1}H-NMR Spectrum in $CDCl_3$

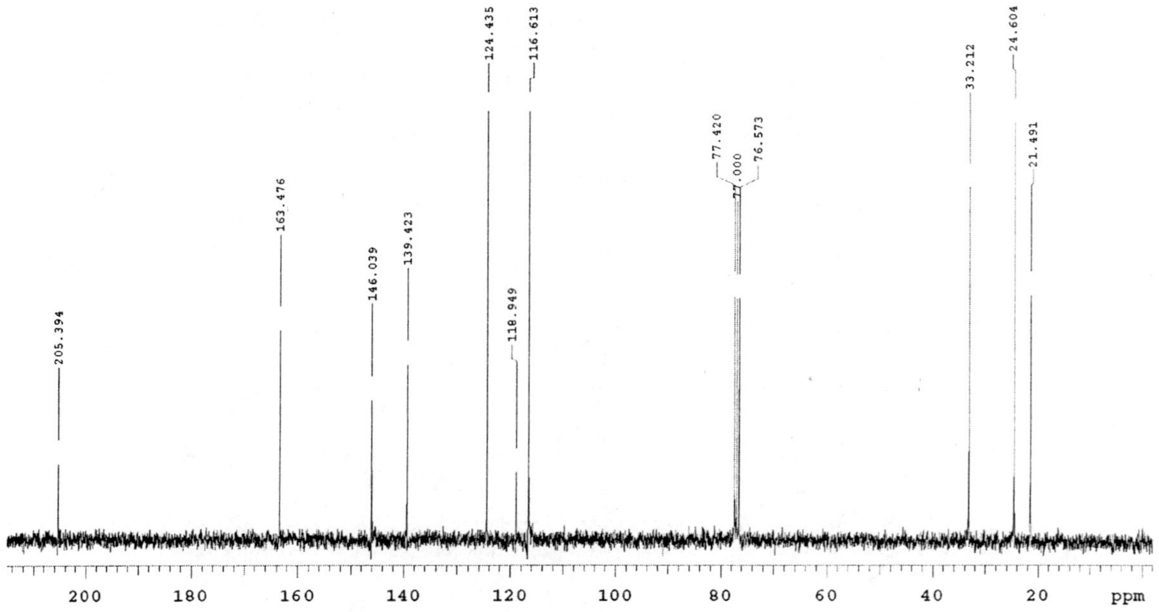

^{13}C-NMR Spectrum in $CDCl_3$

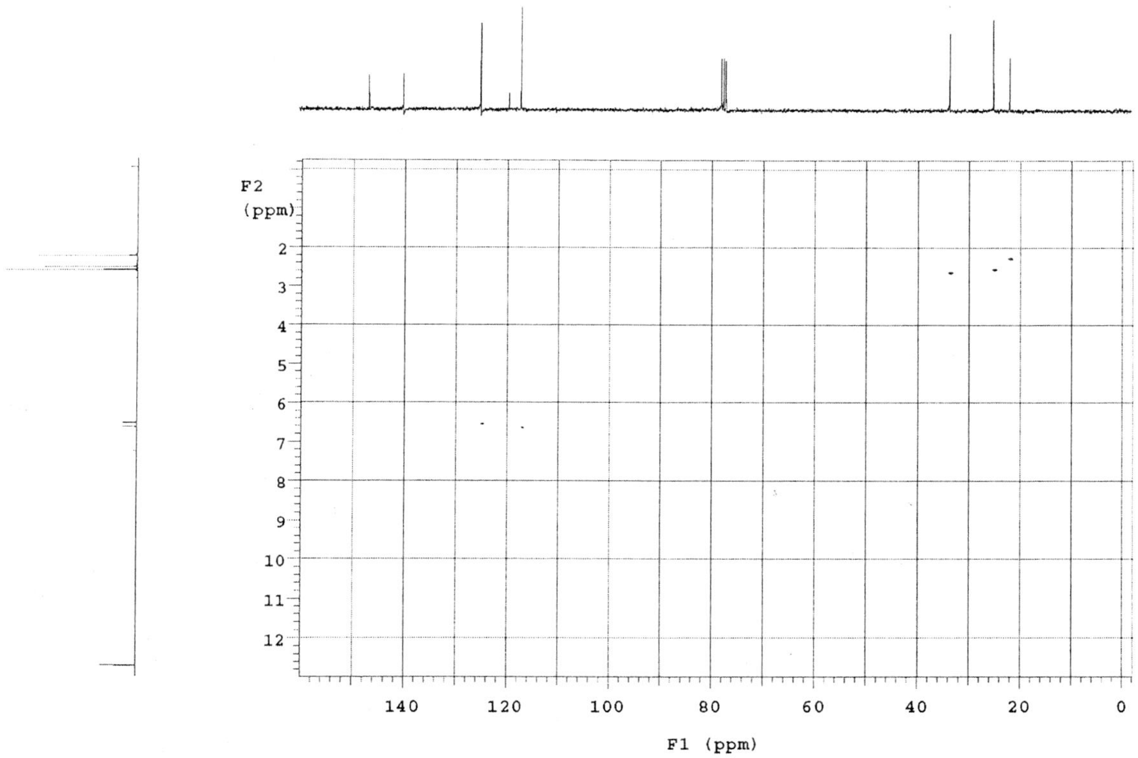

gHSQC Spectrum in $CDCl_3$

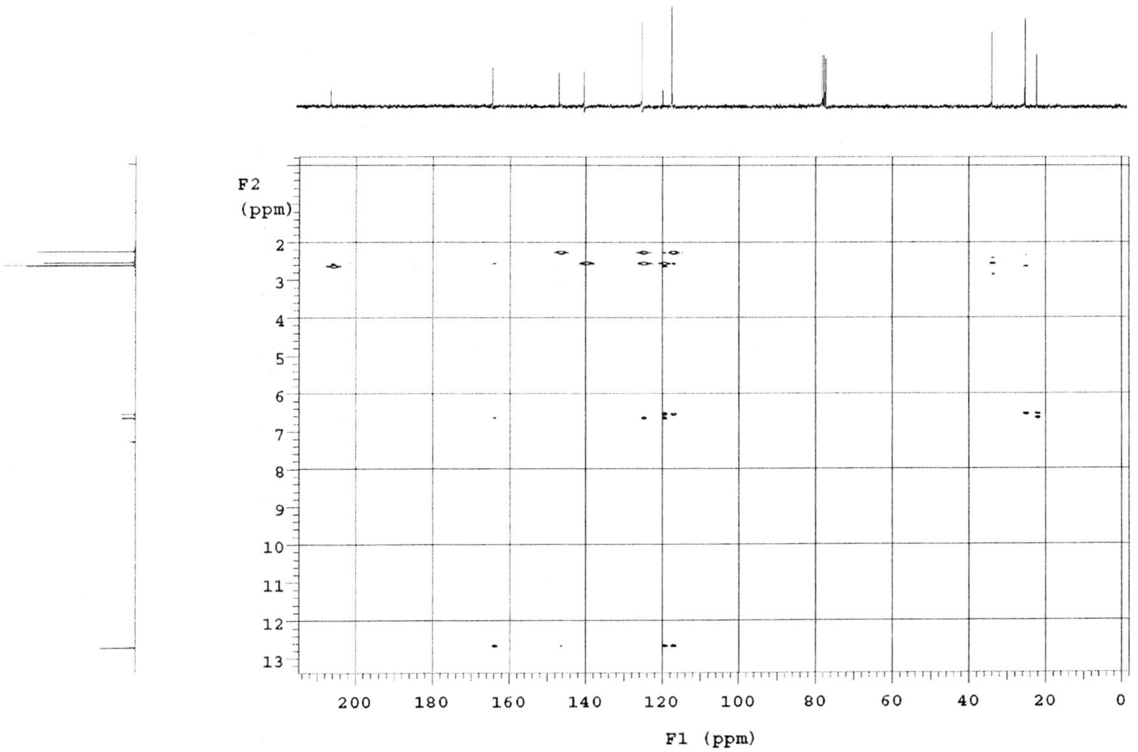

gHMBC Spectrum in $CDCl_3$

Problem 3.04 Given the spectroscopic data below (Proton, Carbon, Edited DEPT, gCOSY, gHSQC, and gHMBC), provide the structure of the molecule. (IR: ~3600 cm^{-1} broad signal)

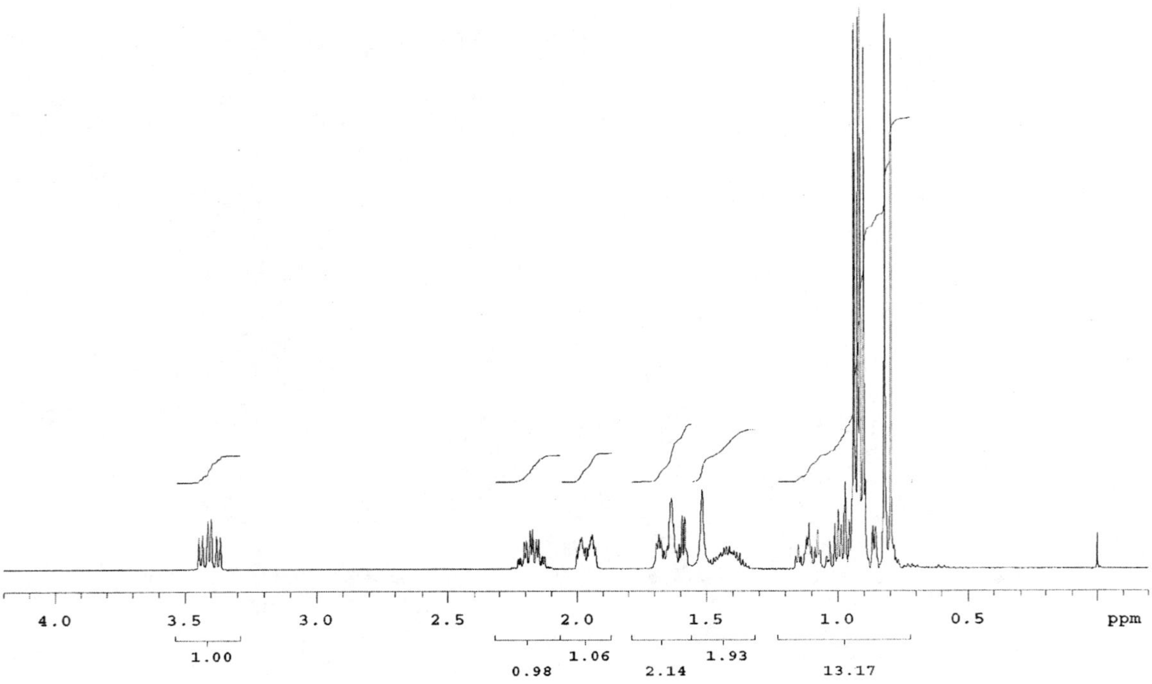

^{1}H-NMR Spectrum in $CDCl_3$

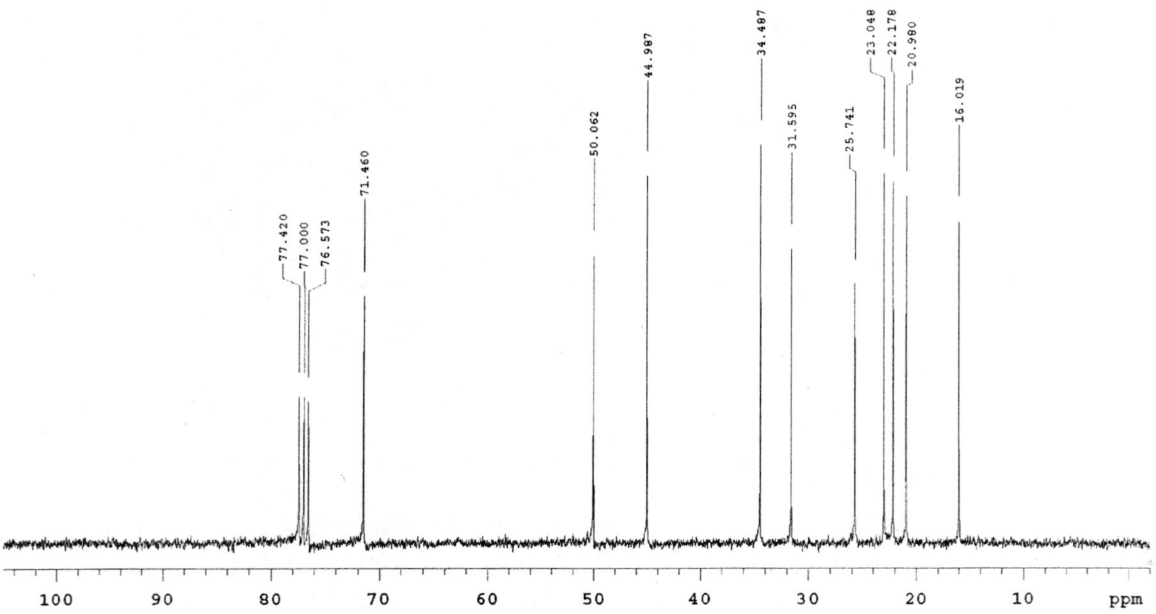

^{13}C-NMR Spectrum in $CDCl_3$

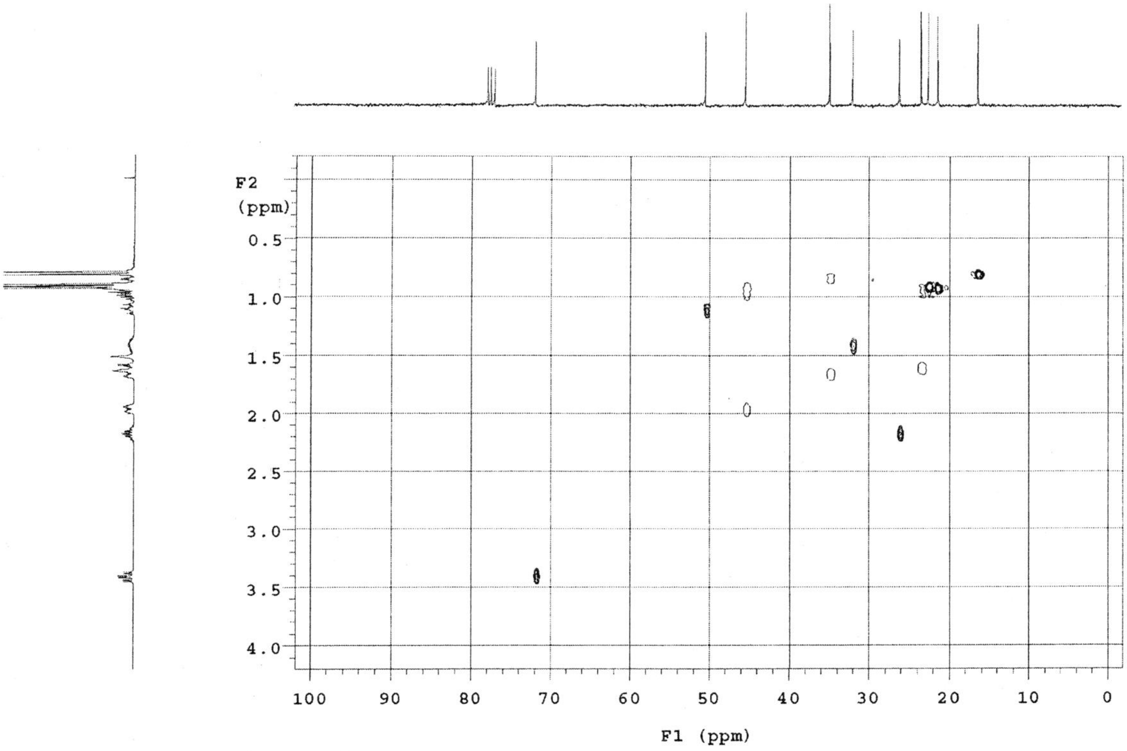

gHSQC Spectrum in $CDCl_3$

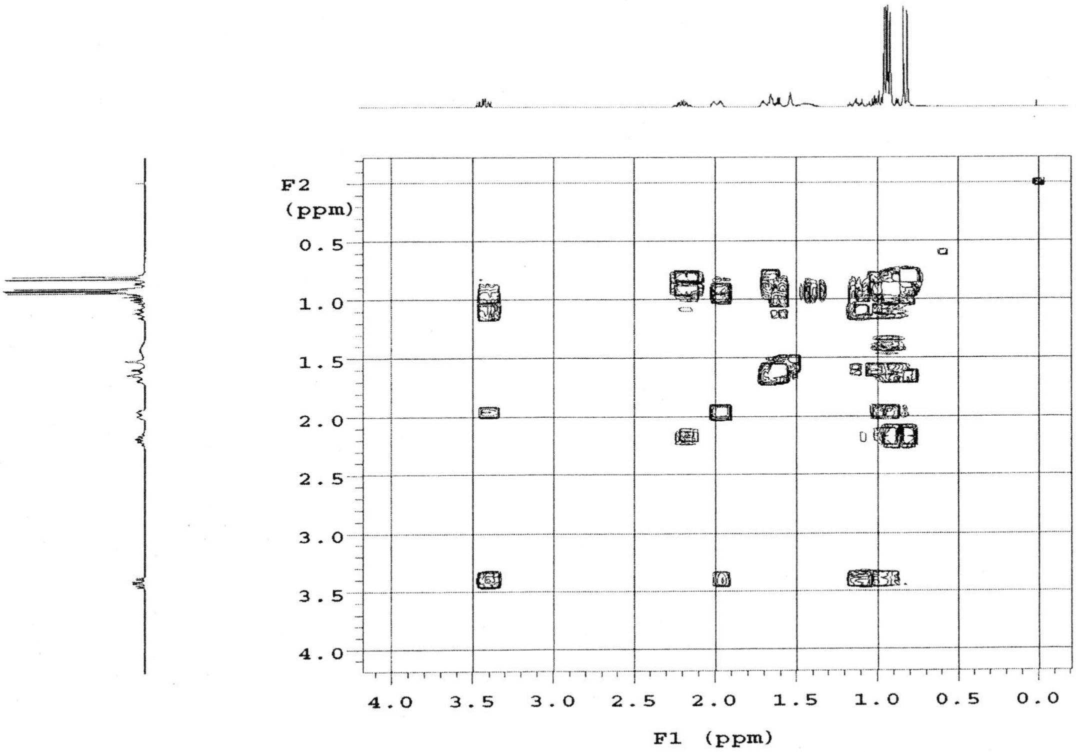

gCOSY Spectrum in $CDCl_3$

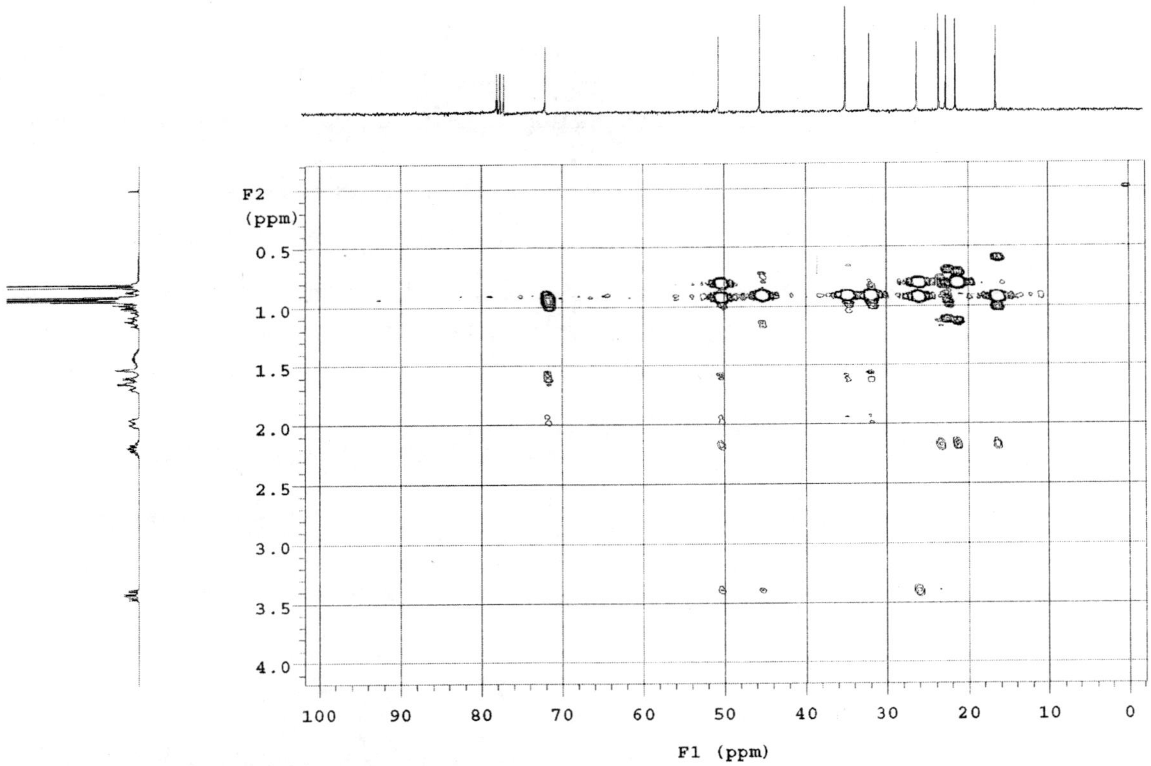

gHMBC Spectrum in $CDCl_3$

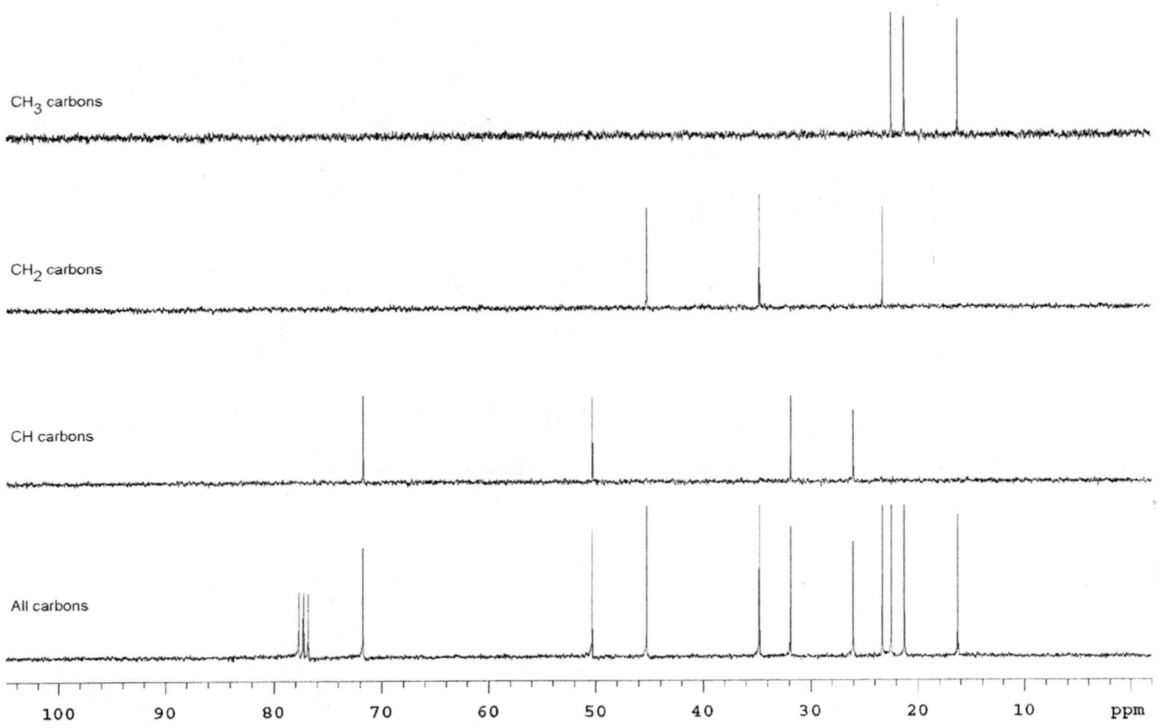

Edited DEPT ^{13}C-NMR Spectrum in $CDCl_3$

Problem 3.05 Given the spectroscopic data below (Proton, Carbon, gCOSY, gHSQC, and gHMBC), provide the structure of the molecule. In the mass spectrum, there are M, M+2, and M+4 peaks in a 100:64:10 ratio.

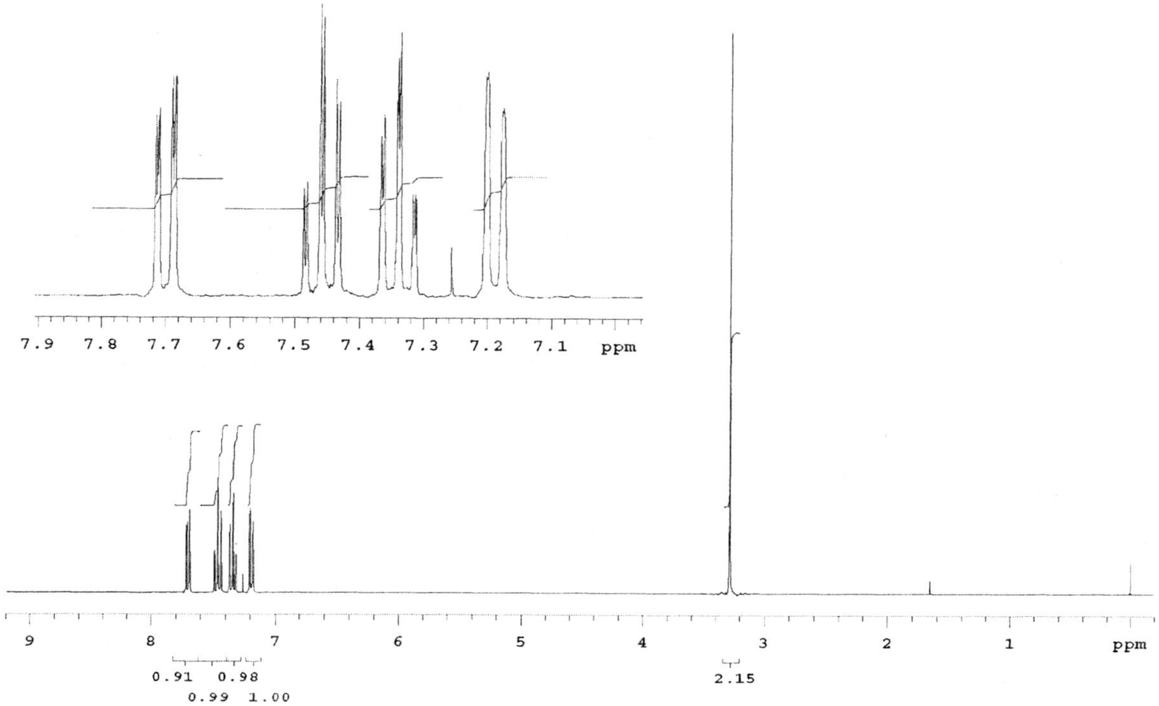

^{1}H-NMR Spectrum in $CDCl_3$

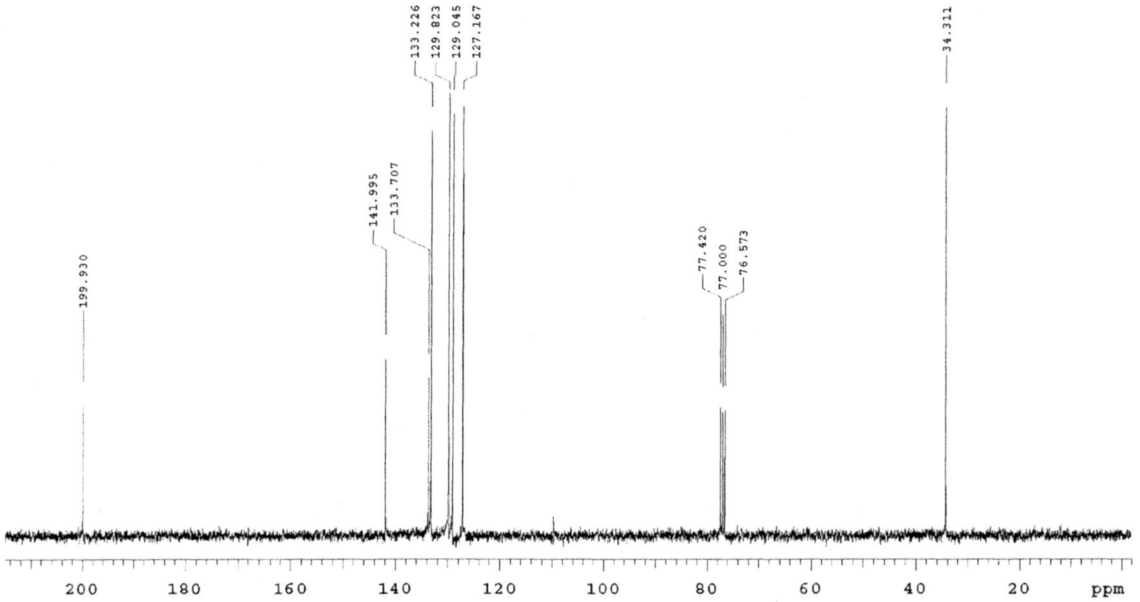

^{13}C-NMR Spectrum in $CDCl_3$

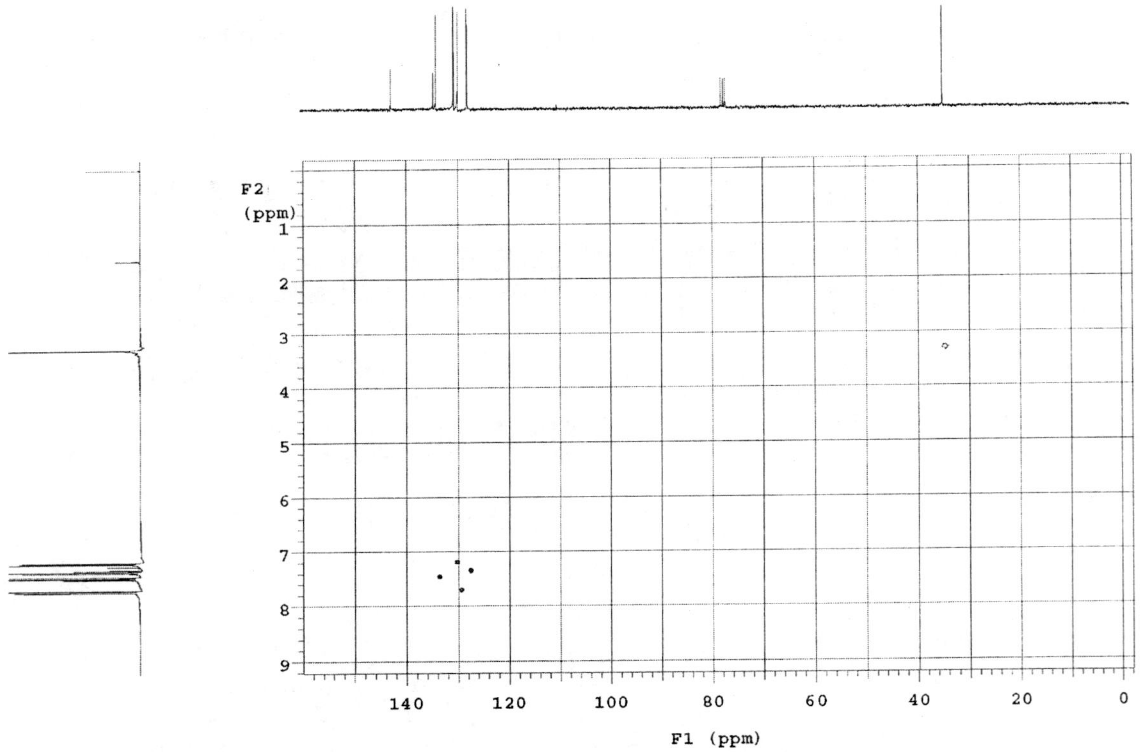

gHSQC Spectrum in $CDCl_3$

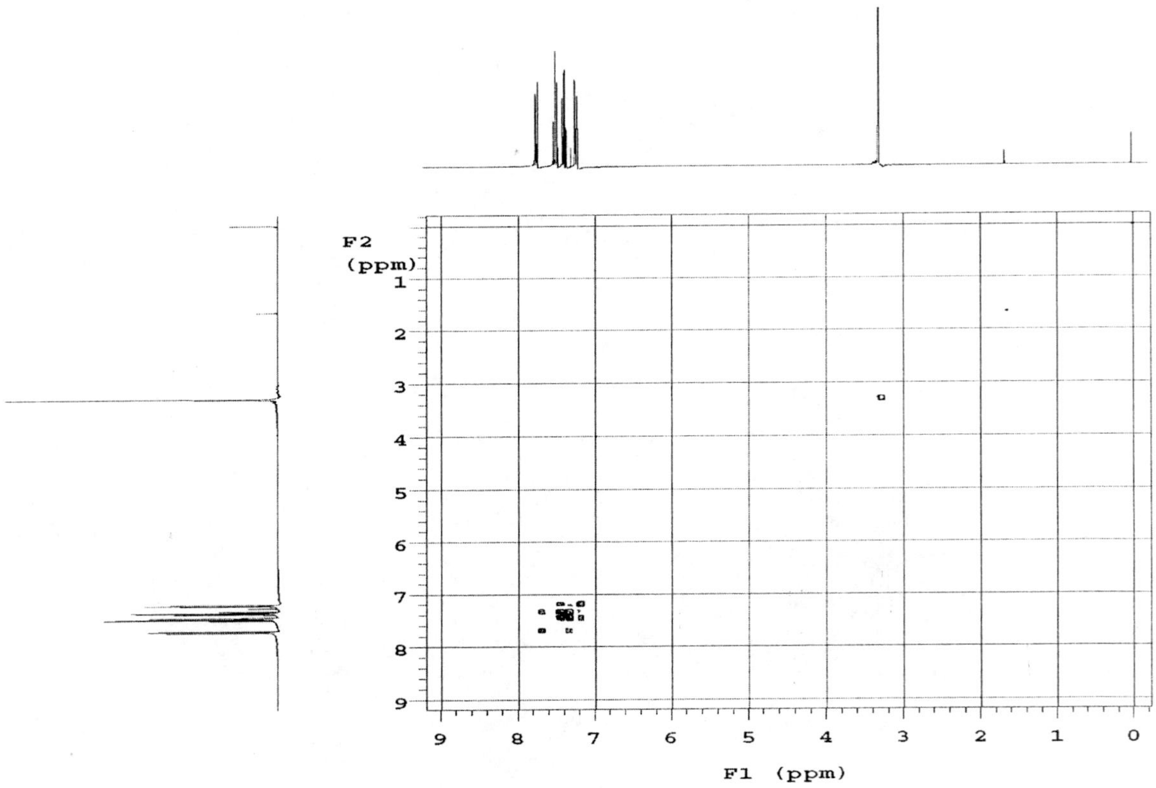

gCOSY Spectrum in $CDCl_3$

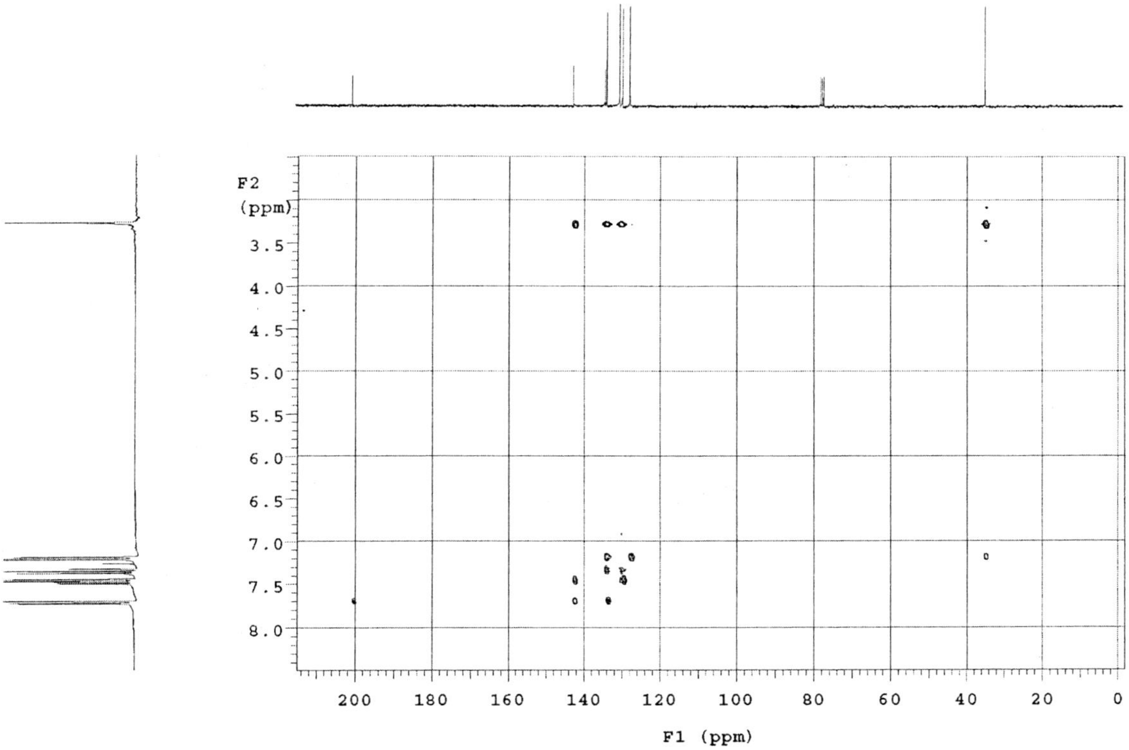

gHMBC Spectrum in $CDCl_3$

Problem 3.06 Given the spectroscopic data below (Proton, Carbon, gHSQC, gCOSY, and gHMBC), provide the structure of the molecule.

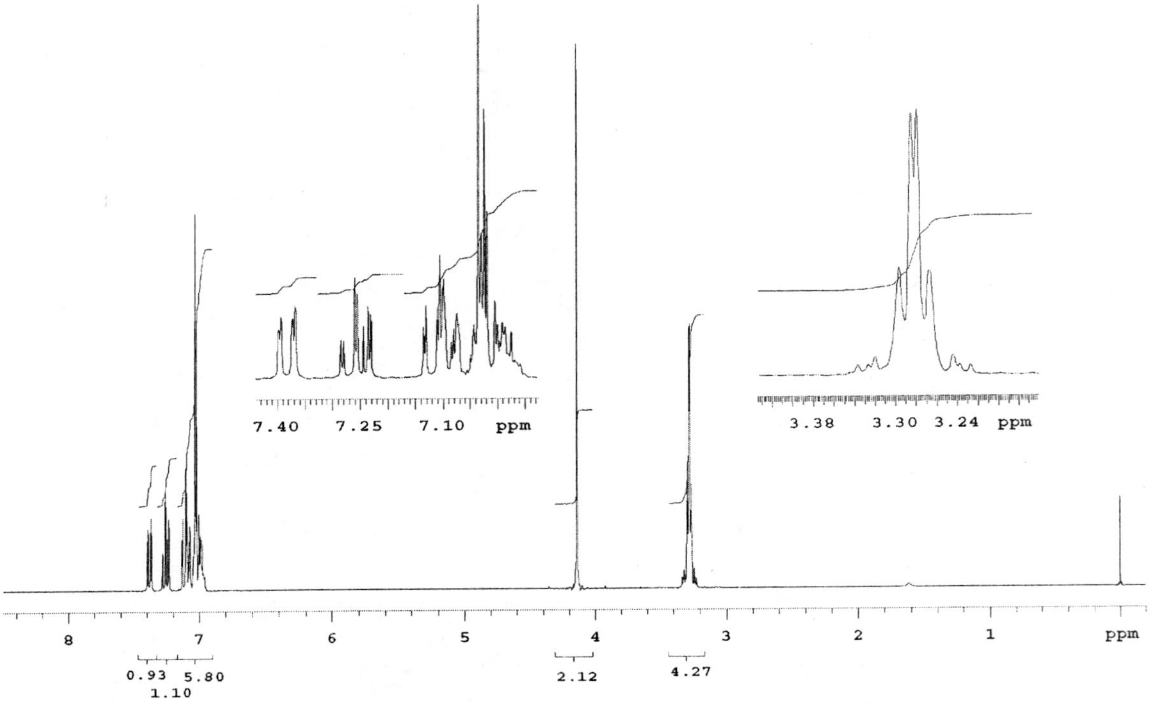

^{1}H-NMR Spectrum in $CDCl_3$

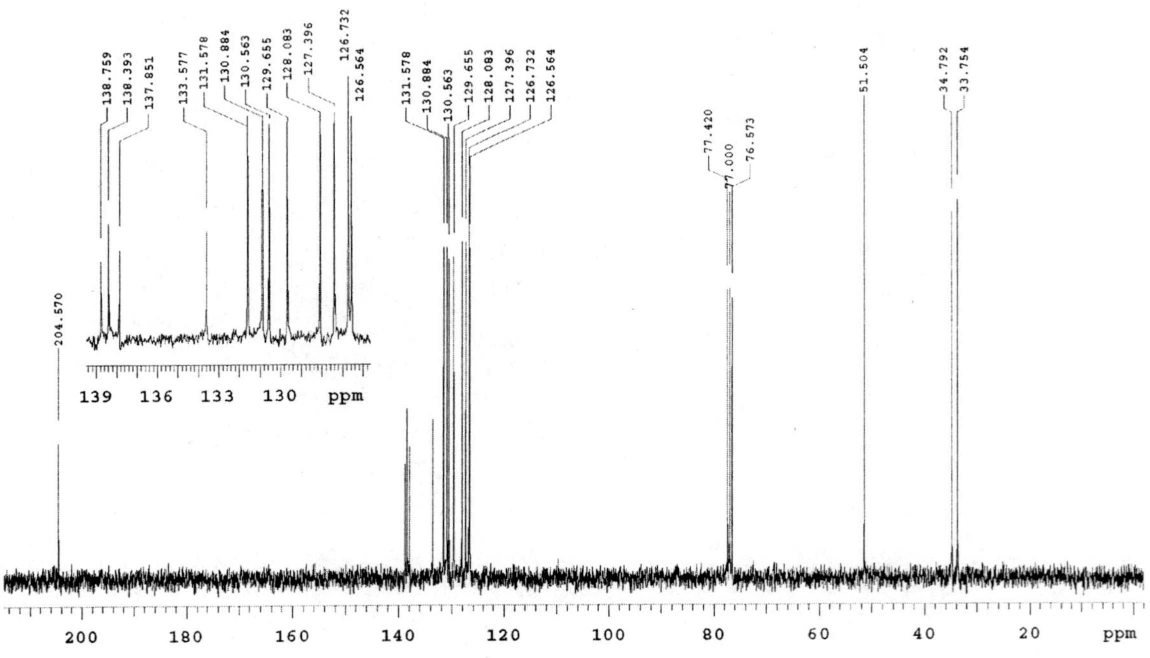

^{13}C-NMR Spectrum in $CDCl_3$

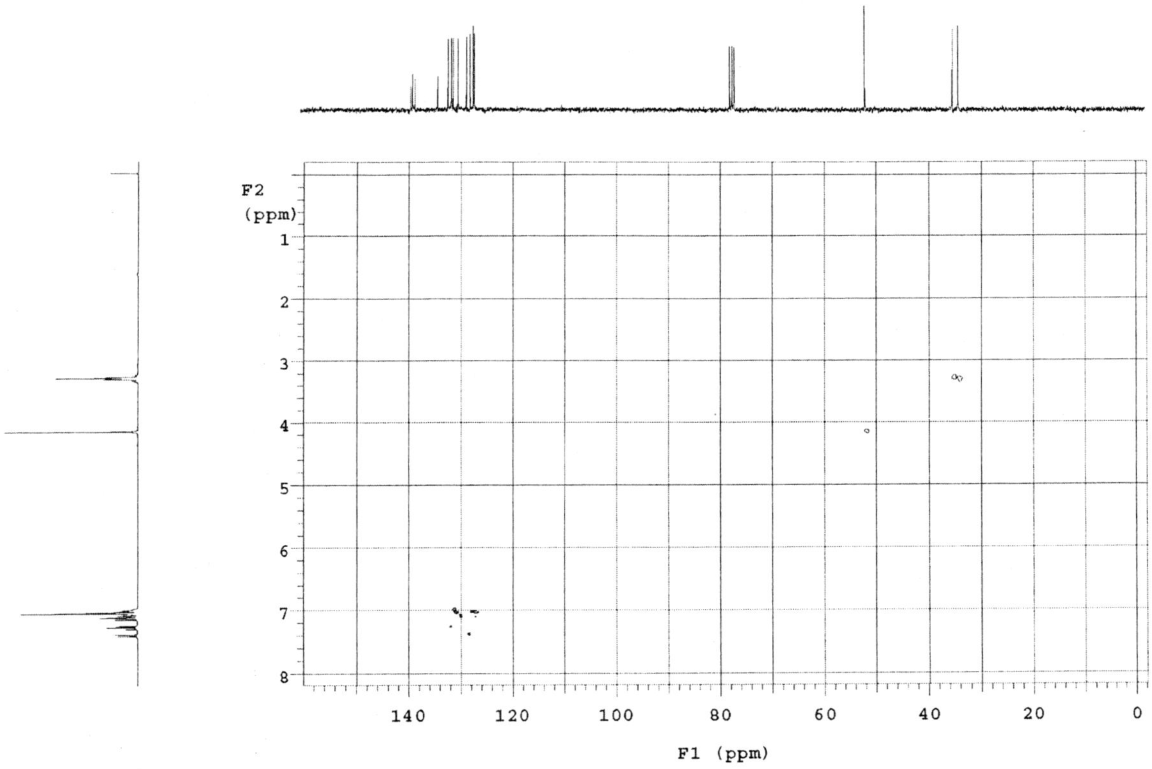

gHSQC Spectrum in $CDCl_3$

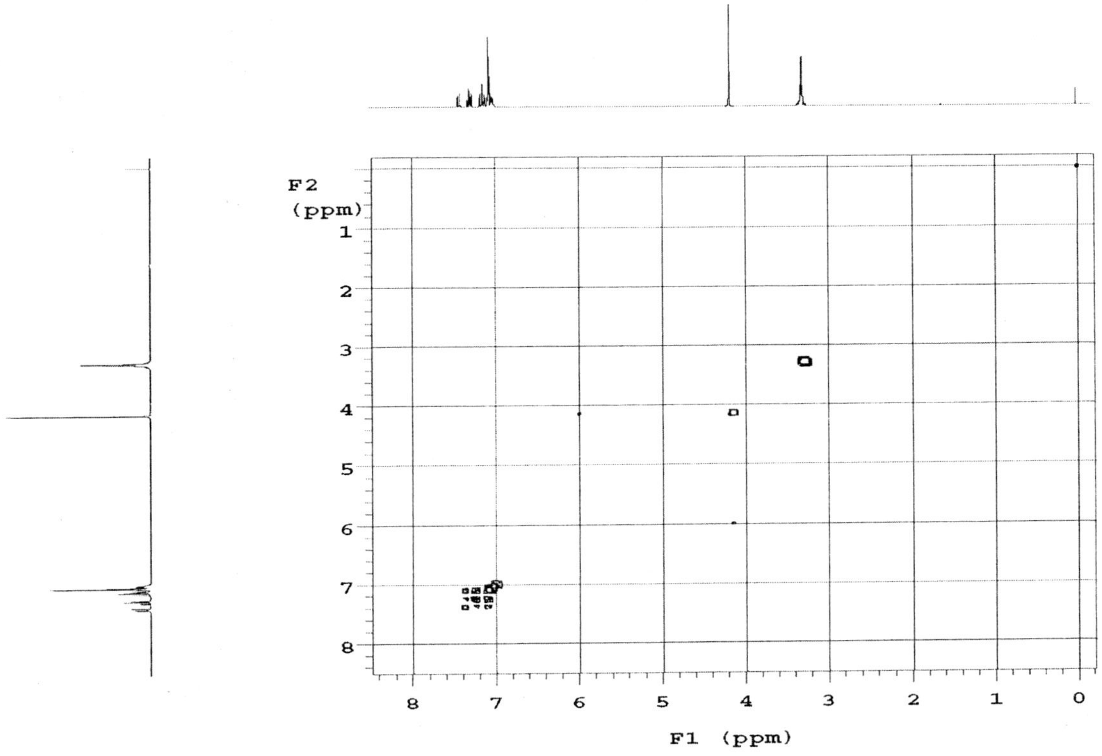

gCOSY Spectrum in $CDCl_3$

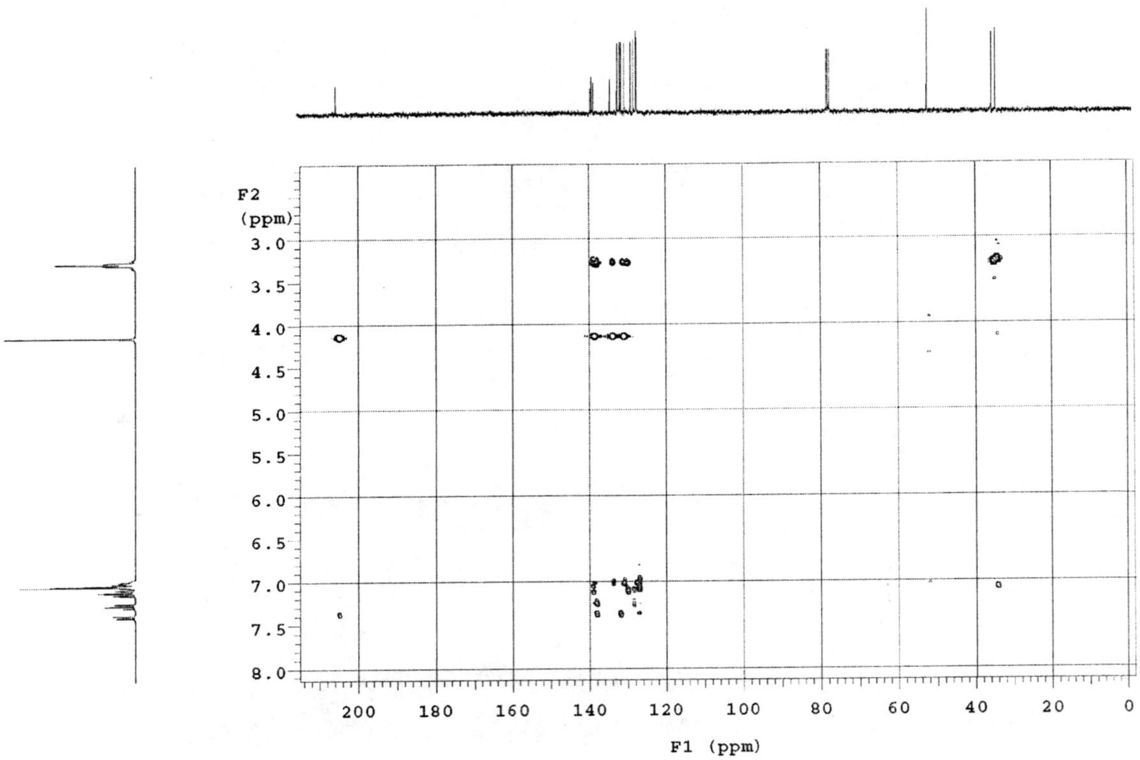

gHMBC Spectrum in $CDCl_3$

Problem 3.07 Given the spectroscopic data below (Proton, Carbon, gHSQC, gCOSY, and gHMBC), provide the structure of the molecule.

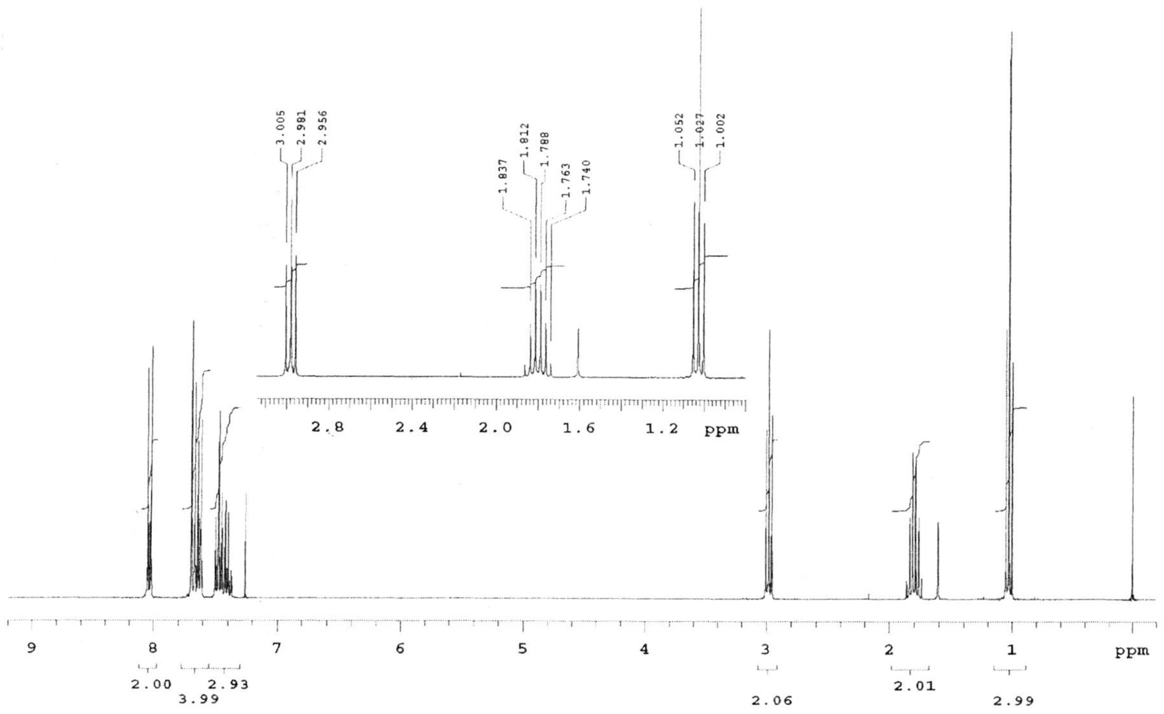

^{1}H-NMR Spectrum in $CDCl_3$

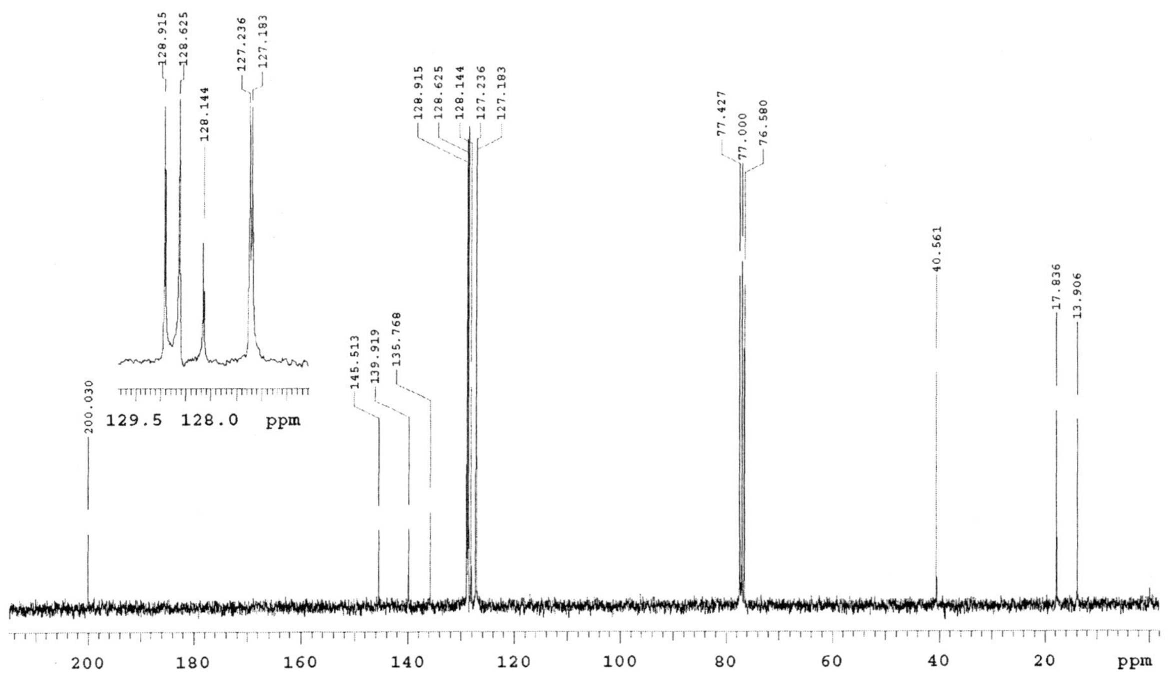

^{13}C-NMR Spectrum in $CDCl_3$

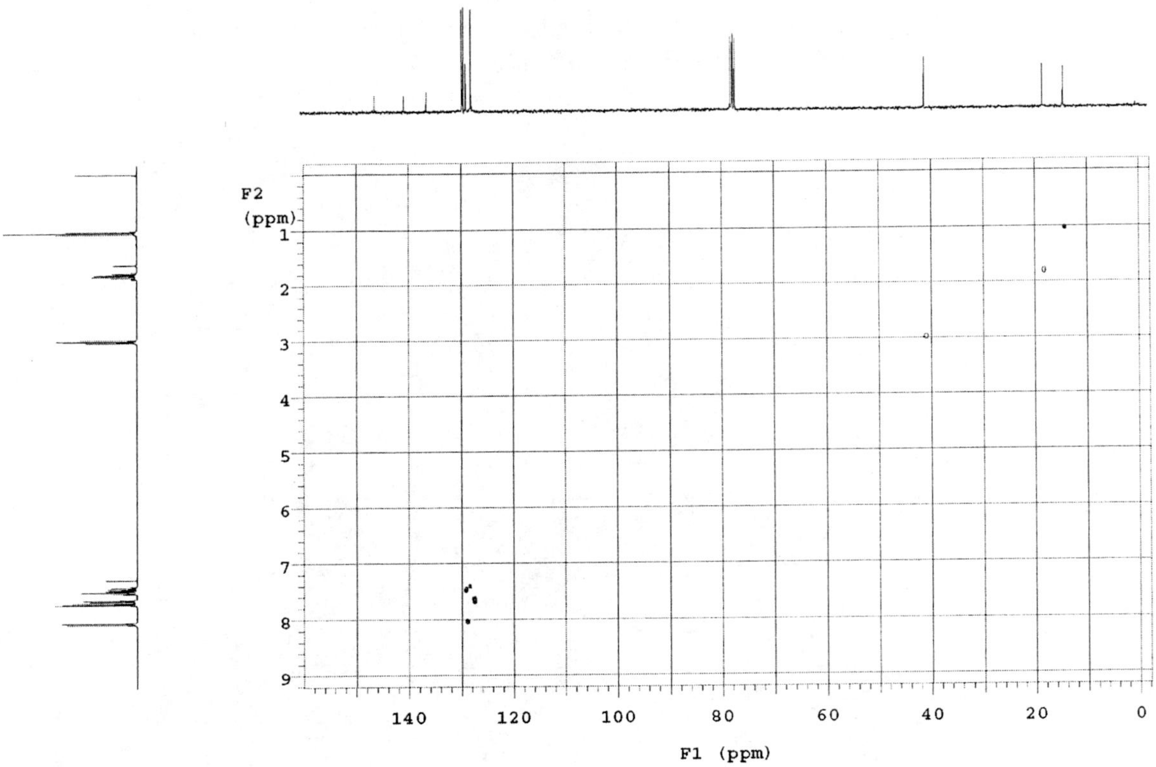

gHSQC Spectrum in $CDCl_3$

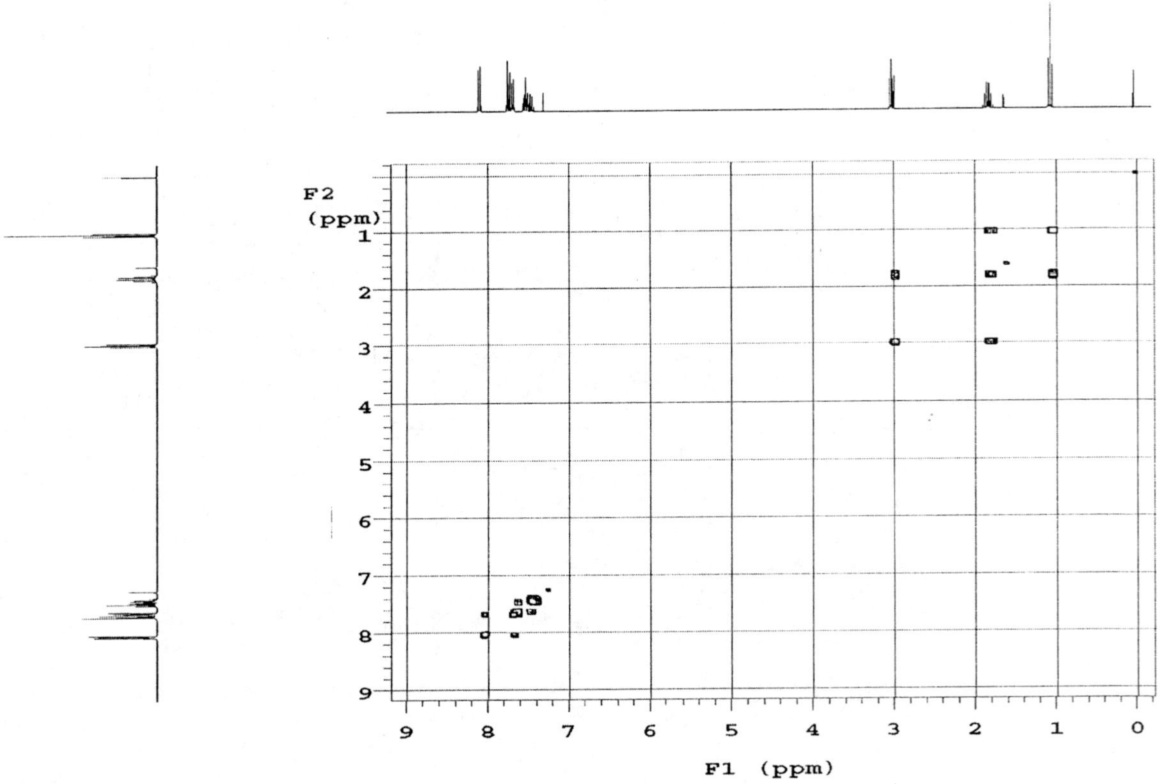

gCOSY Spectrum in $CDCl_3$

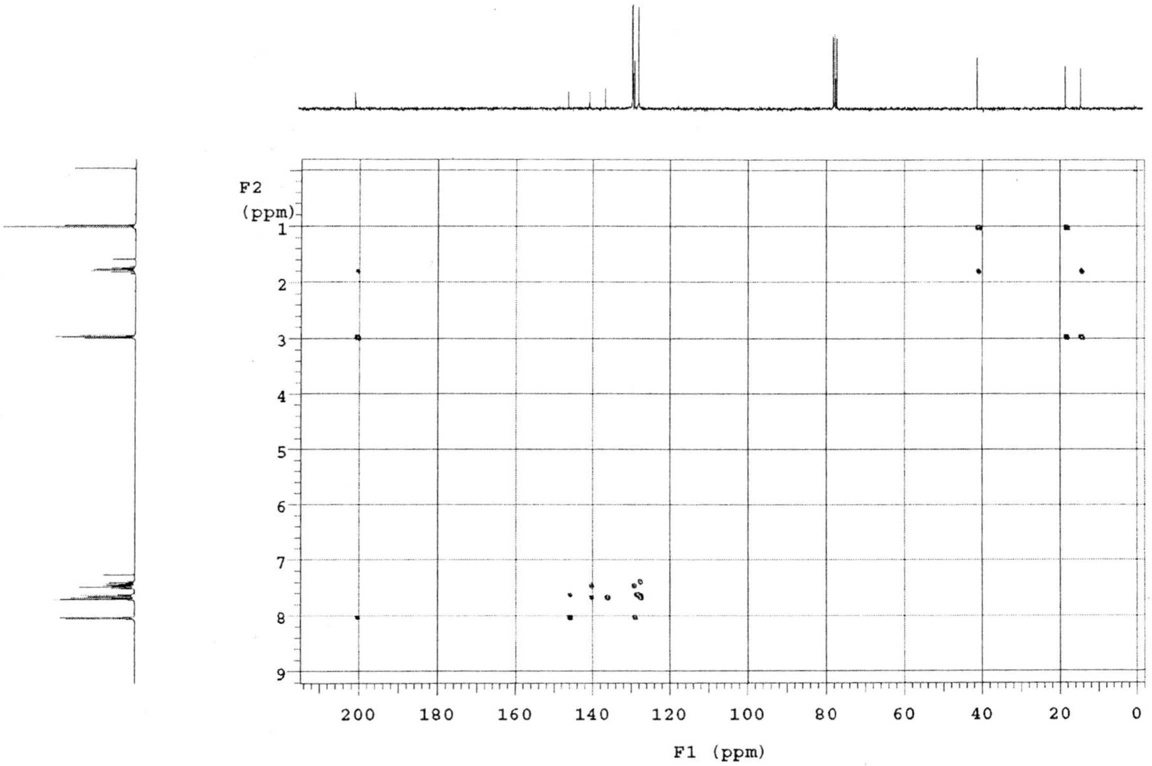

gHMBC Spectrum in $CDCl_3$

Problem 3.08 Given the spectroscopic data below (Proton, Carbon, gHSQC, gCOSY, and gHMBC), provide the structure of the molecule.

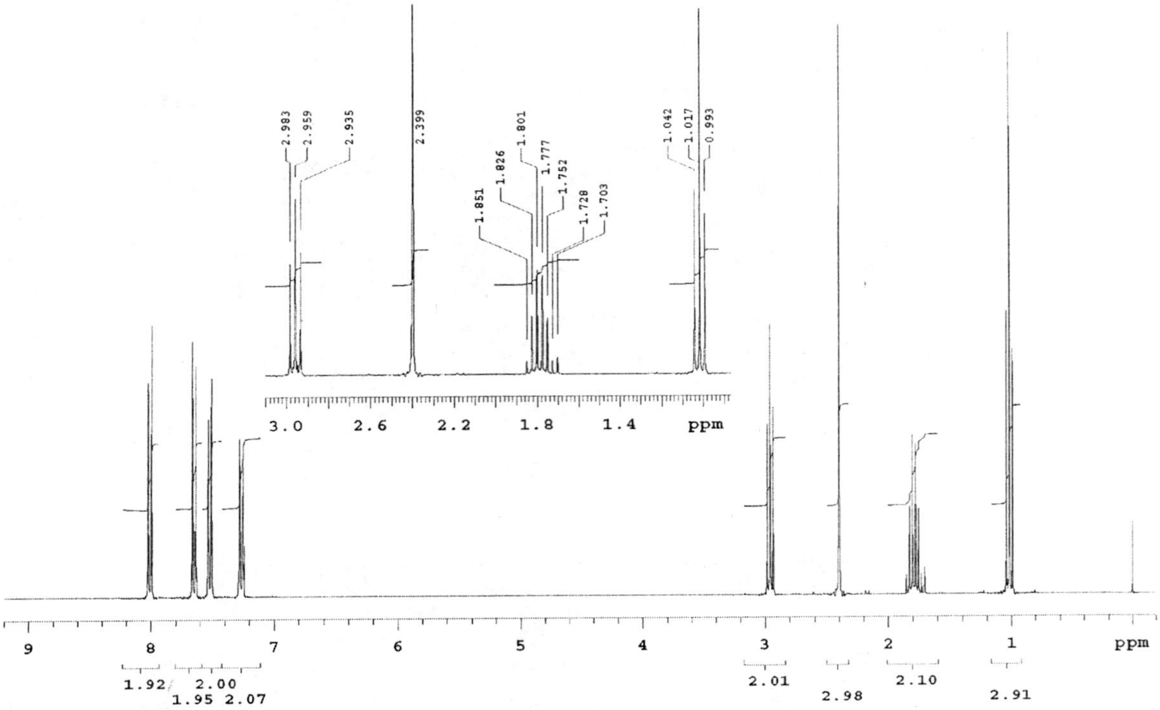

^{1}H-NMR Spectrum in $CDCl_3$

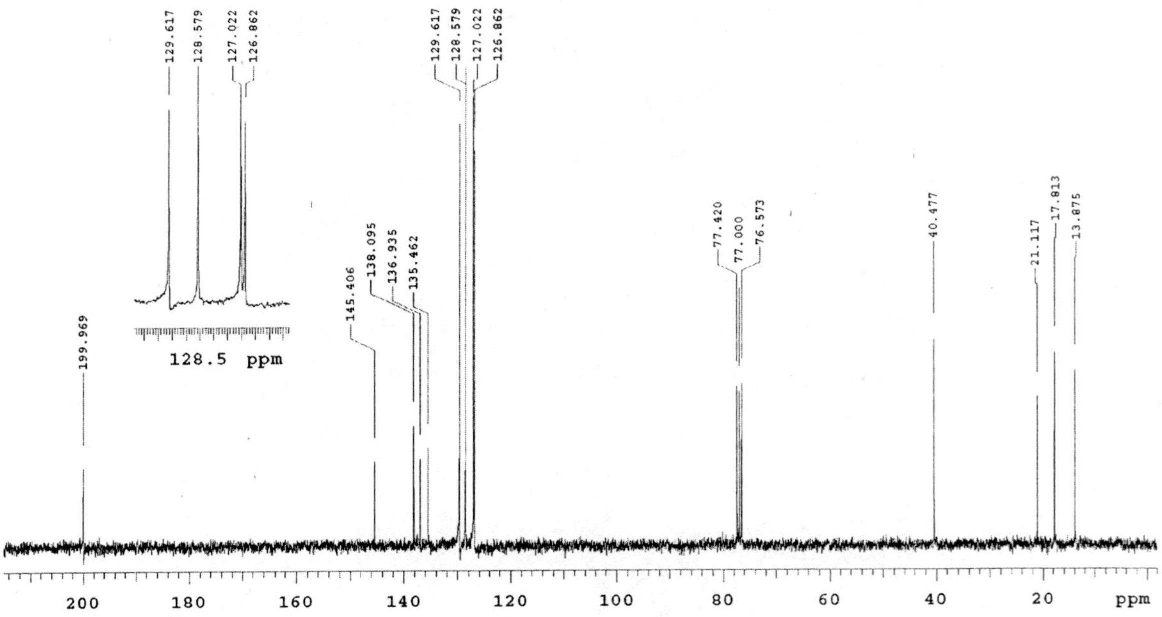

^{13}C-NMR Spectrum in $CDCl_3$

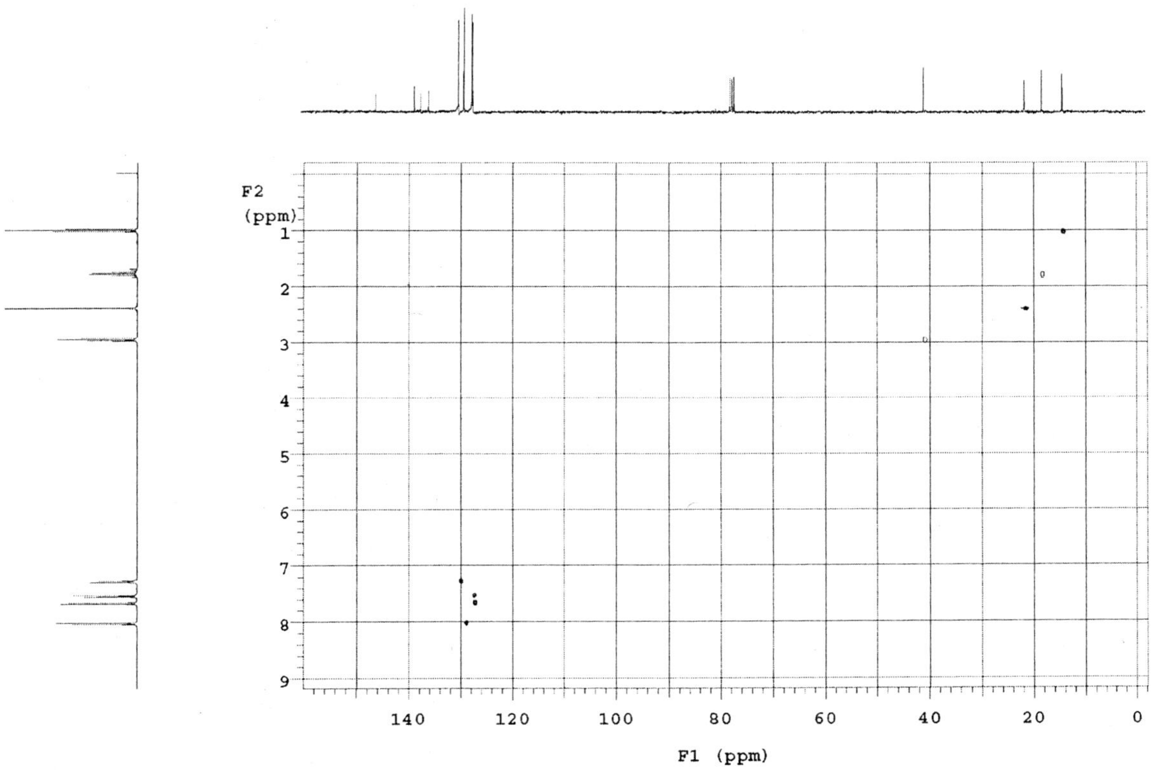

gHSQC Spectrum in $CDCl_3$

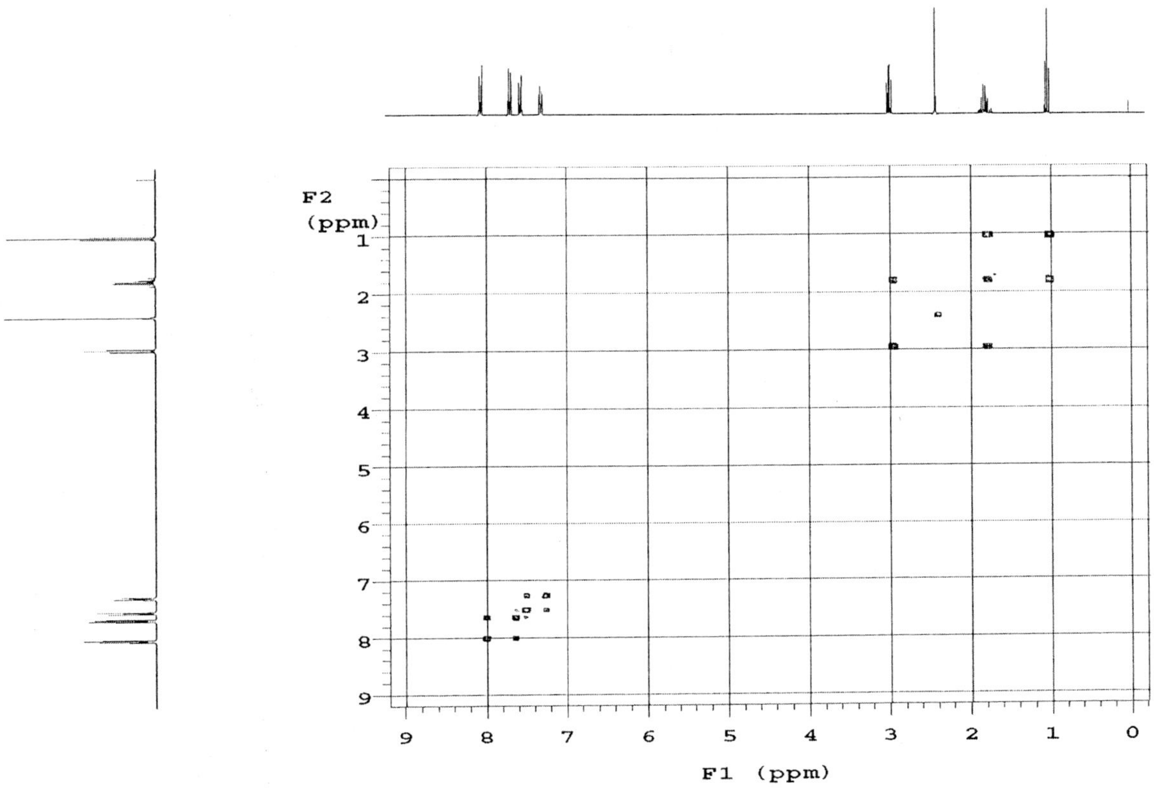

gCOSY Spectrum in $CDCl_3$

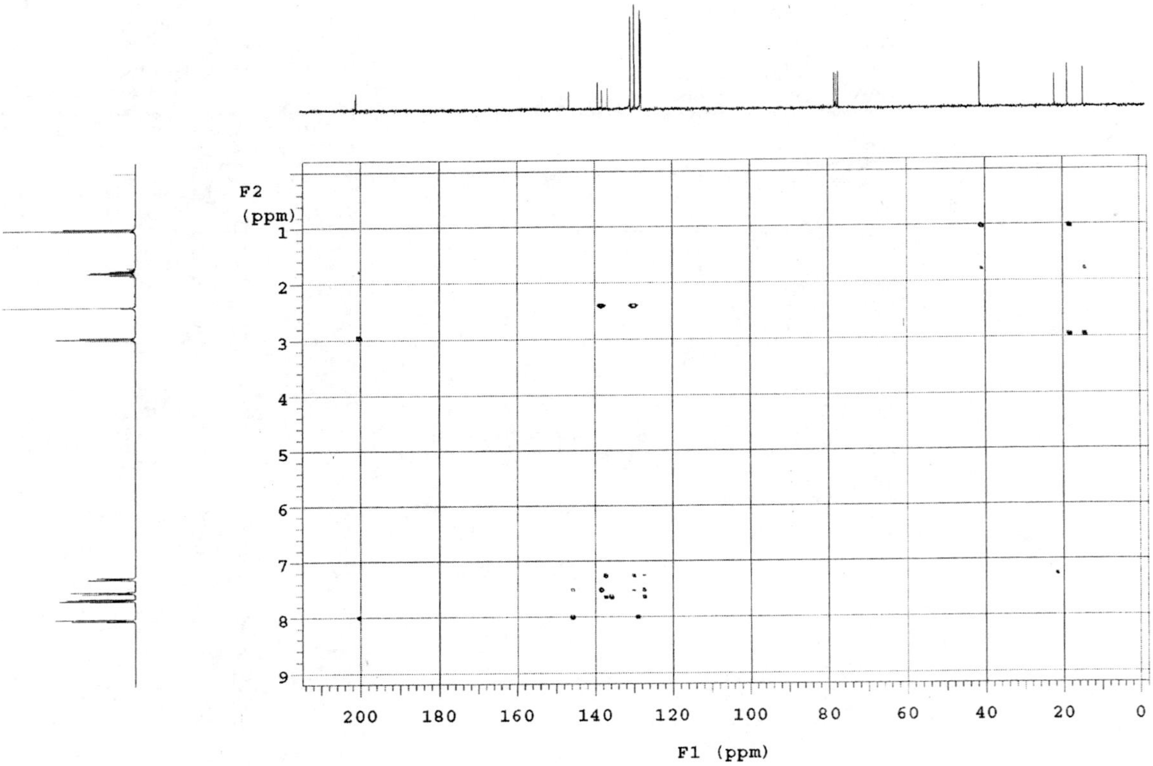

gHMBC Spectrum in $CDCl_3$

Problem 3.09 Given the spectroscopic data below (Proton, Carbon, gHSQC, gCOSY, and gHMBC), provide the structure of the molecule.

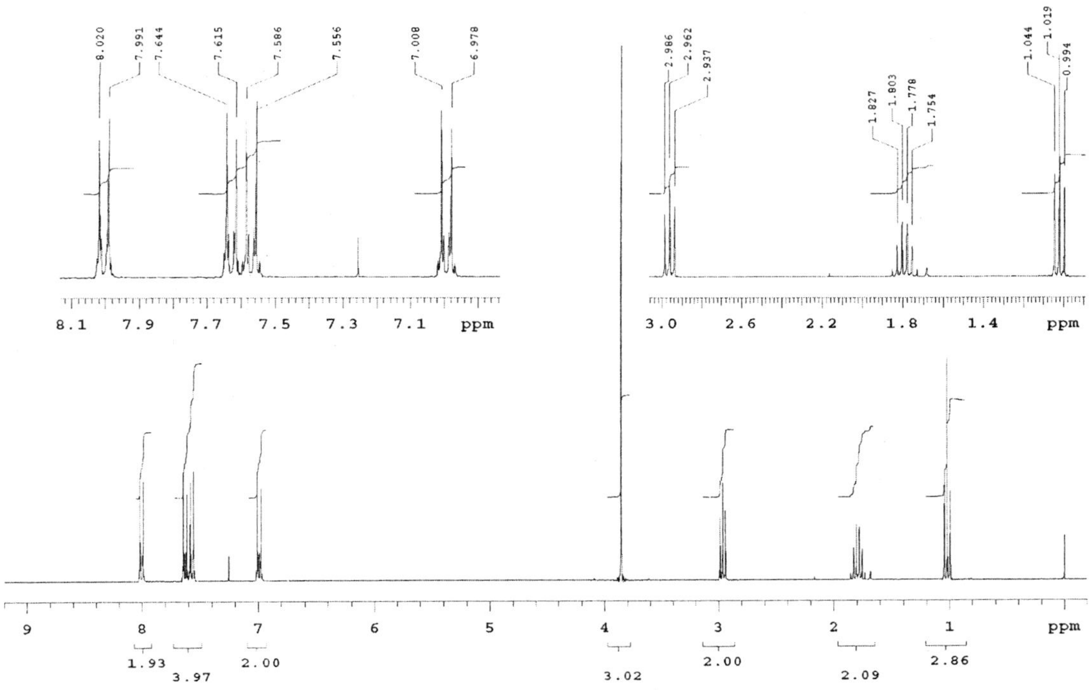

^{1}H-NMR Spectrum in $CDCl_3$

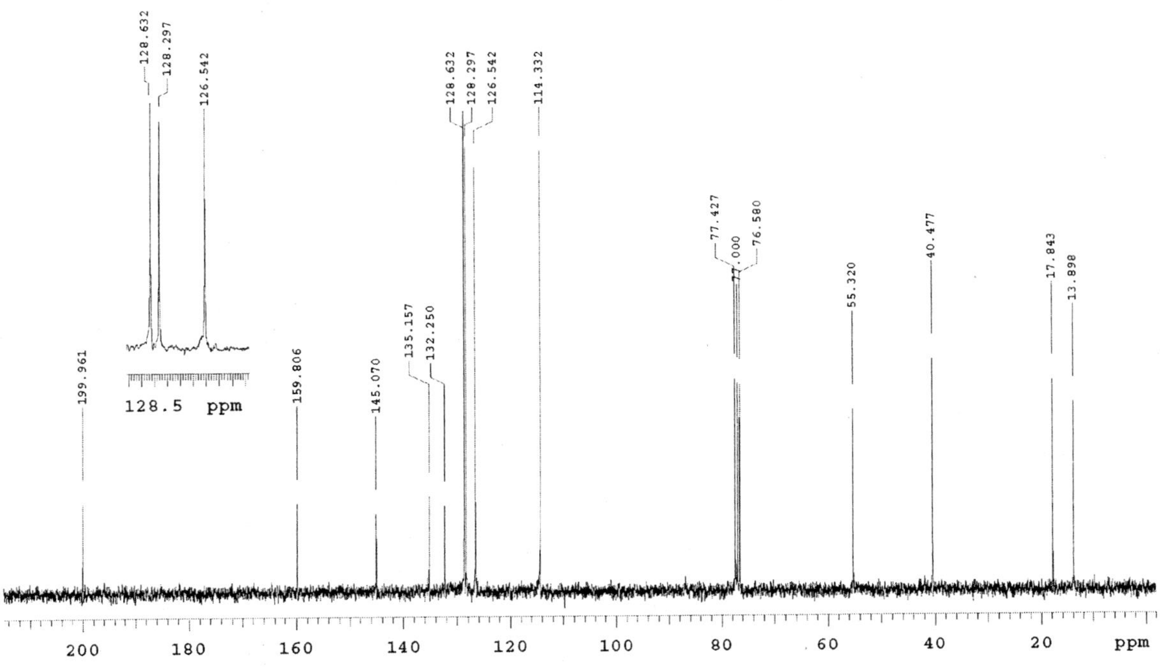

^{13}C-NMR Spectrum in $CDCl_3$

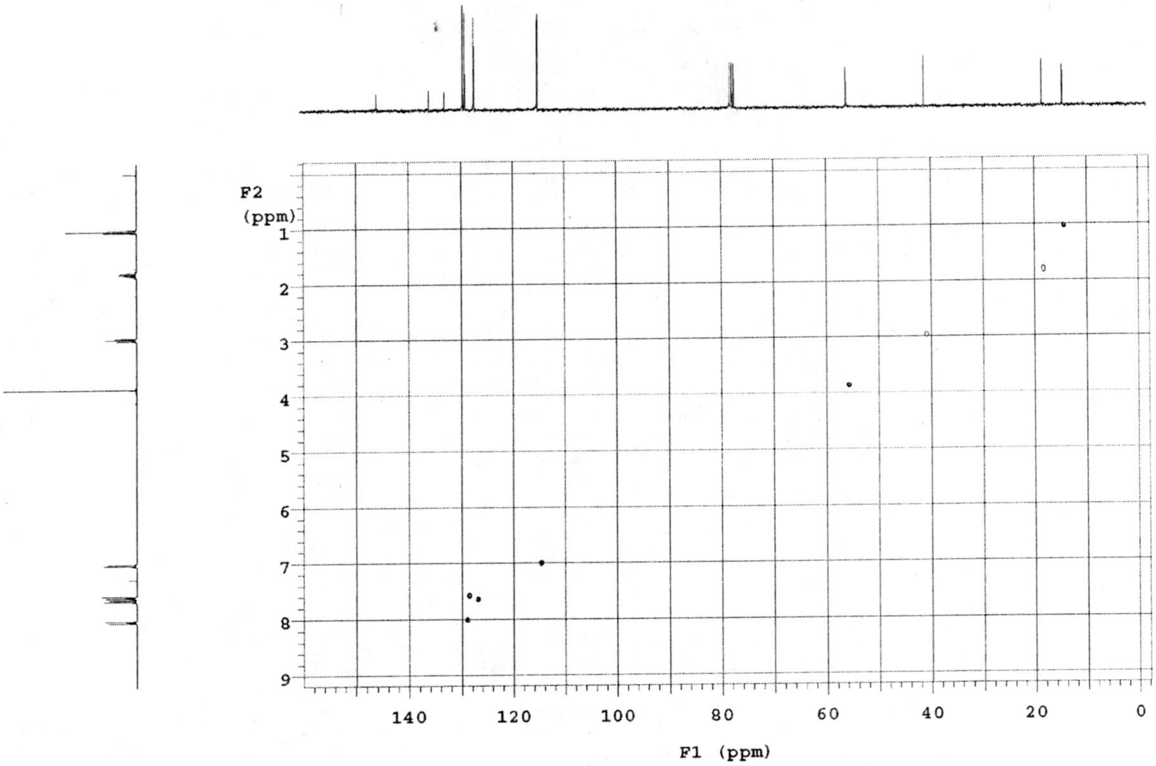

gHSQC Spectrum in $CDCl_3$

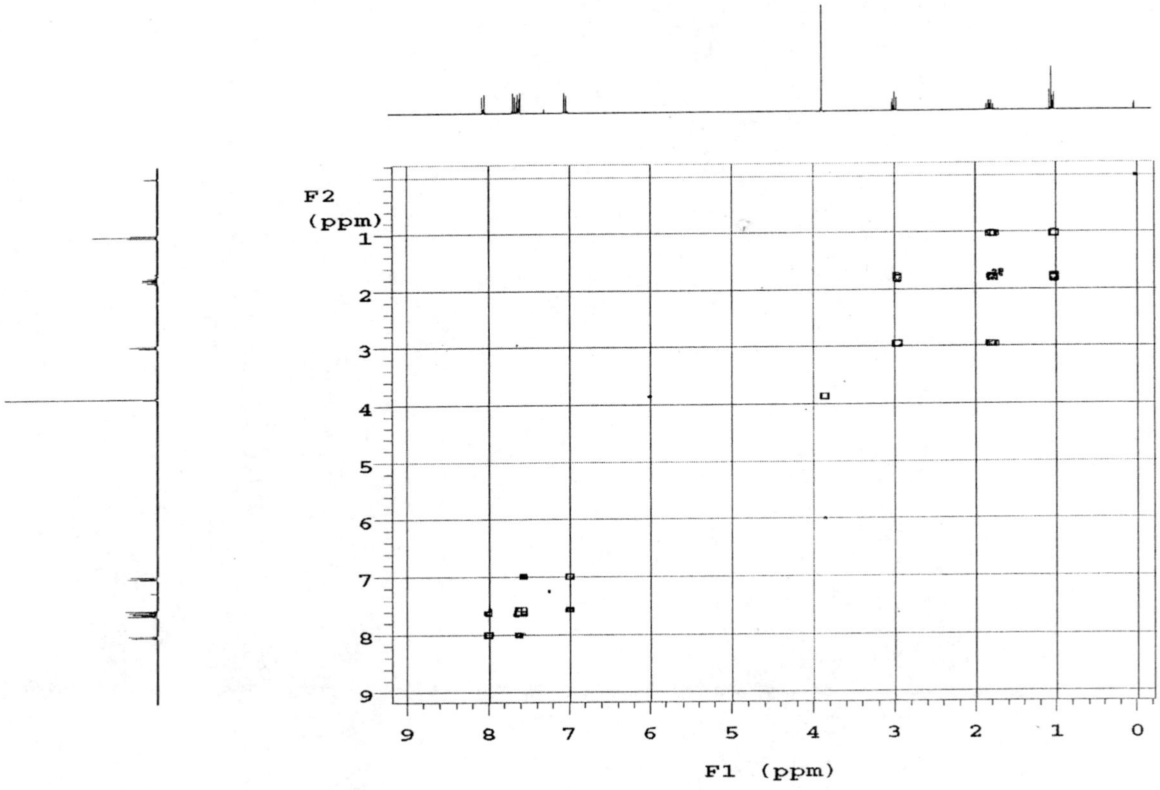

gCOSY Spectrum in $CDCl_3$

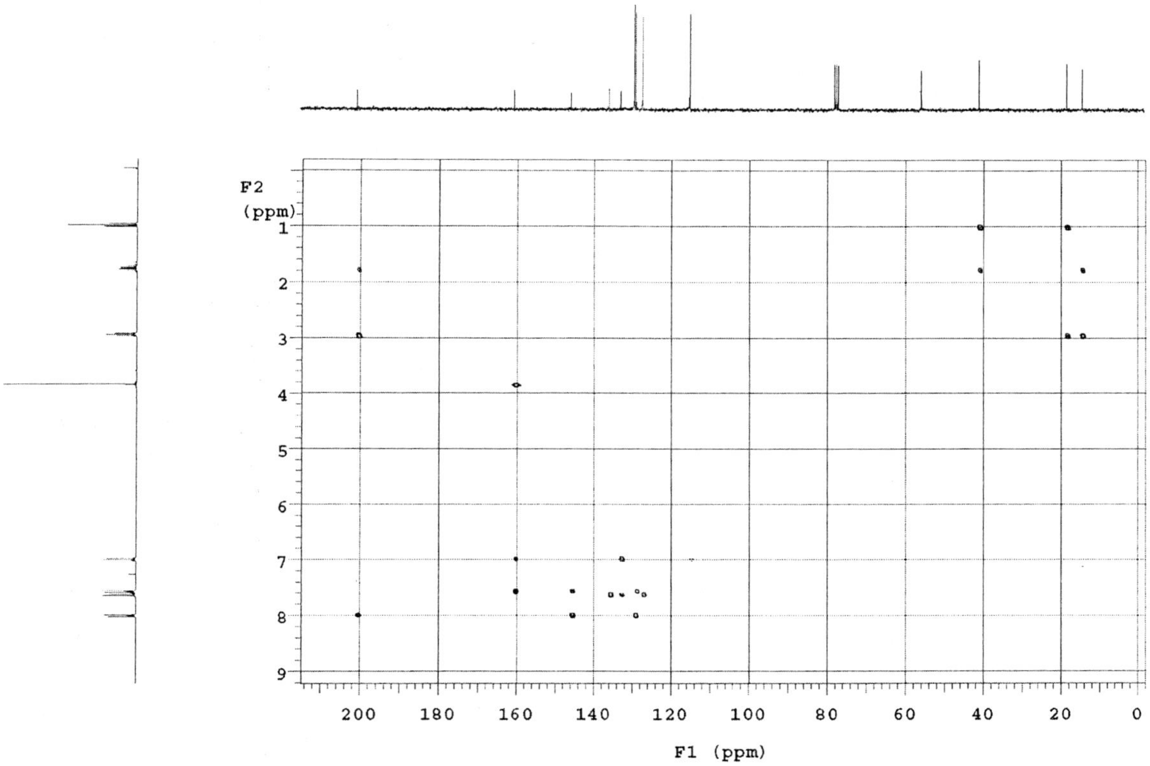

gHMBC Spectrum in $CDCl_3$

Problem 3.10 Given the spectroscopic data below (Proton, Carbon, gHSQC, gCOSY, and gHMBC), provide the structure of the molecule. (MW: 239 g/mol)

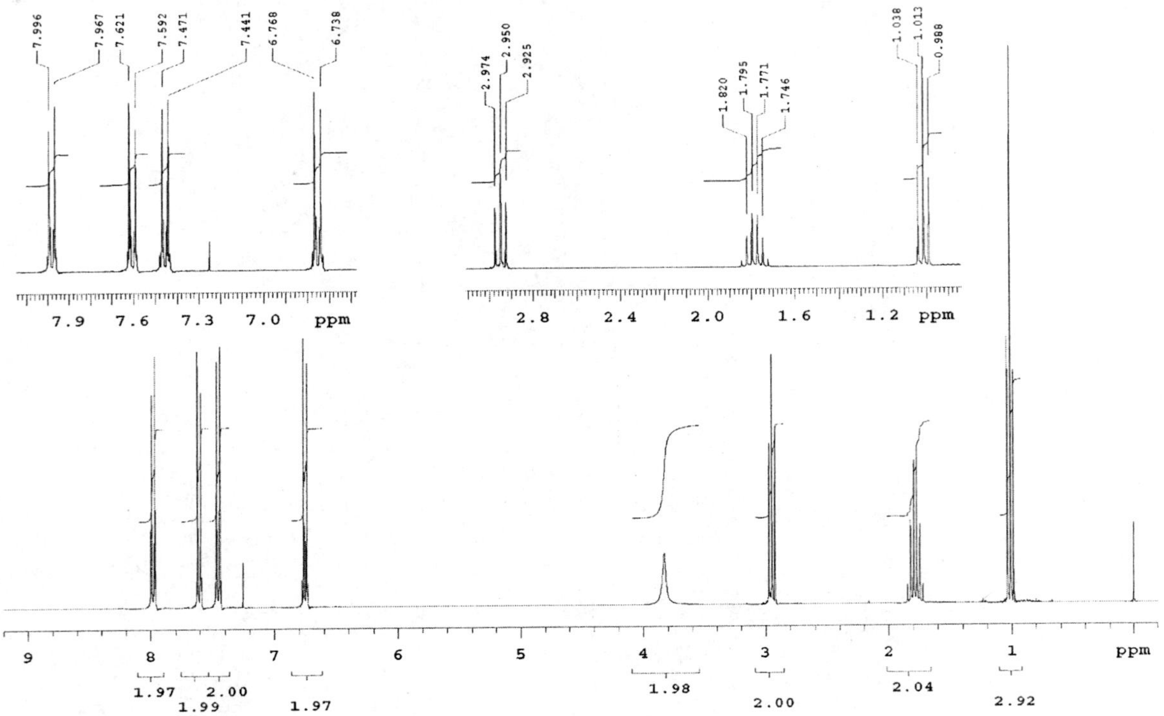

^{1}H-NMR Spectrum in $CDCl_3$

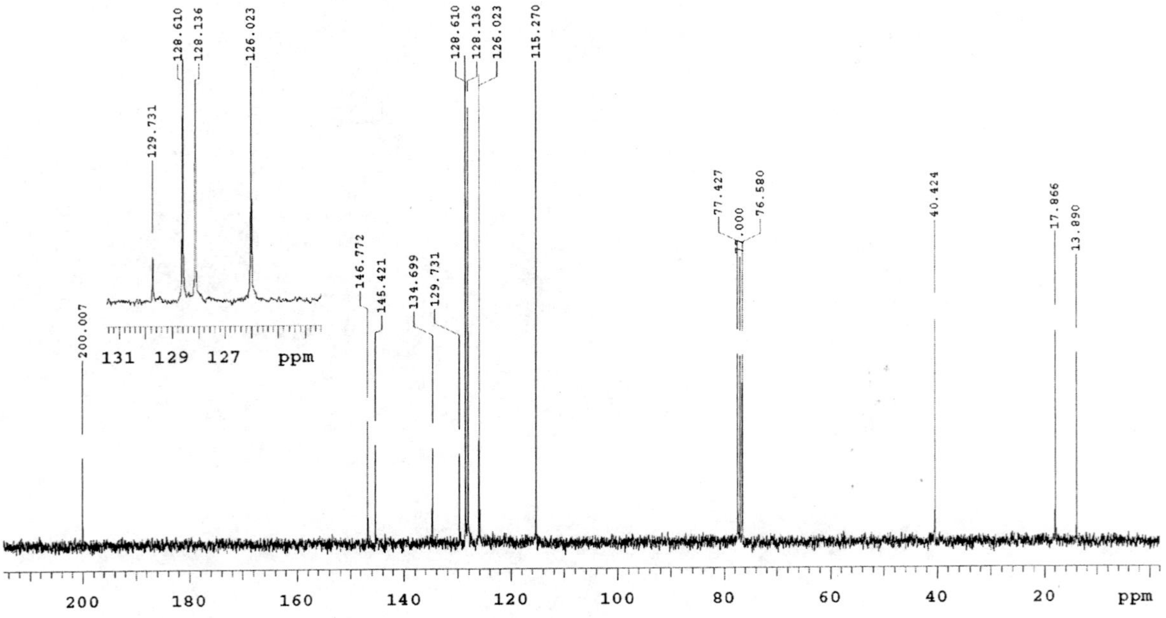

^{13}C-NMR Spectrum in $CDCl_3$

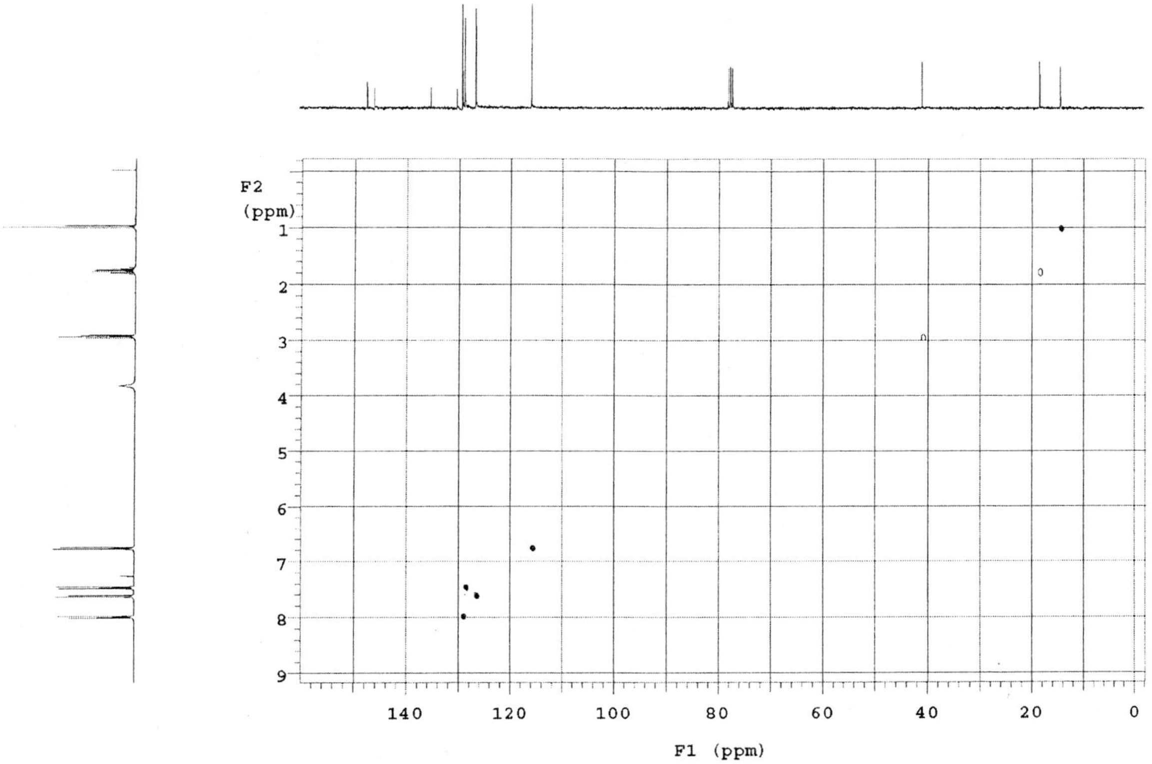

gHSQC Spectrum in $CDCl_3$

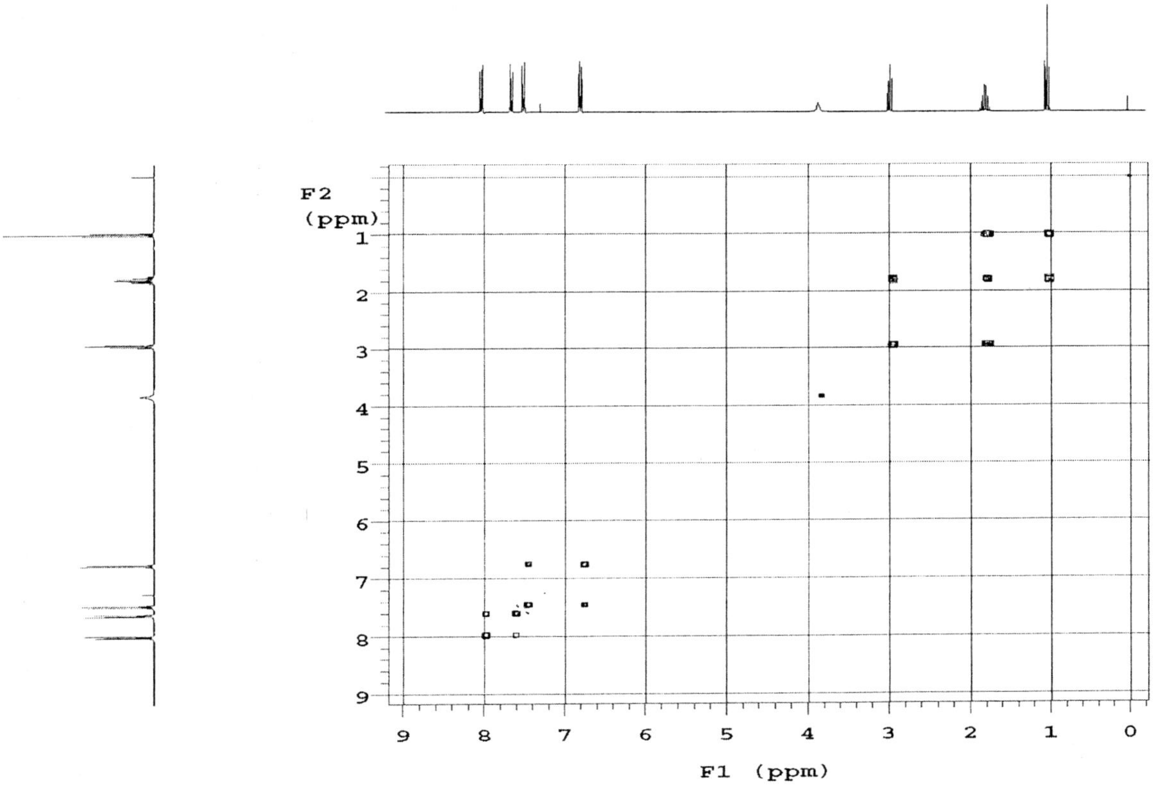

gCOSY Spectrum in $CDCl_3$

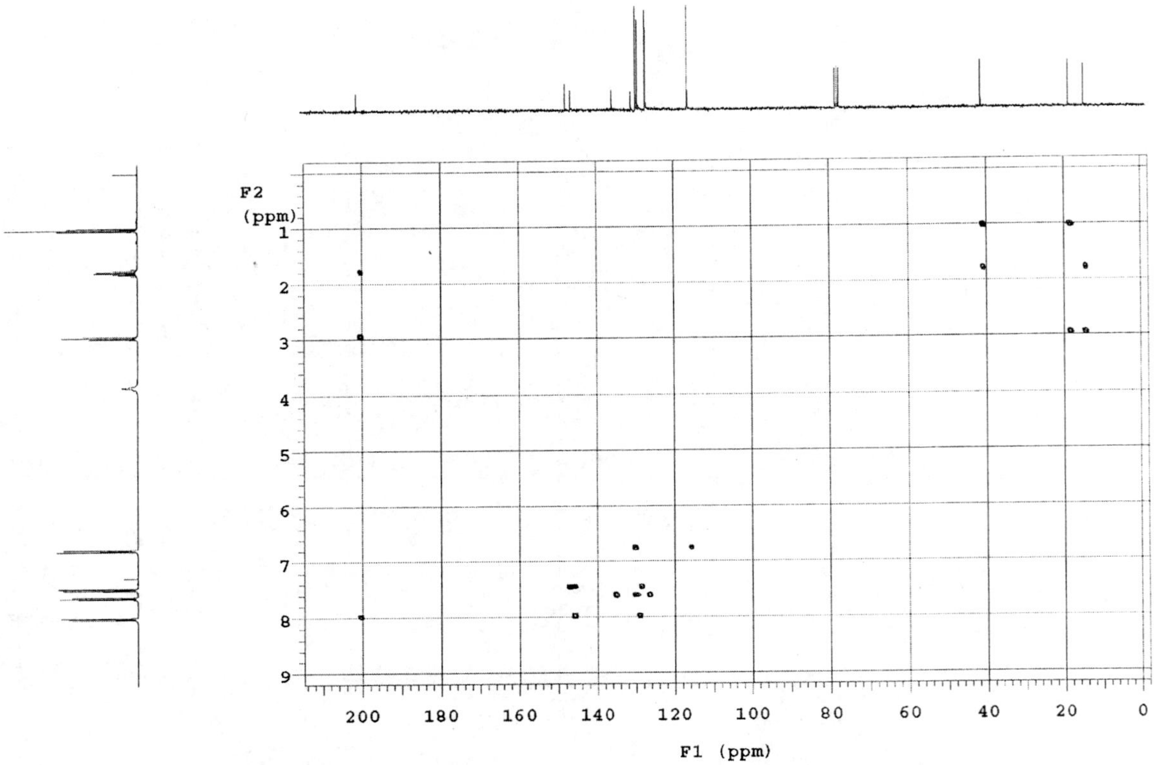

gHMBC Spectrum in $CDCl_3$

Problem 3.11 Given the spectroscopic data below (Proton, Carbon, gHMQC, gCOSY, and gHMBC), provide the structure of the molecule. (MW = 281 g/mol)

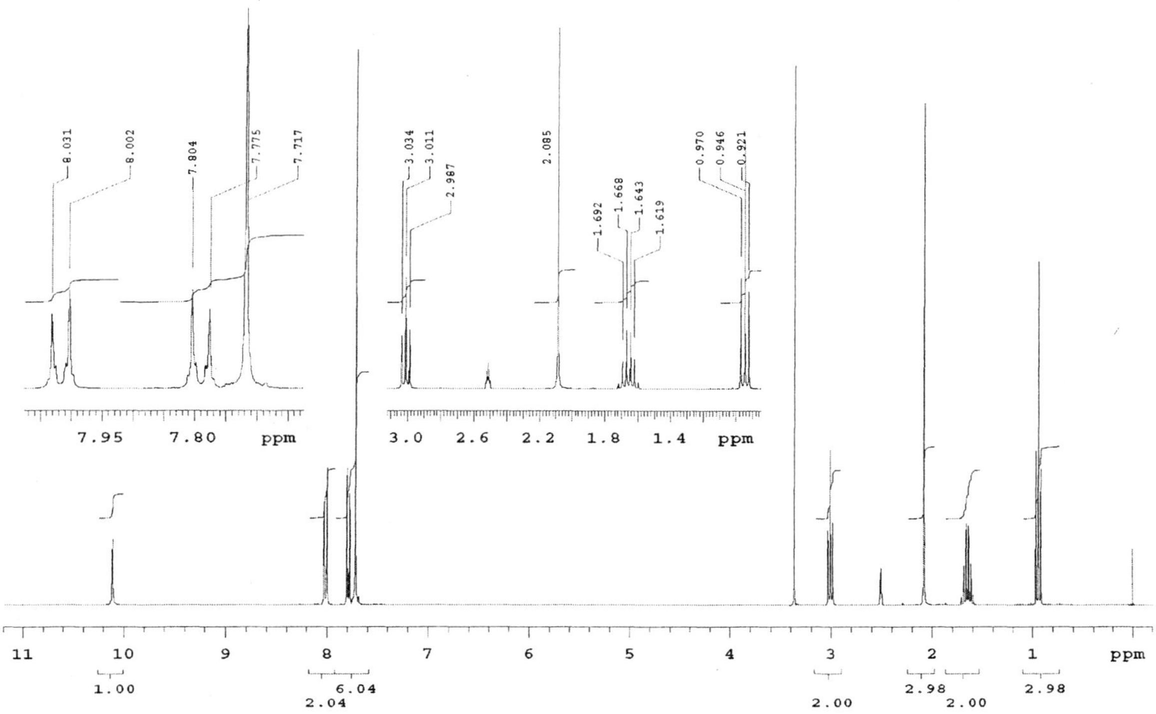

^{1}H-NMR Spectrum in DMSO-d_6

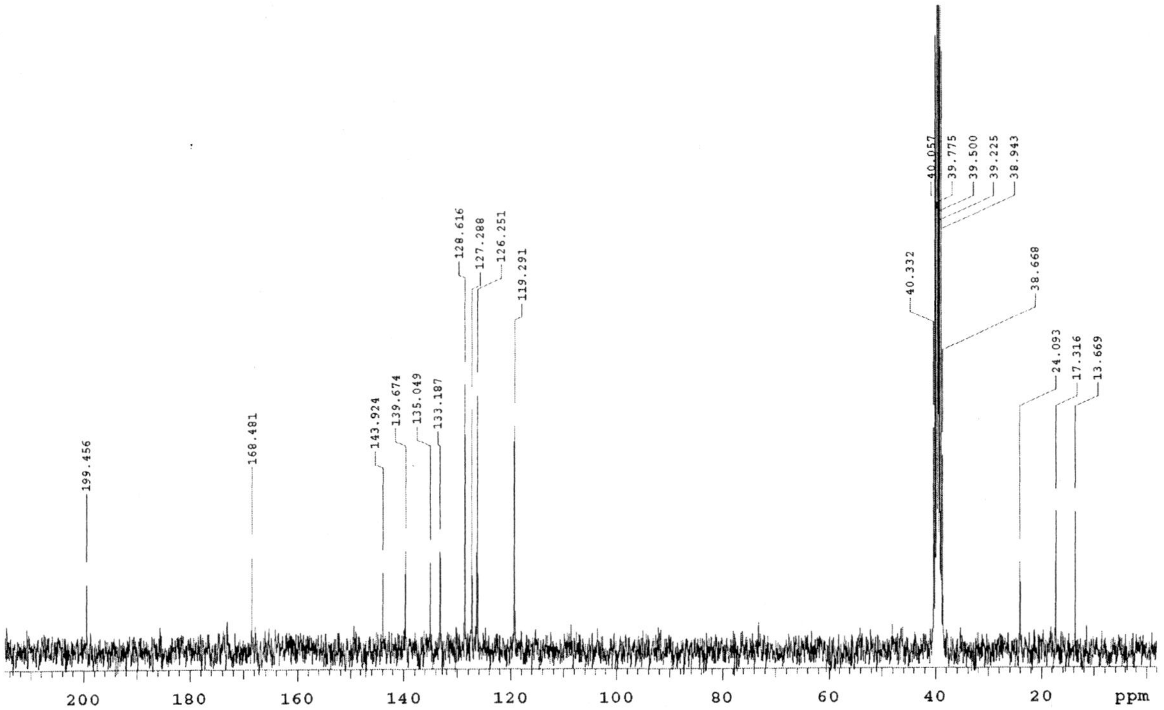

^{13}C-NMR Spectrum in DMSO-d_6

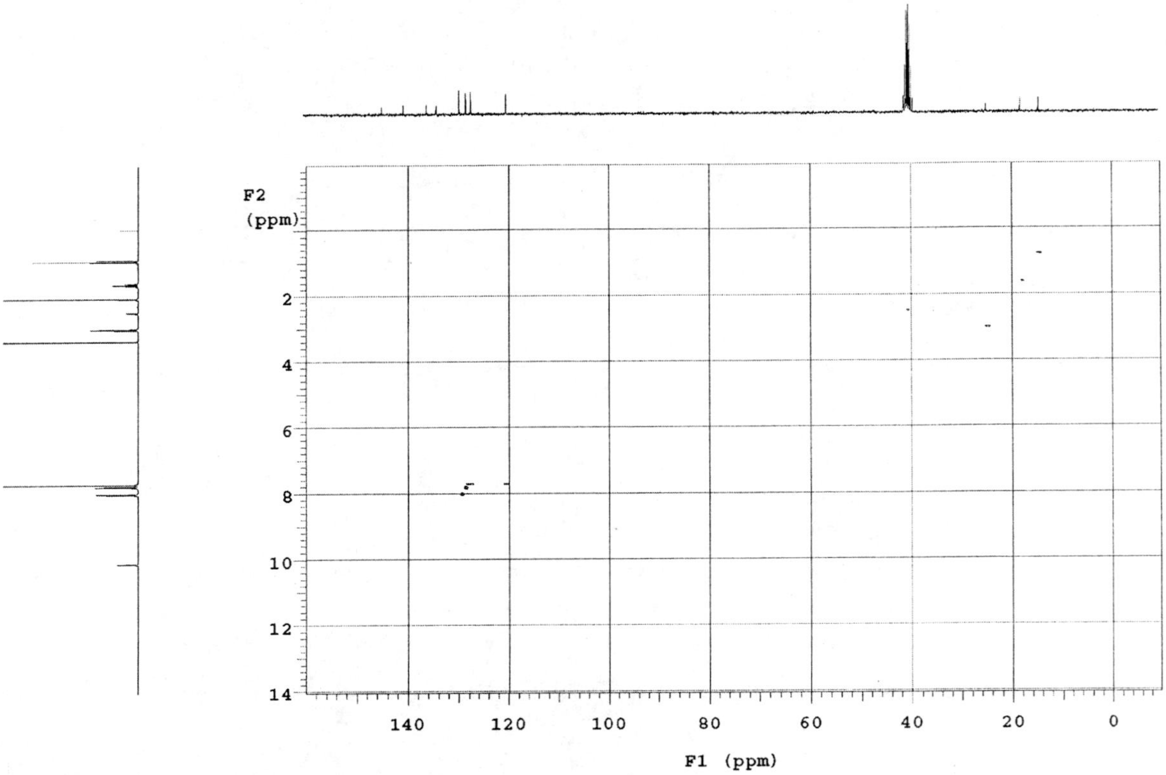

gHMQC Spectrum in DMSO-d_6

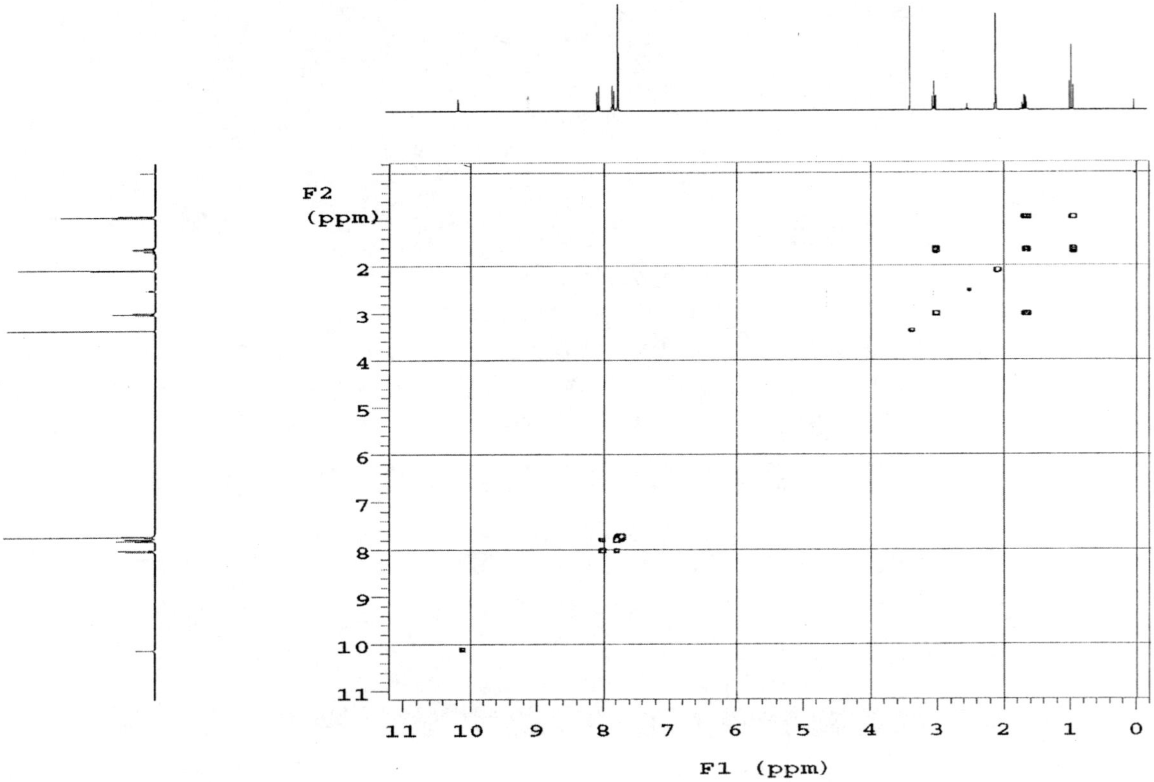

gCOSY Spectrum in DMSO-d_6

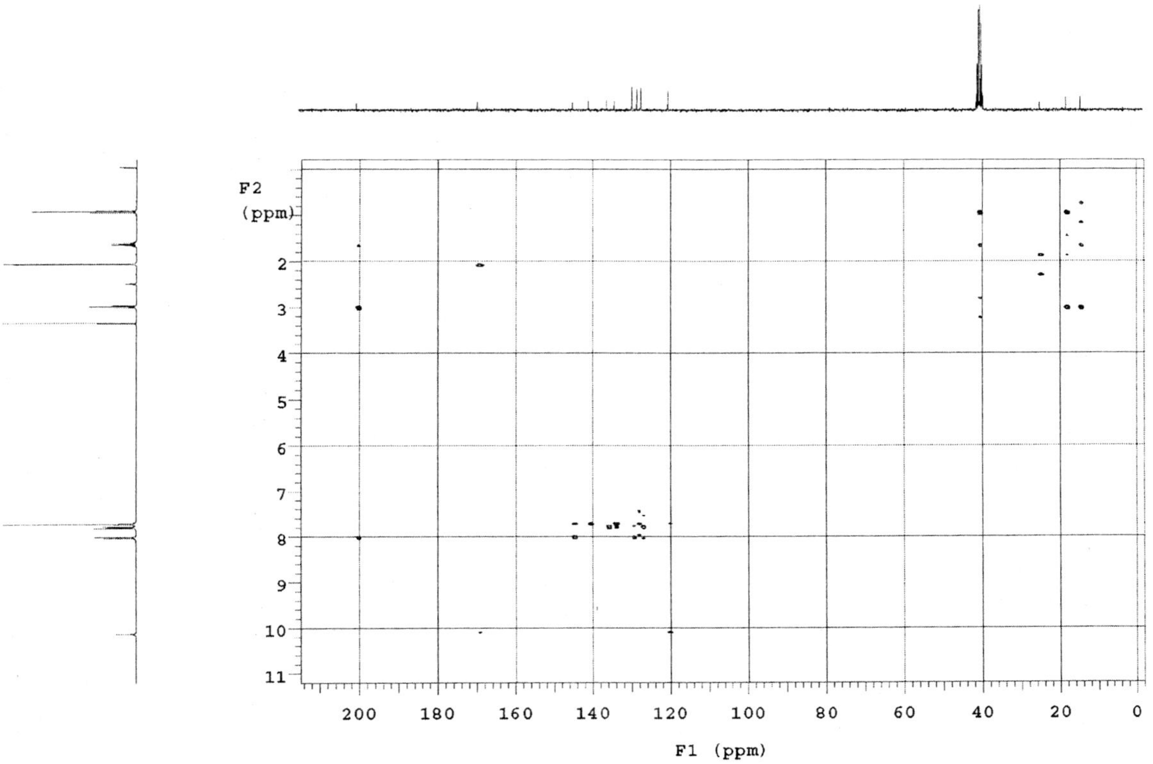

gHMBC Spectrum in DMSO-d_6

Problem 3.12 Given the spectroscopic data below (Proton, Carbon, gHSQC, gCOSY, and gHMBC), provide the structure of the molecule. (C: 81.84%; H: 6.14%; N: 5.65%; IR: 2240-2220 cm^{-1})

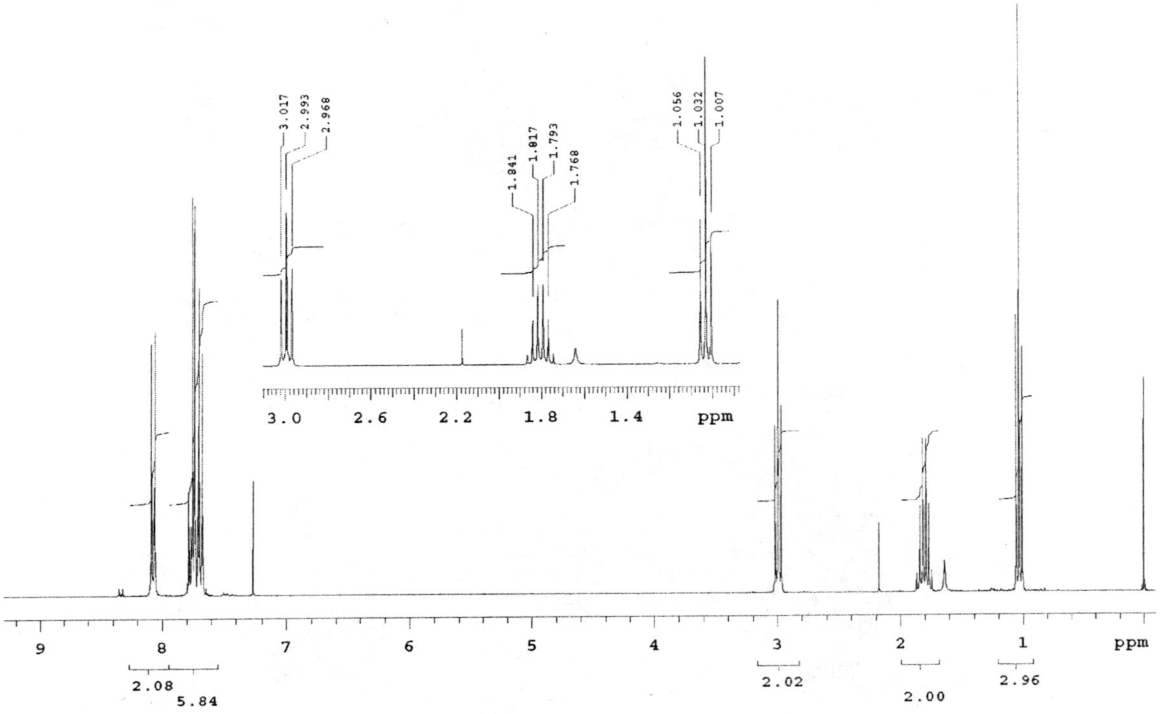

^{1}H-NMR Spectrum in $CDCl_3$

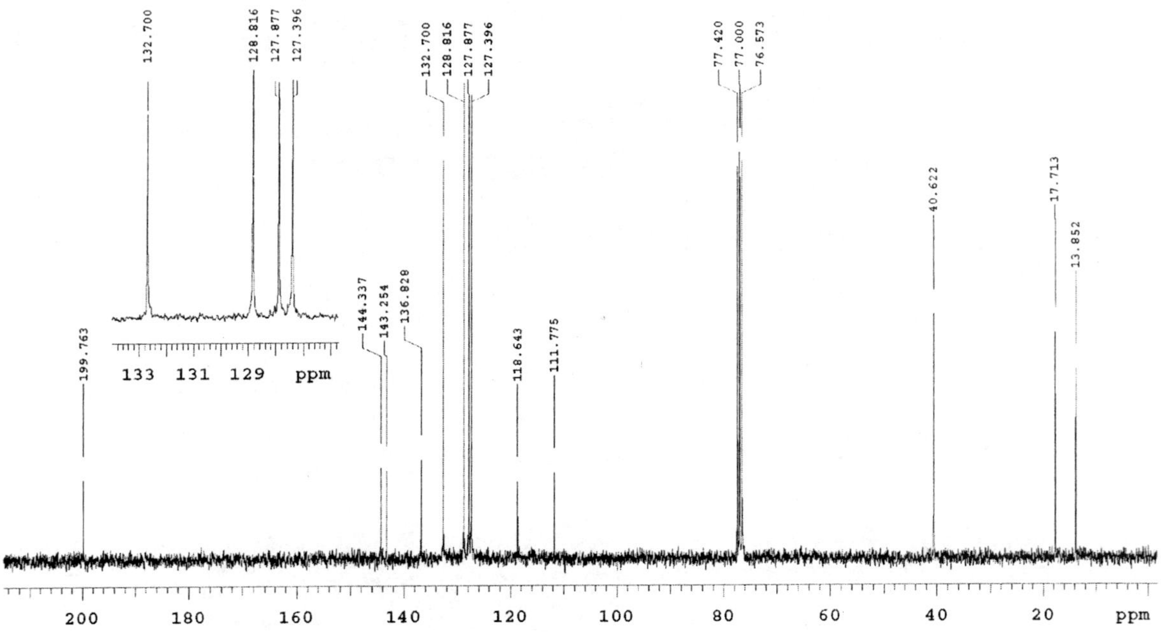

^{13}C-NMR Spectrum in $CDCl_3$

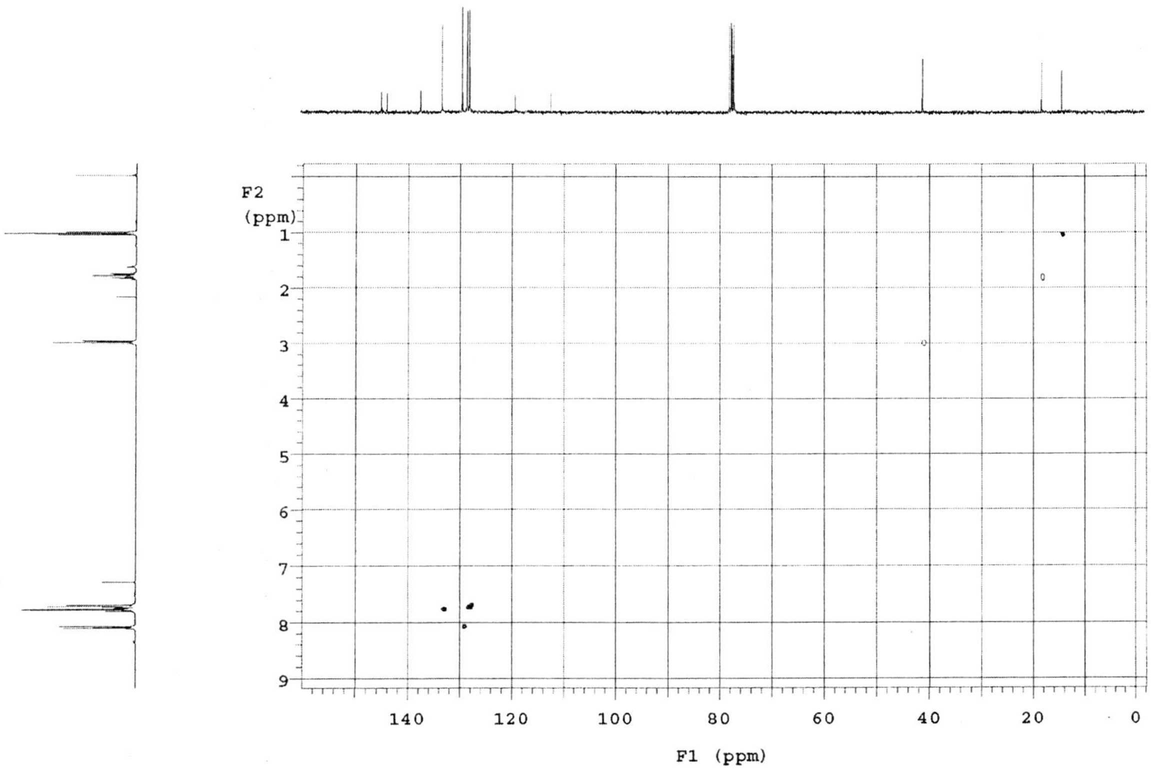

gHSQC Spectrum in $CDCl_3$

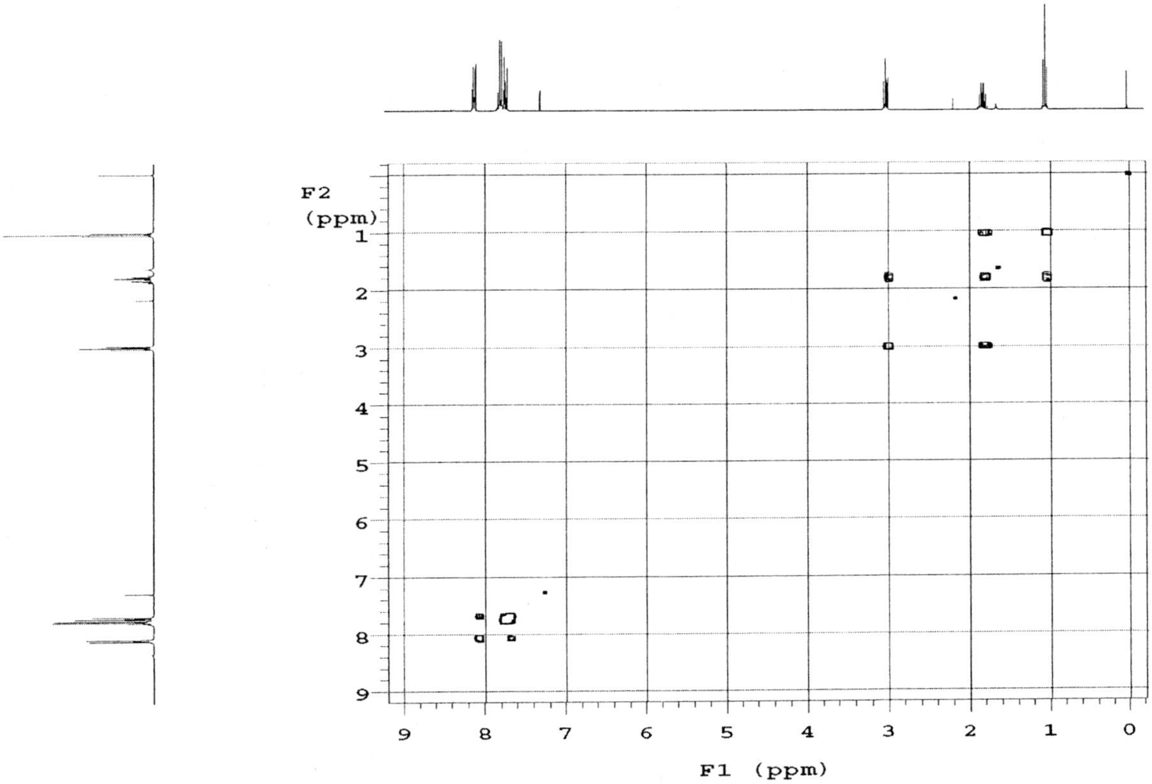

gCOSY Spectrum in $CDCl_3$

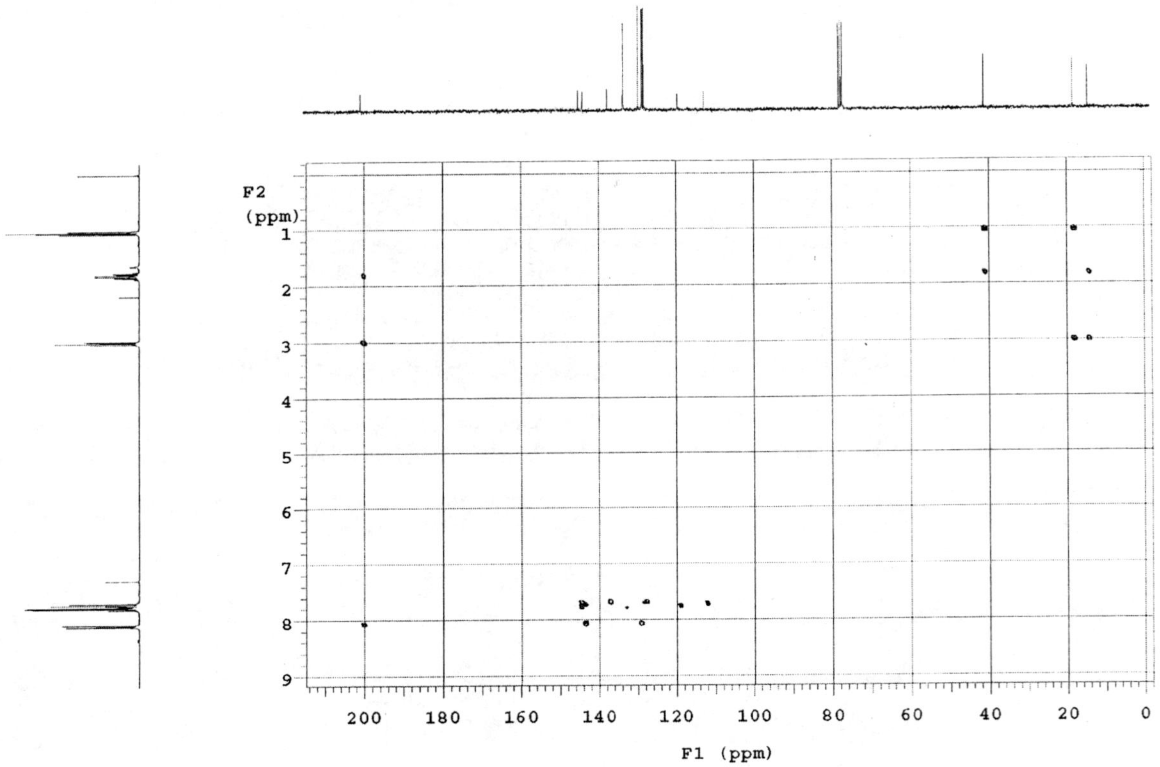

gHMBC Spectrum in $CDCl_3$

Problem 3.13 Given the spectroscopic data below (Proton, Carbon, gHSQC, and gHMBC), provide the structure of the molecule. By vapor pressure osmometry, the molecular weight was determined to be 356 ± 10% g/mol.

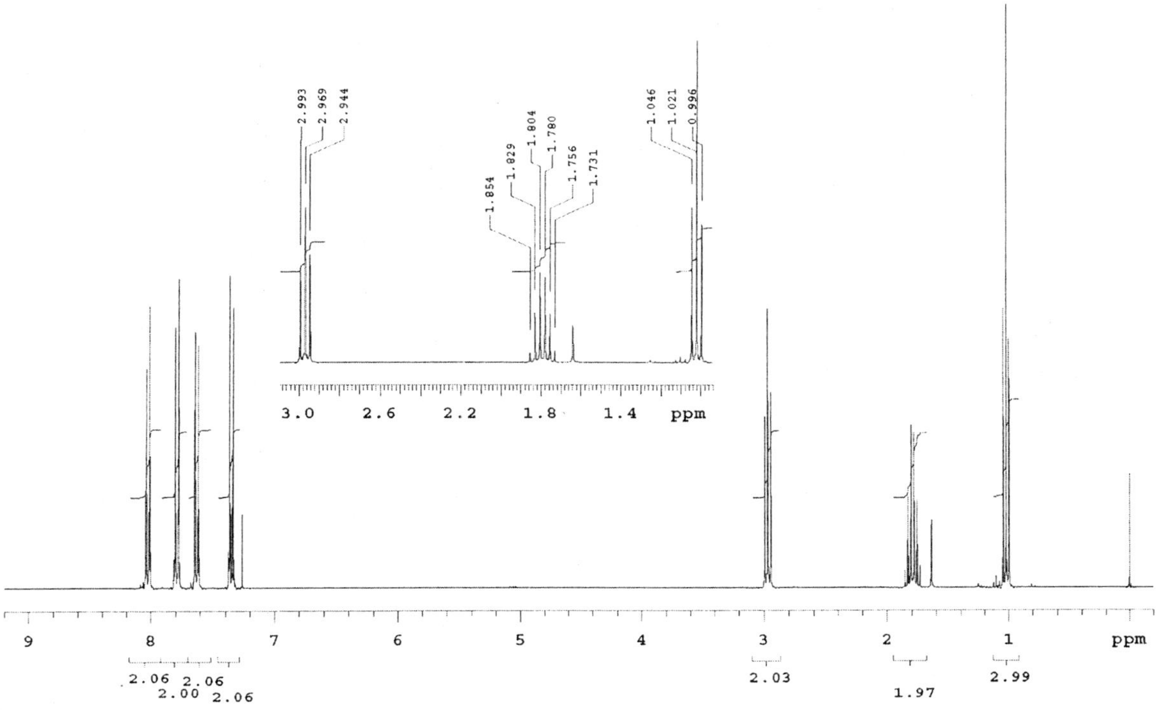

^{1}H-NMR Spectrum in $CDCl_3$

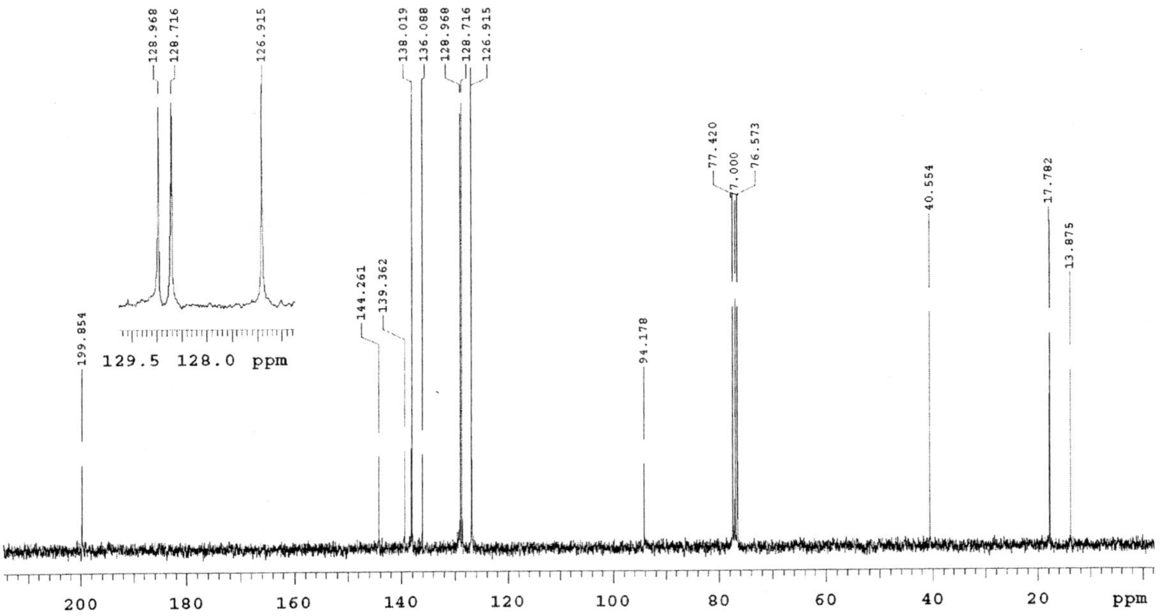

^{13}C-NMR Spectrum in $CDCl_3$

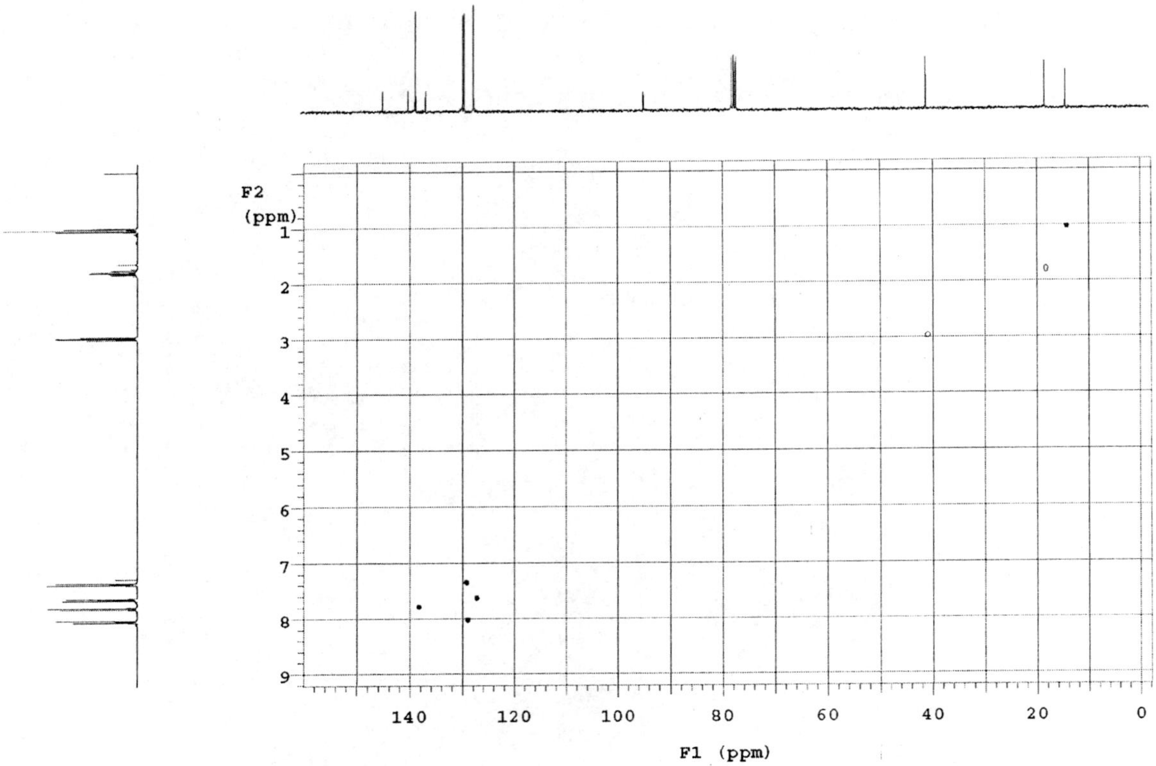

gHSQC Spectrum in $CDCl_3$

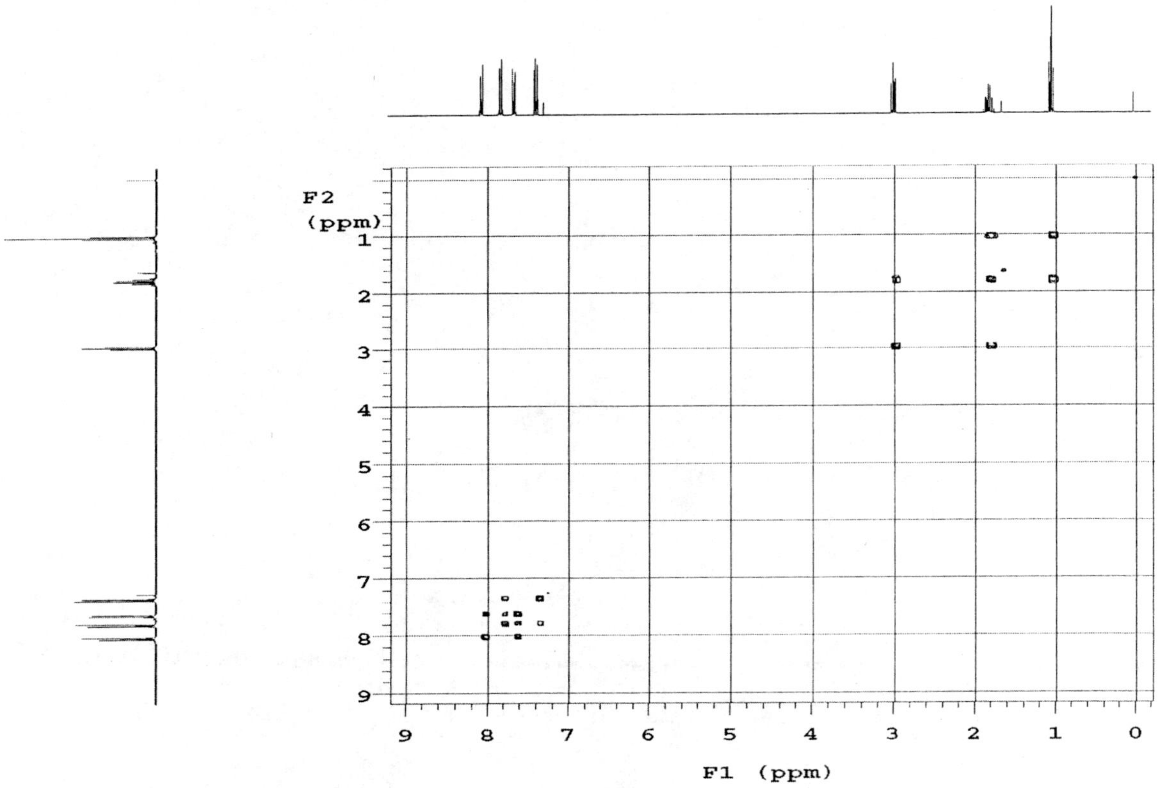

gCOSY Spectrum in $CDCl_3$

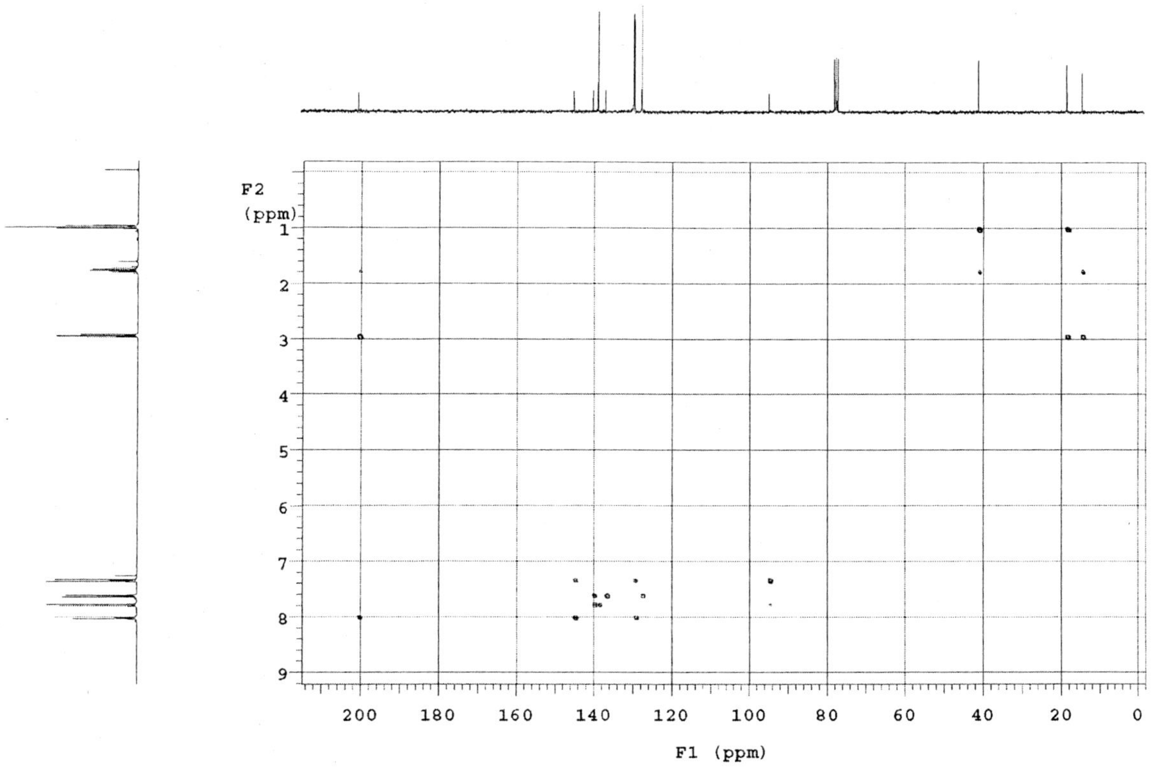

gHMBC Spectrum in $CDCl_3$

Problem 3.14 Given the spectroscopic data below (Proton, Carbon, gHSQC, gCOSY, and gHMBC), provide the structure of the molecule. The position of the signal at ~5 ppm is concentration and temperature dependent.

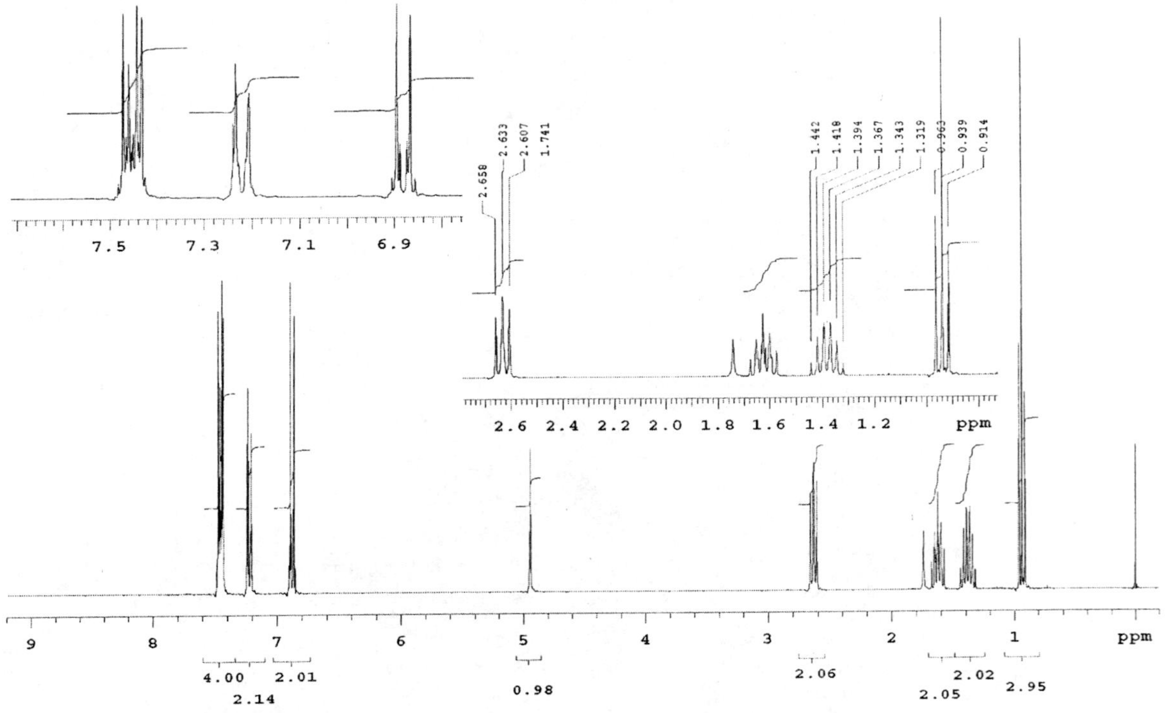

^{1}H-NMR Spectrum in $CDCl_3$

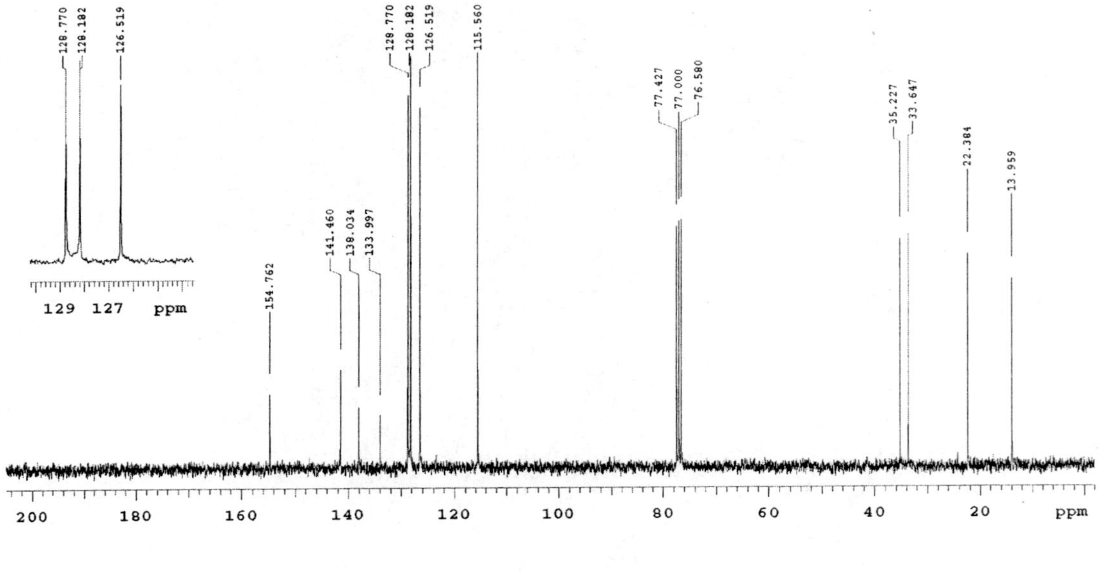

^{13}C-NMR Spectrum in $CDCl_3$

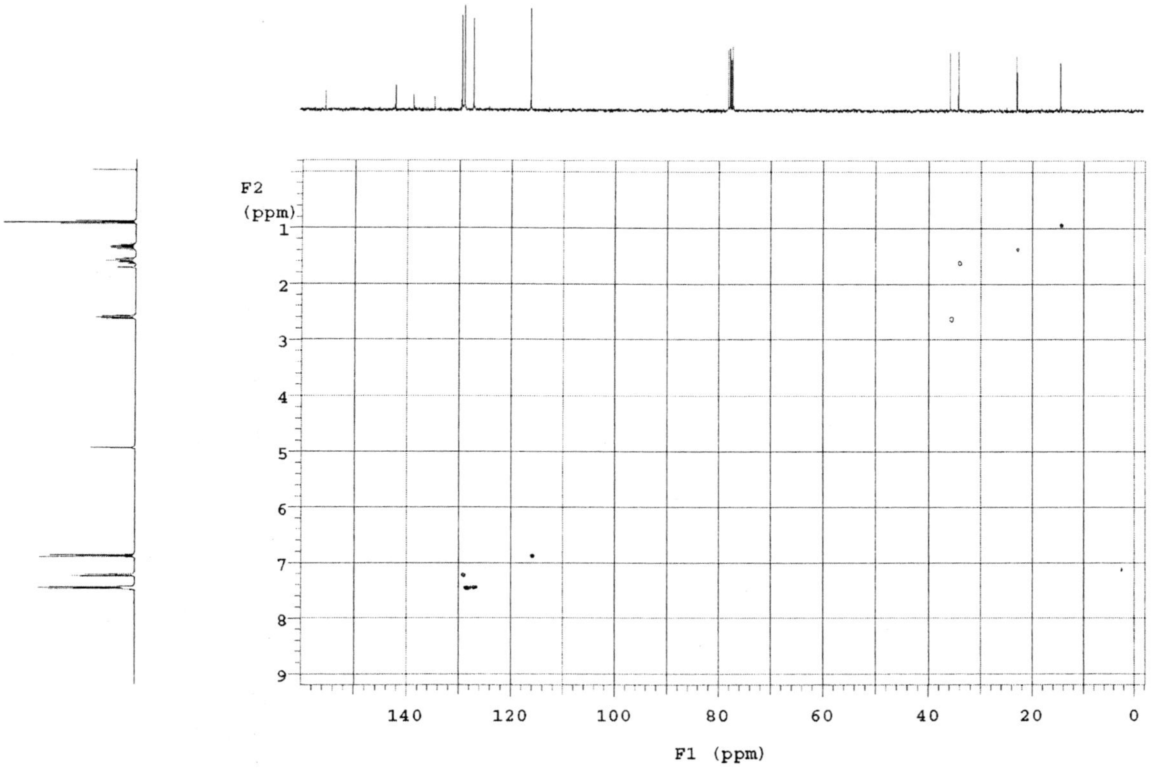

gHSQC Spectrum in $CDCl_3$

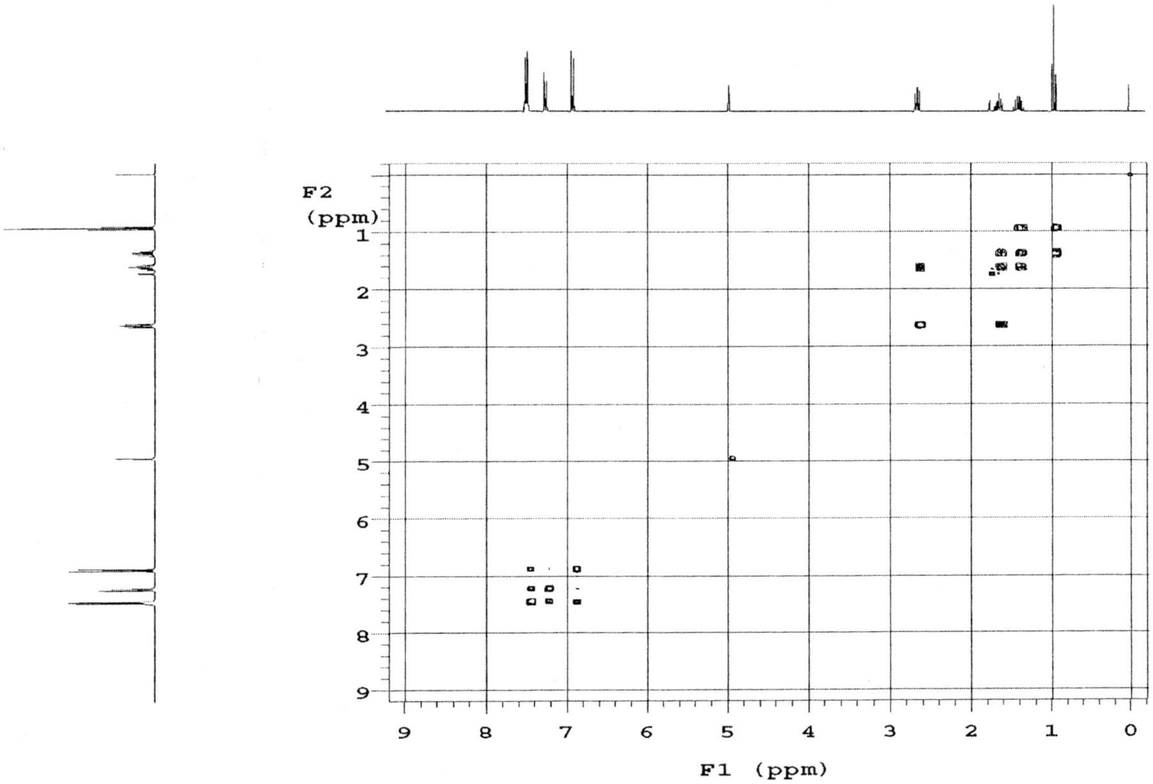

gCOSY Spectrum in $CDCl_3$

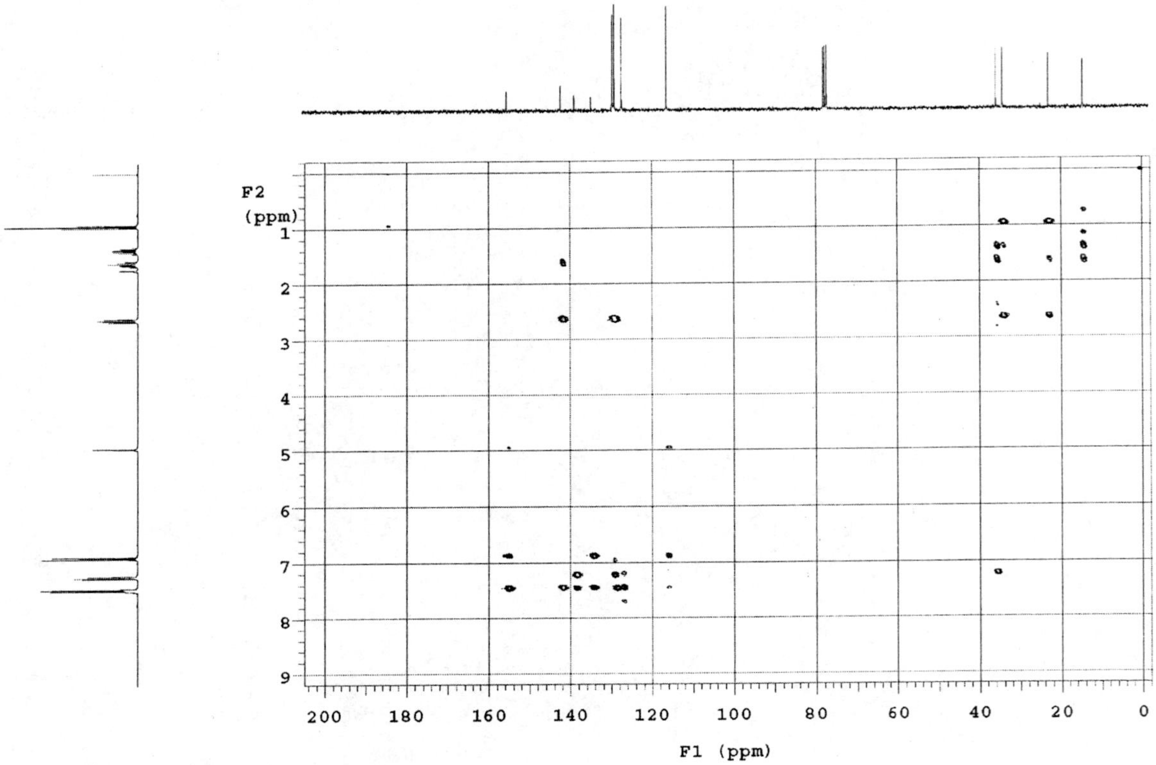

gHMBC Spectrum in $CDCl_3$

Problem 3.15 Given the spectroscopic data below (Proton, Carbon, gHSQC, gCOSY, and gHMBC), provide the structure of the molecule. (C: 71.37%; H: 5.55%; N: 5.14% & IR: 1550 cm^{-1})

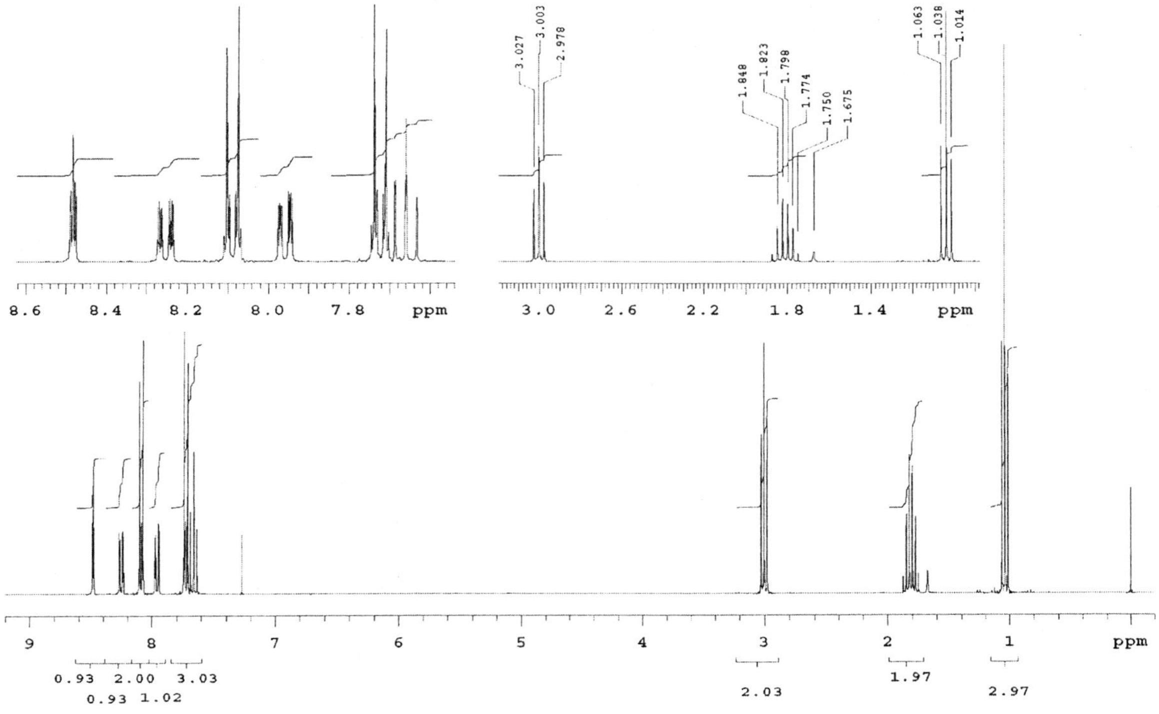

^{1}H-NMR Spectrum in $CDCl_3$

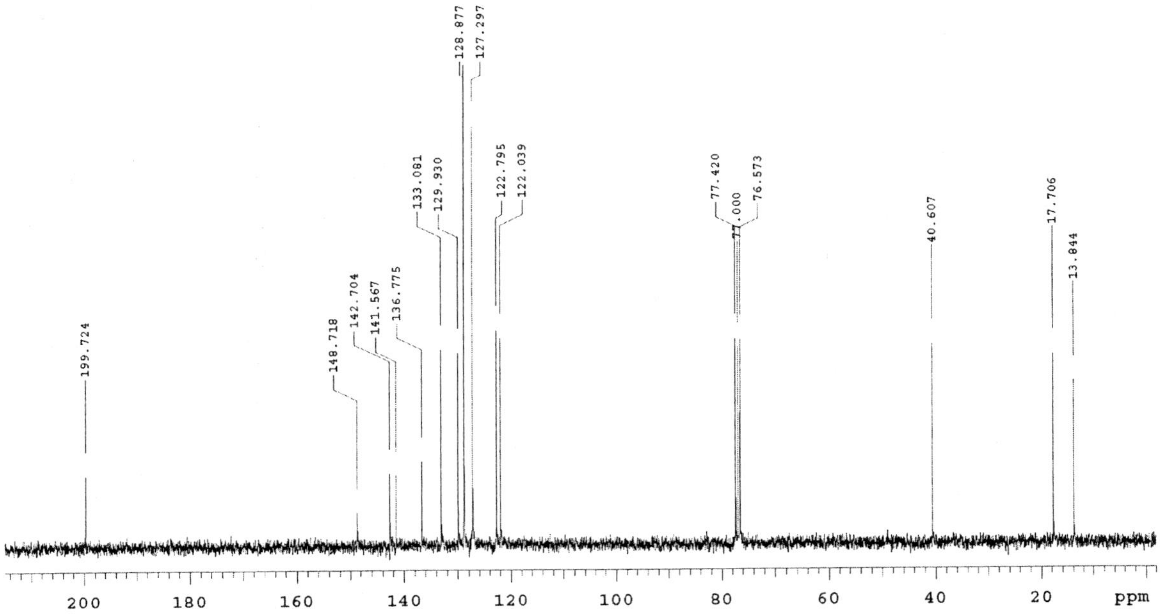

^{13}C-NMR Spectrum in $CDCl_3$

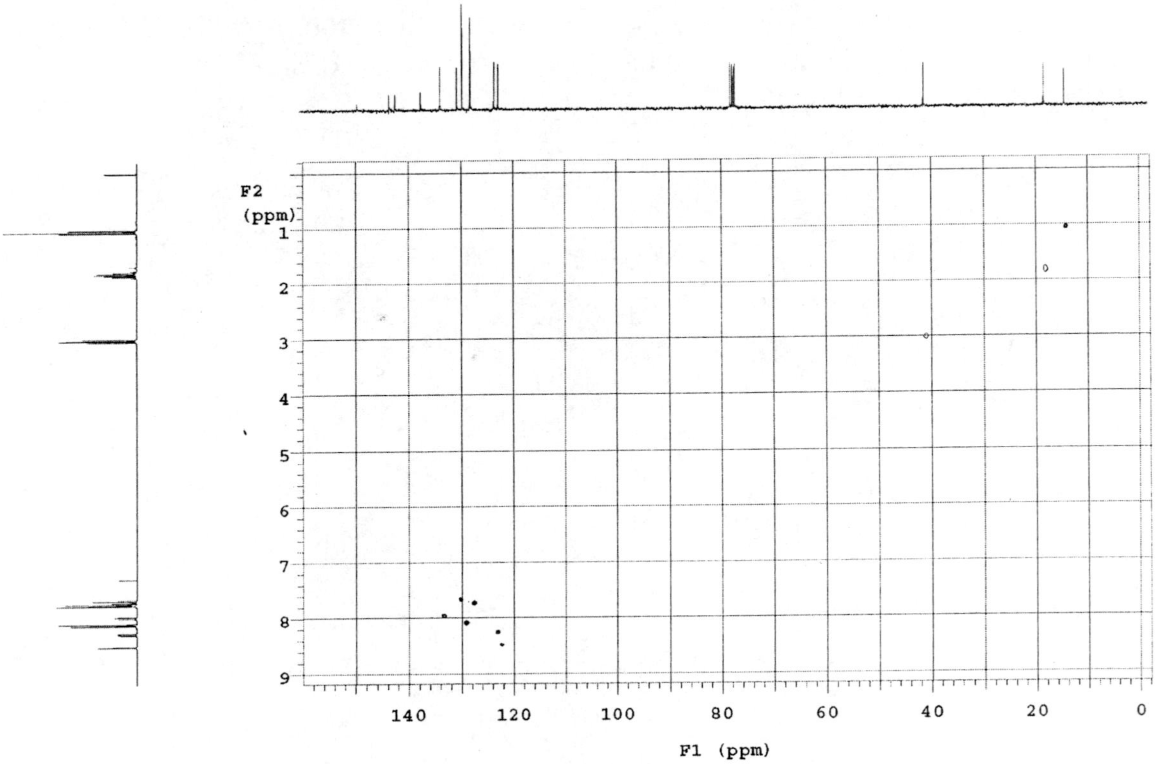

gHSQC Spectrum in $CDCl_3$

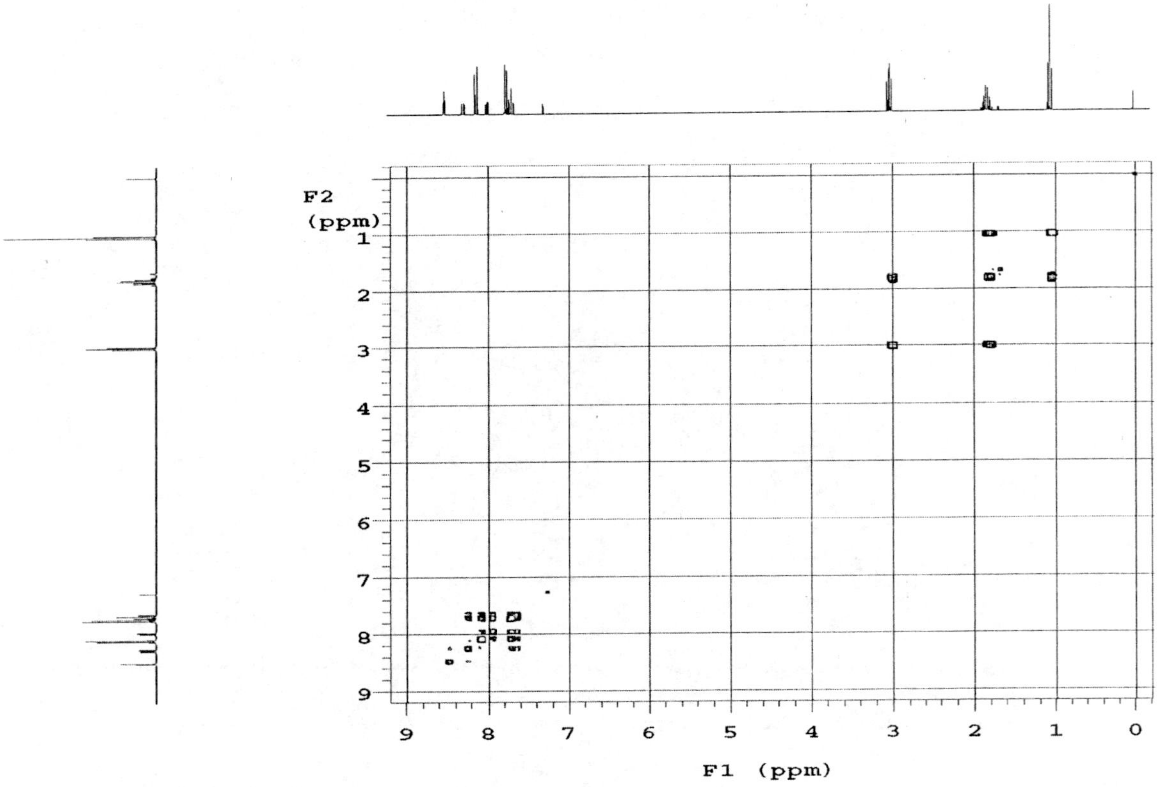

gCOSY Spectrum in $CDCl_3$

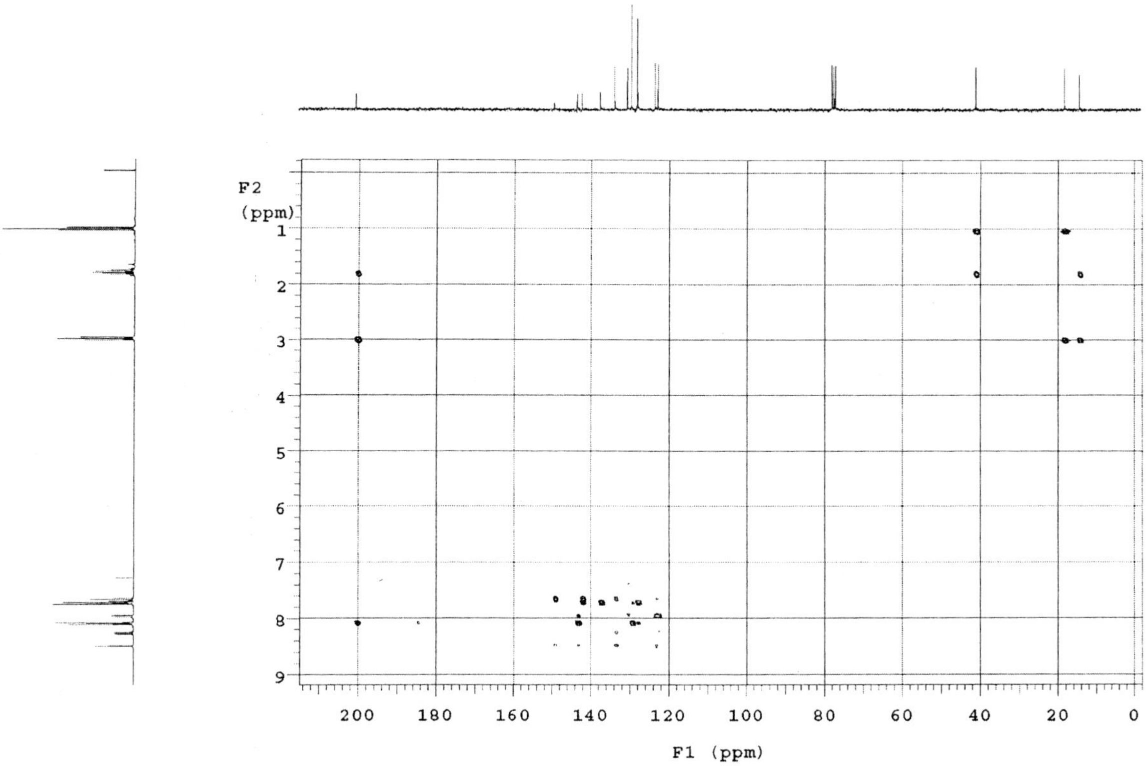

gHMBC Spectrum in $CDCl_3$

Problem 3.16 Given the spectroscopic data below (Proton, Carbon, gHSQC, gCOSY, and gHMBC), provide the structure of the molecule. ($C_{16}H_{17}NO$)

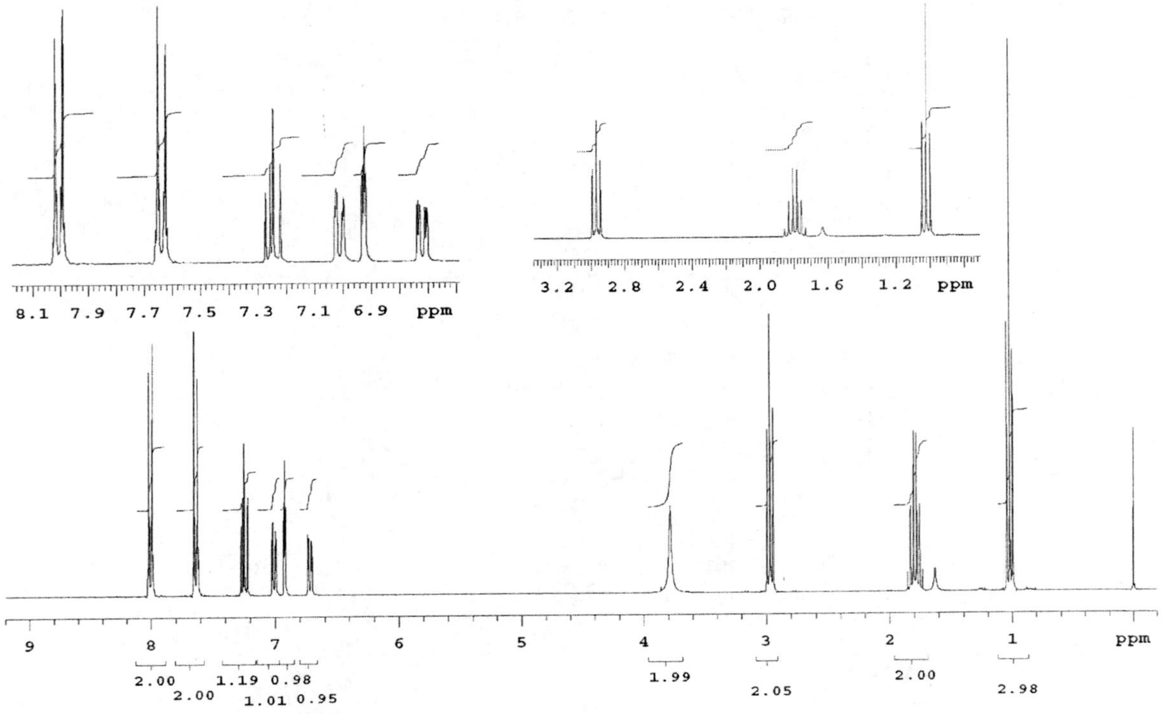

^{1}H-NMR Spectrum in $CDCl_3$

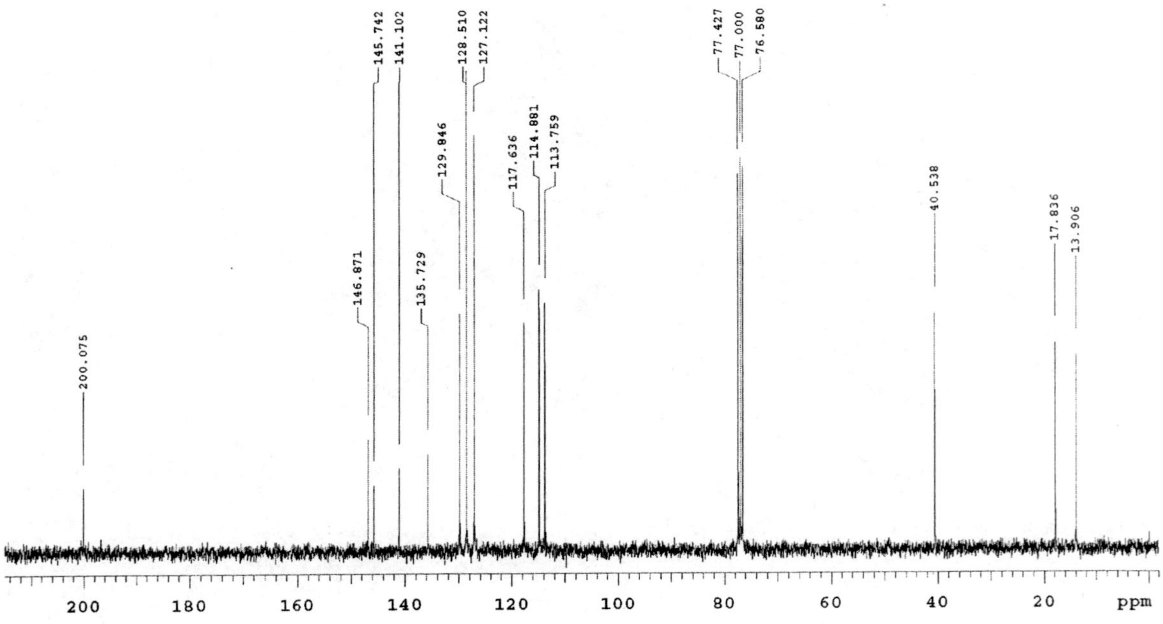

^{13}C-NMR Spectrum in $CDCl_3$

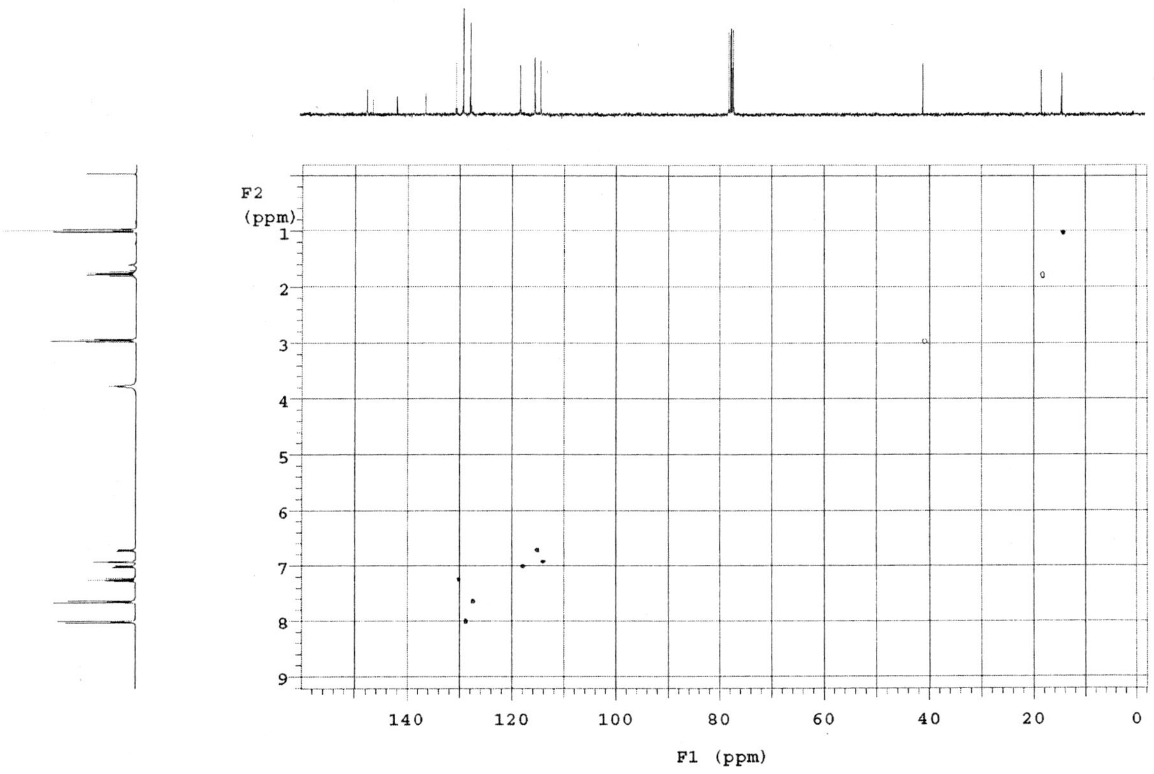

gHSQC Spectrum in $CDCl_3$

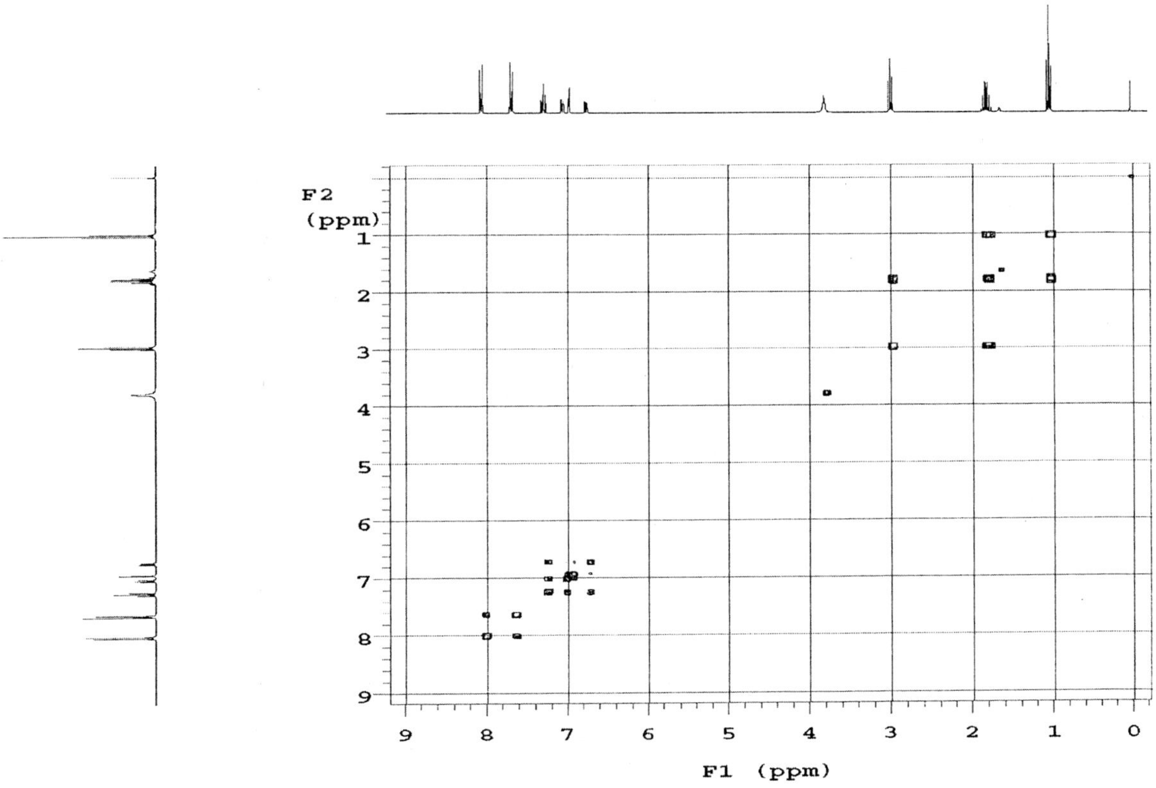

gCOSY Spectrum in $CDCl_3$

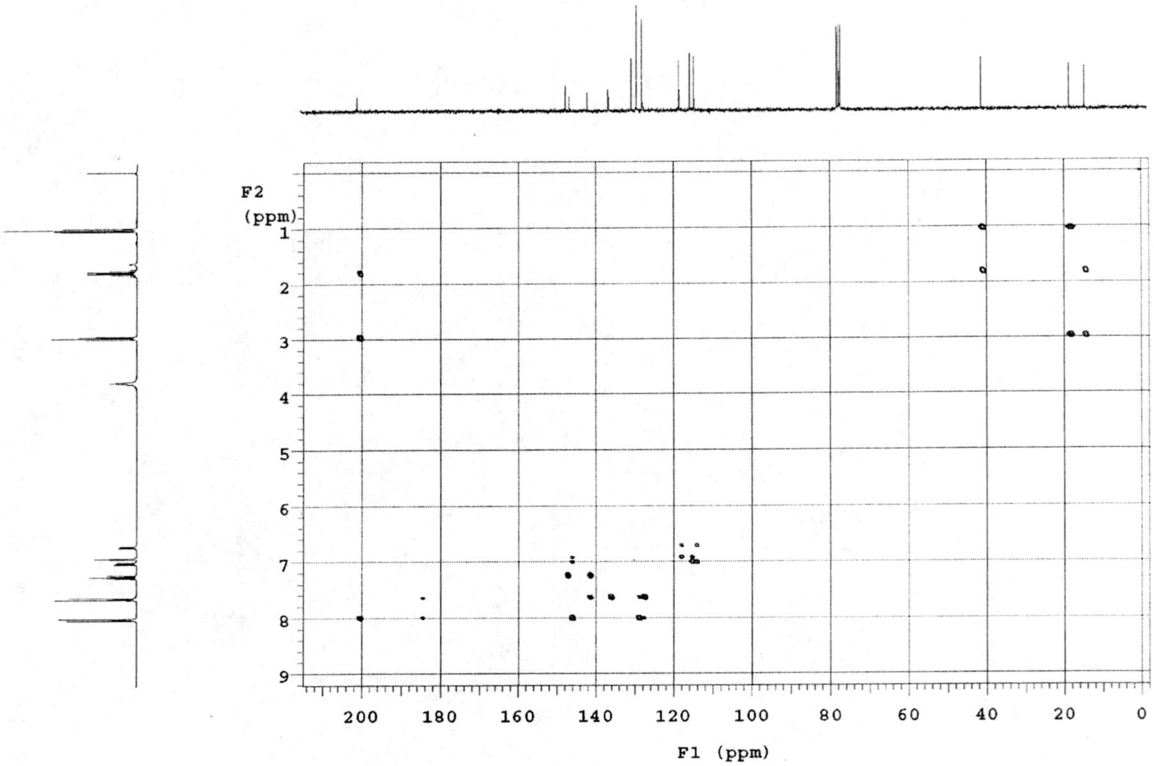

gHMBC Spectrum in $CDCl_3$

Problem 3.17 Given the spectroscopic data below (Proton, Carbon, gHSQC, gCOSY, and gHMBC), provide the structure of the molecule. (MW: 191 g/mol)

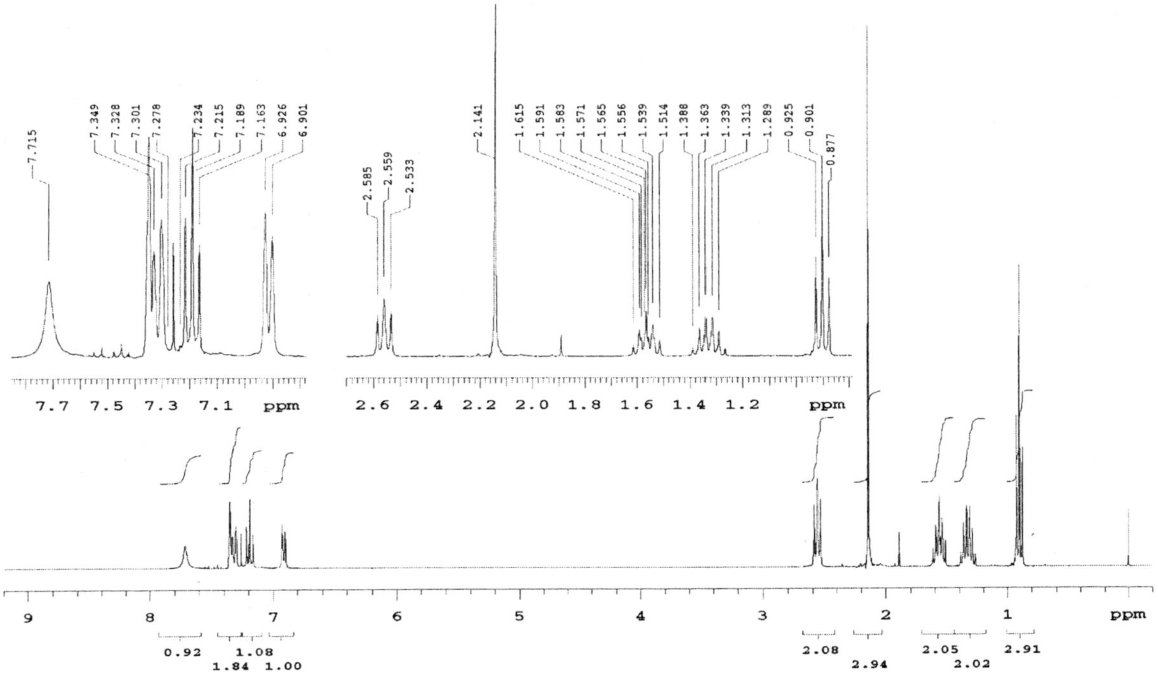

^{1}H-NMR Spectrum in $CDCl_3$

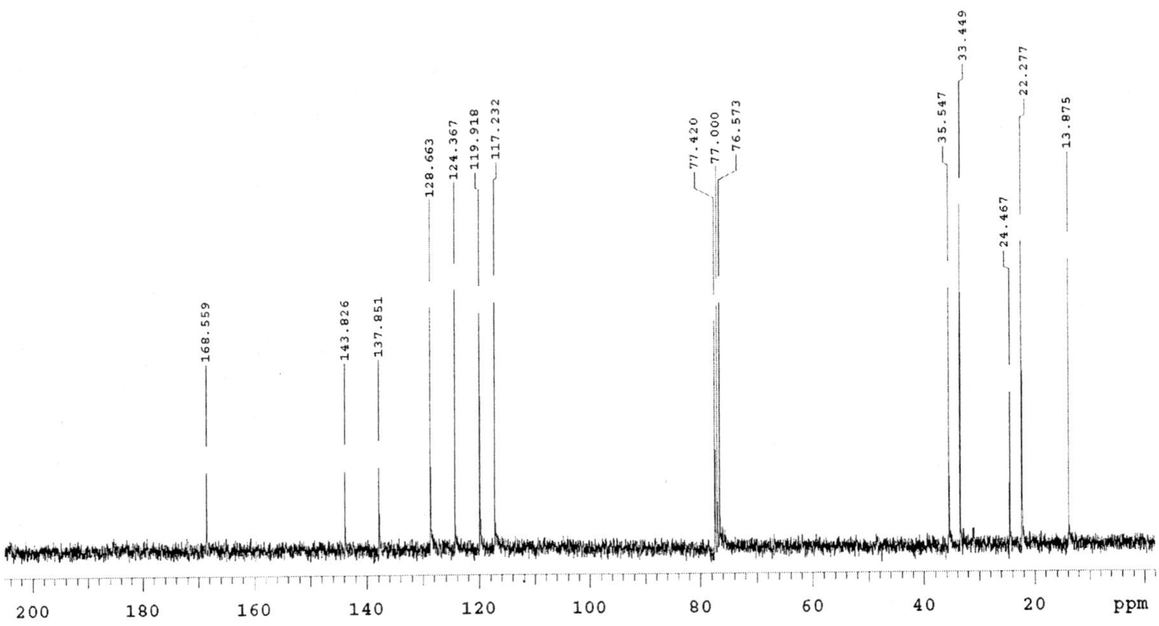

^{13}C-NMR Spectrum in $CDCl_3$

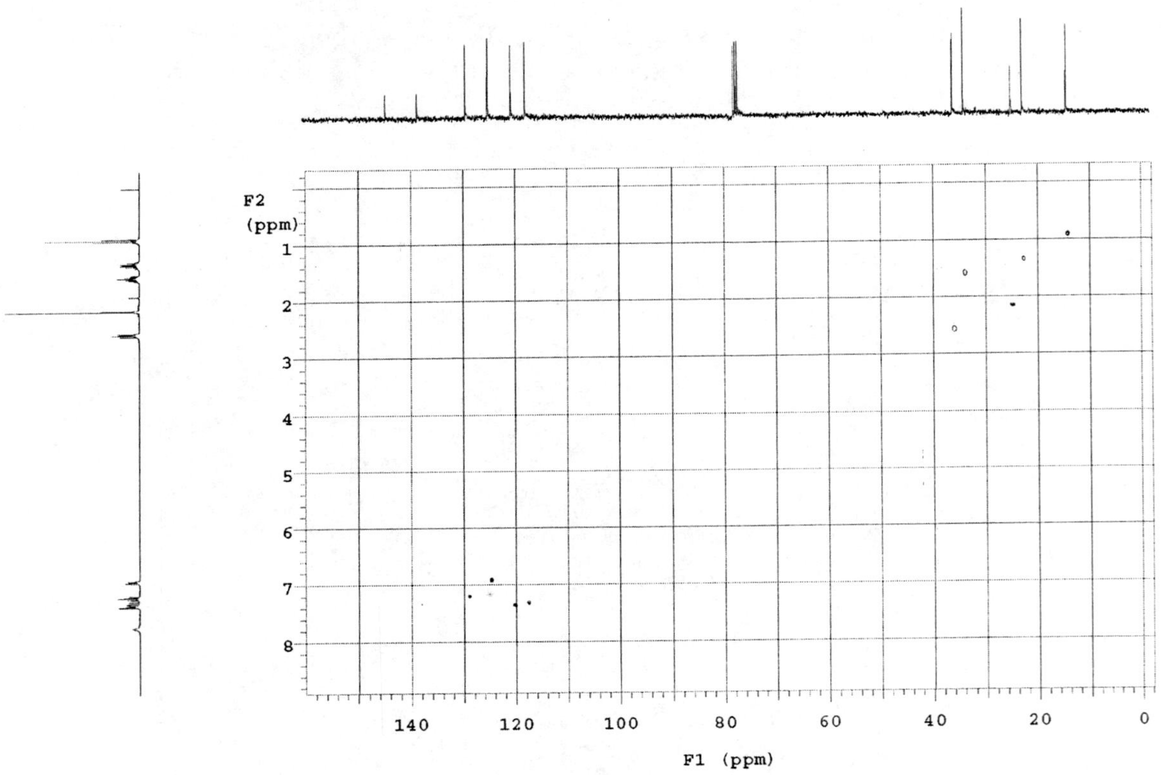

gHSQC Spectrum in $CDCl_3$

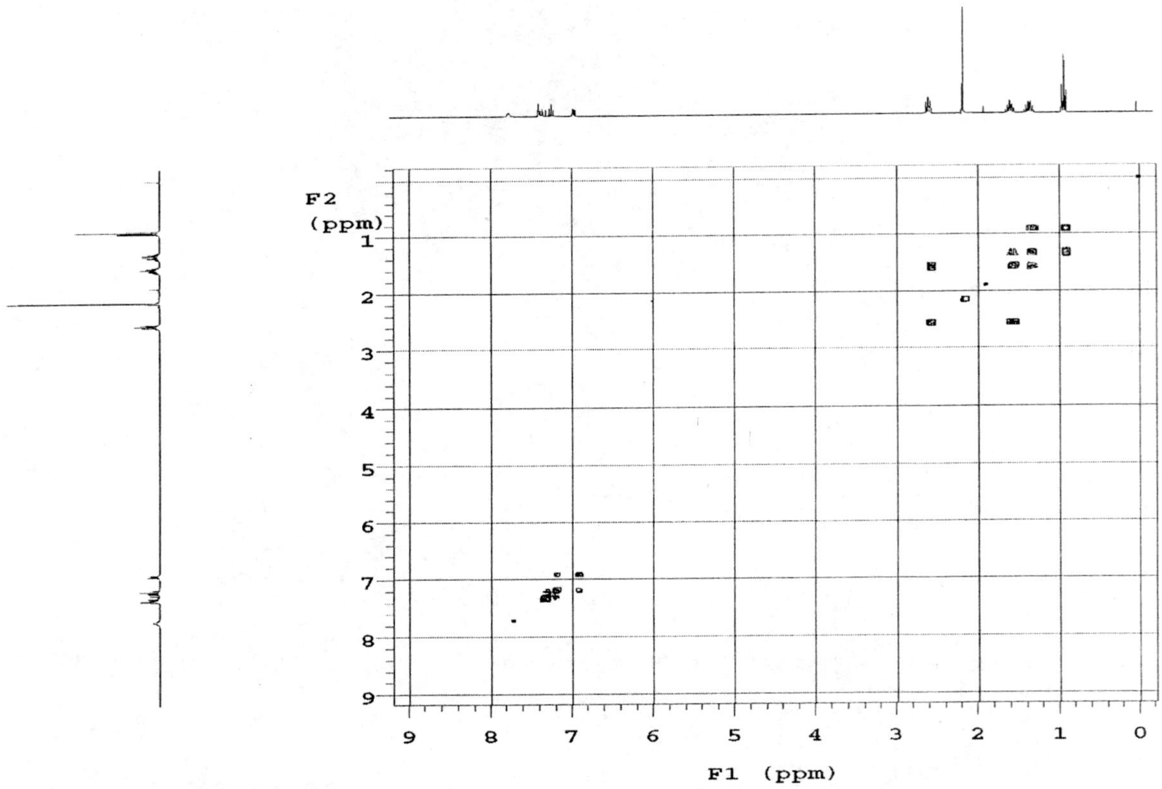

gCOSY Spectrum in $CDCl_3$

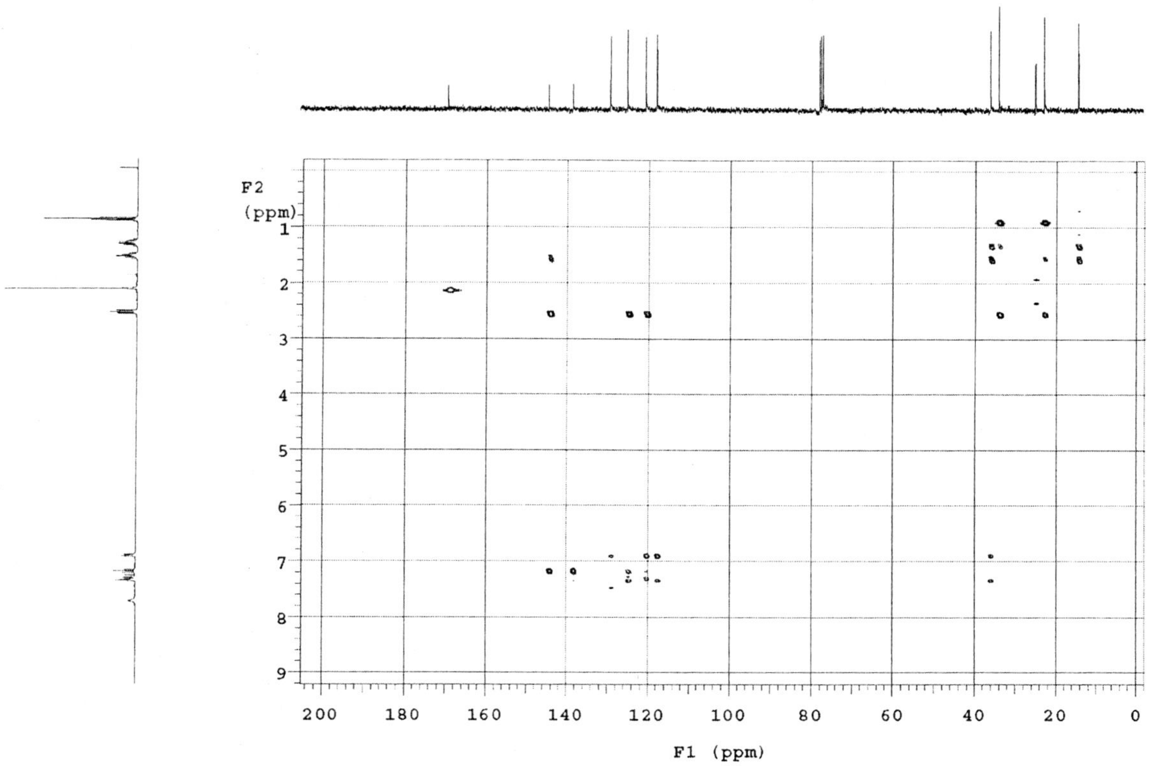

gHMBC Spectrum in $CDCl_3$

Problem 3.18 Given the spectroscopic data below (Proton, Carbon, Edited DEPT, gHSQC, gCOSY and gHMBC), provide the structure of the molecule. ($C_{10}H_{16}O$)

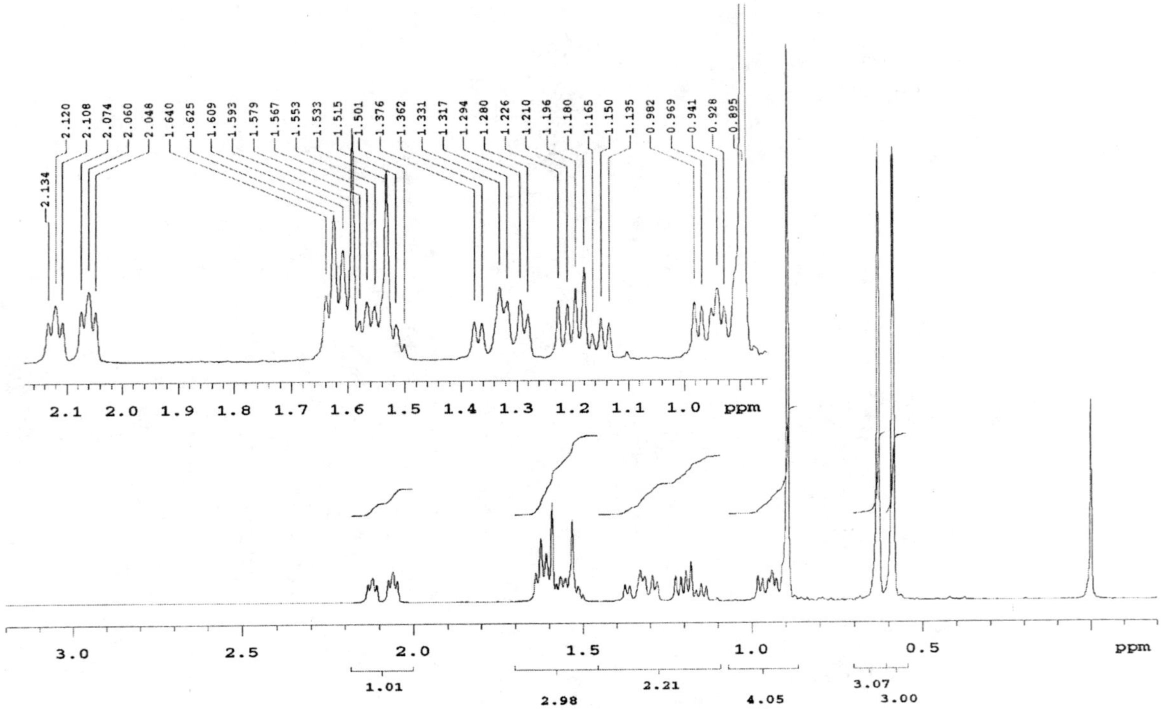

^{1}H-NMR Spectrum in benzene-d_6.

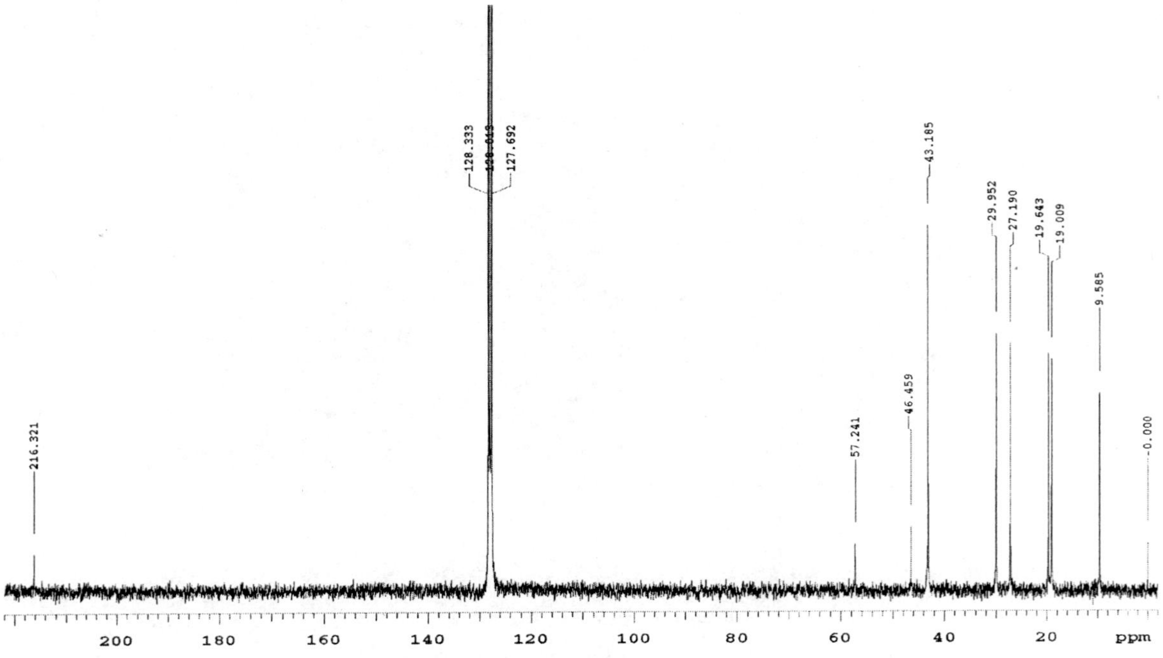

^{13}C-NMR Spectrum in benzene-d_6.

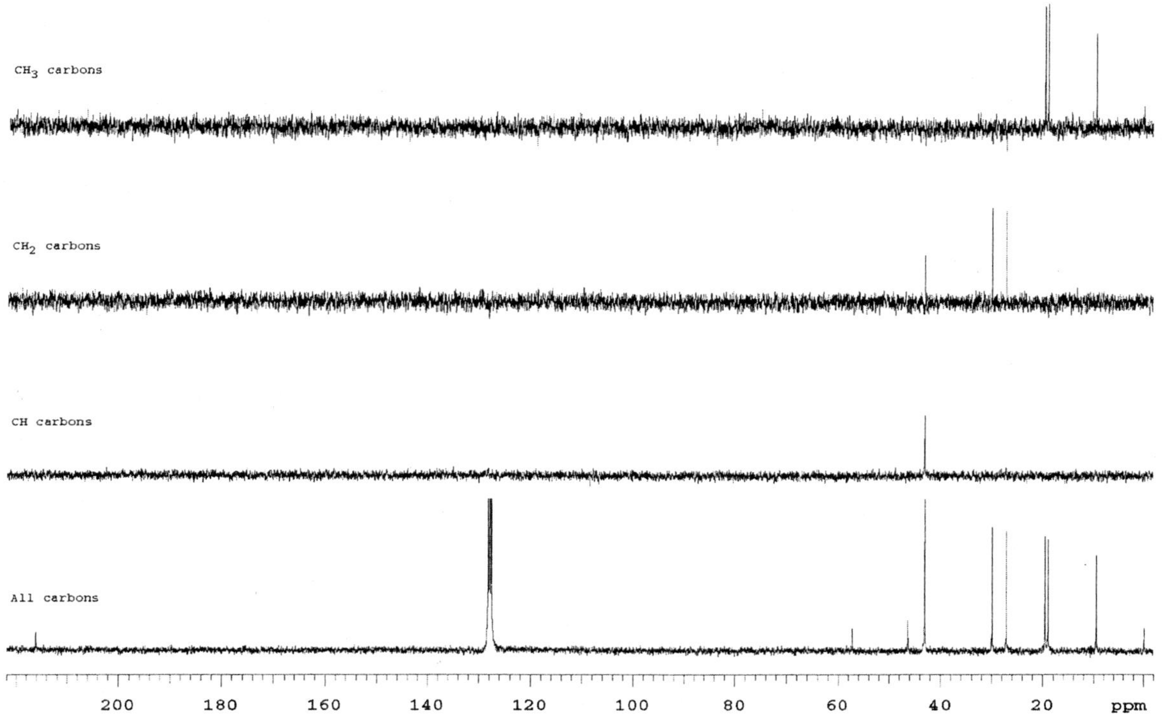

Edited DEPT ^{13}C-NMR Spectrum in benzene-d_6.

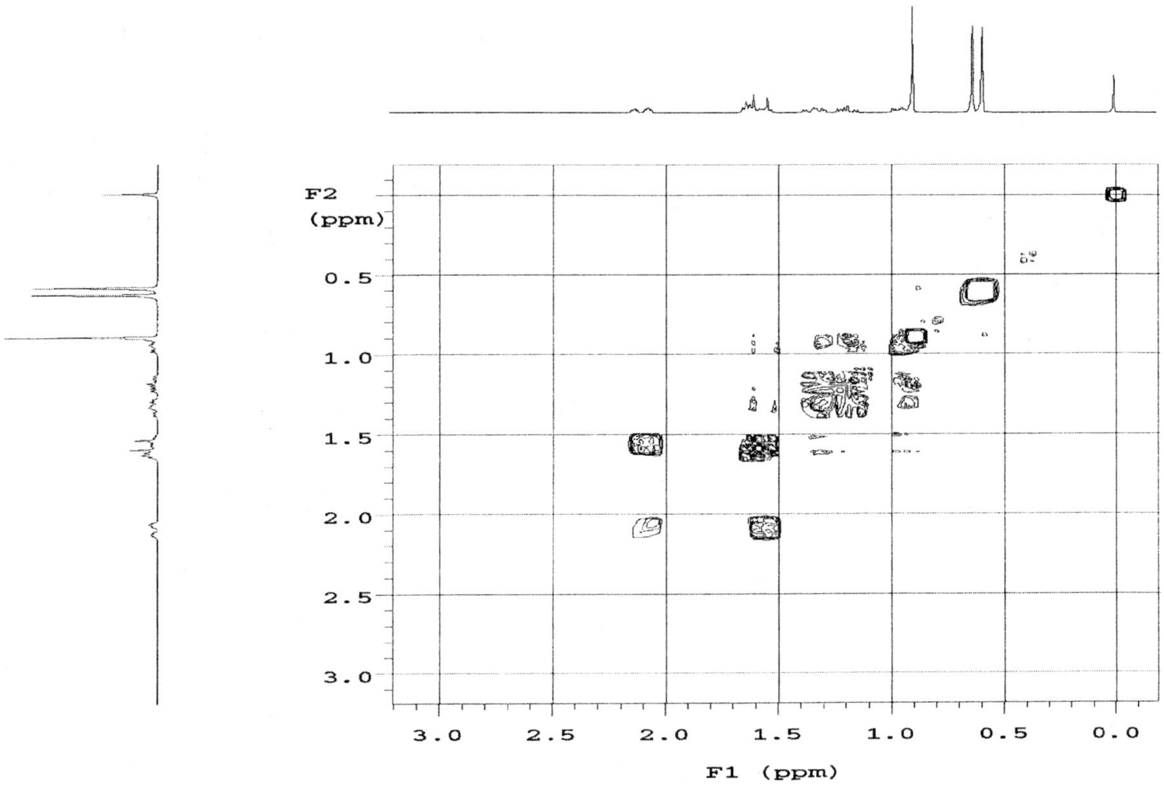

gCOSY Spectrum in benzene-d_6.

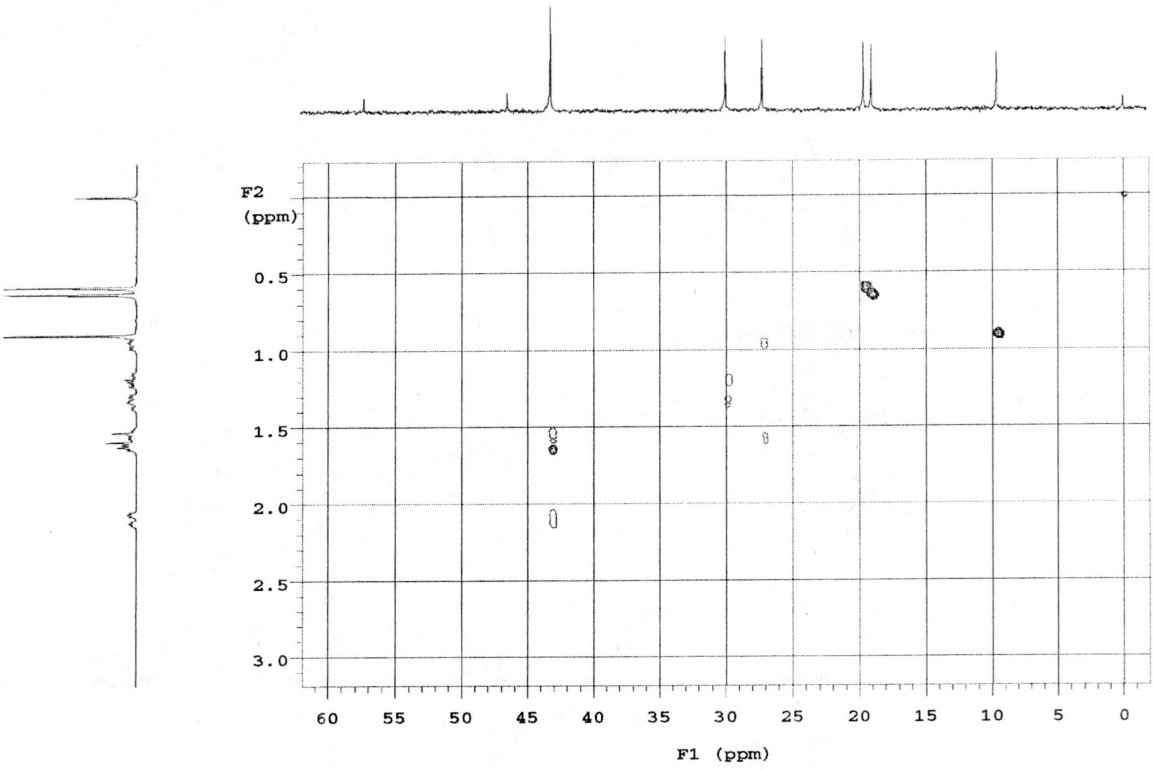

gHSQC Spectrum in benzene-d_6.

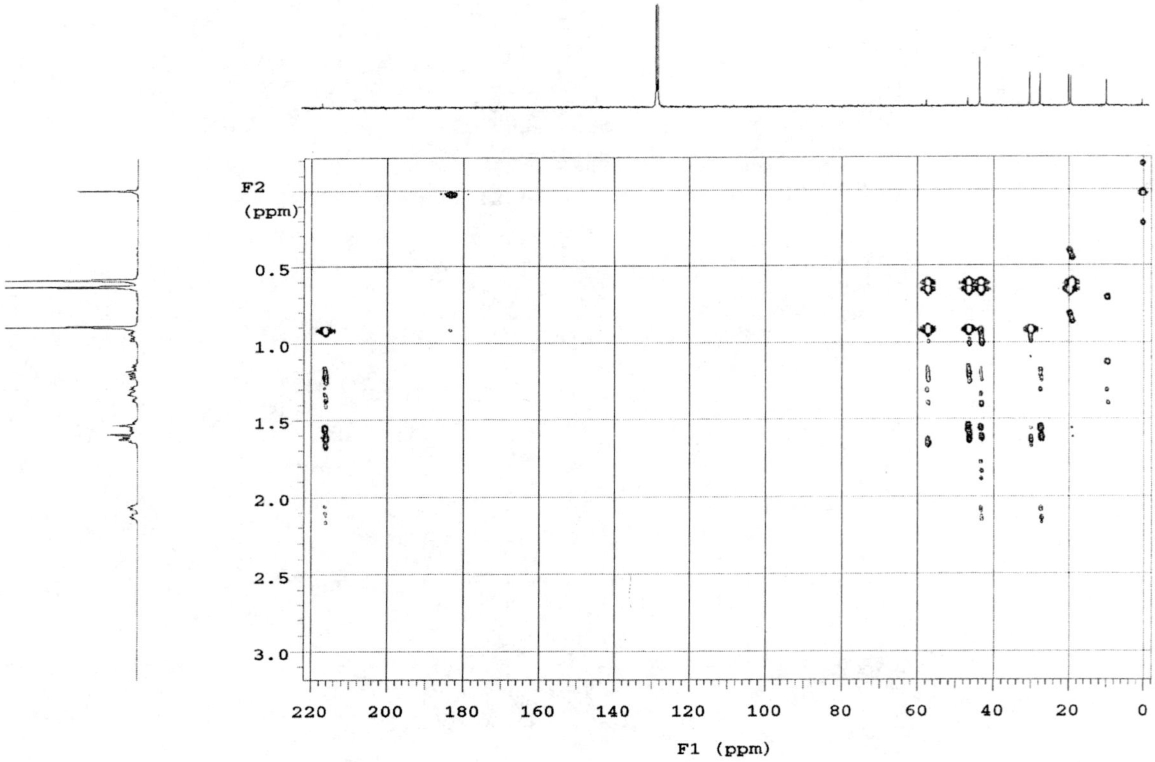

gHMBC Spectrum in benzene-d_6.

Problem 3.19 Given the spectroscopic data below (Proton, Carbon, Edited DEPT, gHSQC, and gCOSY), provide the structure of the molecule.

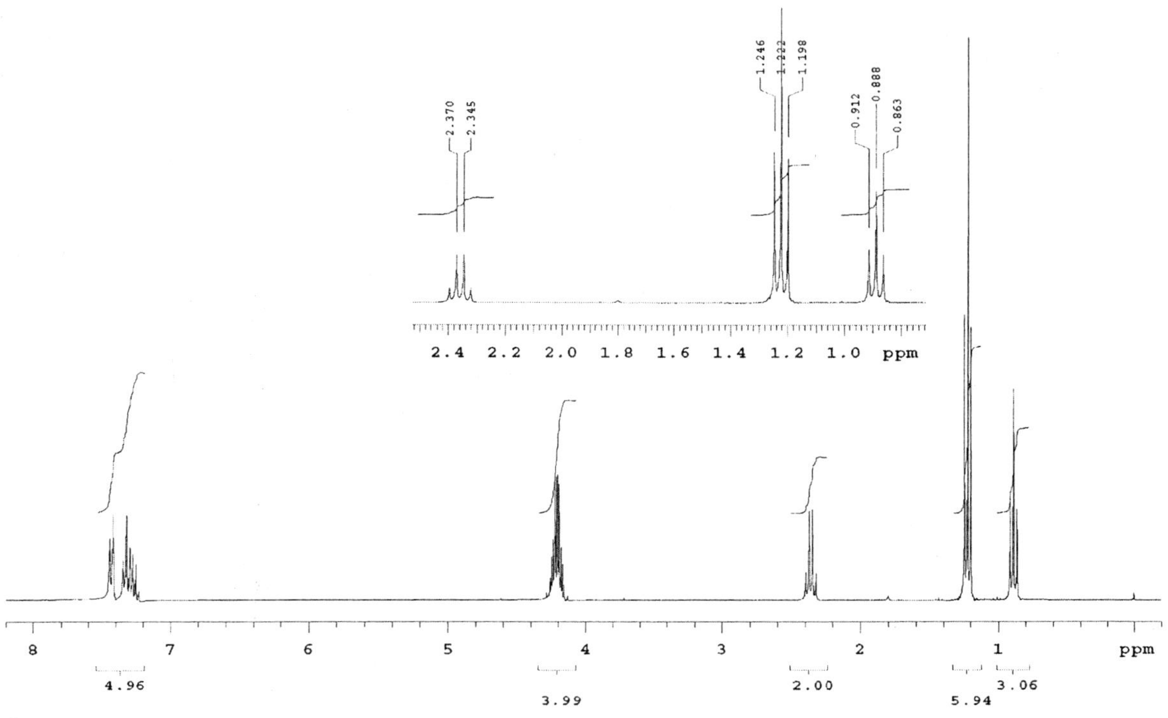

^{1}H-NMR Spectrum in $CDCl_3$

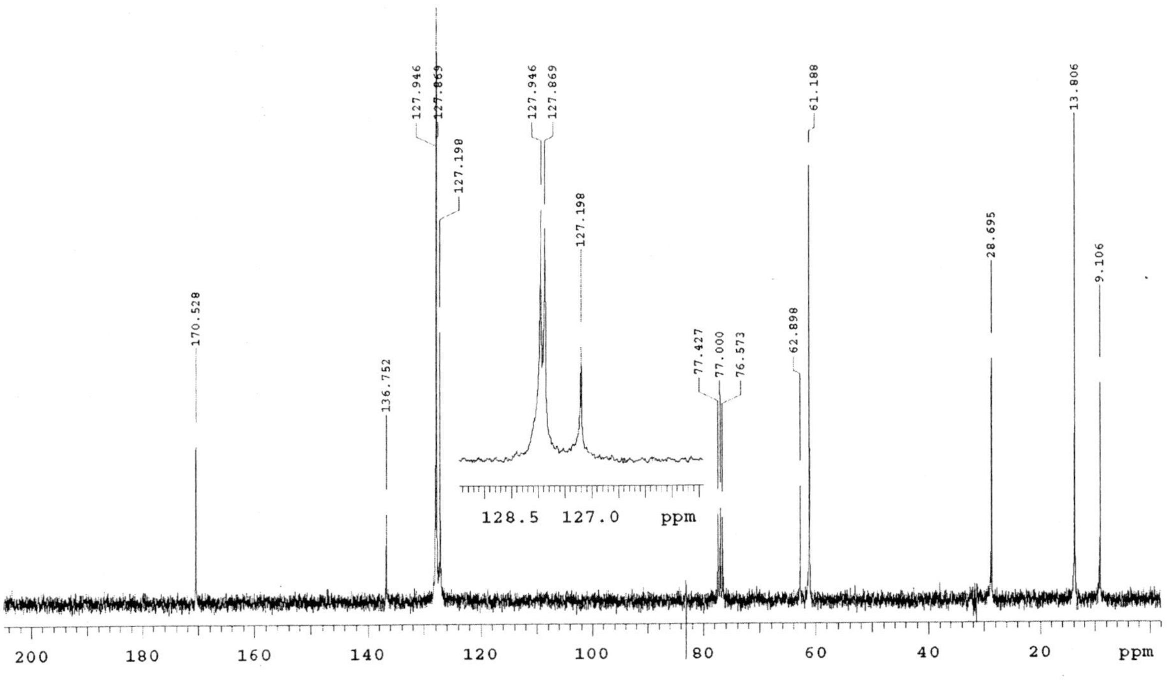

^{13}C-NMR Spectrum in $CDCl_3$

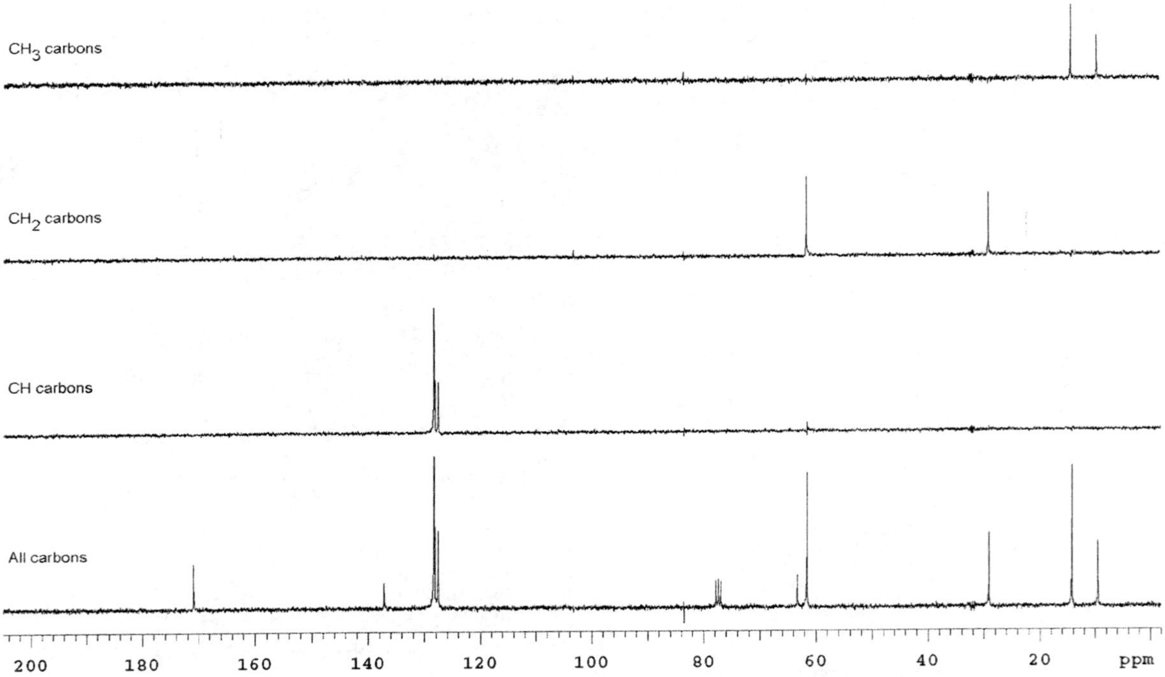

Edited DEPT ^{13}C-NMR Spectrum in $CDCl_3$

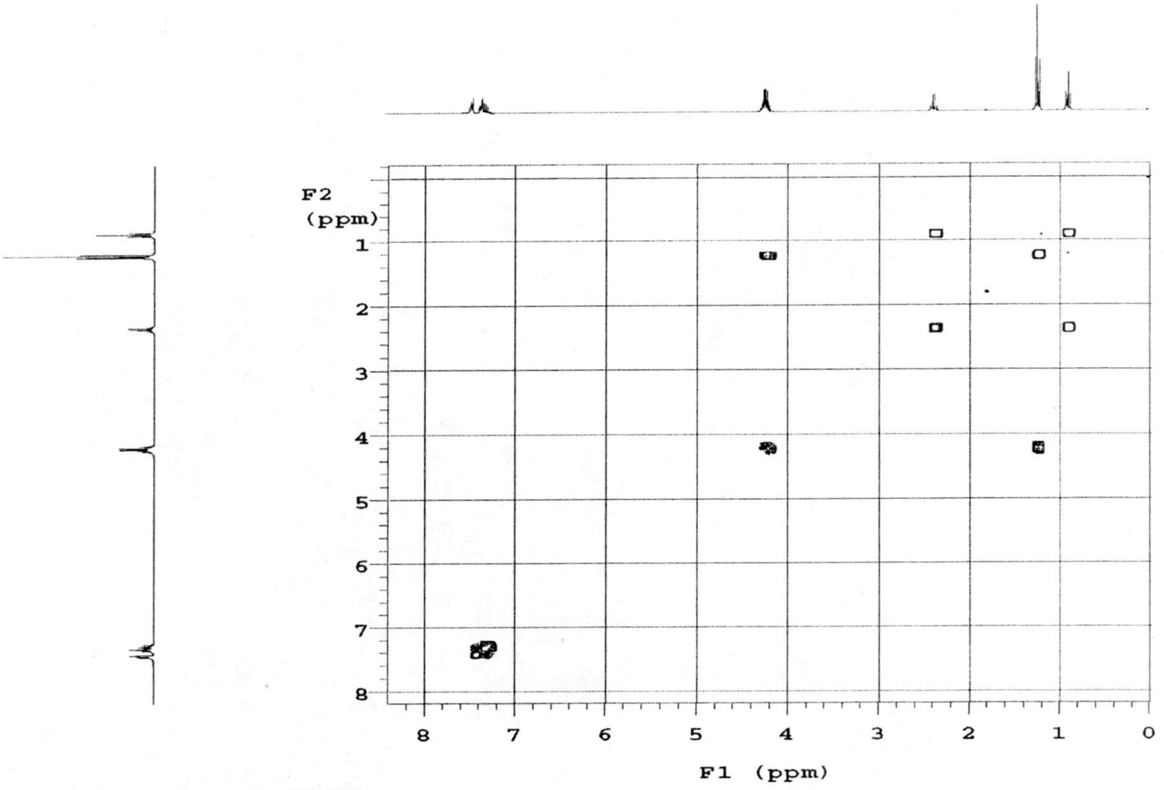

gCOSY Spectrum in $CDCl_3$

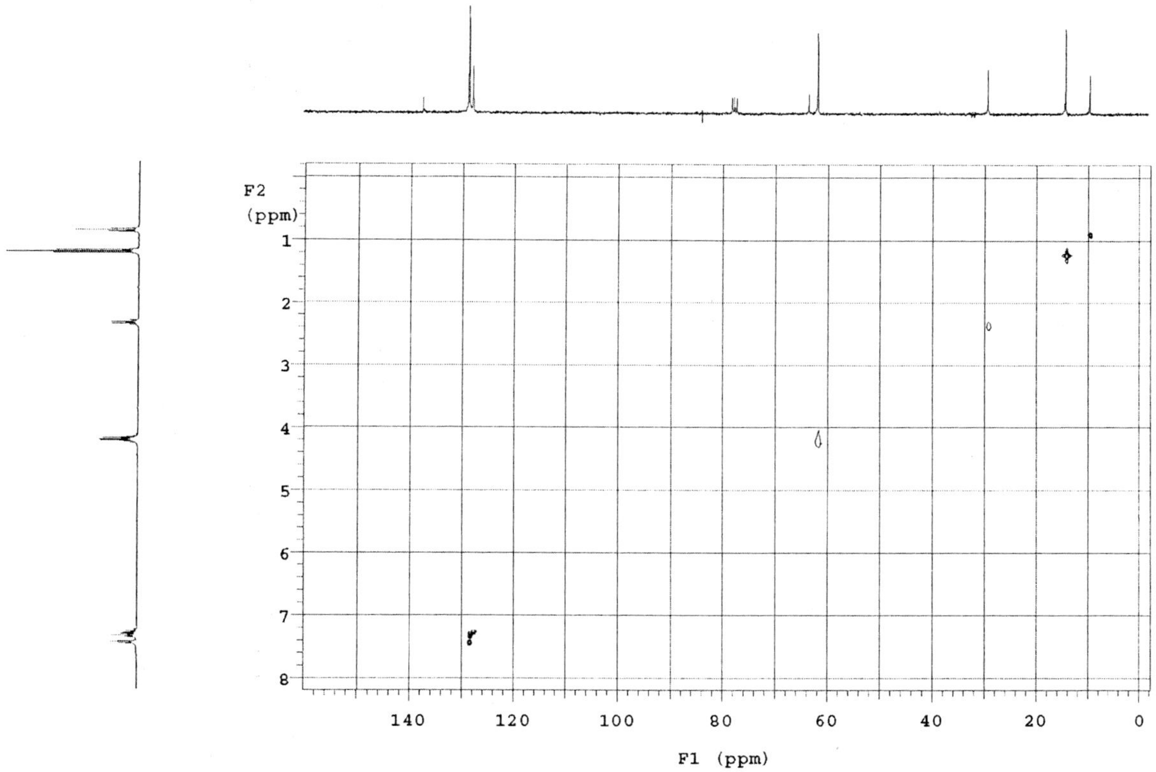

gHSQC Spectrum in $CDCl_3$

Problem 3.20 Given the spectroscopic data below (Proton, Carbon, Edited DEPT, gHSQC, and gCOSY), provide the structure of the molecule. ($C_{10}H_{15}NO$)

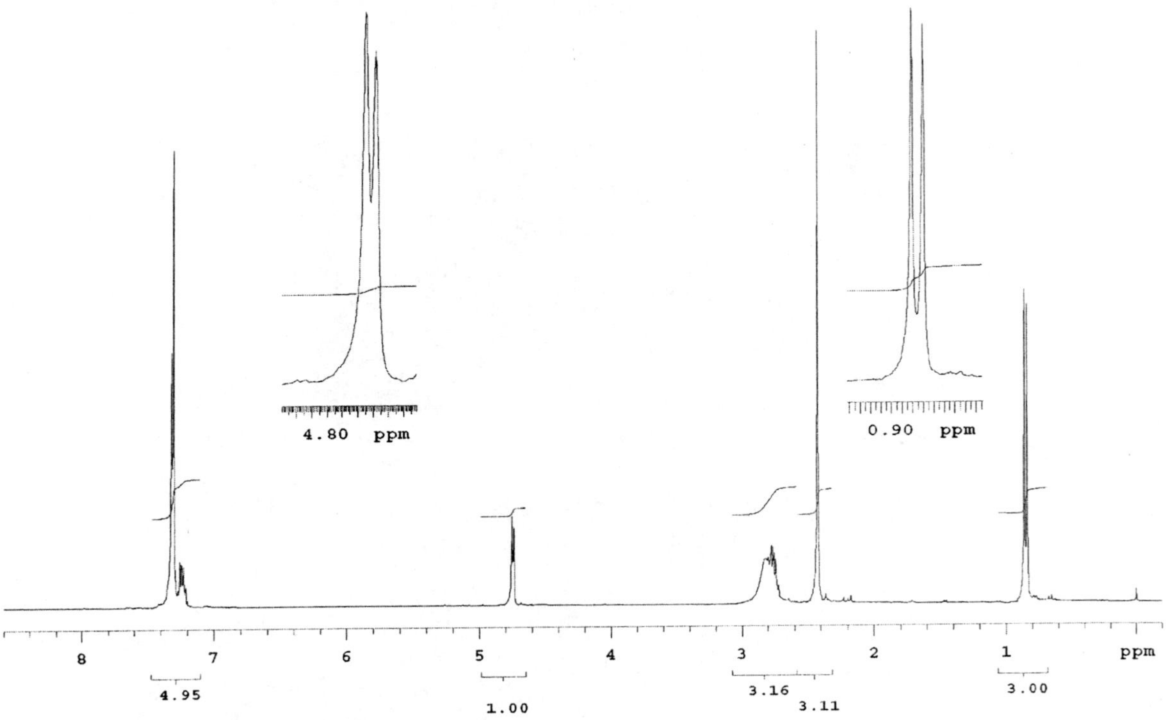

^{1}H-NMR Spectrum in $CDCl_3$

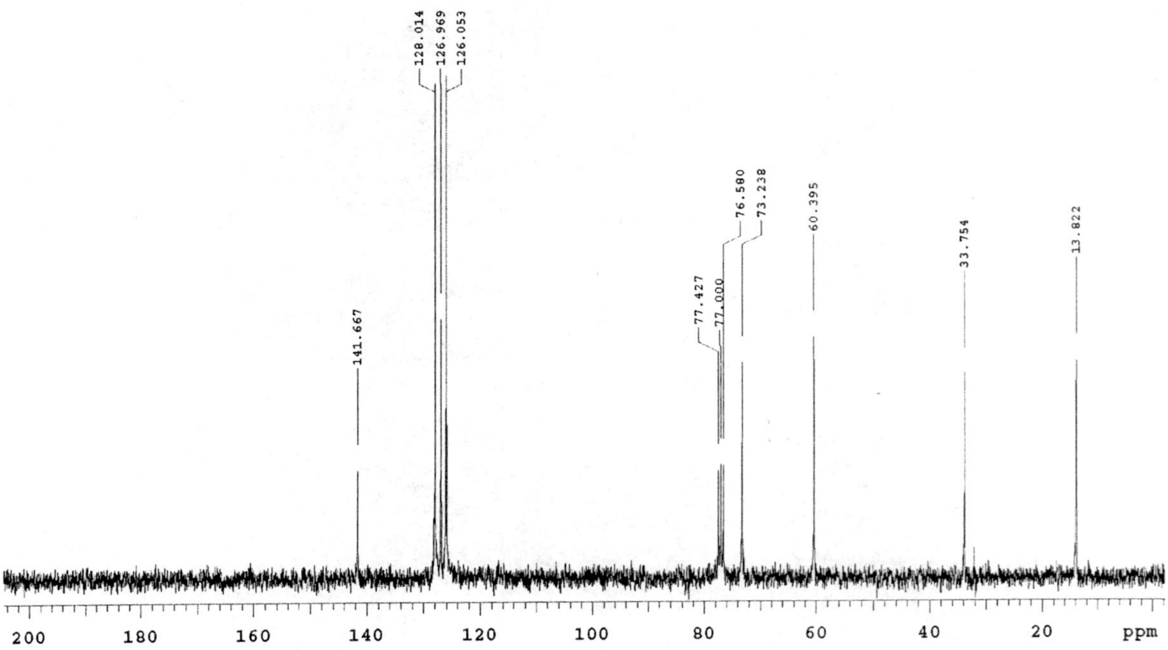

^{13}C-NMR Spectrum in $CDCl_3$

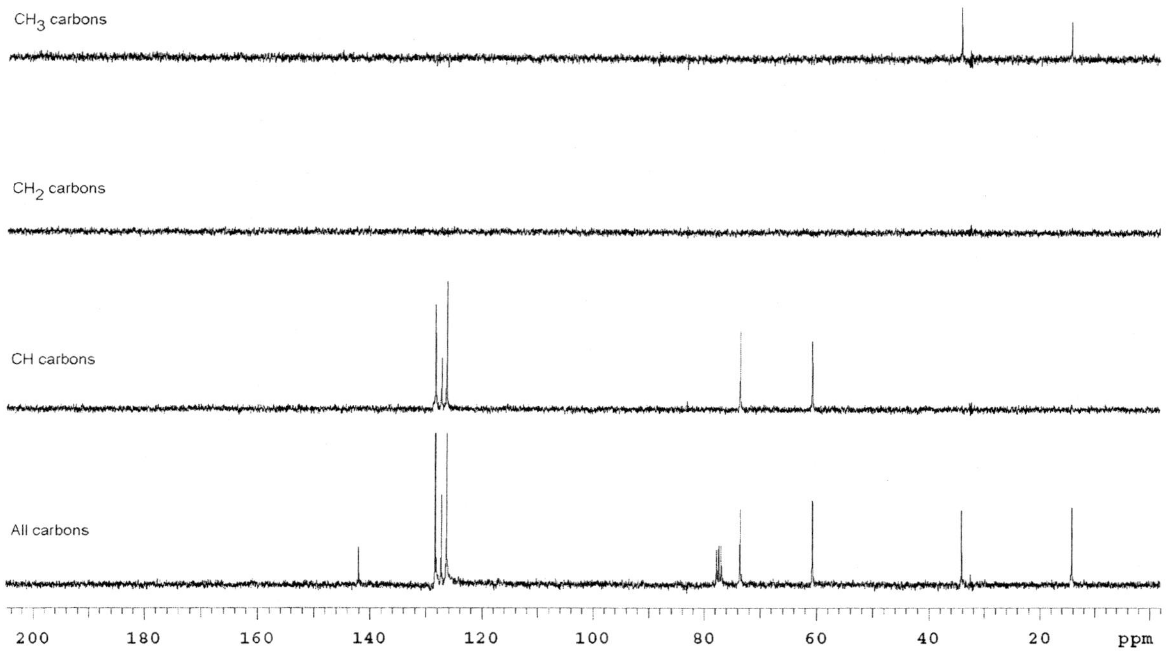

Edited DEPT ^{13}C-NMR Spectrum in $CDCl_3$

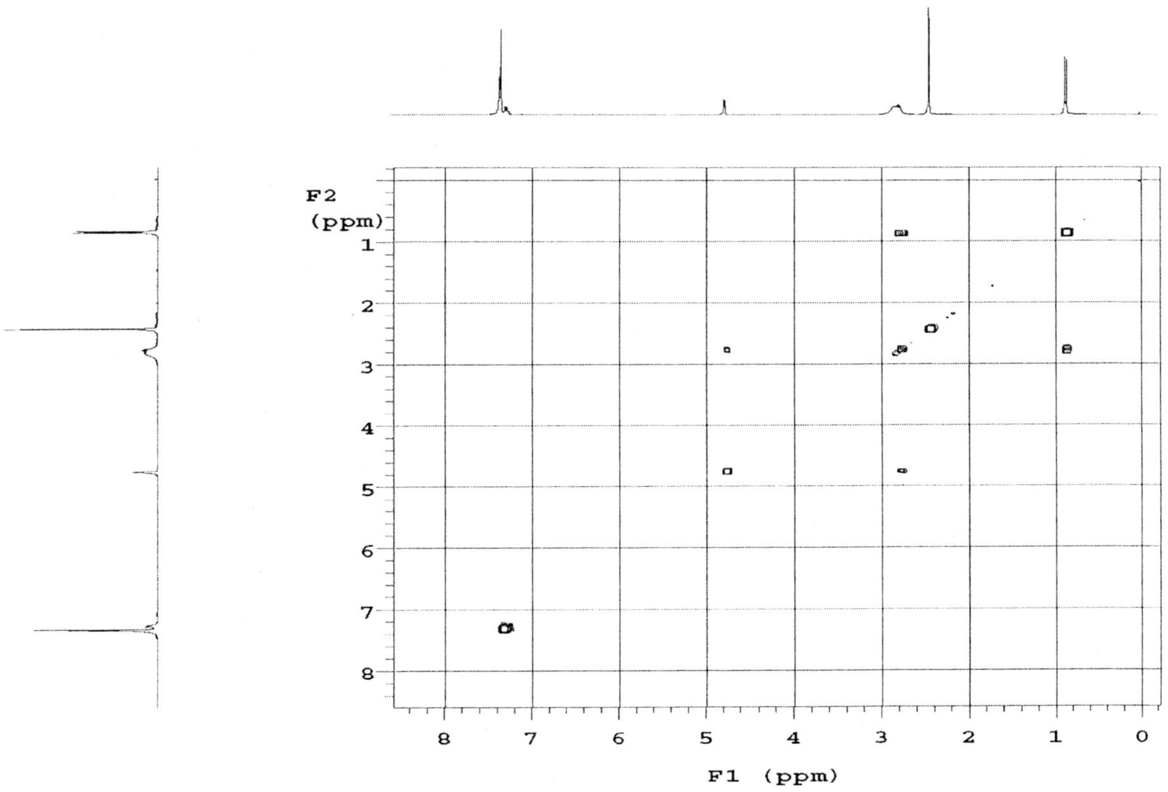

gCOSY Spectrum in $CDCl_3$

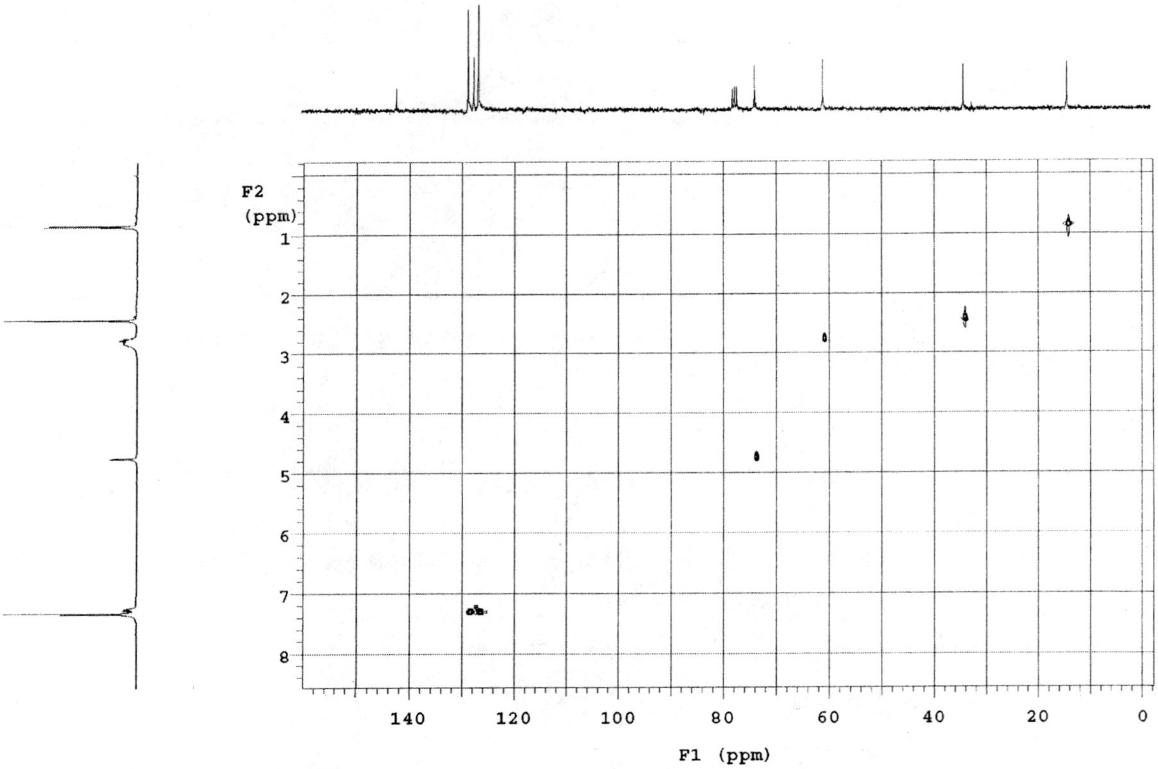

gHSQC Spectrum in $CDCl_3$

Problem 3.21 Given the spectroscopic data below (Proton, Carbon, Edited DEPT, gHSQC, and gCOSY), provide the structure of the molecule. (HR-MS: 268.1576 amu)

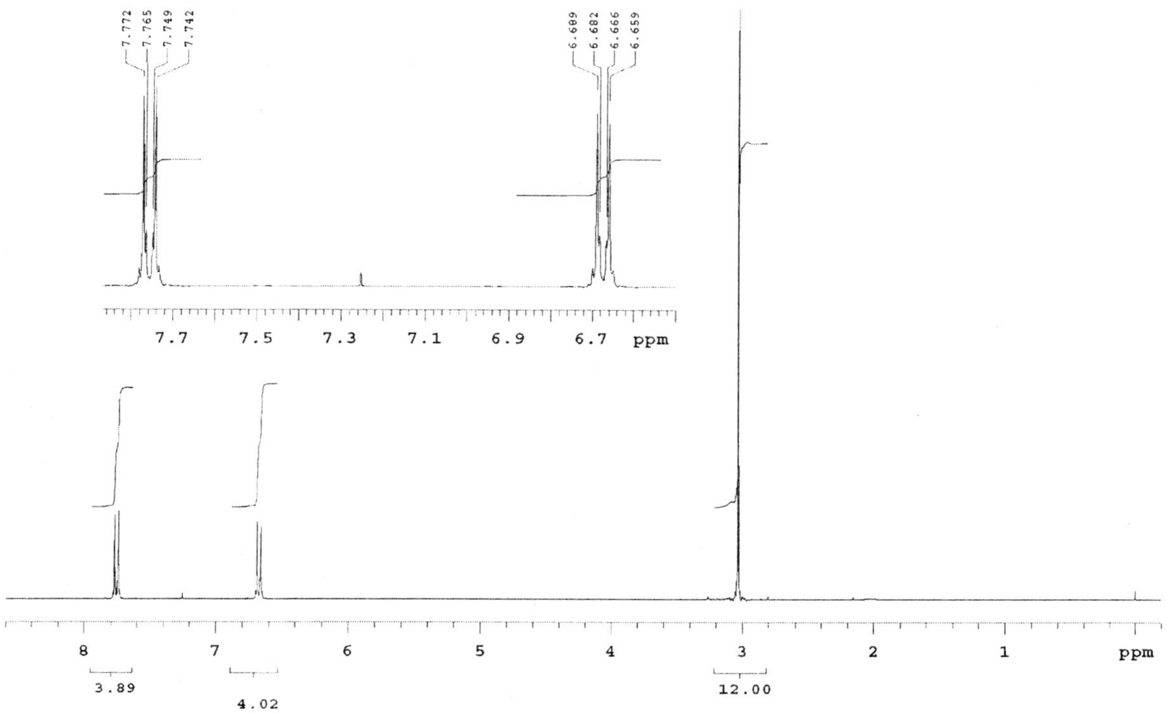

^{1}H-NMR Spectrum in $CDCl_3$

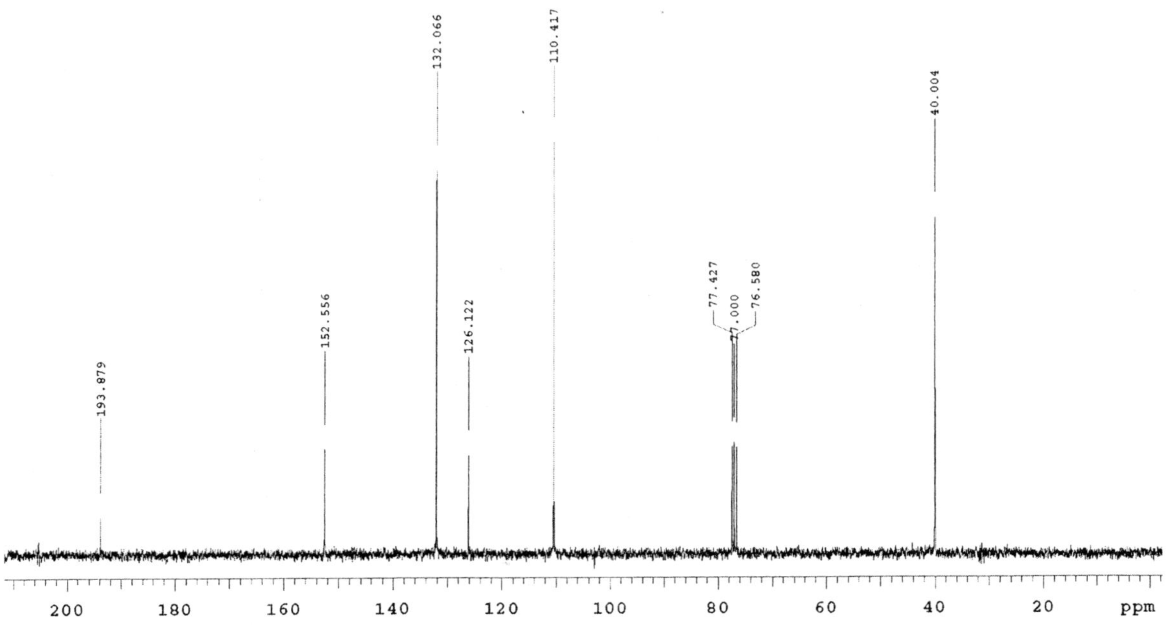

^{13}C-NMR Spectrum in $CDCl_3$

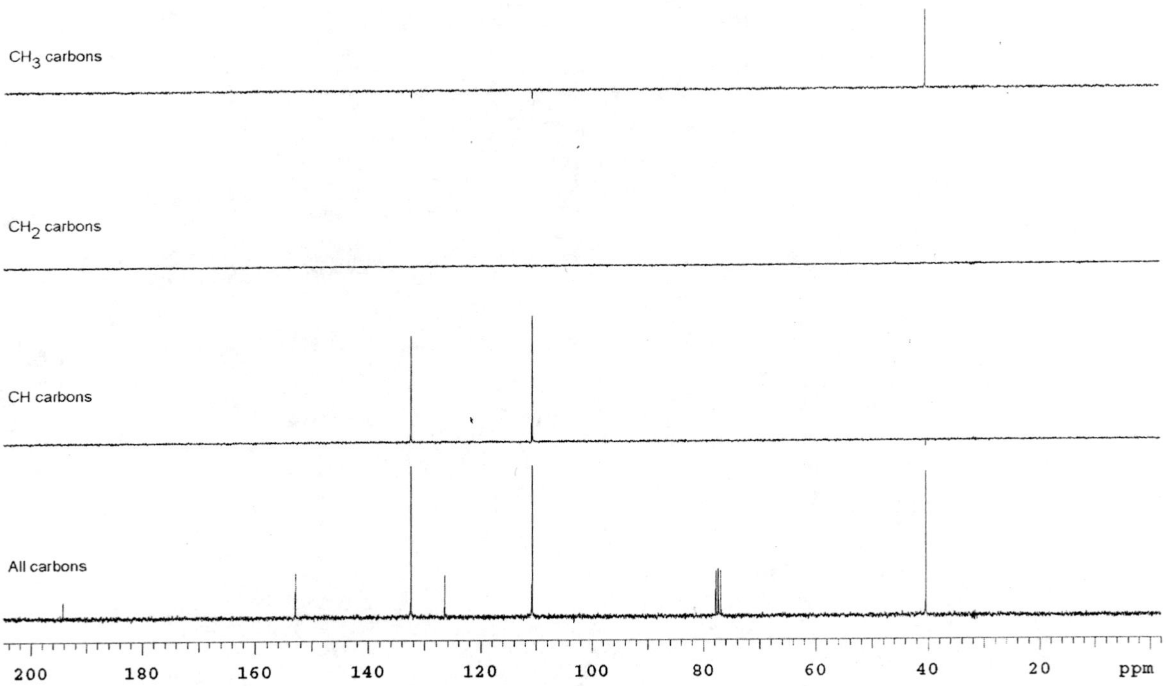

Edited DEPT ^{13}C-NMR Spectrum in $CDCl_3$

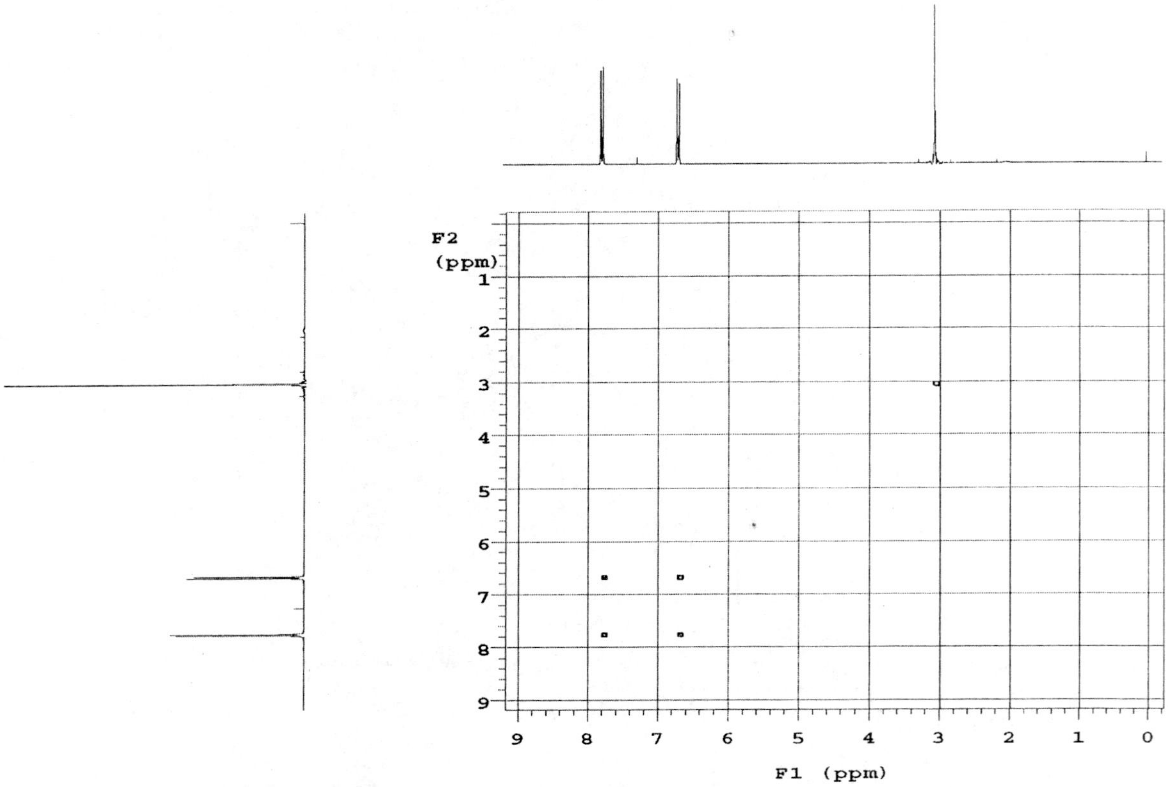

gCOSY Spectrum in $CDCl_3$

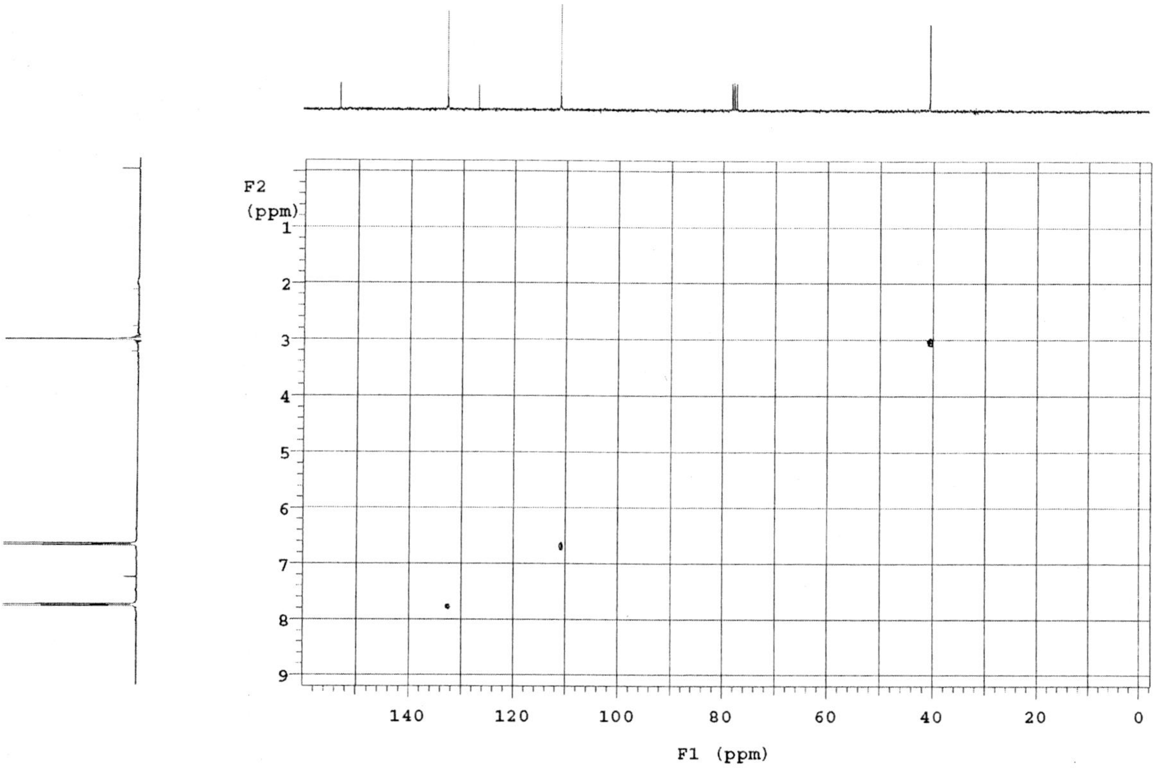

gHSQC Spectrum in $CDCl_3$

Problem 3.22 Given the spectroscopic data below (Proton, Carbon, Edited DEPT, gHSQC, gCOSY, and gHMBC), provide the structure of the molecule.

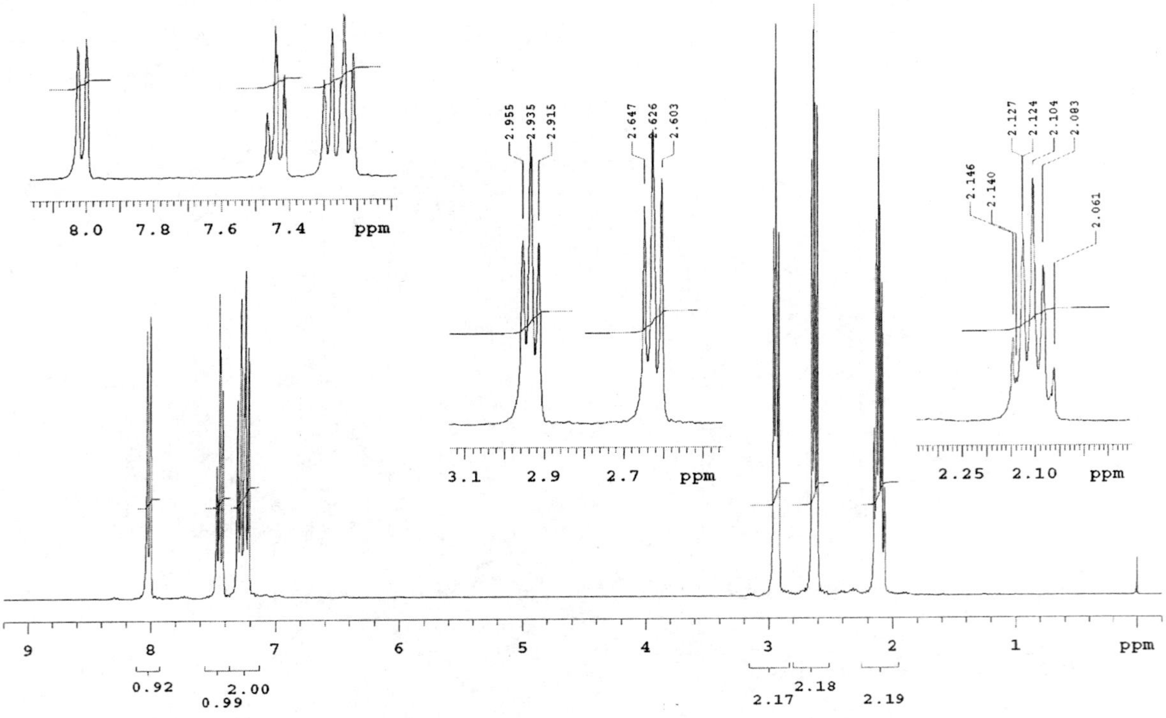

^{1}H-NMR Spectrum in $CDCl_3$

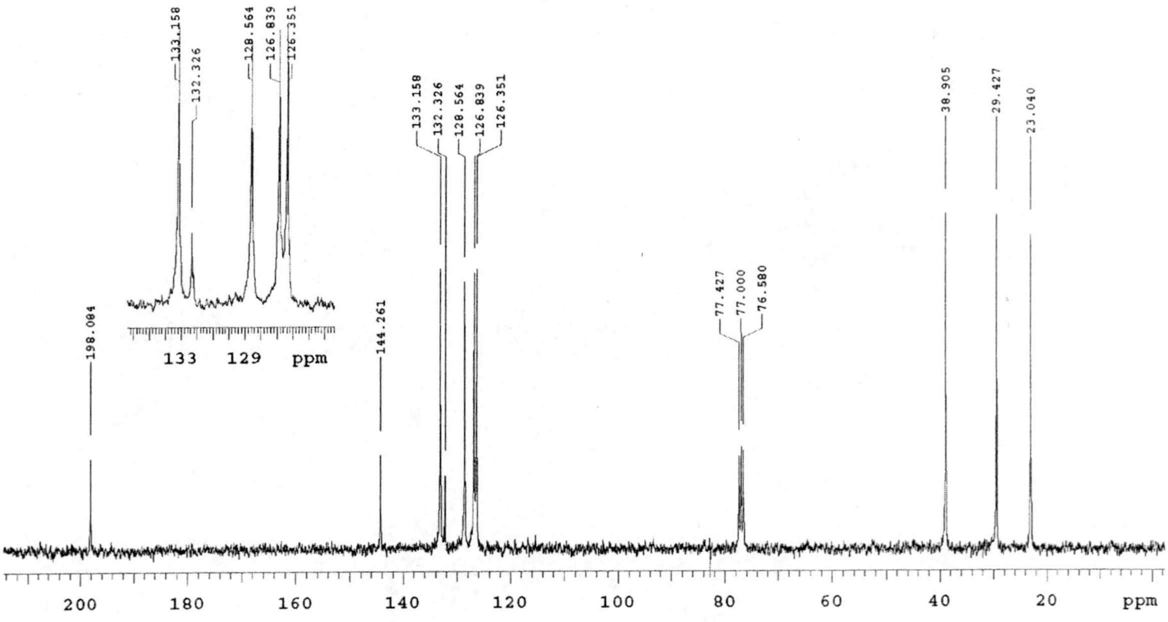

^{13}C-NMR Spectrum in $CDCl_3$

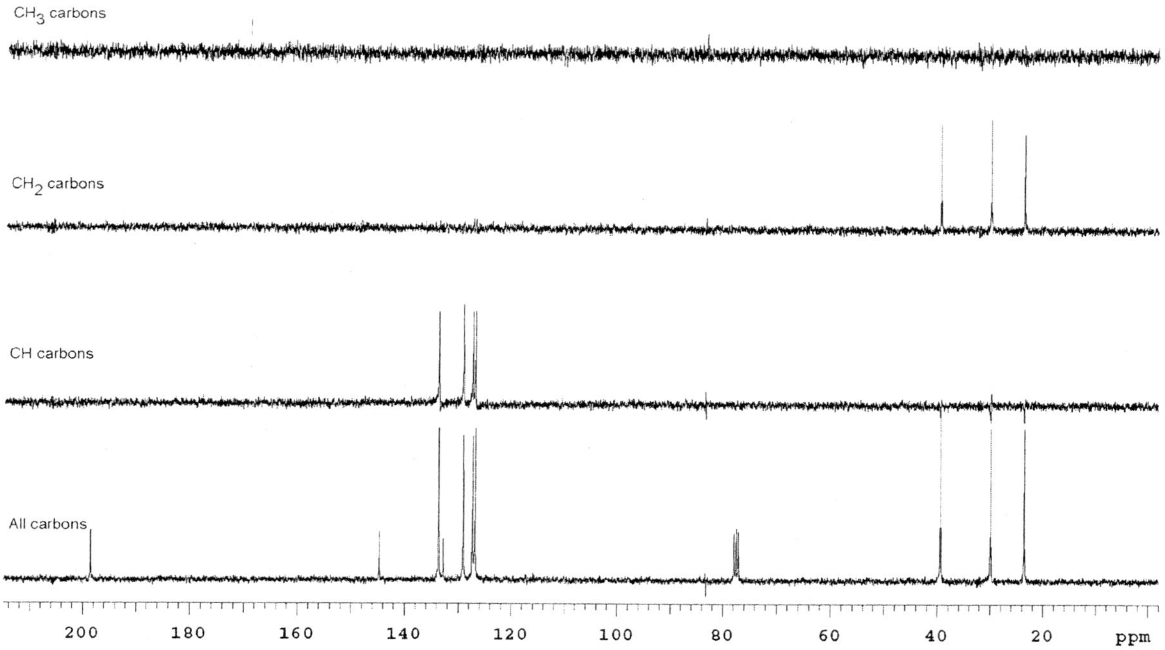

Edited DEPT ^{13}C-NMR Spectrum in $CDCl_3$

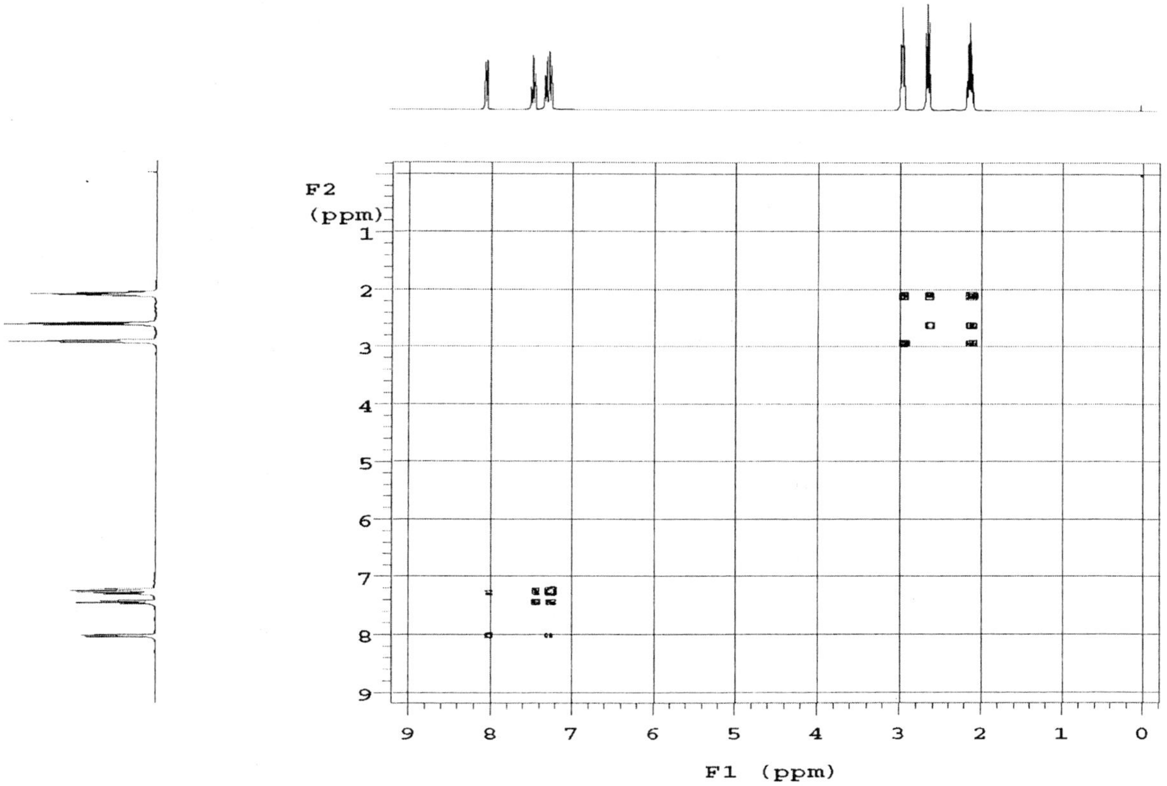

gCOSY Spectrum in $CDCl_3$

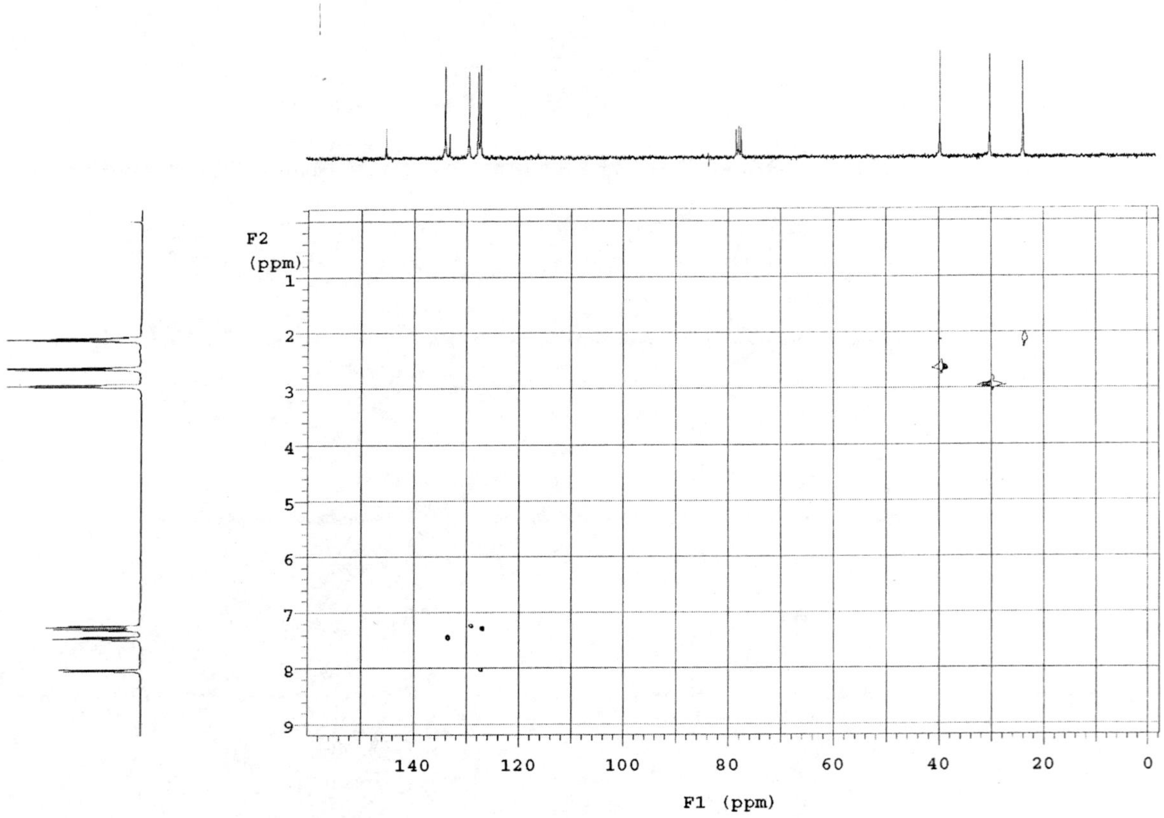

gHSQC Spectrum in $CDCl_3$

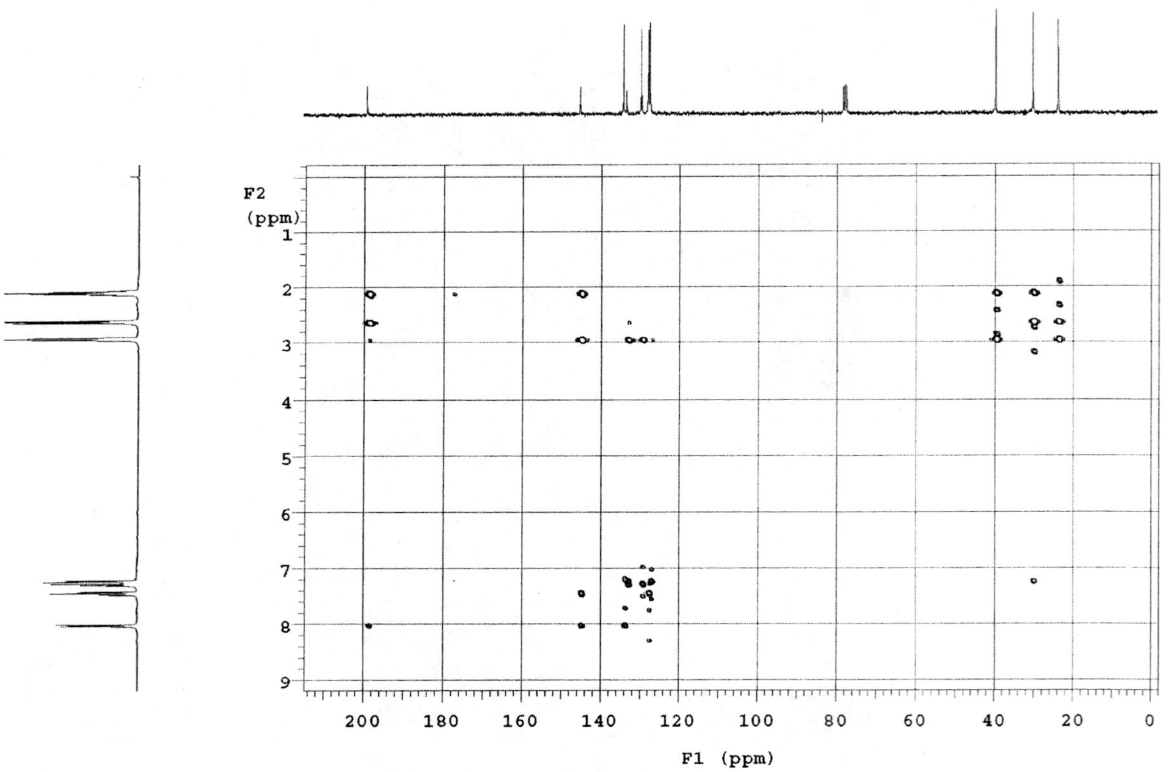

gHMBC Spectrum in $CDCl_3$

Problem 3.23 Given the spectroscopic data below (Proton, Carbon, Edited DEPT, gHSQC, gCOSY and gHMBC), provide the structure of the molecule. (MW: 143 g/mol)

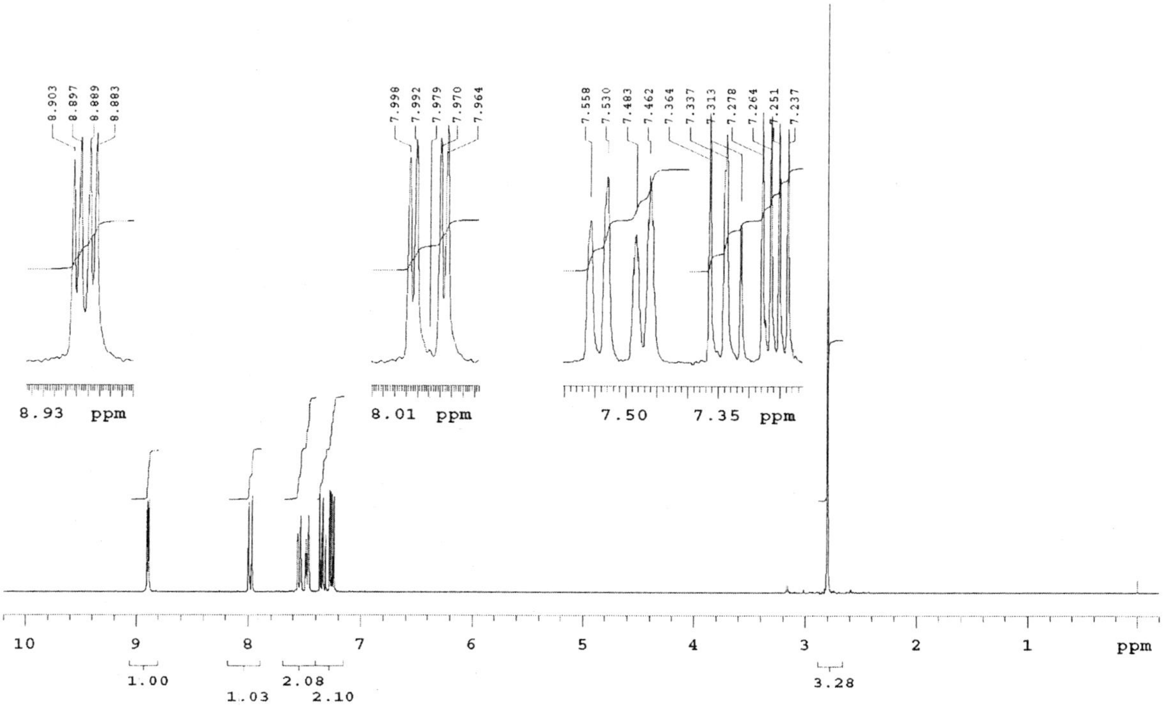

^{1}H-NMR Spectrum in $CDCl_3$

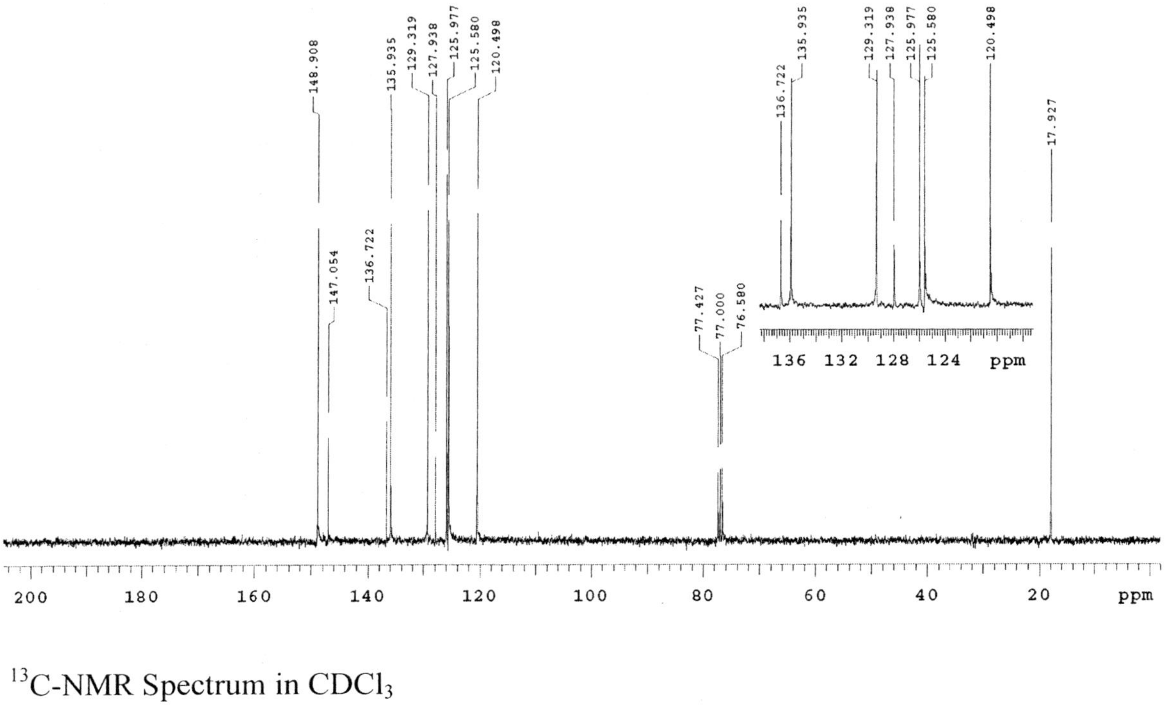

^{13}C-NMR Spectrum in $CDCl_3$

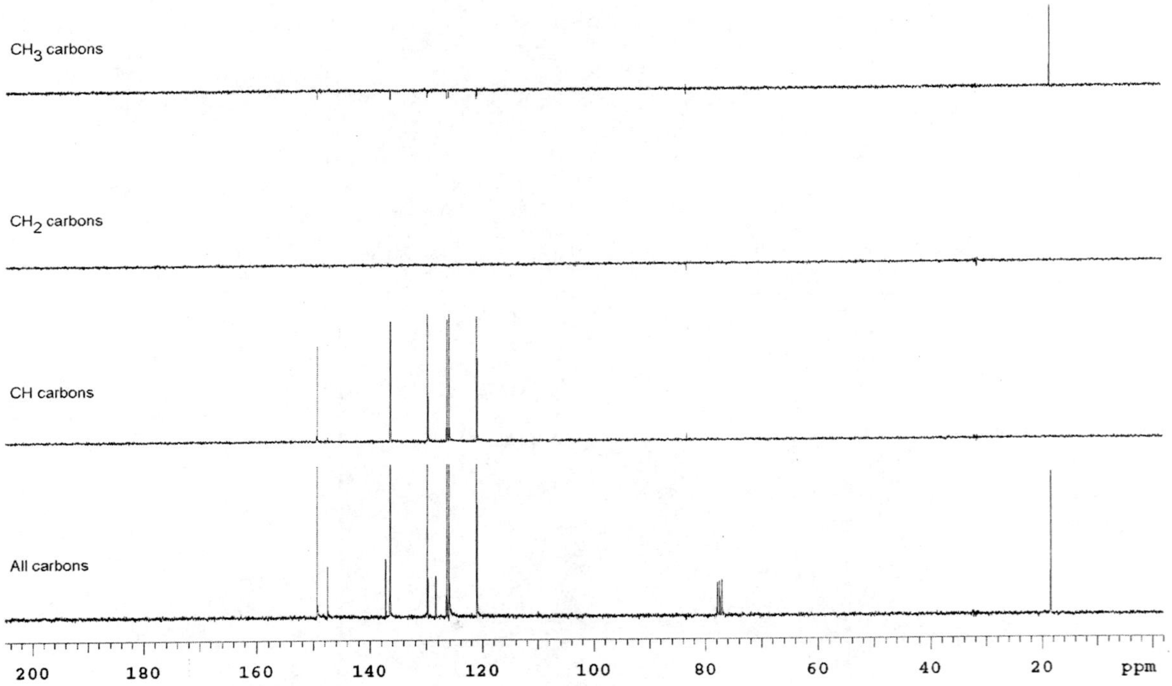

Edited DEPT ^{13}C-NMR Spectrum in $CDCl_3$

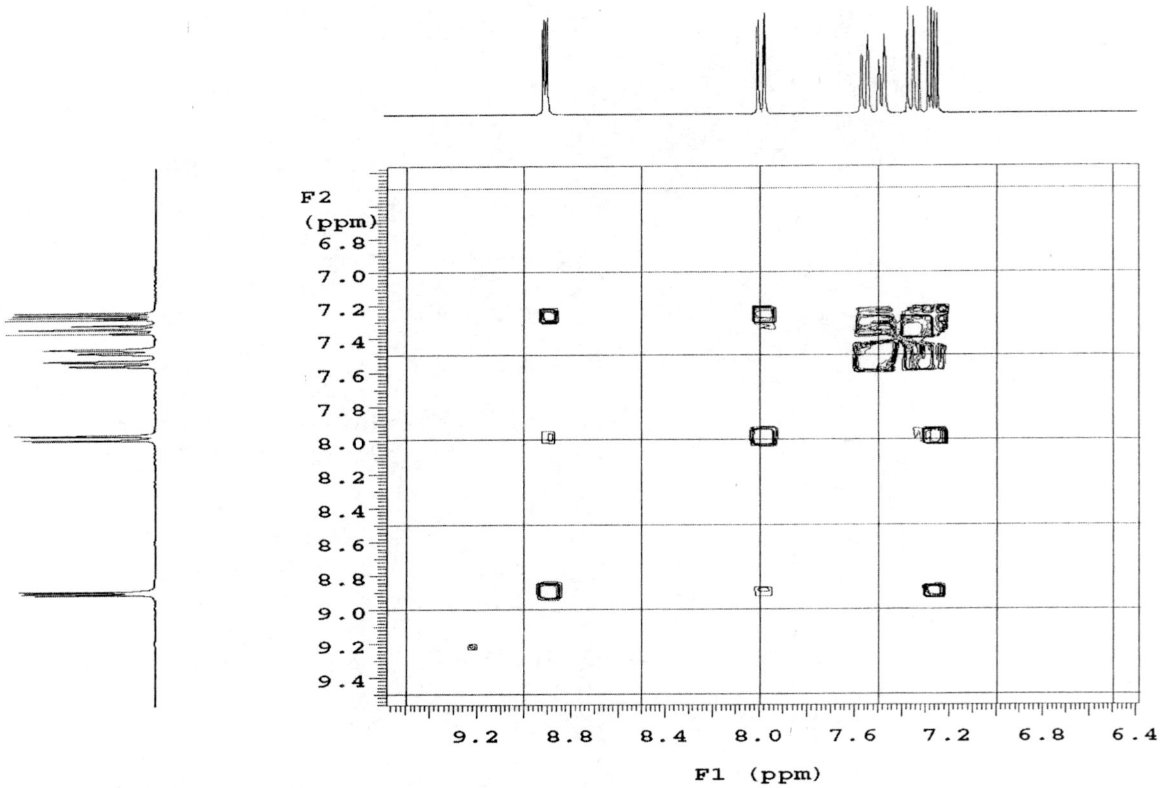

gCOSY Spectrum in $CDCl_3$

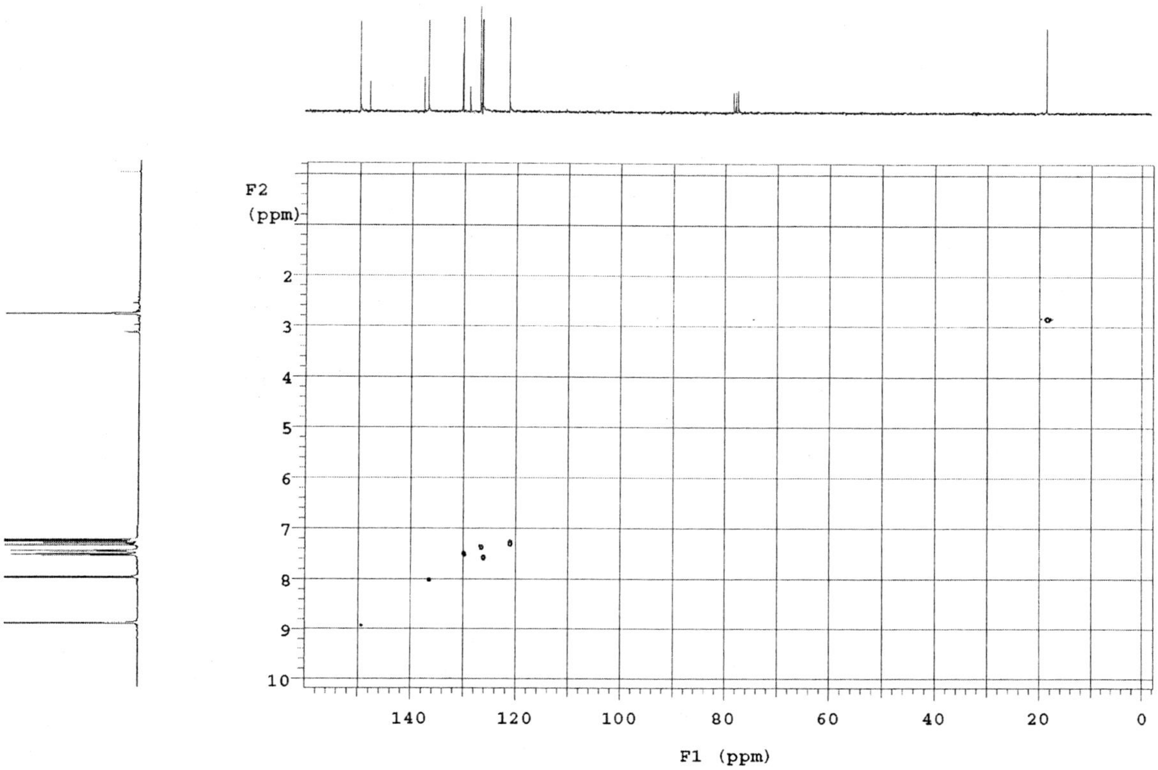

gHSQC Spectrum in $CDCl_3$

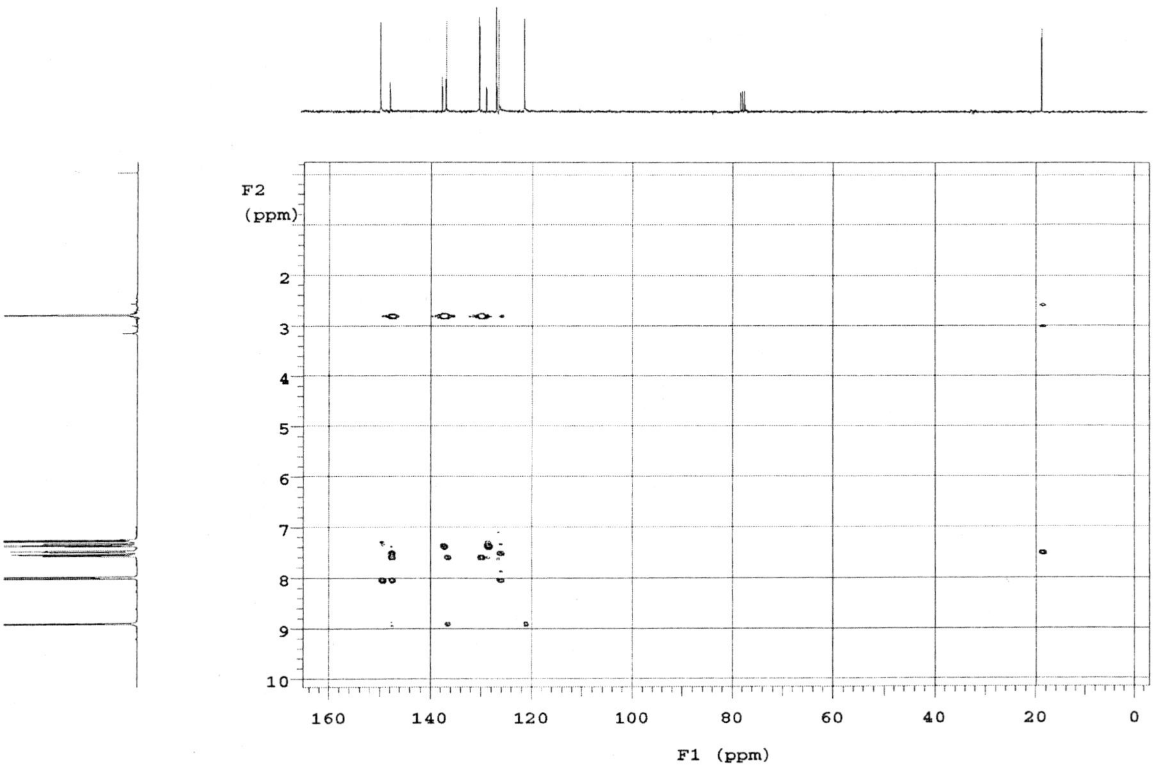

gHMBC Spectrum in $CDCl_3$

Problem 3.24 Given the spectroscopic data below (Proton, Carbon, Edited DEPT, gHSQC, gCOSY and gHMBC), provide the structure of the molecule. (MW: 143 g/mol)

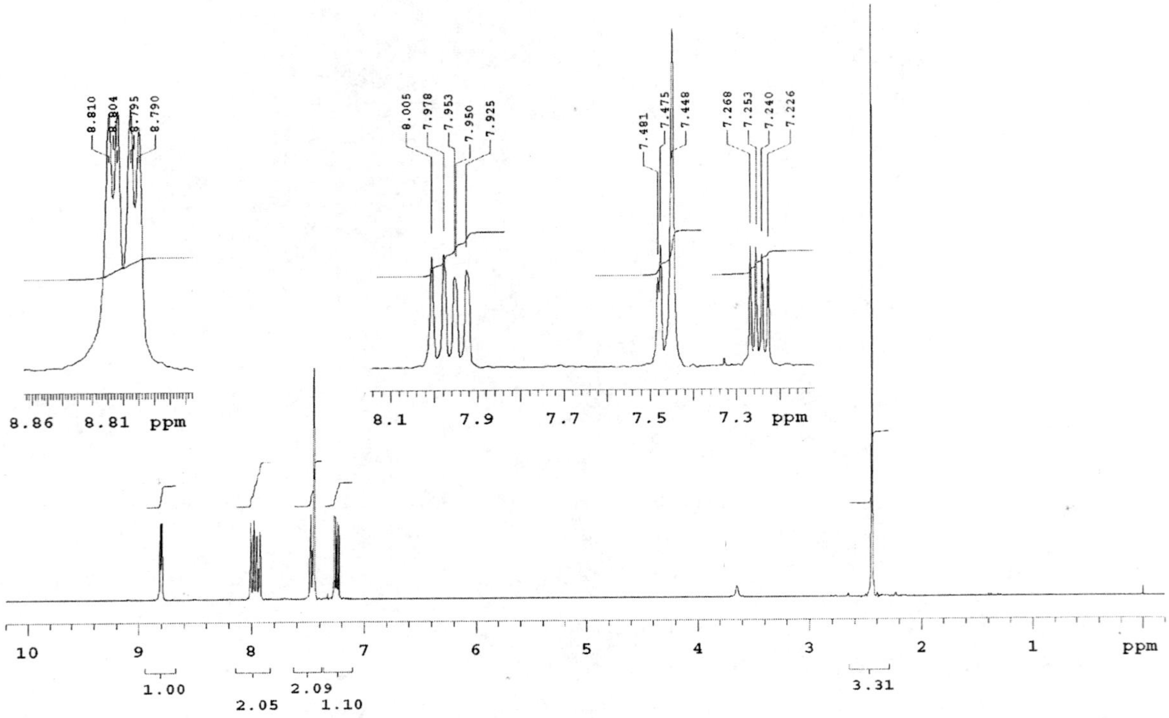

^{1}H-NMR Spectrum in $CDCl_3$

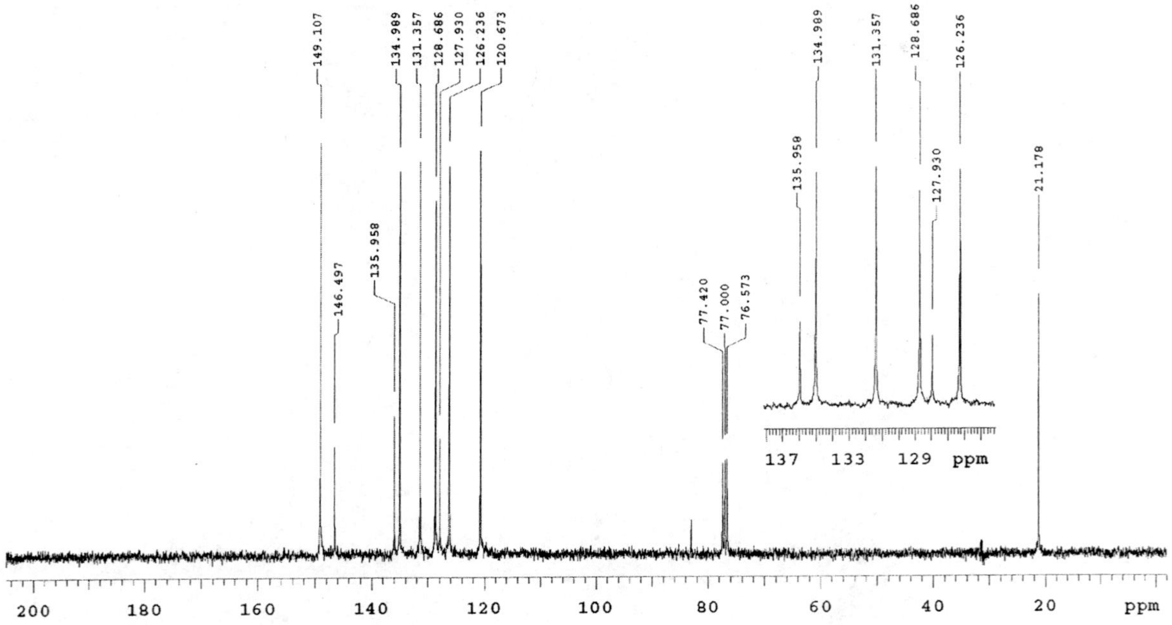

^{13}C-NMR Spectrum in $CDCl_3$

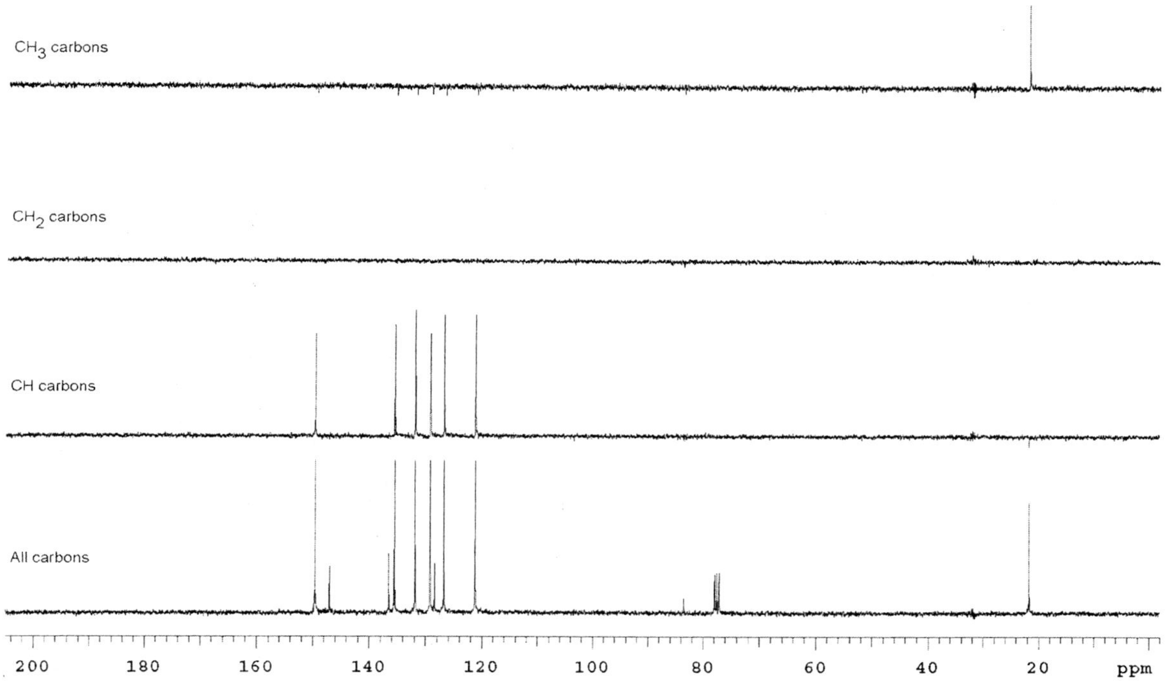

Edited DEPT ^{13}C-NMR Spectrum in $CDCl_3$

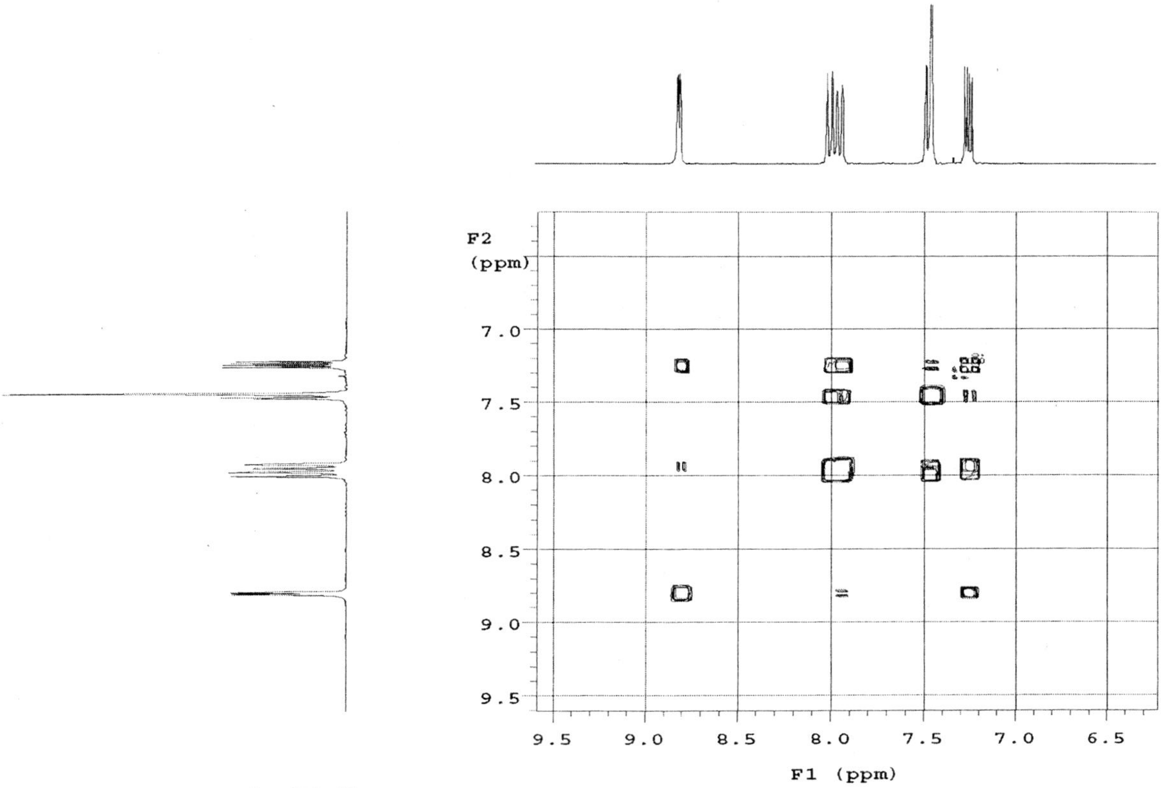

gCOSY Spectrum in $CDCl_3$

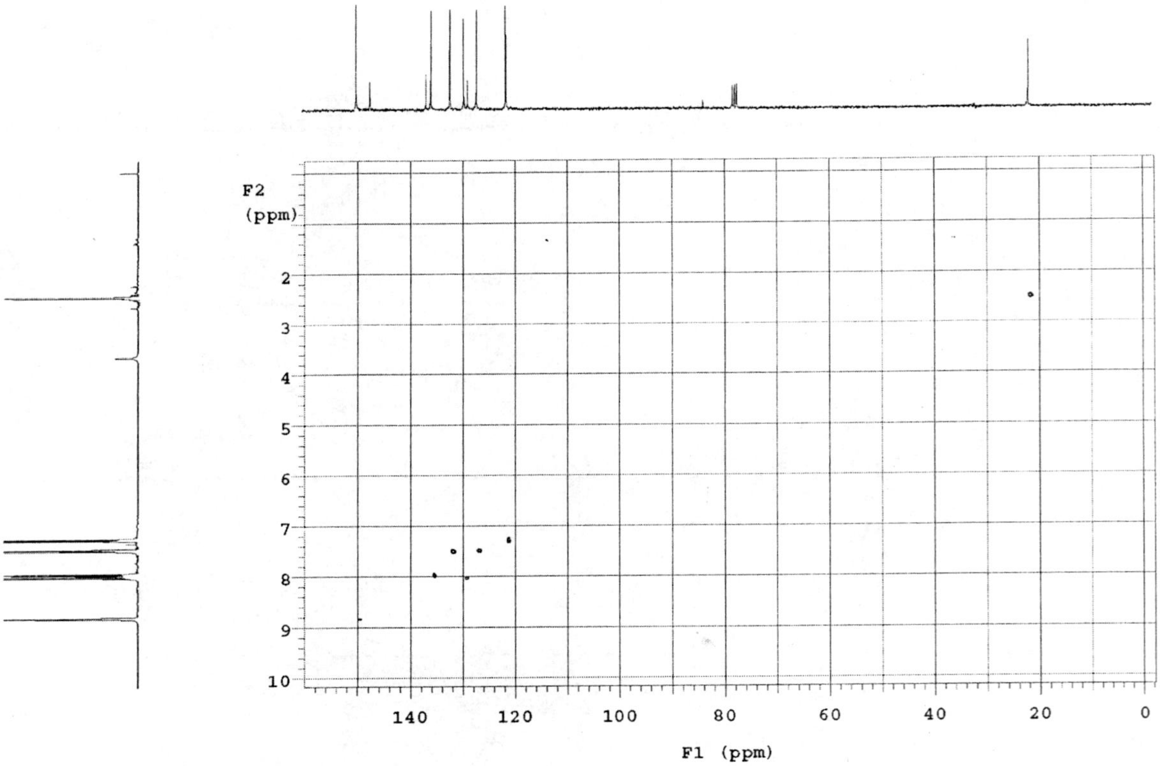

gHSQC Spectrum in $CDCl_3$

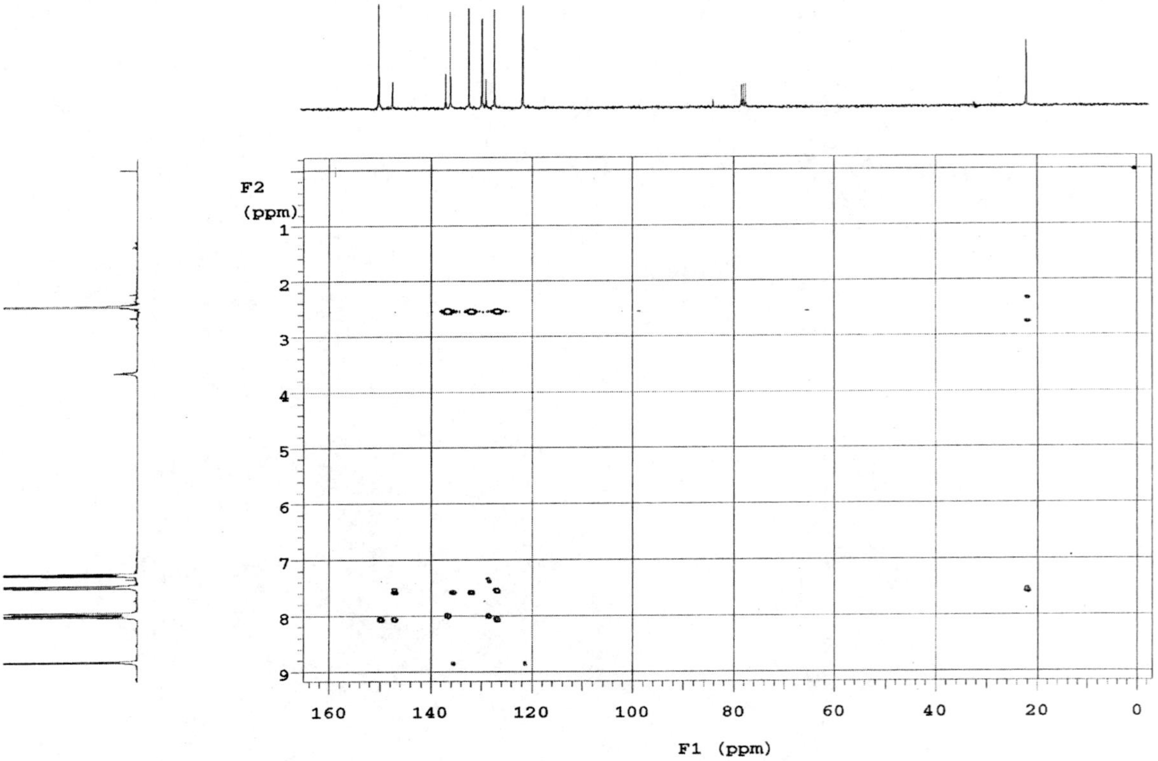

gHMBC Spectrum in $CDCl_3$

Problem 3.25 Given the spectroscopic data below (Proton, Carbon, Edited DEPT, gHSQC, and gCOSY), provide the structure of the molecule.

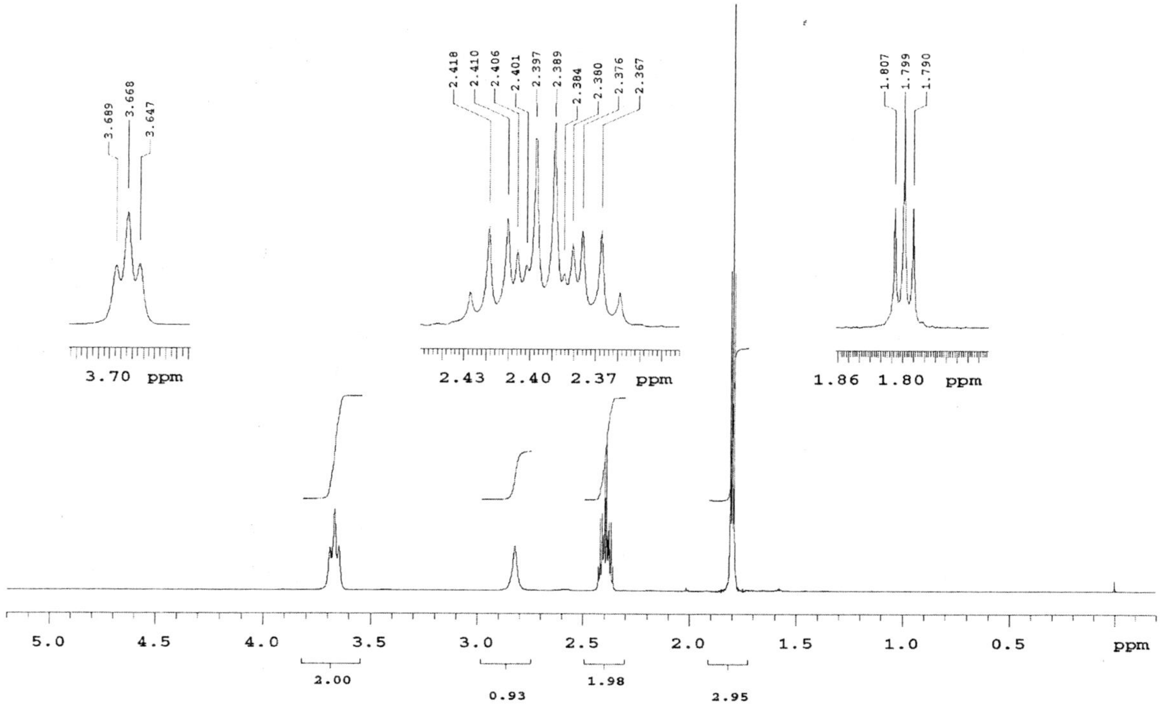

^{1}H-NMR Spectrum in $CDCl_3$

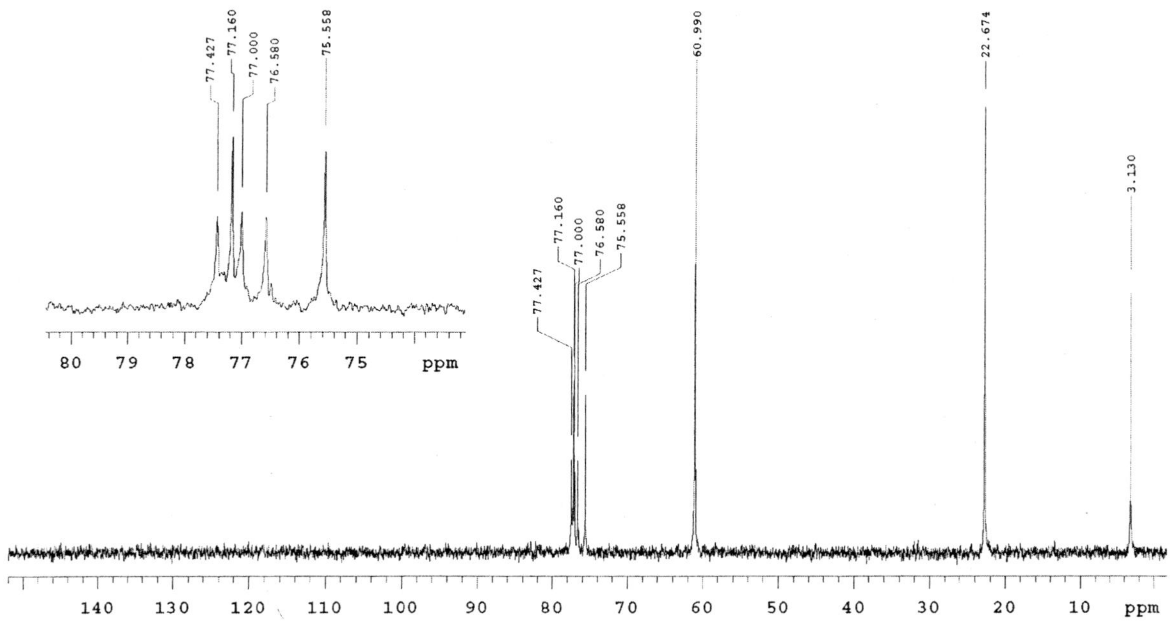

^{13}C-NMR Spectrum in $CDCl_3$

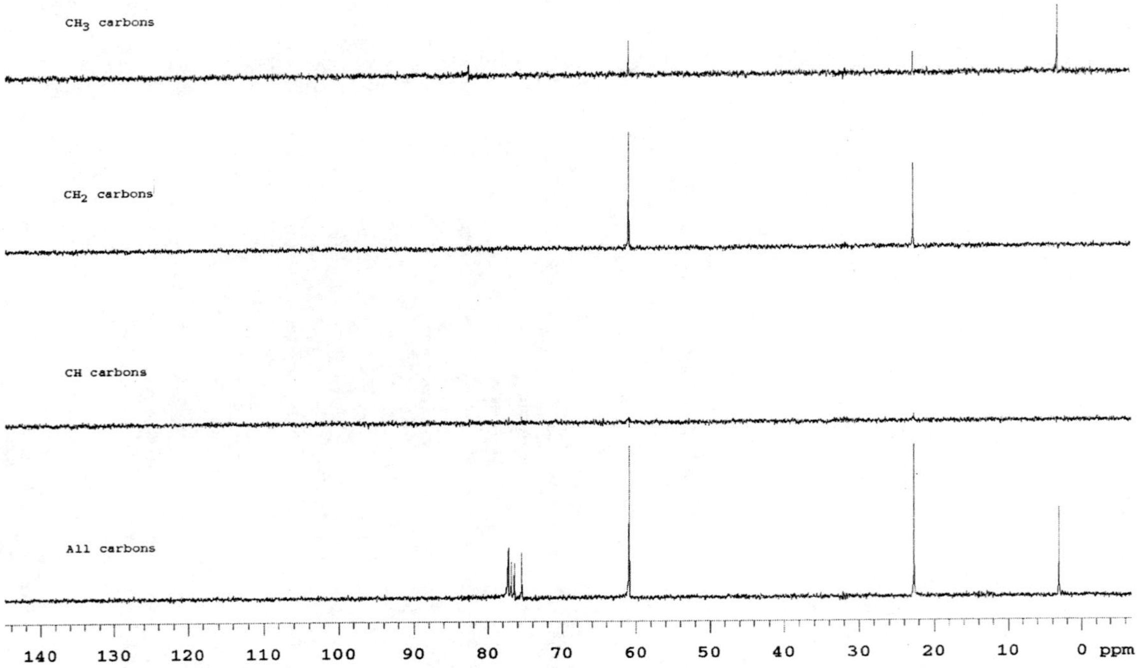

Edited DEPT ^{13}C-NMR Spectrum in $CDCl_3$

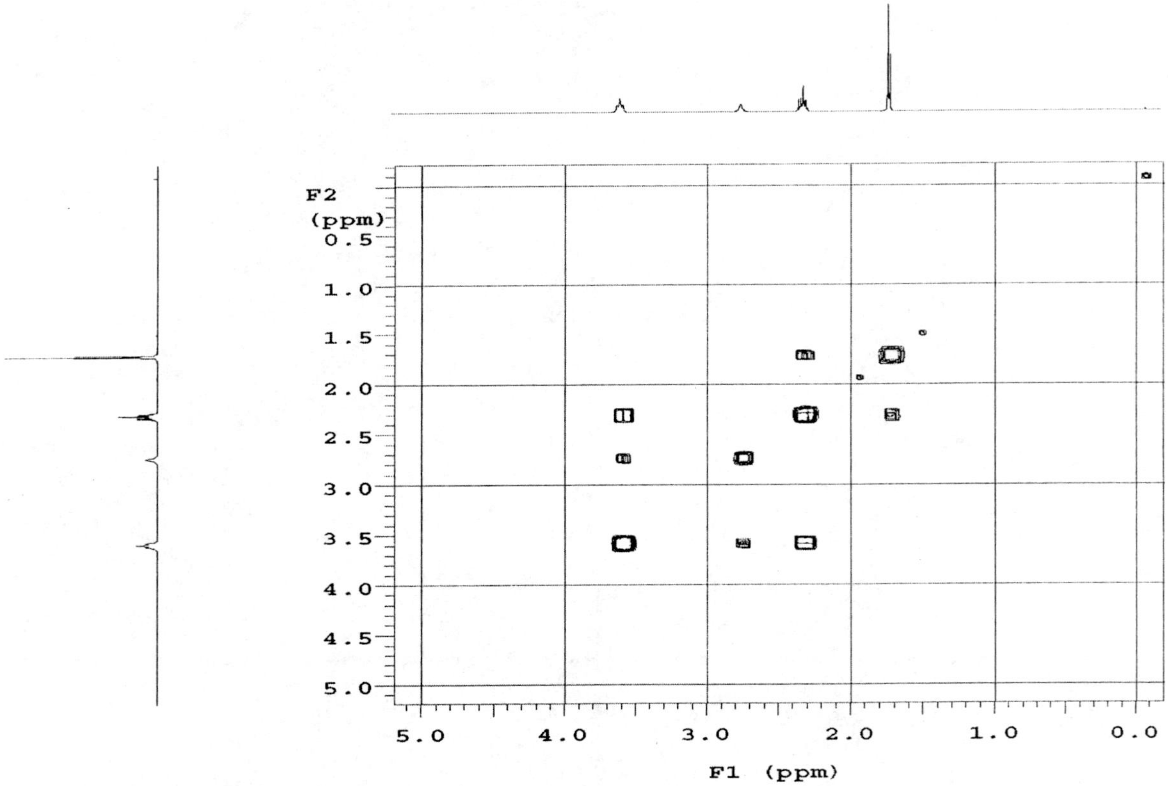

gCOSY Spectrum in $CDCl_3$

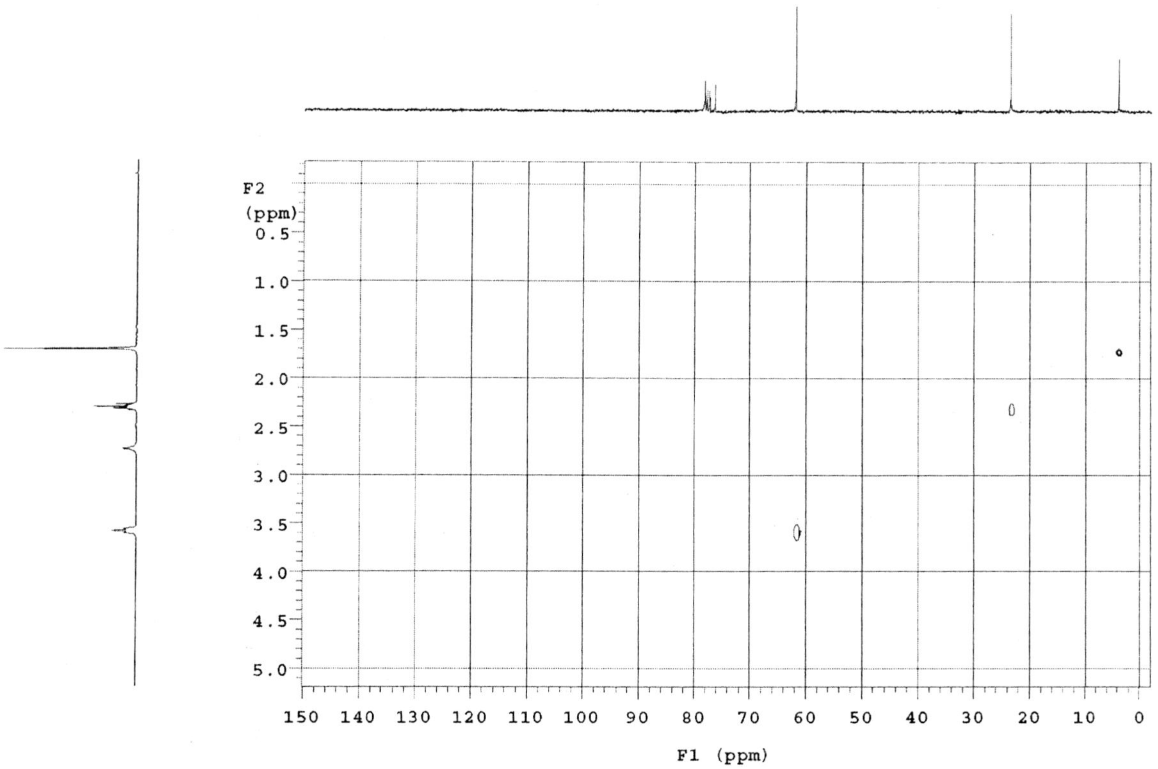

gHSQC Spectrum in $CDCl_3$

Problem 3.26 Given the spectroscopic data below (Proton, Carbon, Edited DEPT, gHSQC, and gCOSY), provide the structure of the molecule. (MW: 144 g/mol)

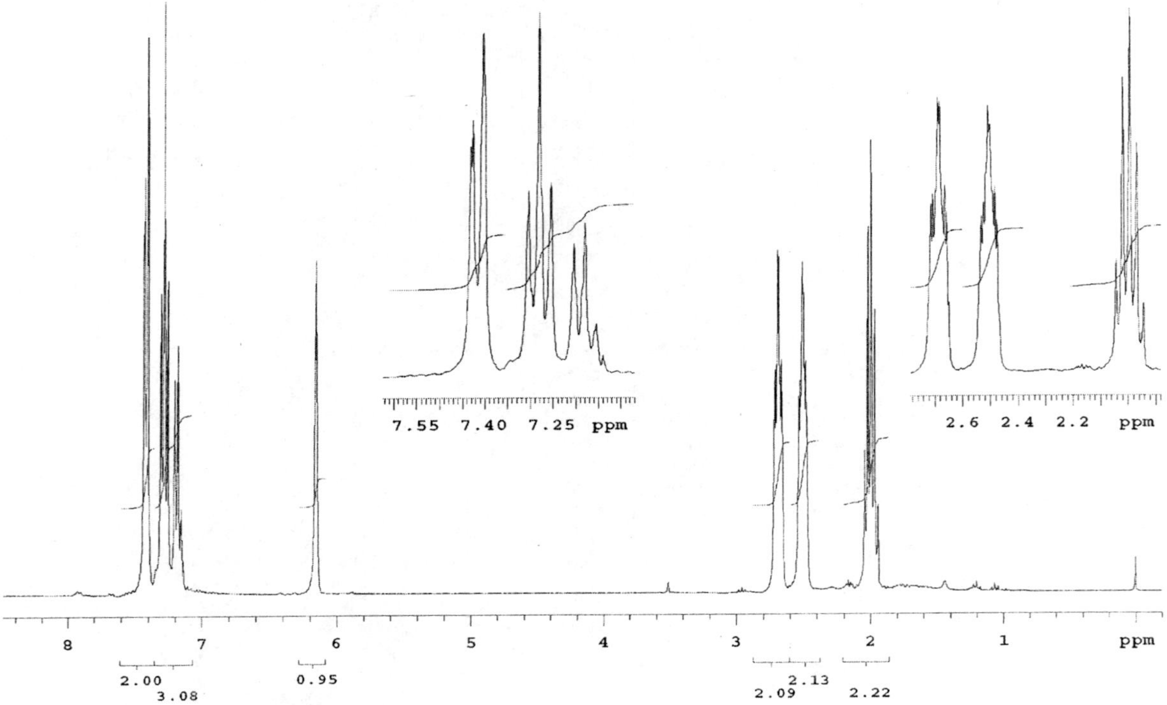

^{1}H-NMR Spectrum in $CDCl_3$

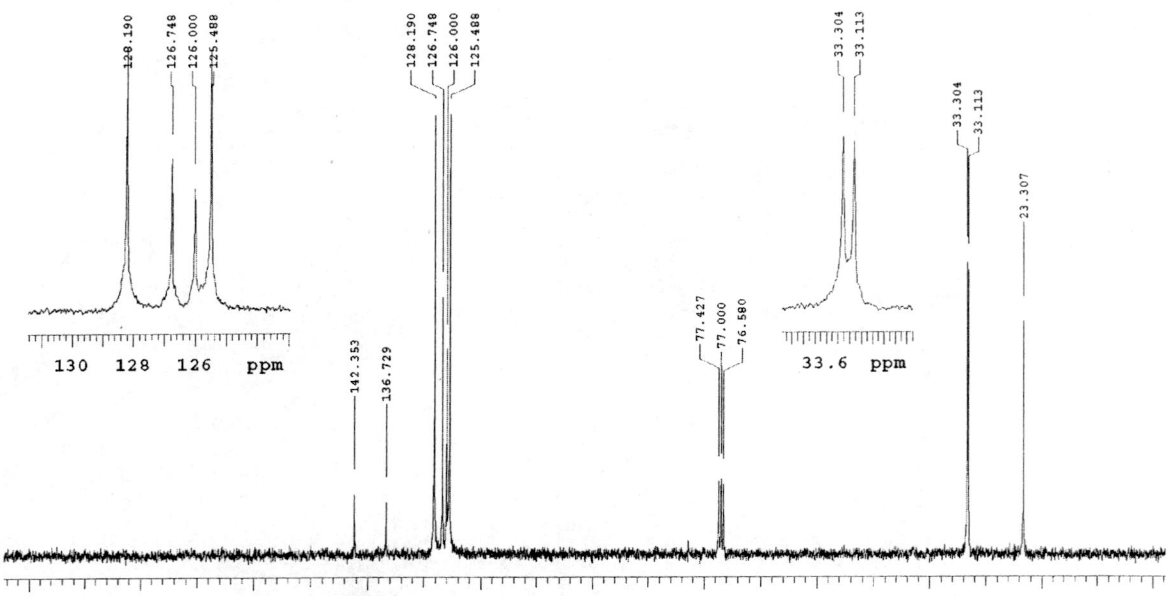

^{13}C-NMR Spectrum in $CDCl_3$

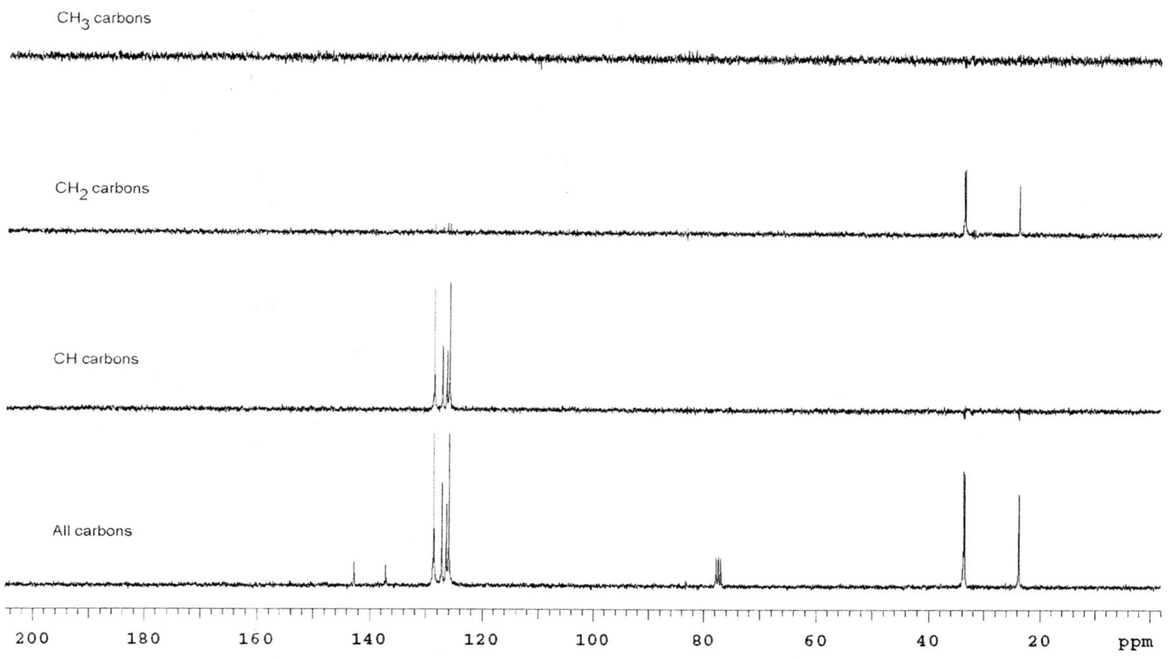

Edited DEPT ^{13}C-NMR Spectrum in $CDCl_3$

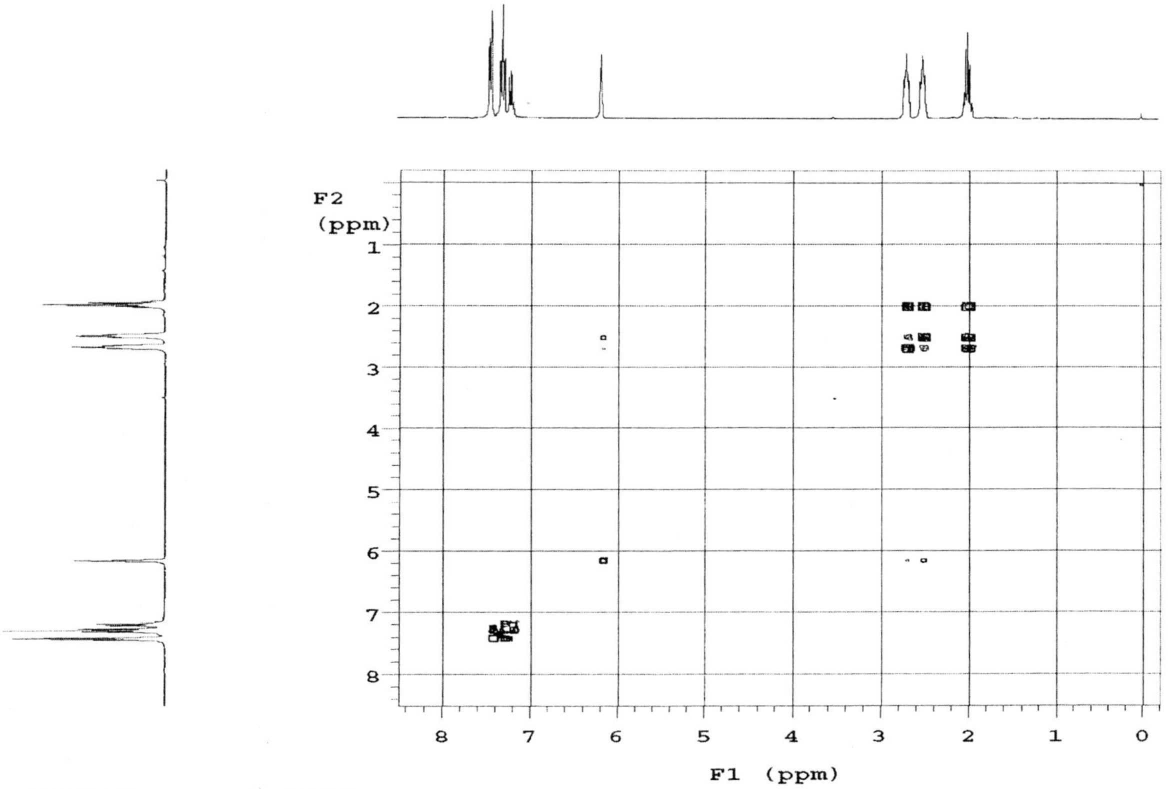

gCOSY Spectrum in $CDCl_3$

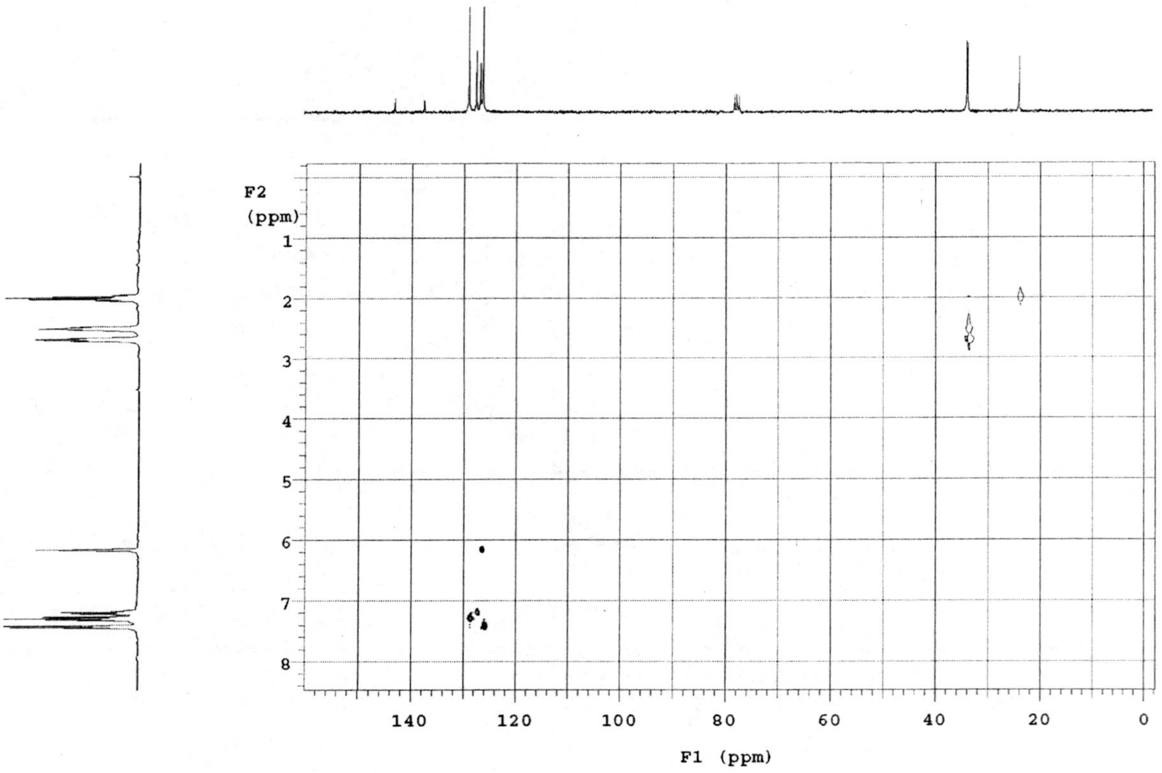

gHSQC Spectrum in $CDCl_3$

Problem 3.27 Given the spectroscopic data below (Proton, Carbon, Edited DEPT, gHSQC, and gCOSY), provide the structure of the molecule.

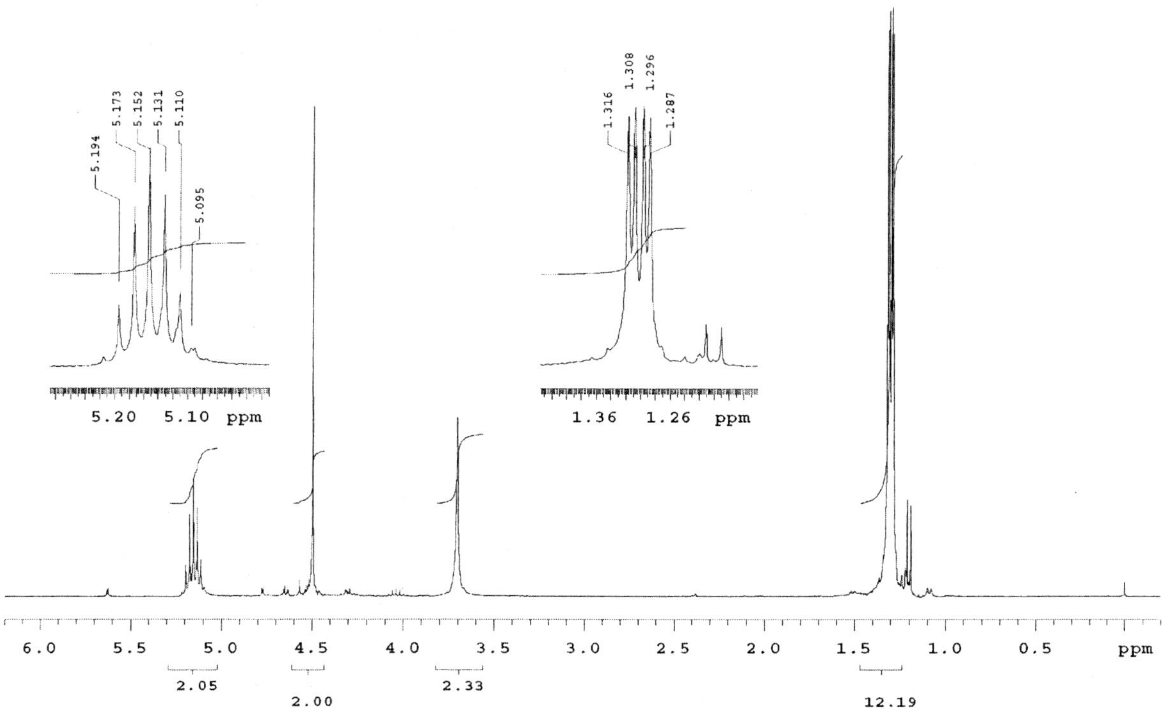

^{1}H-NMR Spectrum in $CDCl_3$

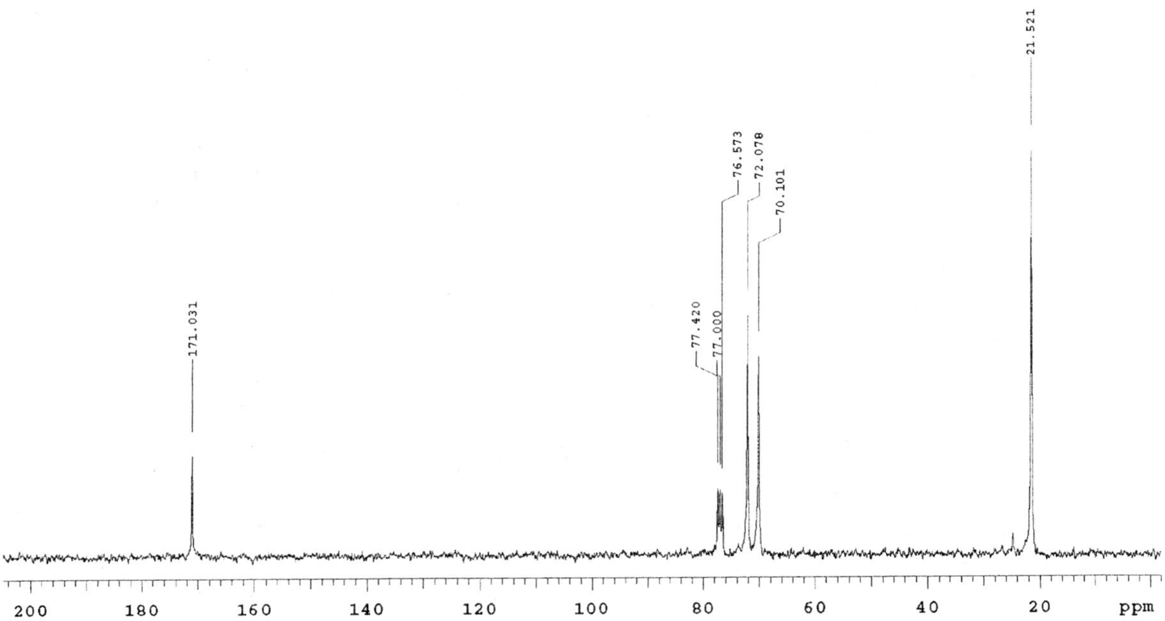

^{13}C-NMR Spectrum in $CDCl_3$

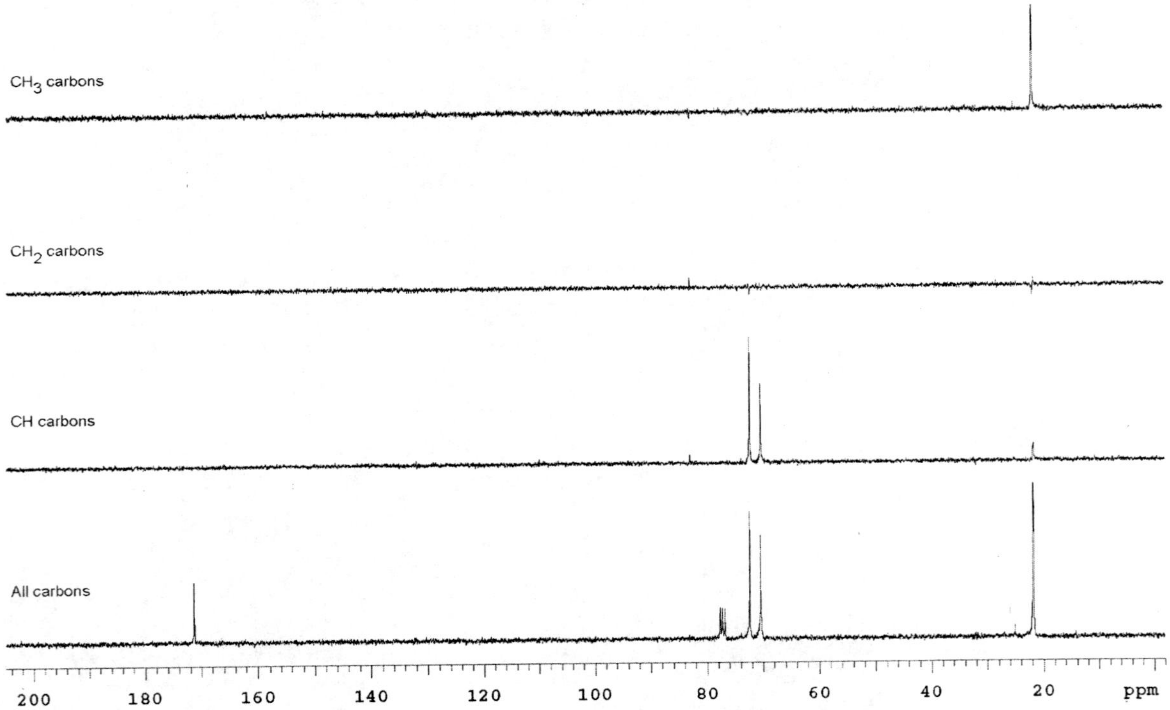

Edited DEPT ^{13}C-NMR Spectrum in $CDCl_3$

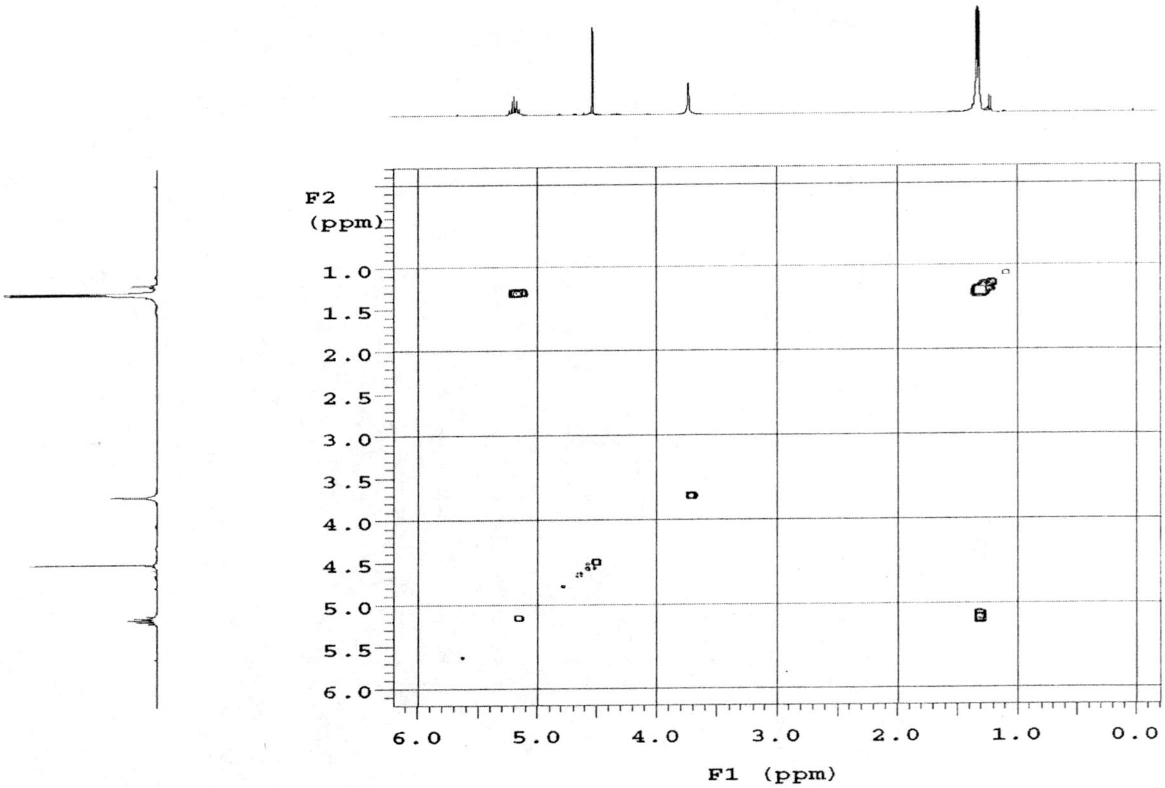

gCOSY Spectrum in $CDCl_3$

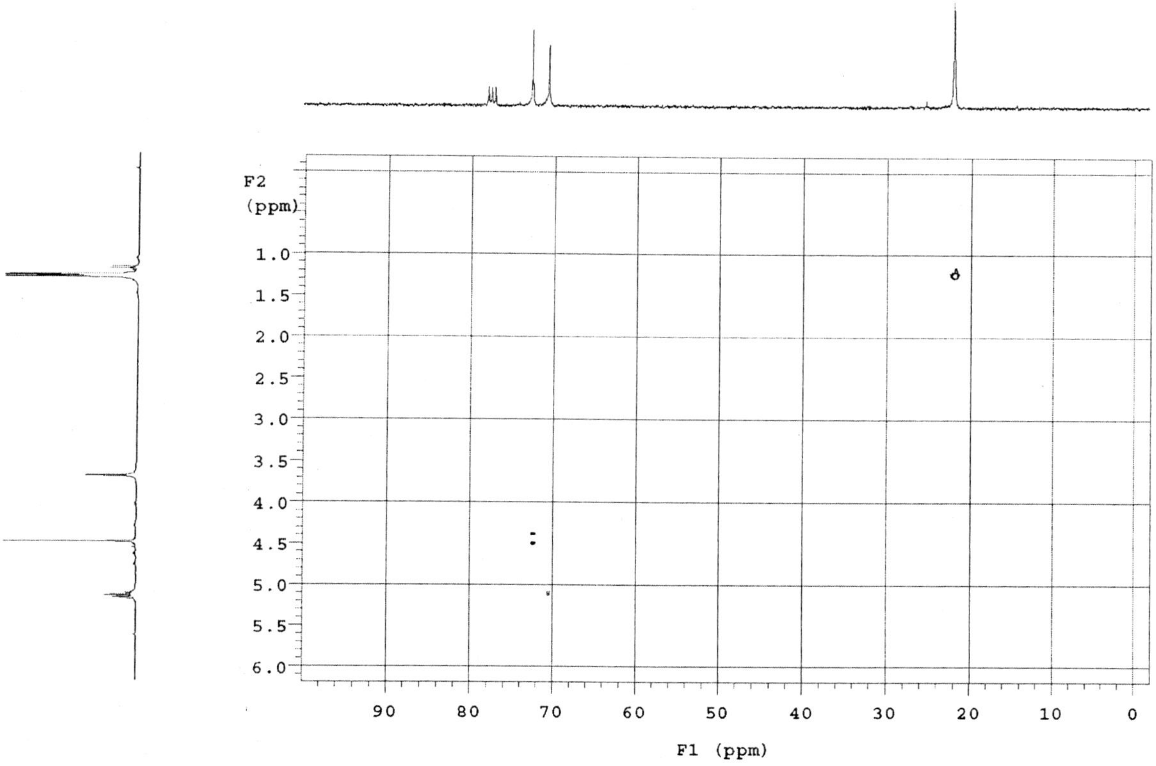

gHSQC Spectrum in $CDCl_3$

Problem 3.28 Given the spectroscopic data below (Proton, Carbon, gHSQC, and Fluorine), provide the structure of the molecule. (MW 174 & 176 g/mol) The carbon signal centered at ~159 ppm has no correlation in the gHSQC spectrum.

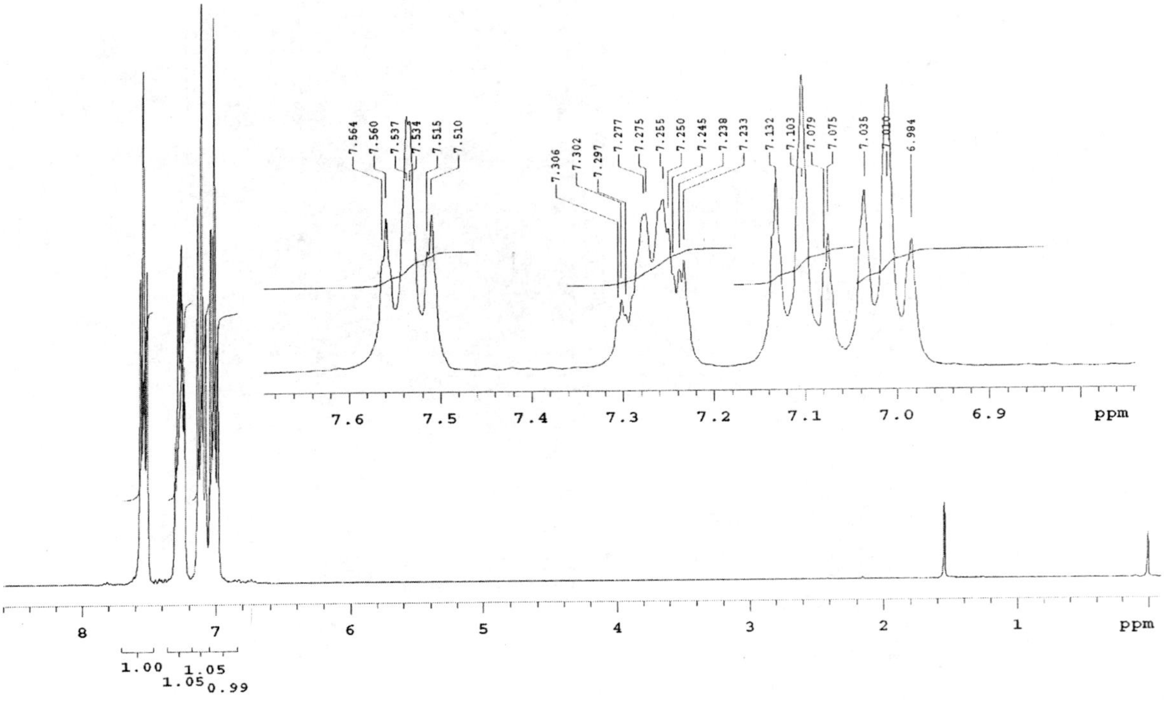

^{1}H-NMR Spectrum in $CDCl_3$

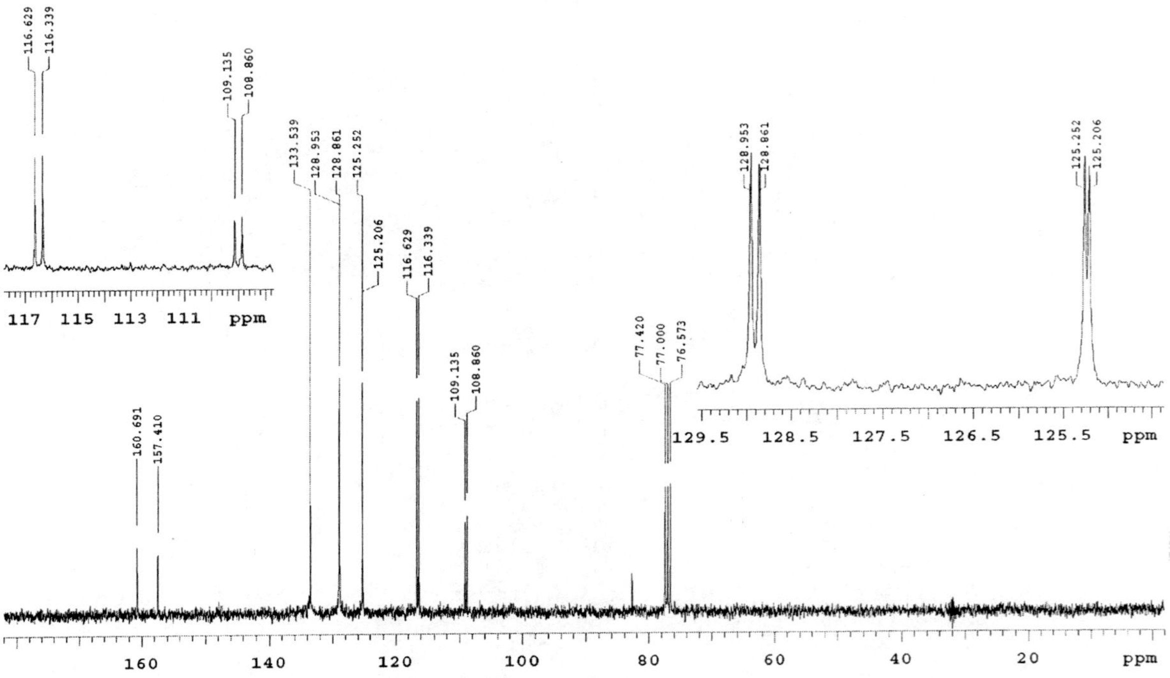

^{13}C-NMR Spectrum in $CDCl_3$

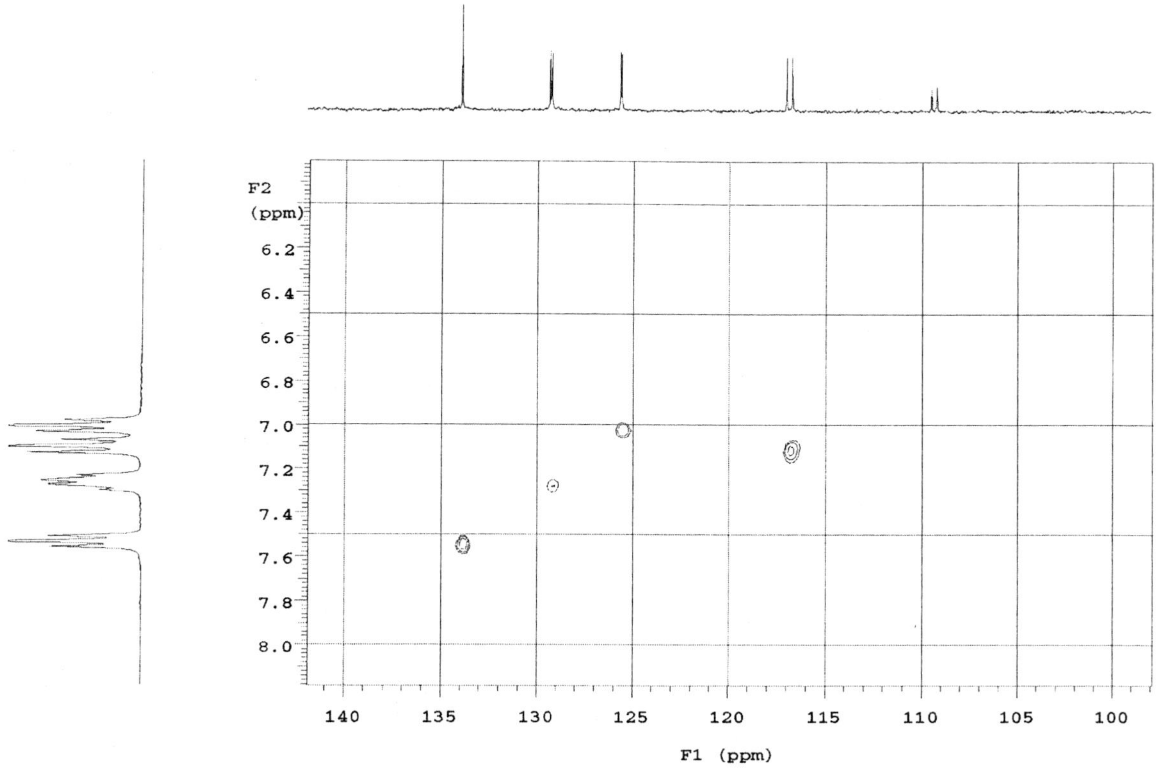

gHSQC Spectrum in $CDCl_3$

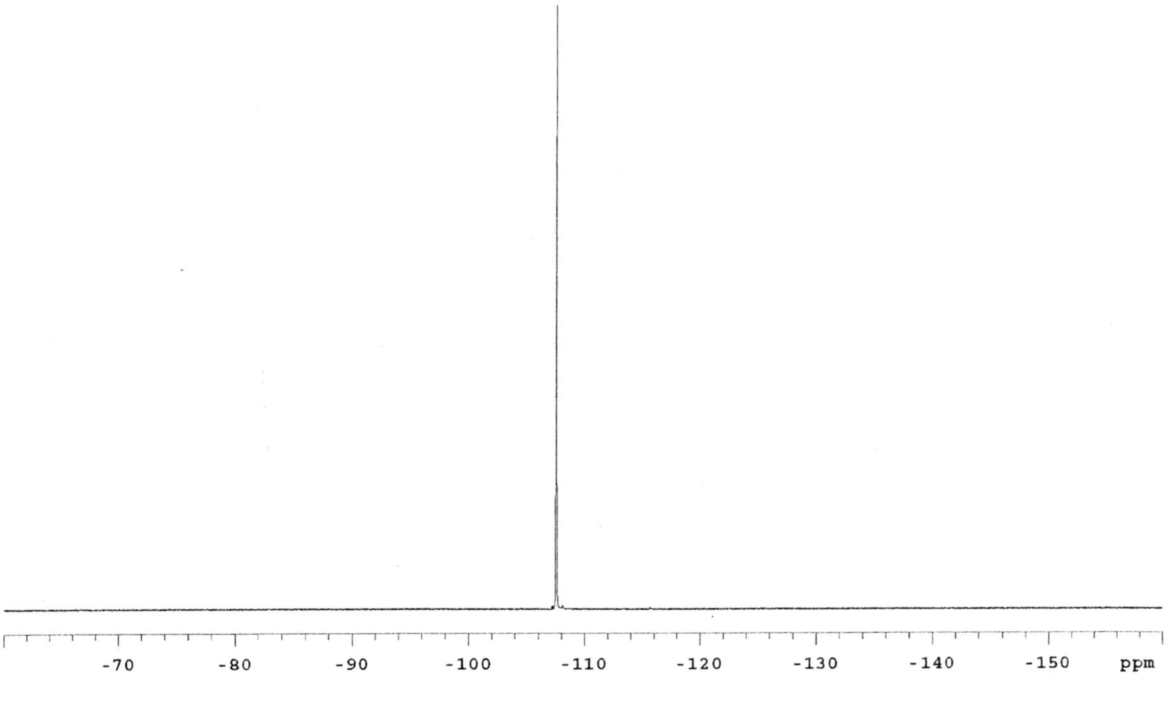

^{19}F-NMR Spectrum in $CDCl_3$

Problem 3.29 Given the spectroscopic data below (Proton, Carbon, gHSQC, and Fluorine), provide the structure of the molecule. Hint: Isomer of 3.28. The carbon signal centered at ~162 ppm has no correlation in the gHSQC spectrum.

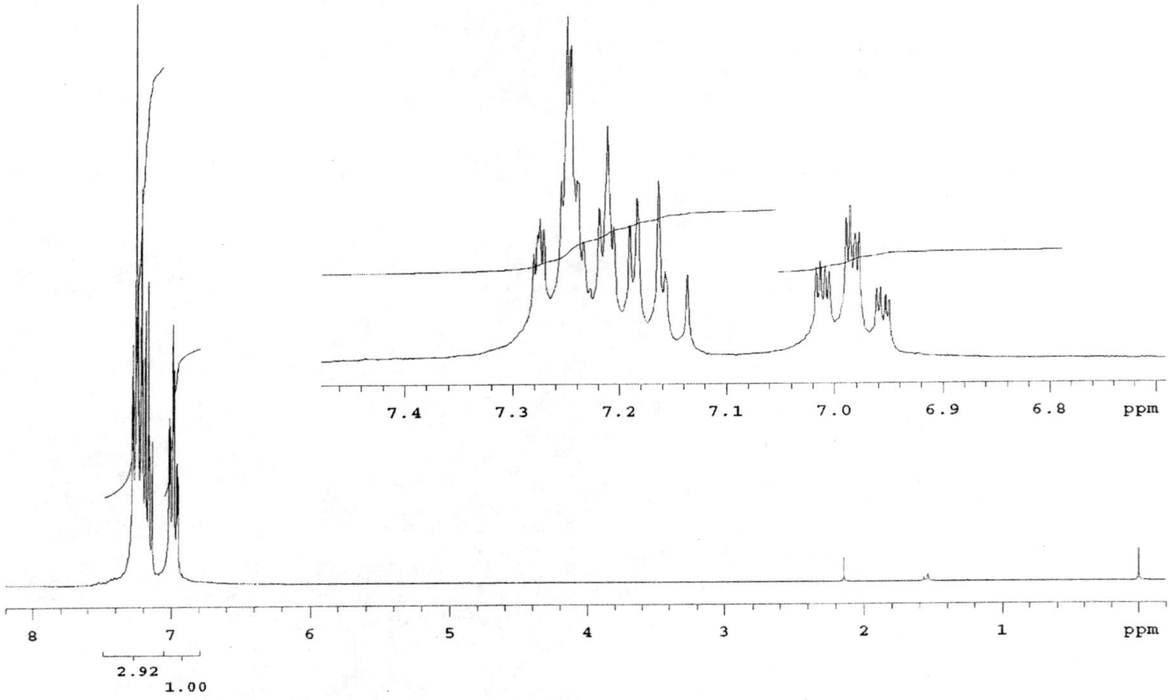

^{1}H-NMR Spectrum in $CDCl_3$

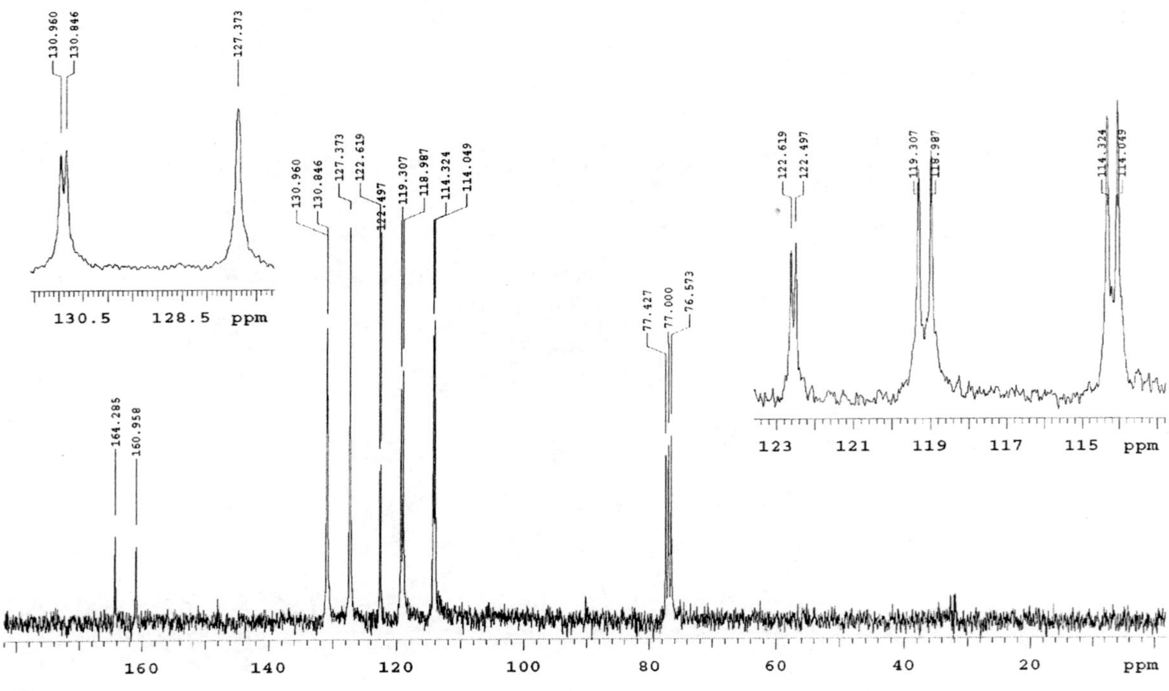

^{13}C-NMR Spectrum in $CDCl_3$

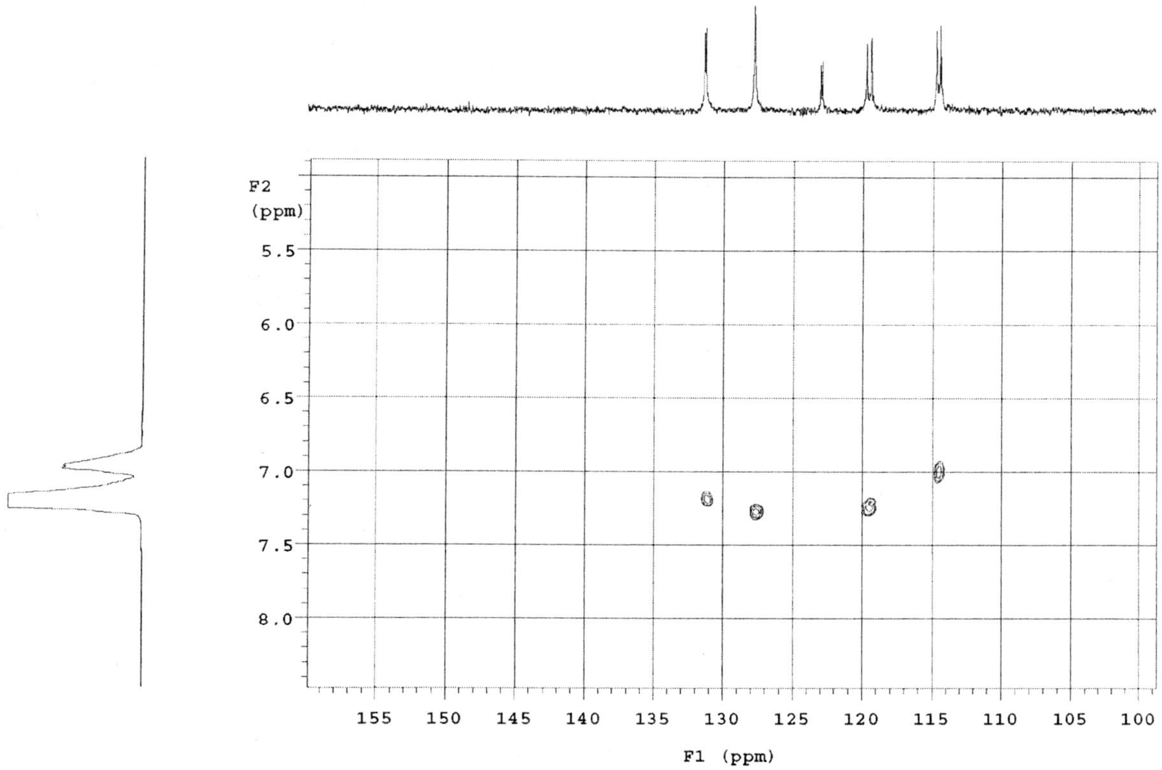

gHSQC Spectrum in $CDCl_3$

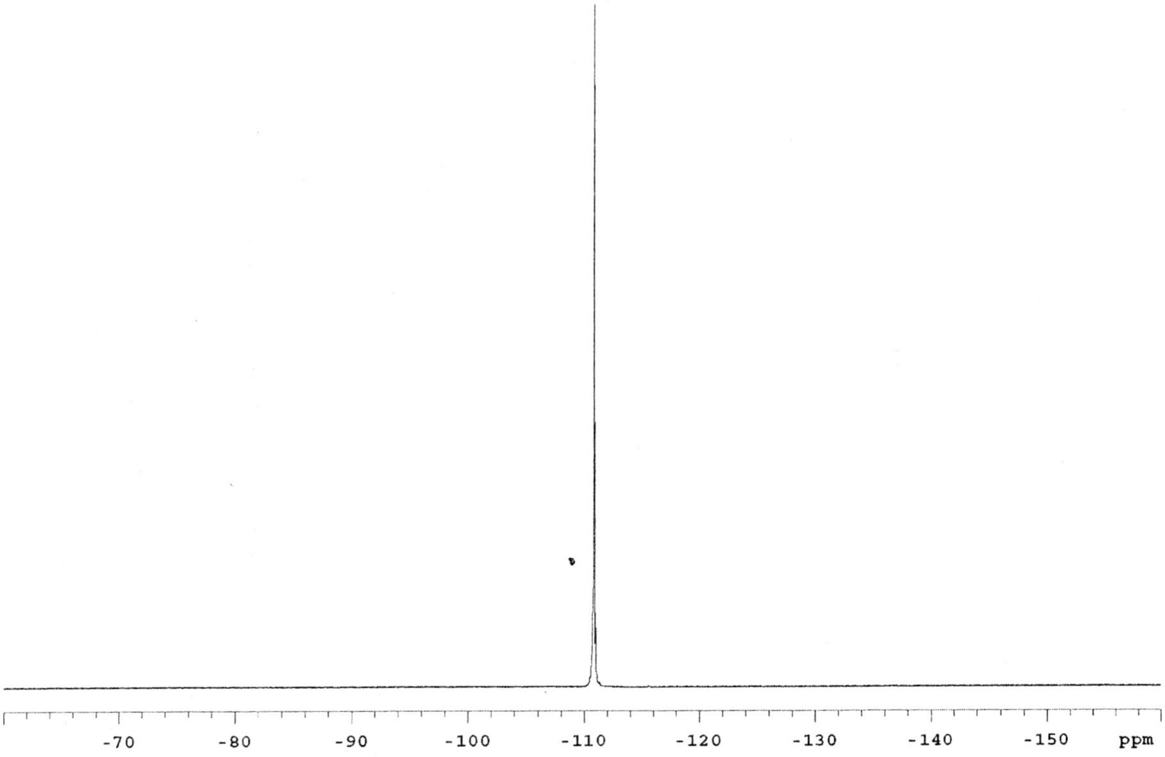

^{19}F-NMR Spectrum in $CDCl_3$

Problem 3.30 Given the spectroscopic data below (Proton, Carbon, Edited DEPT, gCOSY, gHSQC, and Fluorine), provide the structure of the molecule. (IR: 700 & 830 cm^{-1}) Caution: Contains acetone impurity. In the gHSQC spectrum, no correlations exist for the signals located beyond 160 ppm.

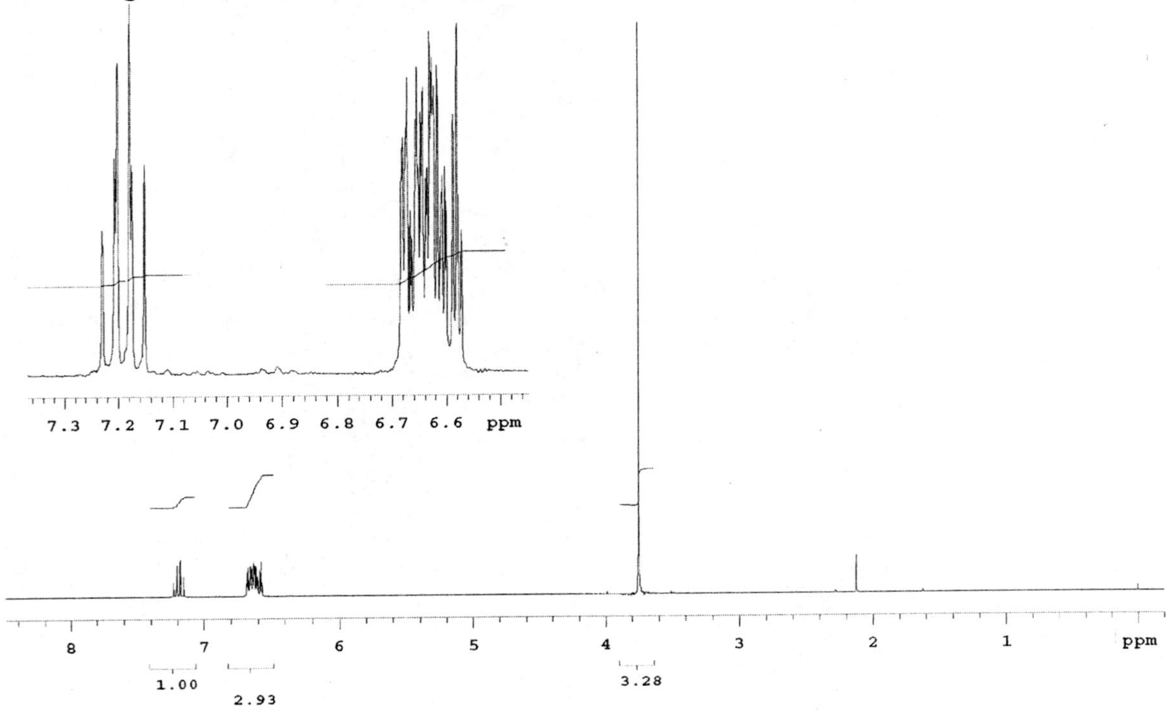

^{1}H-NMR Spectrum in $CDCl_3$

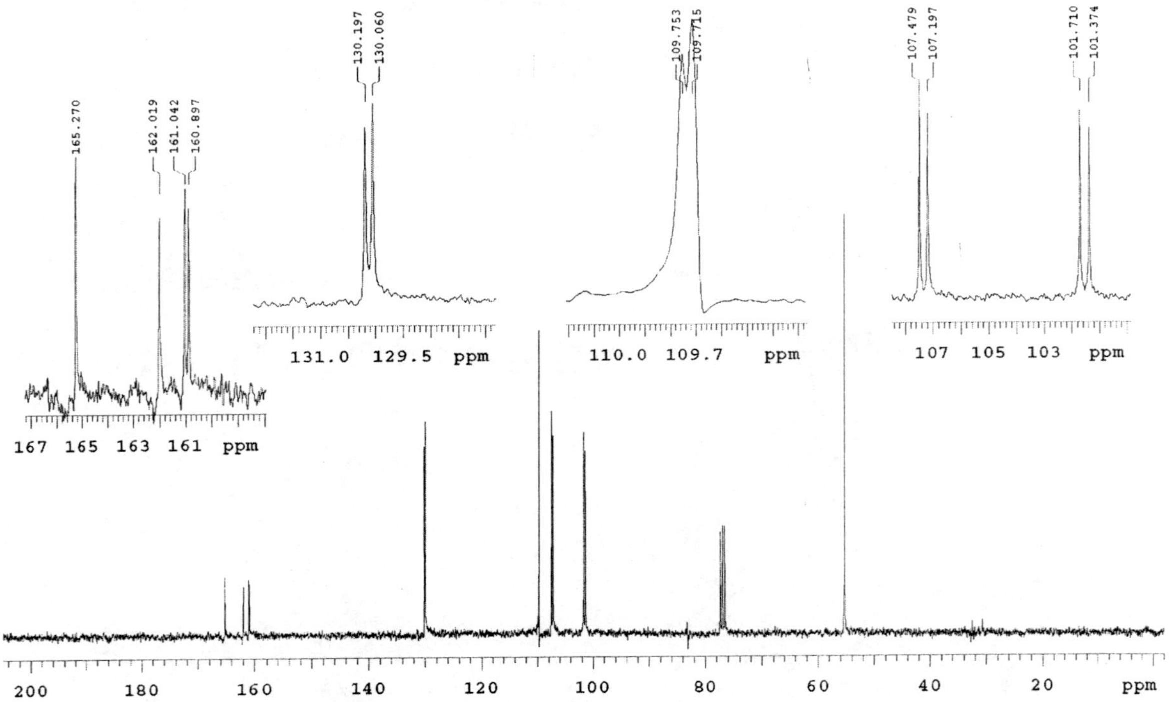

^{13}C-NMR Spectrum in $CDCl_3$

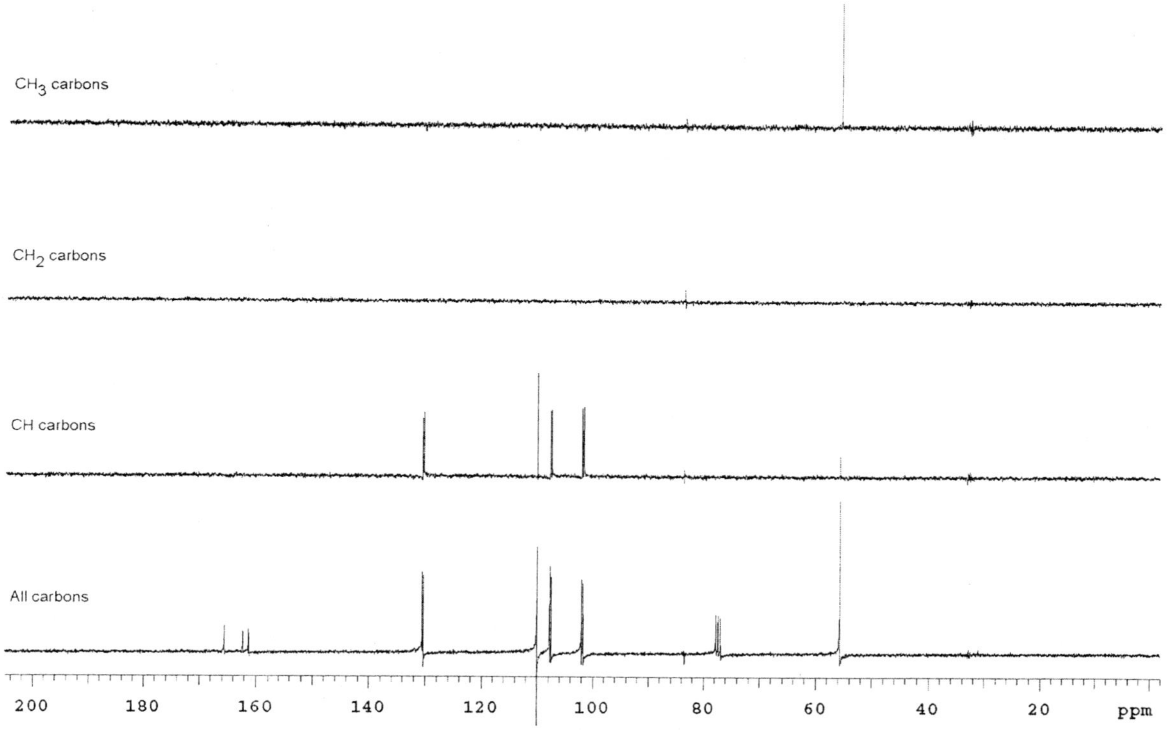

Edited DEPT ^{13}C-NMR Spectrum in $CDCl_3$

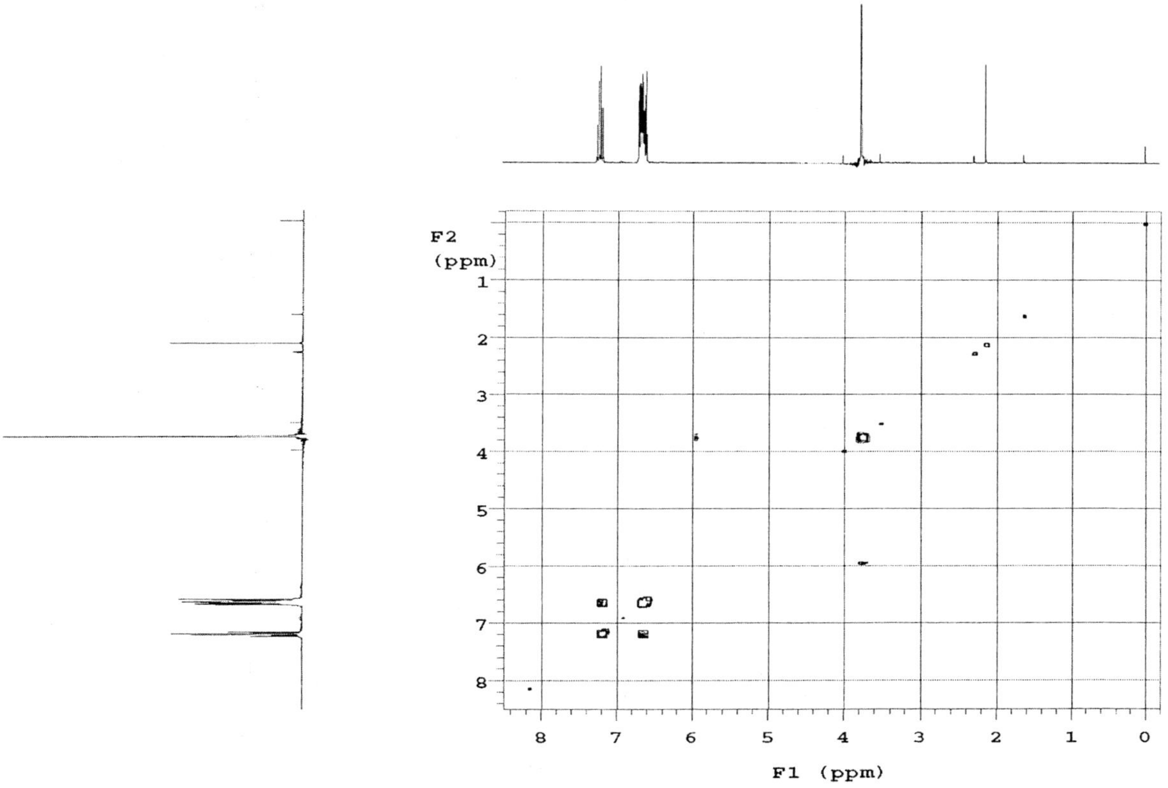

gCOSY Spectrum in $CDCl_3$

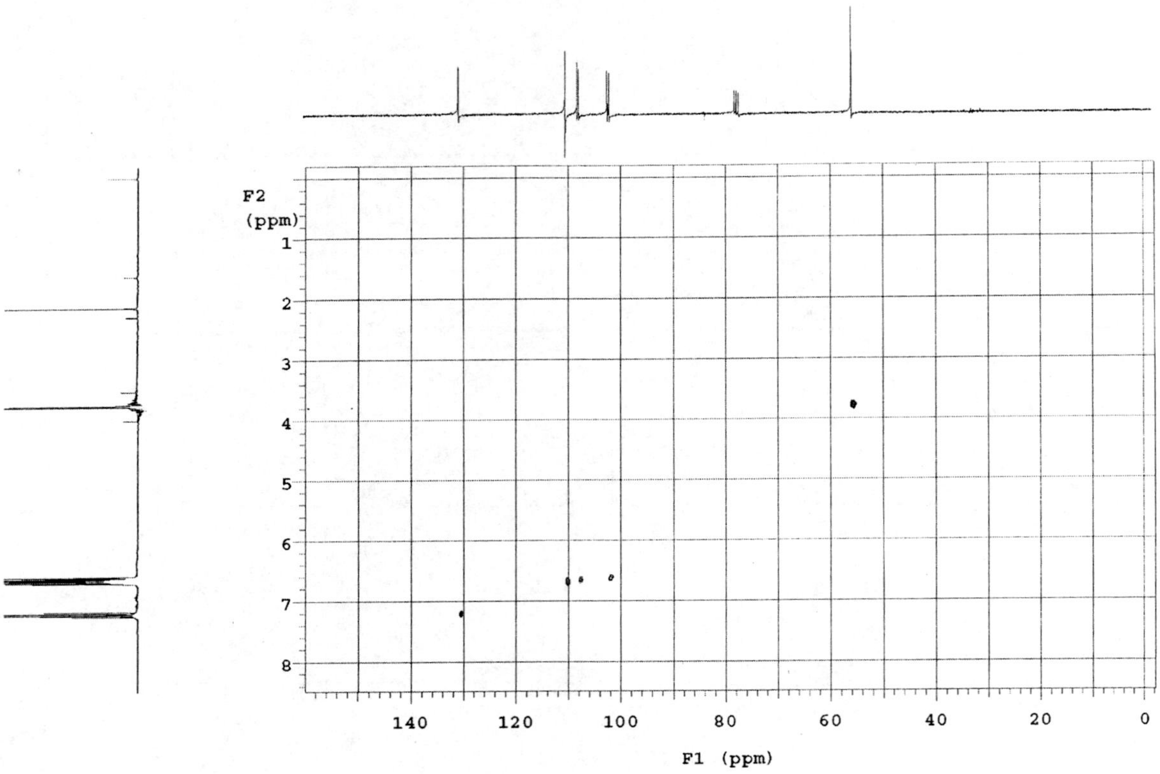

gHSQC Spectrum in $CDCl_3$

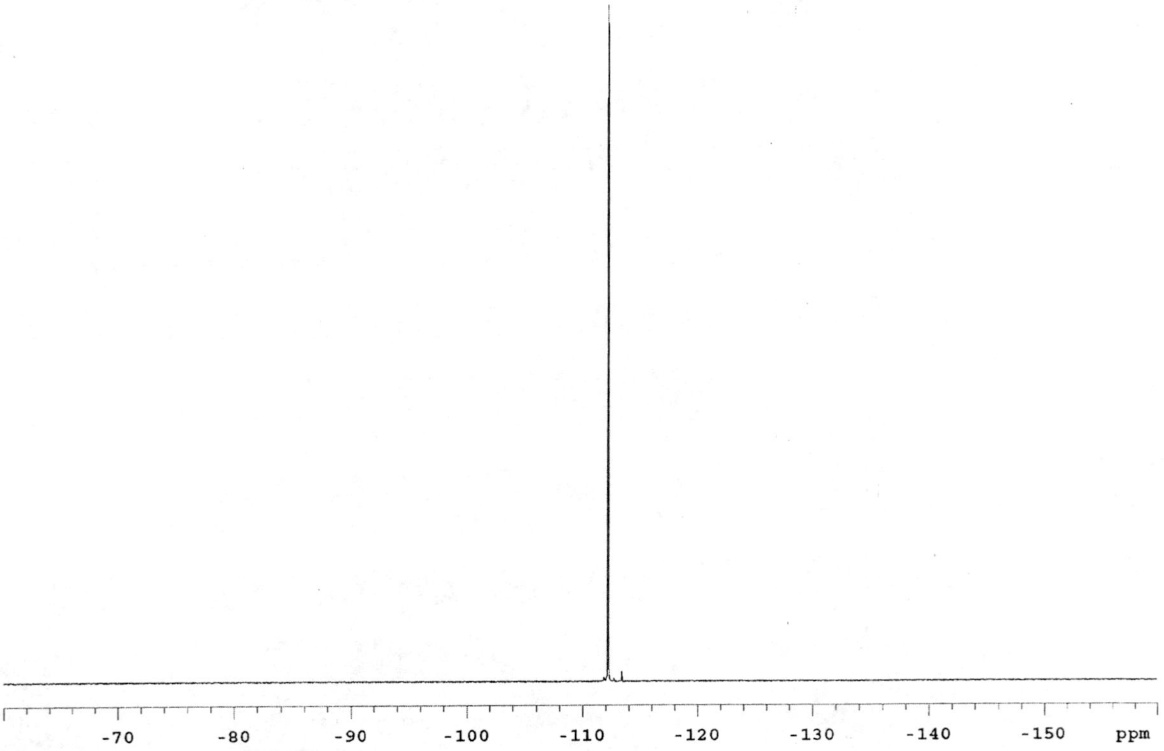

Fluorine Spectrum in $CDCl_3$

Problem 3.31 Given the spectroscopic data below (Proton, Carbon, gHSQC, and gCOSY, gHMBC), provide the structure of the molecule. (MW: 226 g/mol)

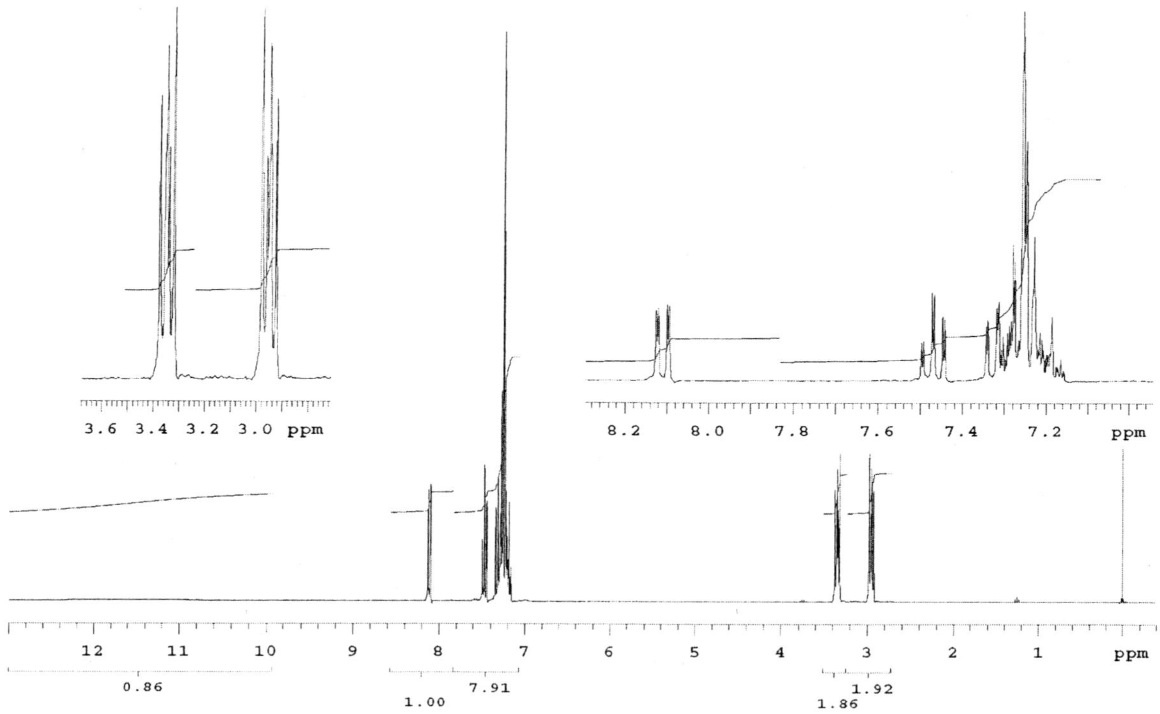

^{1}H-NMR Spectrum in $CDCl_3$

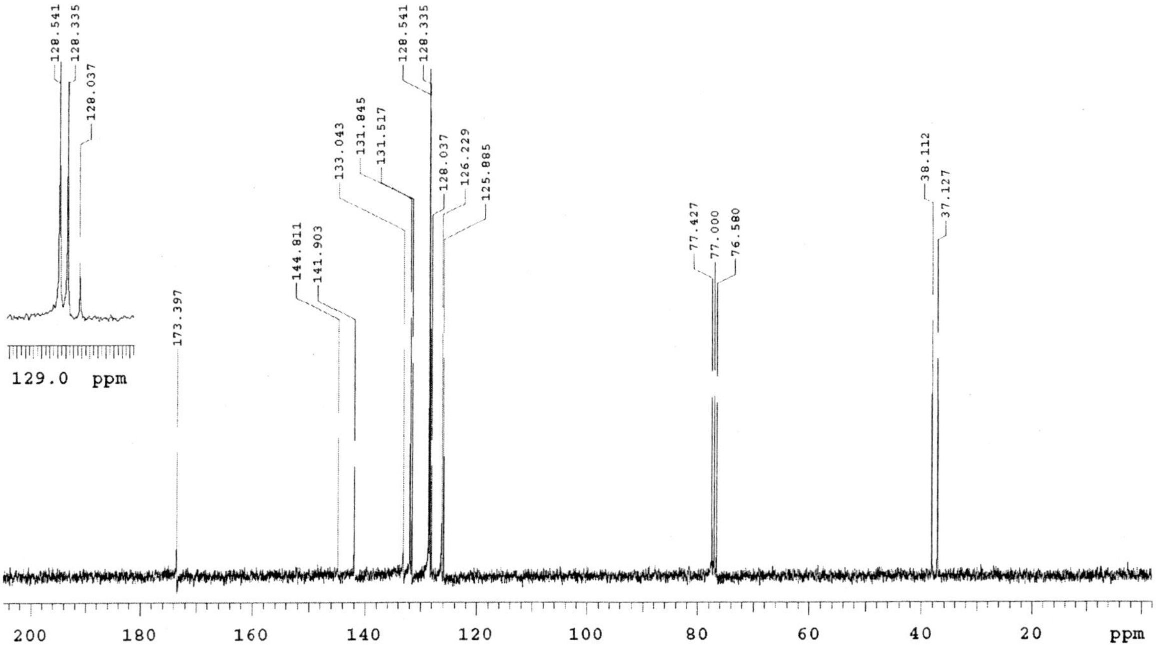

^{13}C-NMR Spectrum in $CDCl_3$

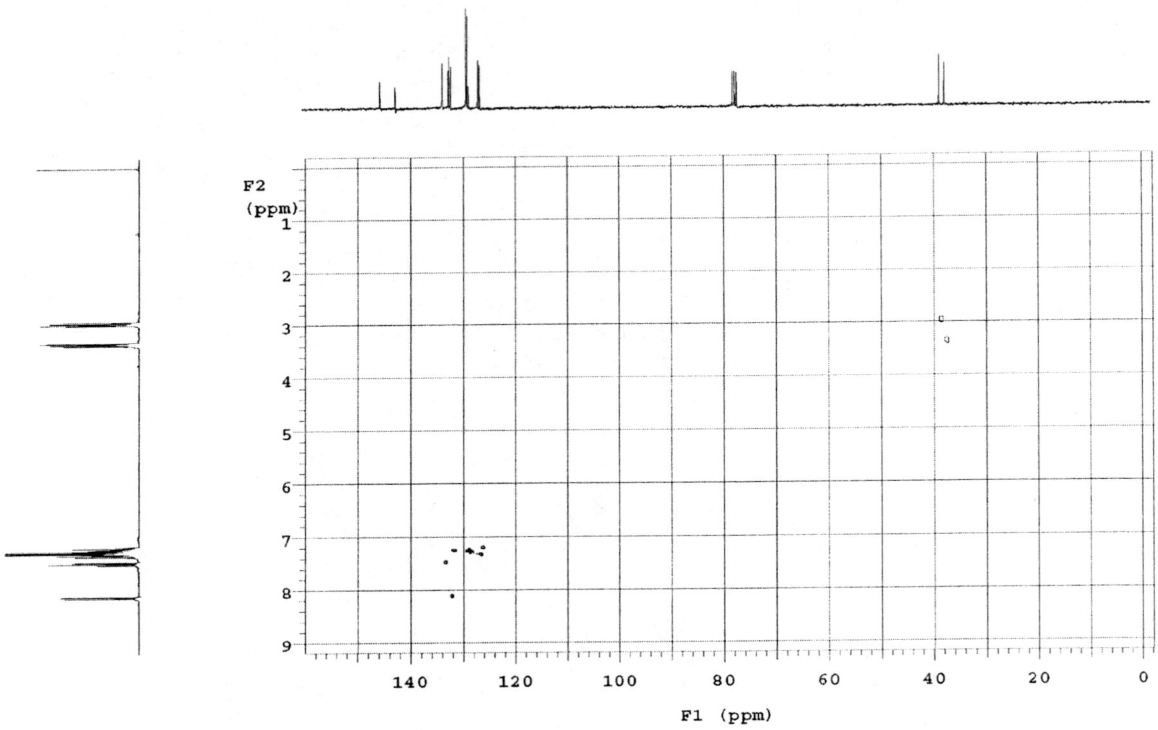

gHSQC Spectrum in $CDCl_3$

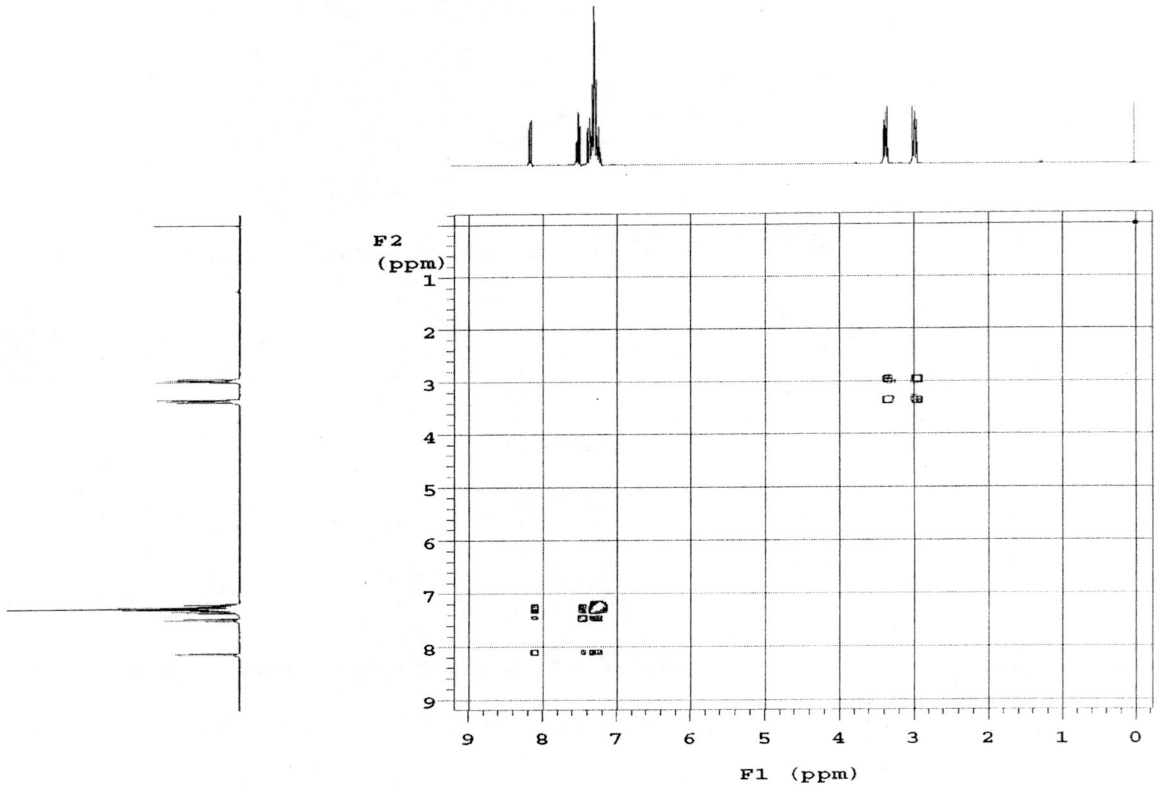

gCOSY Spectrum in $CDCl_3$

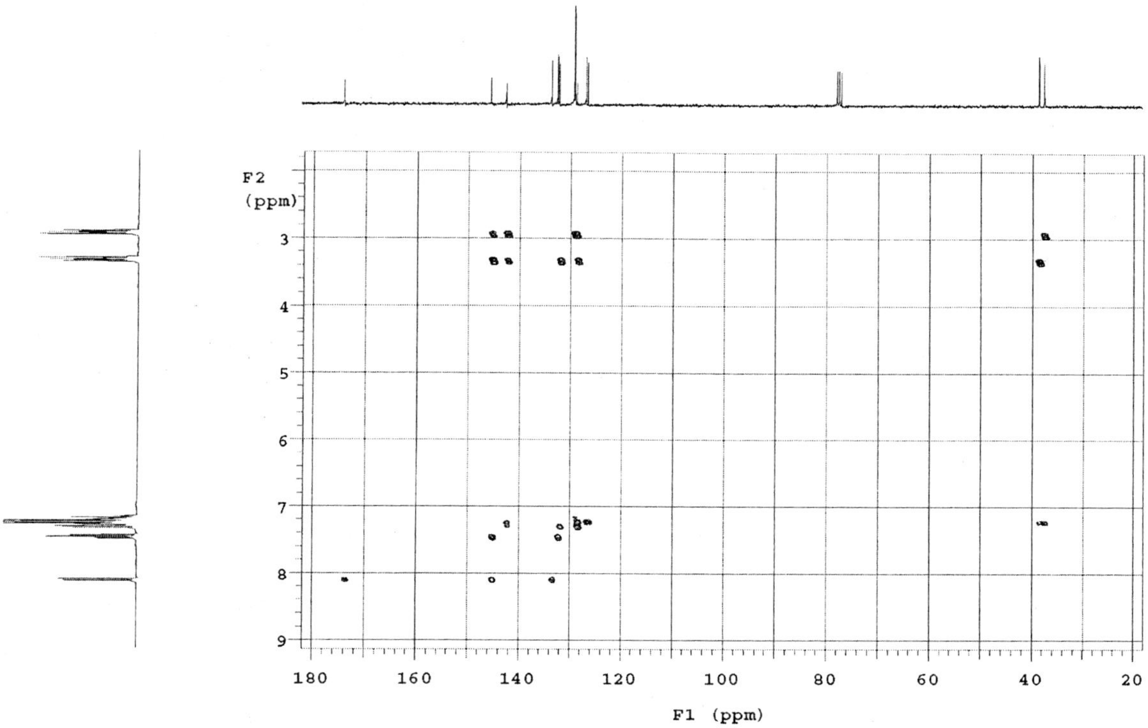

gHMBC Spectrum in $CDCl_3$

Problem 3.32 Given the spectroscopic data below (Proton, Carbon, gHSQC, and gCOSY), provide the structure of the molecule. (MW: 125 g/mol)

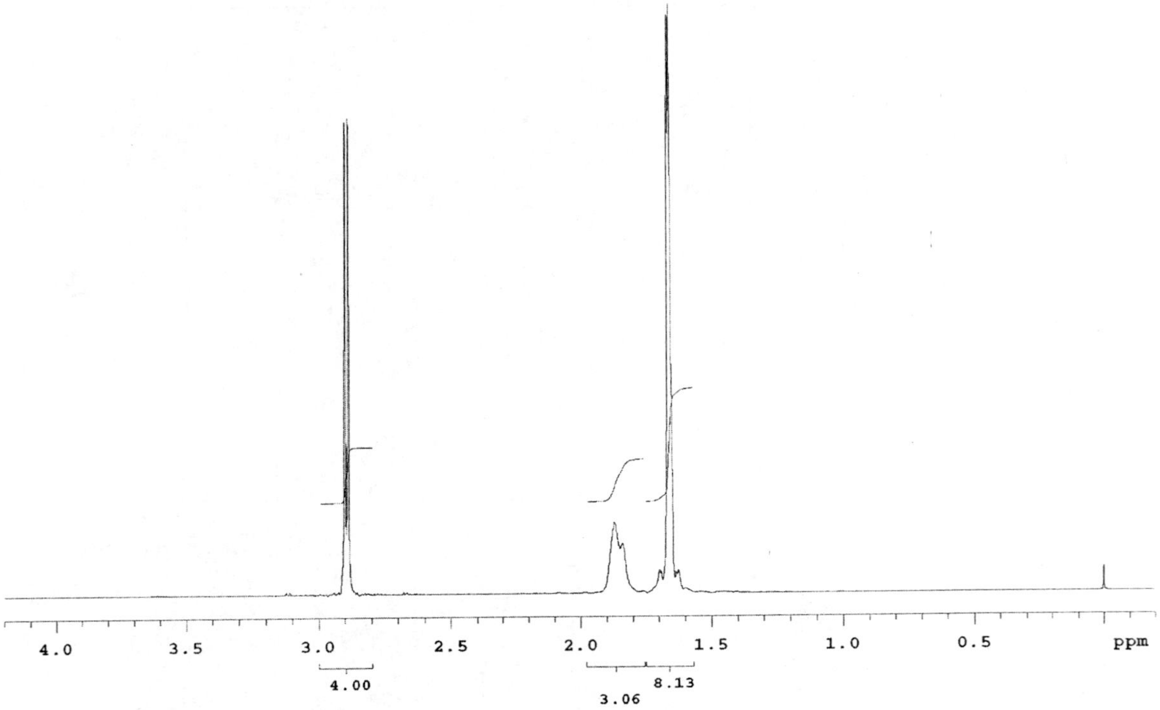

^{1}H-NMR Spectrum in $CDCl_3$

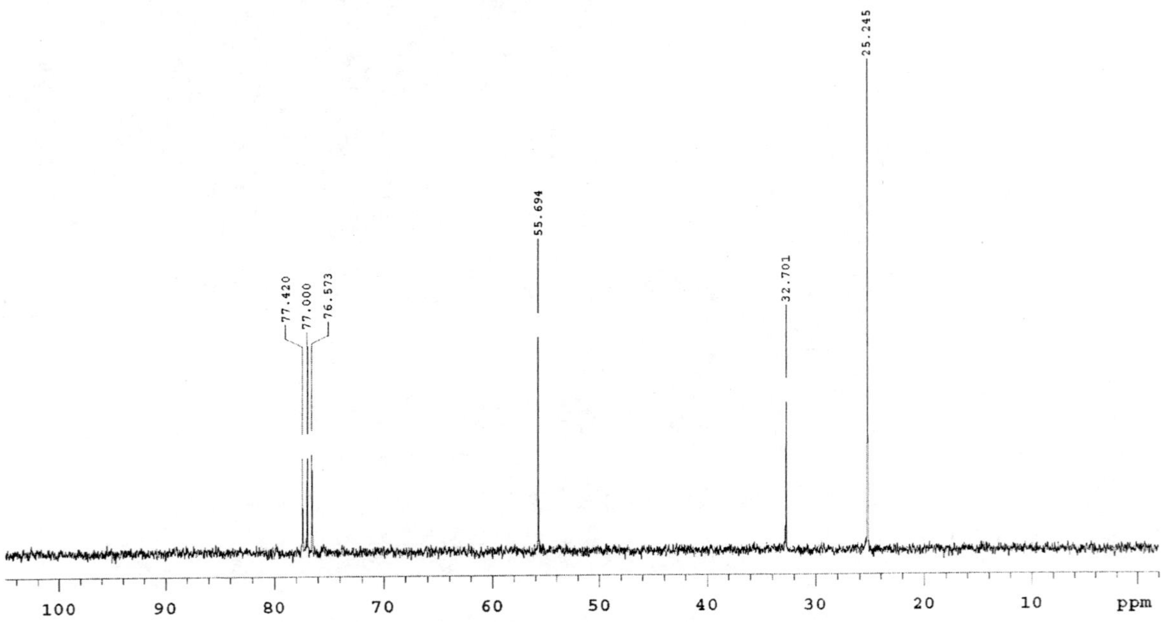

^{13}C-NMR Spectrum in $CDCl_3$

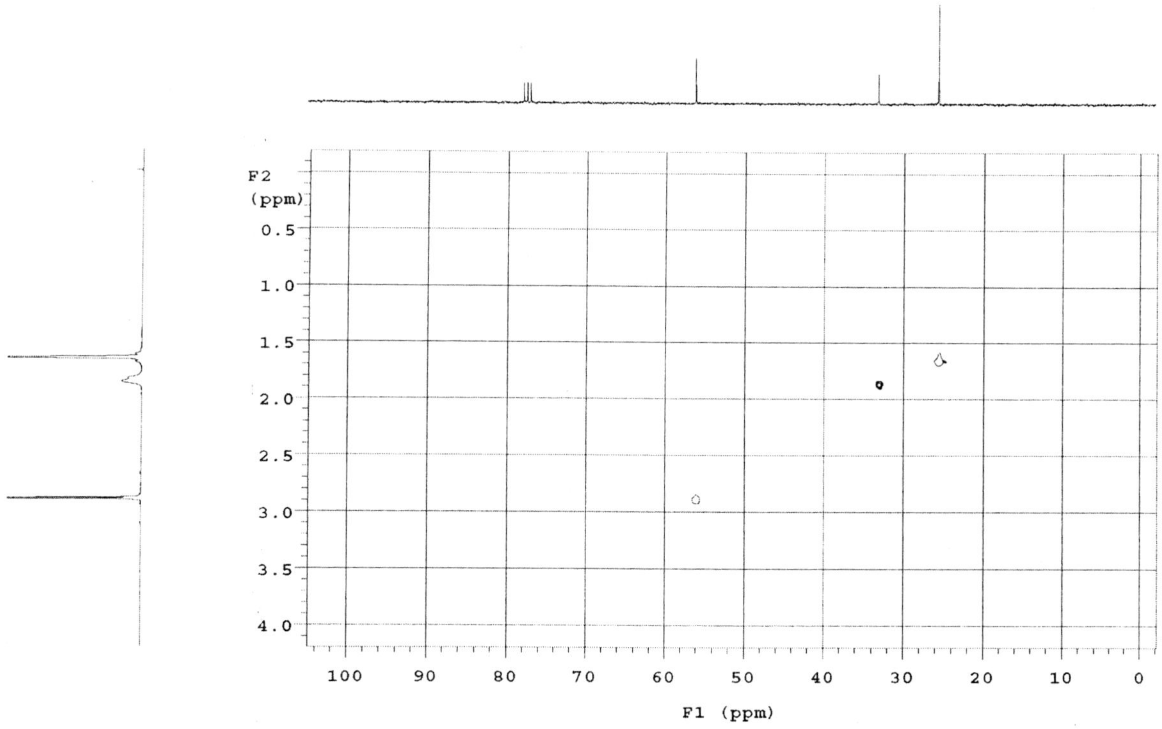

gHSQC Spectrum in $CDCl_3$

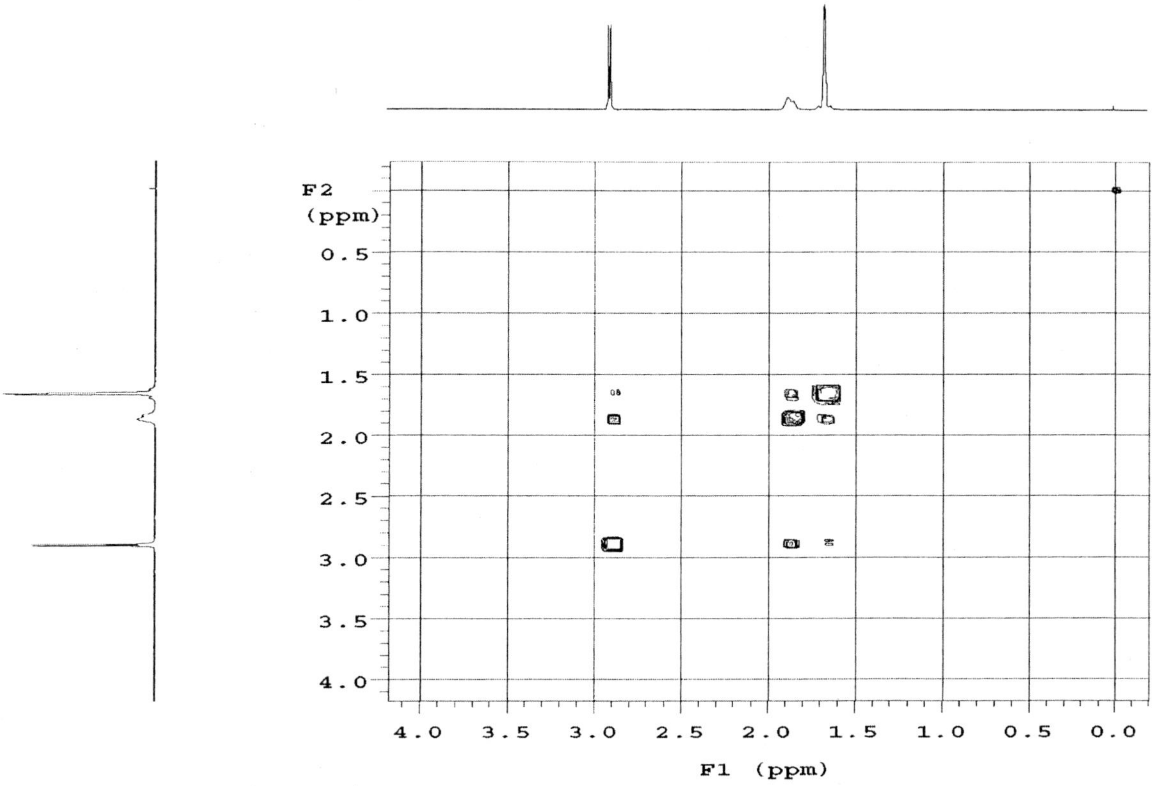

gCOSY Spectrum in $CDCl_3$

Problem 3.33 Given the spectroscopic data below (Proton, Carbon, gHSQC, and gCOSY), provide the structure of the molecule. (MW: 128 g/mol)

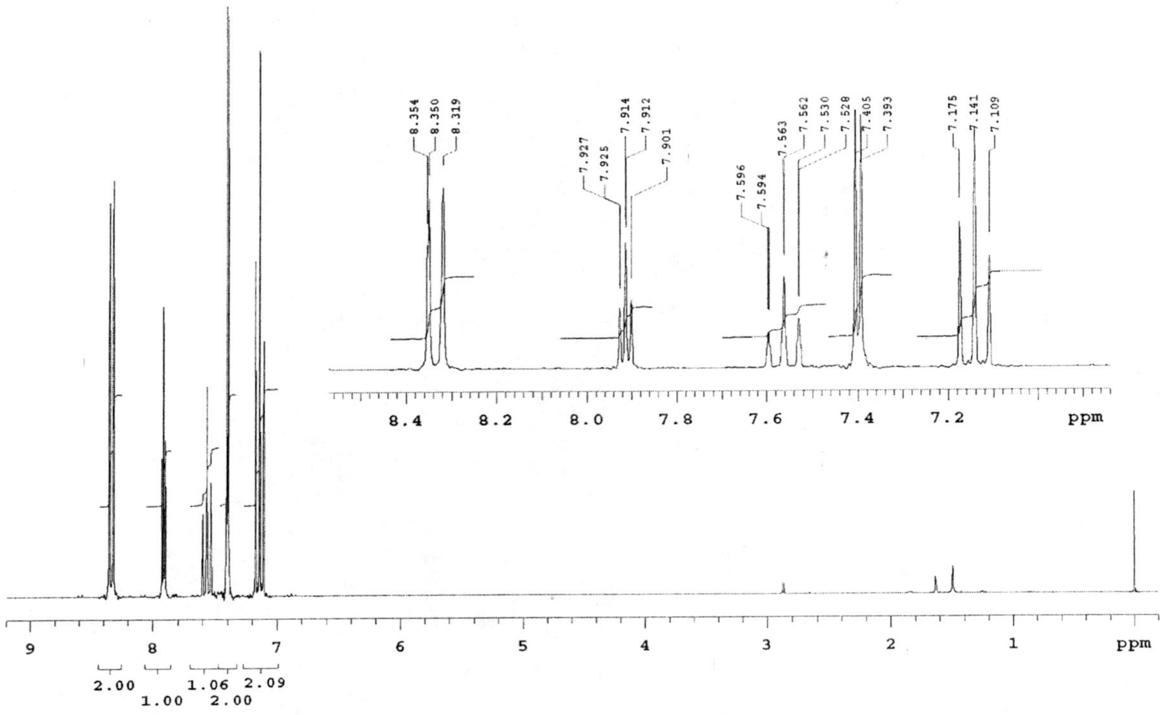

^{1}H-NMR Spectrum in $CDCl_3$

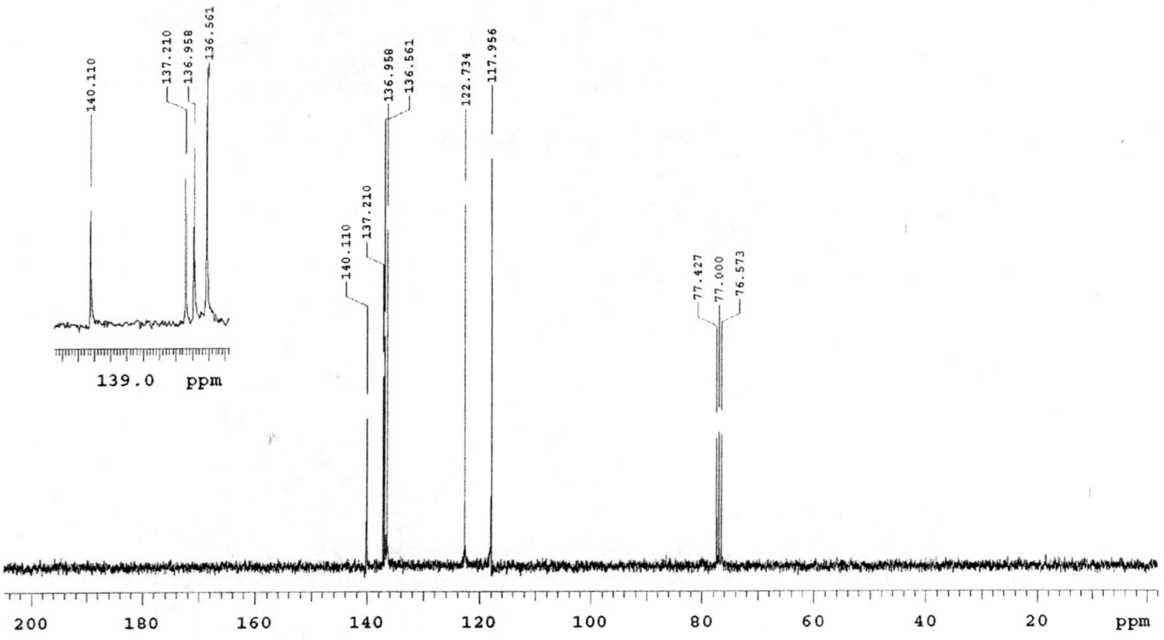

^{13}C-NMR Spectrum in $CDCl_3$

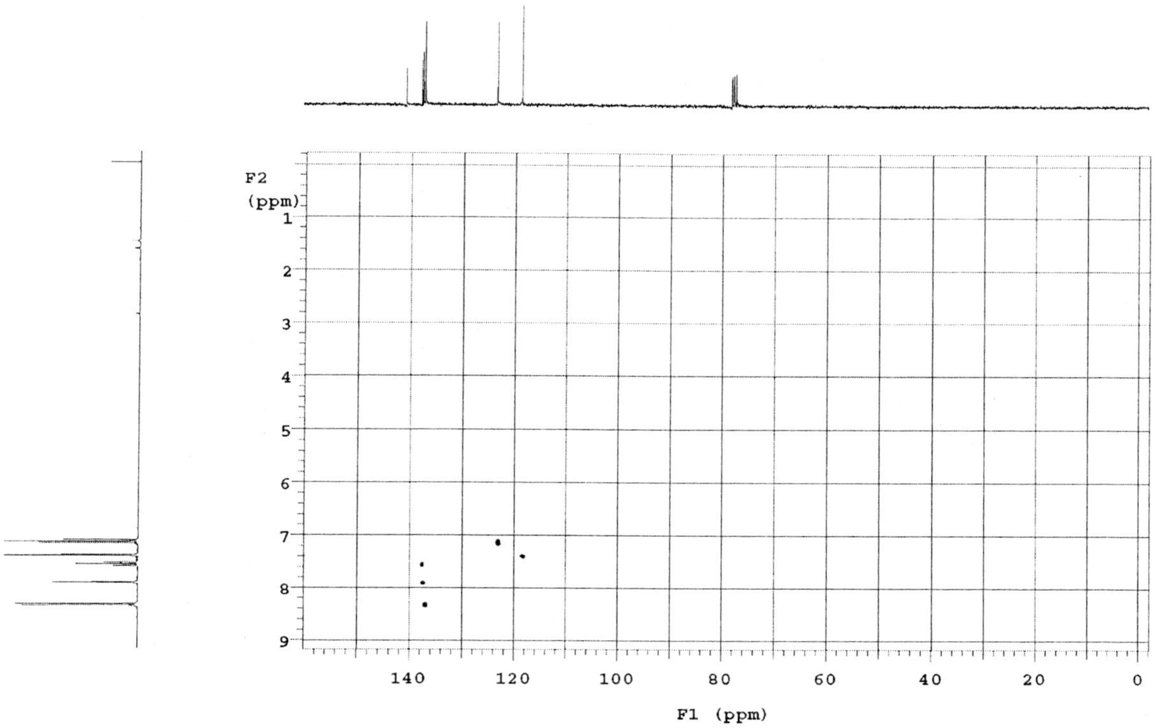

gHSQC Spectrum in $CDCl_3$

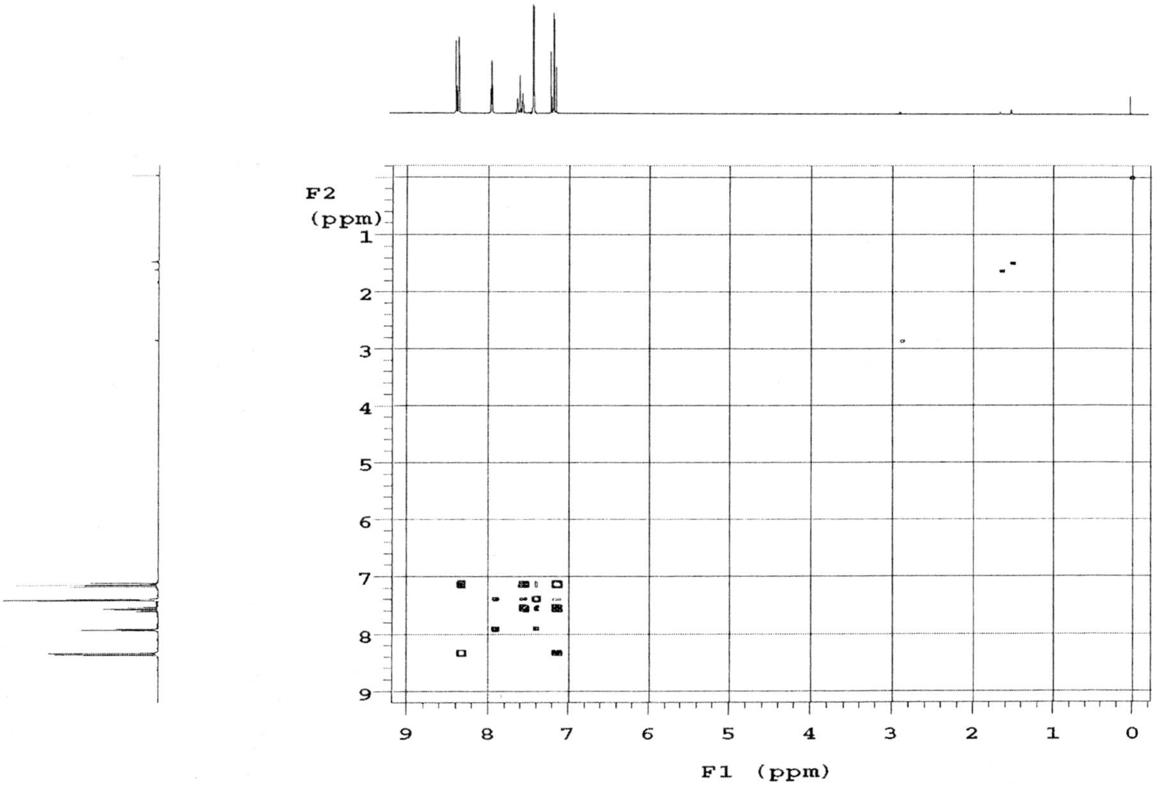

gCOSY Spectrum in $CDCl_3$

Problem 3.34 Given the spectroscopic data below (Proton, Carbon, gHSQC, gCOSY, and gHMBC), provide the structure of the molecule. (HR-MS: 184.1000 amu)

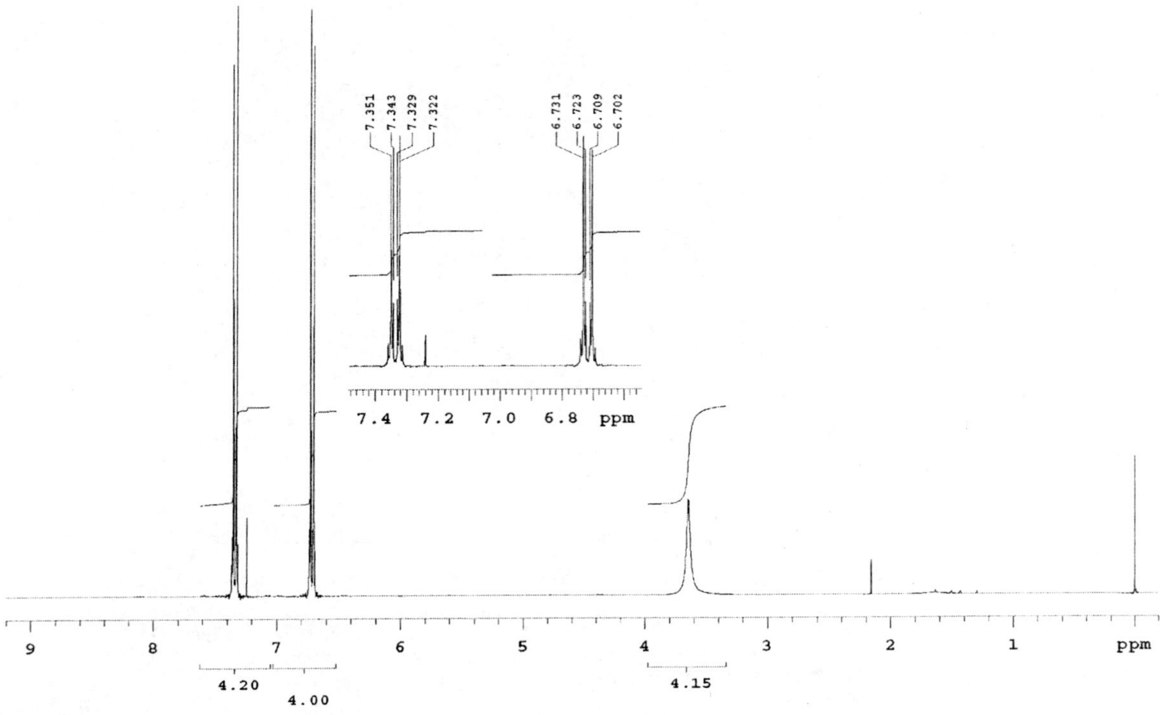

^{1}H-NMR Spectrum in $CDCl_3$

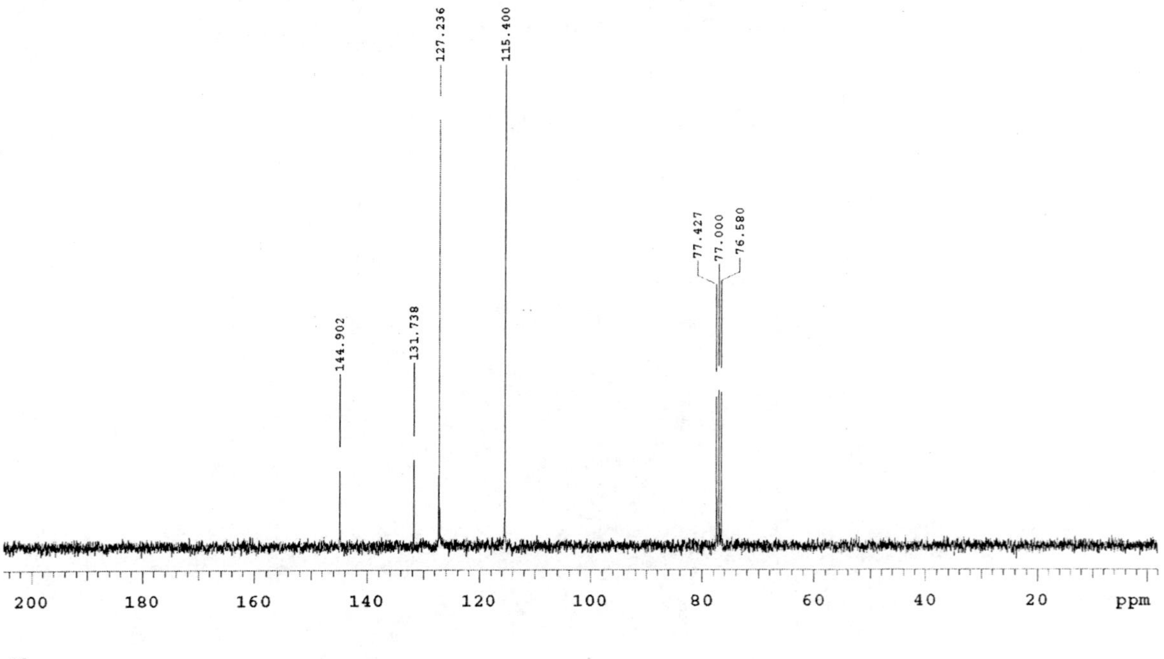

^{13}C-NMR Spectrum in $CDCl_3$

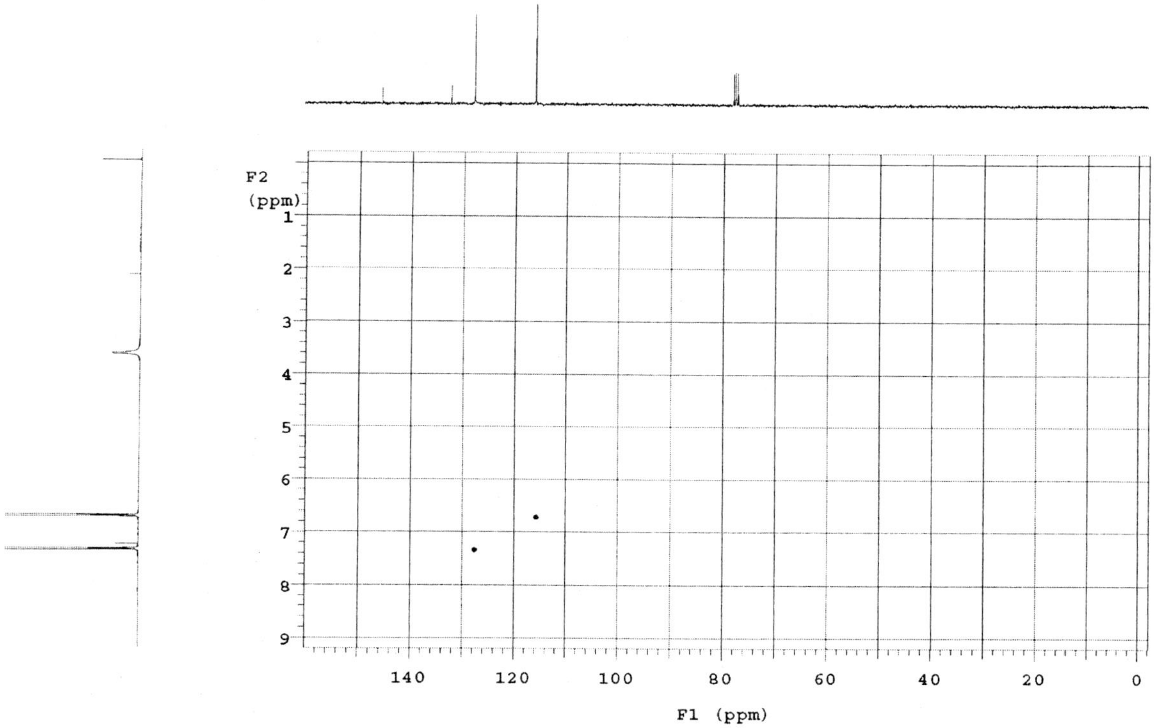

gHSQC Spectrum in $CDCl_3$

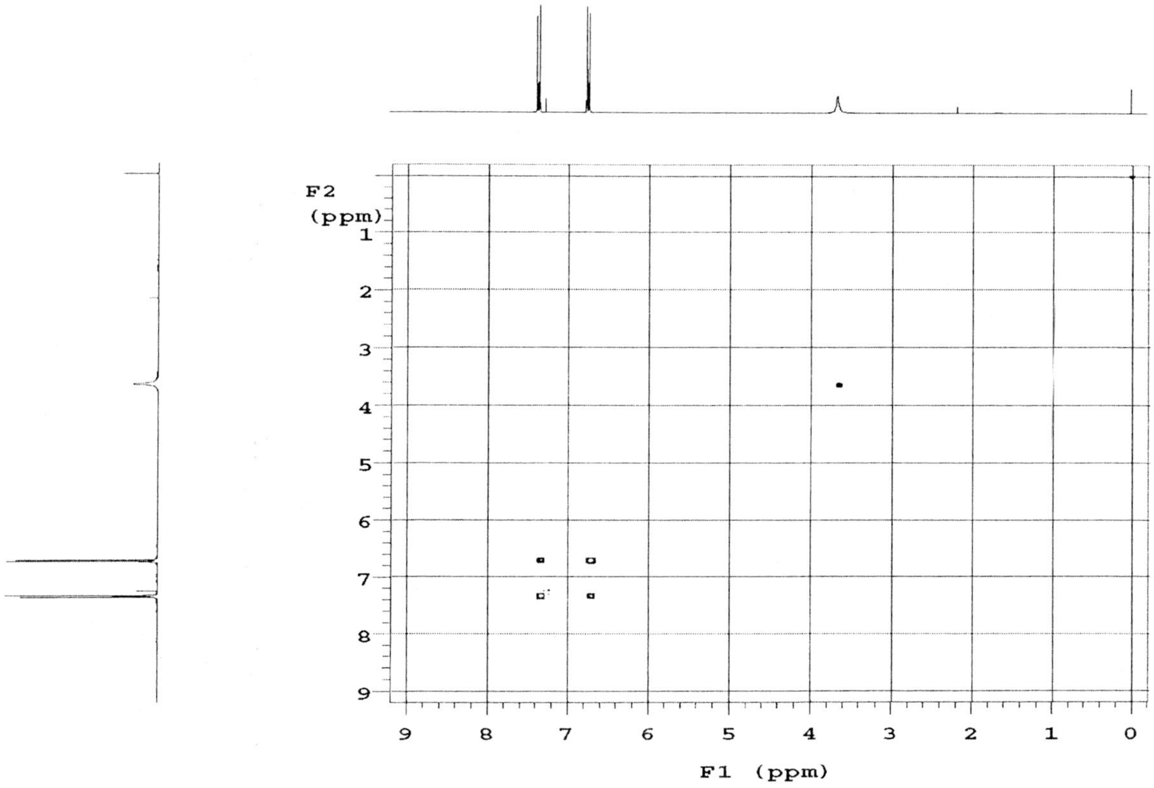

gCOSY Spectrum in $CDCl_3$

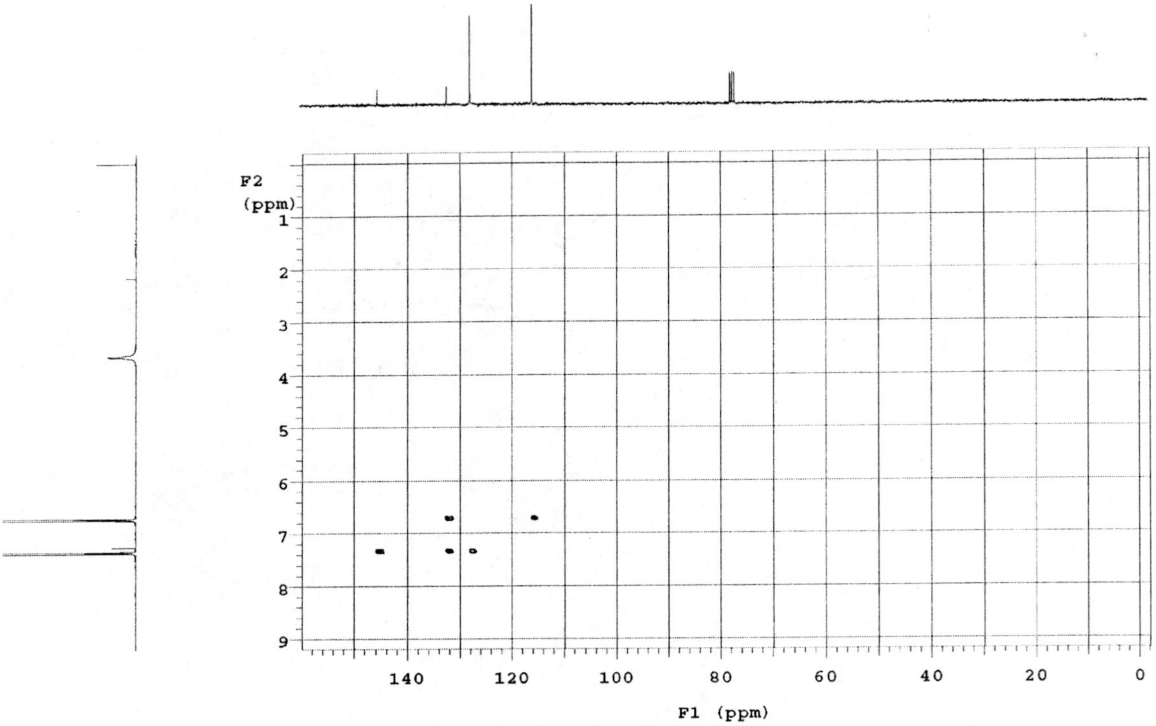

gHMBC Spectrum in $CDCl_3$

Problem 3.35 Given the spectroscopic data below (Proton, Carbon, gHSQC, and gCOSY), provide the structure of the molecule. (HR-MS: 156.0687 amu)

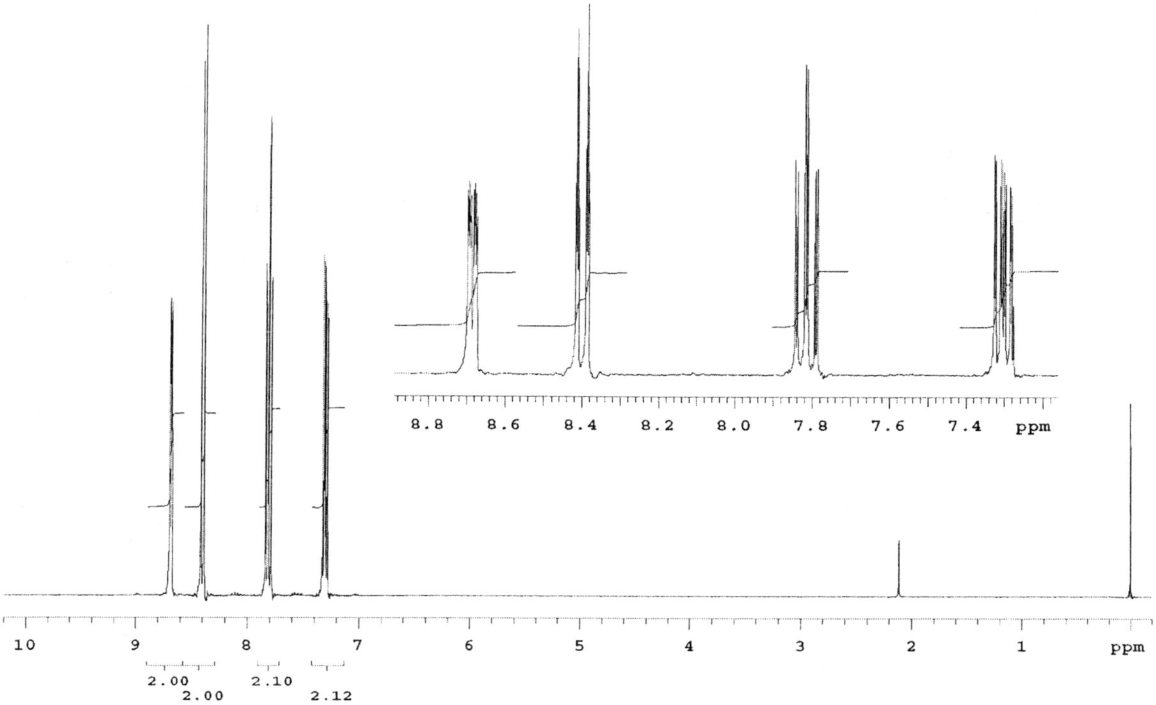

^{1}H-NMR Spectrum in $CDCl_3$

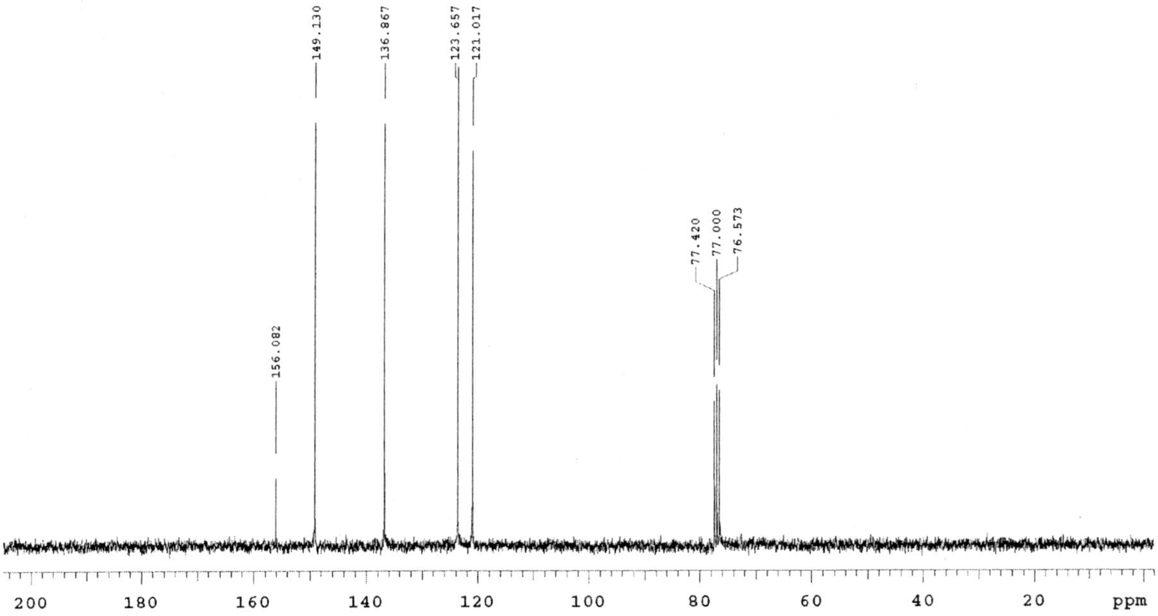

^{13}C-NMR Spectrum in $CDCl_3$

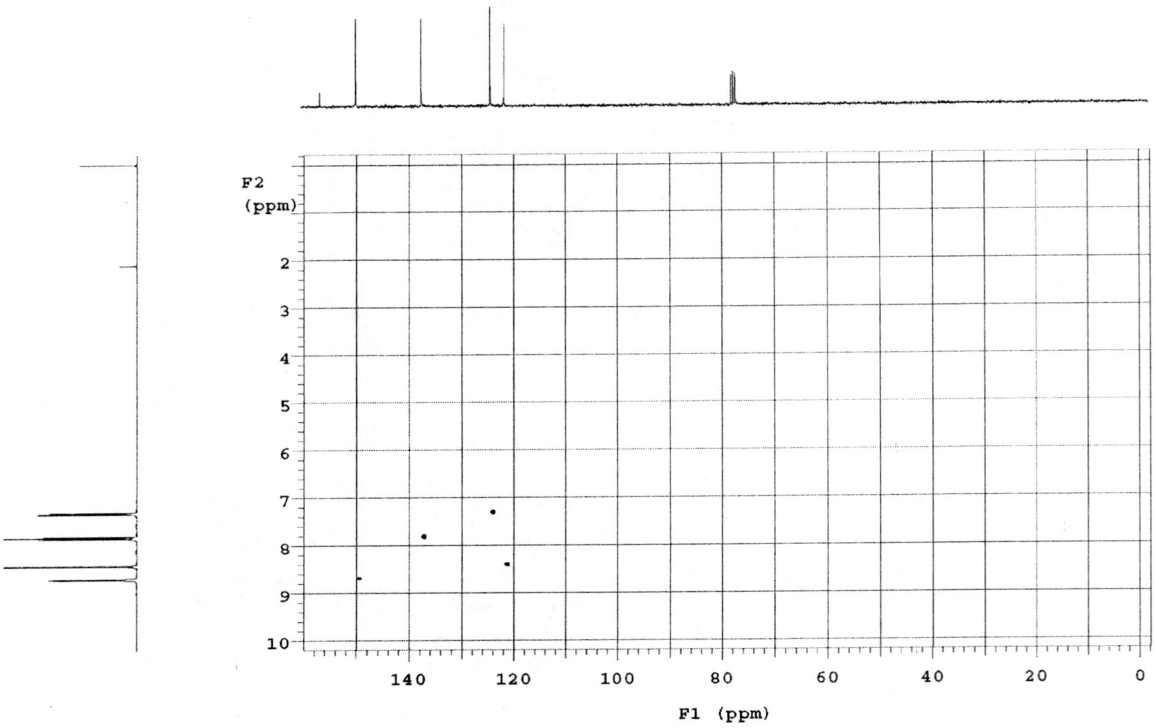

gHSQC Spectrum in $CDCl_3$

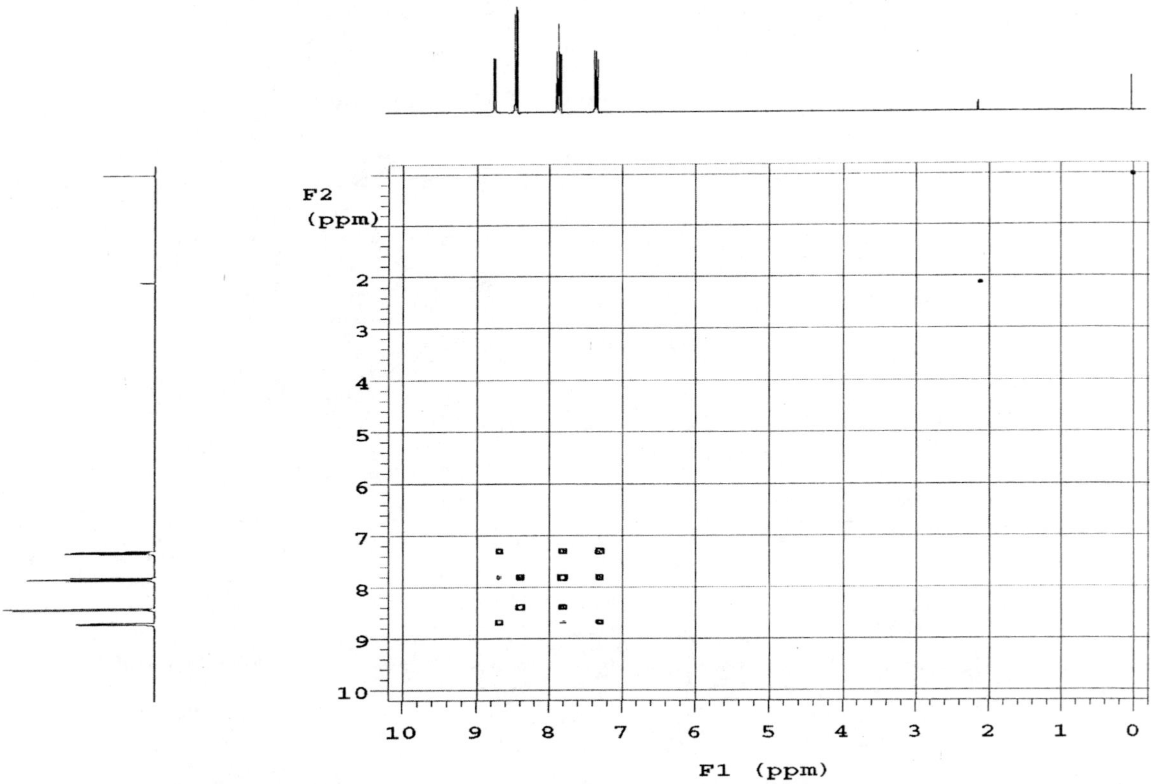

gCOSY Spectrum in $CDCl_3$

Problem 3.36 Given the spectroscopic data below (Proton, Carbon, gHSQC, gCOSY, and gHMBC), provide the structure of the molecule. (HR-MS: 281.1627 amu)

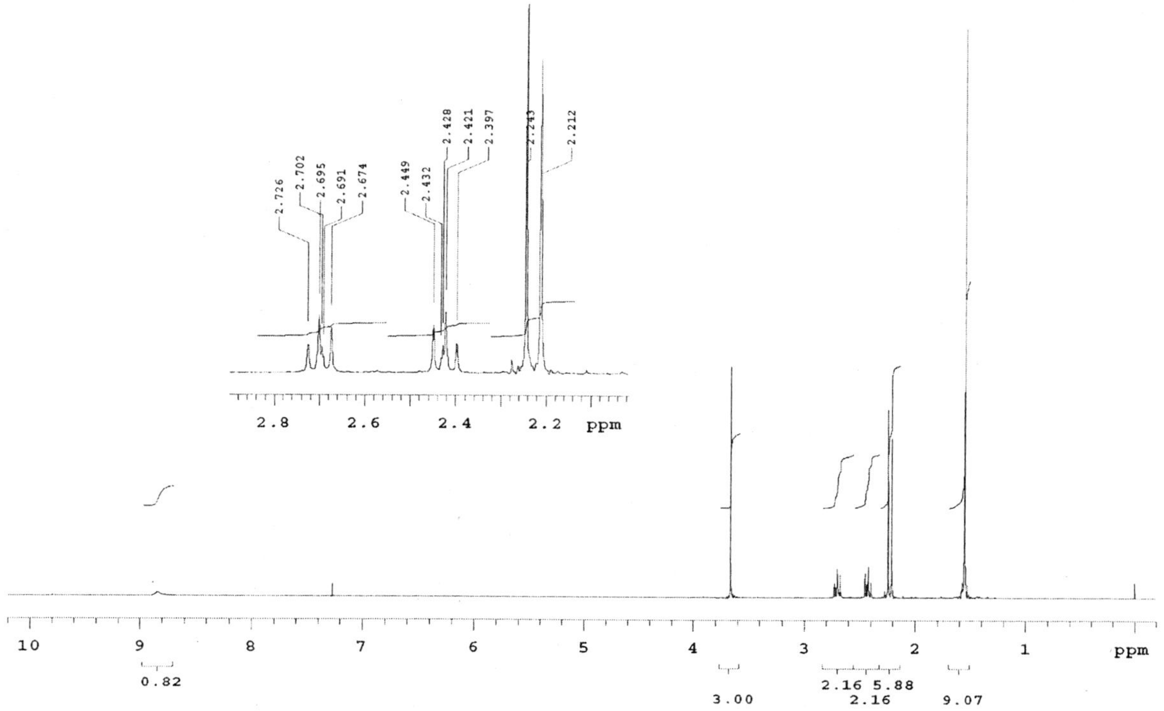

^{1}H-NMR Spectrum in $CDCl_3$

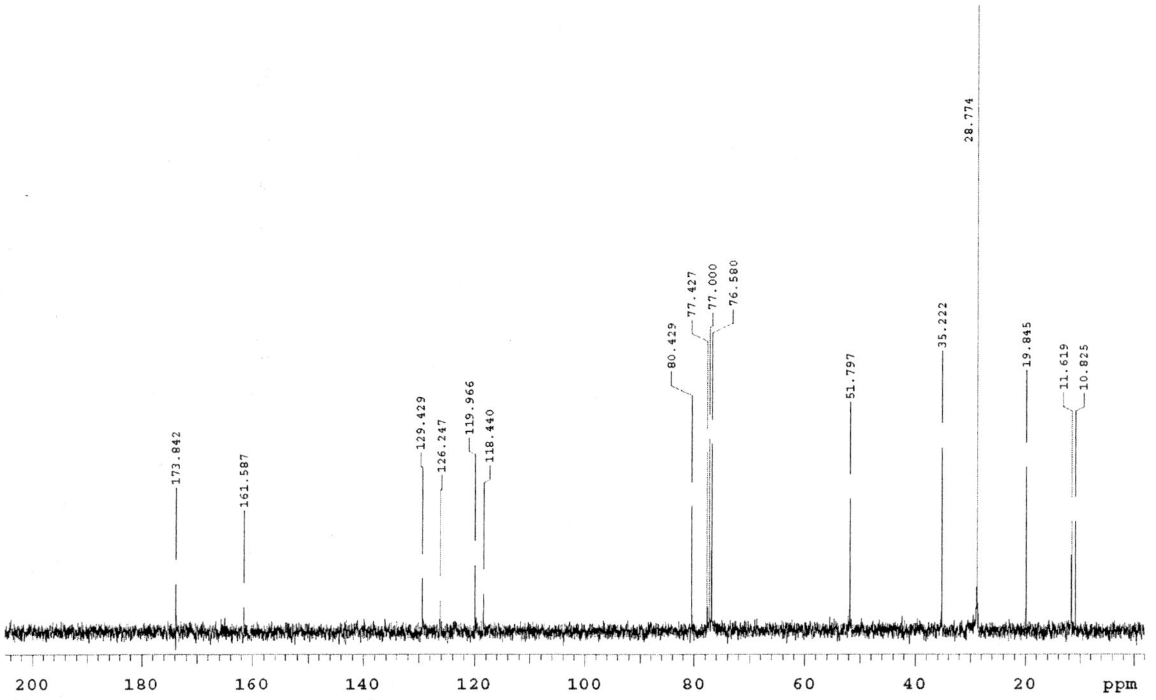

^{13}C-NMR Spectrum in $CDCl_3$

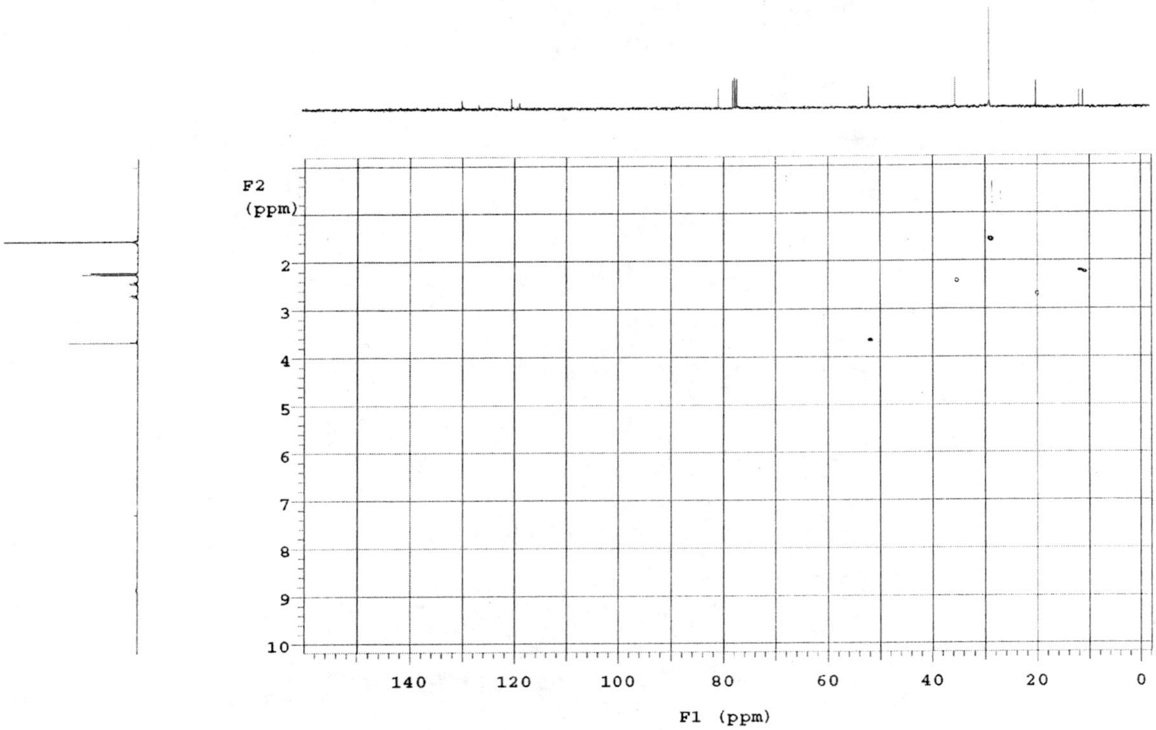

gHSQC Spectrum in $CDCl_3$

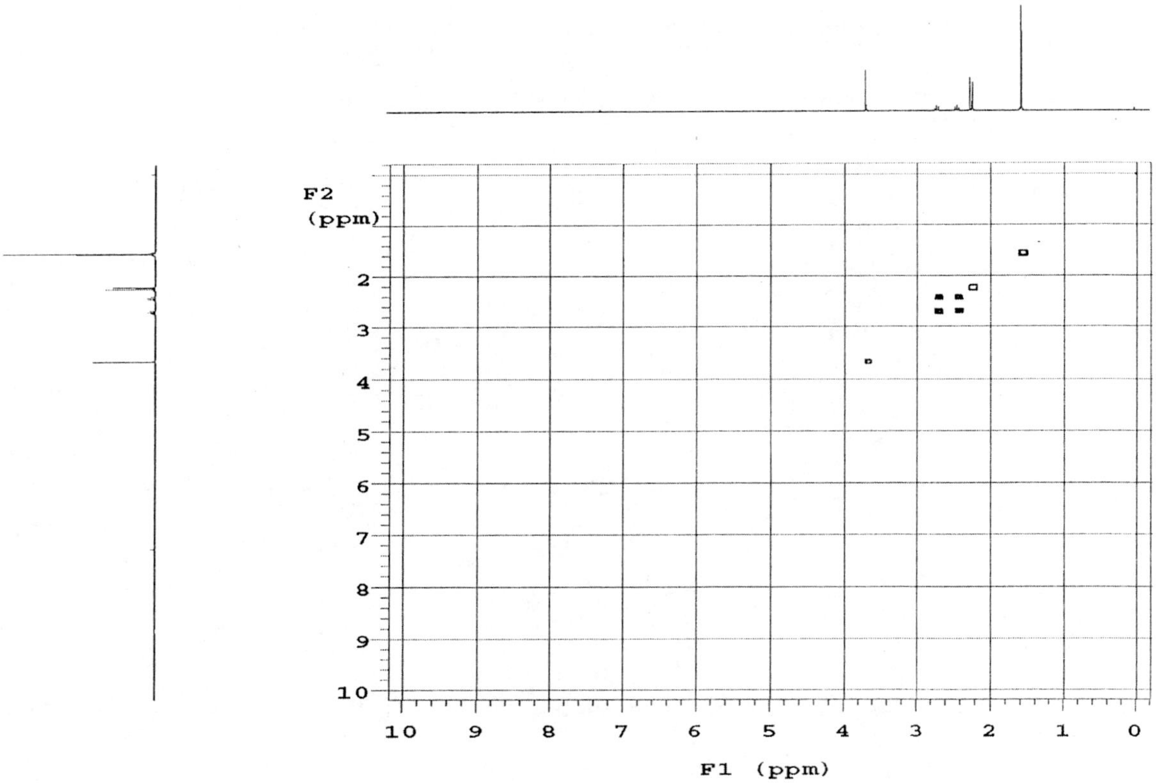

gCOSY Spectrum in $CDCl_3$

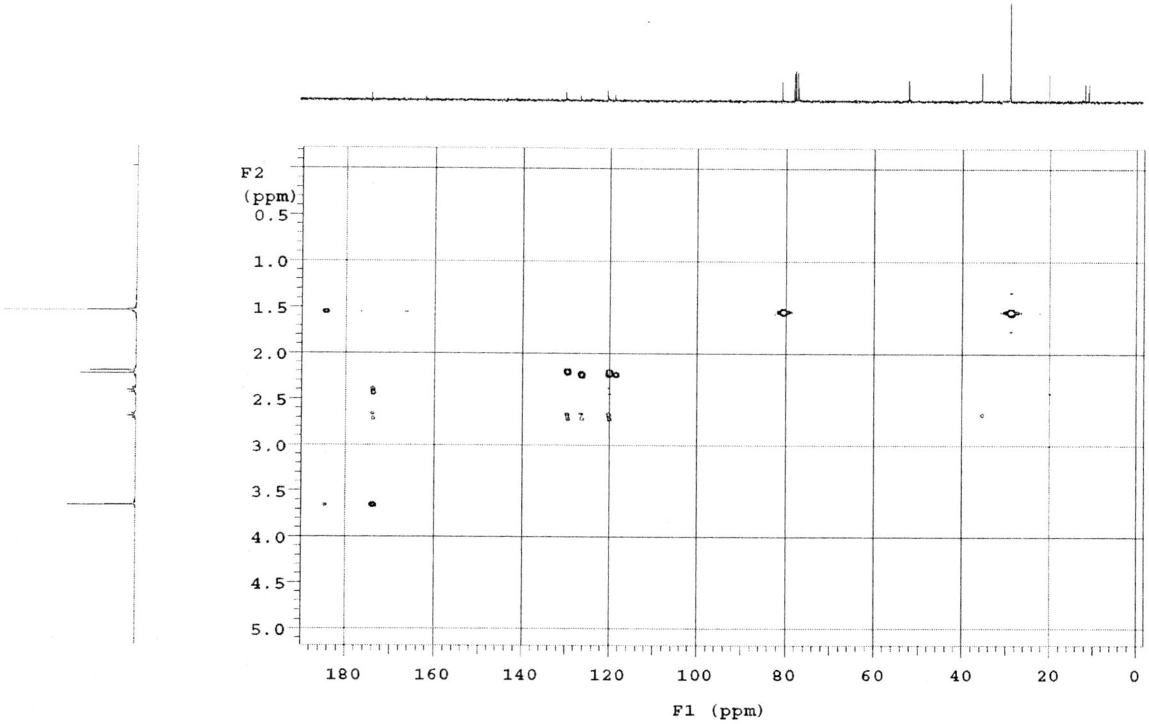

gHMBC Spectrum in $CDCl_3$

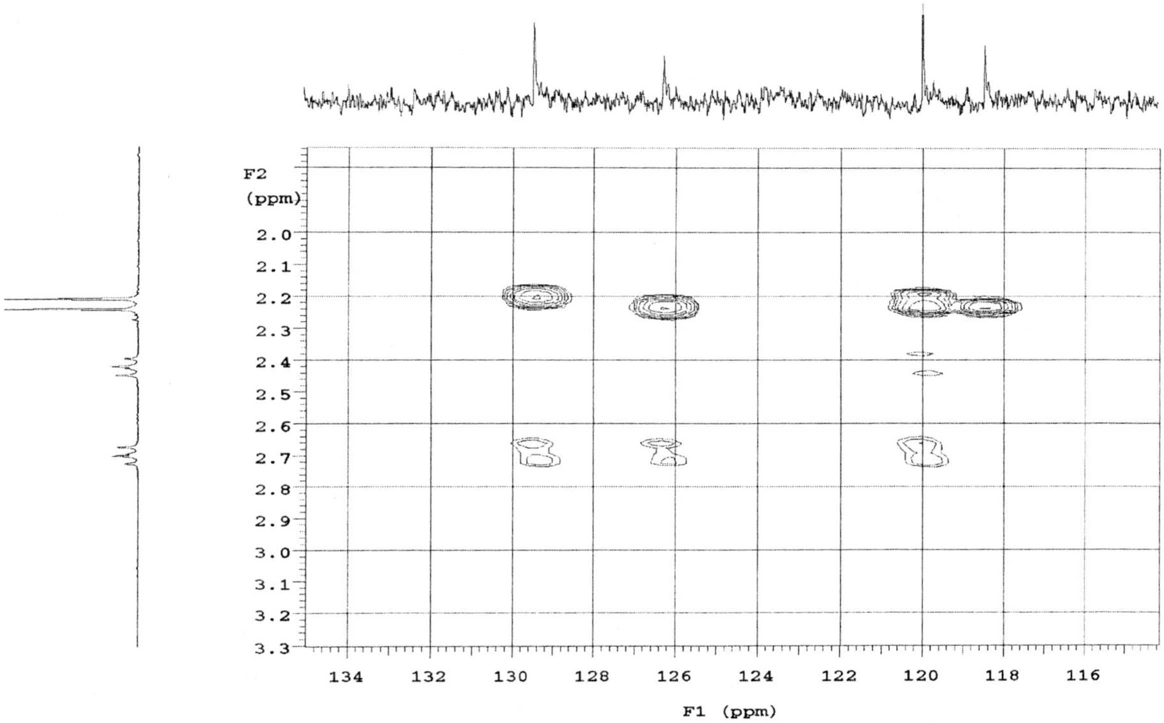

Expansion of gHMBC Spectrum in $CDCl_3$

Problem 3.37 Given the spectroscopic data below (Proton, Carbon, gHSQC, gCOSY, and gHMBC), provide the structure of the molecule. ($C_{13}H_{19}NO_4$)

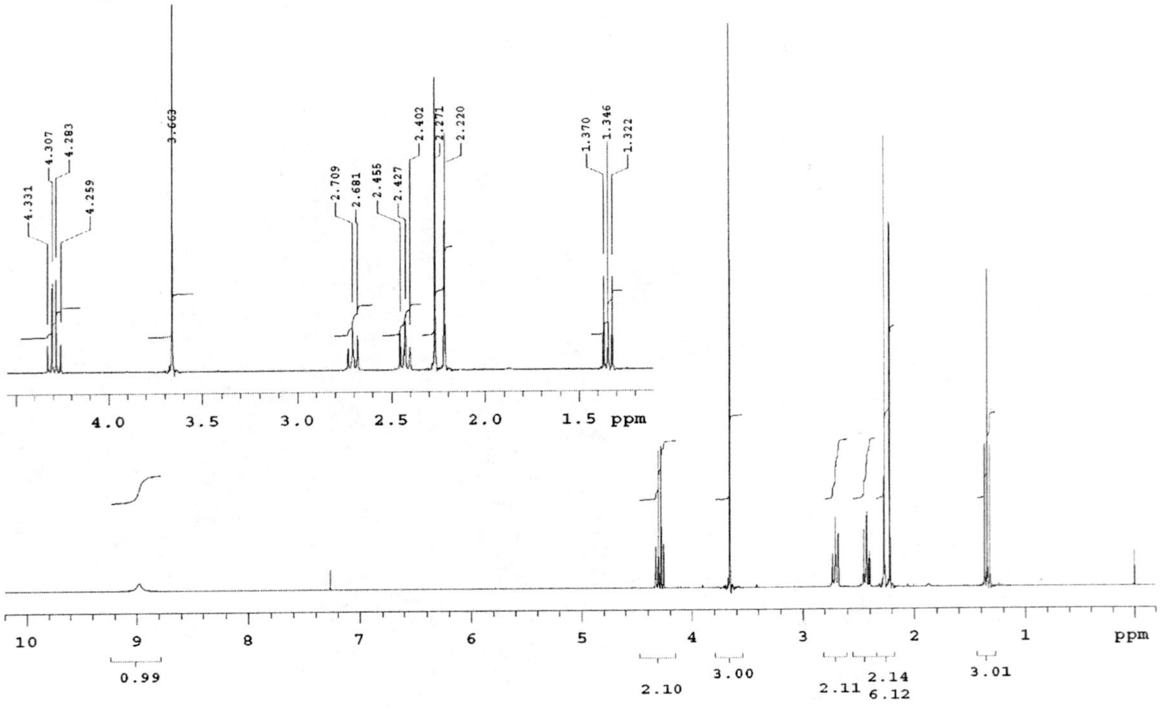

^{1}H-NMR Spectrum in $CDCl_3$

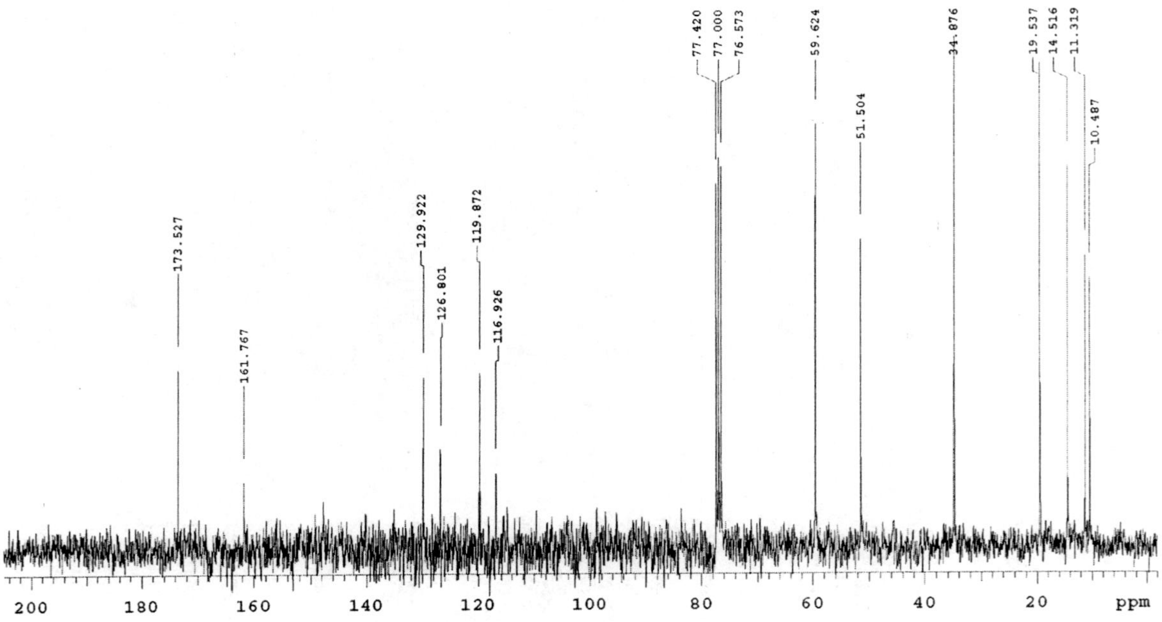

^{13}C-NMR Spectrum in $CDCl_3$

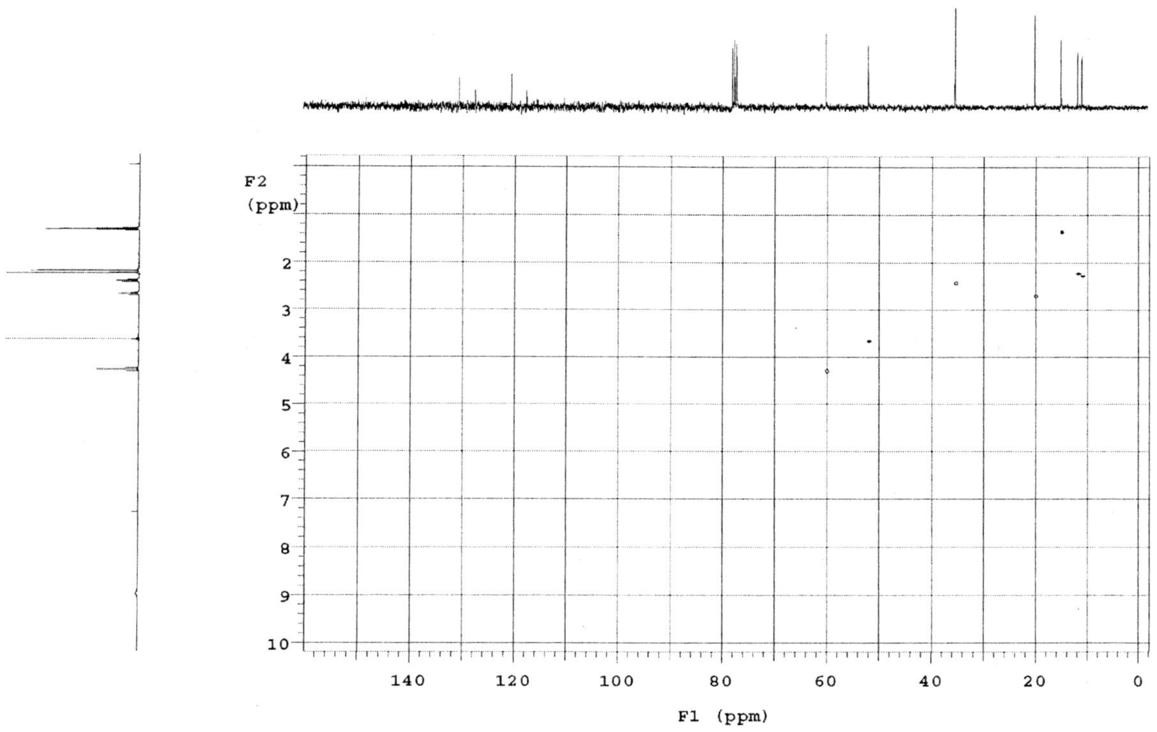

gHSQC Spectrum in $CDCl_3$

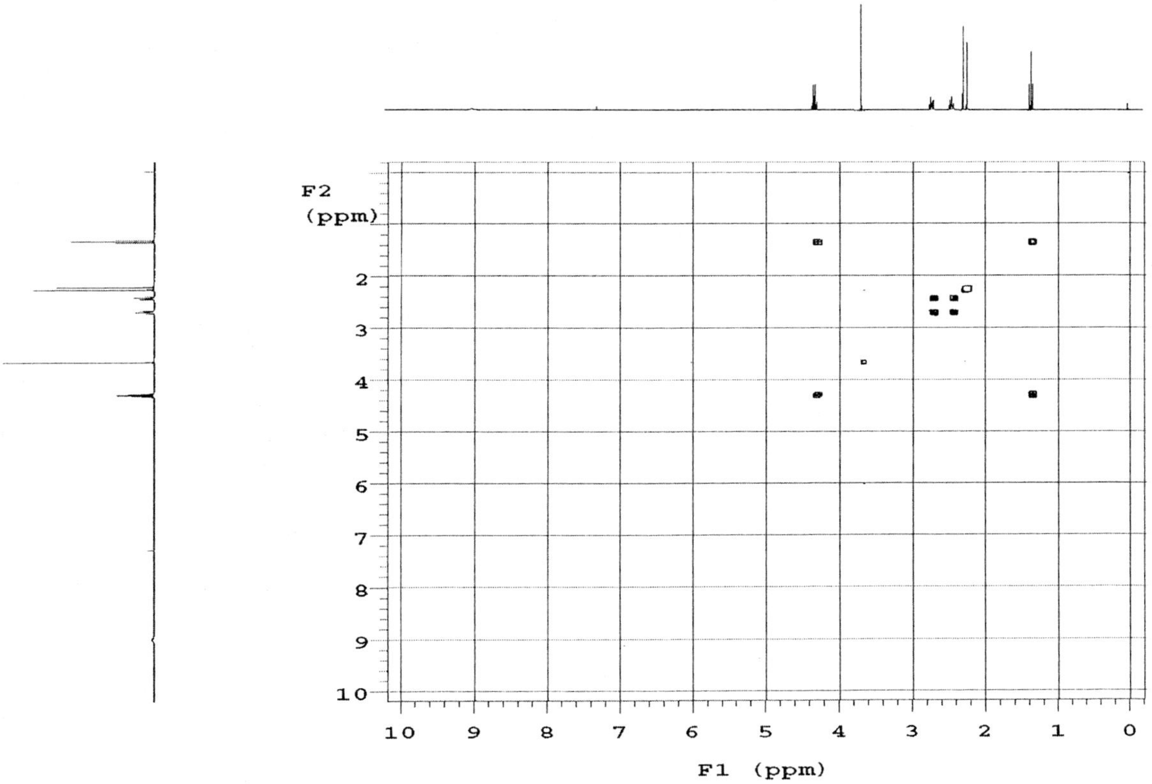

gCOSY Spectrum in $CDCl_3$

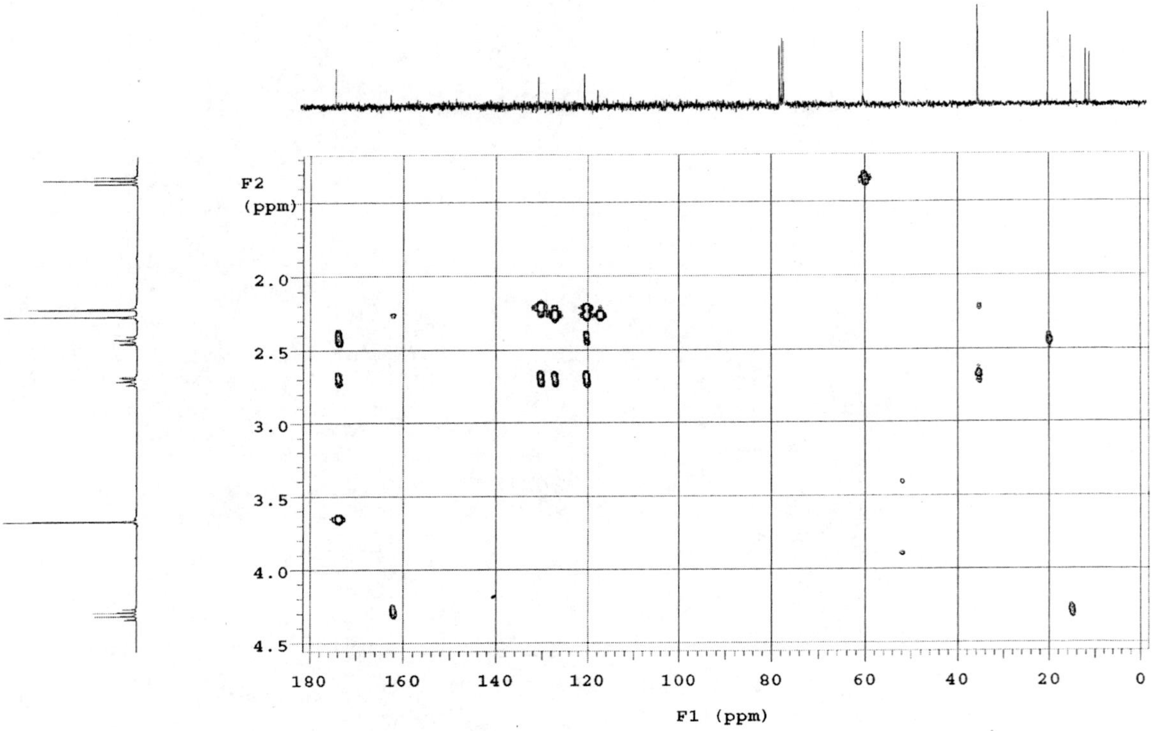

gHMBC Spectrum in $CDCl_3$

Problem 3.38 Given the spectroscopic data below (Proton, Carbon, gCOSY, gHSQC, and gHMBC), provide the structure of the molecule. (HR-MS: 267.1470 amu)

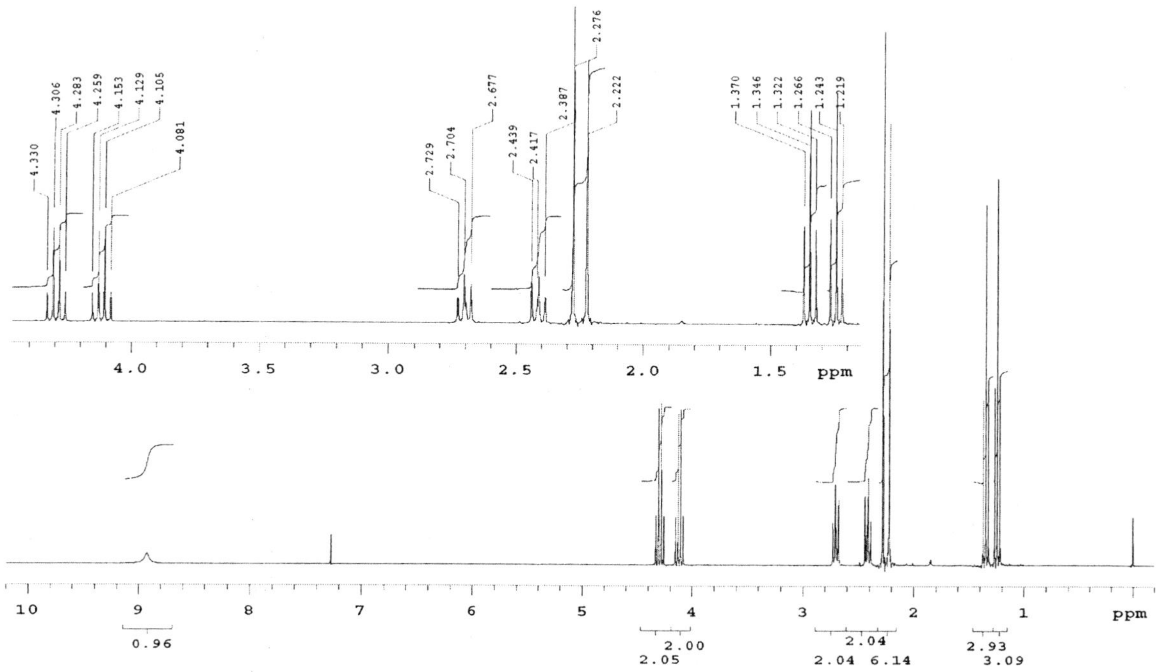

^{1}H-NMR Spectrum in $CDCl_3$

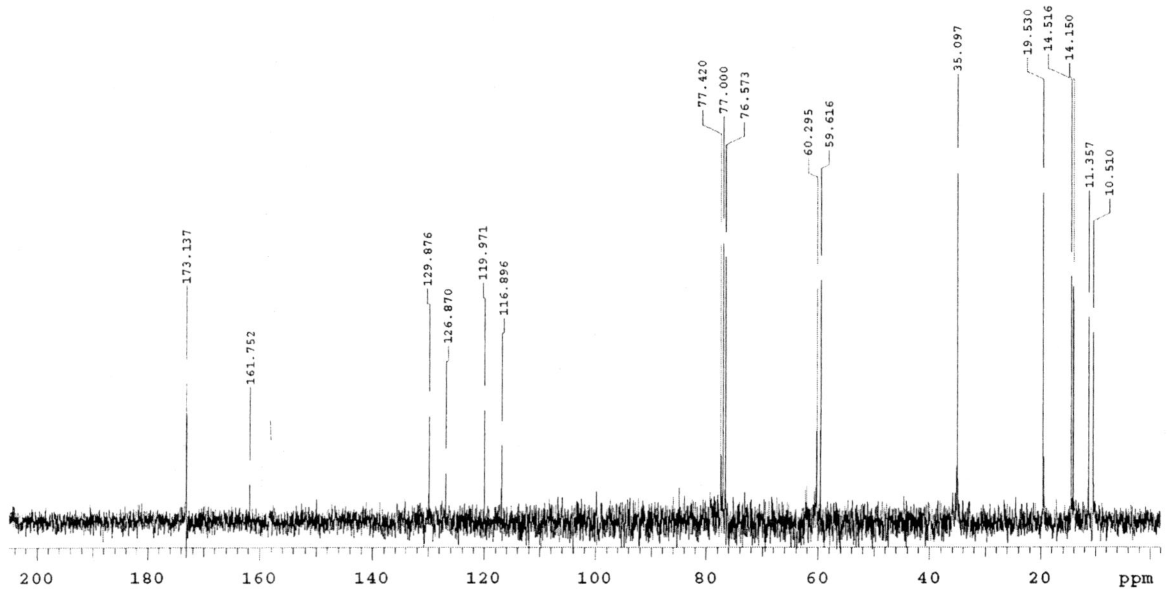

^{13}C-NMR Spectrum in $CDCl_3$

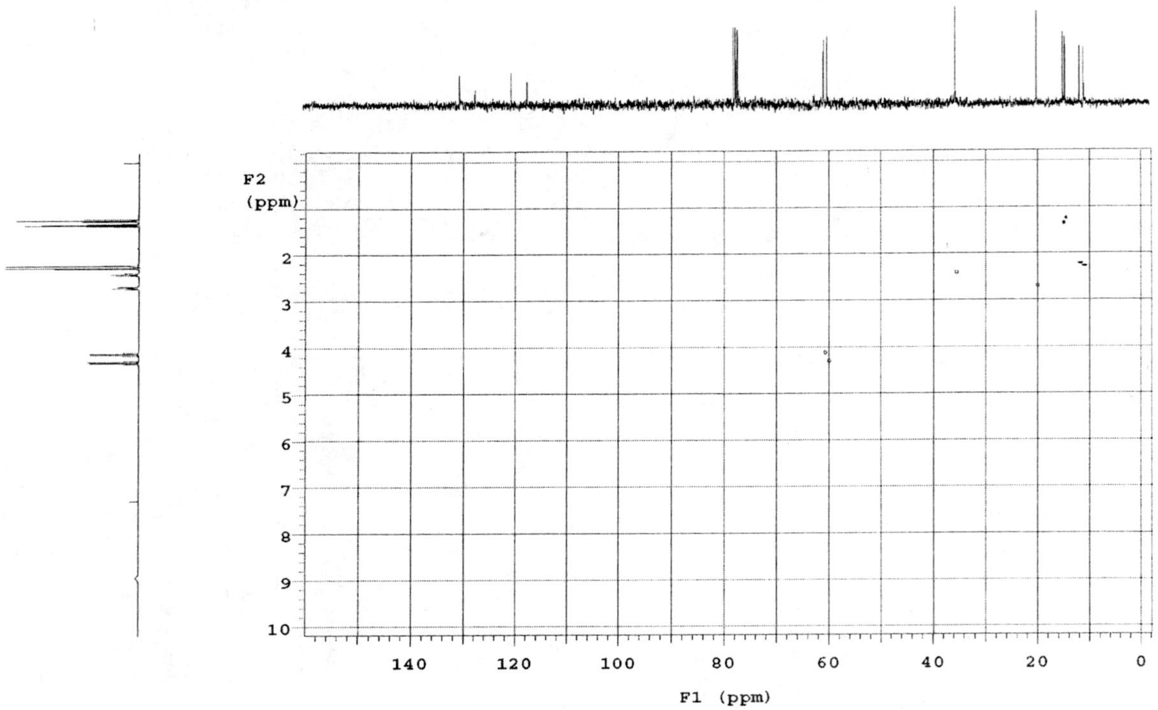

gHSQC Spectrum in $CDCl_3$

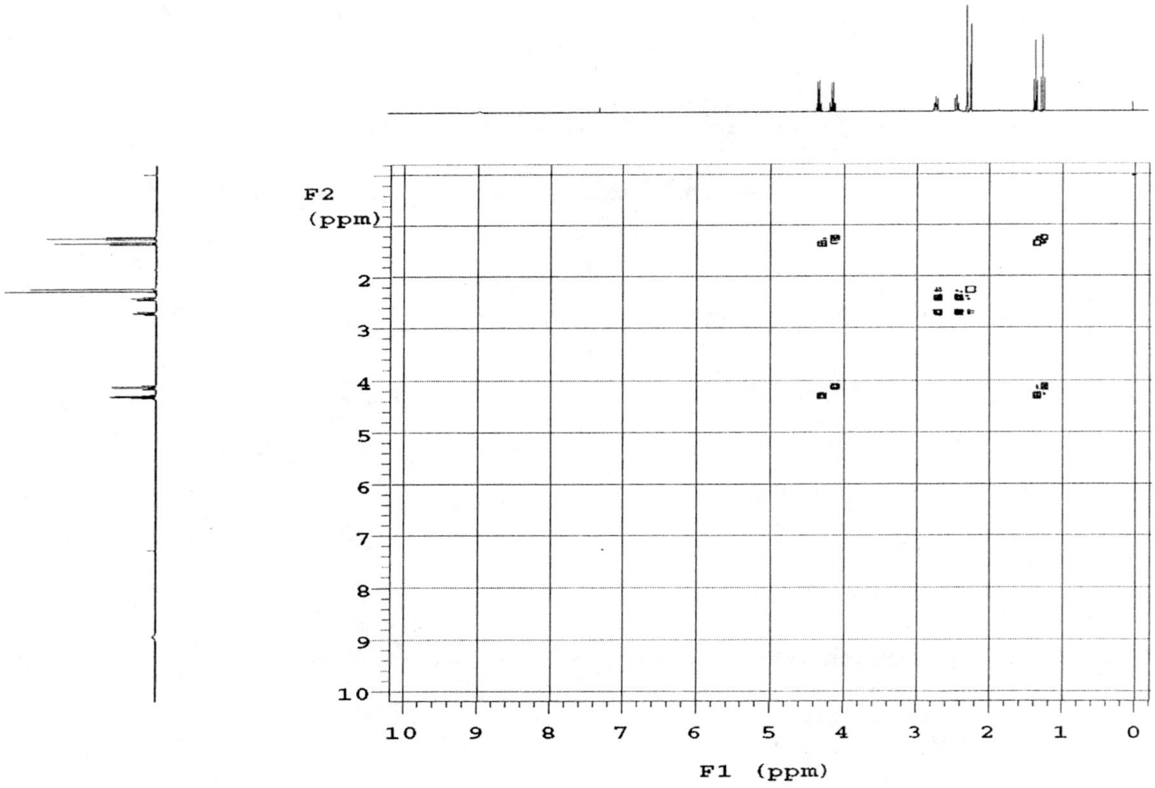

gCOSY Spectrum in $CDCl_3$

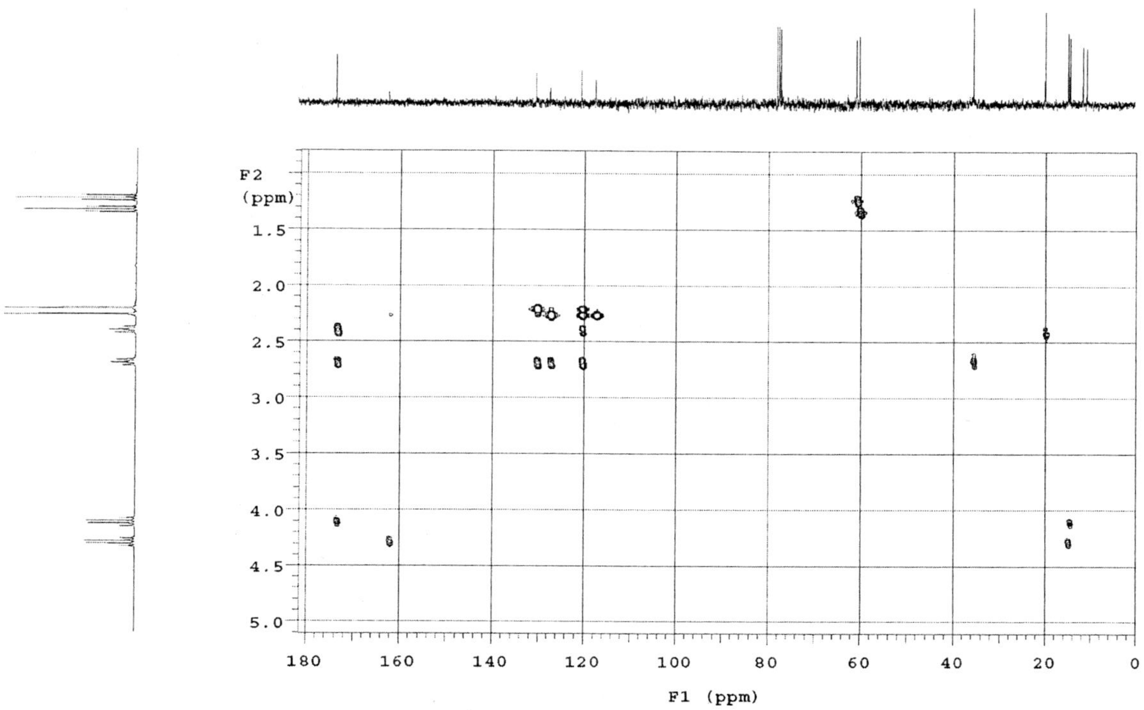

gHMBC Spectrum in $CDCl_3$

Problem 3.39 Given the spectroscopic data below (Proton, Carbon, gHSQC, gCOSY, and gHMBC), provide the structure of the molecule. (MW: 168 g/mol)

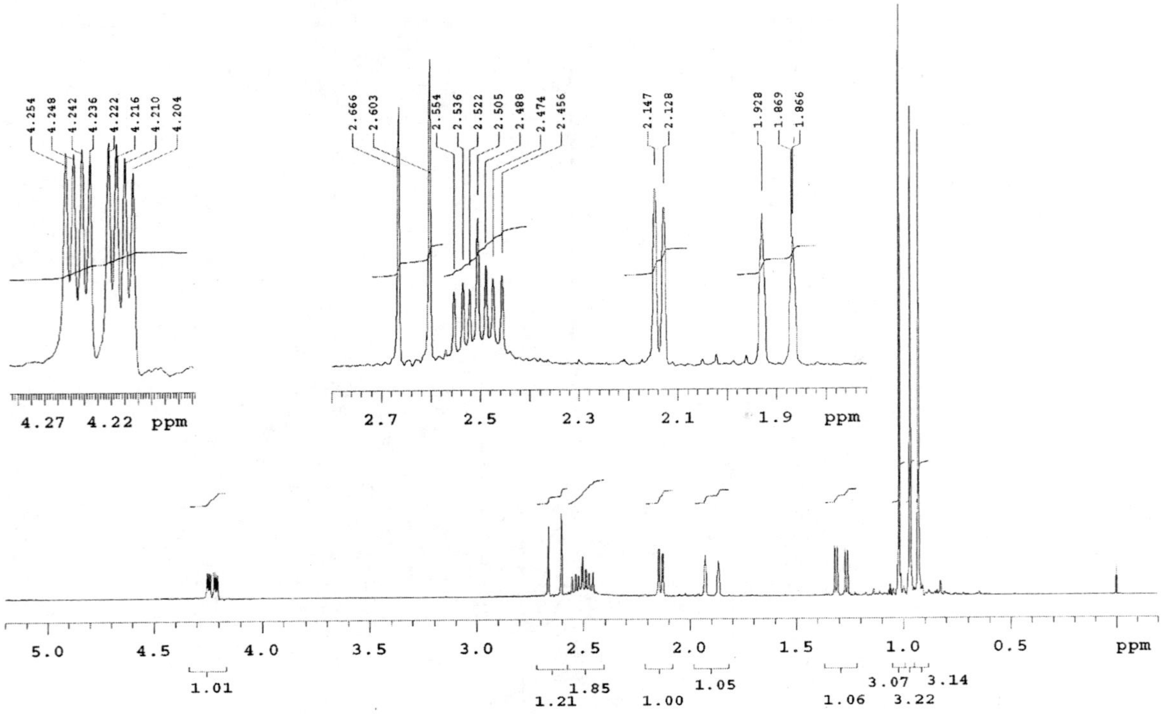

^{1}H-NMR Spectrum in $CDCl_3$

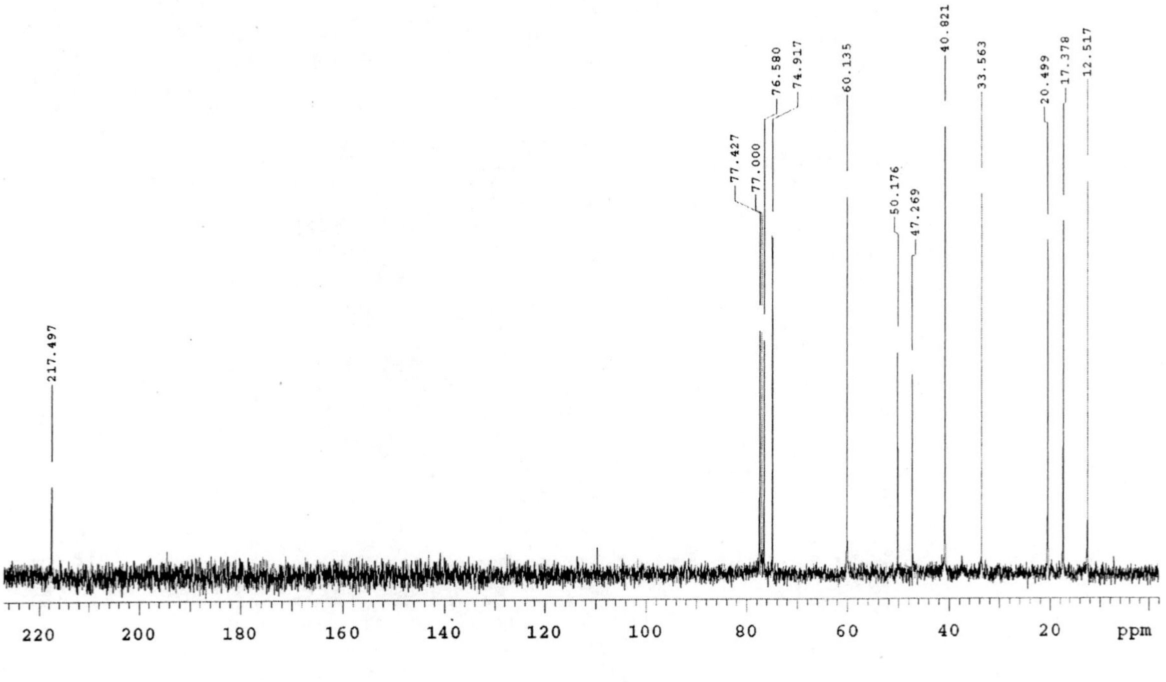

^{13}C-NMR Spectrum in $CDCl_3$

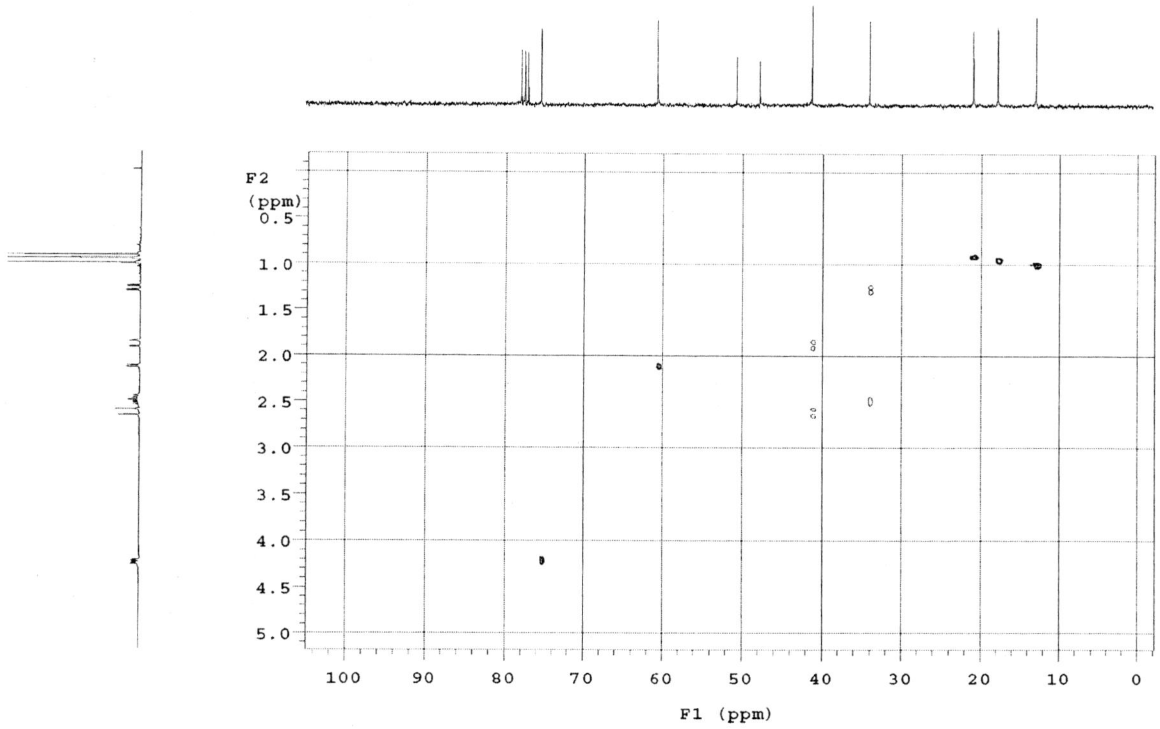

gHSQC Spectrum in $CDCl_3$

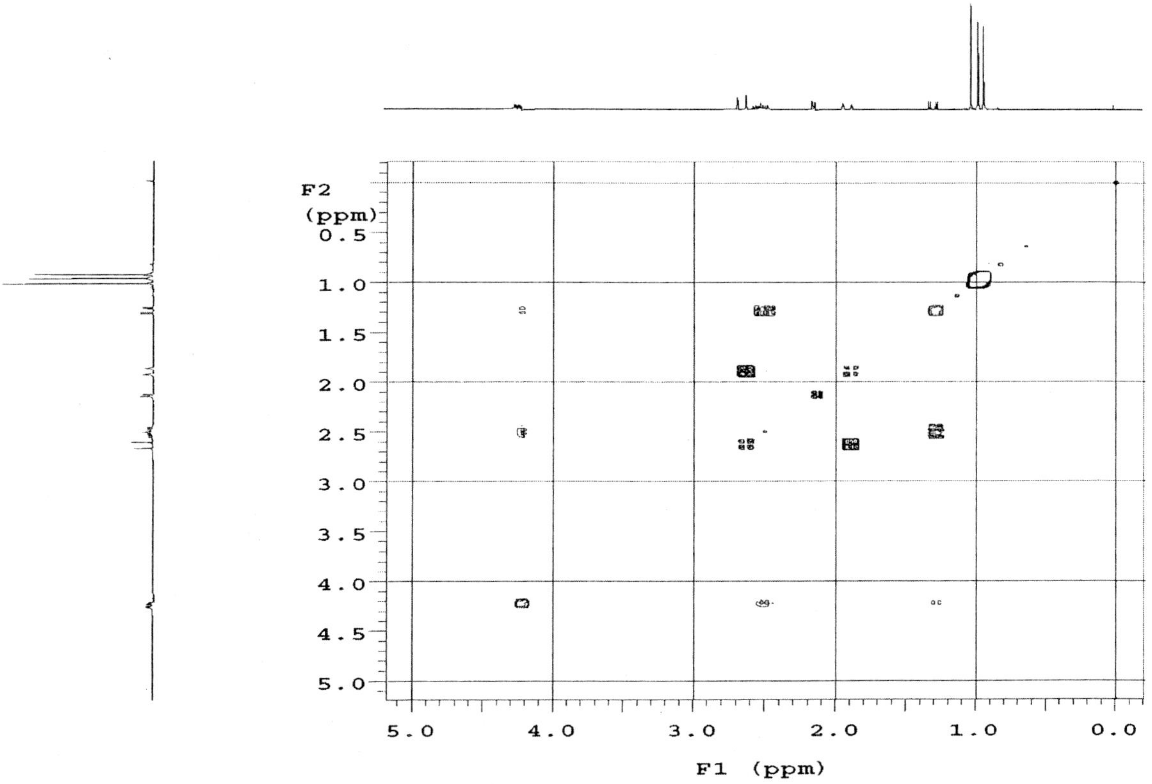

gCOSY Spectrum in $CDCl_3$

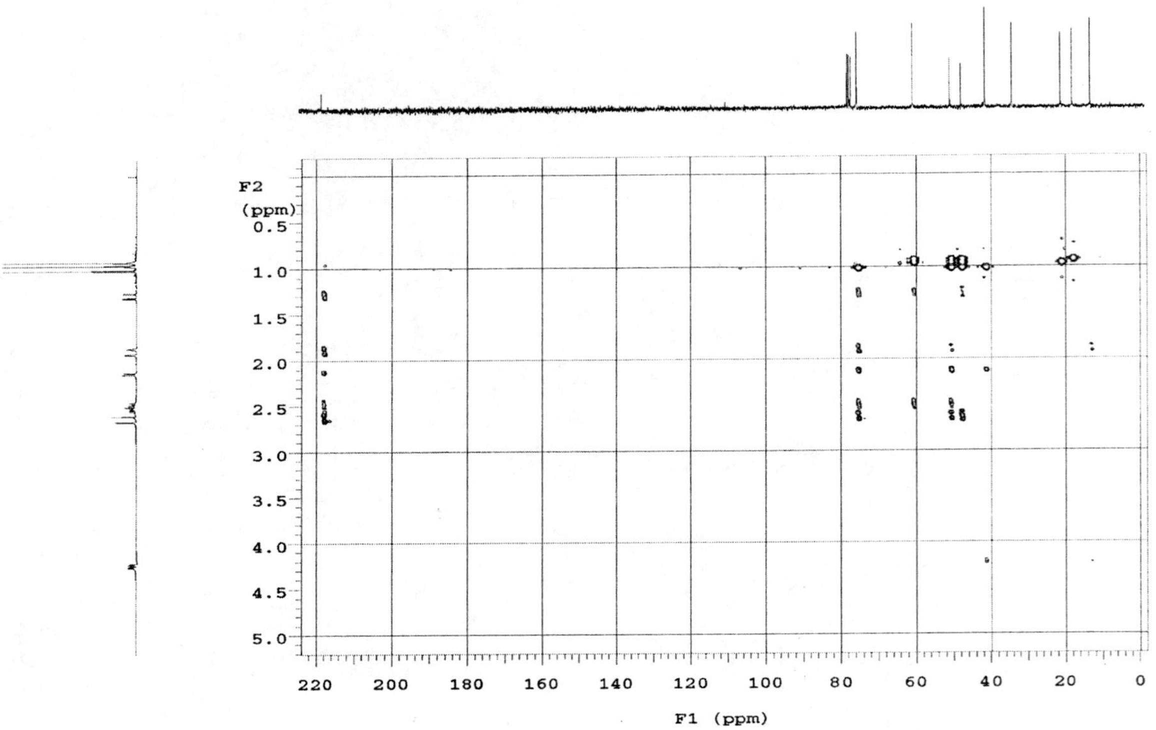

gHMBC Spectrum in $CDCl_3$

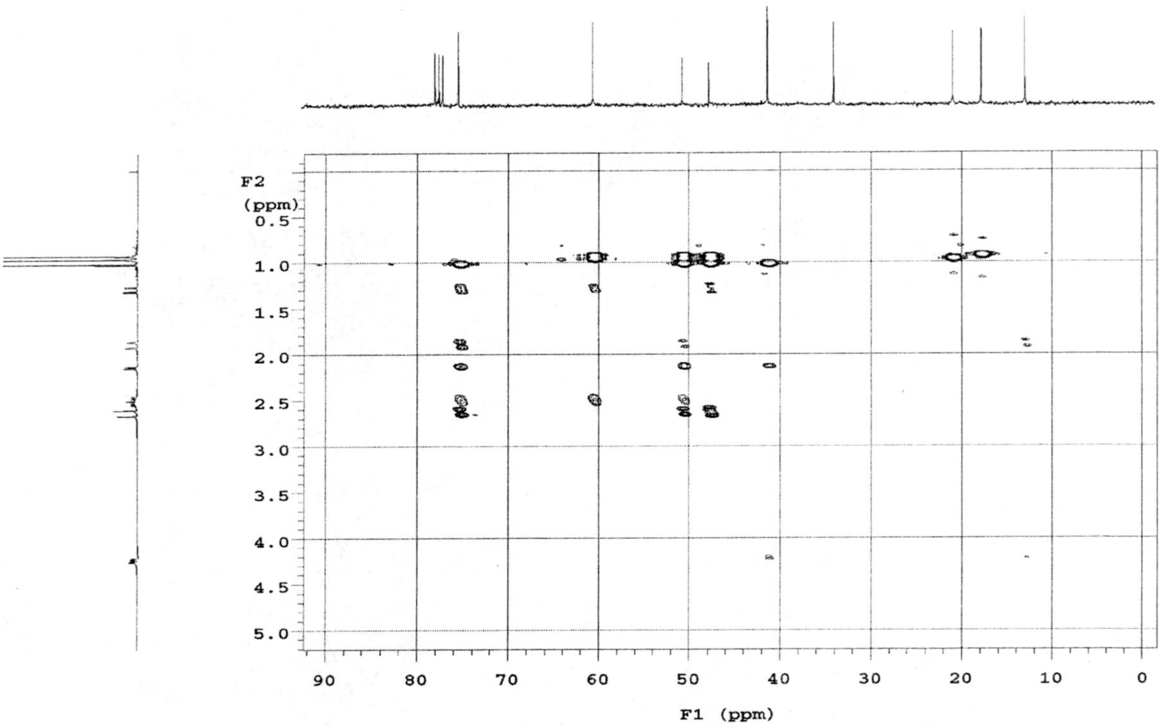

Expansion of gHMBC Spectrum in $CDCl_3$

Problem 3.40 Given the spectroscopic data below (Proton, Carbon, Edited DEPT, gHSQC, and gCOSY), provide the structure of the molecule. (MW: 124 g/mol)

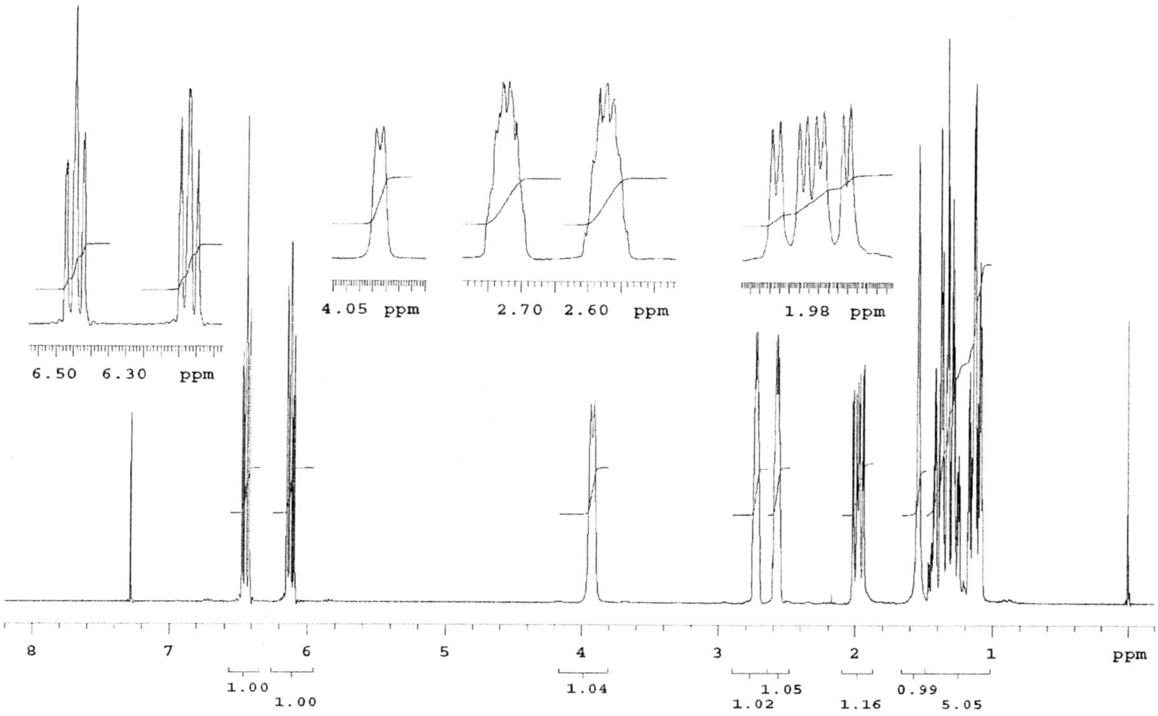

^{1}H-NMR Spectrum in $CDCl_3$

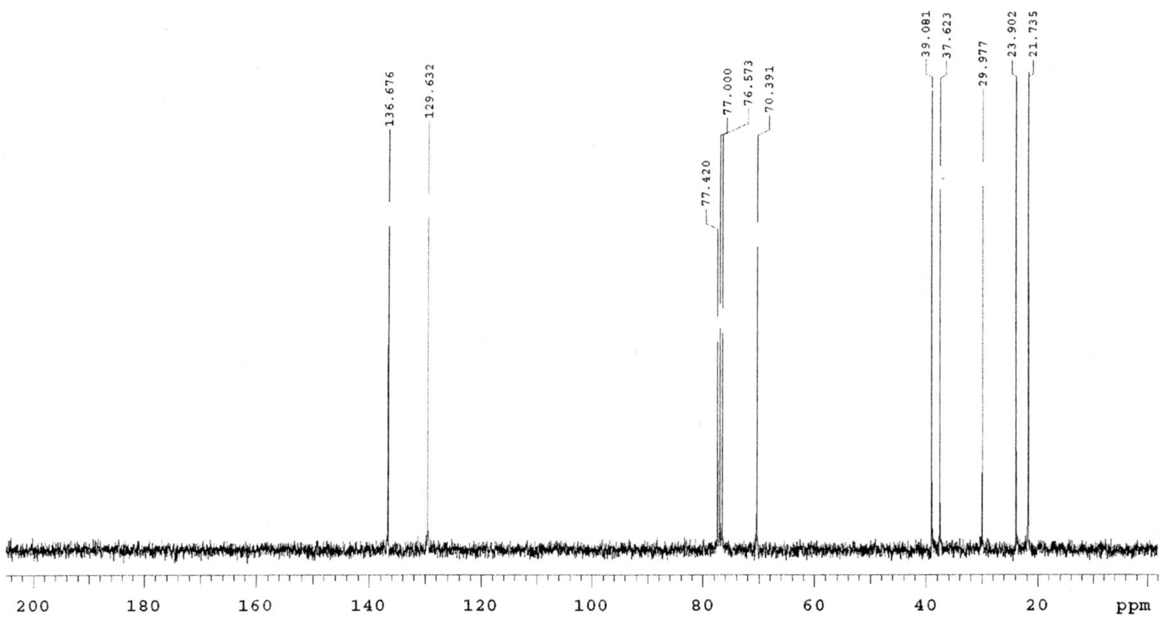

^{13}C-NMR Spectrum in $CDCl_3$

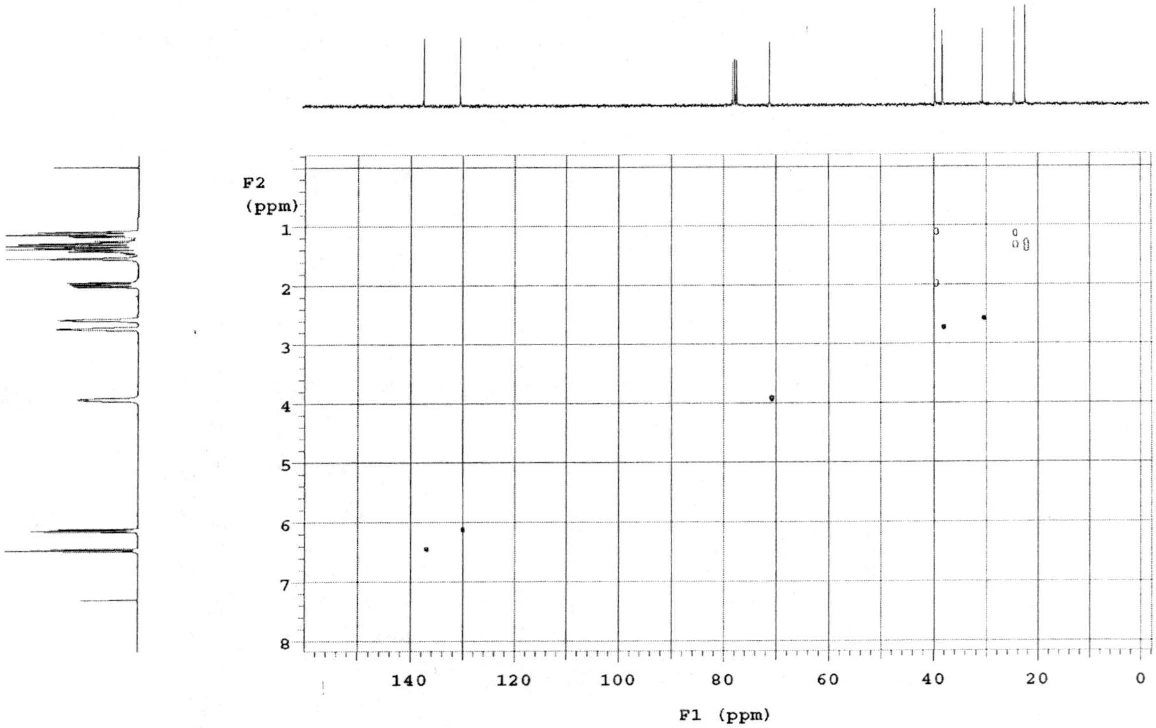

gHSQC Spectrum in $CDCl_3$

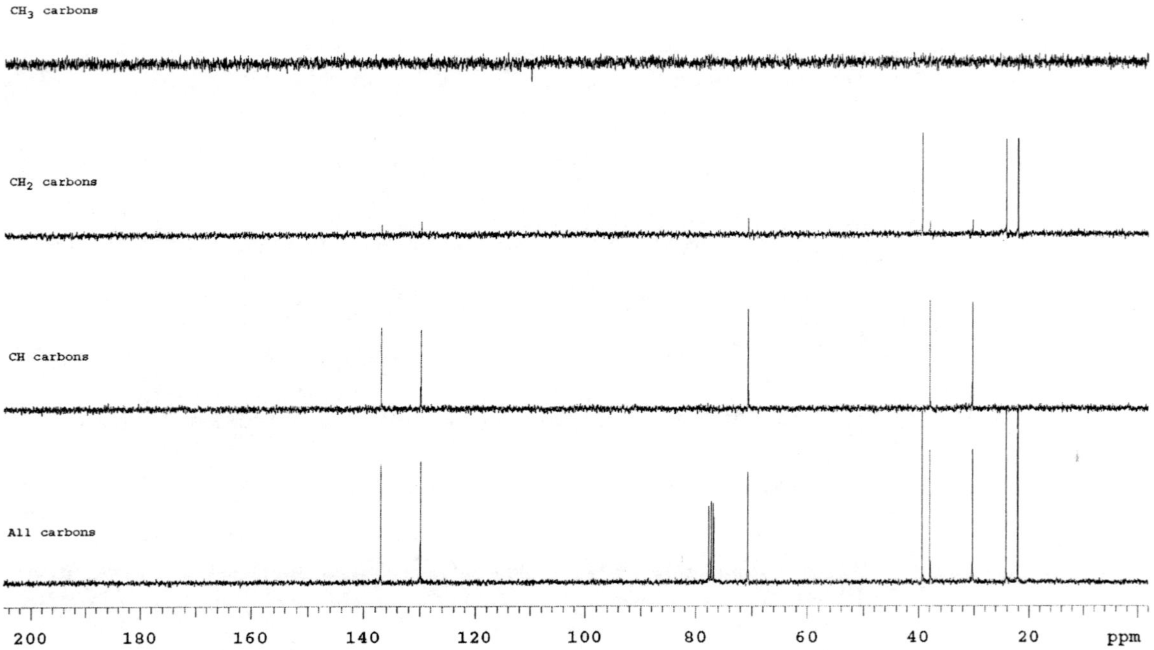

Edited DEPT ^{13}C-NMR Spectrum in $CDCl_3$

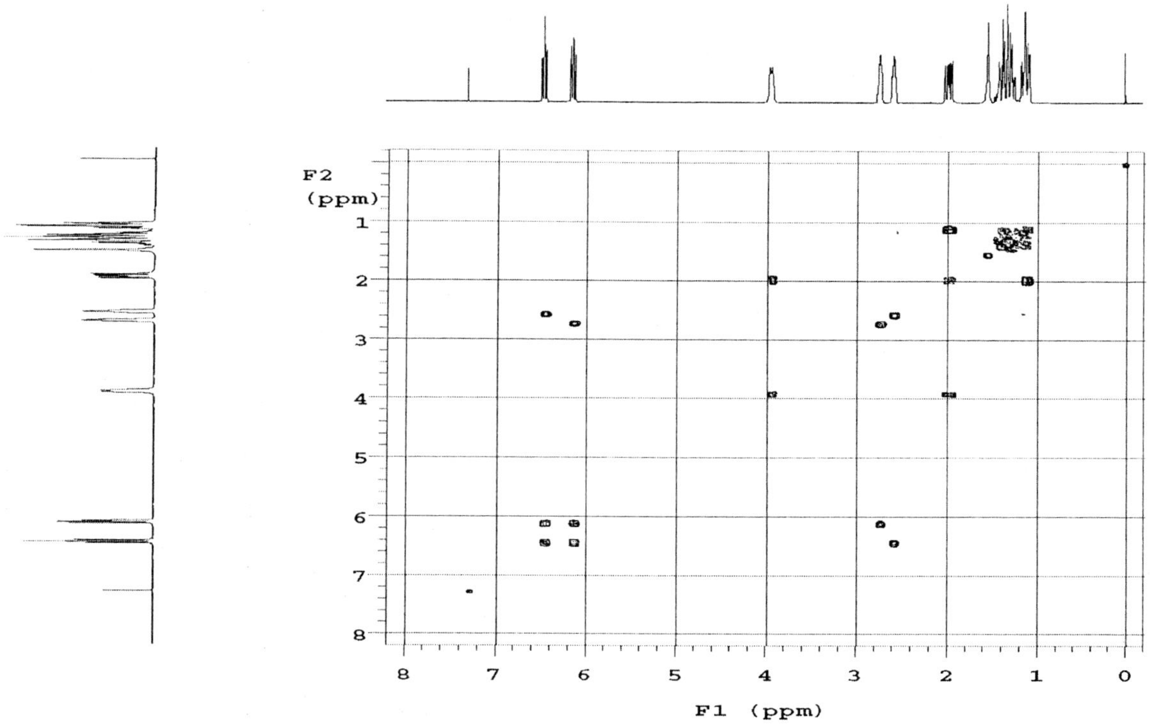

gCOSY Spectrum in $CDCl_3$

Problem 3.41 Given the spectroscopic data below (Proton, Carbon, gHSQC, gCOSY, and gHMBC), provide the structure of the molecule.

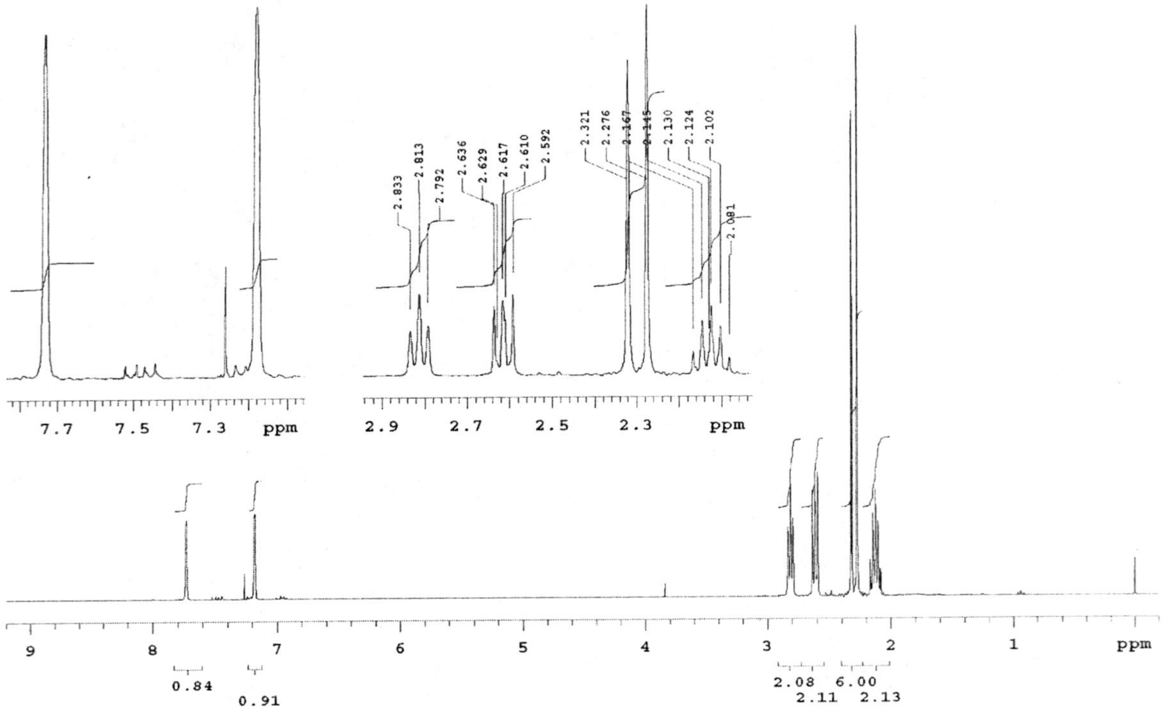

^{1}H-NMR Spectrum in $CDCl_3$

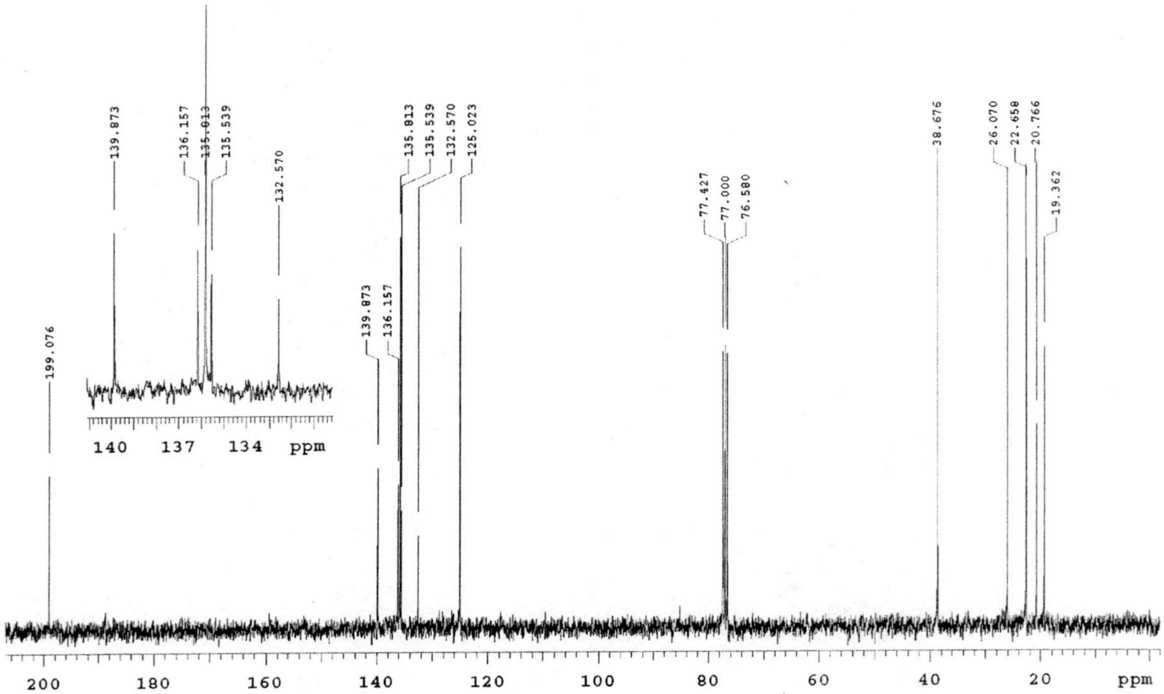

^{13}C-NMR Spectrum in $CDCl_3$

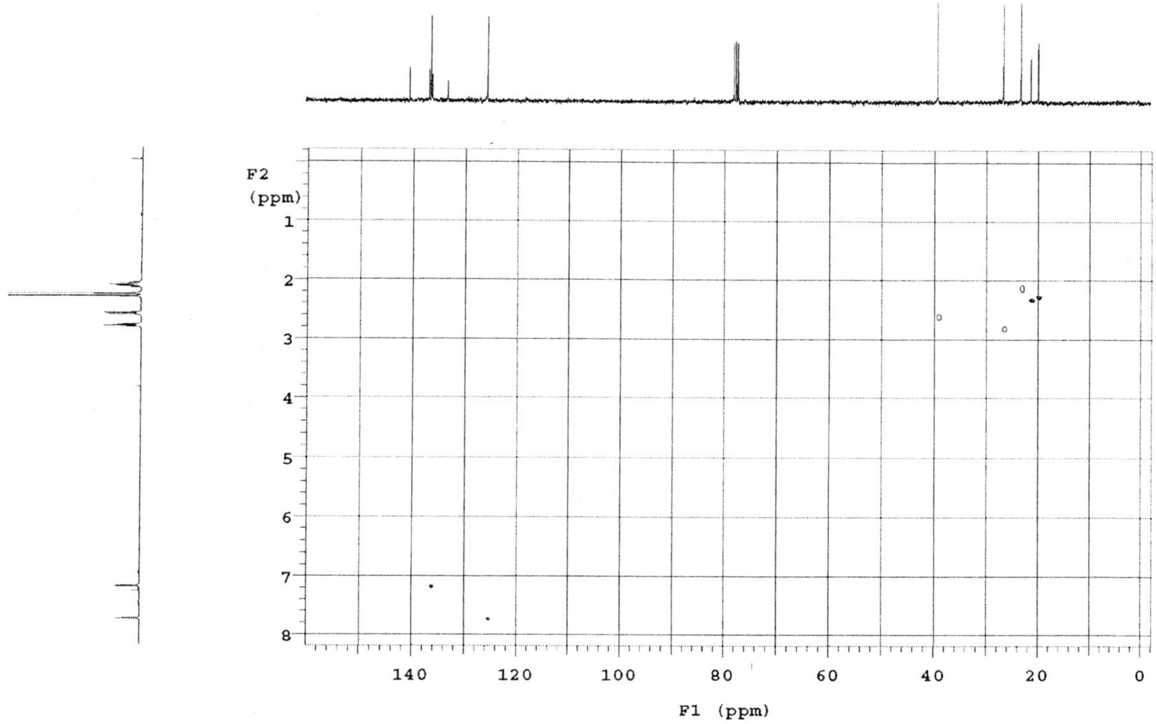

gHSQC Spectrum in $CDCl_3$

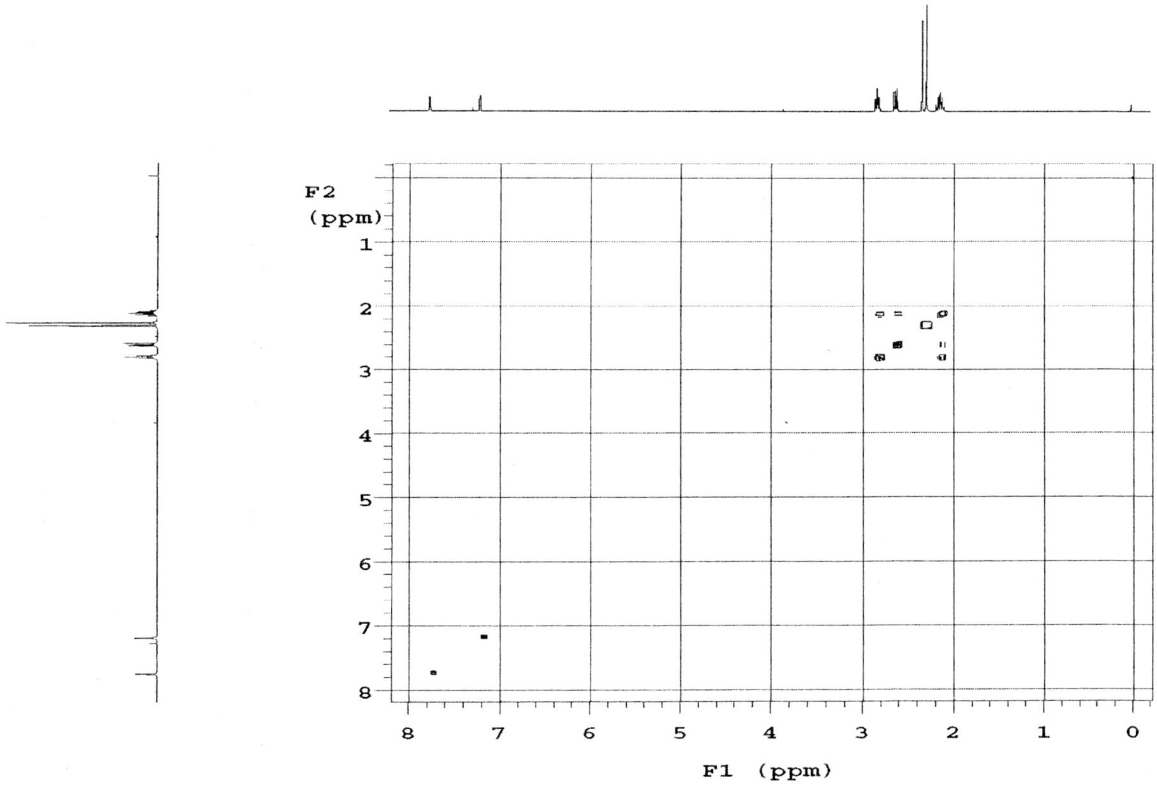

gCOSY Spectrum in $CDCl_3$

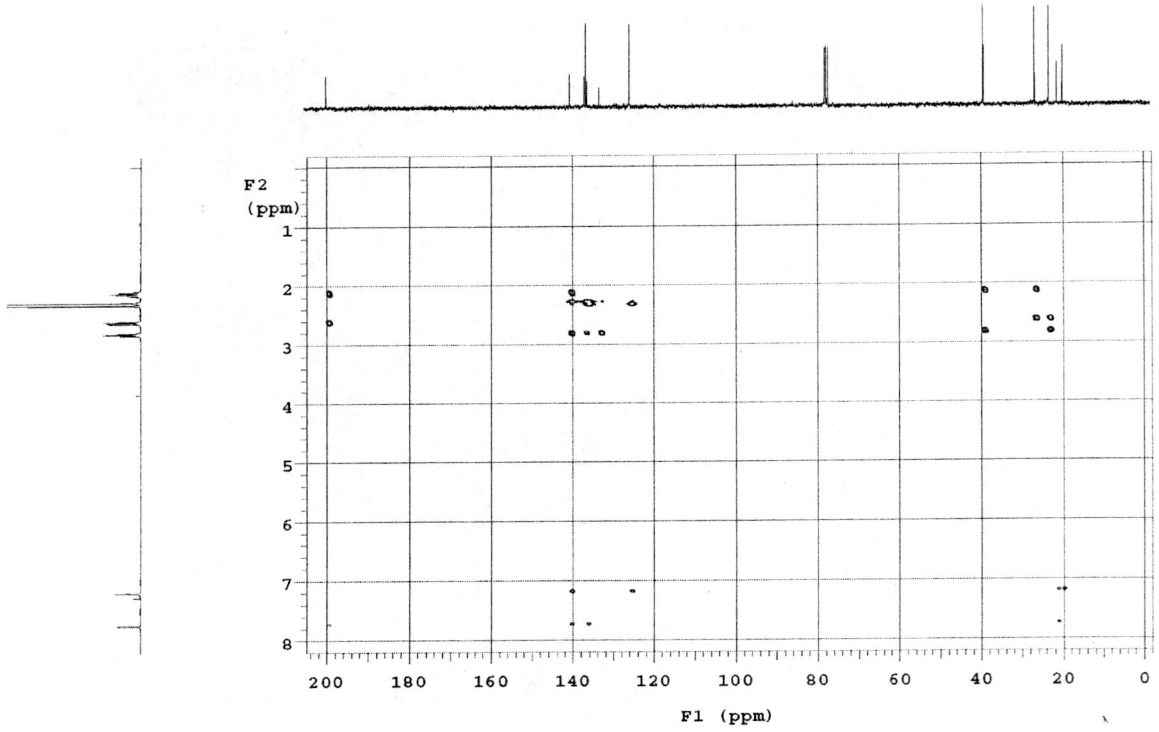

gHMBC Spectrum in $CDCl_3$

Problem 3.42 Given the spectroscopic data below (Proton, Carbon, gHSQC, and gCOSY), provide the structure of this blue colored molecule.

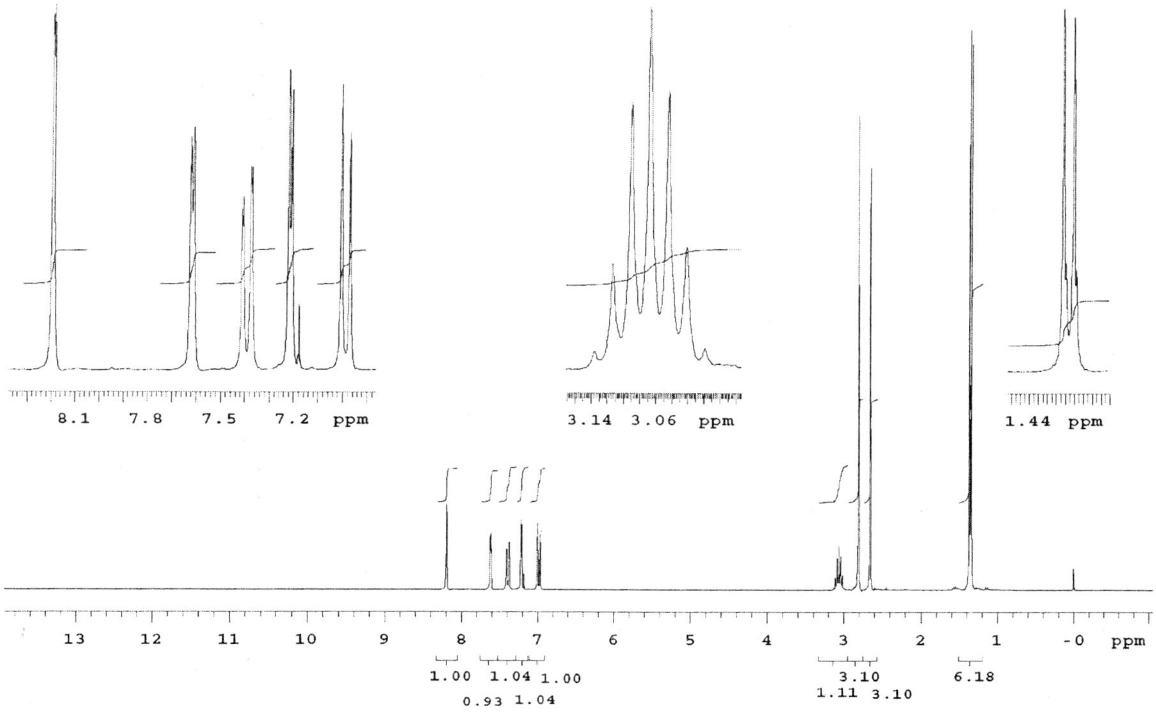

^{1}H-NMR Spectrum in $CDCl_3$

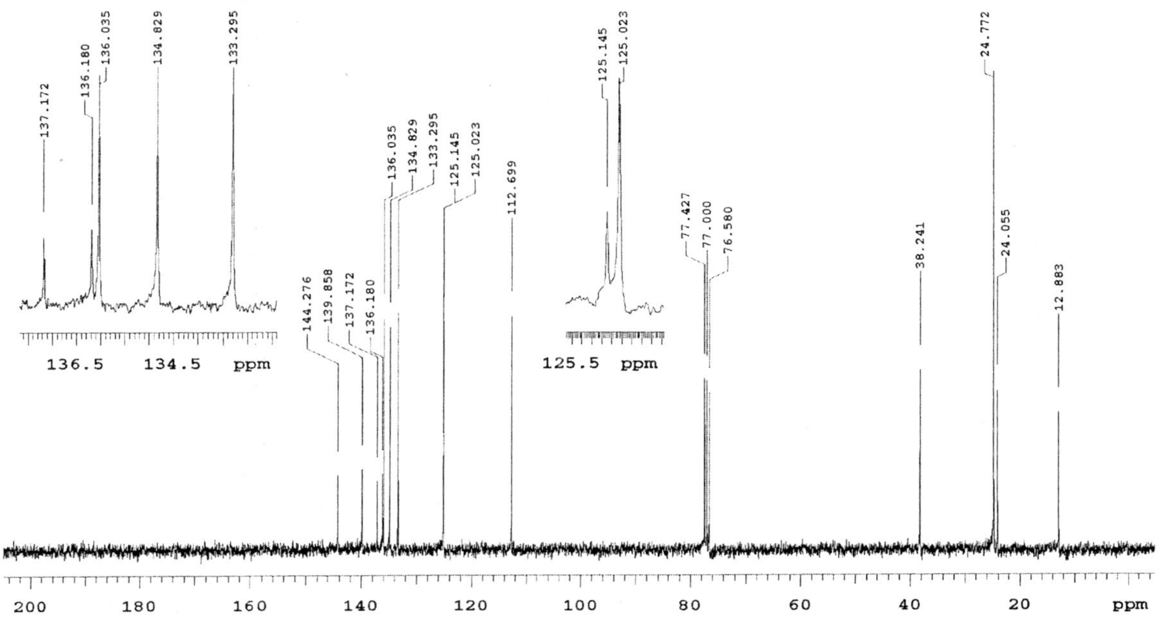

^{13}C-NMR Spectrum in $CDCl_3$

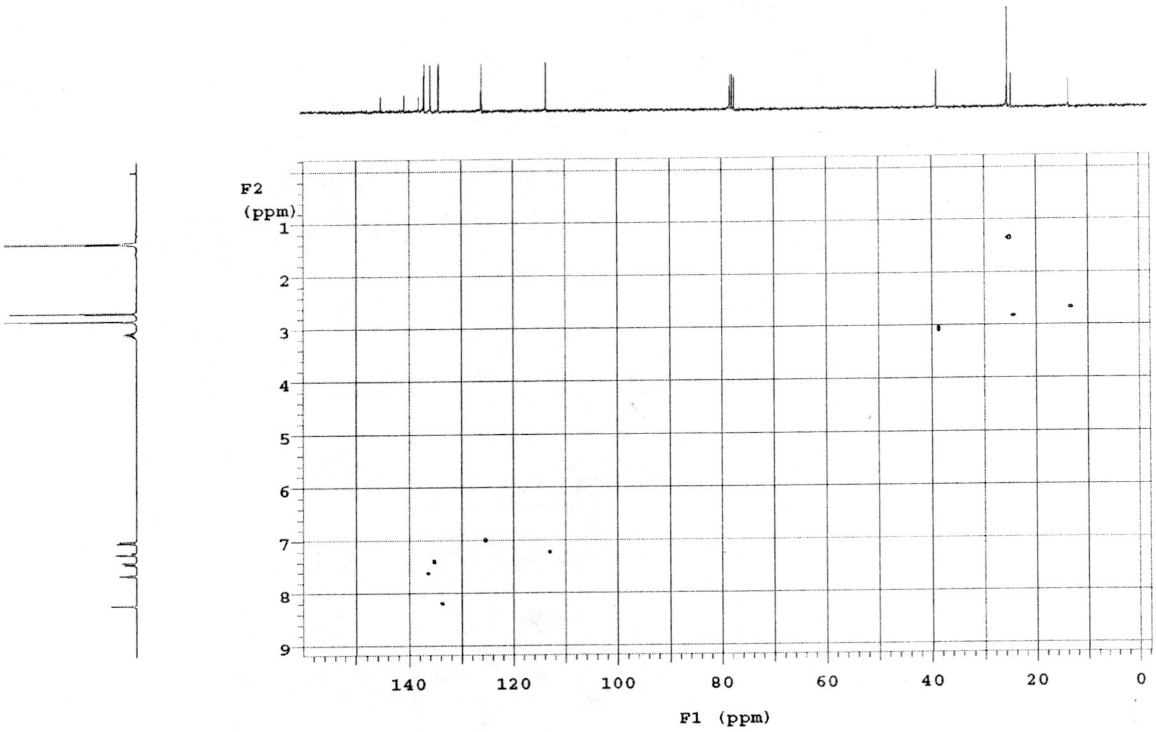

gHSQC Spectrum in $CDCl_3$

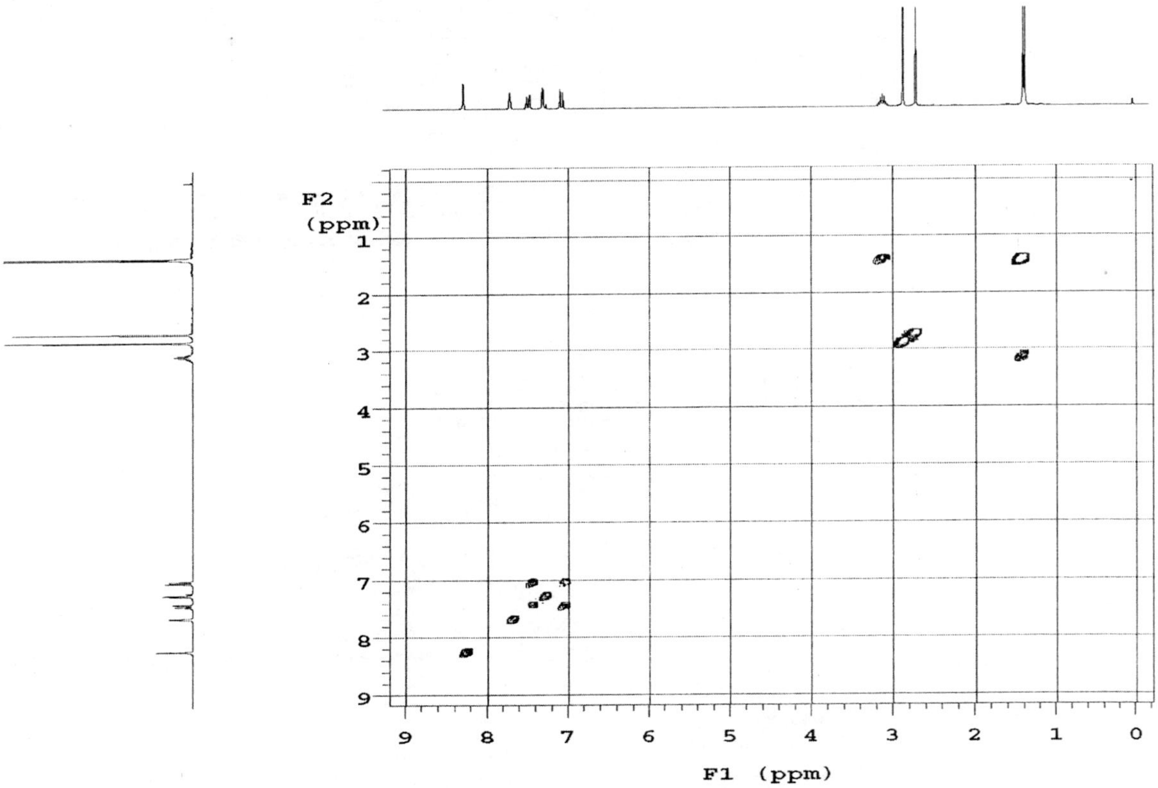

gCOSY Spectrum in $CDCl_3$

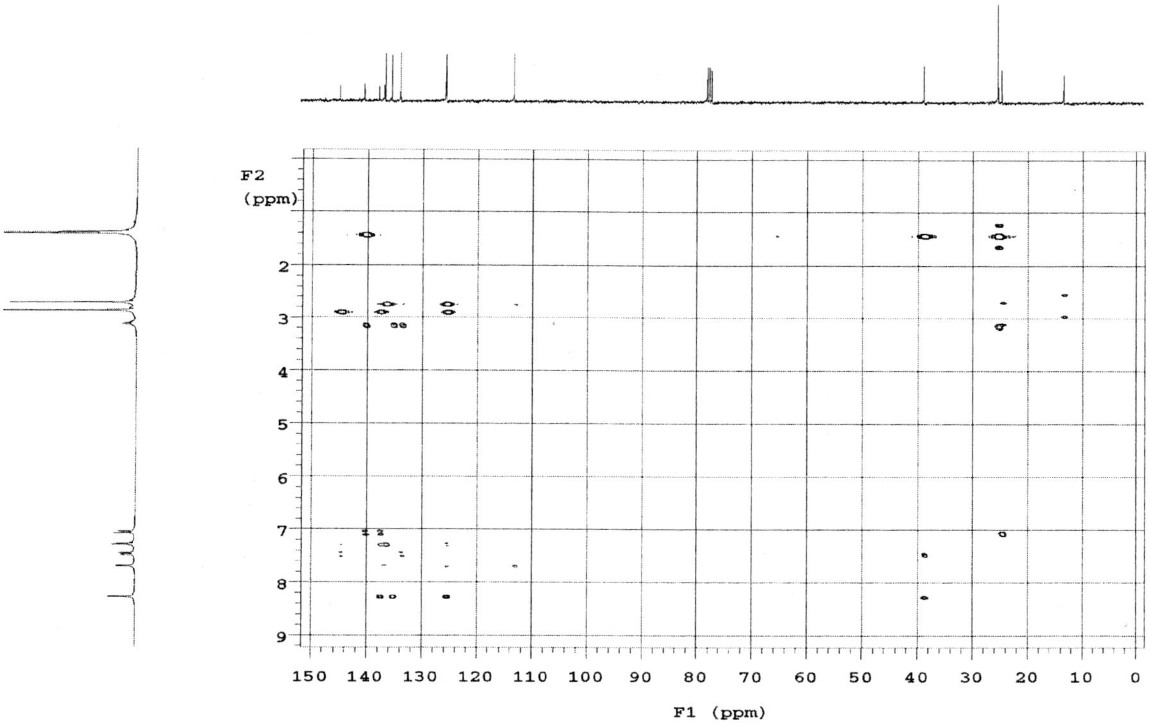

gHMBC Spectrum in $CDCl_3$

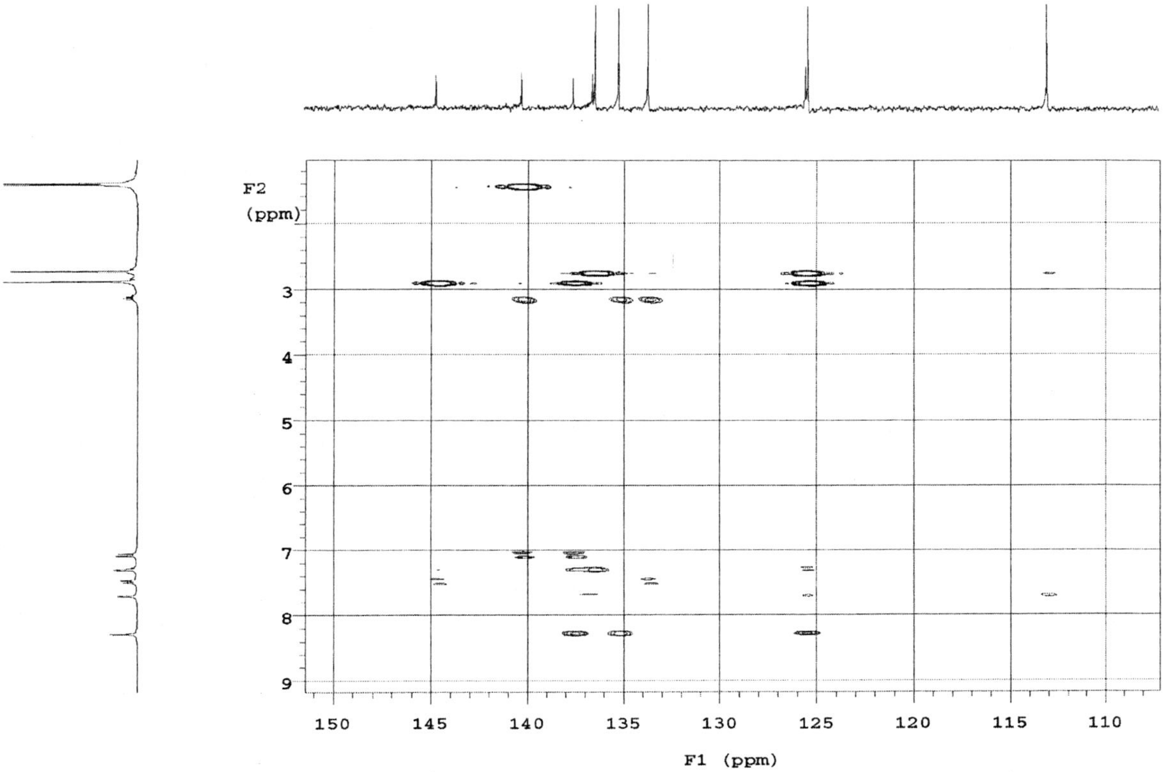

Expansion of gHMBC Spectrum in $CDCl_3$

Problem 3.43 Given the spectroscopic data below (Proton, Carbon, gHSQC, gCOSY, and gHMBC), provide the structure of the molecule. (HR-MS: 166.0994 amu)

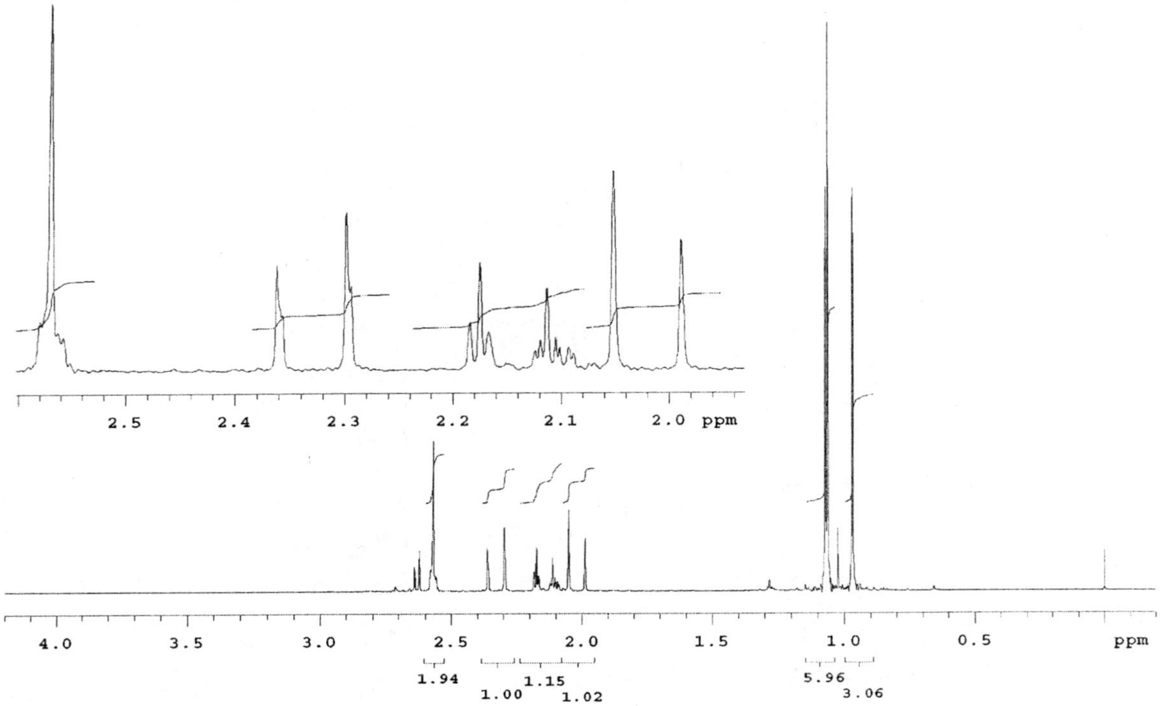

^{1}H-NMR Spectrum in $CDCl_3$

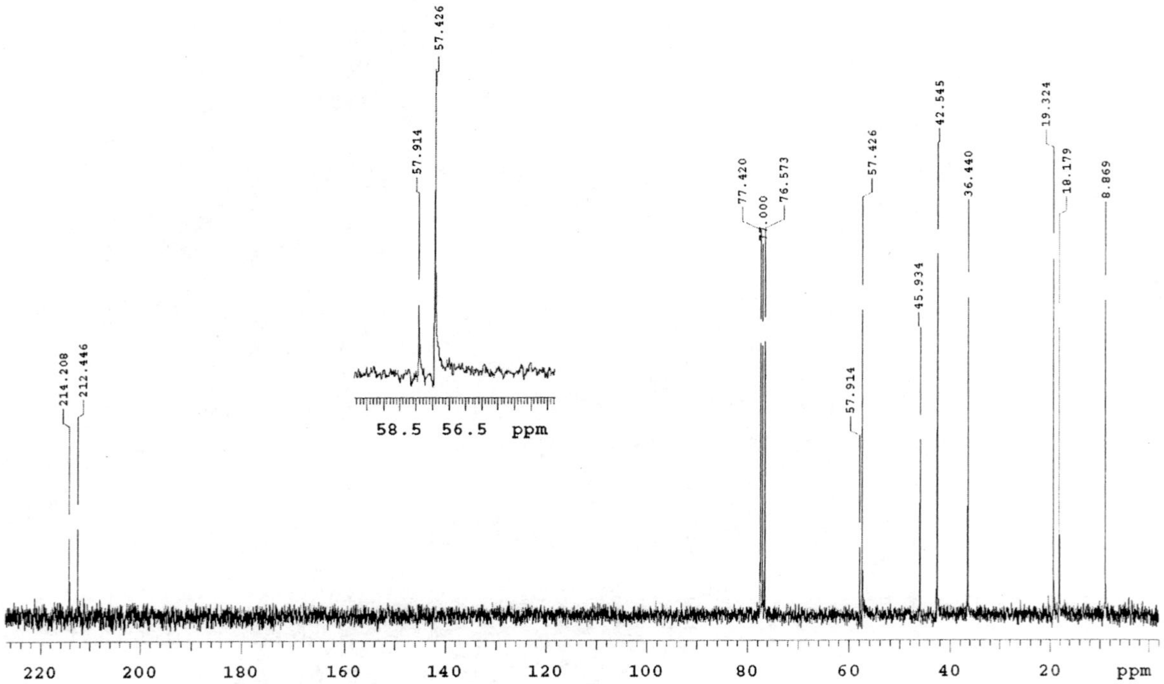

^{13}C-NMR Spectrum in $CDCl_3$

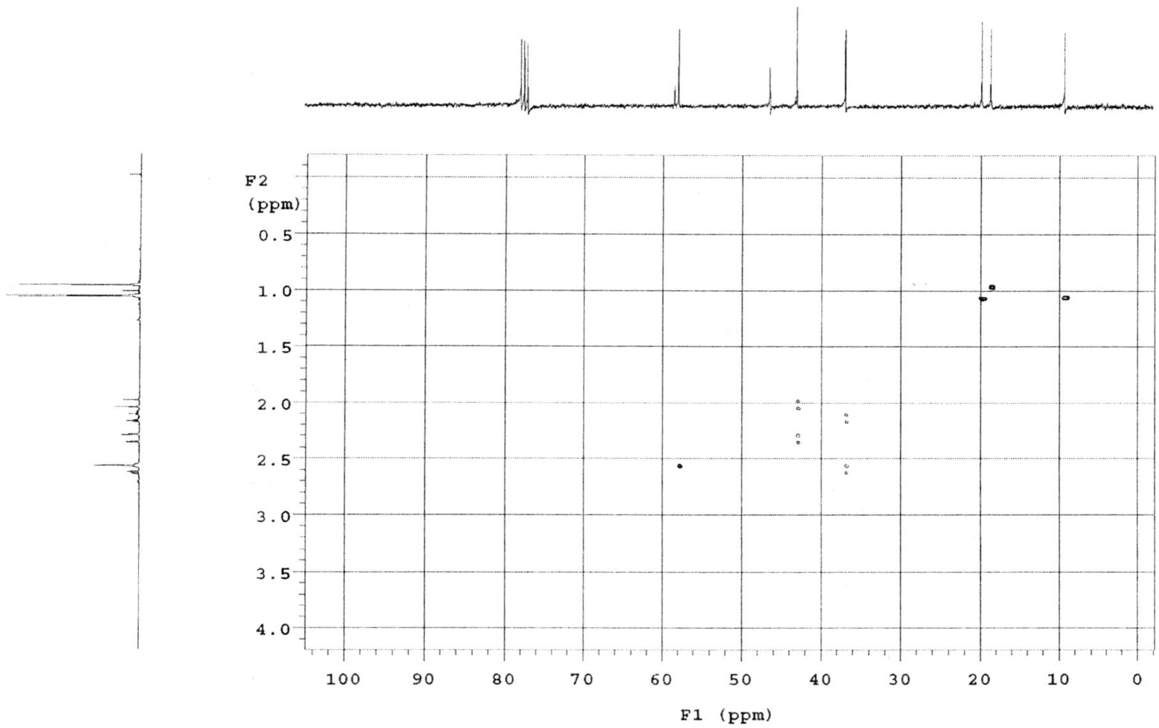

gHSQC Spectrum in $CDCl_3$

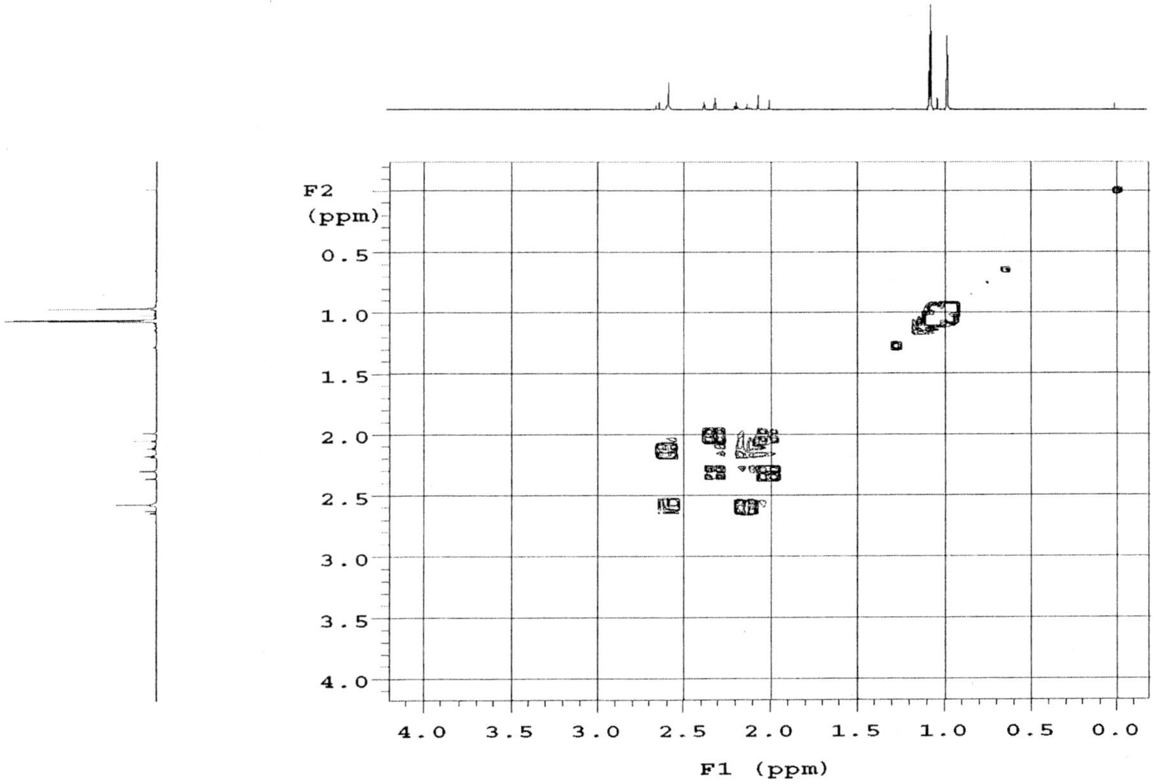

gCOSY Spectrum in $CDCl_3$

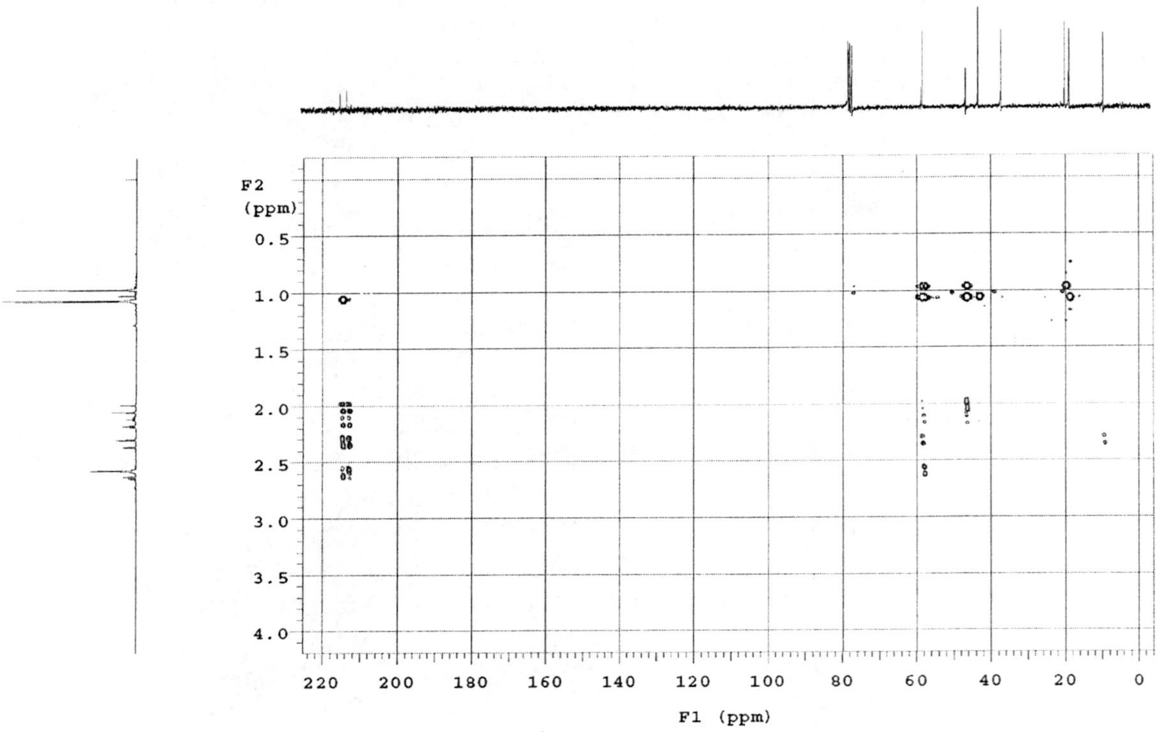

gHMBC Spectrum in $CDCl_3$

Problem 3.44 Given the spectroscopic data below (Proton, Carbon, gHSQC, gCOSY, and gHMBC), provide the structure of the molecule. (HR-MS: 166.0994 amu)

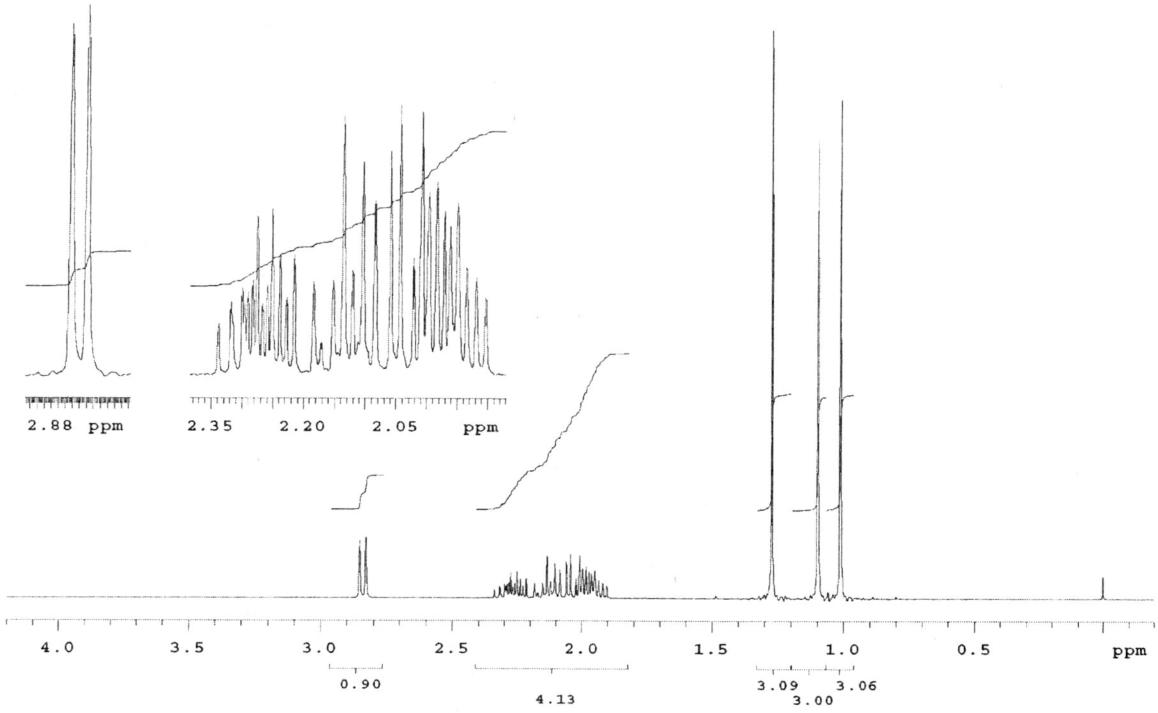

^{1}H-NMR Spectrum in $CDCl_3$

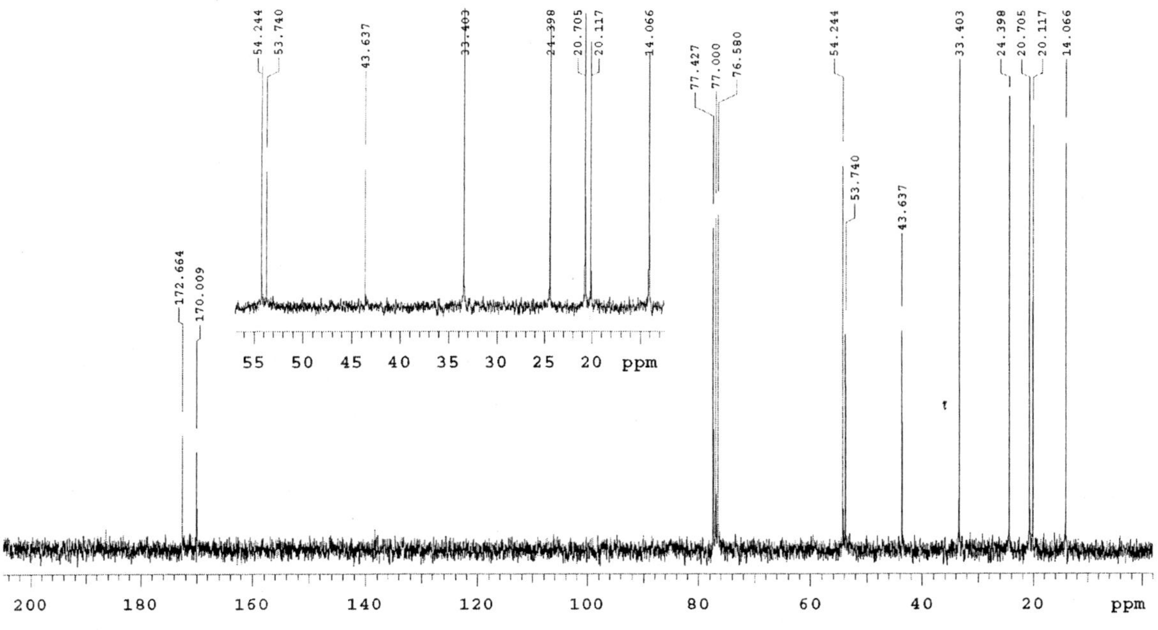

^{13}C-NMR Spectrum in $CDCl_3$

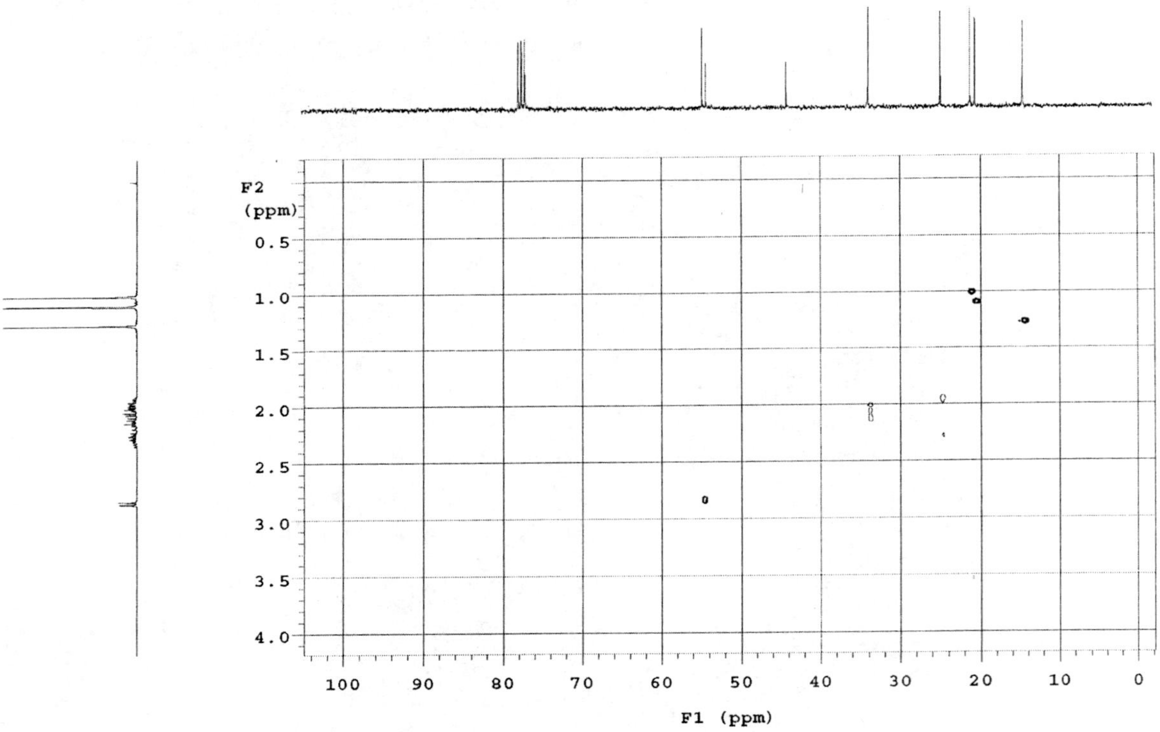

gHSQC Spectrum in $CDCl_3$

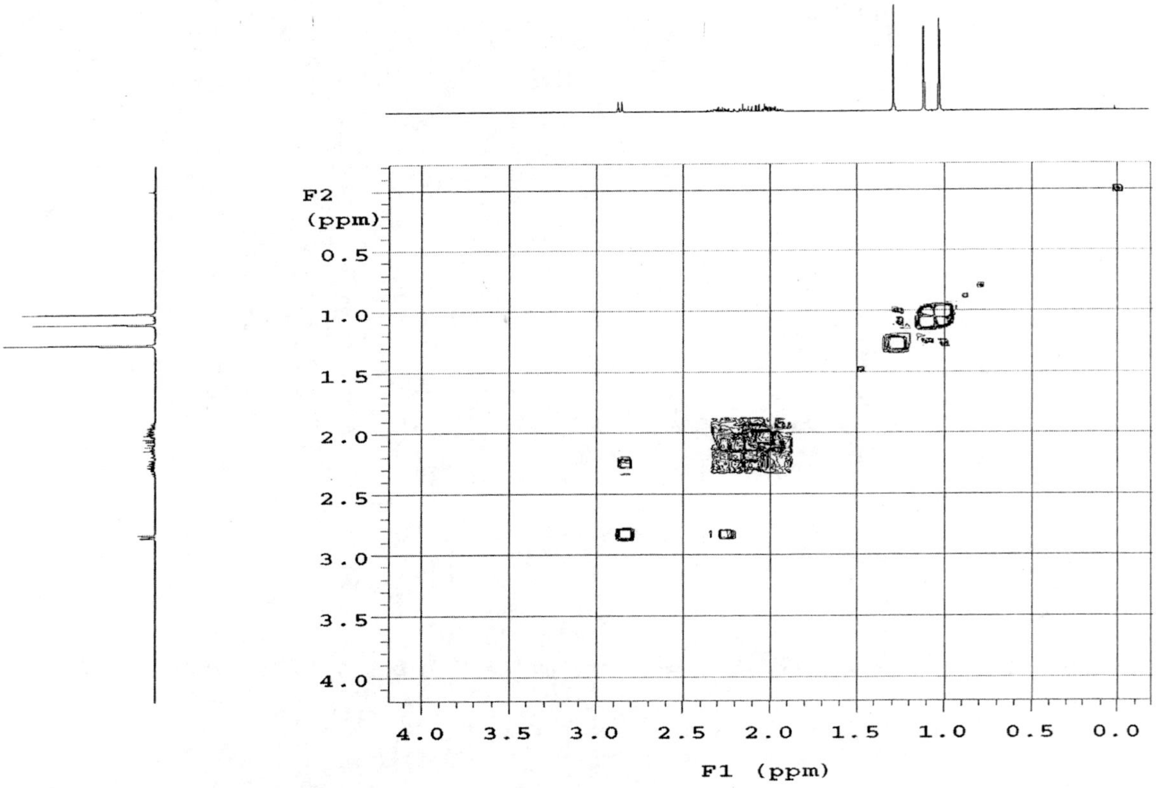

gCOSY Spectrum in $CDCl_3$

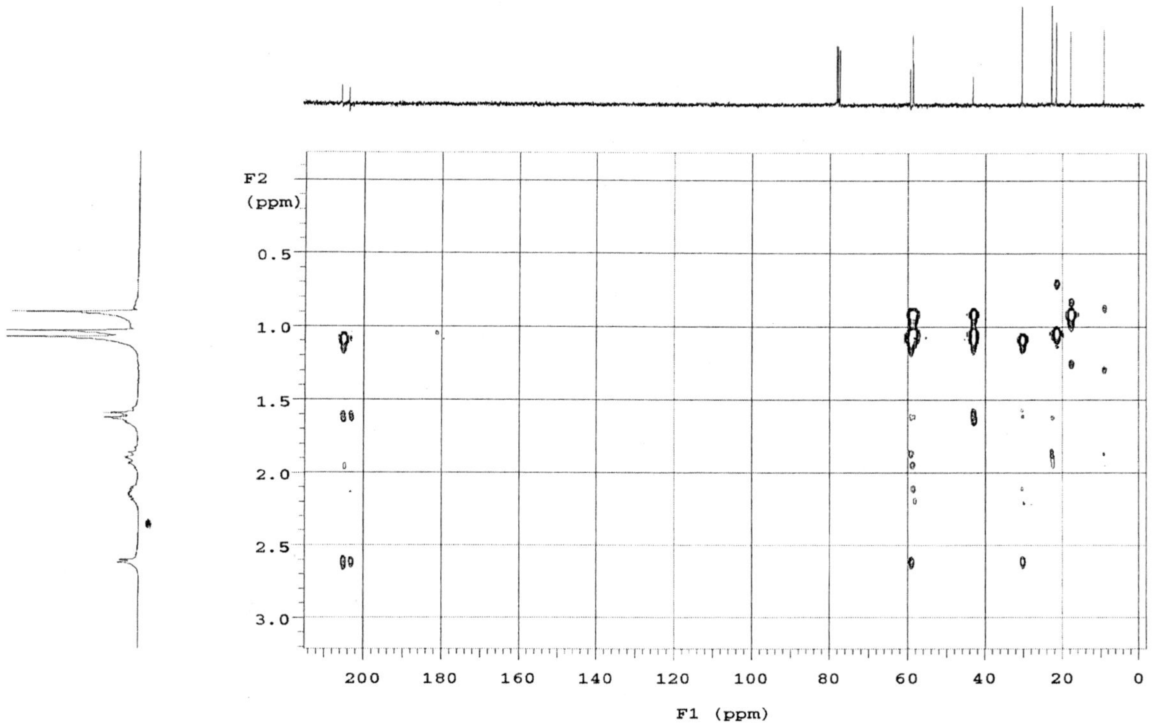

gHMBC Spectrum in $CDCl_3$

Problem 3.45 Given the spectroscopic data below (Proton, Carbon, gHSQC, gCOSY, and gHMBC), provide the structure of the molecule. (MW: 167 g/mol)

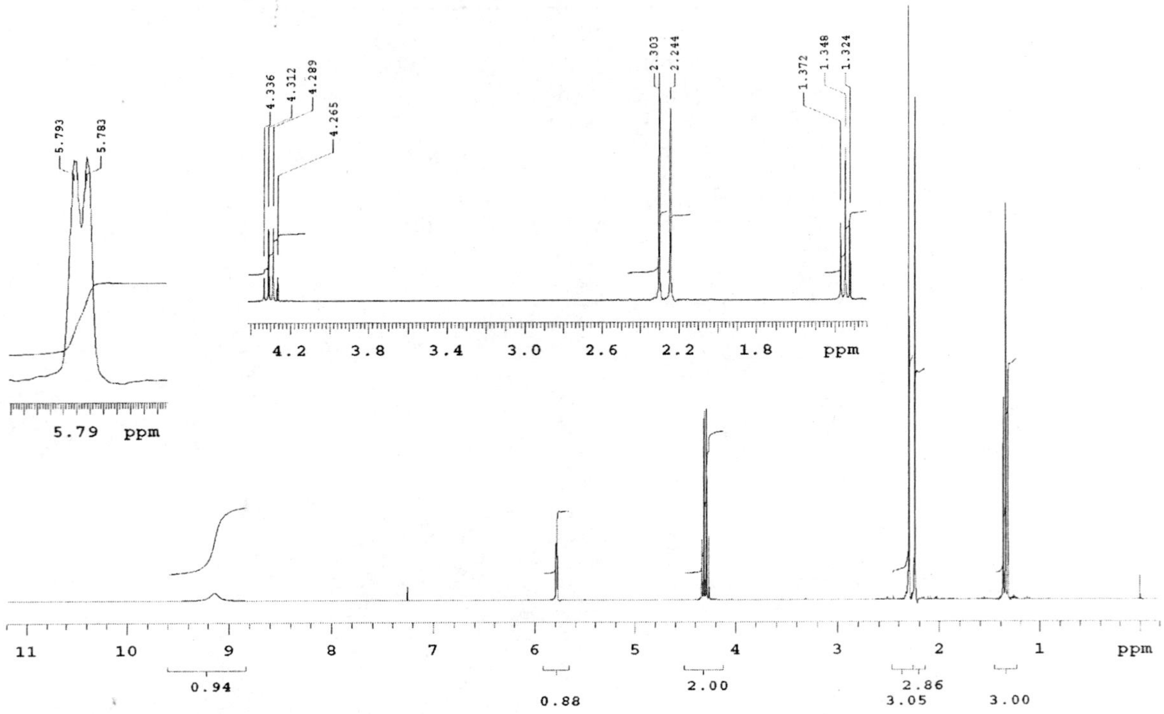

^{1}H-NMR Spectrum in $CDCl_3$

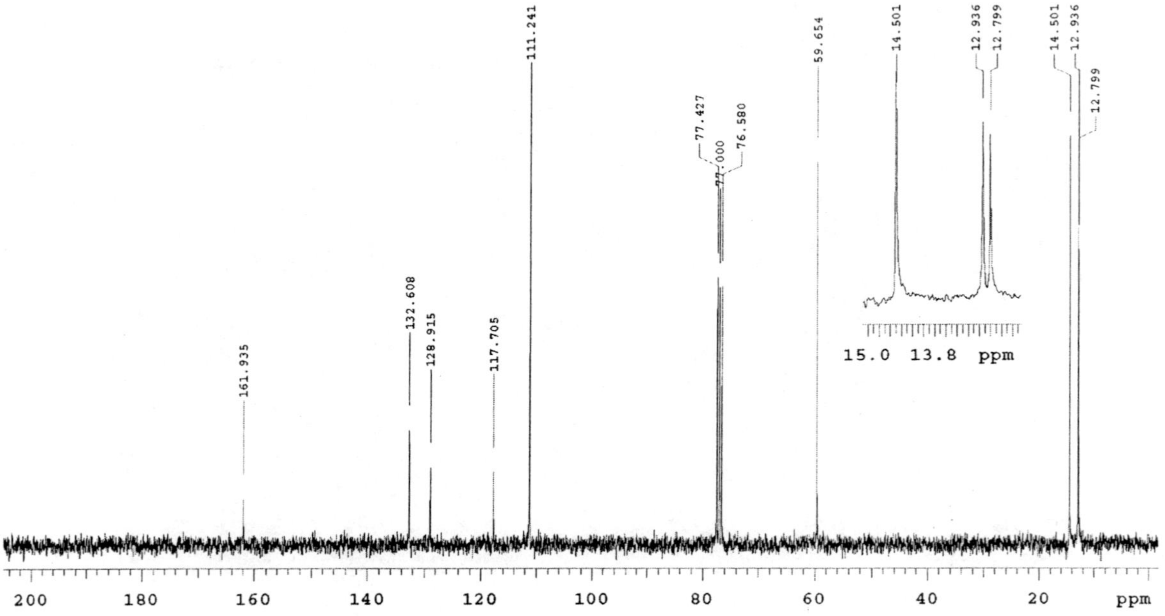

^{13}C-NMR Spectrum in $CDCl_3$

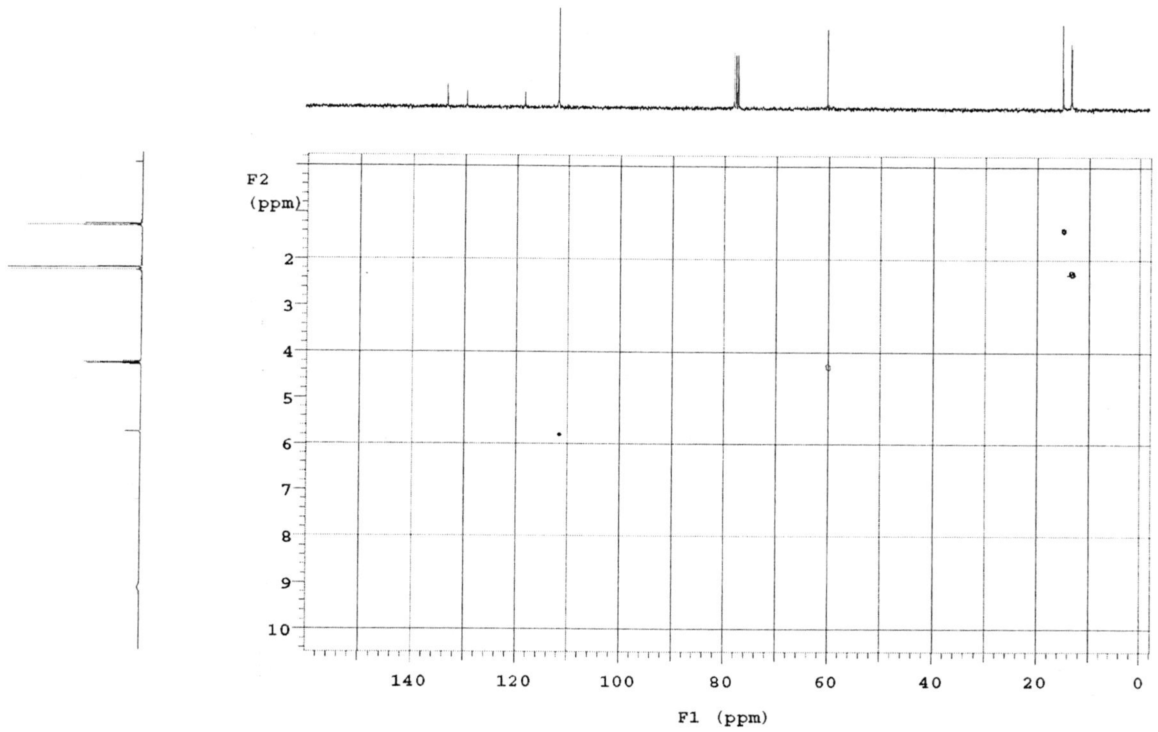

gHSQC Spectrum in $CDCl_3$

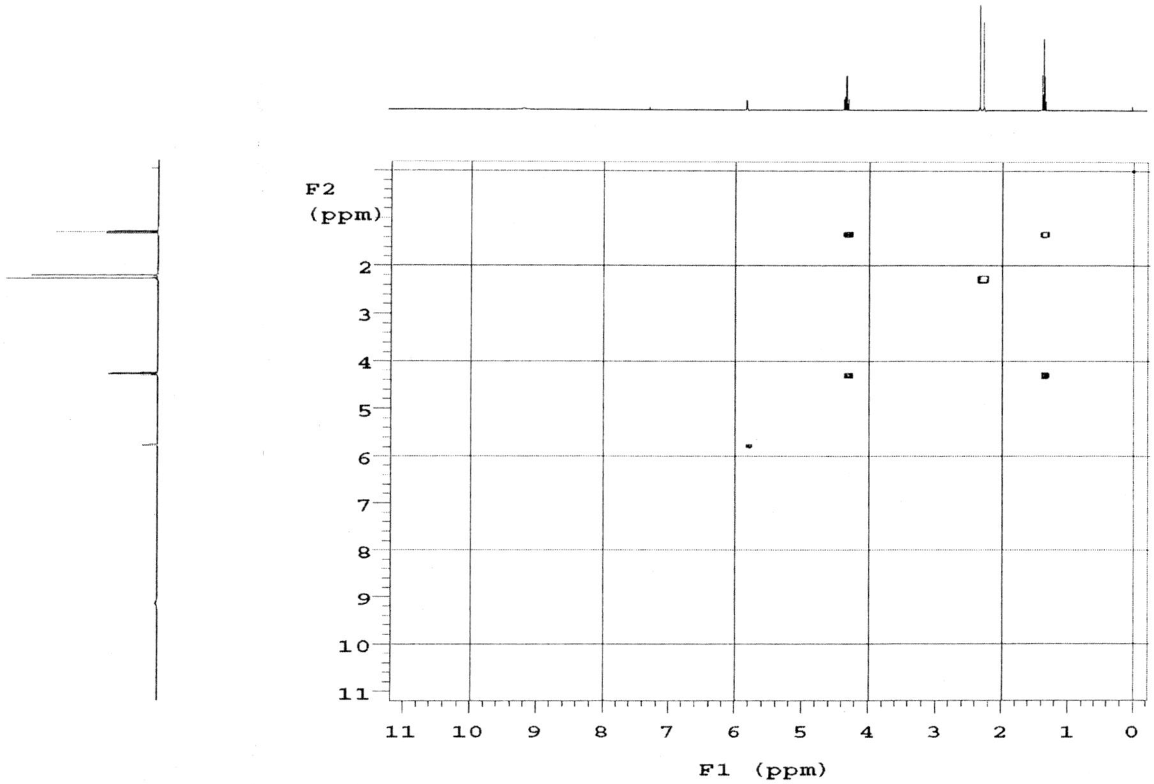

gCOSY Spectrum in $CDCl_3$

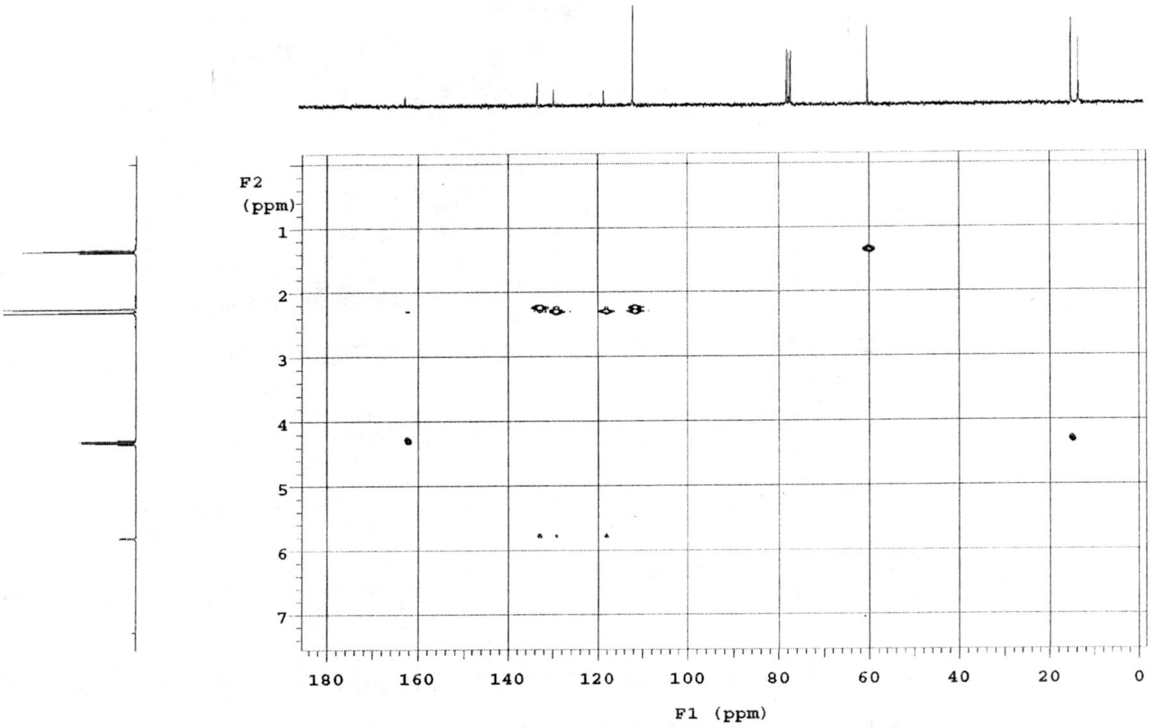

gHMBC Spectrum in $CDCl_3$

Problem 3.46 Given the spectroscopic data below (Proton, Carbon, Edited DEPT, gHSQC, and gCOSY), provide the structure of the molecule.

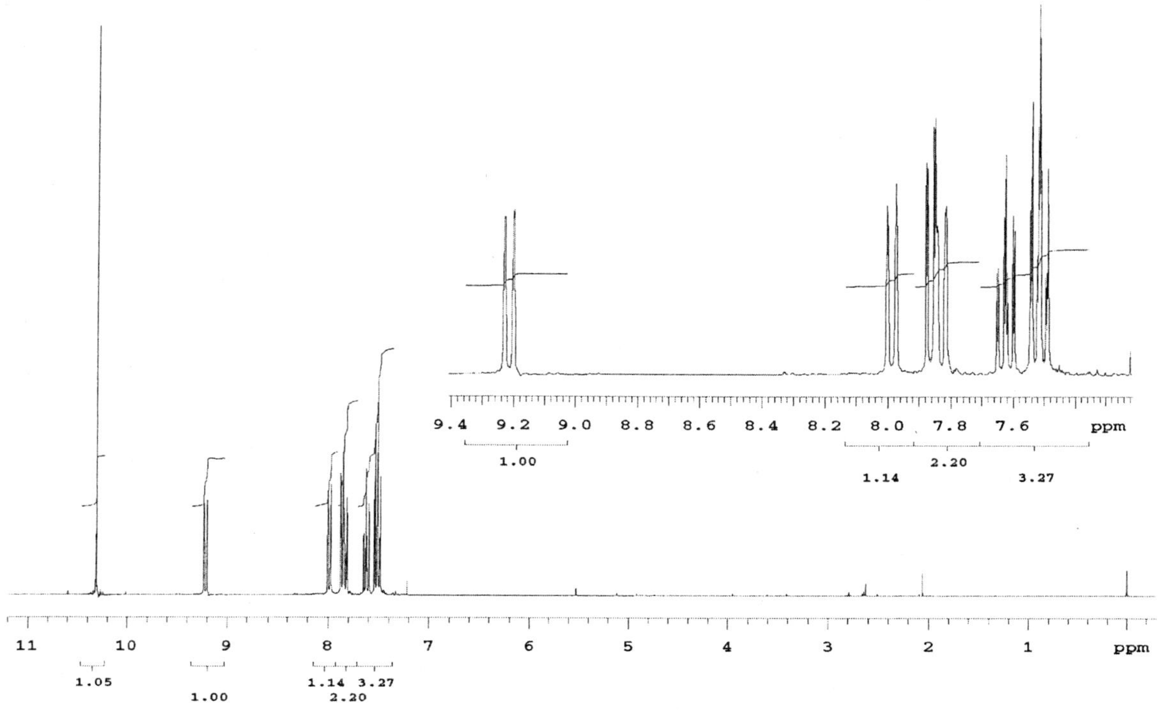

^{1}H-NMR Spectrum in $CDCl_3$

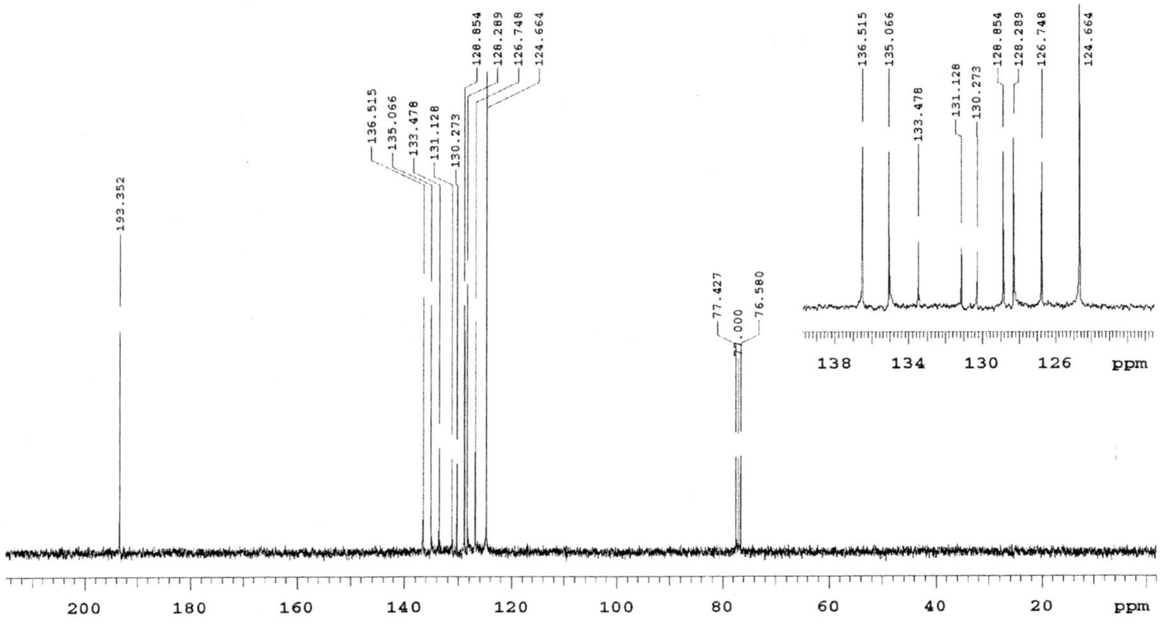

^{13}C-NMR Spectrum in $CDCl_3$

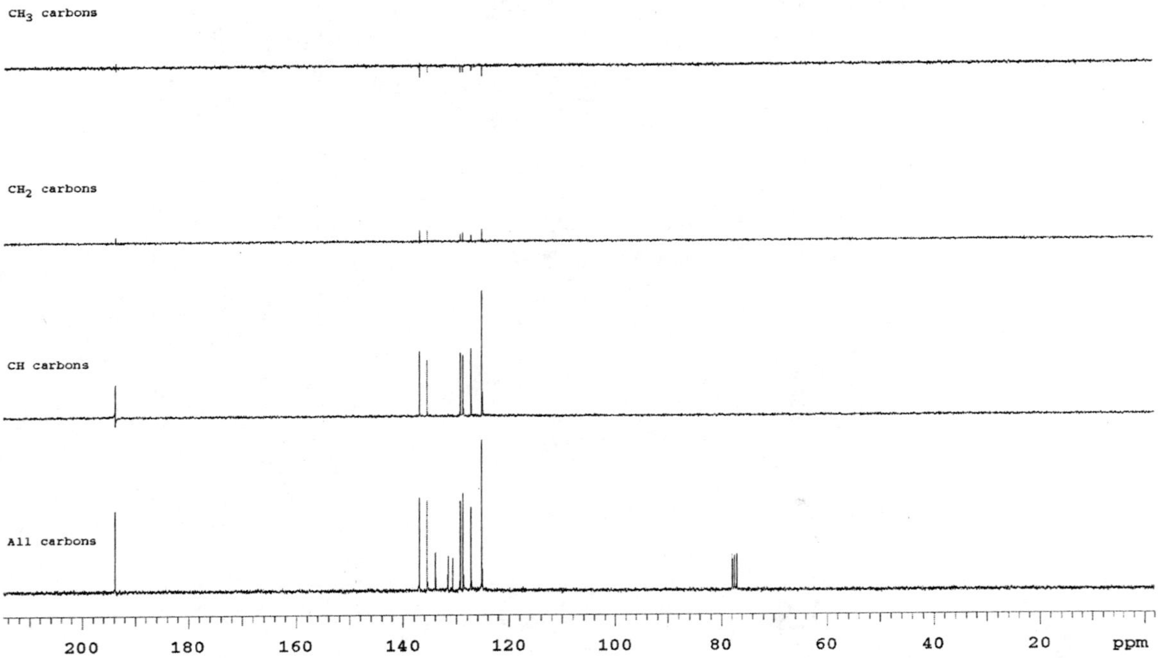

Edited DEPT ^{13}C-NMR Spectrum in $CDCl_3$

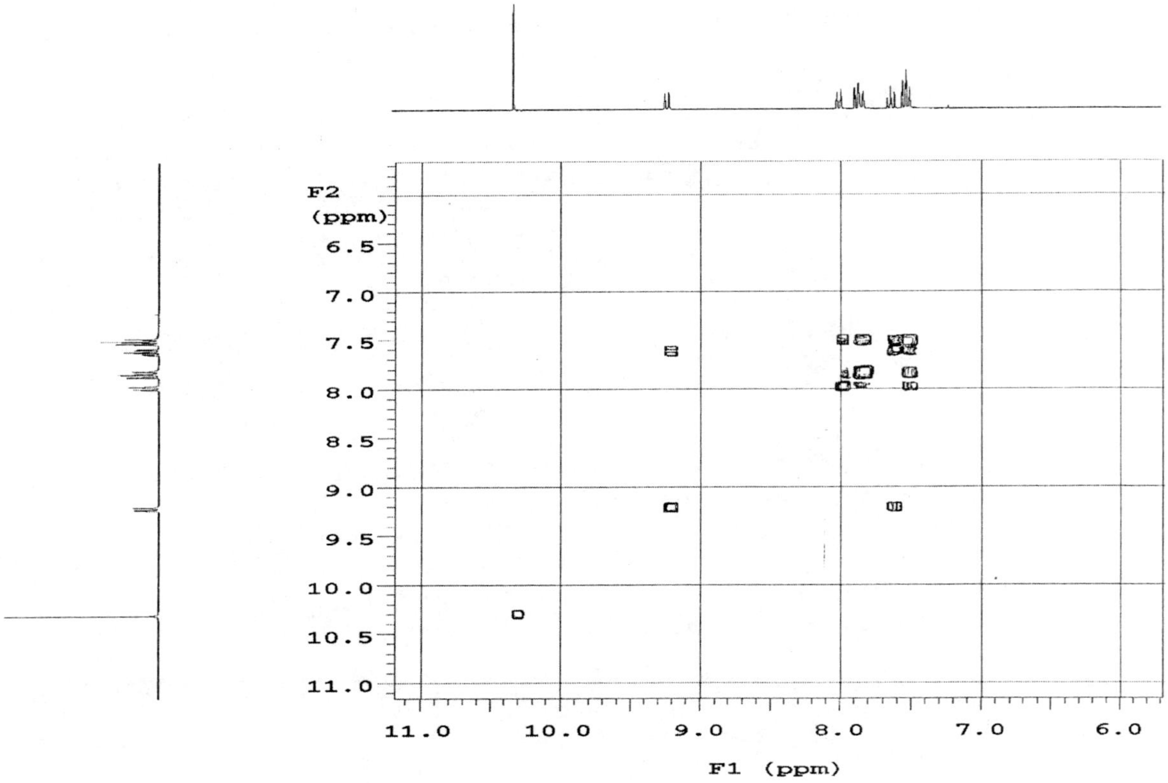

gCOSY Spectrum in $CDCl_3$

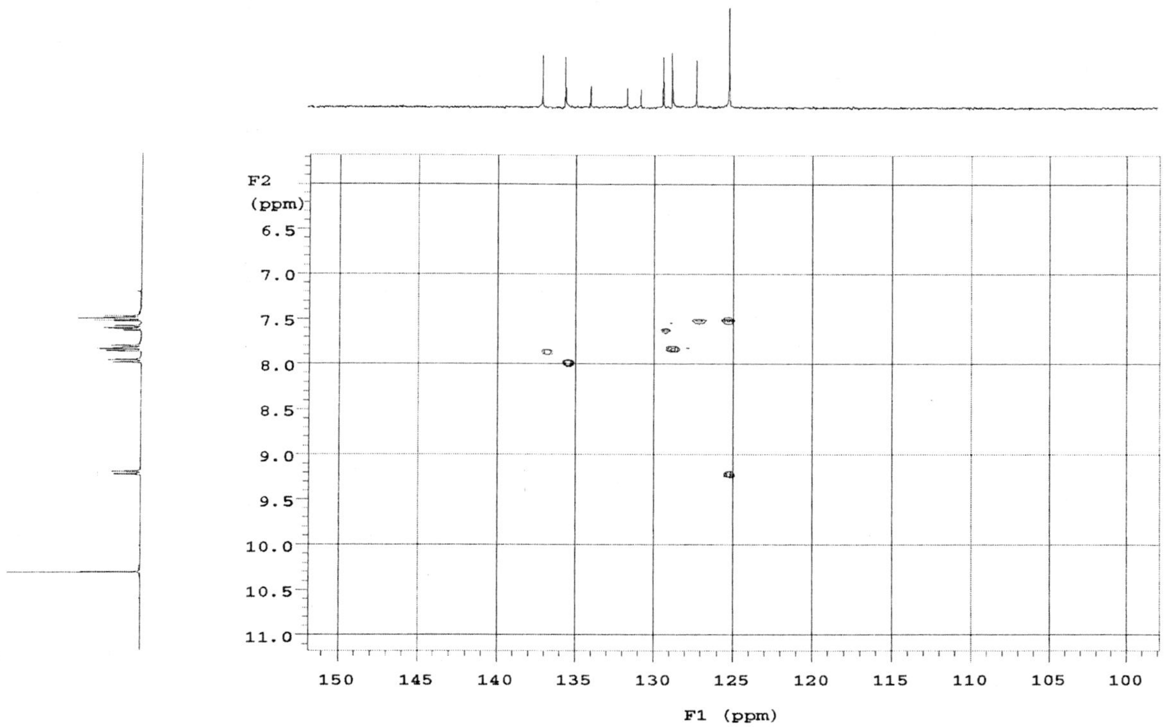

gHSQC Spectrum in $CDCl_3$

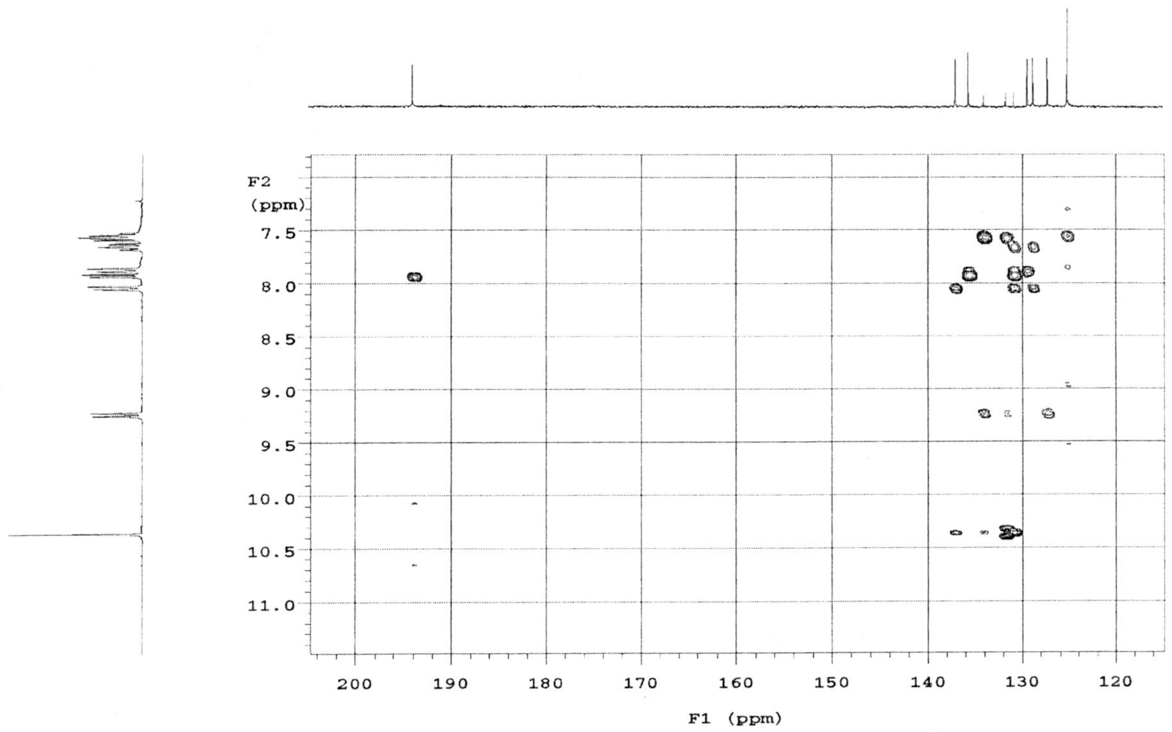

gHMBC Spectrum in $CDCl_3$

Problem 3.47 Given the spectroscopic data below (Proton, Carbon, gHSQC, and gHMBC), provide the structure of the molecule.

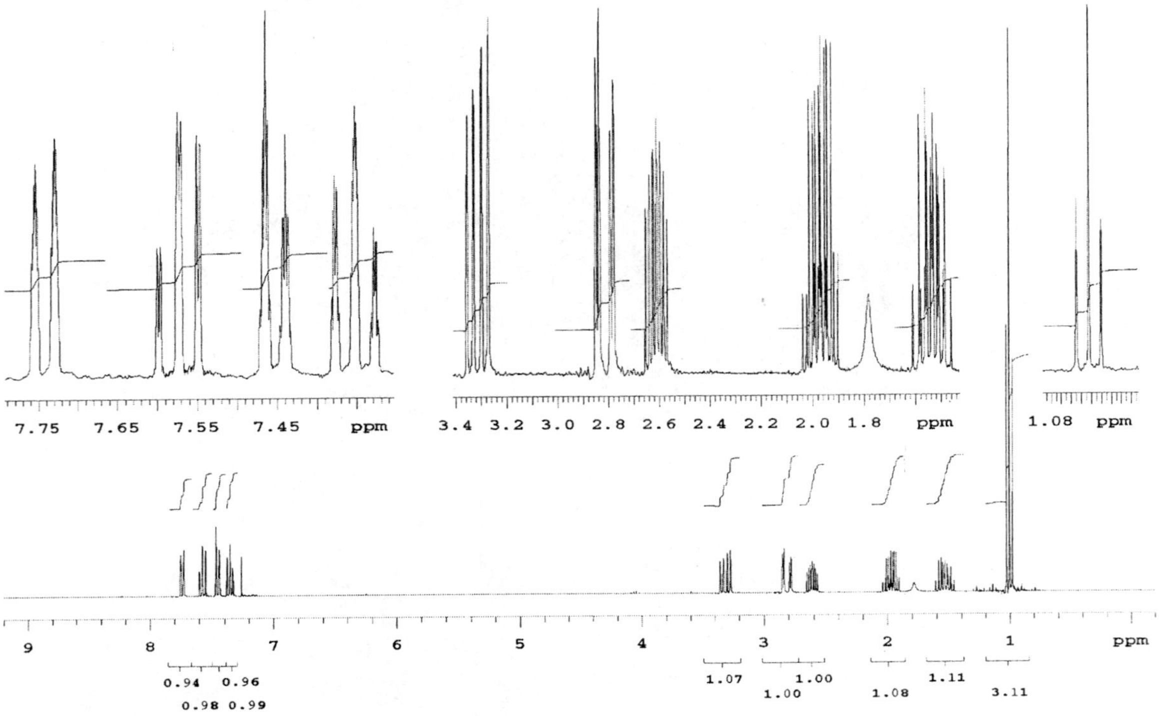

^{1}H-NMR Spectrum in $CDCl_3$

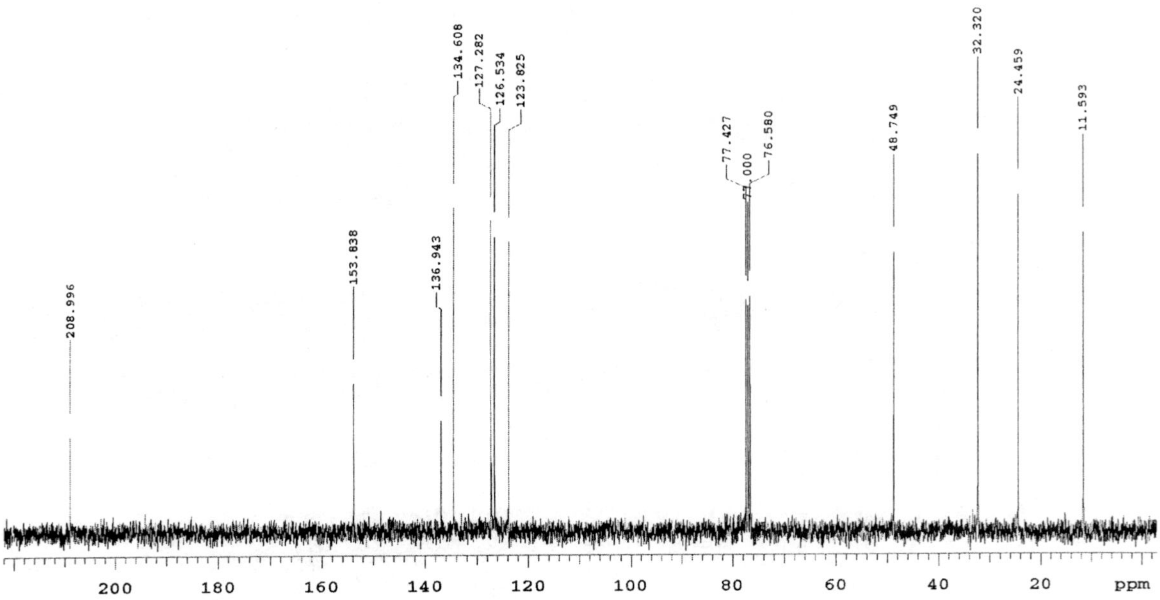

^{13}C-NMR Spectrum in $CDCl_3$

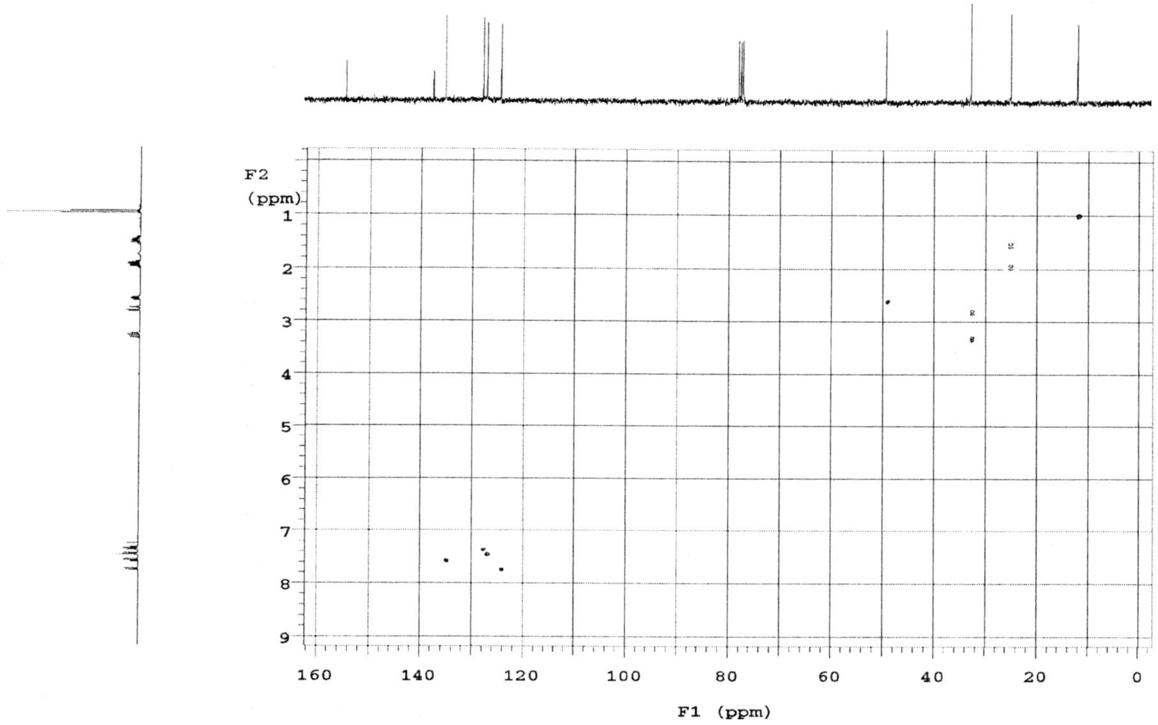

gHSQC Spectrum in $CDCl_3$

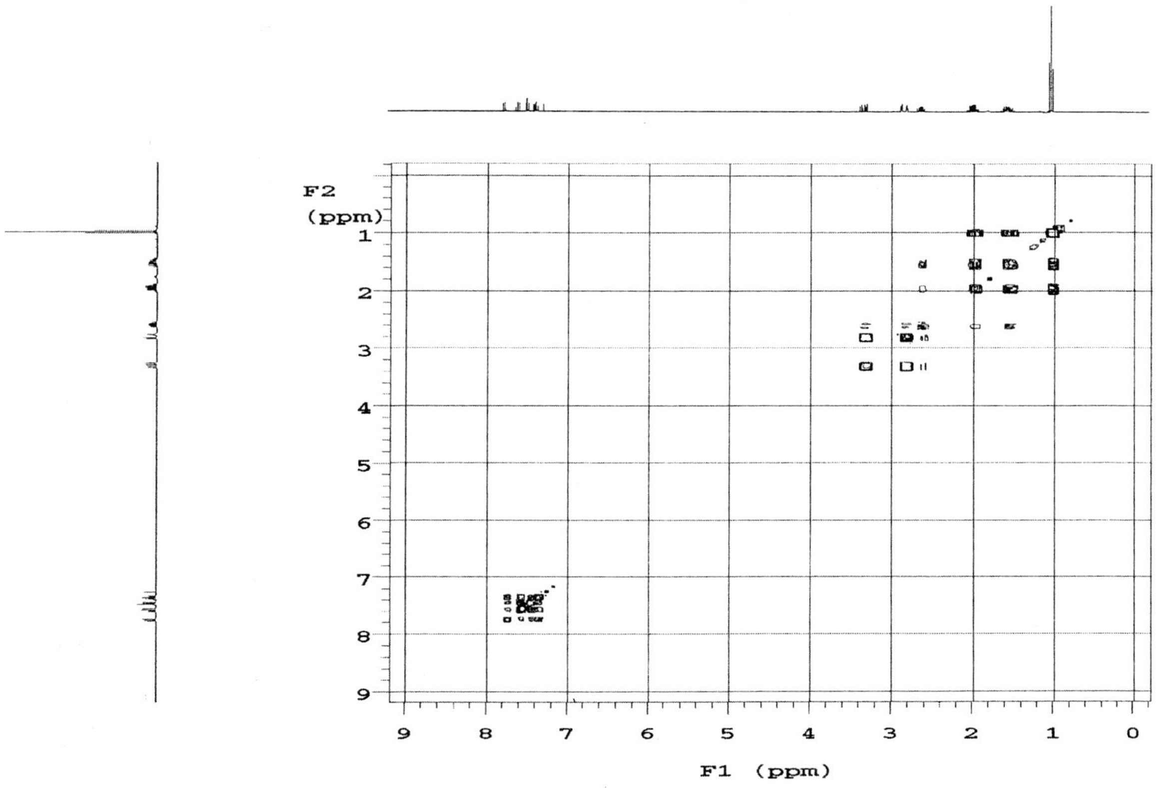

gCOSY Spectrum in $CDCl_3$

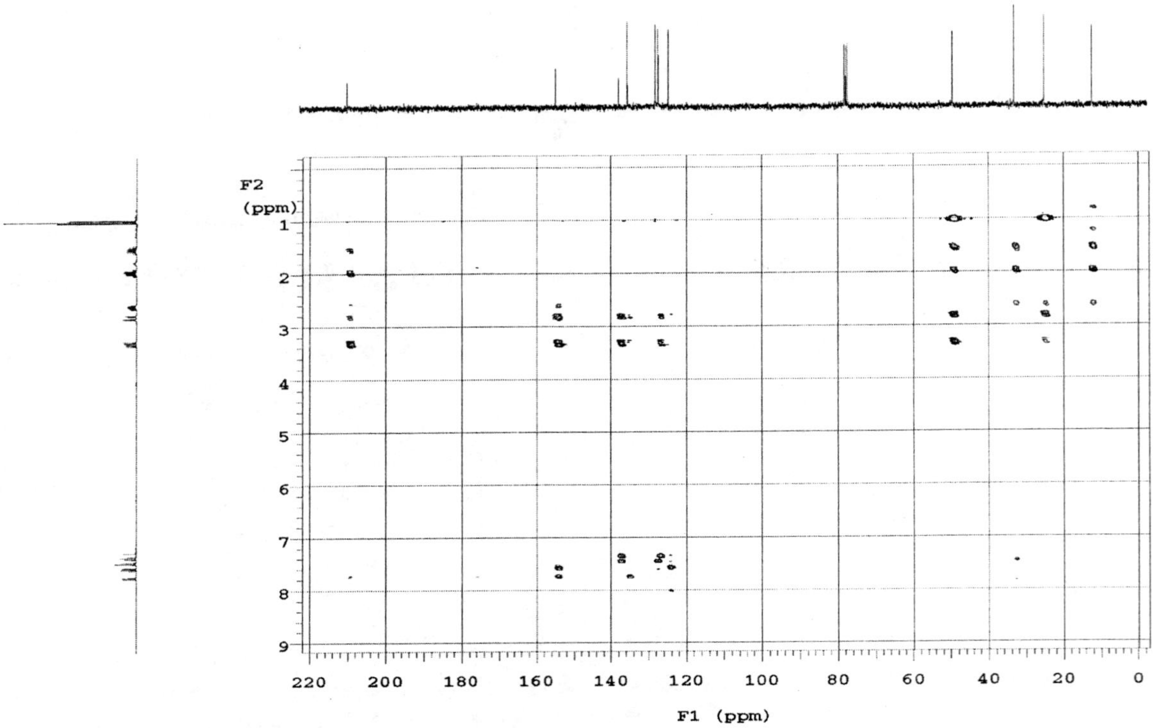

gHMBC Spectrum in $CDCl_3$

Problem 3.48 Given the spectroscopic data below (Proton, Carbon, gHSQC, gCOSY, and Phosphorus), provide the structure of the molecule. ($C_6H_{15}PO_3$) Contains acetone impurity.

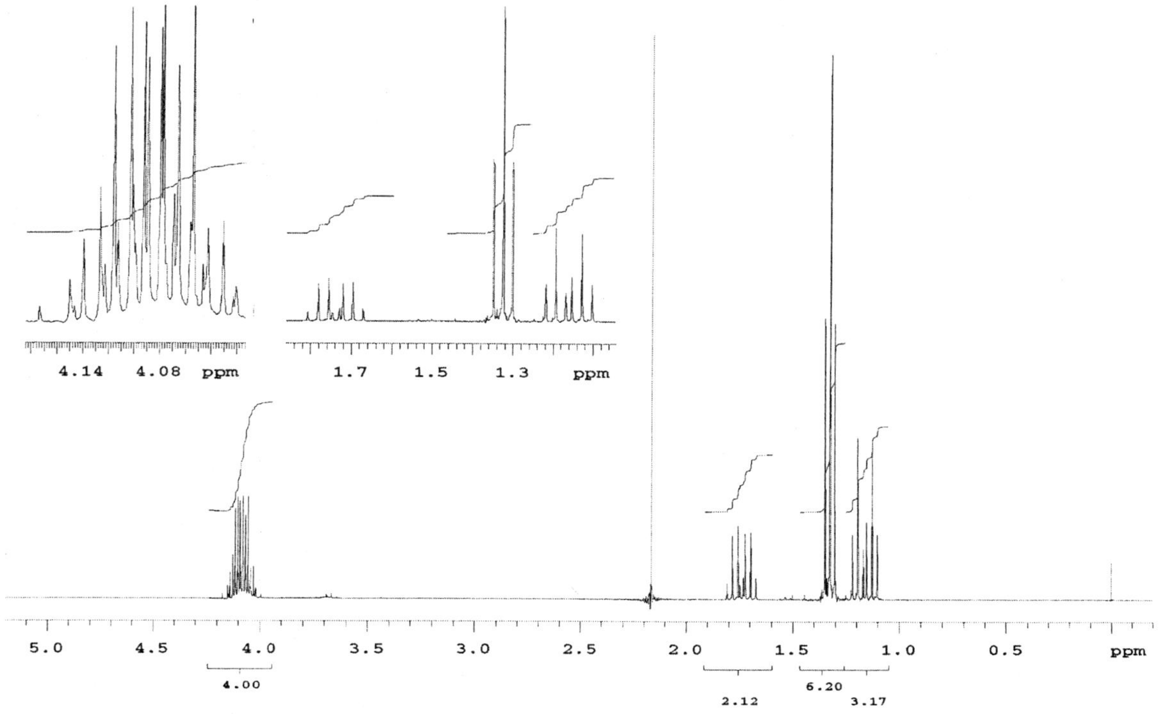

^{1}H-NMR Spectrum in $CDCl_3$

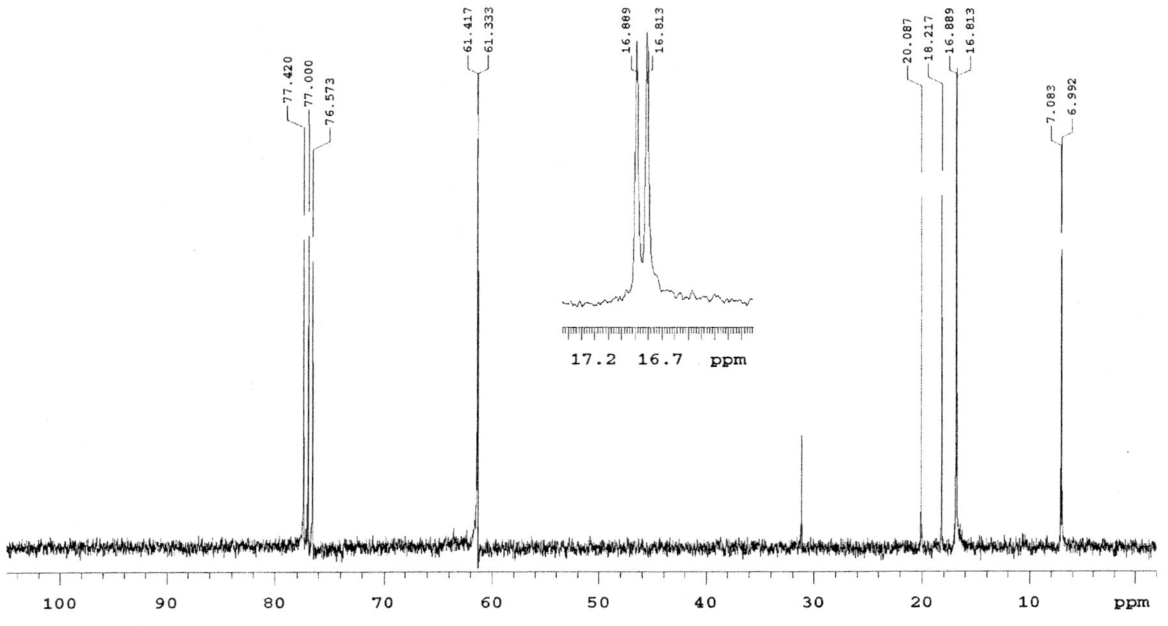

^{13}C-NMR Spectrum in $CDCl_3$

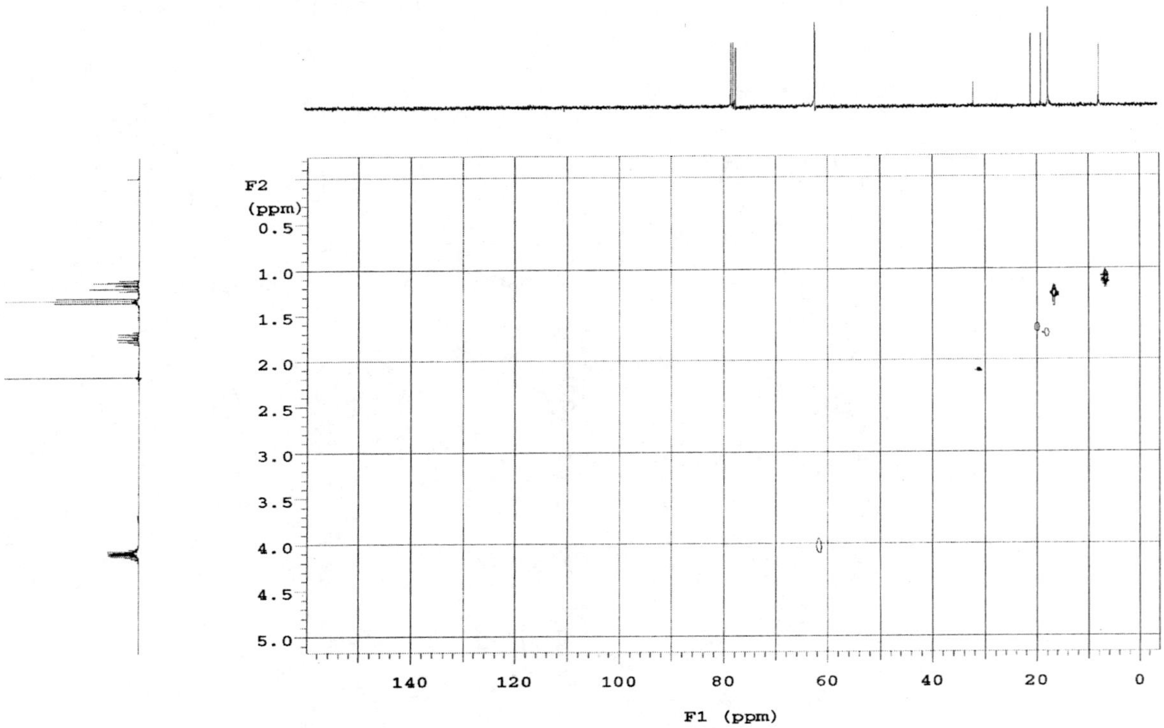

gHSQC Spectrum in $CDCl_3$

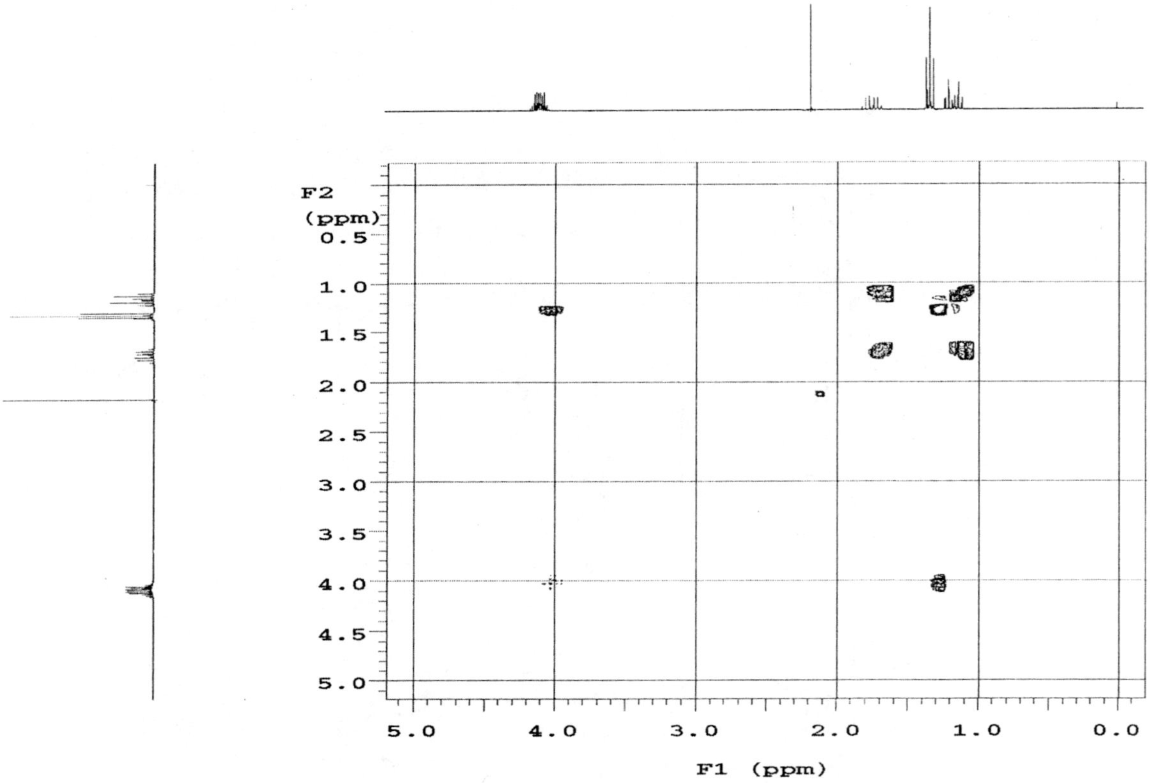

gCOSY Spectrum in $CDCl_3$

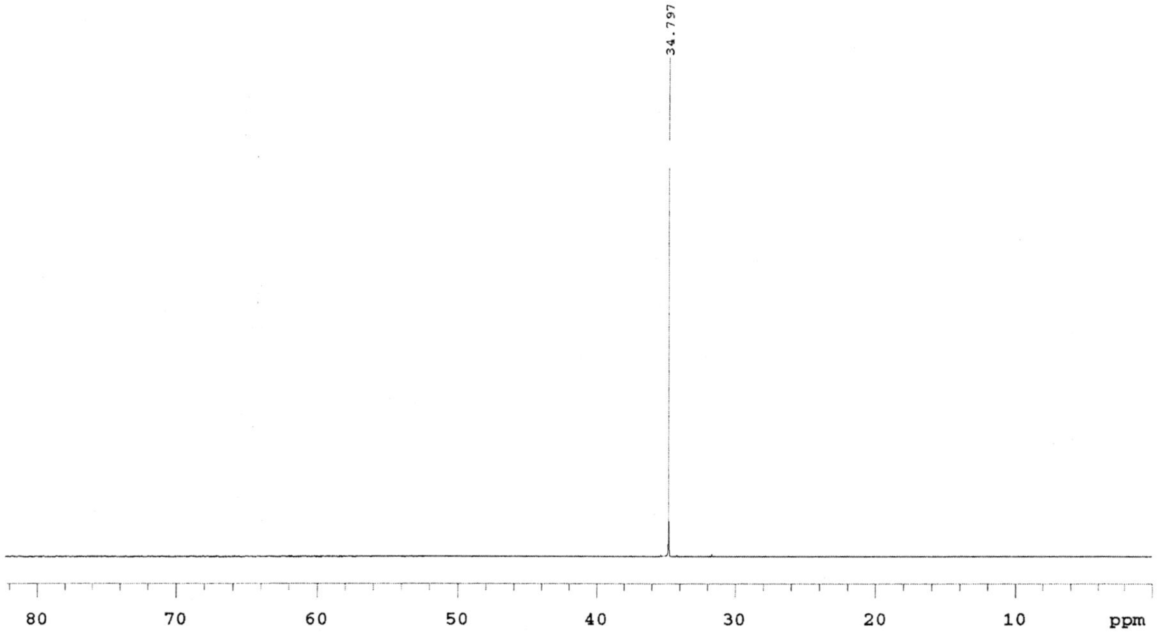

^{31}P-NMR Spectrum in $CDCl_3$

Problem 3.49 Given the spectroscopic data below (Proton, Carbon, Edited DEPT, gHSQC, and gHMBC), provide the structure of the molecule. ($C_{11}H_{11}NO_4$)

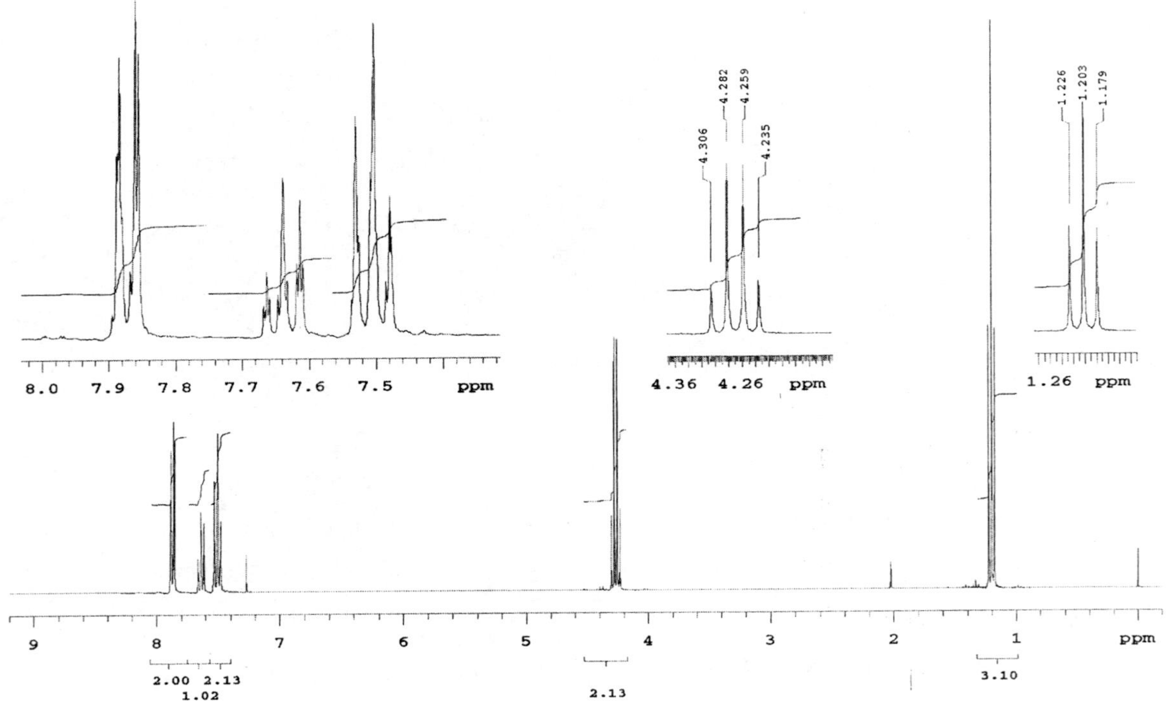

^{1}H-NMR Spectrum in $CDCl_3$

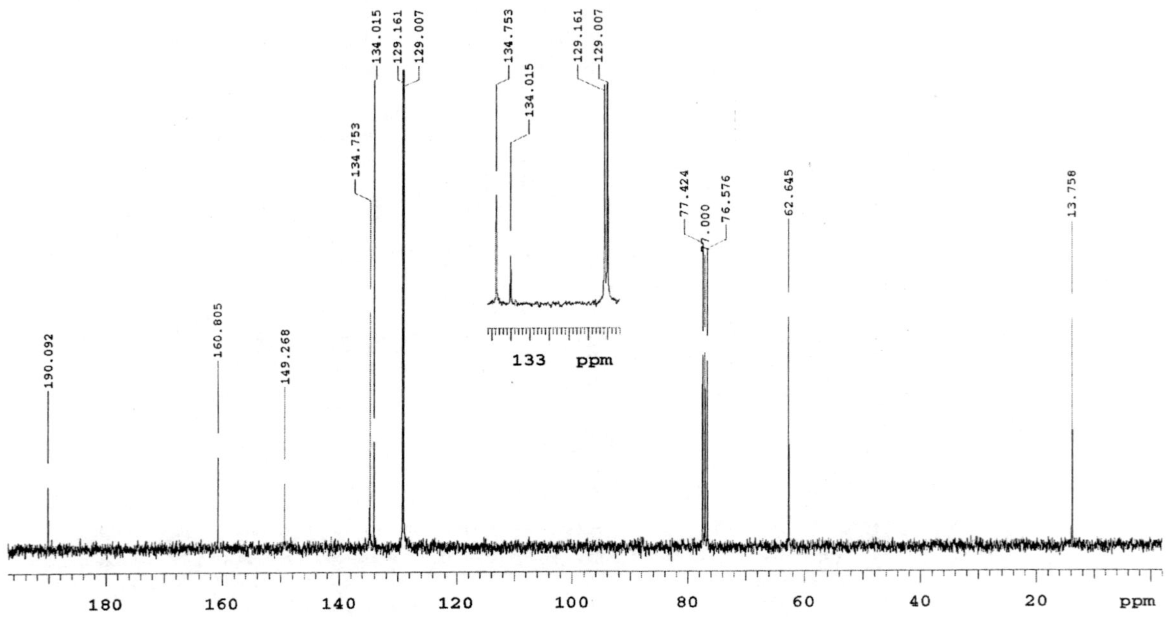

^{13}C-NMR Spectrum in $CDCl_3$

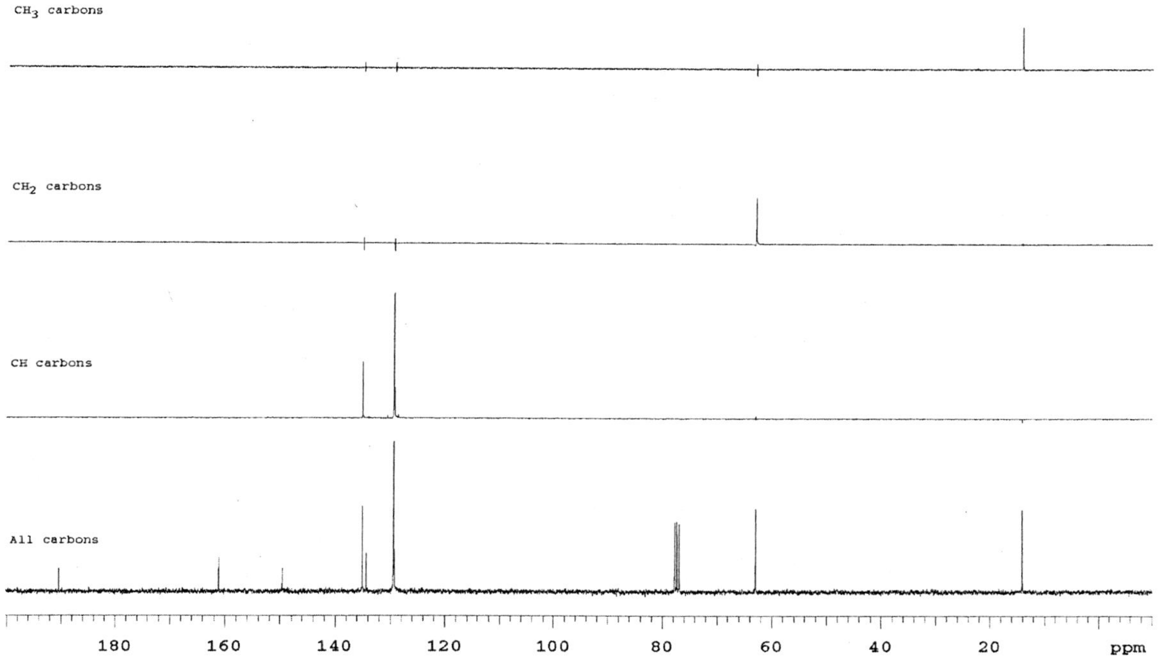

Edited DEPT ^{13}C-NMR Spectrum in $CDCl_3$

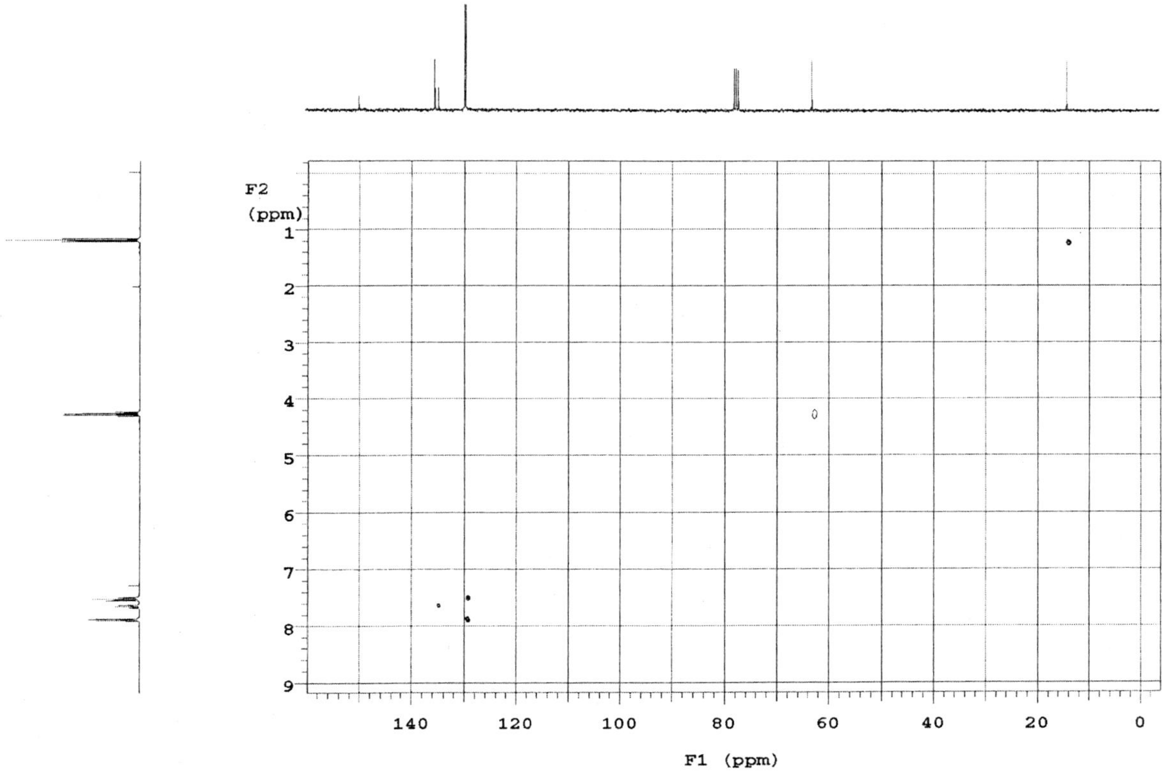

gHSQC Spectrum in $CDCl_3$

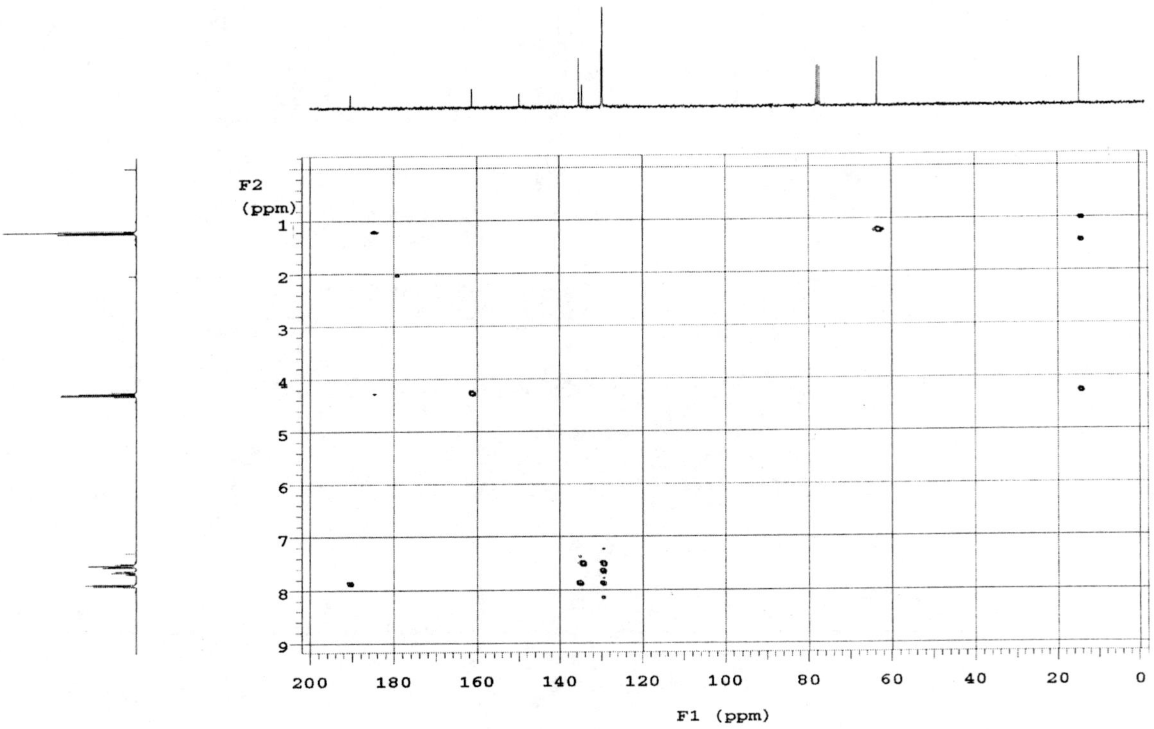

gHMBC Spectrum in $CDCl_3$

Problem 3.50 Given the spectroscopic data below (Proton, Carbon, gHSQC, gCOSY, and gHMBC), provide the structure of the molecule. (MW: 195 g/mol)

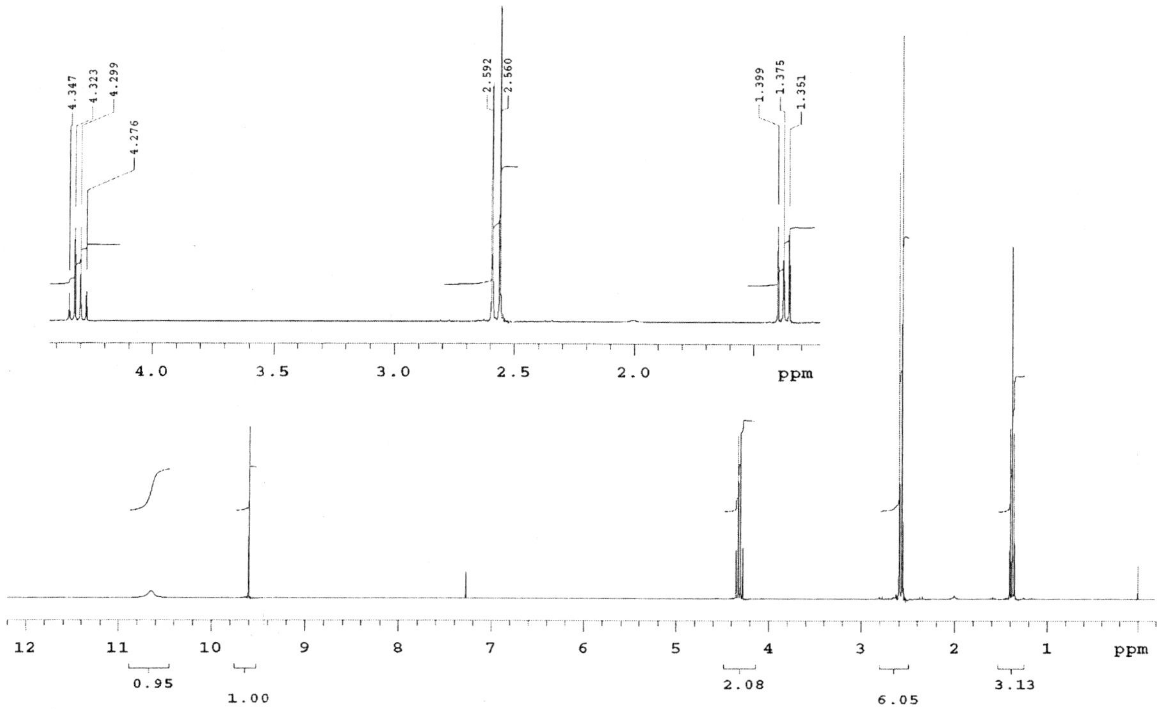

^{1}H-NMR Spectrum in $CDCl_3$

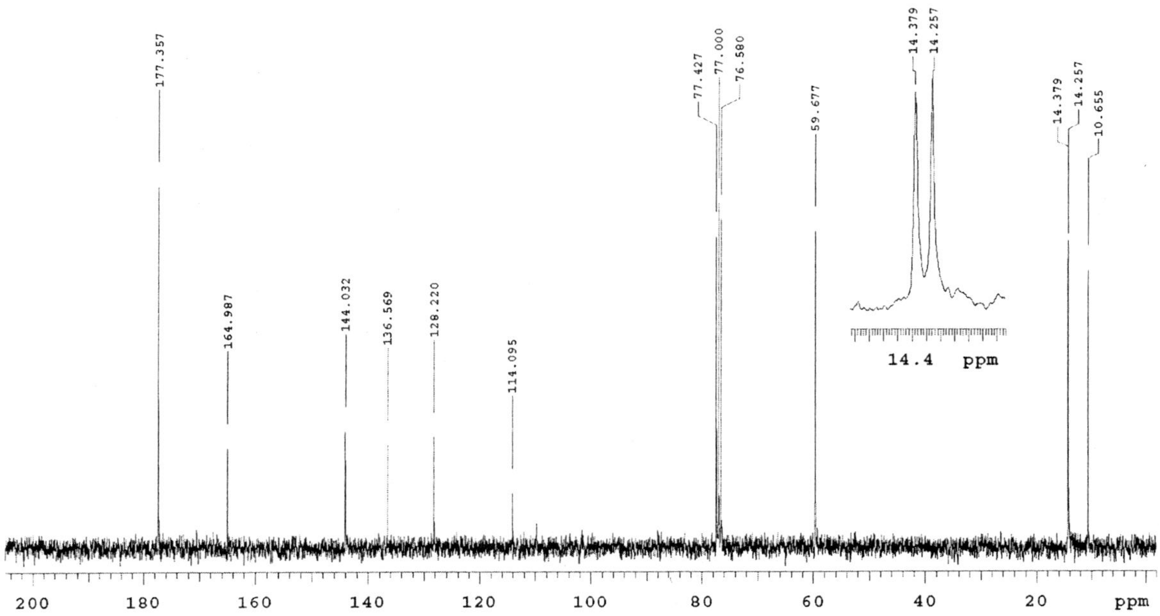

^{13}C-NMR Spectrum in $CDCl_3$

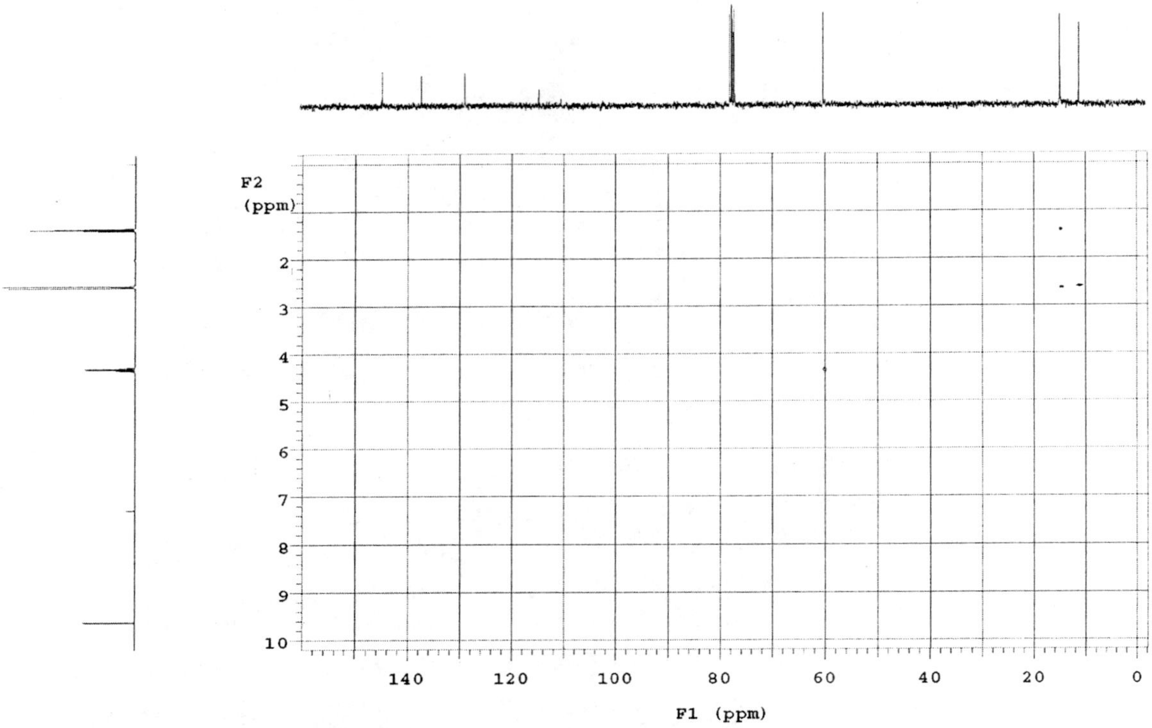

gHSQC Spectrum in $CDCl_3$ (A gHSQC correlation between the ~9.5 ppm signal in the ^{1}H-NMR and the ~177.4 ppm signal in the ^{13}C-NMR was also observed.)

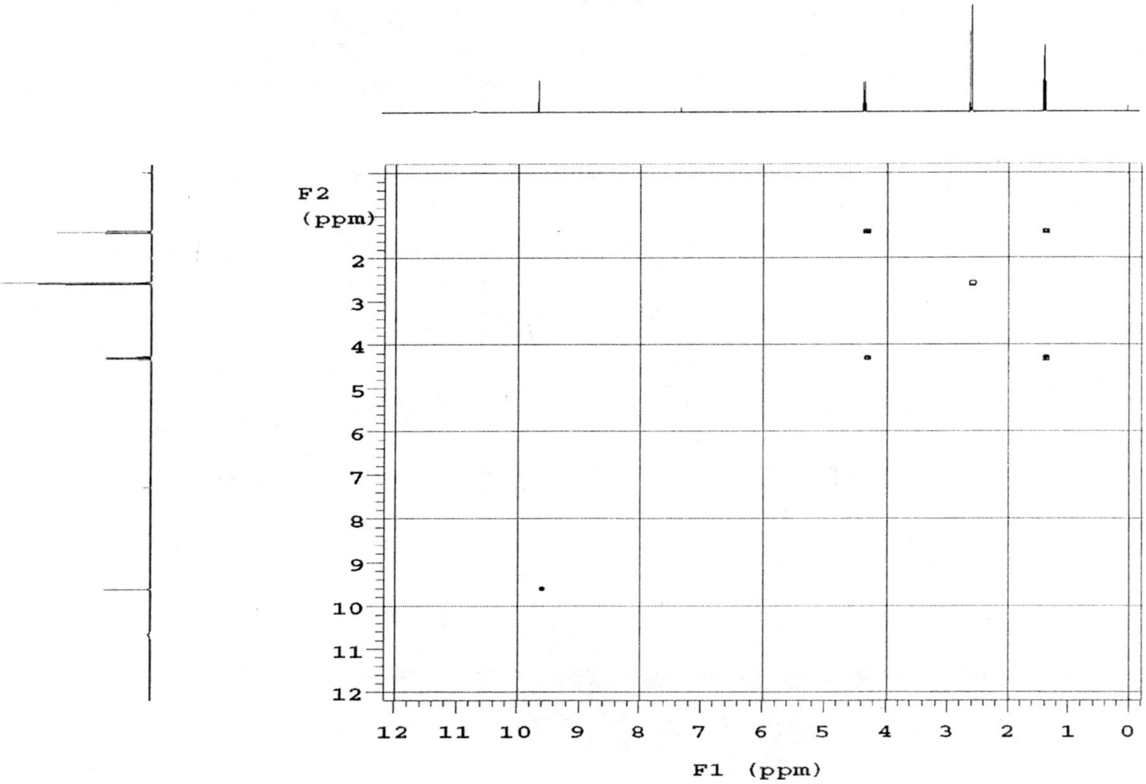

gCOSY Spectrum in $CDCl_3$

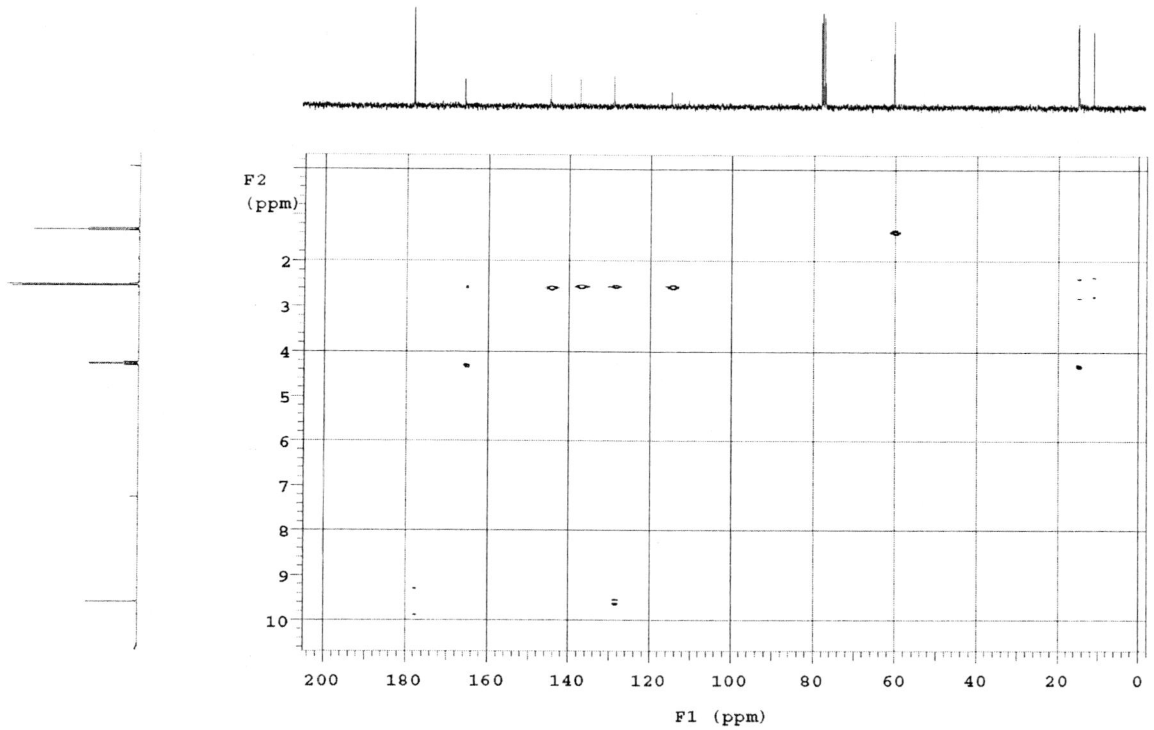

gHMBC Spectrum in $CDCl_3$

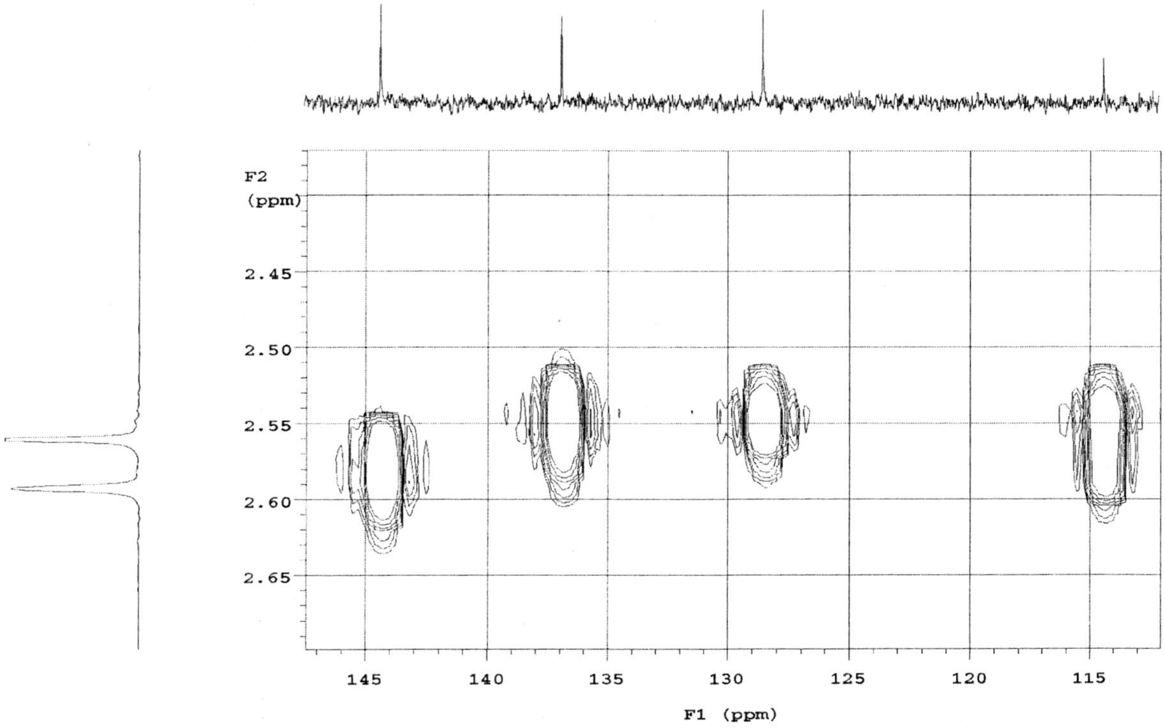

Expansion of gHMBC Spectrum in $CDCl_3$

Chapter 4
More-Challenging Problems

Problem 4.01 Given the spectroscopic data below (Proton, Carbon, gHSQC, and gCOSY), provide the structure of the molecule. (HR-MS: 136.1252 amu)

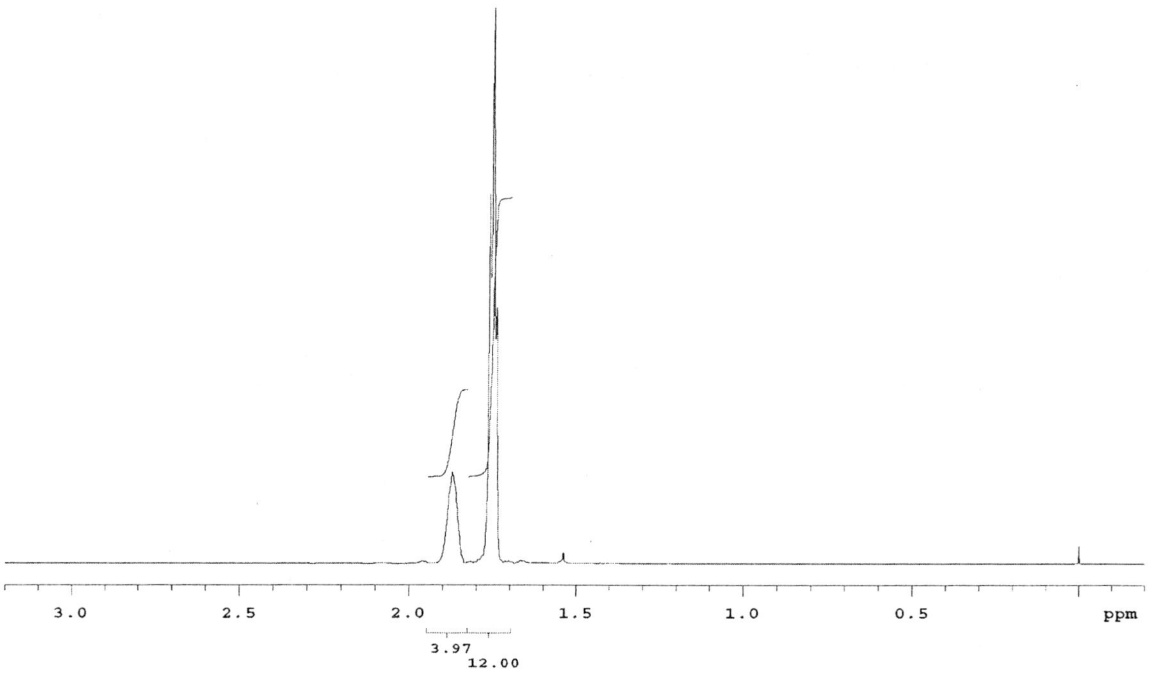

^{1}H-NMR Spectrum in $CDCl_3$

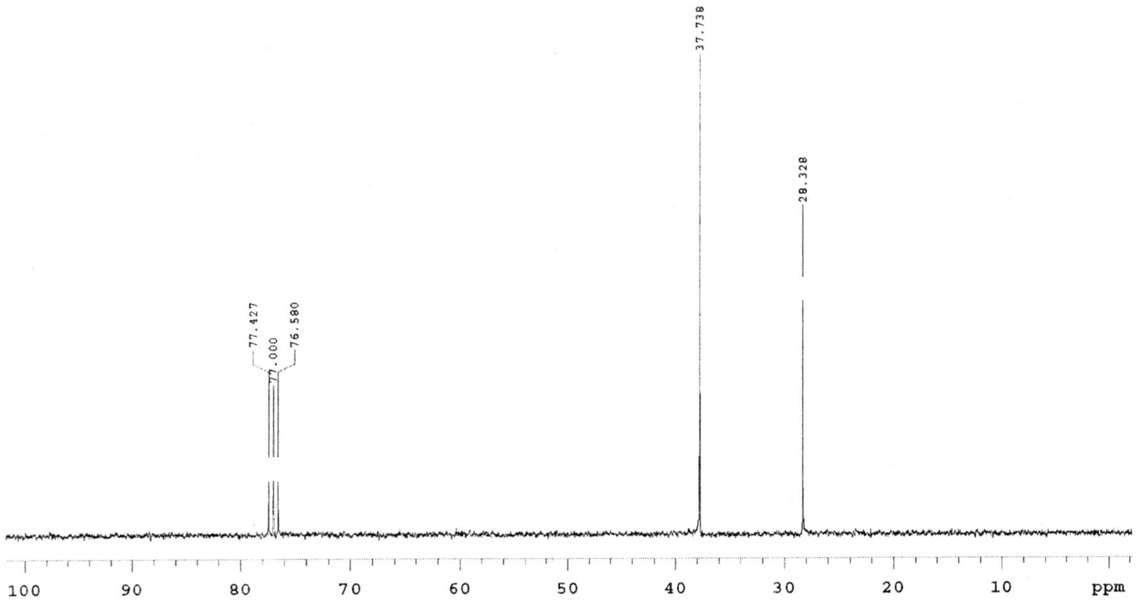

^{13}C-NMR Spectrum in $CDCl_3$

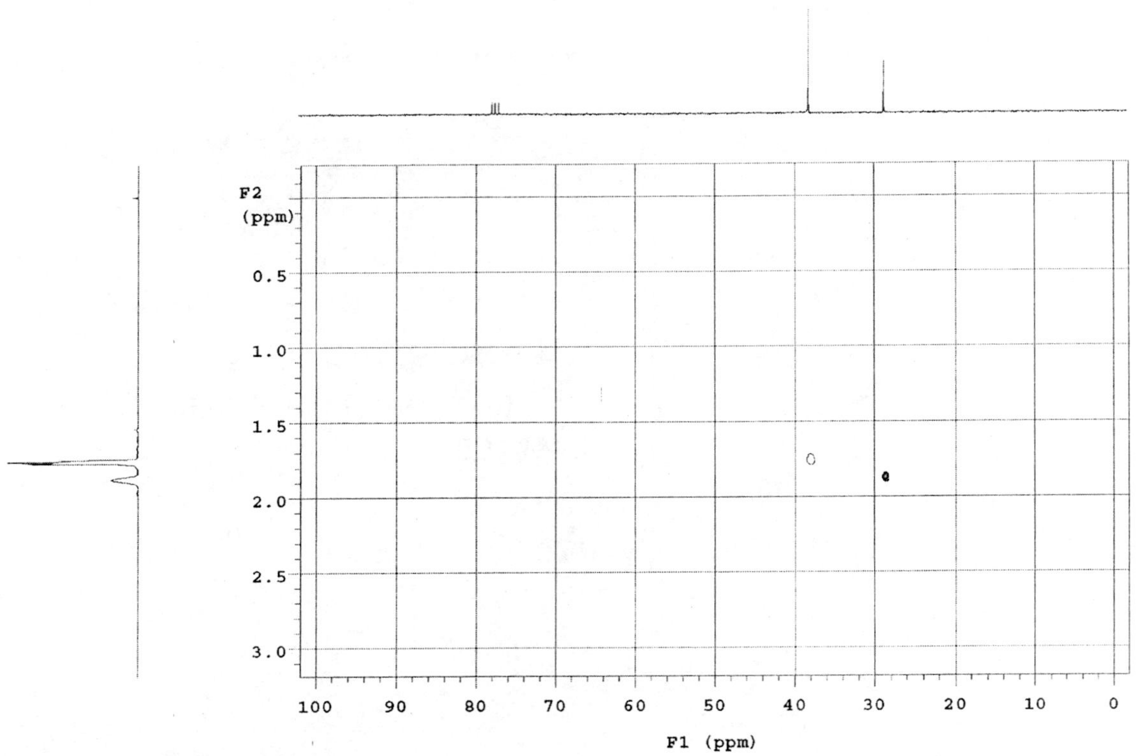

gHSQC Spectrum in $CDCl_3$

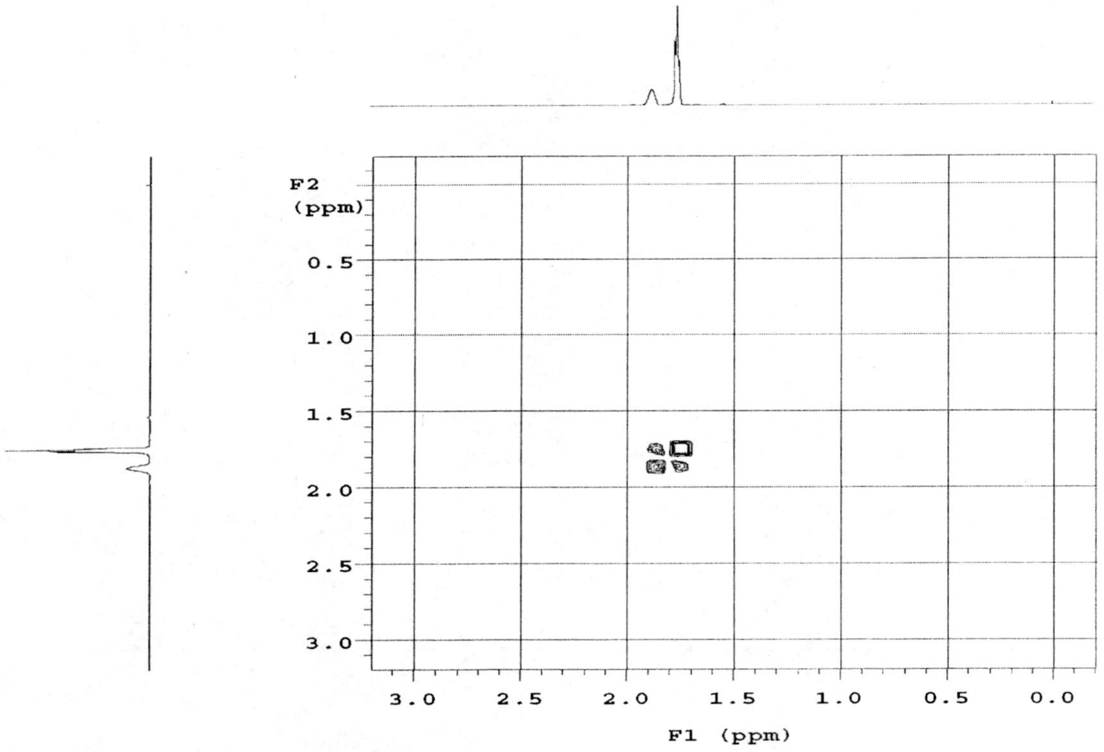

gCOSY Spectrum in $CDCl_3$

Problem 4.02 Given the spectroscopic data below (Proton, Carbon, gCOSY, gHSQC, and gHMBC), provide the structure of the molecule. (MW: 214 g/mol and 216 g/mol: present in approximately equal intensities)

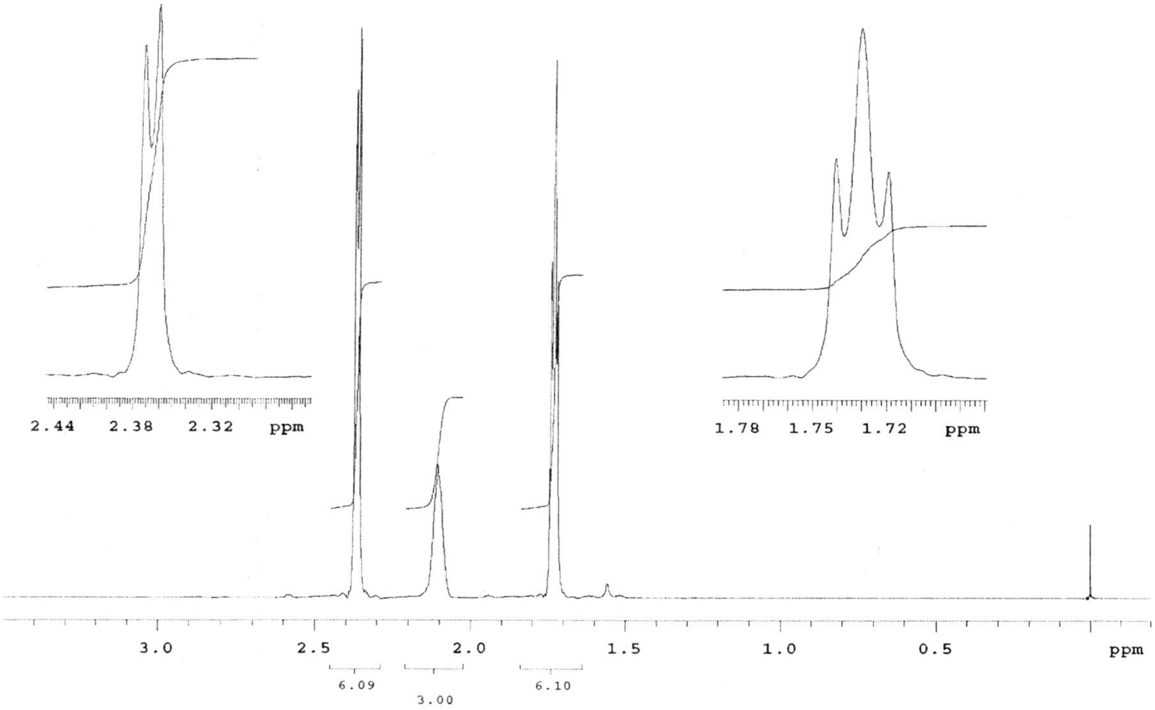

^{1}H-NMR Spectrum in $CDCl_3$

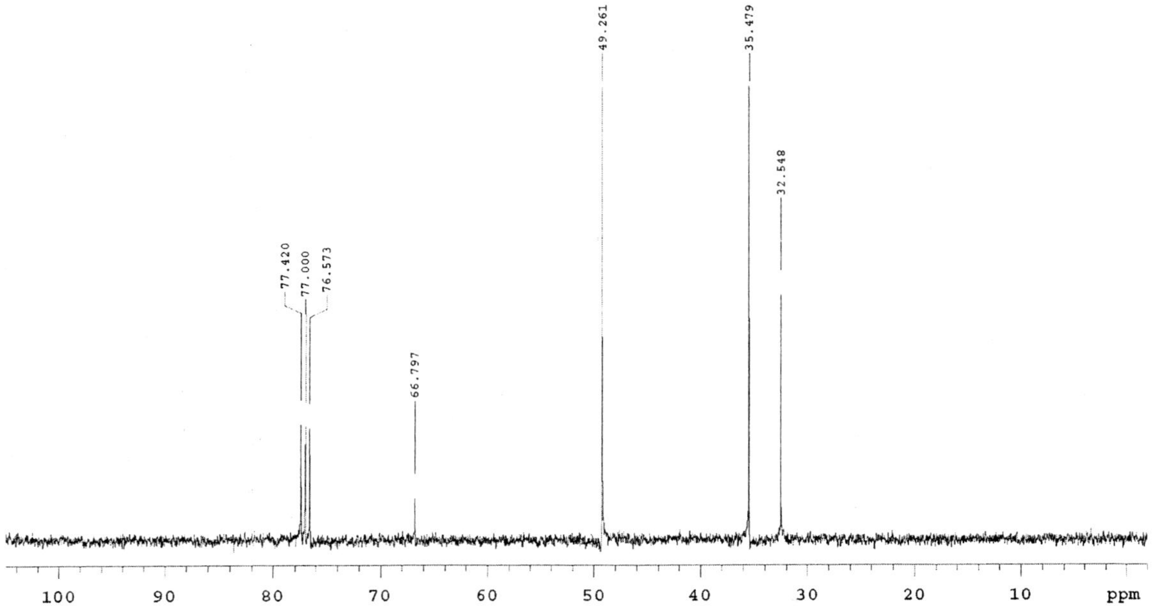

^{13}C-NMR Spectrum in $CDCl_3$

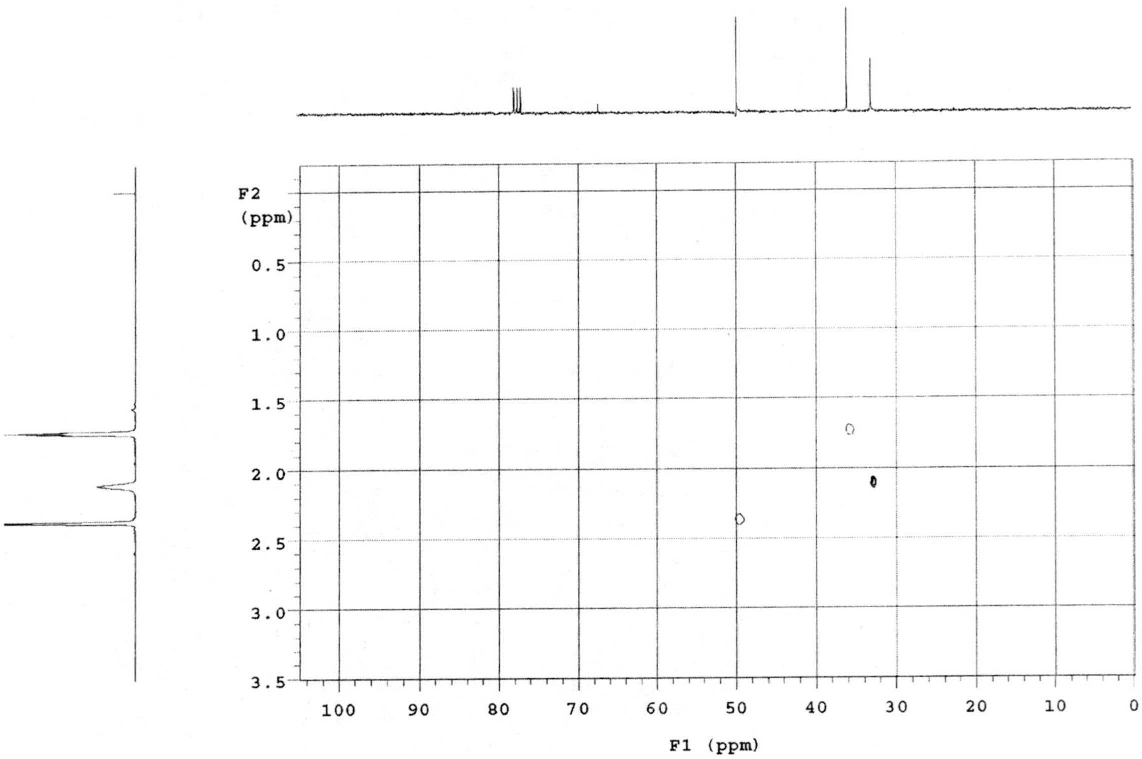

gHSQC Spectrum in $CDCl_3$

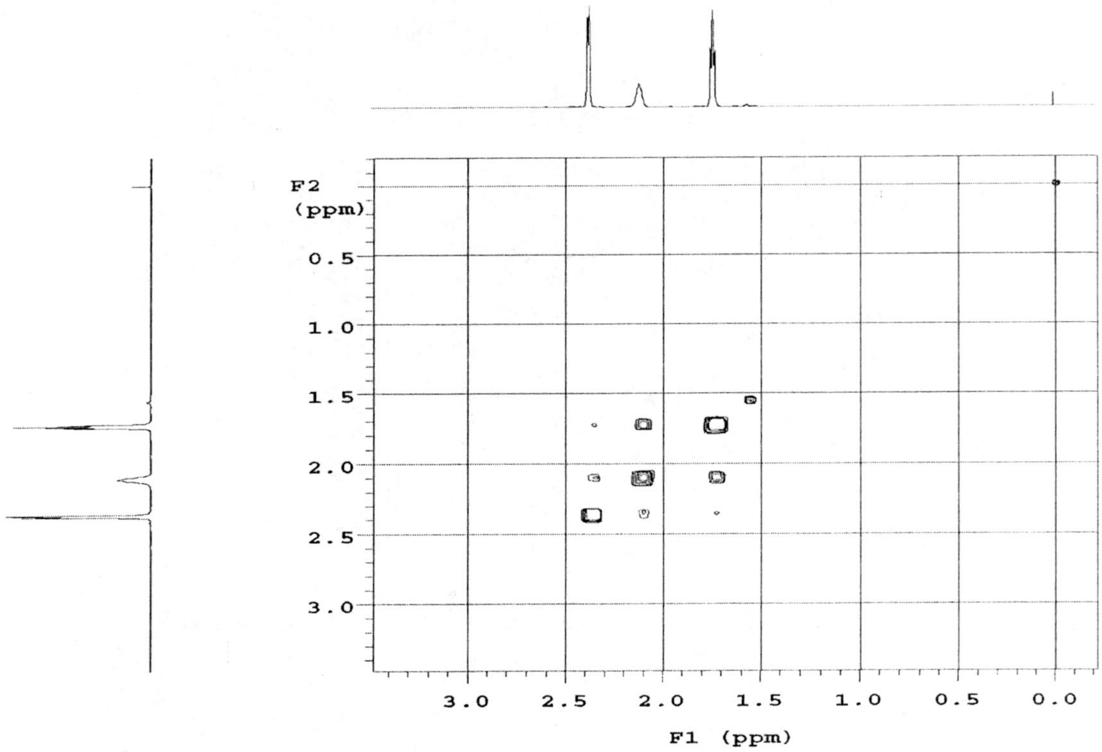

gCOSY Spectrum in $CDCl_3$

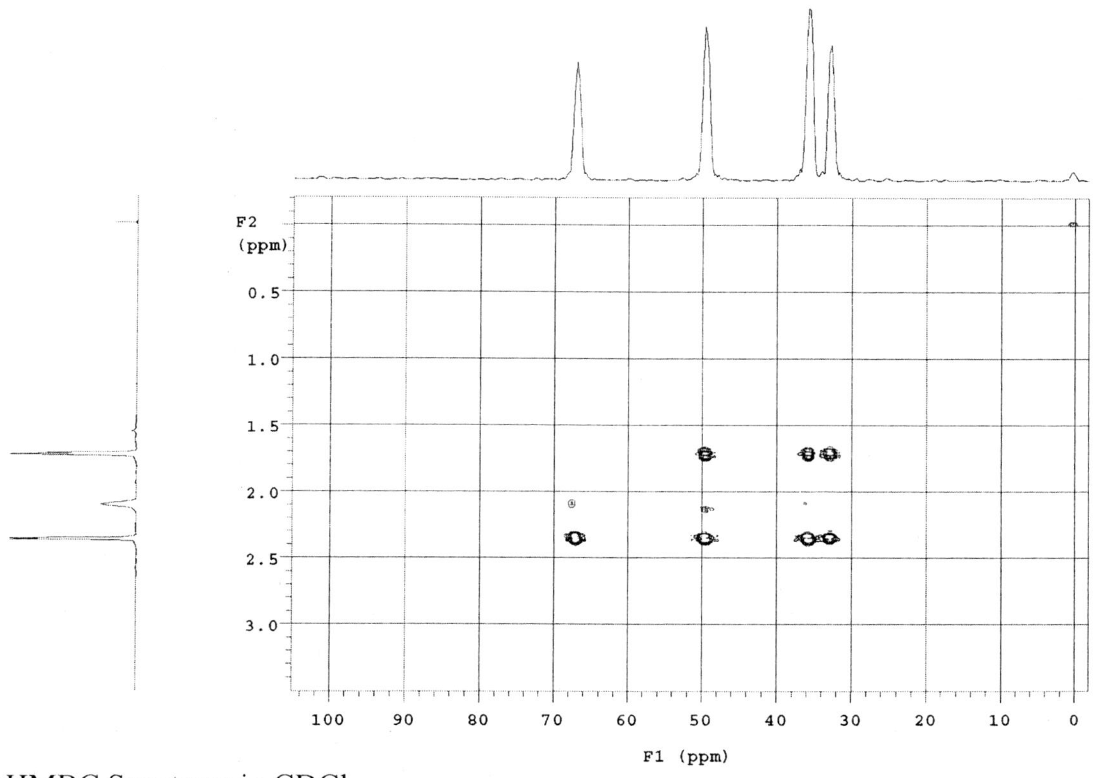

gHMBC Spectrum in $CDCl_3$

Problem 4.03 Given the spectroscopic data below (Proton, Carbon, gHSQC, and gCOSY), provide the structure of the molecule. (MS: 292, 294, 296 m/z in ratio 1:2:1)

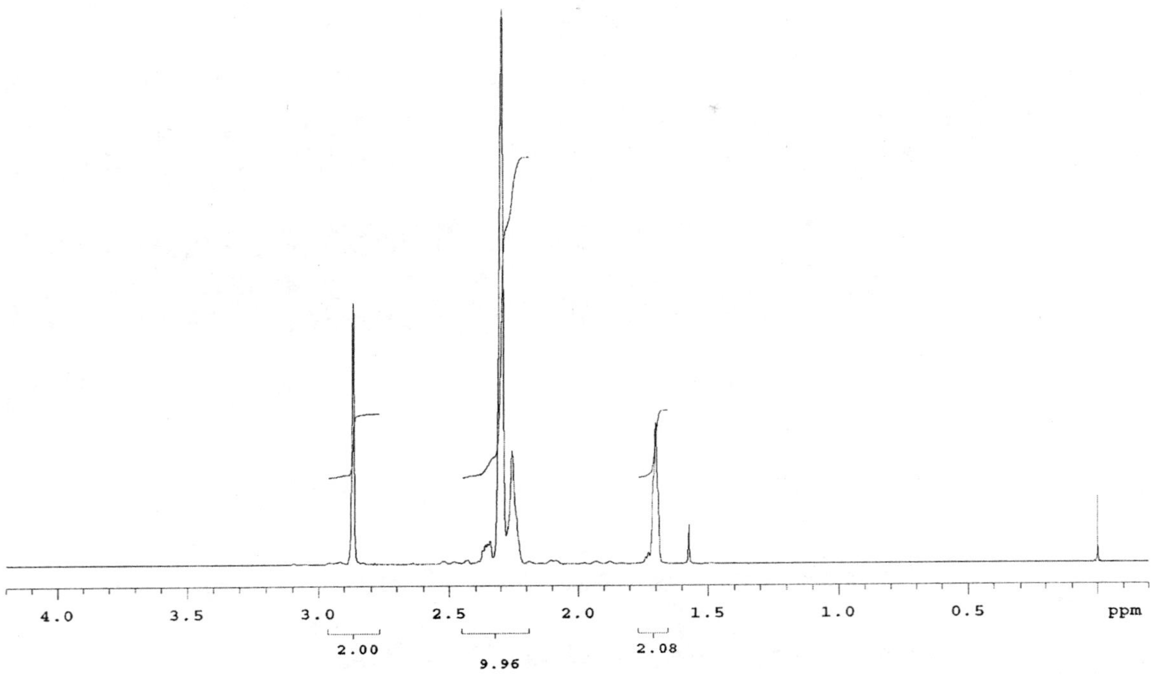

^1H-NMR Spectrum in $CDCl_3$

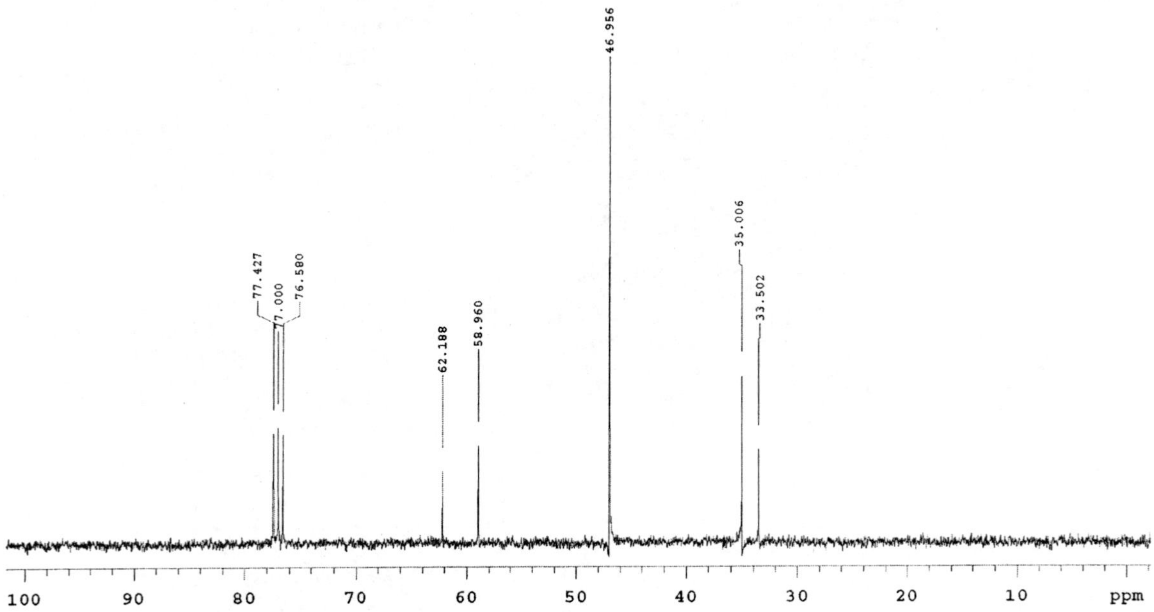

^{13}C-NMR Spectrum in $CDCl_3$

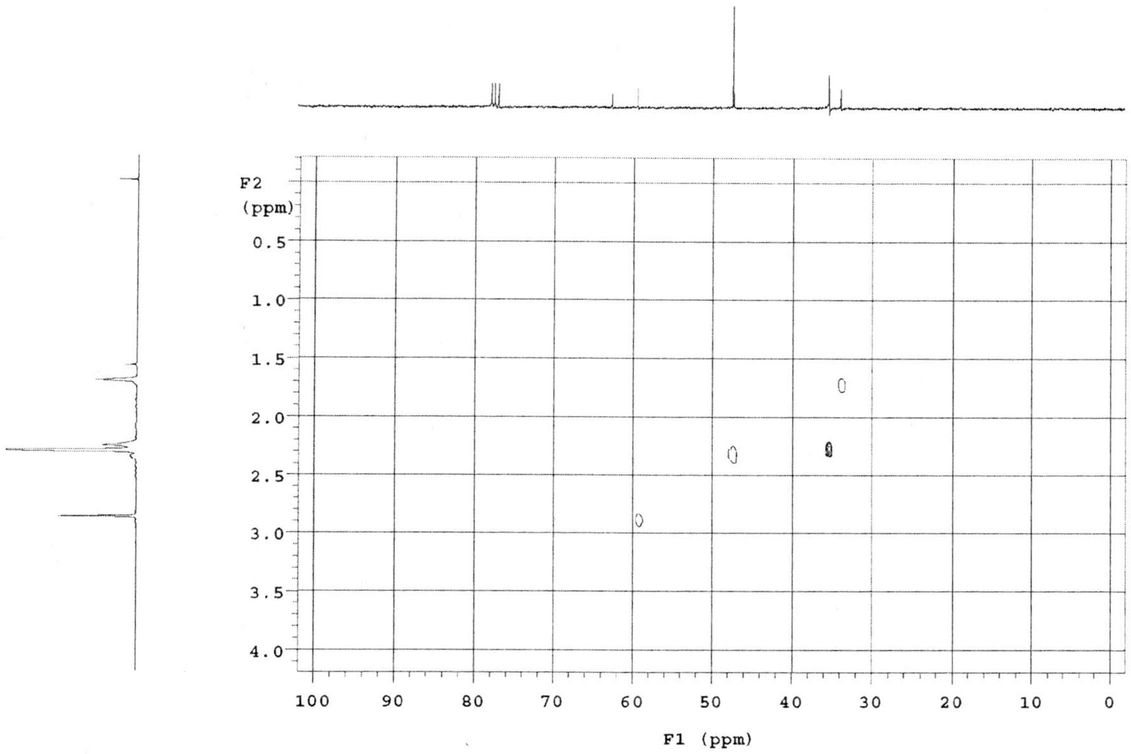

gHSQC Spectrum in $CDCl_3$

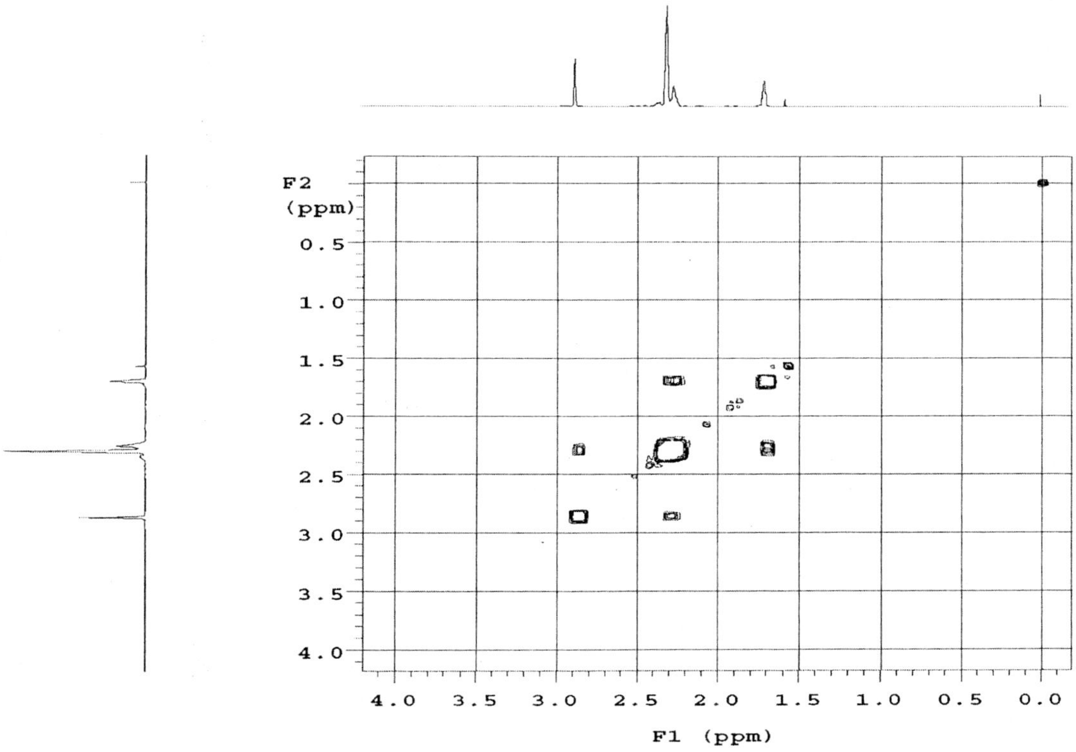

gCOSY Spectrum in $CDCl_3$

Problem 4.04 Given the spectroscopic data below (Proton, Carbon, gHSQC, gCOSY, and gHMBC), provide the structure of the molecule. The peak at ~2.2 ppm disappears upon addition of a small amount of D_2O.

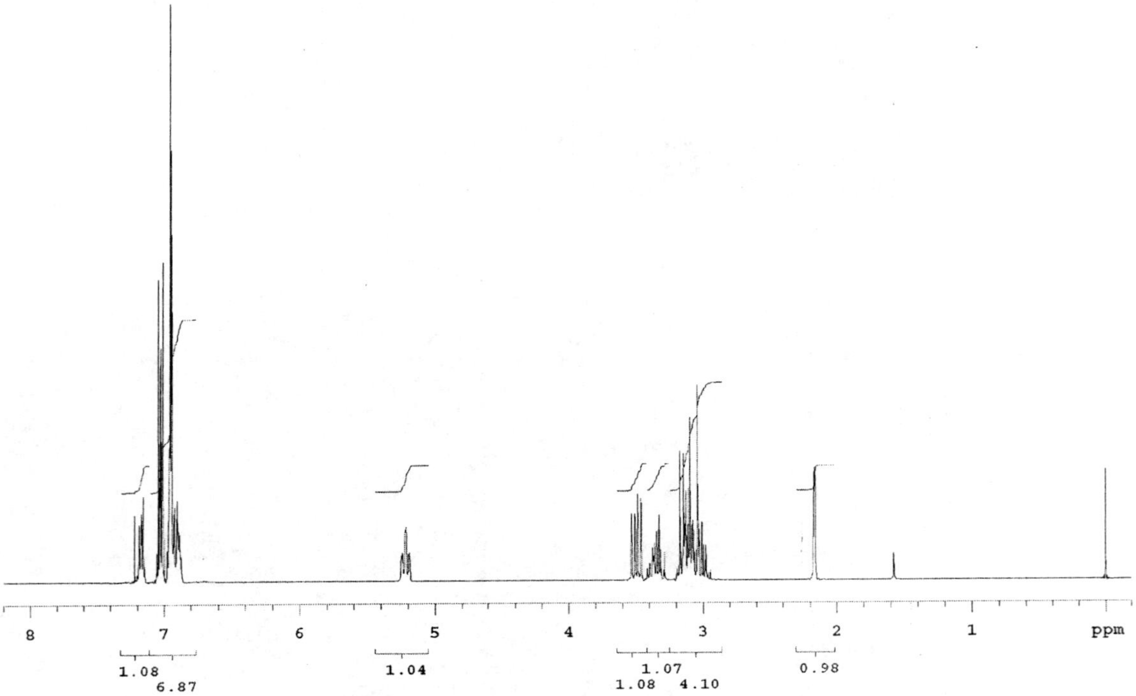

^{1}H-NMR Spectrum in $CDCl_3$

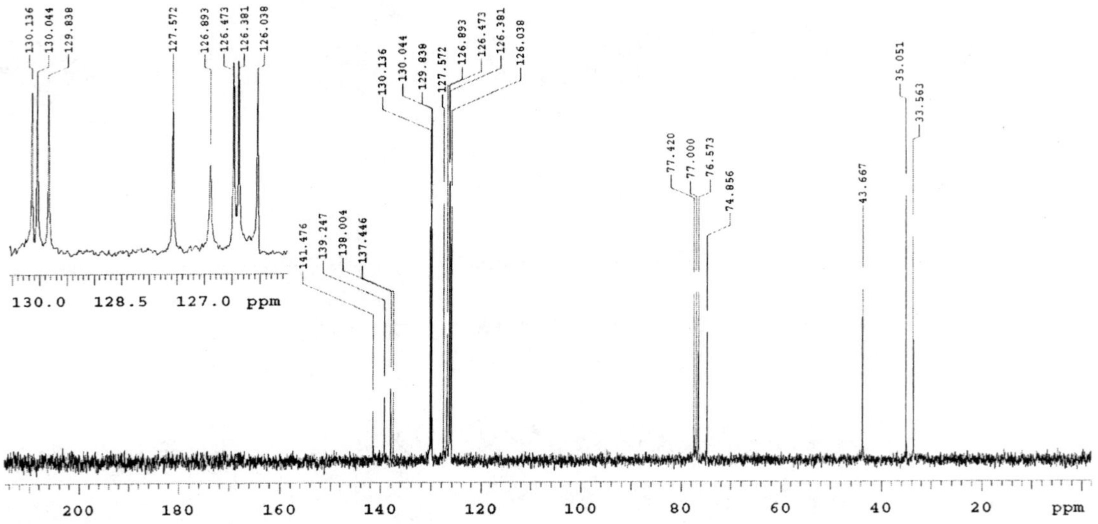

^{13}C-NMR Spectrum in $CDCl_3$

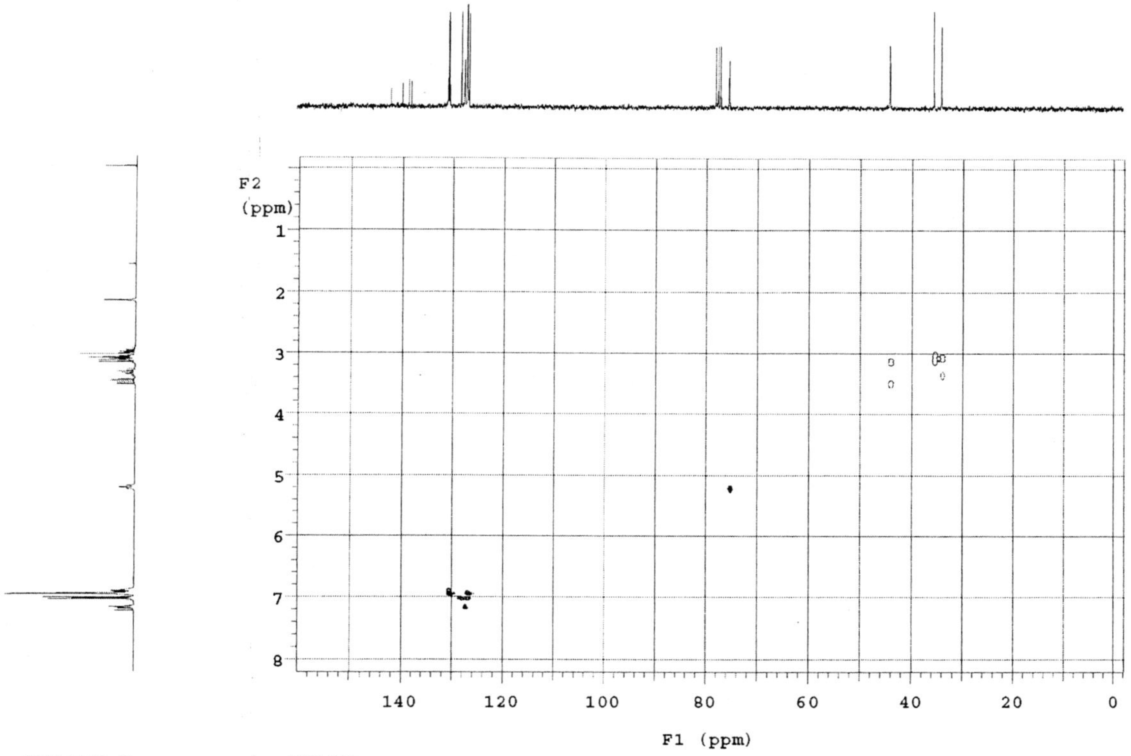

gHSQC Spectrum in $CDCl_3$

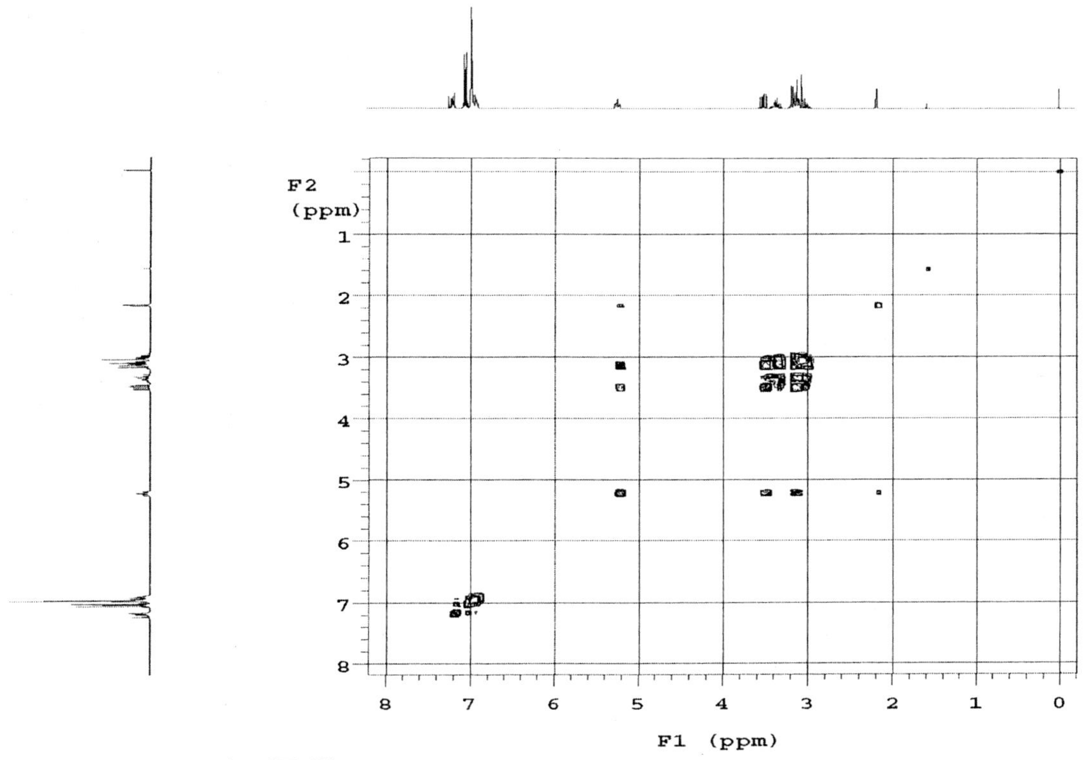

gCOSY Spectrum in $CDCl_3$

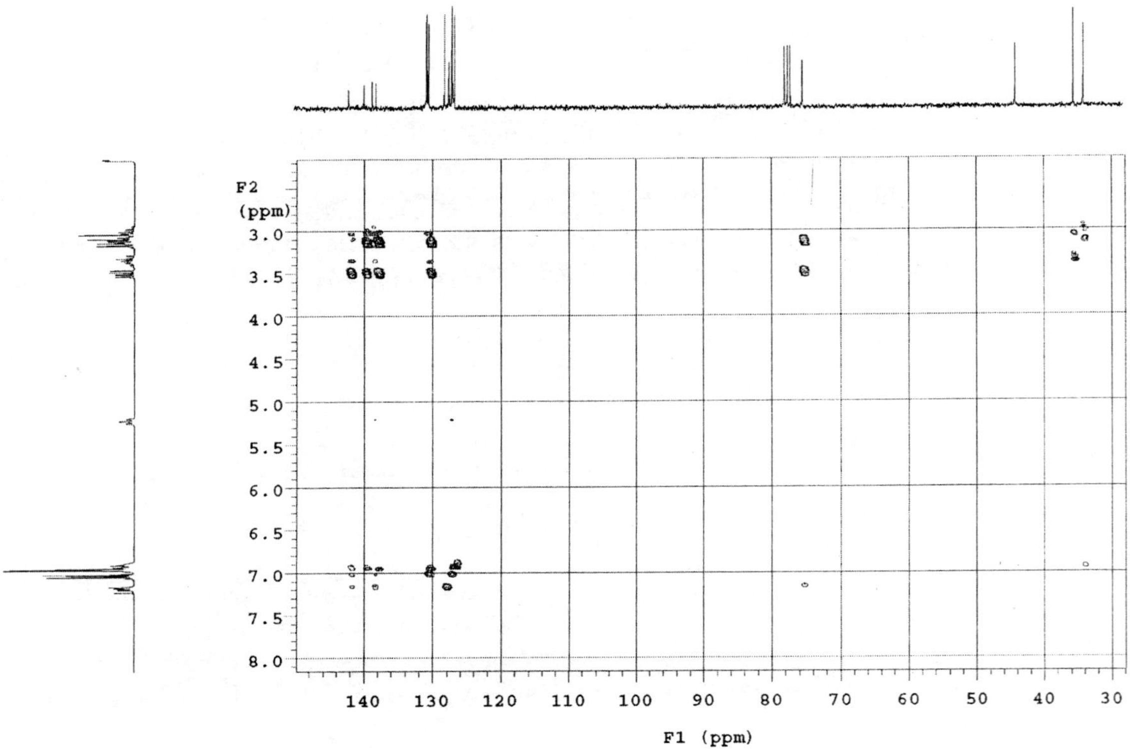

gHMBC Spectrum in $CDCl_3$

Problem 4.05 Given the spectroscopic data below (Proton, Carbon, gHSQC, gCOSY, and gHMBC), provide the structure of the molecule. (HR-MS: 238.1358 amu), {^{1}H, ^{1}H} Nuclear Overhauser effect correlation was observed between the signals at 2.57 and 2.90 ppm.

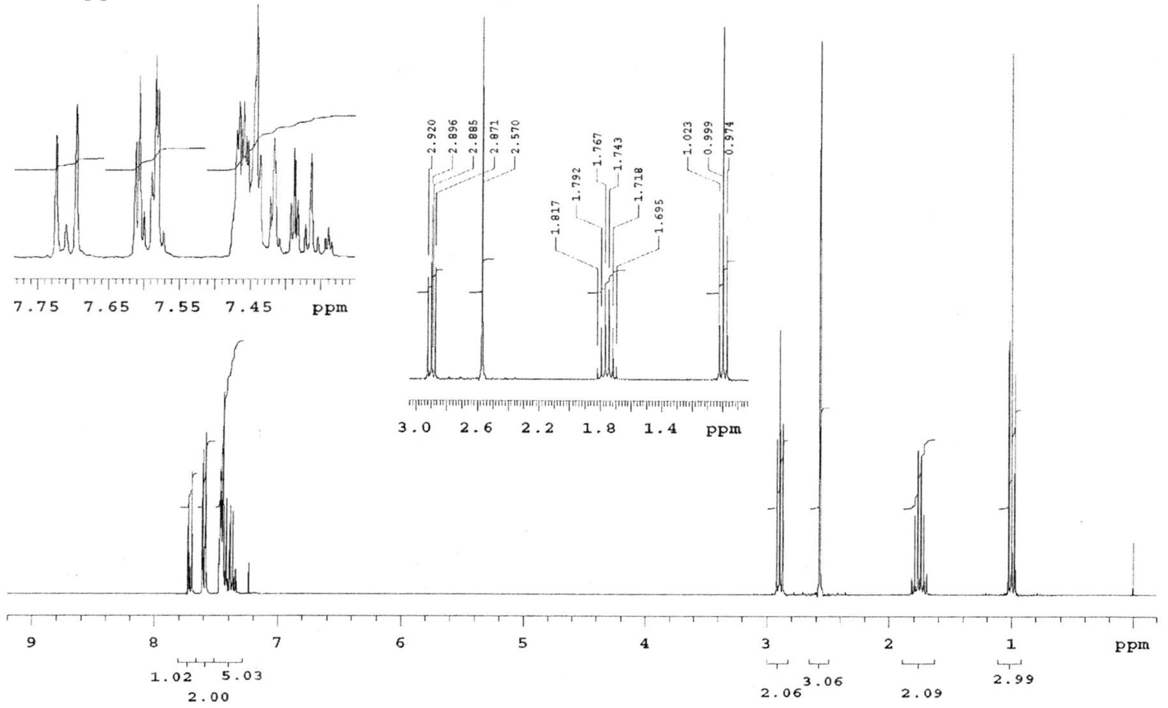

^{1}H-NMR Spectrum in $CDCl_3$

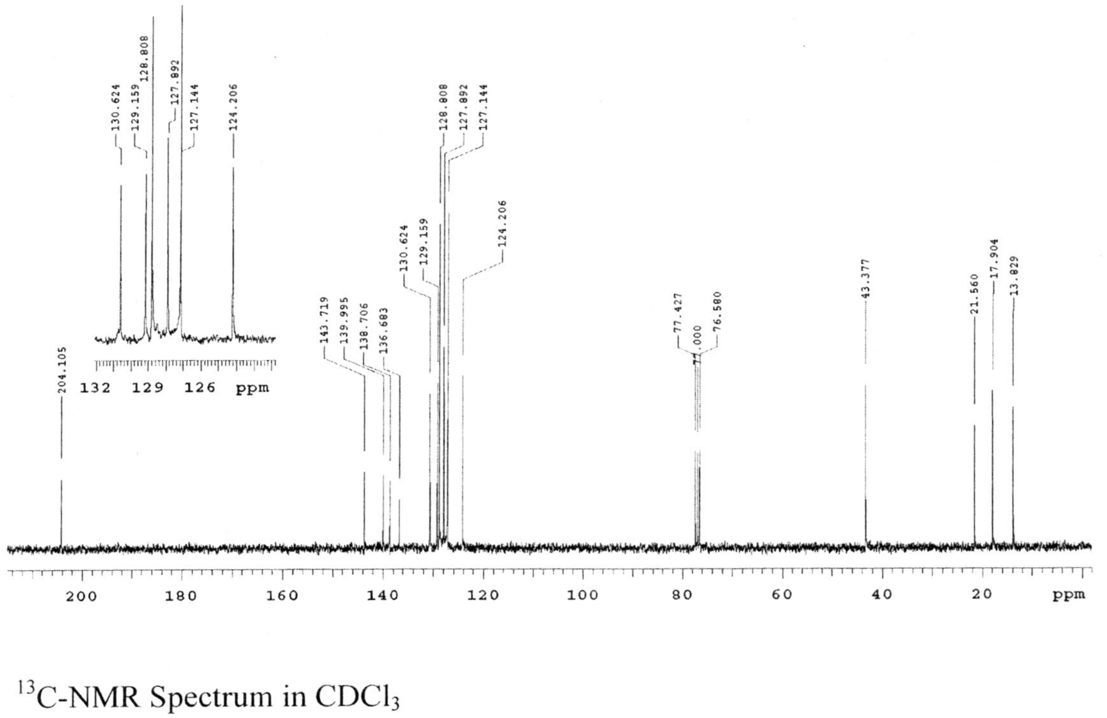

^{13}C-NMR Spectrum in $CDCl_3$

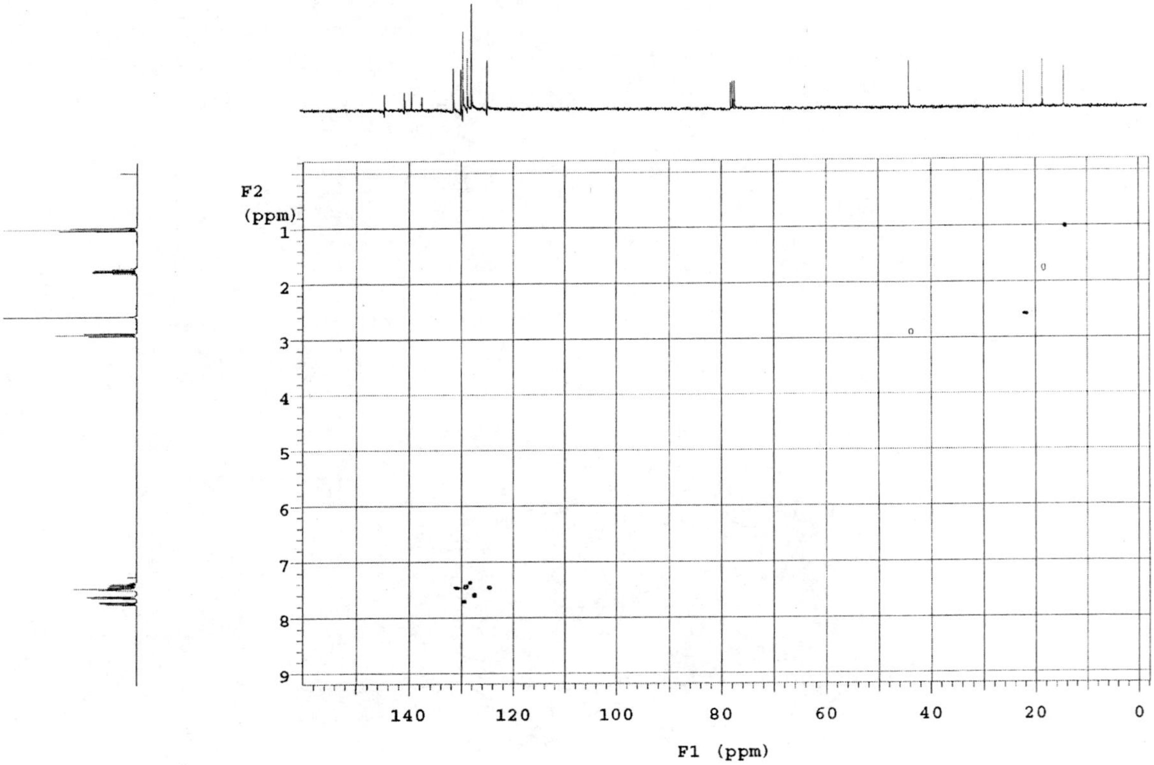

gHSQC Spectrum in $CDCl_3$

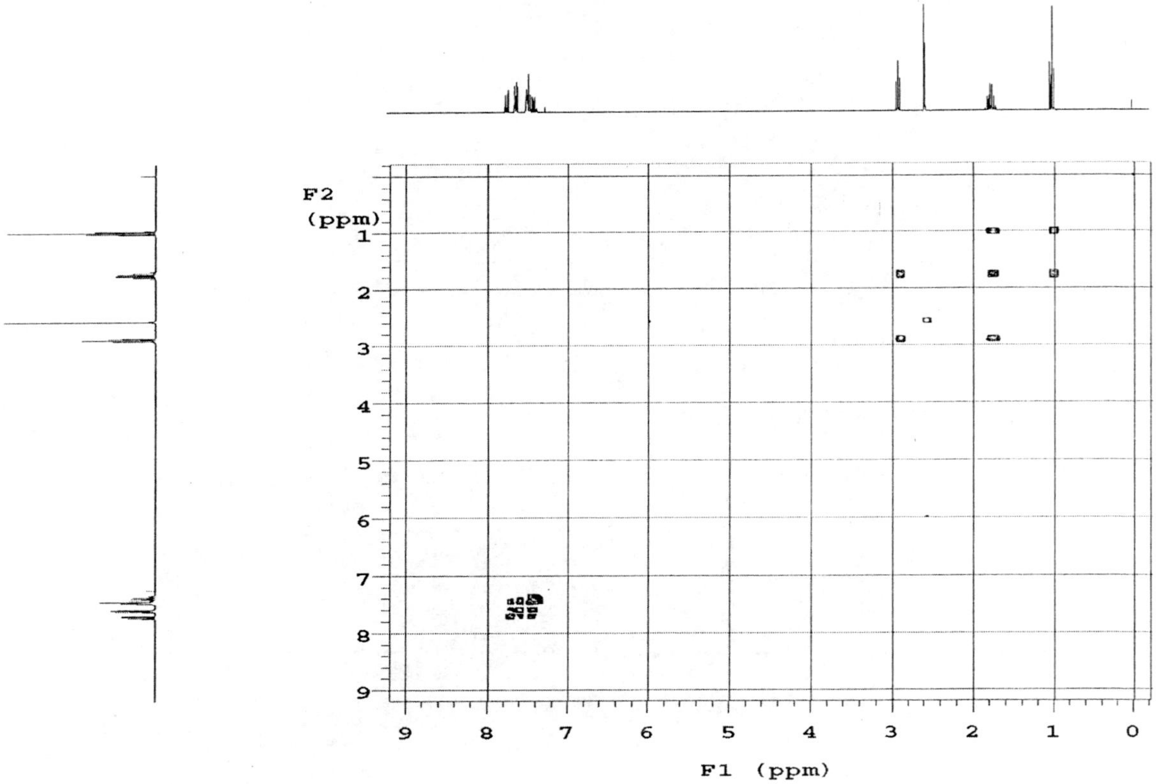

gCOSY Spectrum in $CDCl_3$

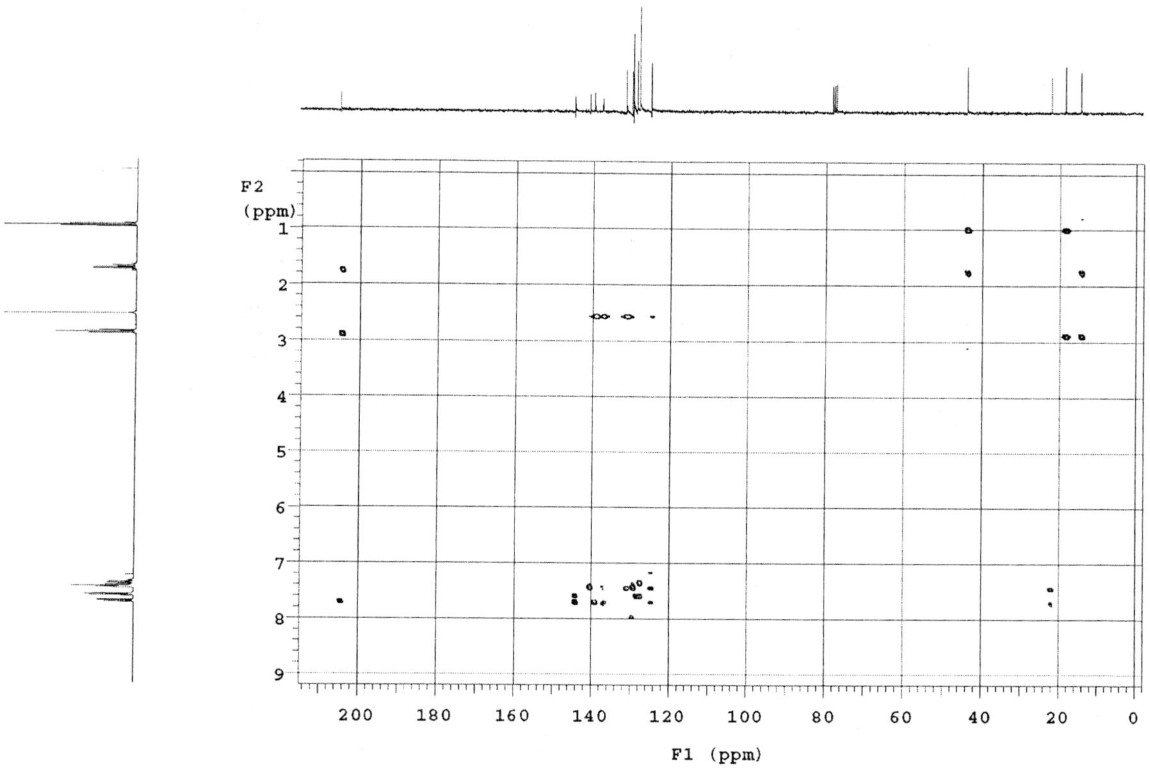

gHMBC Spectrum in $CDCl_3$

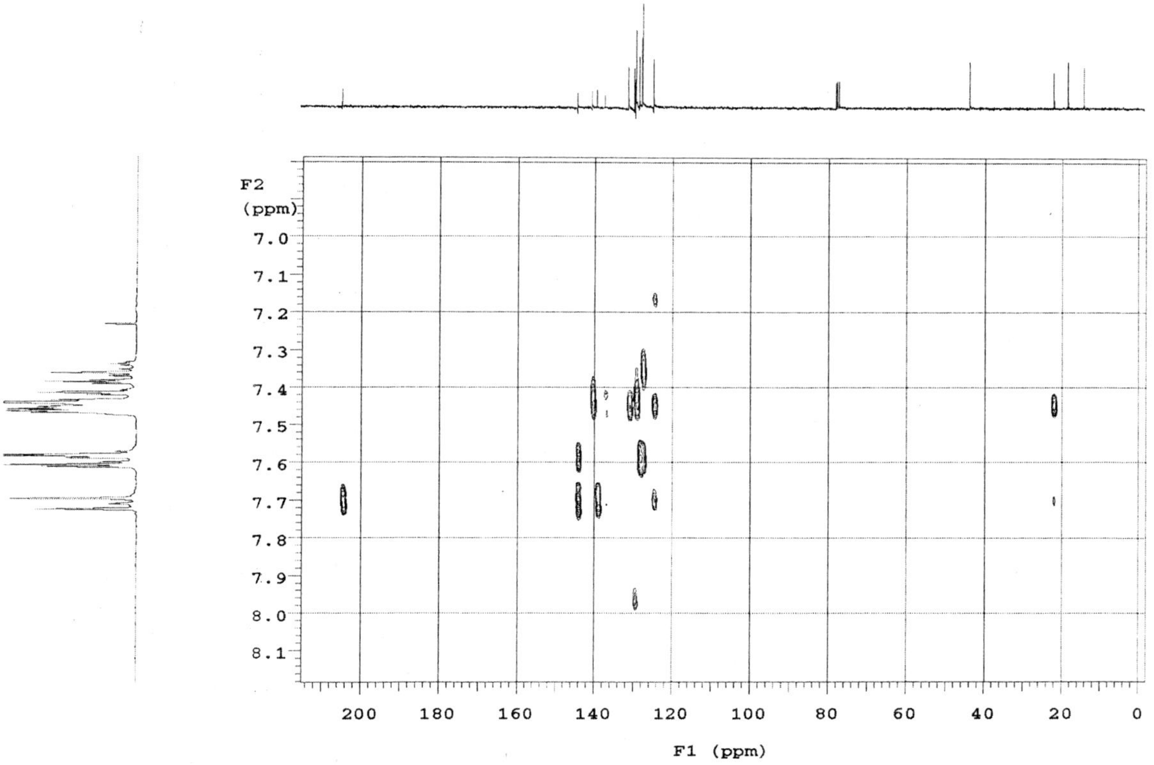

Expansion of gHMBC Spectrum in $CDCl_3$

Problem 4.06 Given the spectroscopic data below (Proton, Carbon, Edited DEPT, gCOSY, and gHSQC), provide the structure of the molecule. (HR-MS: 234.1256 amu)

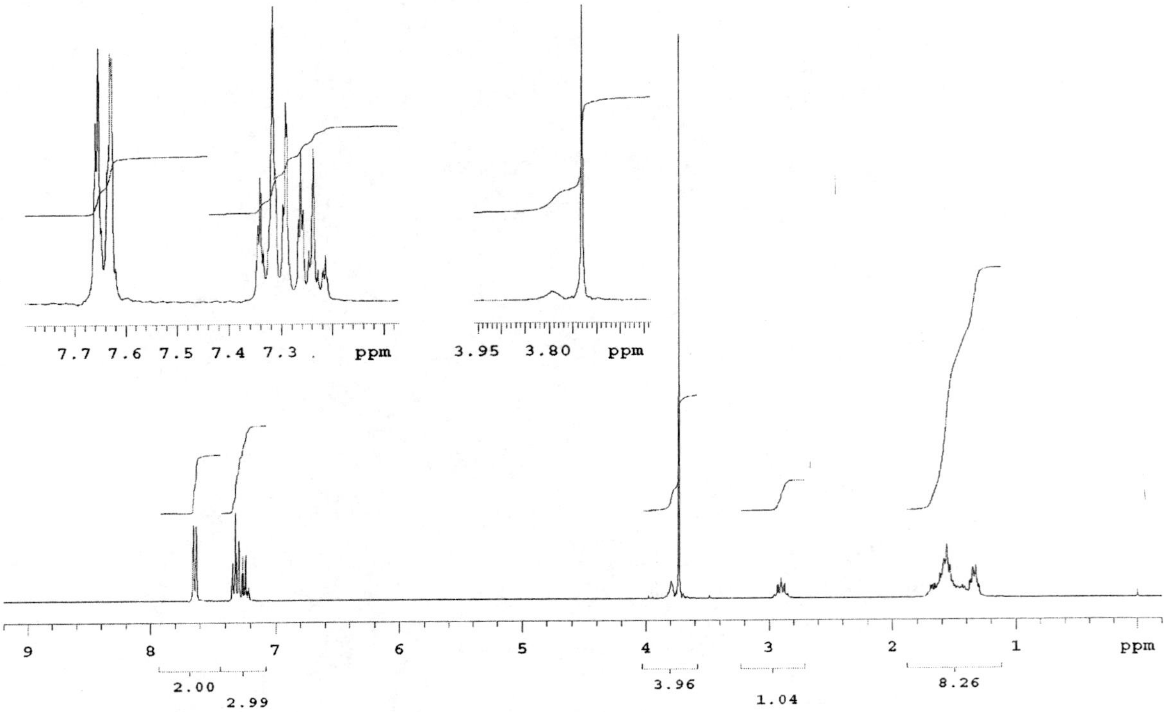

^{1}H-NMR Spectrum in $CDCl_3$

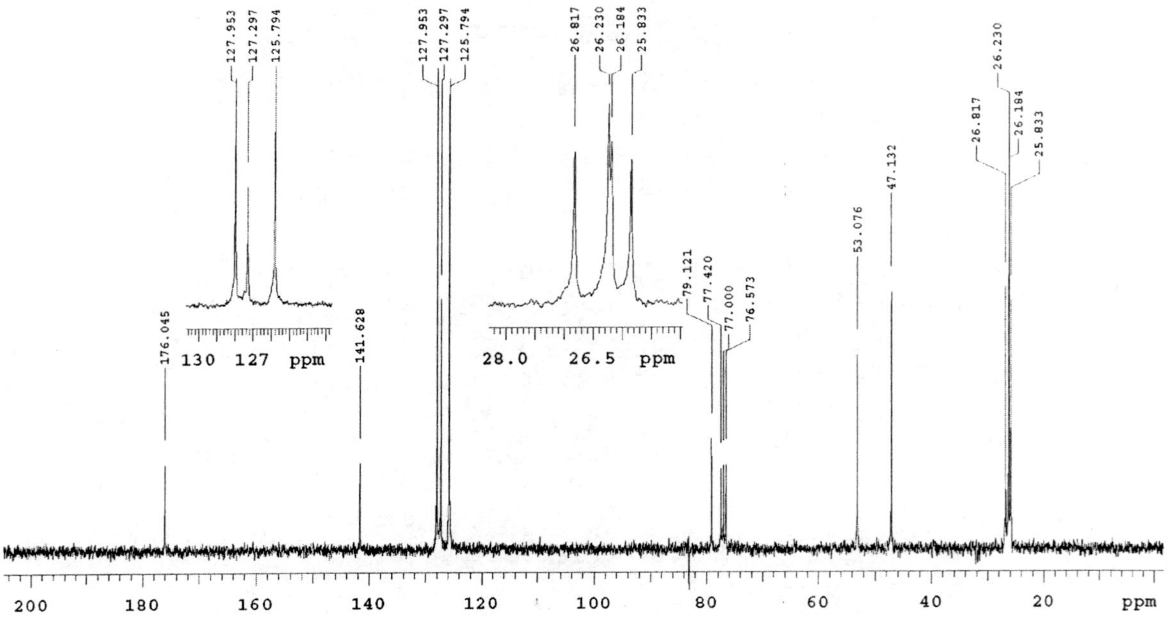

^{13}C-NMR Spectrum in $CDCl_3$

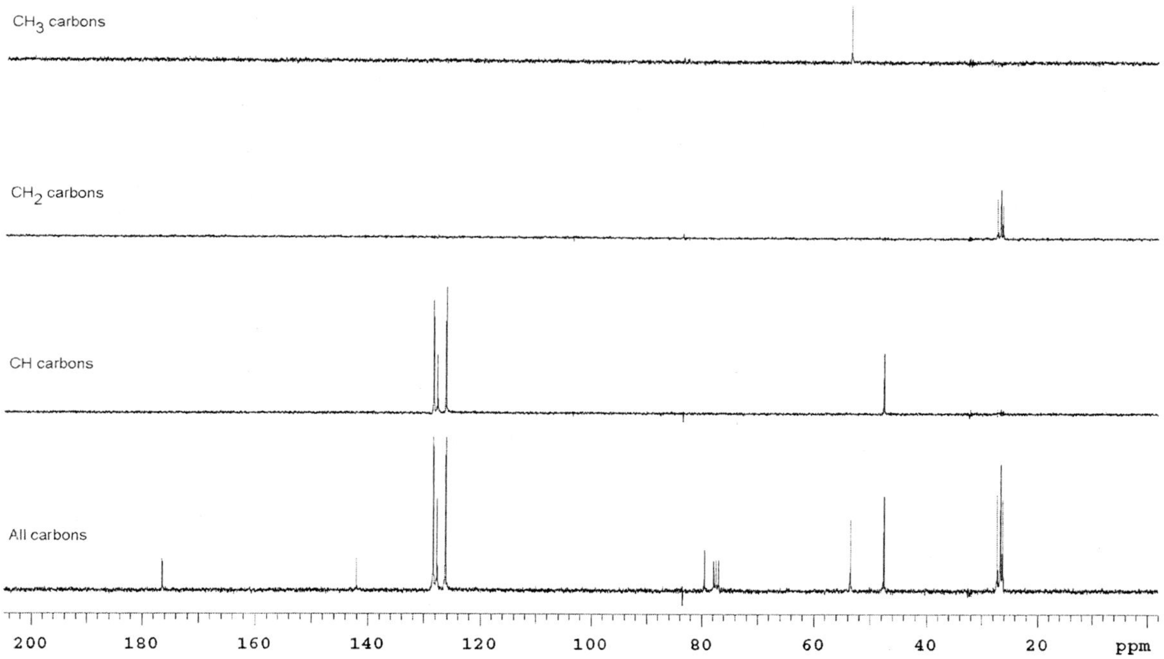

Edited DEPT ^{13}C-NMR Spectrum in $CDCl_3$

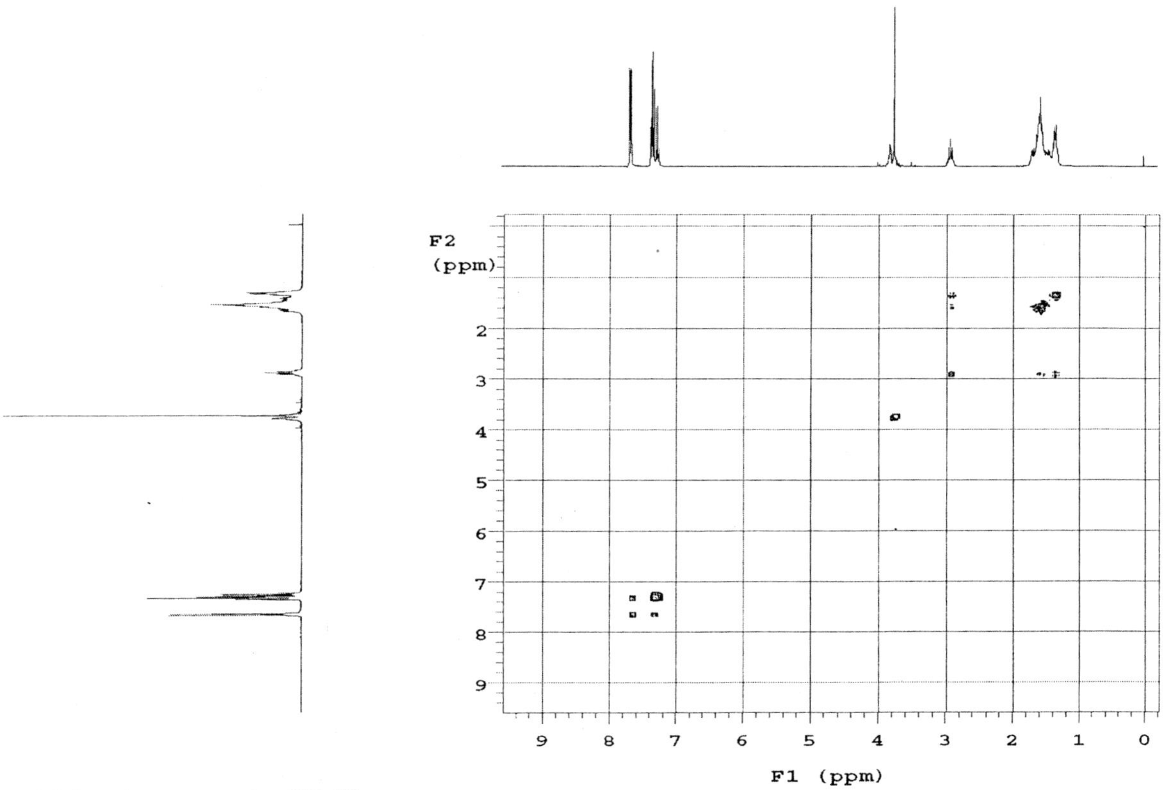

gCOSY Spectrum in $CDCl_3$

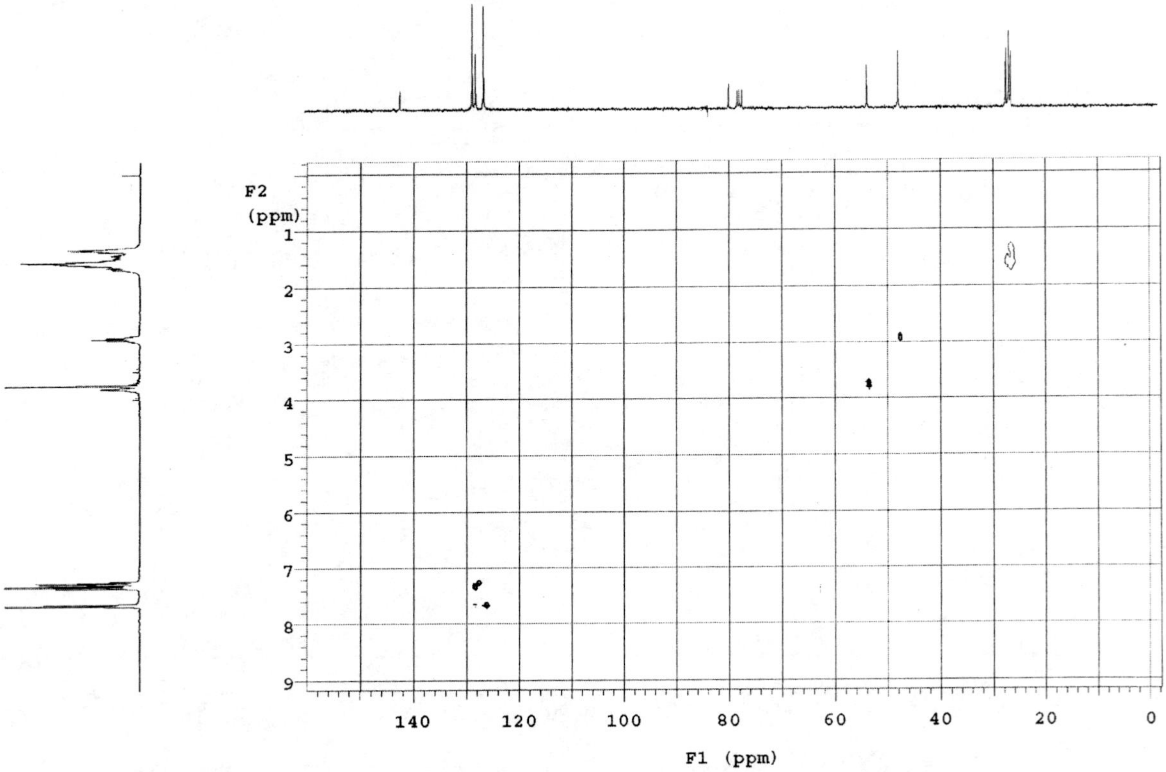

gHSQC Spectrum in $CDCl_3$

Problem 4.07 Given the spectroscopic data below (Proton, Carbon, Edited DEPT, gCOSY, gHSQC, and gHMBC), provide the structure of the molecule. HINT: Problems **4.07** and **4.08** share the same carbon skeleton.

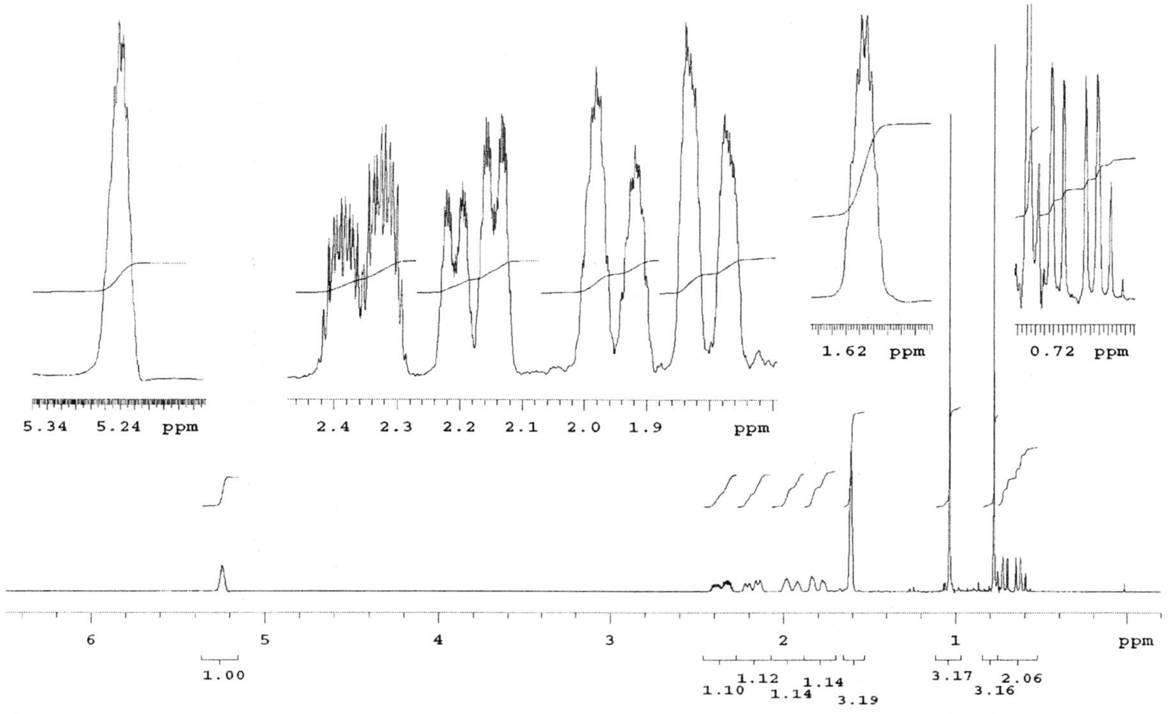

^{1}H-NMR Spectrum in $CDCl_3$

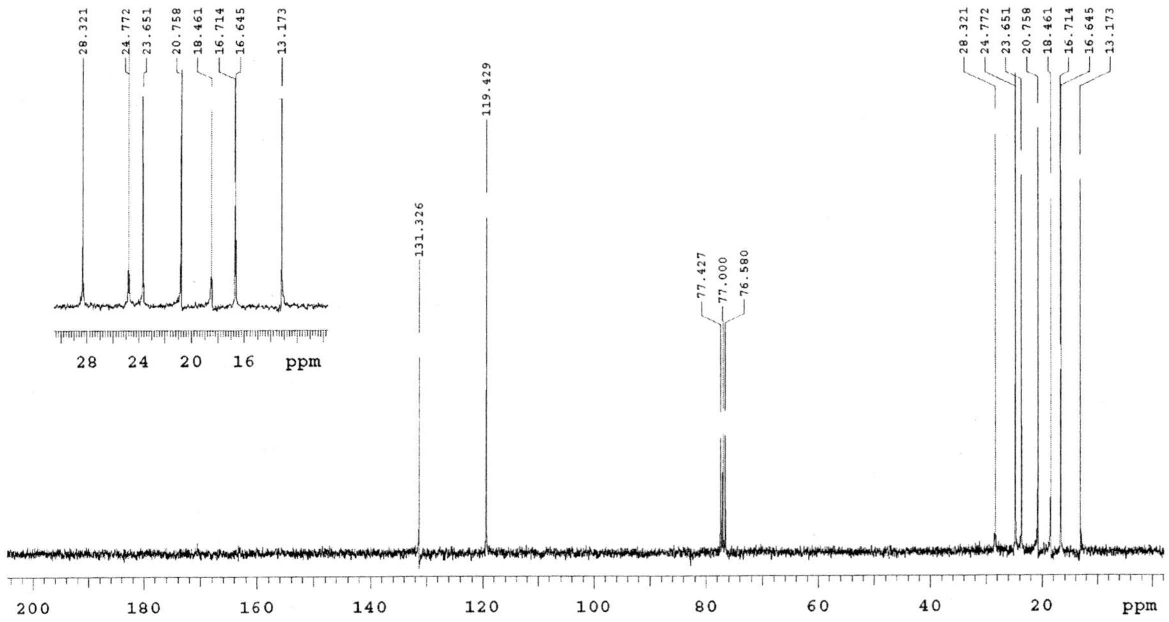

^{13}C-NMR Spectrum in $CDCl_3$

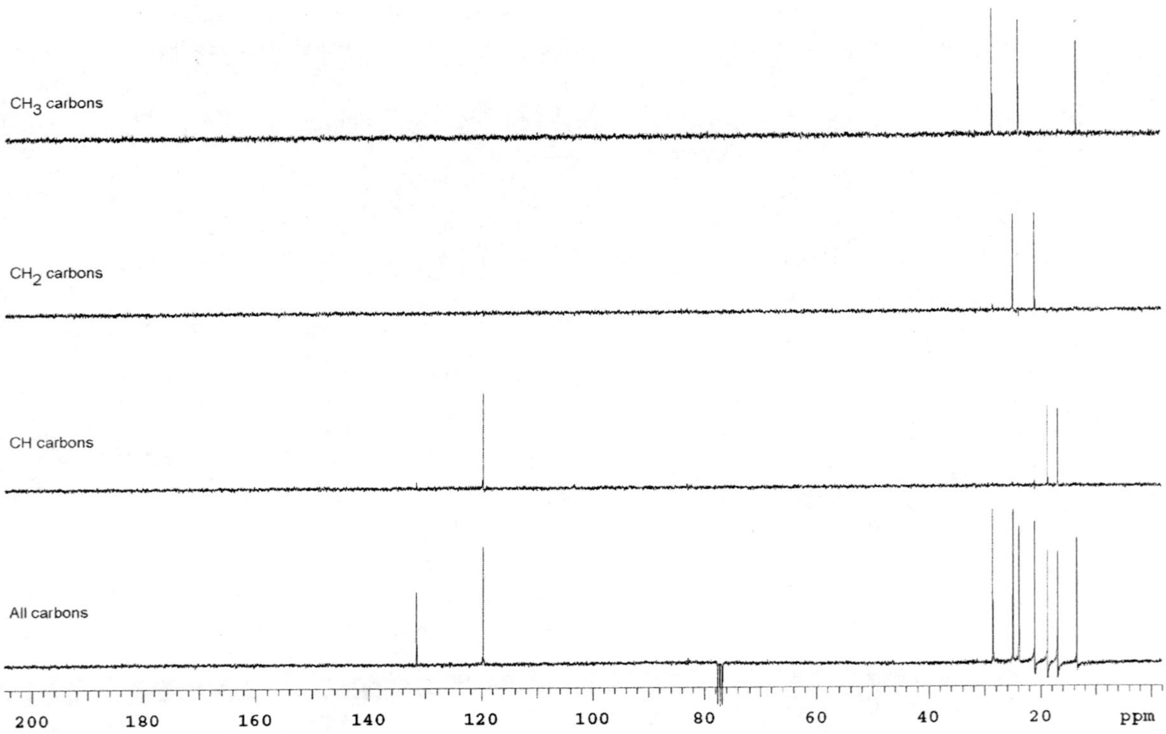

Edited DEPT ^{13}C-NMR Spectrum in $CDCl_3$

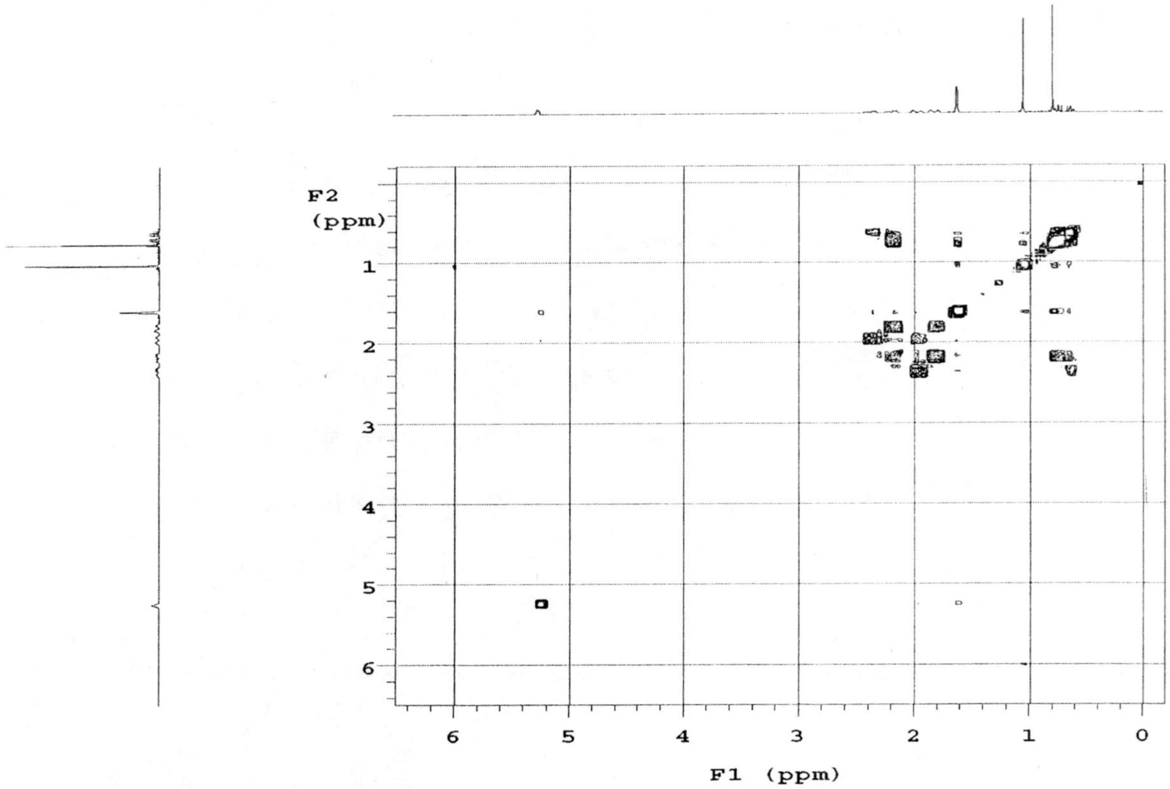

gCOSY Spectrum in $CDCl_3$

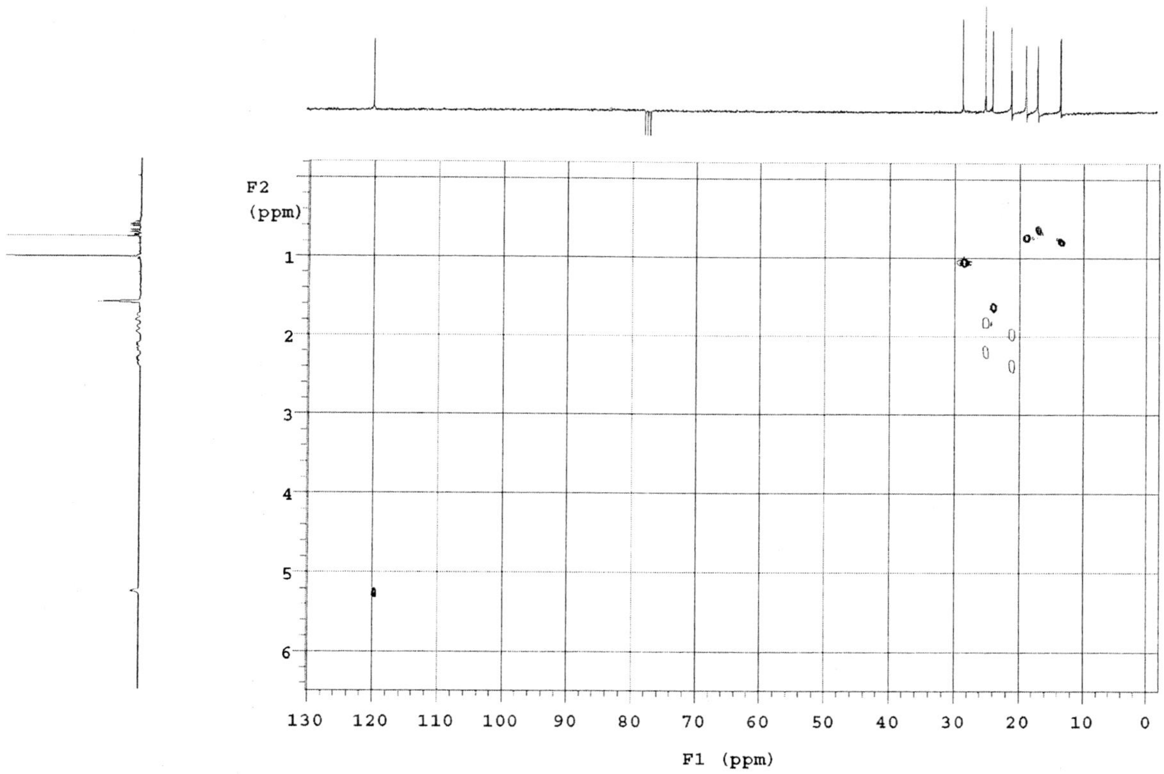

gHSQC Spectrum in $CDCl_3$

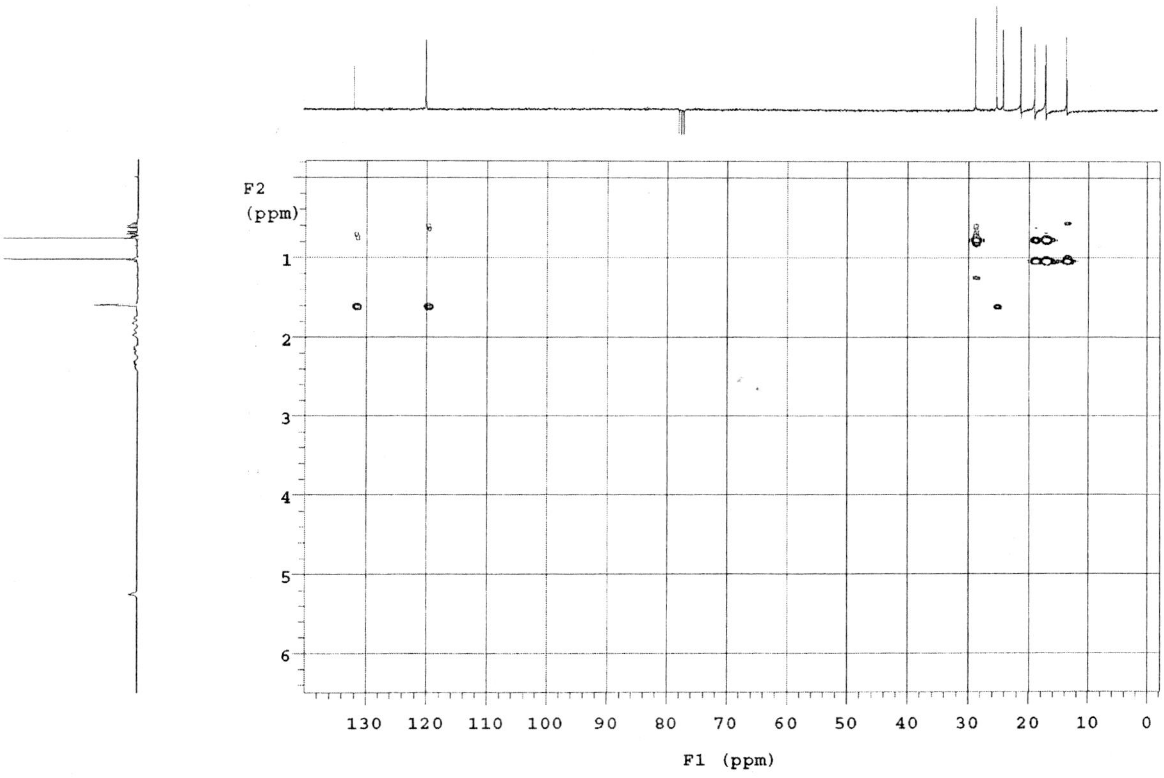

gHMBC Spectrum in $CDCl_3$

Problem 4.08 Given the spectroscopic data below (Proton, Carbon, Edited DEPT, gCOSY, gHSQC, and gHMBC), provide the structure of the molecule. HINT: Problems **4.07** and **4.08** share the same carbon skeleton.

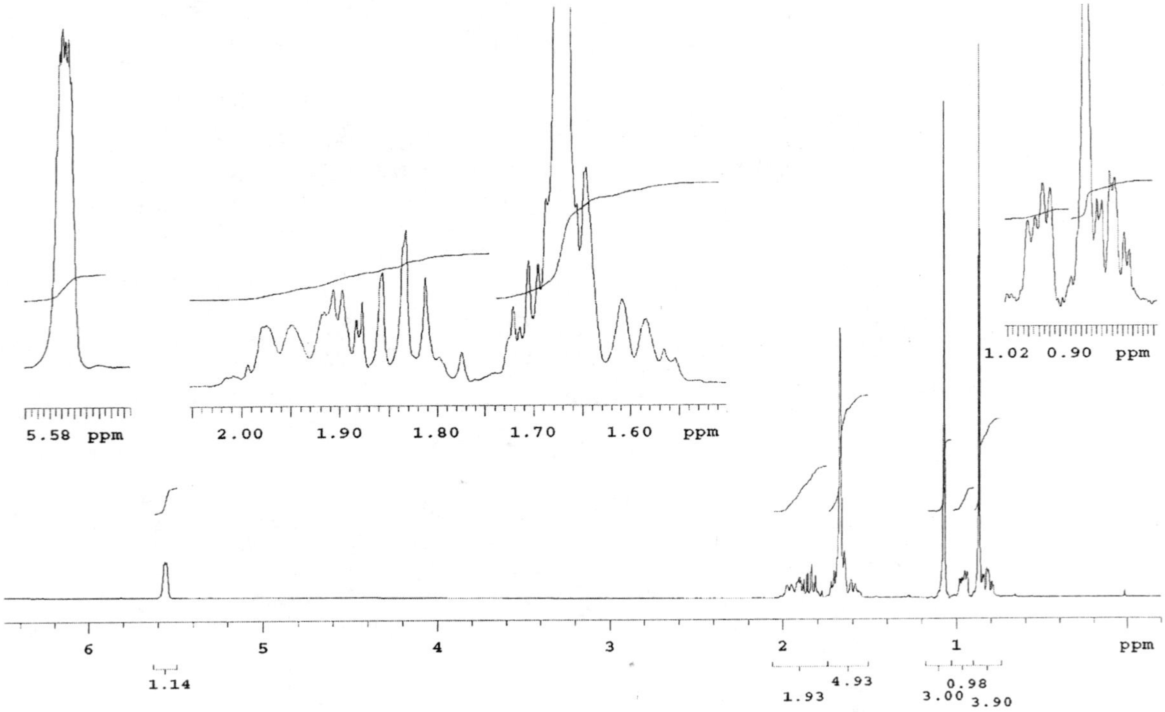

^{1}H-NMR Spectrum in $CDCl_3$

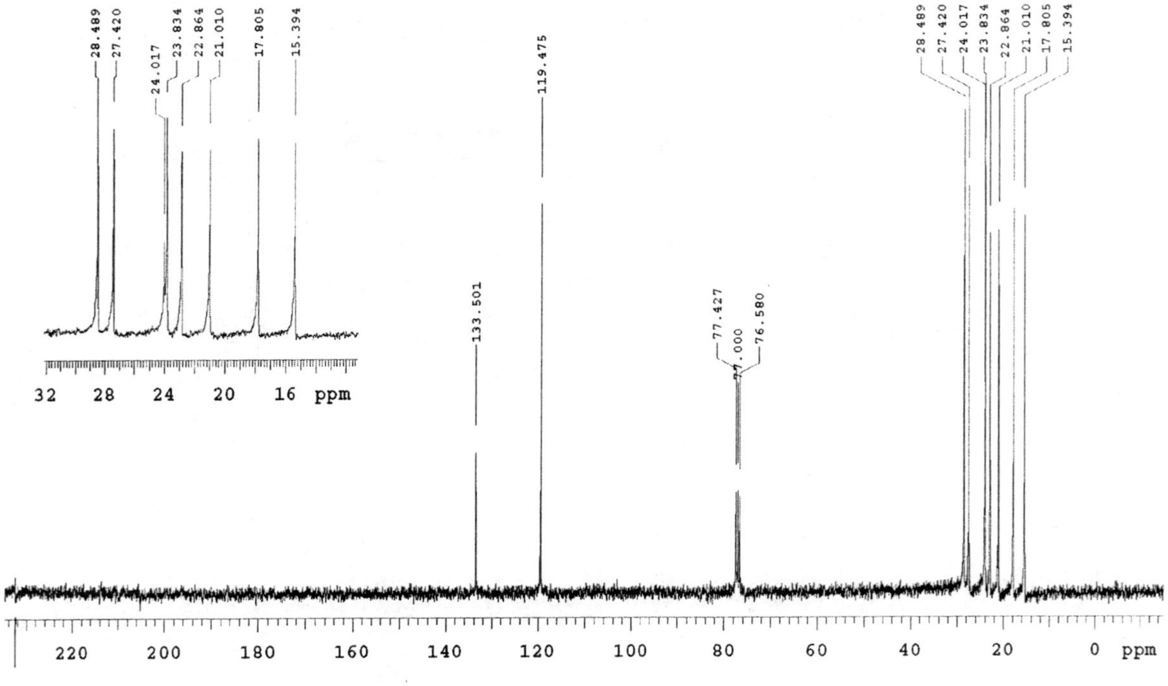

^{13}C-NMR Spectrum in $CDCl_3$

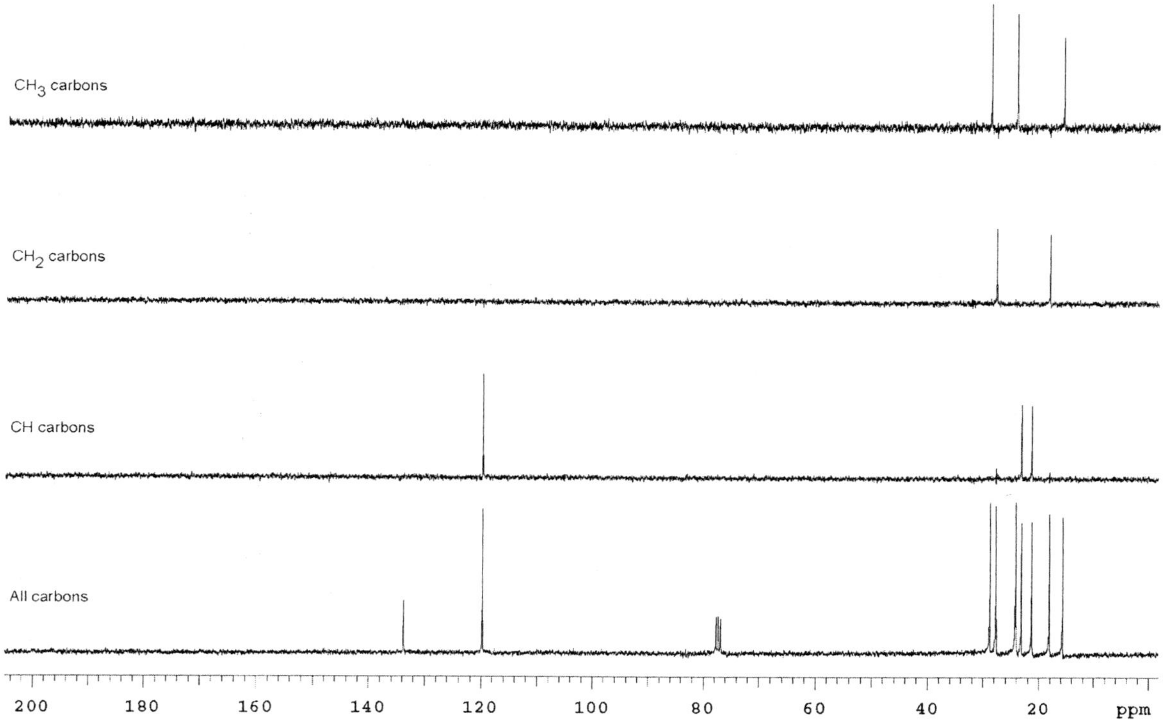

Edited DEPT ^{13}C-NMR Spectrum in $CDCl_3$

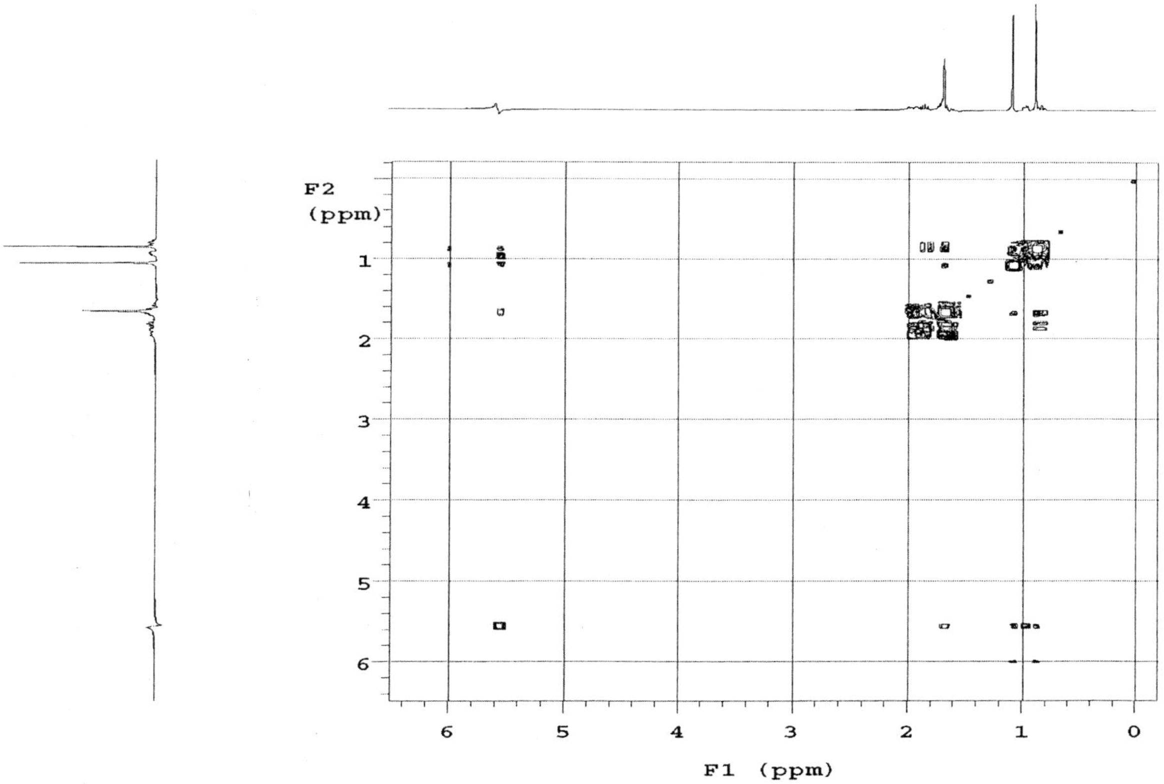

gCOSY Spectrum in $CDCl_3$

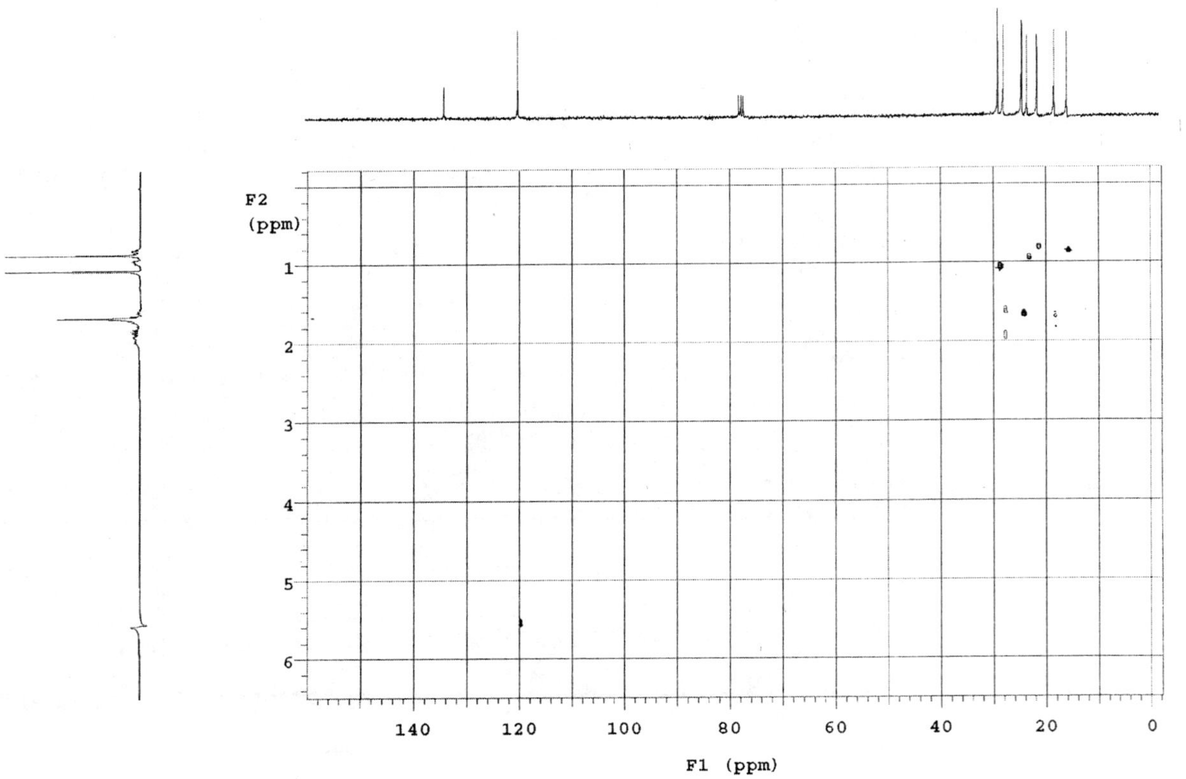

gHSQC Spectrum in $CDCl_3$

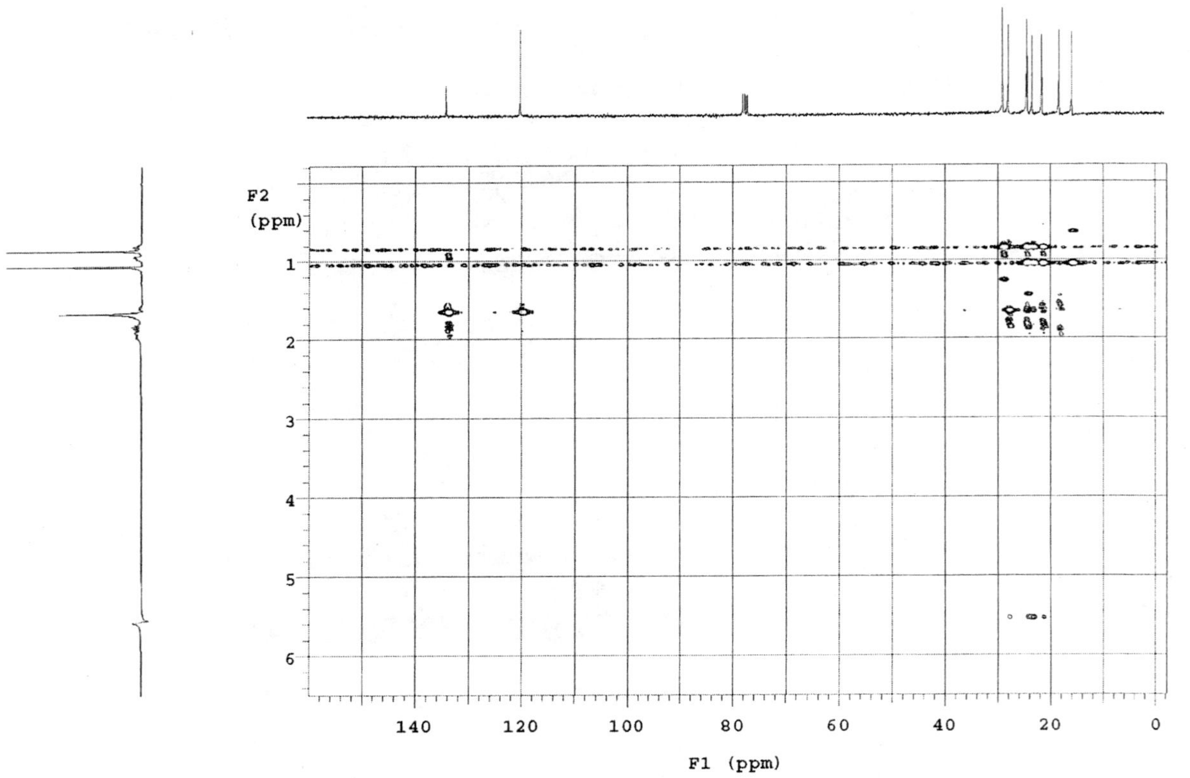

gHMBC Spectrum in $CDCl_3$

Problem 4.09 Given the spectroscopic data below (Proton, Carbon, Edited DEPT, gHSQC, and gHMBC), provide the structure of the molecule. HINT: This compound is related to sample **4.10**.

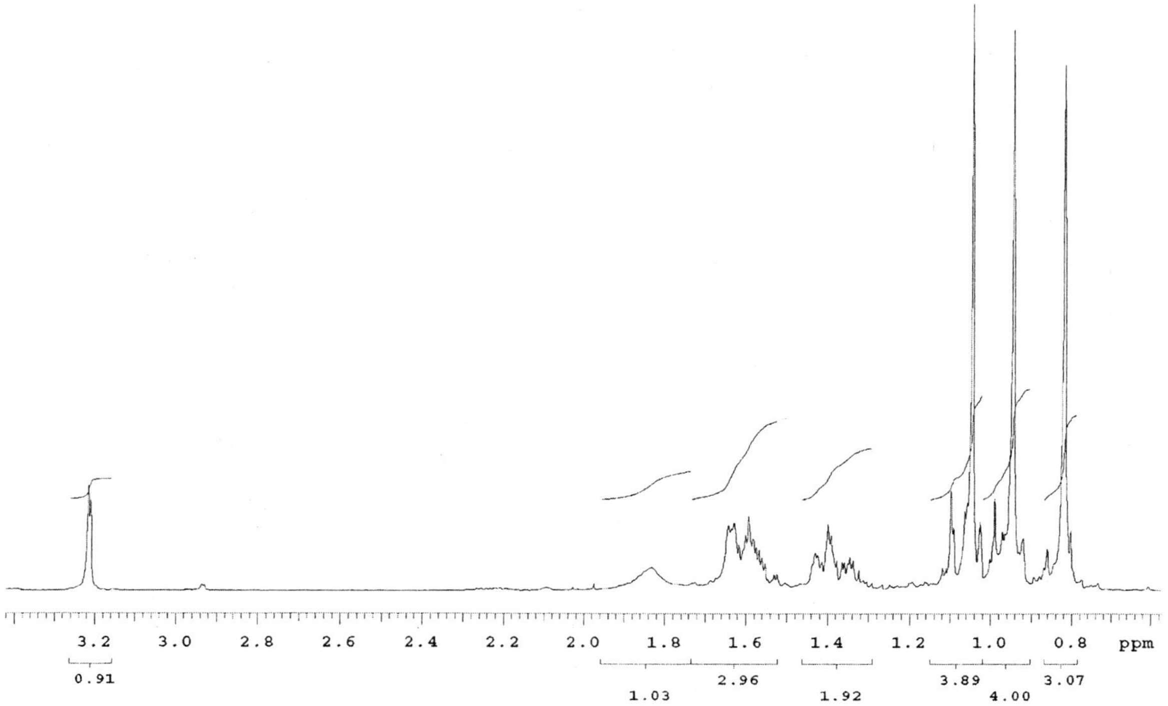

^{1}H-NMR Spectrum in $CDCl_3$

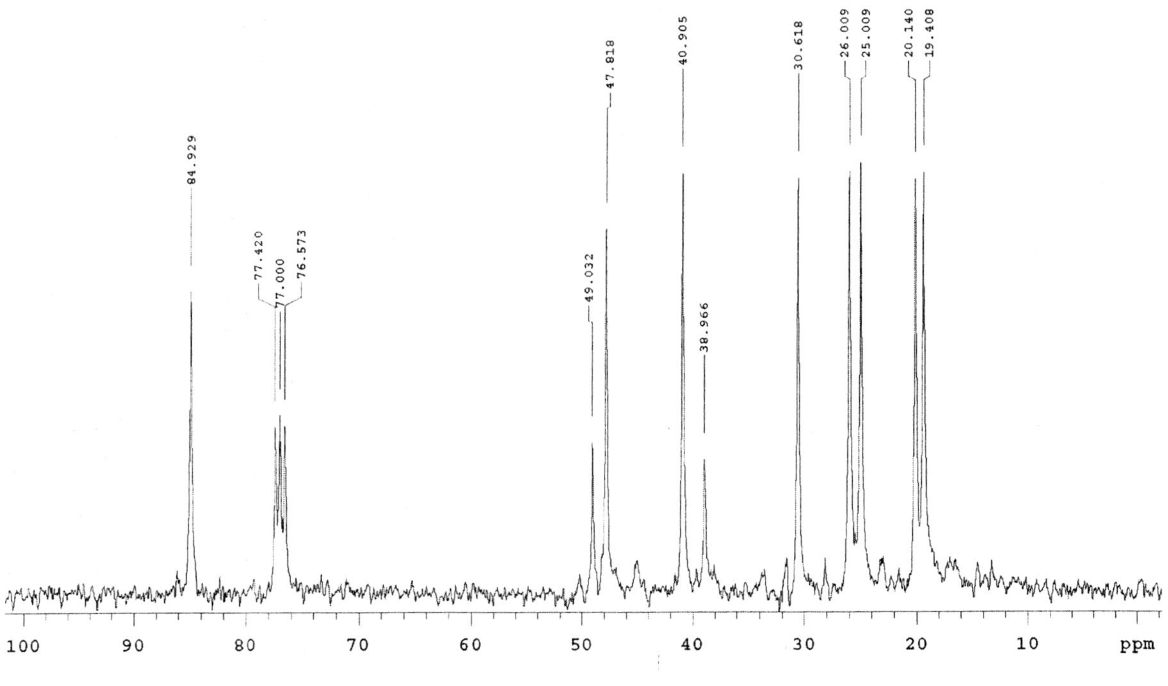

^{13}C-NMR Spectrum in $CDCl_3$

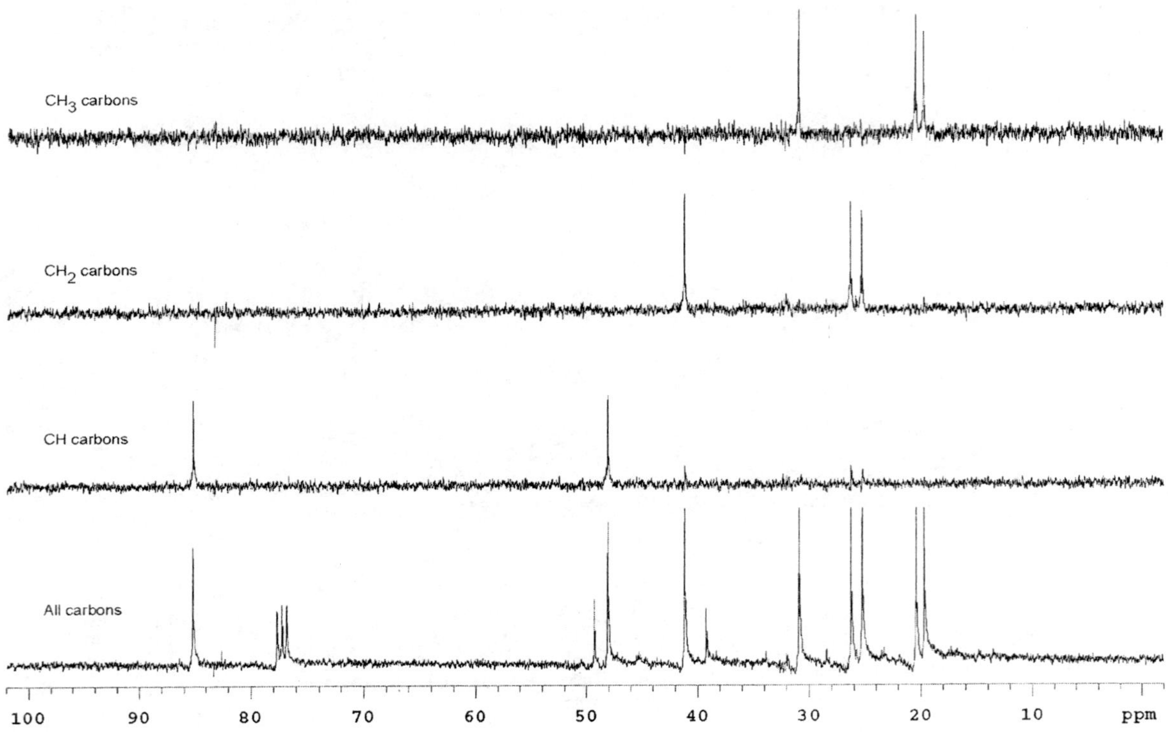

Edited DEPT ^{13}C-NMR Spectrum in $CDCl_3$

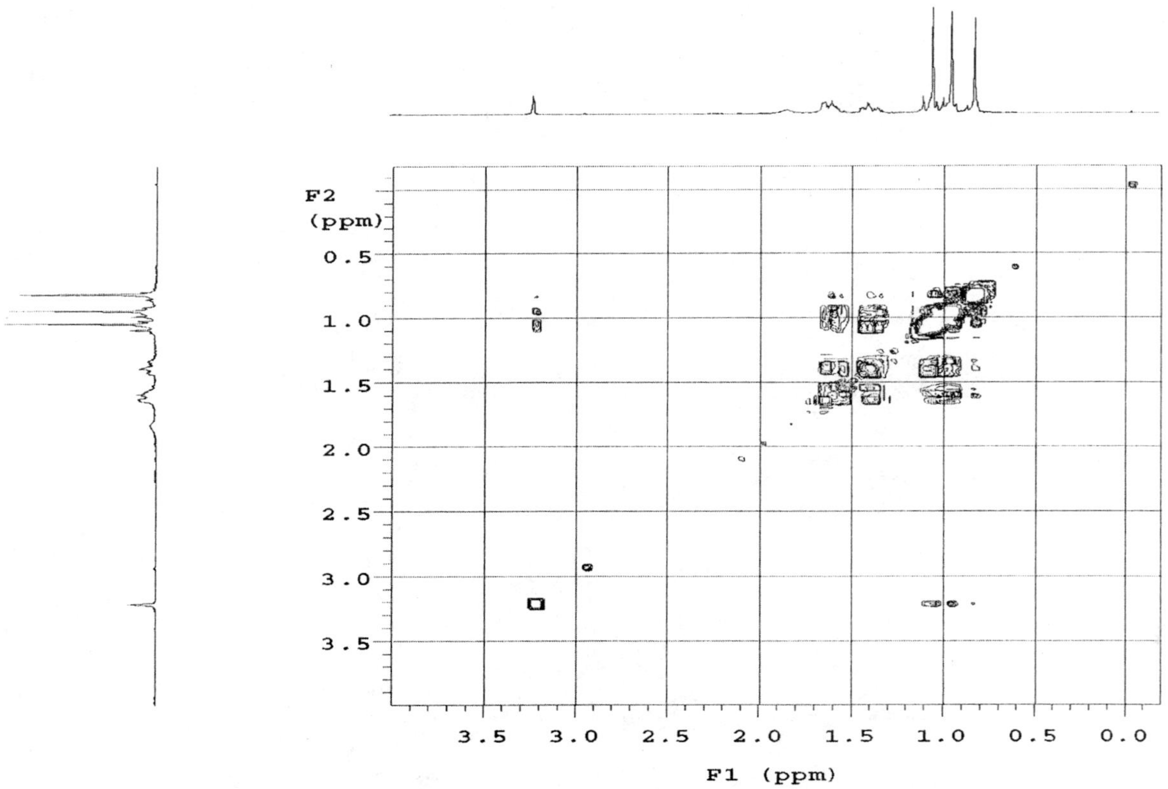

gCOSY Spectrum in $CDCl_3$

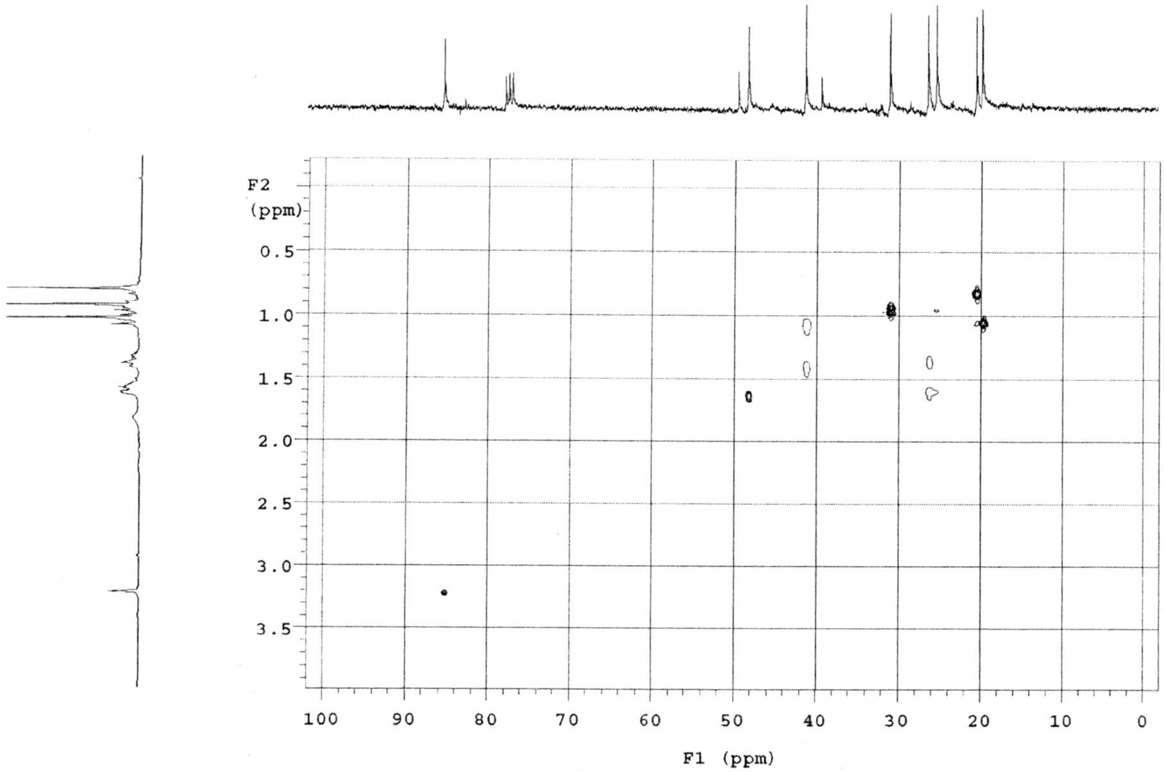

gHSQC Spectrum in $CDCl_3$

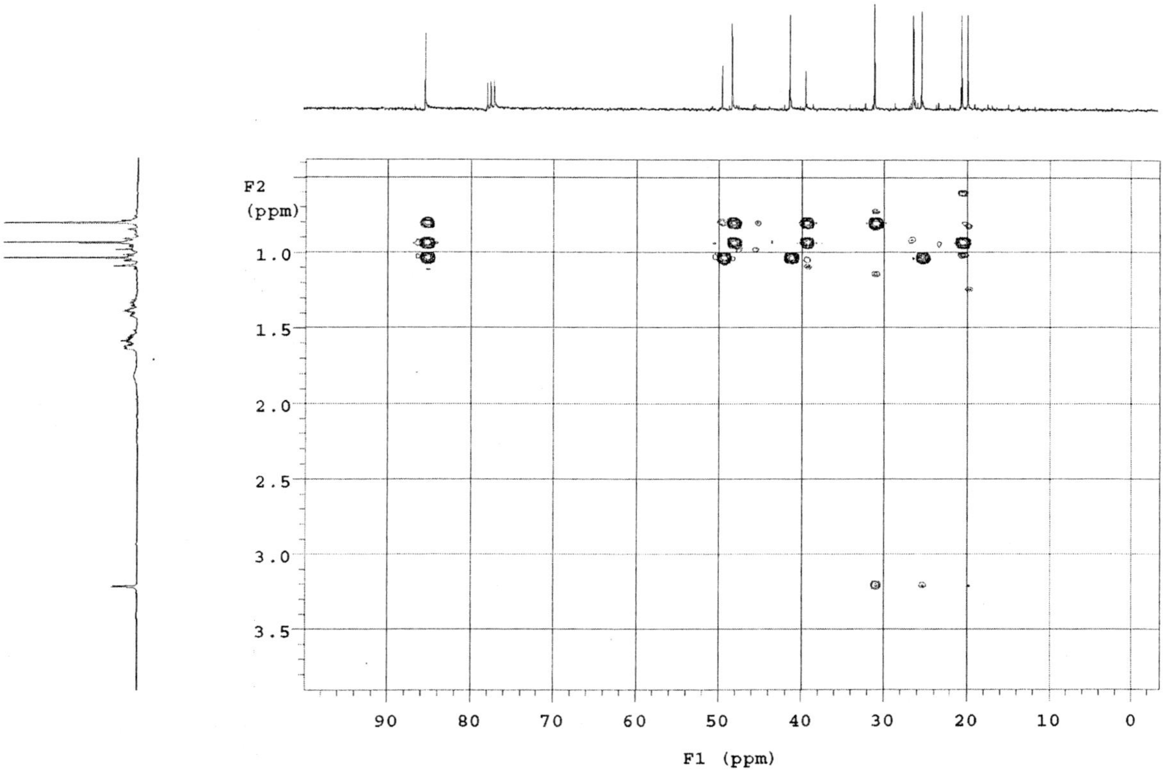

gHMBC Spectrum in $CDCl_3$

Problem 4.10 Given the spectroscopic data below (Proton, Carbon, Edited DEPT, gCOSY, and gHSQC), provide the structurre of the molecule. The sample can be readily derived by simple chemical reaction of **4.09**.

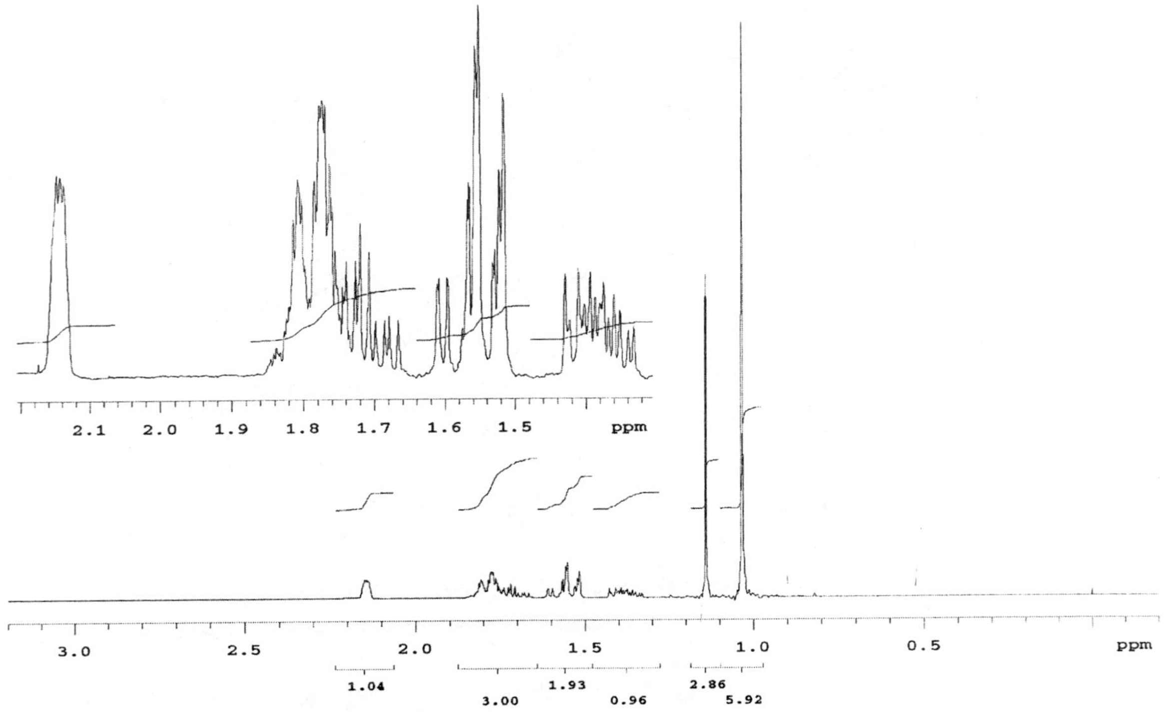

^{1}H-NMR Spectrum in $CDCl_3$

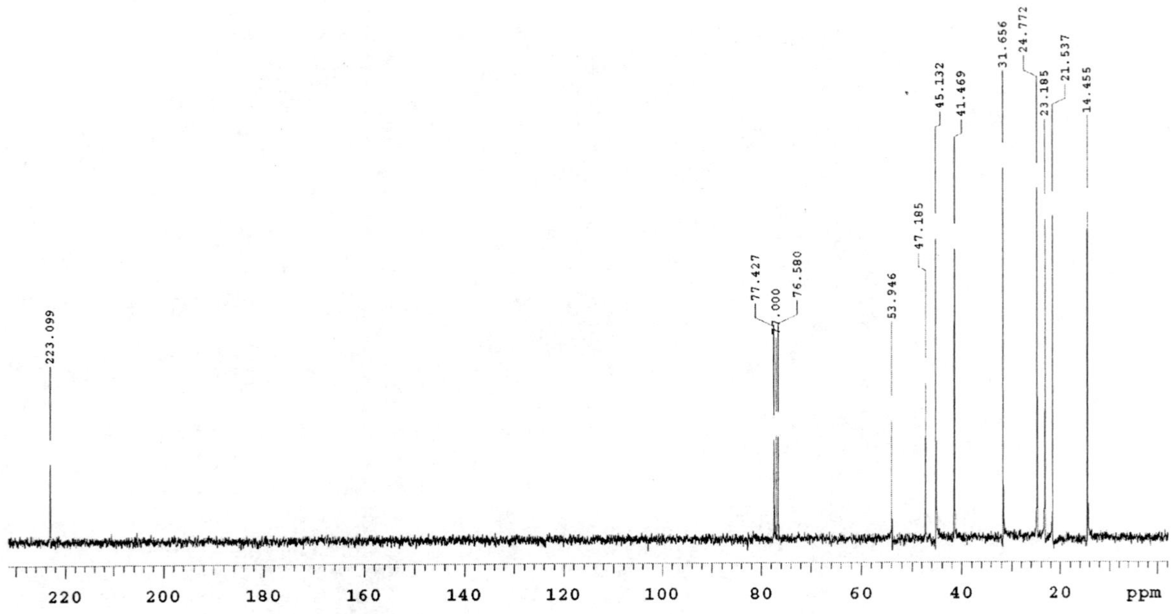

^{13}C-NMR Spectrum in $CDCl_3$

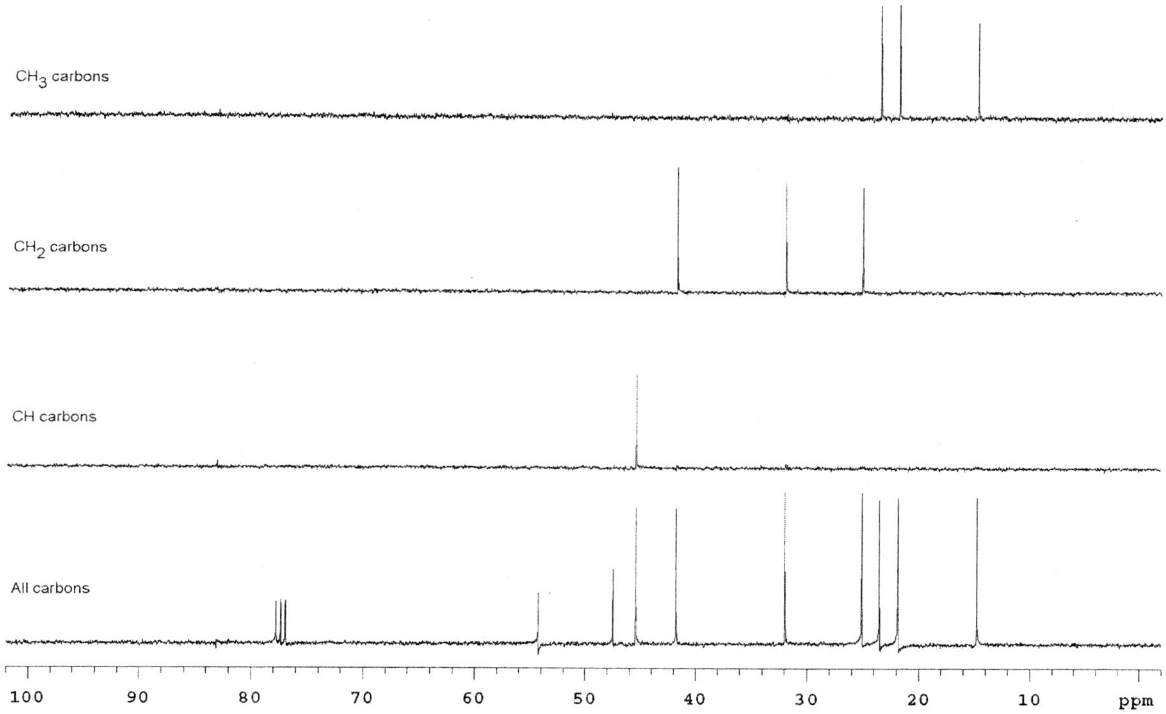

Edited DEPT ^{13}C-NMR Spectrum in $CDCl_3$

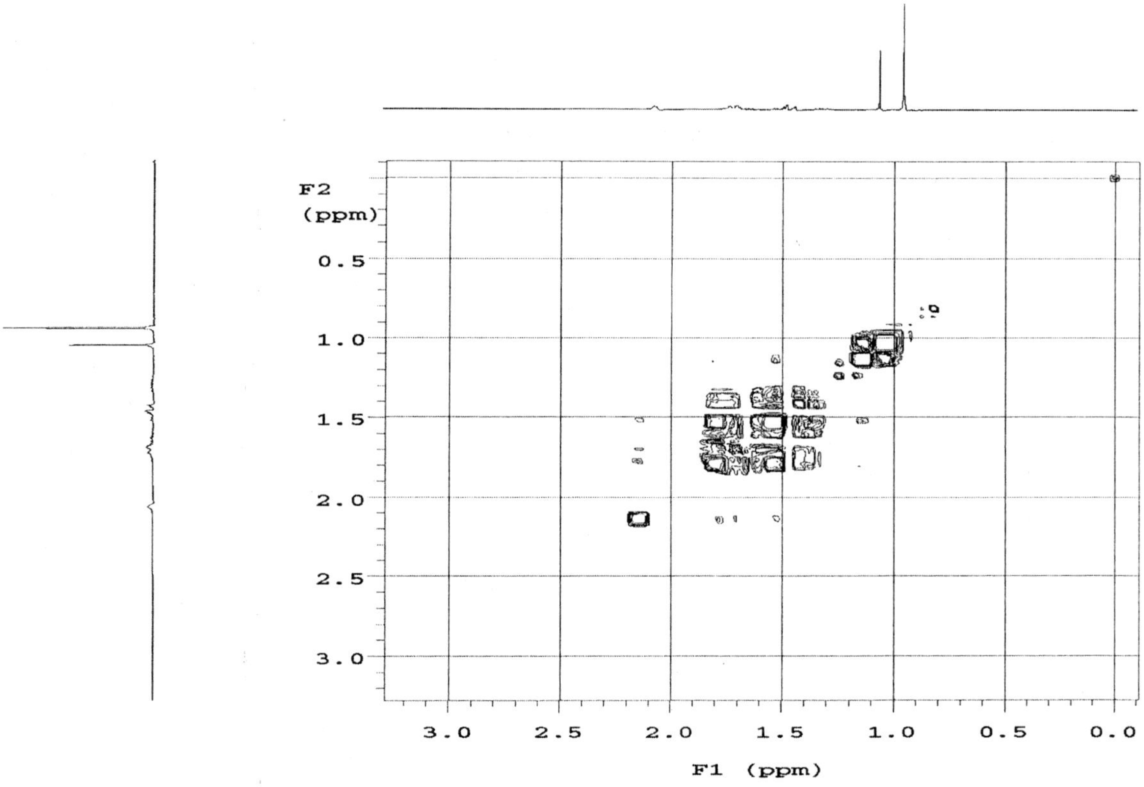

gCOSY Spectrum in $CDCl_3$

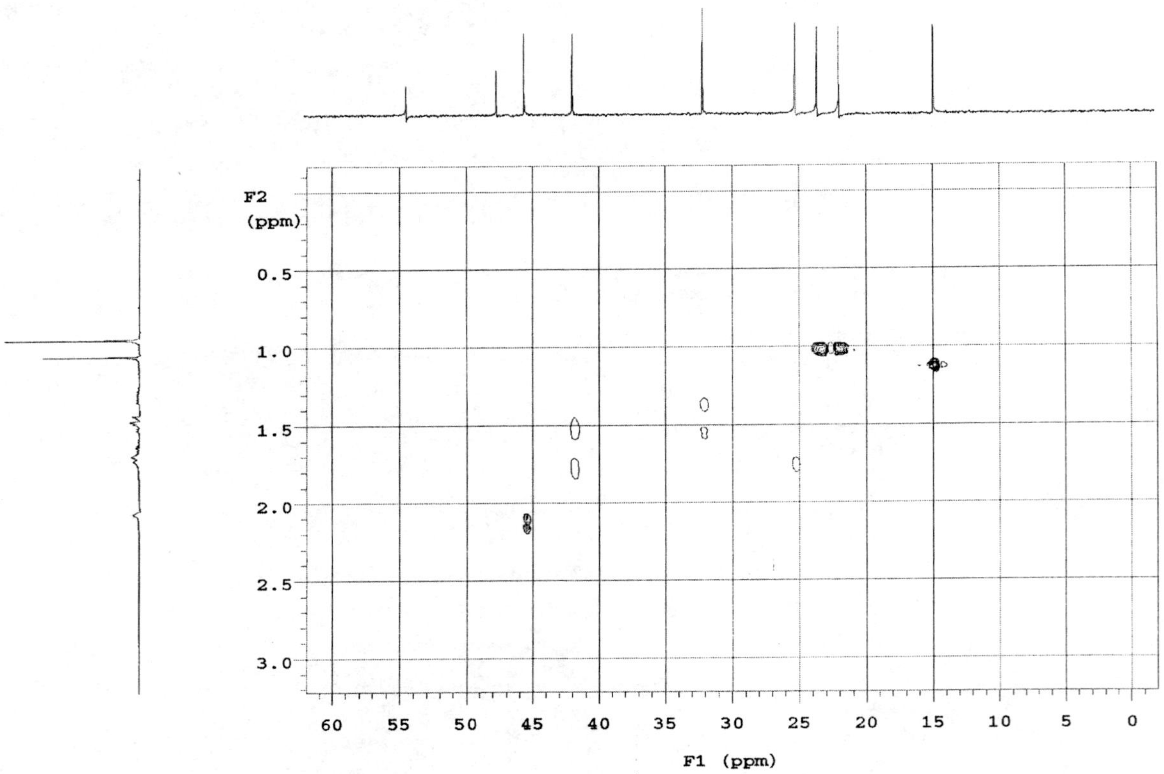

gHSQC Spectrum in $CDCl_3$

Problem 4.11 Given the spectroscopic data below (Proton, Carbon, Edited DEPT, gCOSY, gHSQC, and gHMBC), provide the structure of the molecule. (IR: 3400 cm^{-1}, broad)

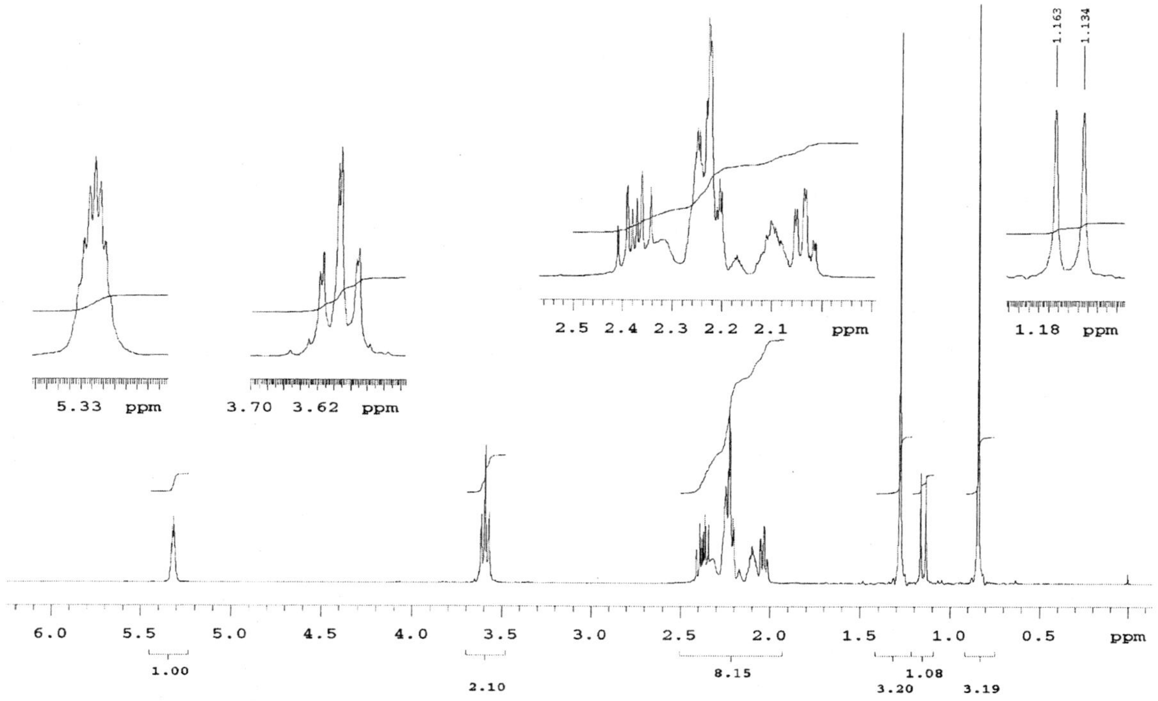

^{1}H-NMR Spectrum in $CDCl_3$

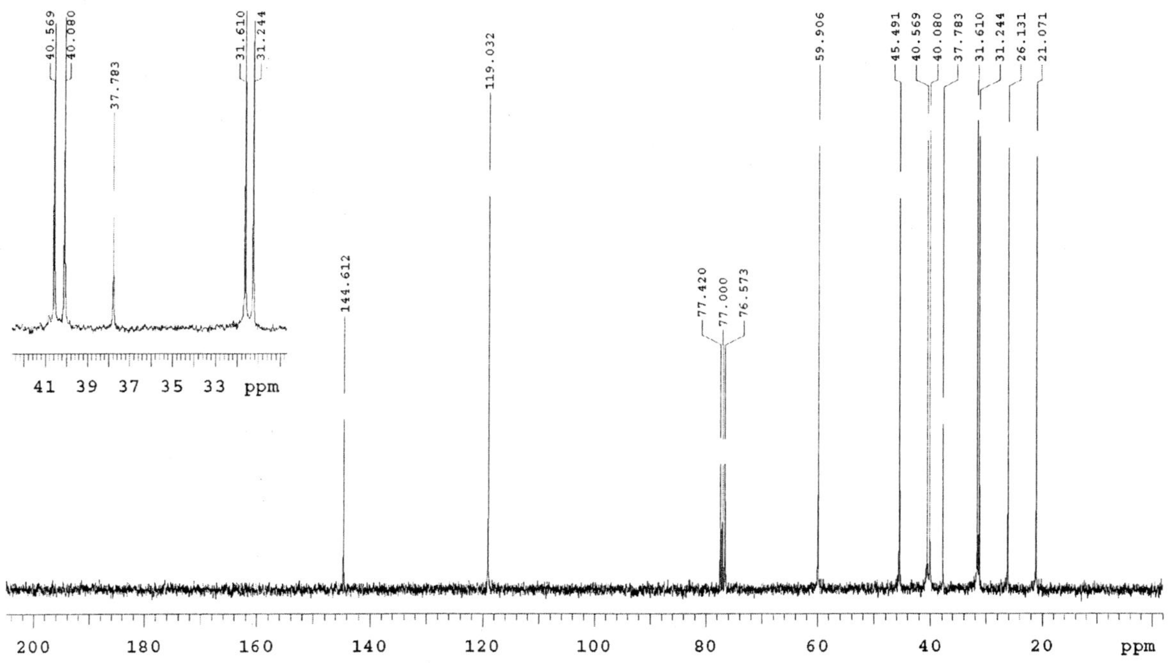

^{13}C-NMR Spectrum in $CDCl_3$

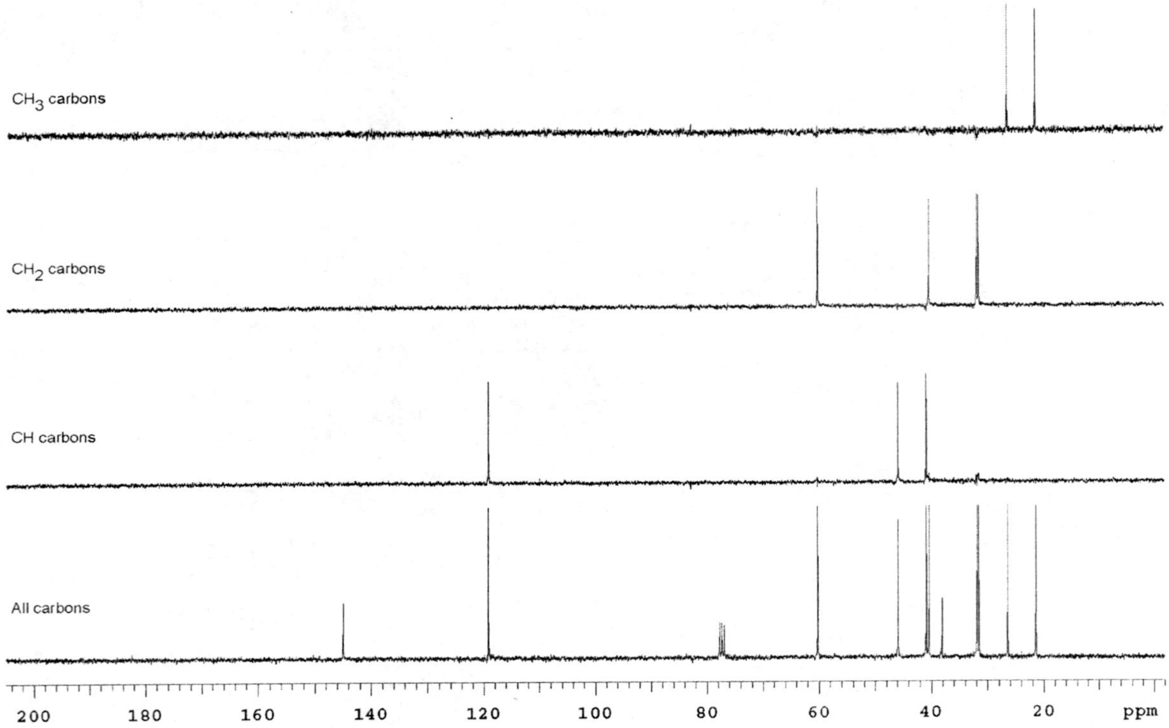

Edited DEPT ^{13}C-NMR Spectrum in $CDCl_3$

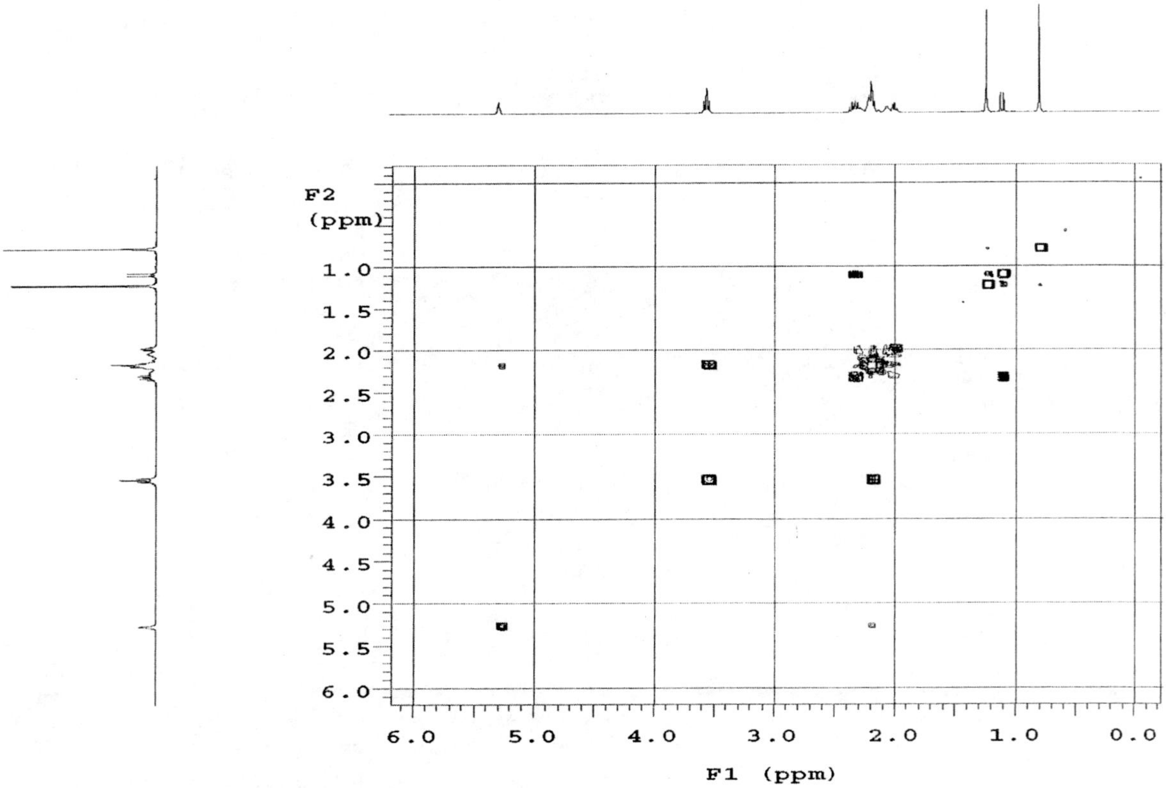

gCOSY Spectrum in $CDCl_3$

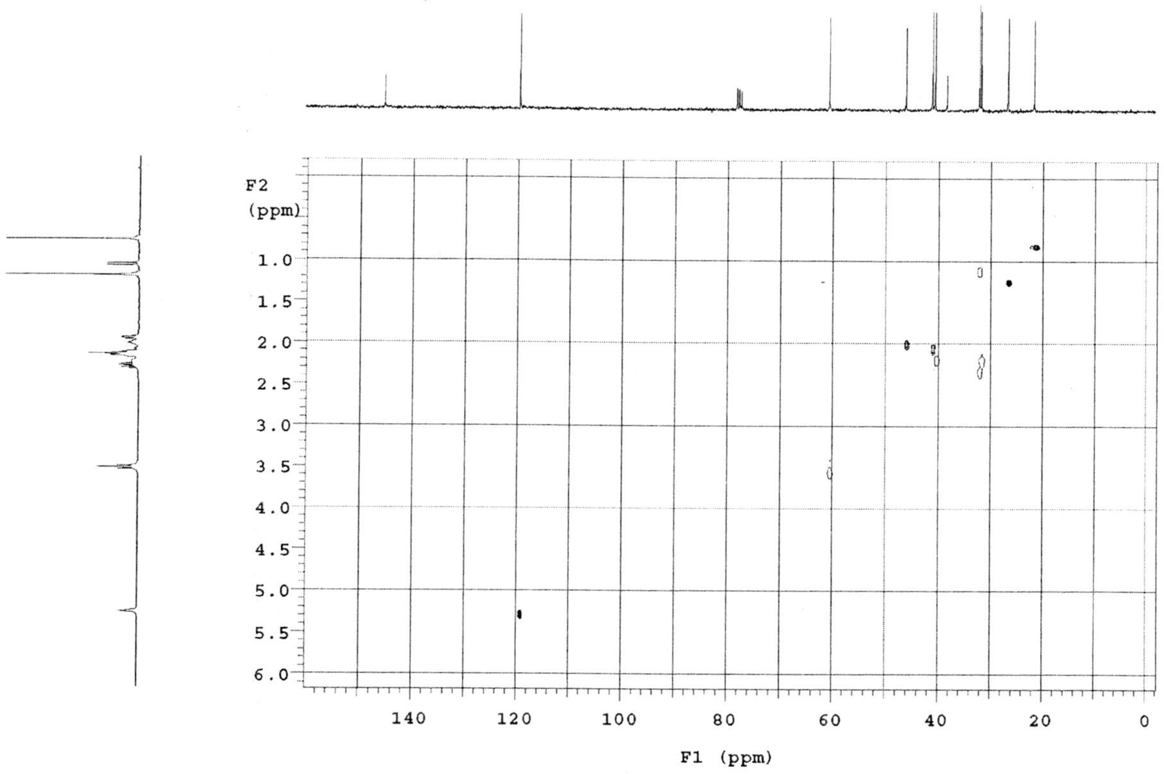

gHSQC Spectrum in $CDCl_3$

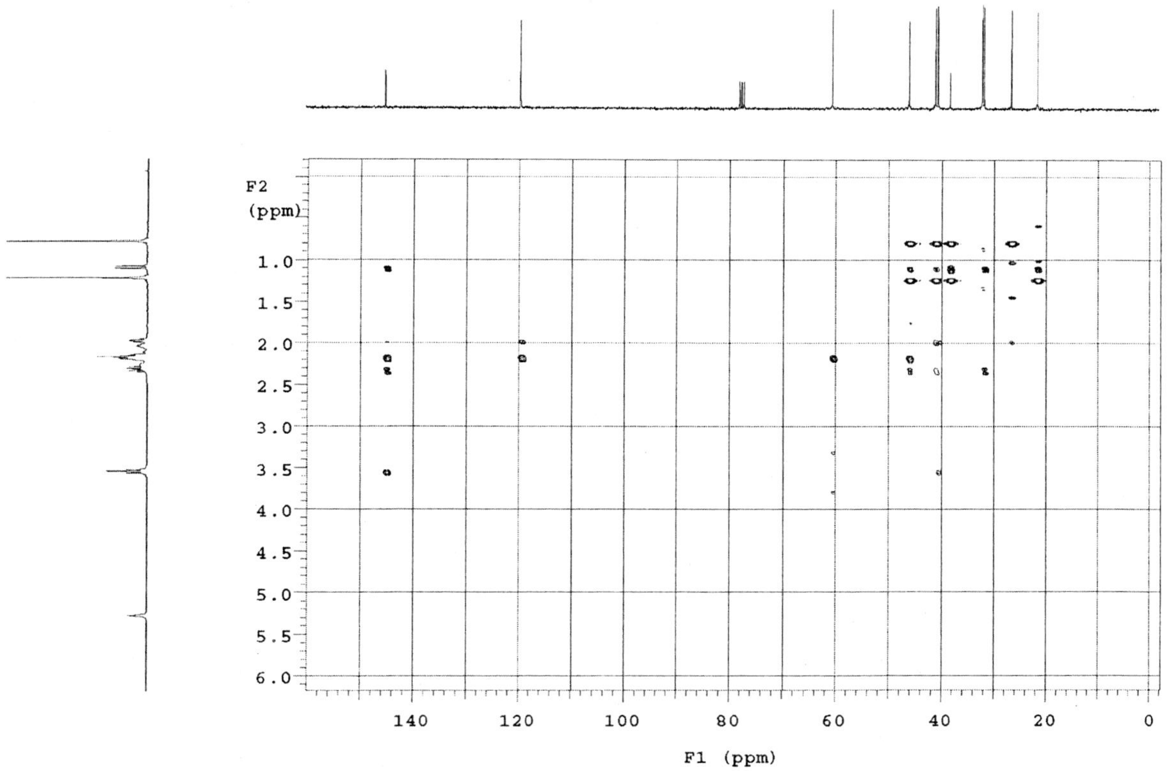

gHMBC Spectrum in $CDCl_3$

Problem 4.12 Given the spectroscopic data below (Proton, Carbon, Edited DEPT, gCOSY, gHSQC, and gHMBC), provide the structure of the molecule.

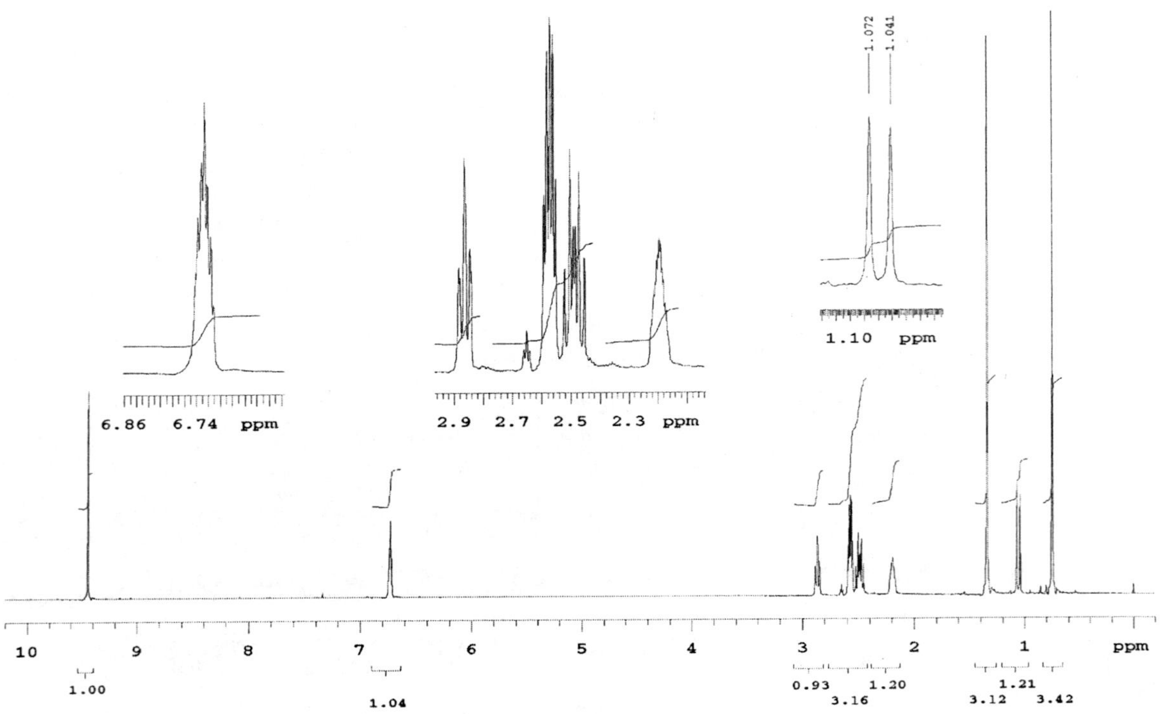

^{1}H-NMR Spectrum in $CDCl_3$

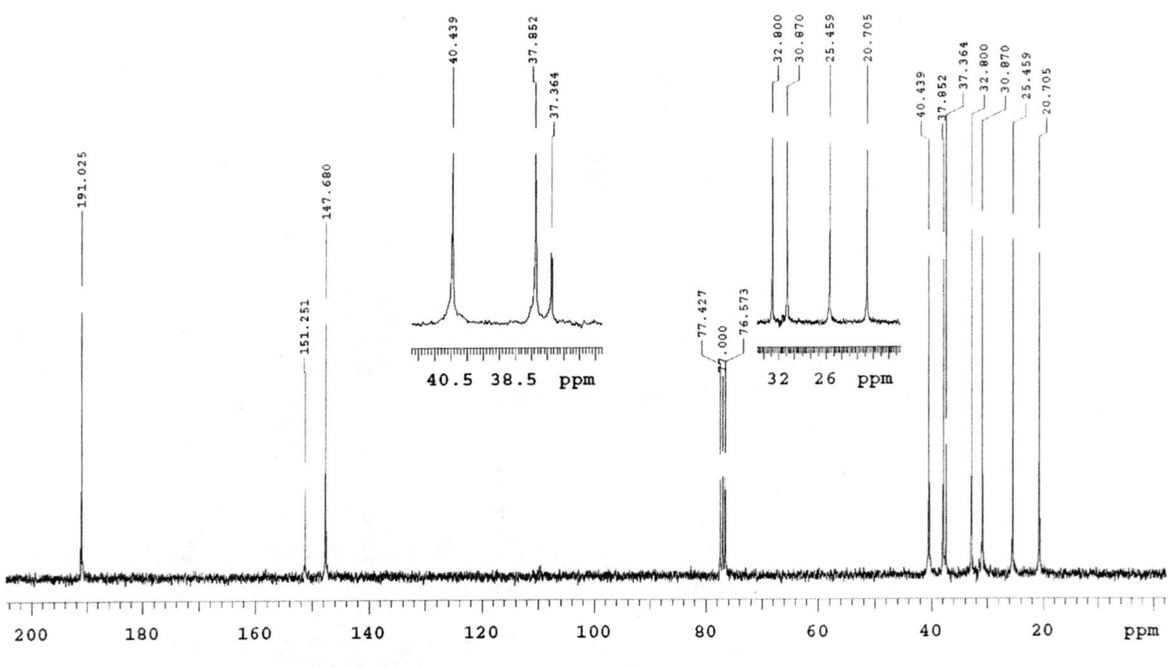

^{13}C-NMR Spectrum in $CDCl_3$

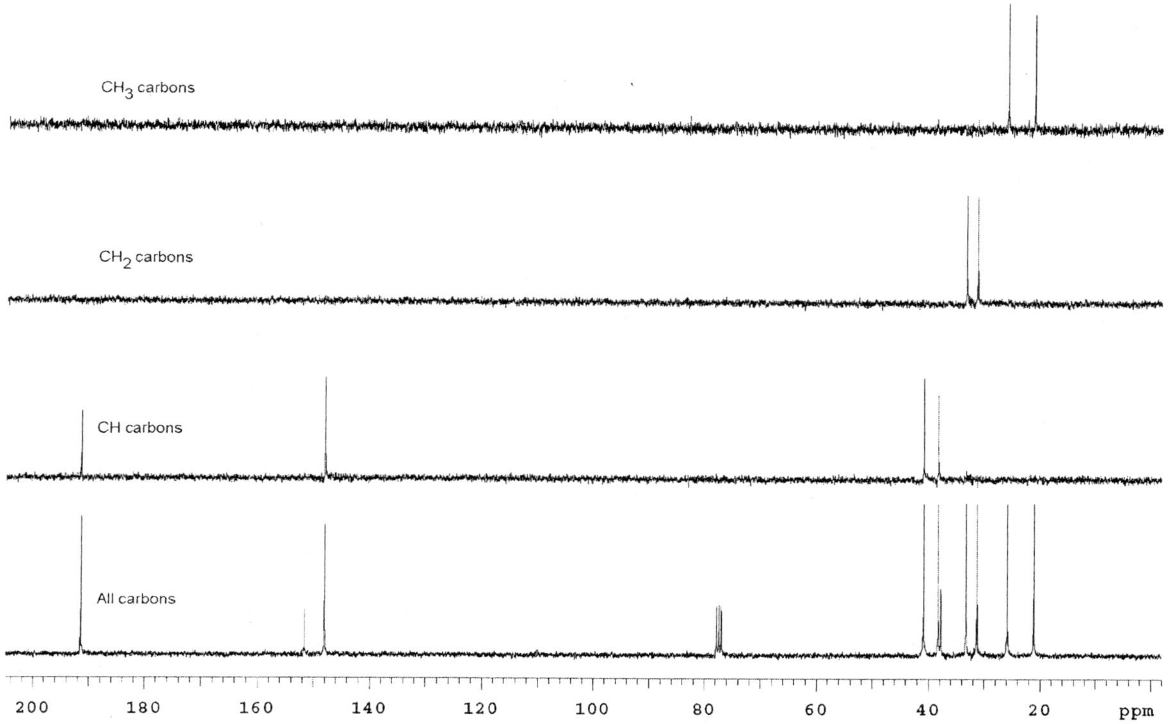

Edited DEPT ^{13}C-NMR Spectrum in $CDCl_3$

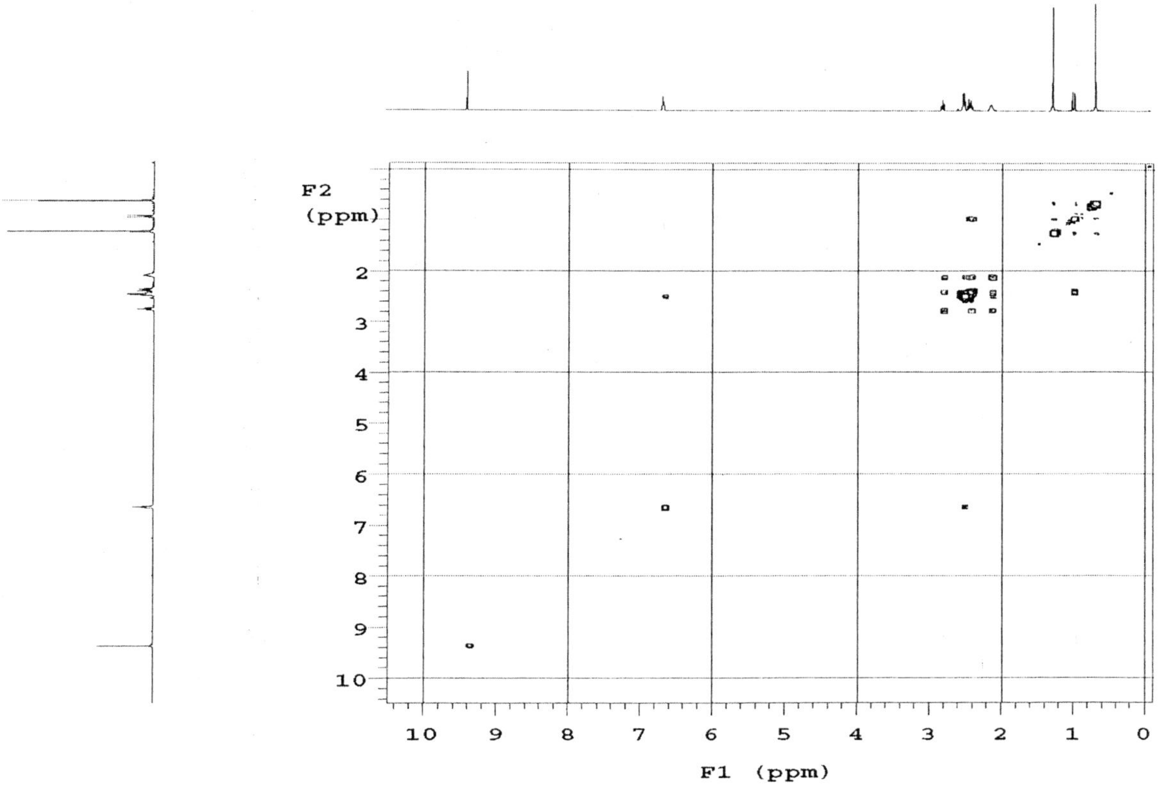

gCOSY Spectrum in $CDCl_3$

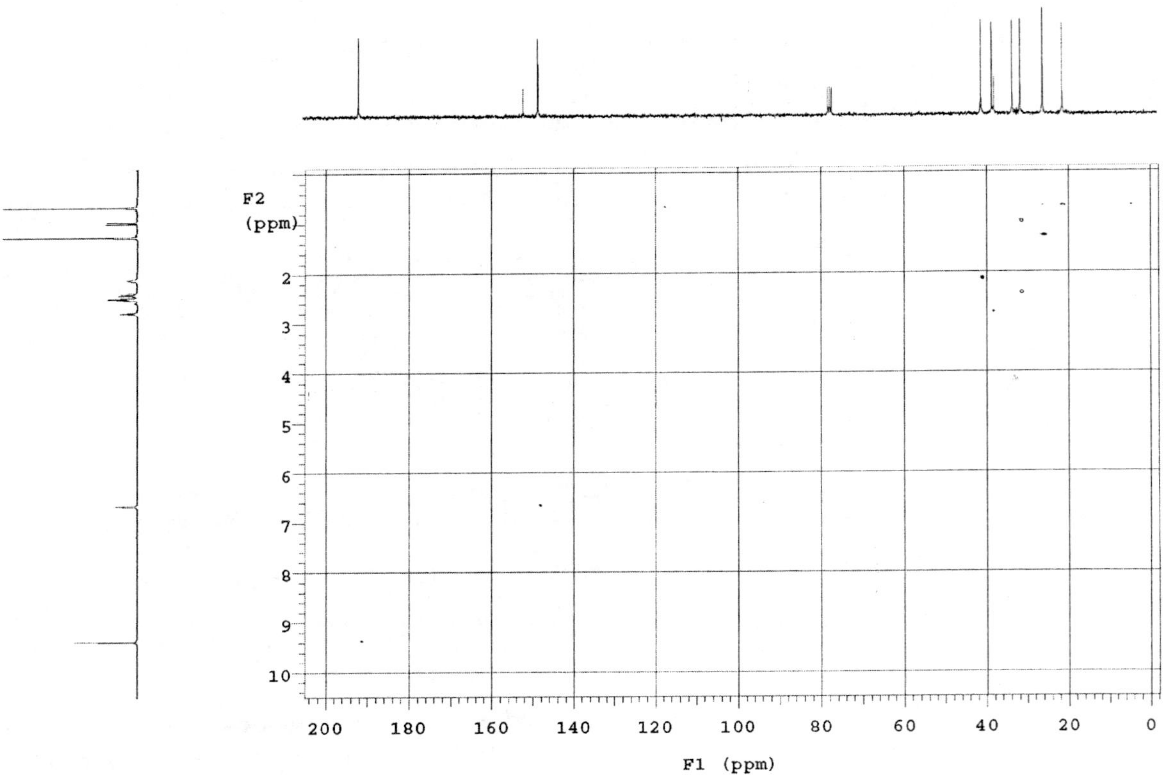

gHSQC Spectrum in $CDCl_3$

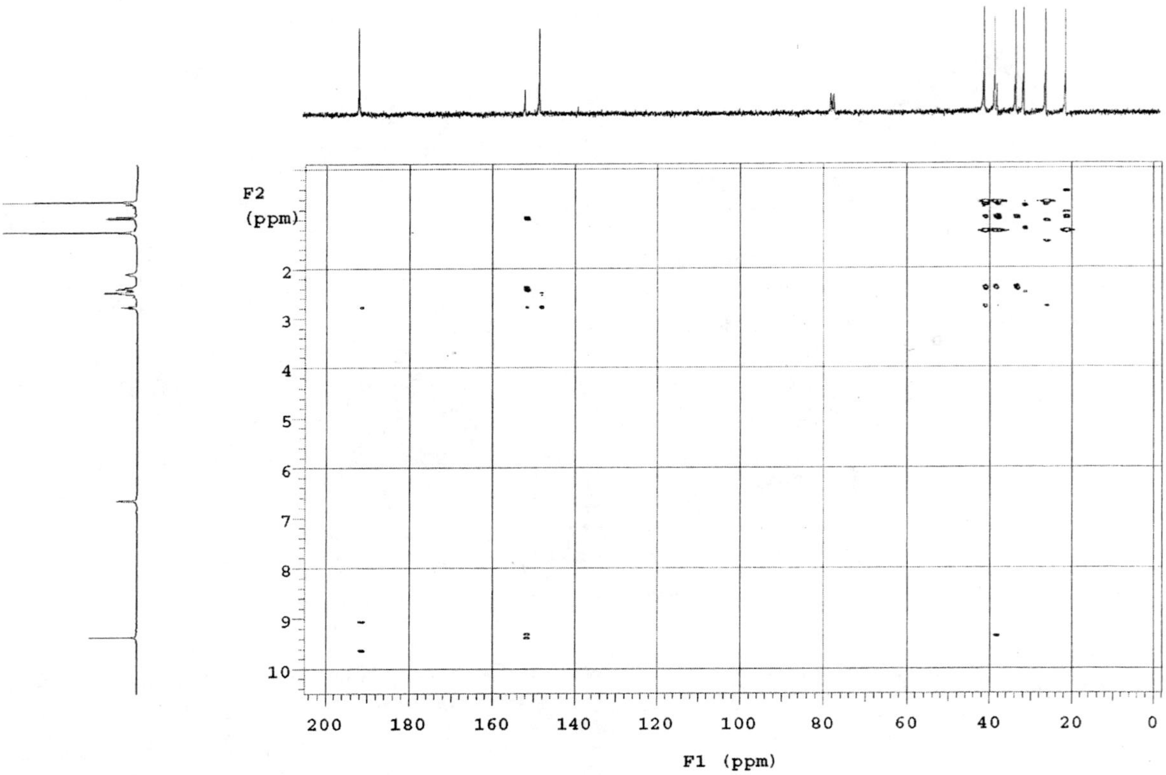

gHMBC Spectrum in $CDCl_3$

Problem 4.13 Given the spectroscopic data below (Proton, Carbon, Edited DEPT, gCOSY, and gHMQC), provide the structure of the molecule. ($C_7H_{16}O_3$)

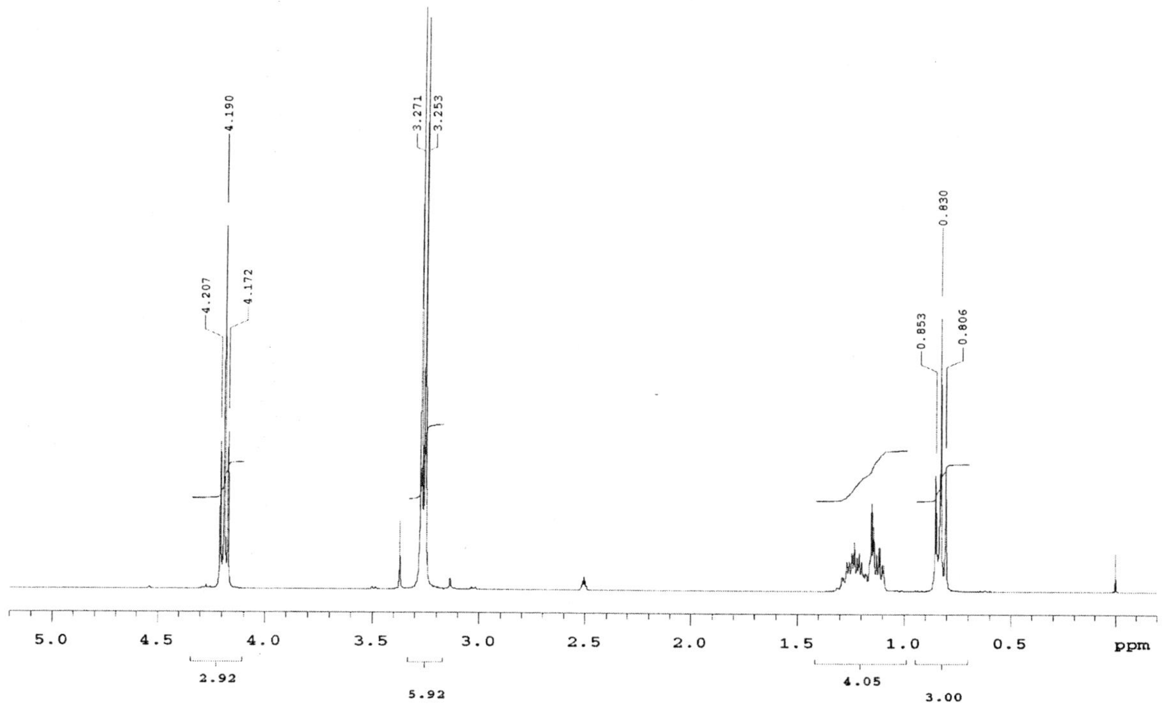

^{1}H-NMR Spectrum in DMSO-d_6

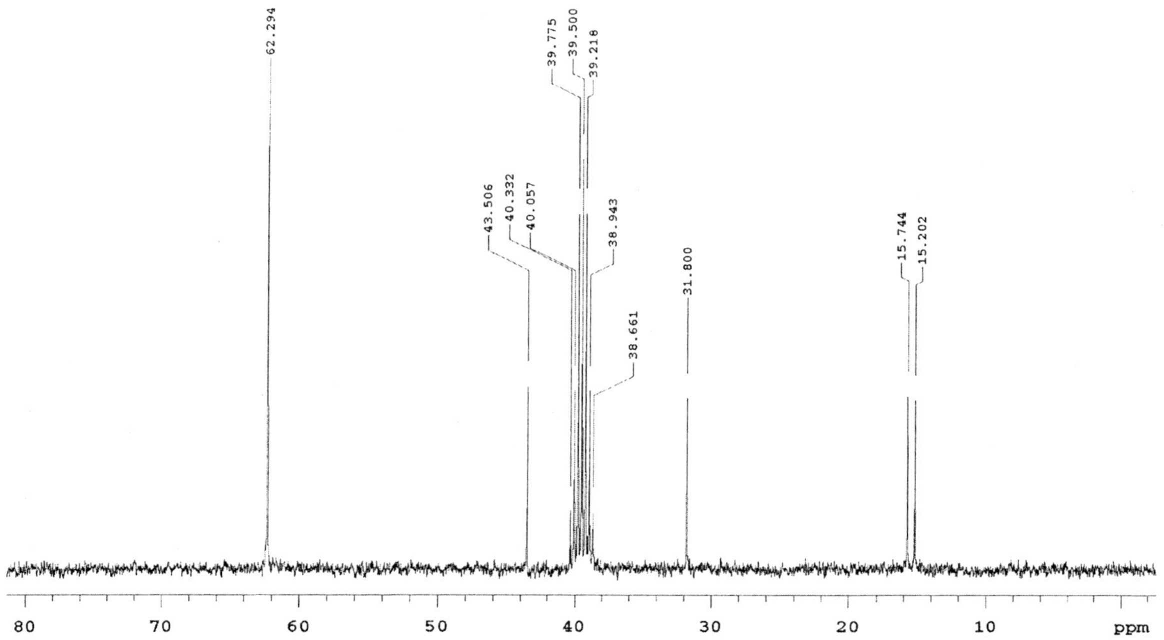

^{13}C-NMR Spectrum in DMSO-d_6

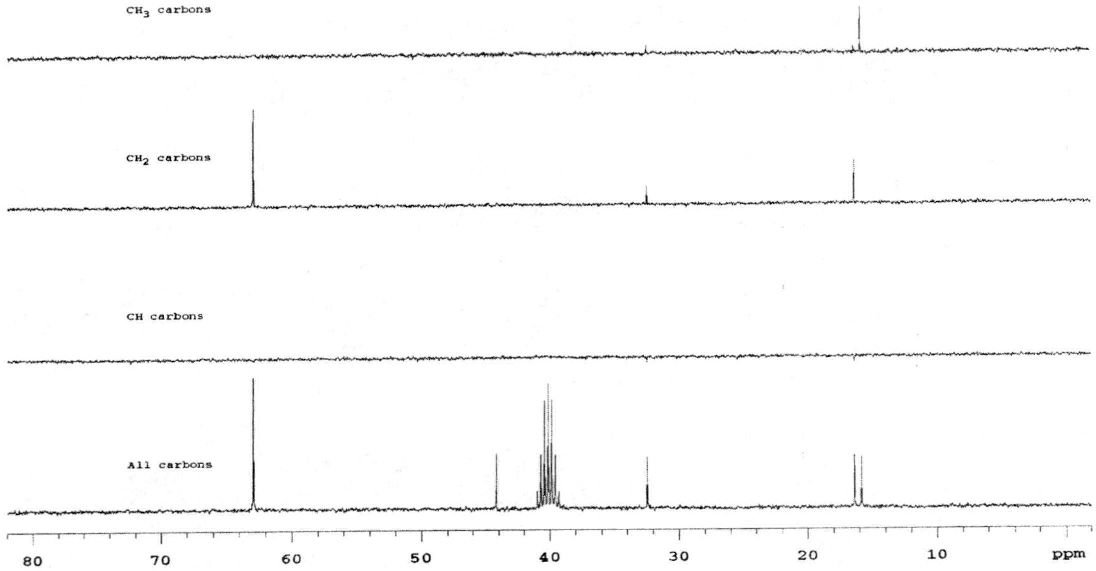

Edited DEPT ^{13}C-NMR Spectrum in DMSO-d_6

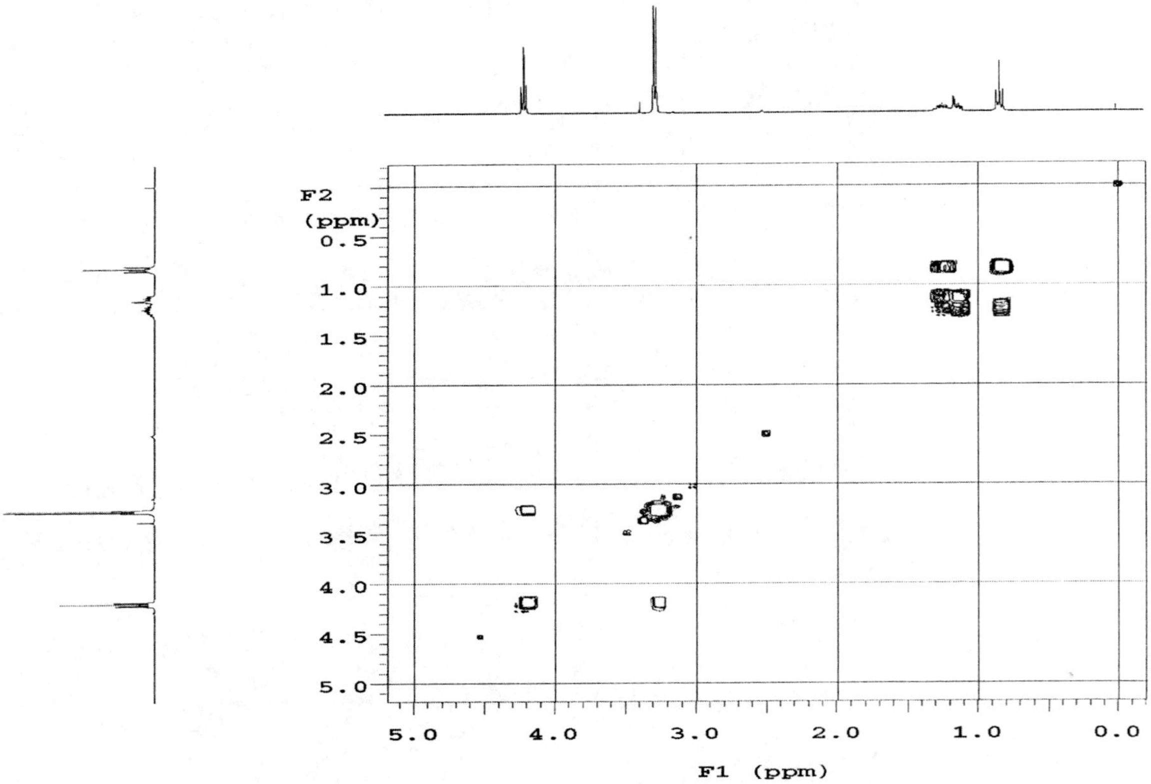

gCOSY Spectrum in DMSO-d_6

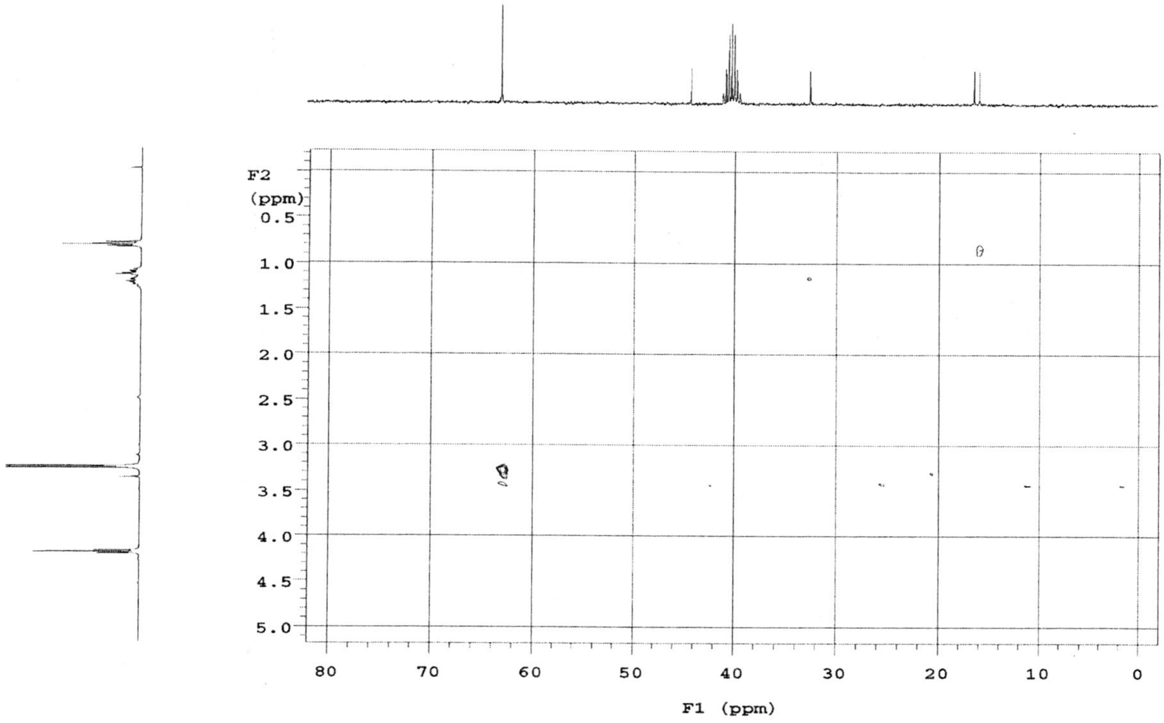

gHMQC Spectrum in DMSO-d_6

Problem 4.14 Given the spectroscopic data below (Proton, Carbon, gHSQC, gCOSY, and gHMBC), provide the structure of the molecule. (IR: 3400 cm^{-1})

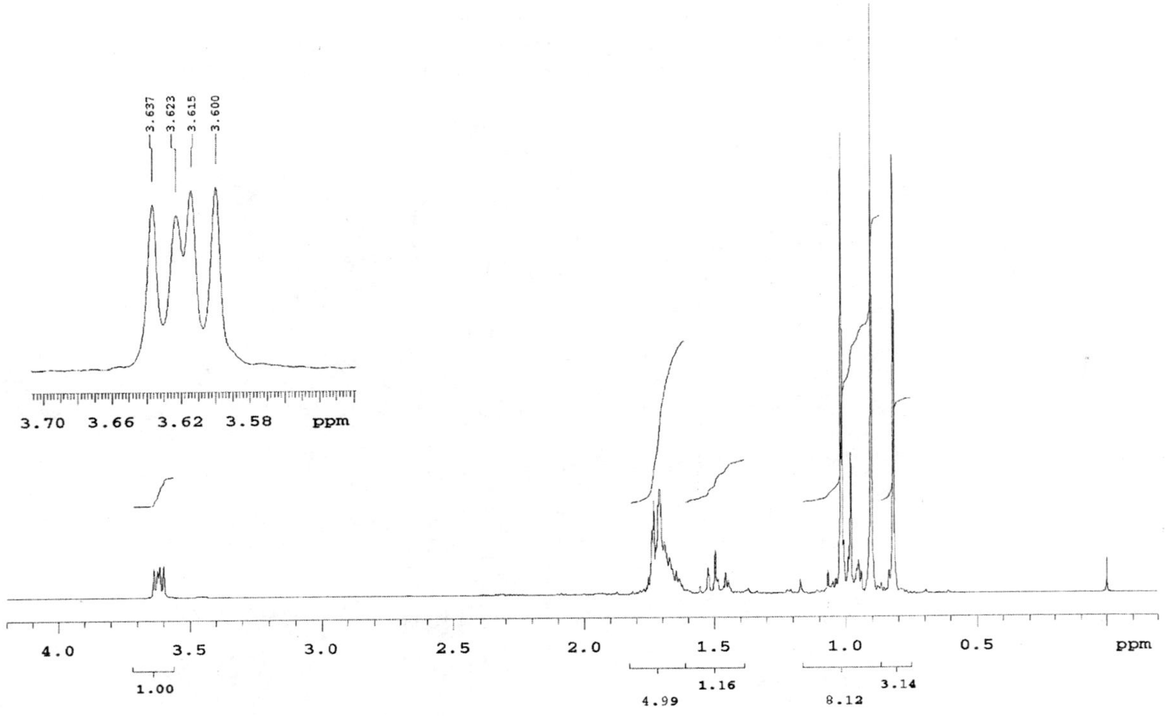

^{1}H-NMR Spectrum in $CDCl_3$

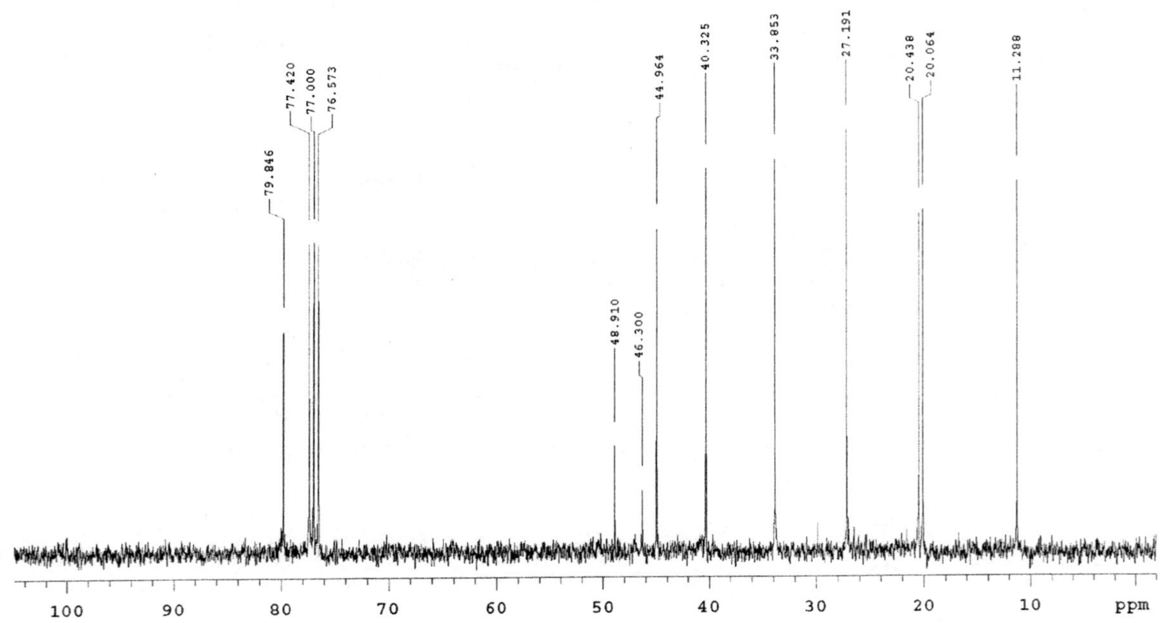

^{13}C-NMR Spectrum in $CDCl_3$

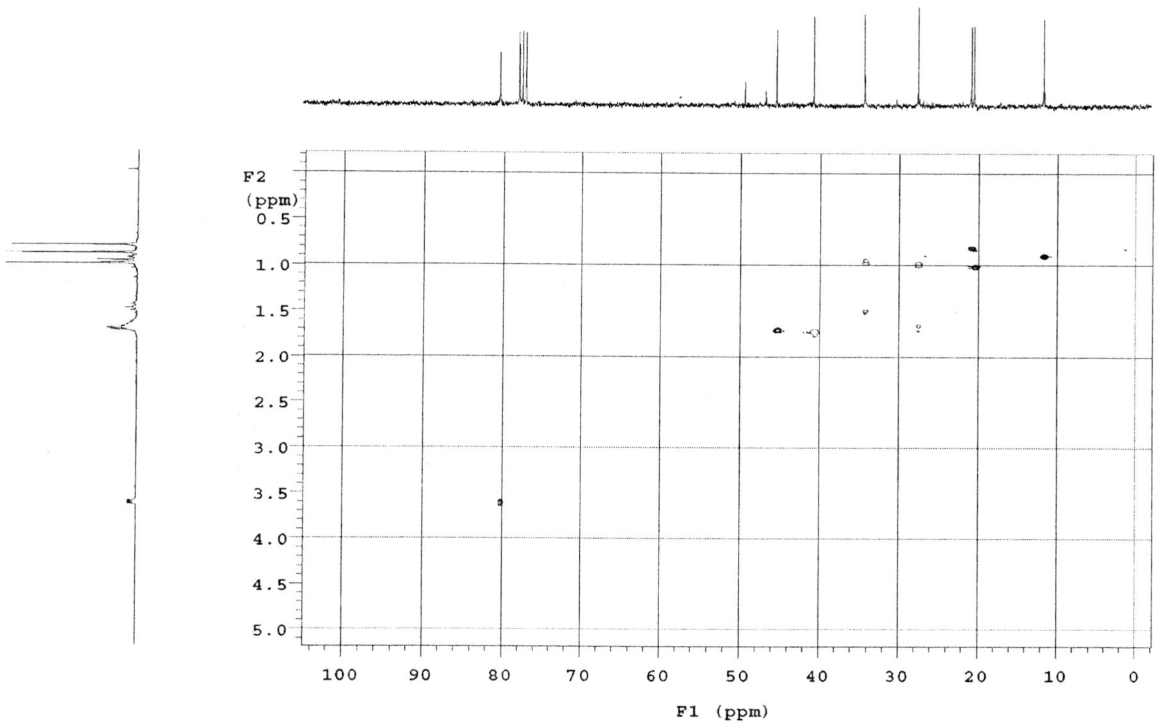

gHSQC Spectrum in $CDCl_3$

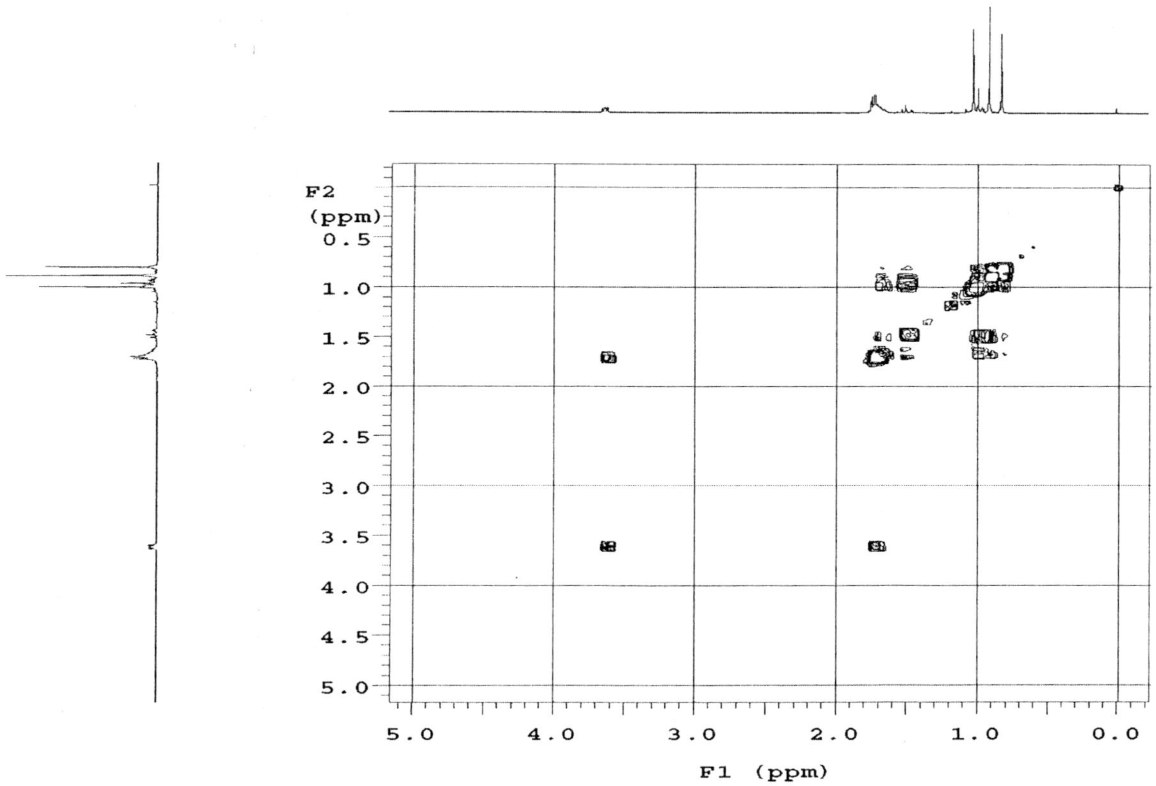

gCOSY Spectrum in $CDCl_3$

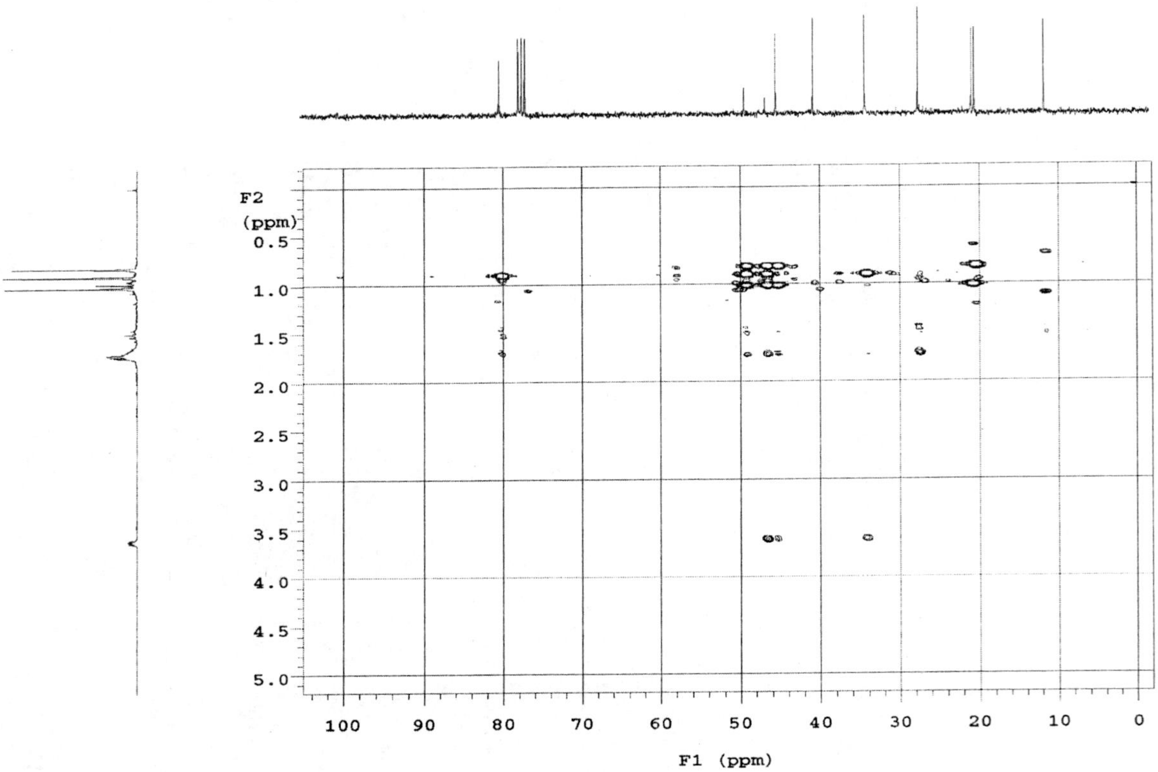

gHMBC Spectrum in $CDCl_3$

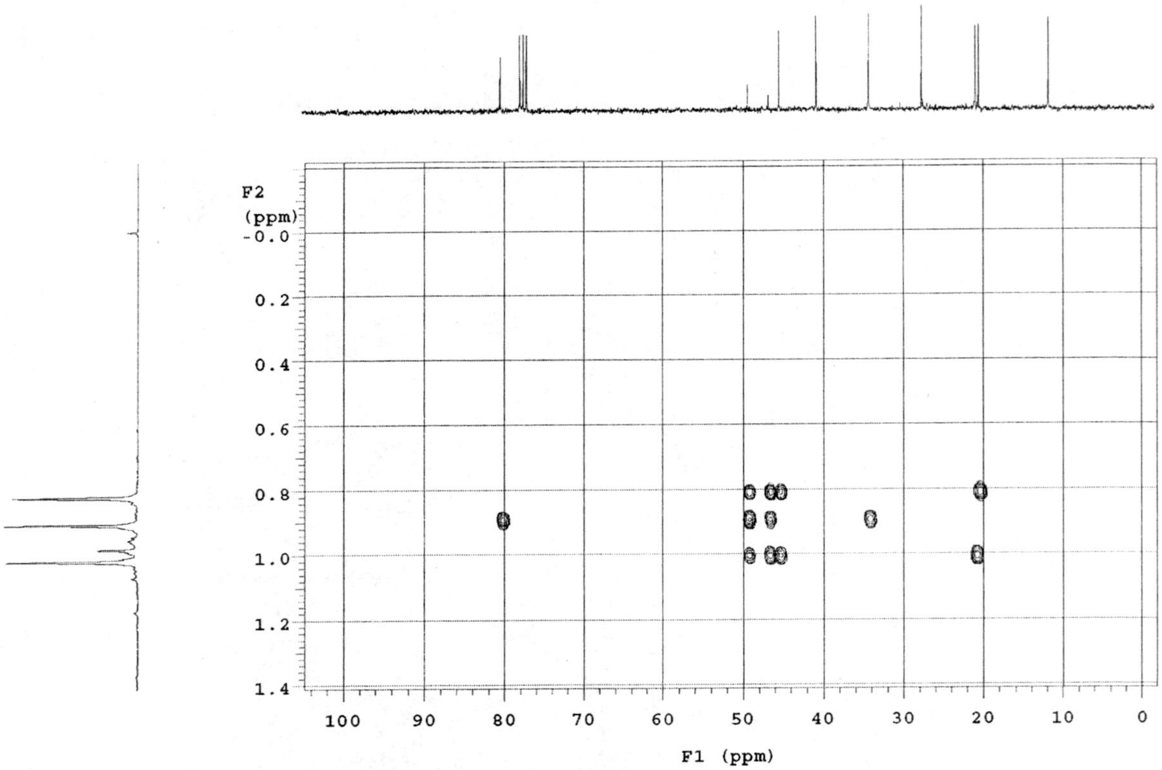

Expansion of gHMBC Spectrum in $CDCl_3$

Problem 4.15 Given the spectroscopic data below (Proton, Carbon, gCOSY, gHSQC, and gHMBC), provide the structure of the molecule.

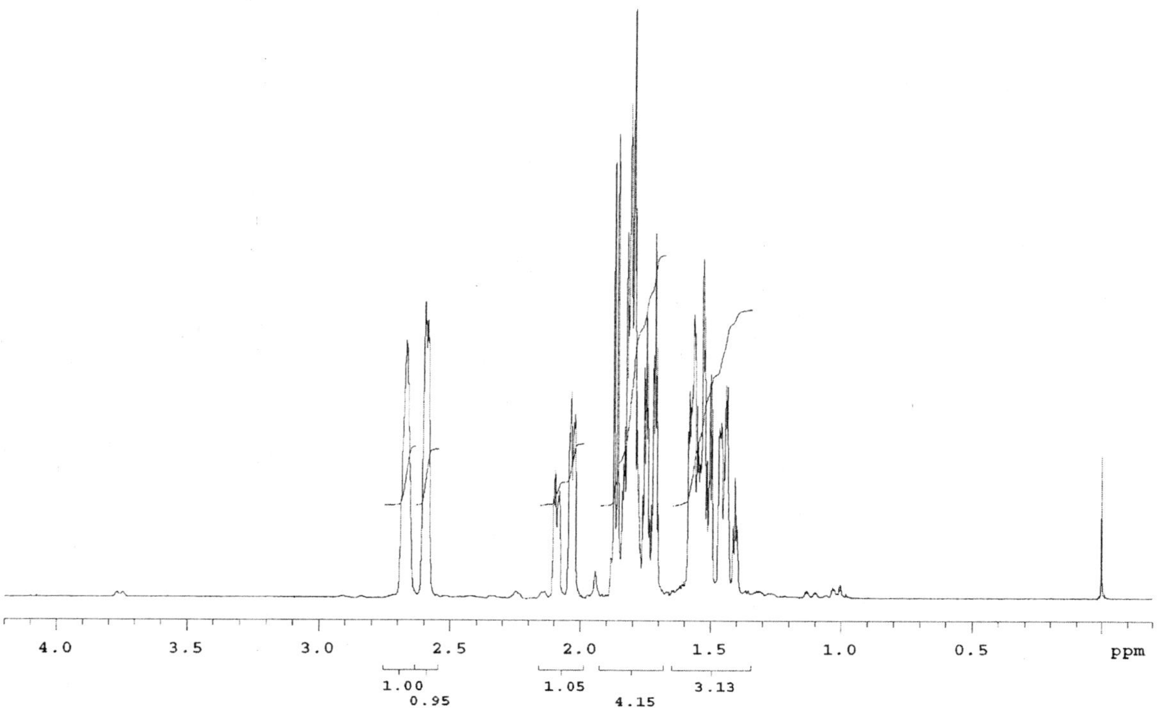

^{1}H-NMR Spectrum in $CDCl_3$

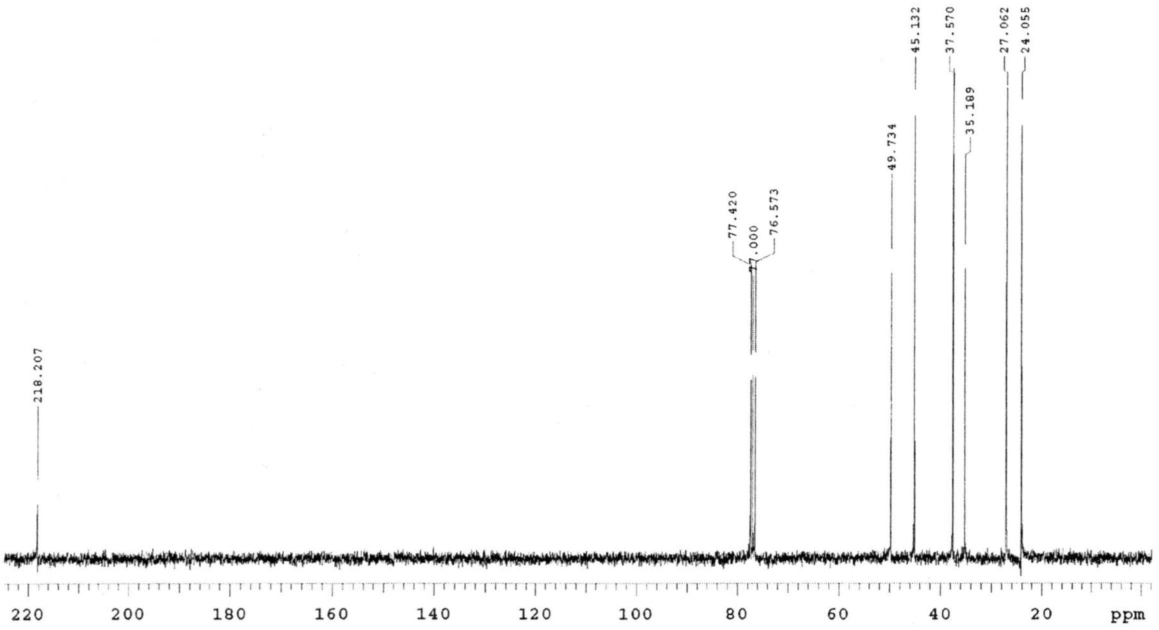

^{13}C-NMR Spectrum in $CDCl_3$

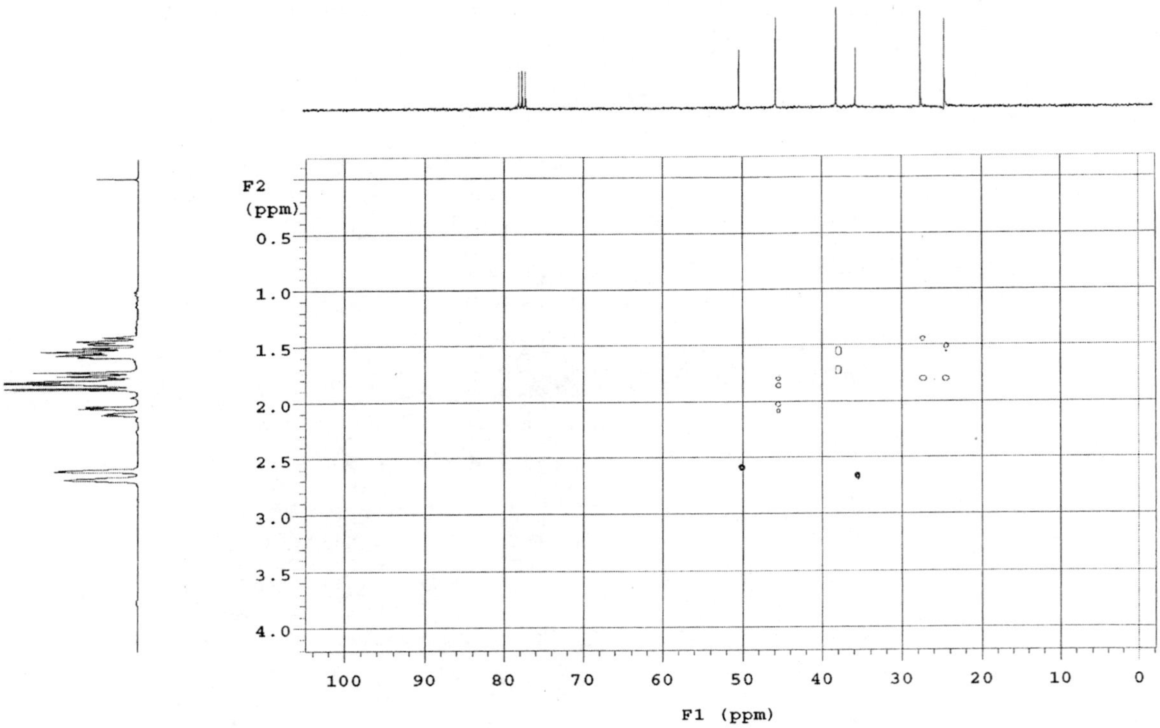

gHSQC Spectrum in $CDCl_3$

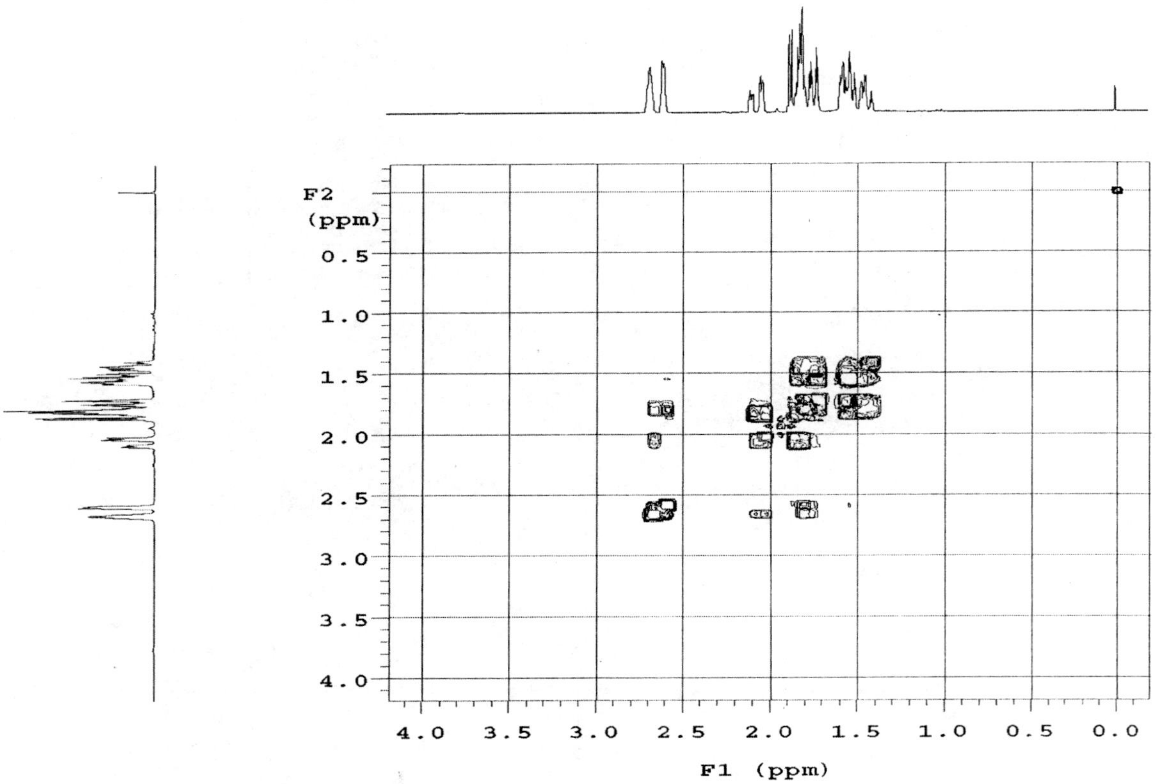

gCOSY Spectrum in $CDCl_3$

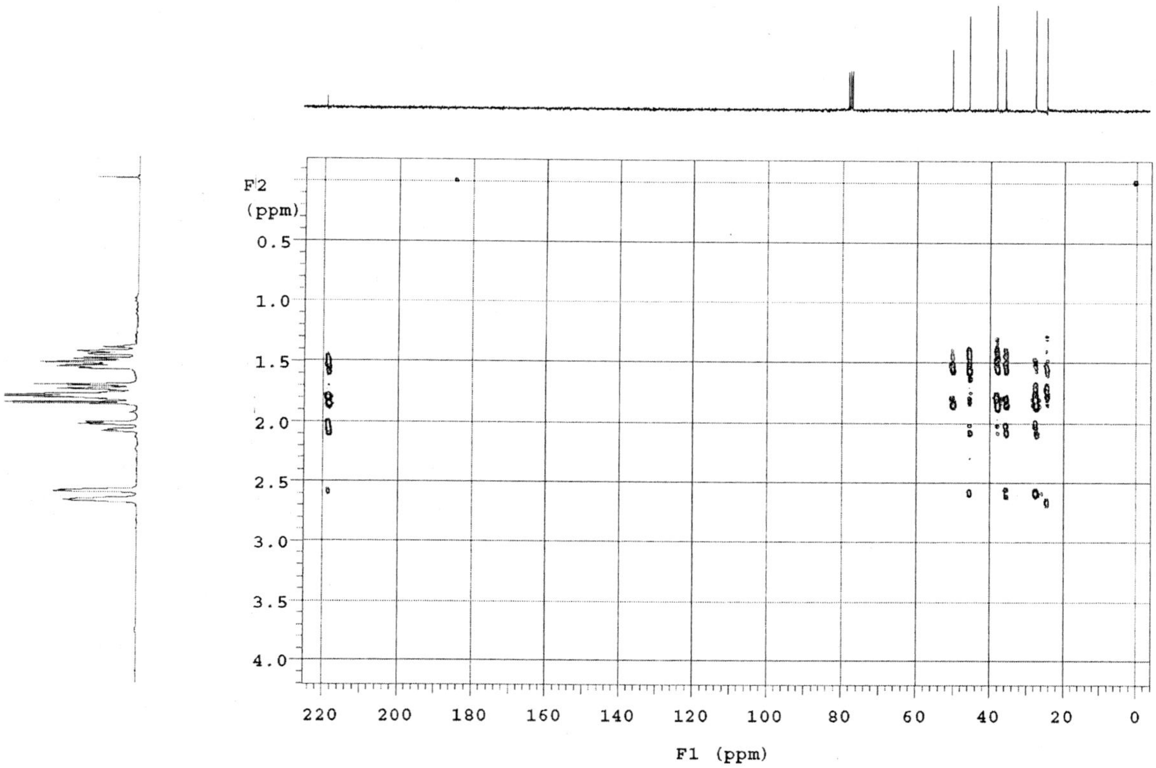

gHMBC Spectrum in $CDCl_3$

Problem 4.16 Given the spectroscopic data below (Proton, Carbon, gHSQC, and gHMBC), provide the structure of the molecule. (HR-MS: 138.1045 amu) Contains small amount of an epimer, note 3.85 and 4.45 signals as well as the additional set of small signals in ^{13}C spectrum.

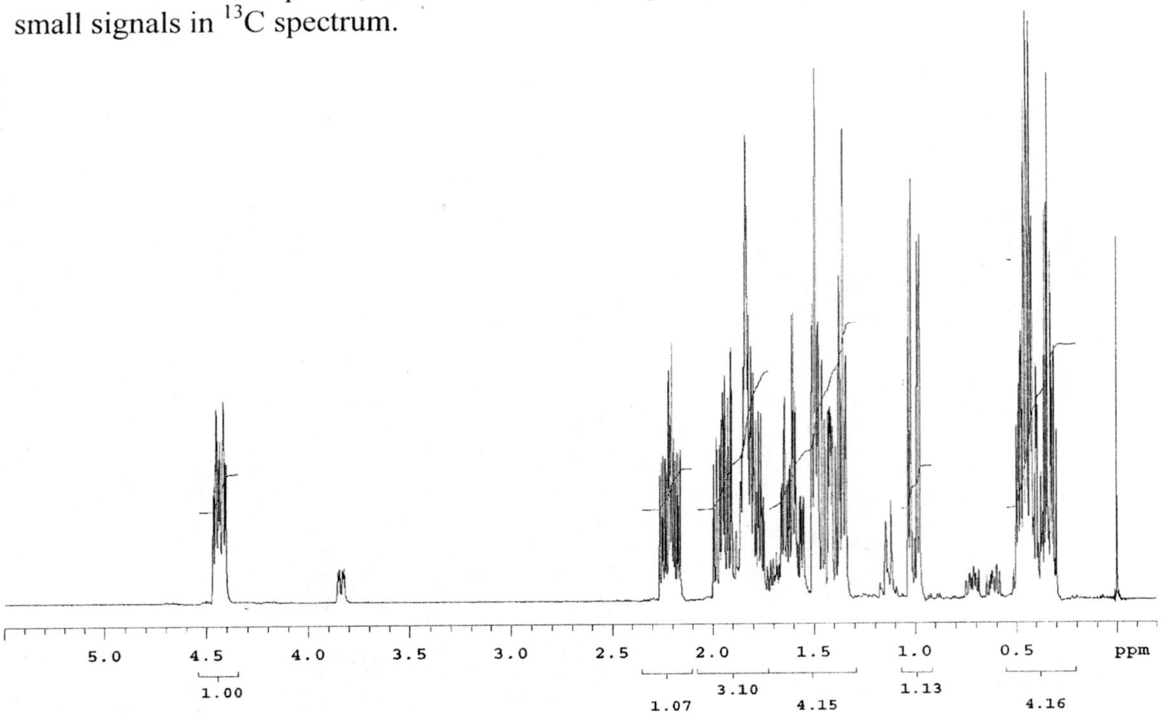

^{1}H-NMR Spectrum in $CDCl_3$

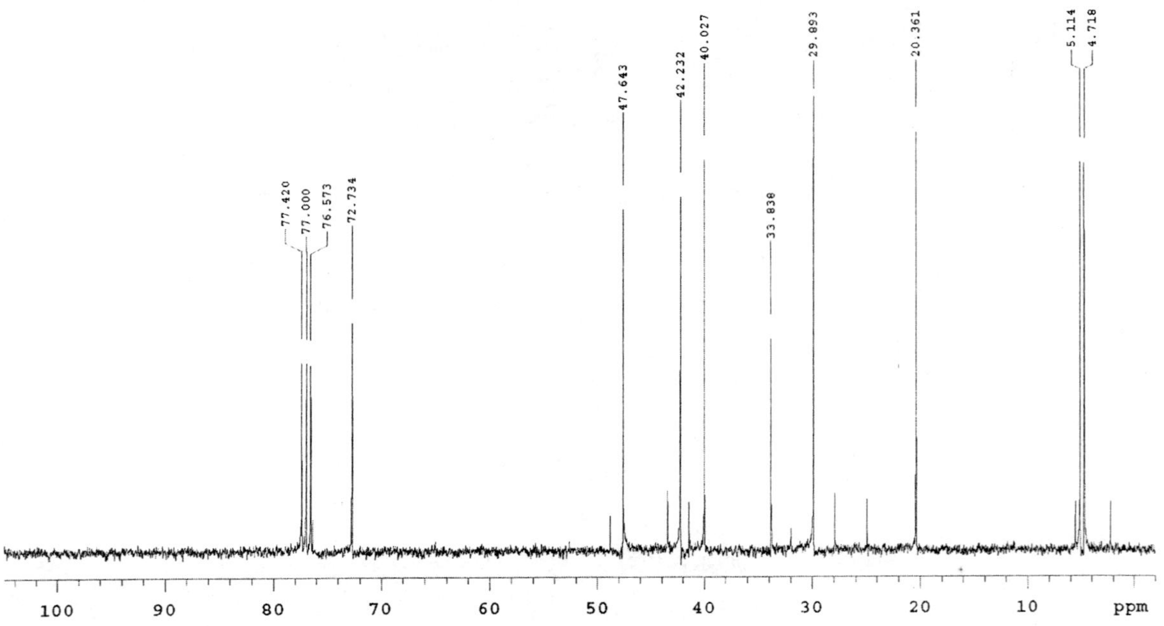

^{13}C-NMR Spectrum in $CDCl_3$

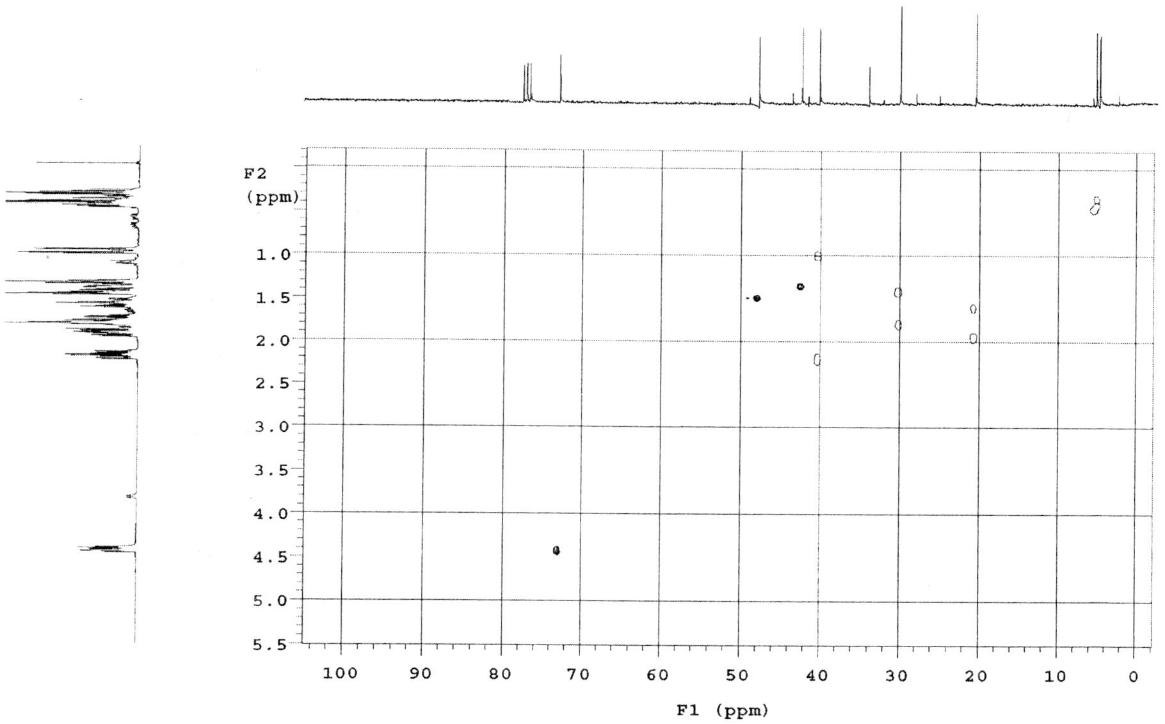

gHSQC Spectrum in $CDCl_3$

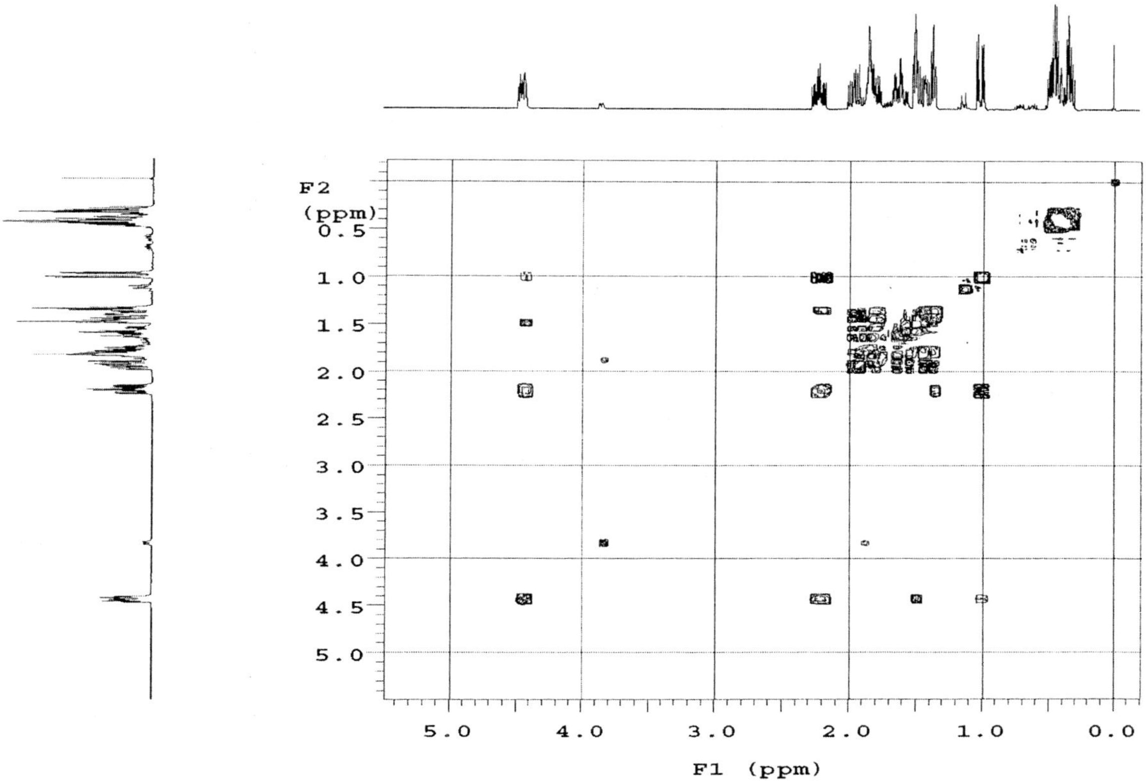

gCOSY Spectrum in $CDCl_3$

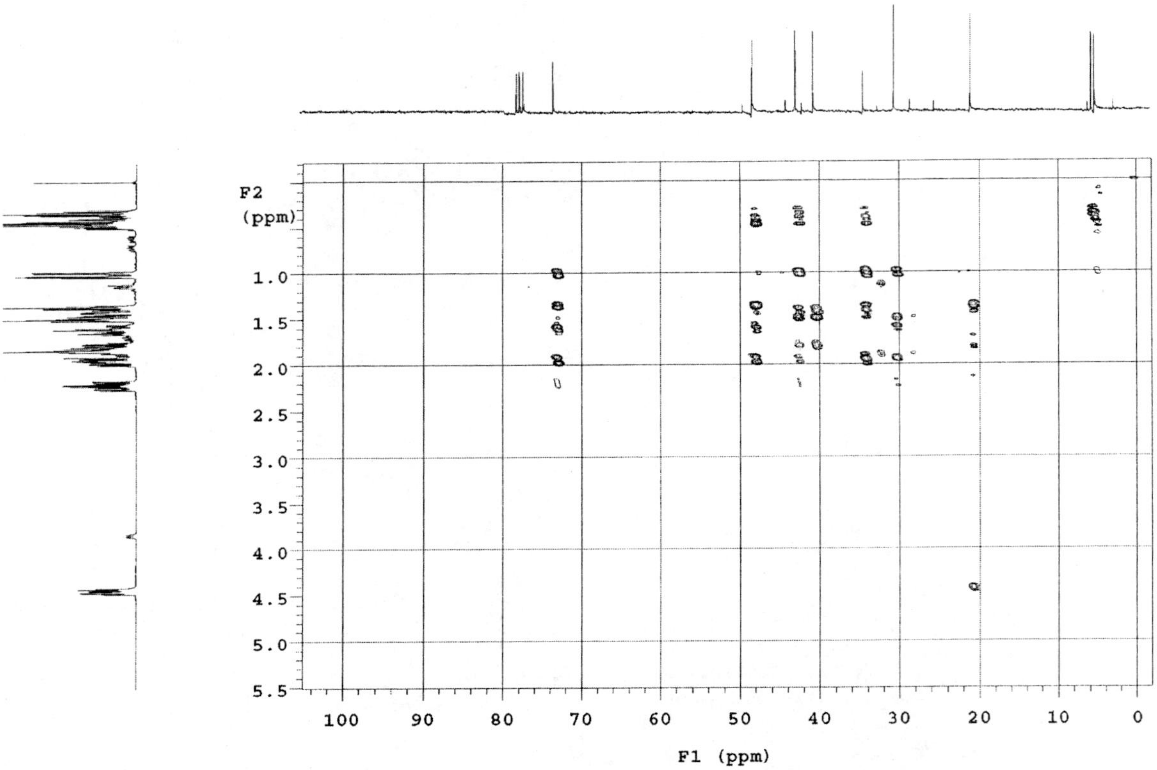

gHMBC Spectrum in $CDCl_3$

Problem 4.17 Given the spectroscopic data below (Proton, Carbon, gHSQC, and gCOSY), provide the structure of the camphor derivative. ($C_{17}H_{20}O$)

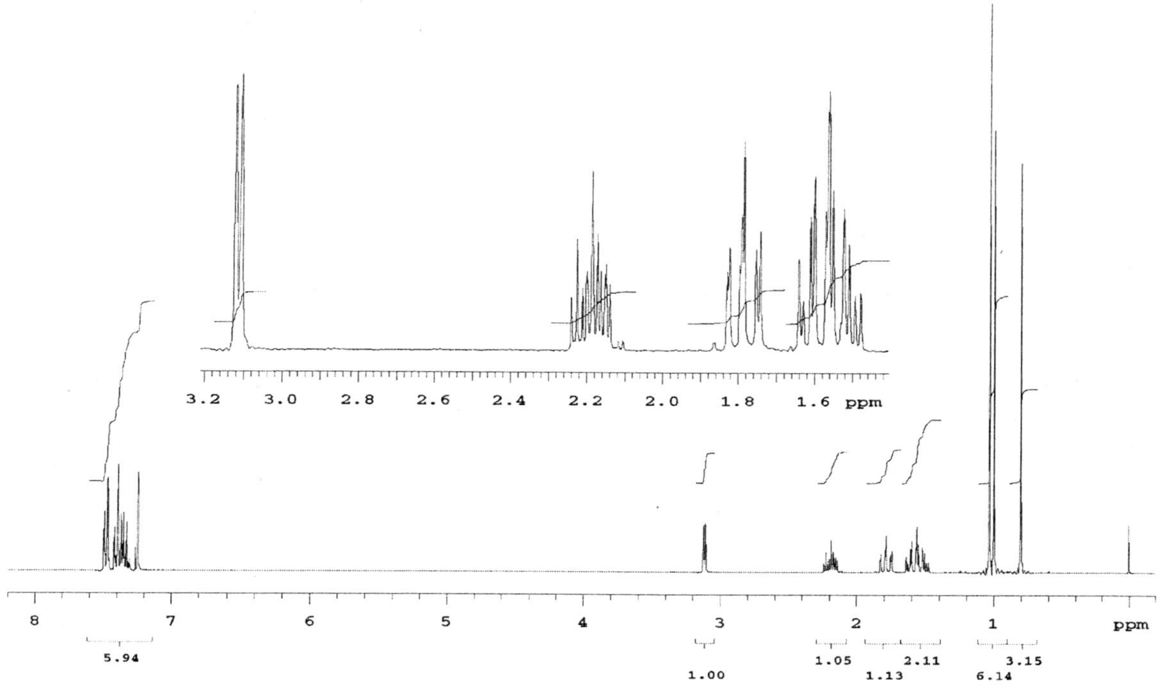

^{1}H-NMR Spectrum in $CDCl_3$

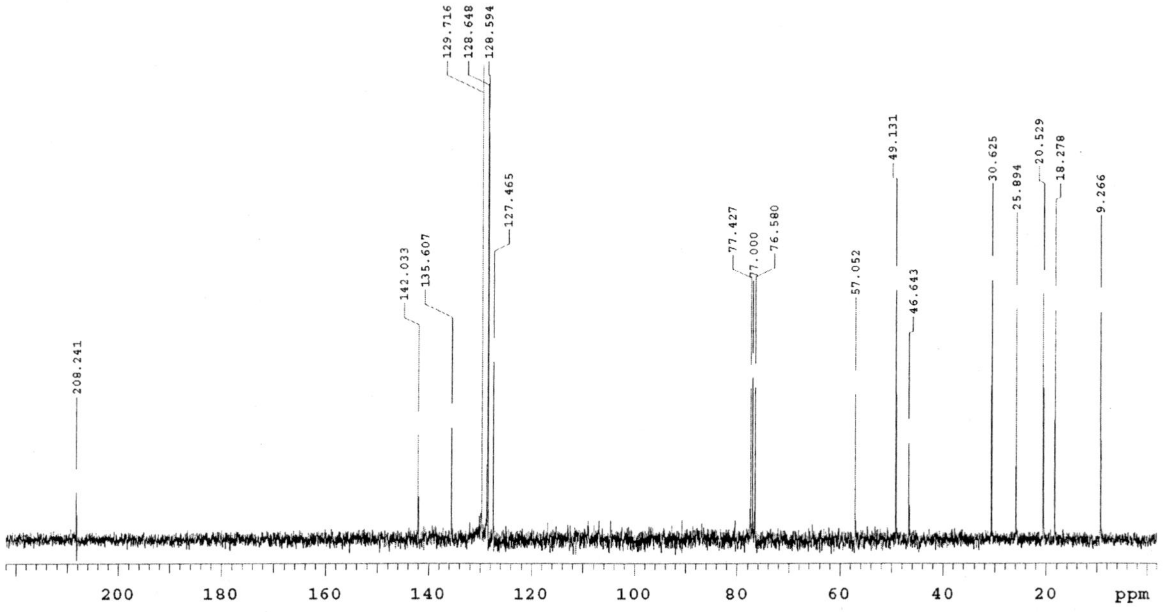

^{13}C-NMR Spectrum in $CDCl_3$

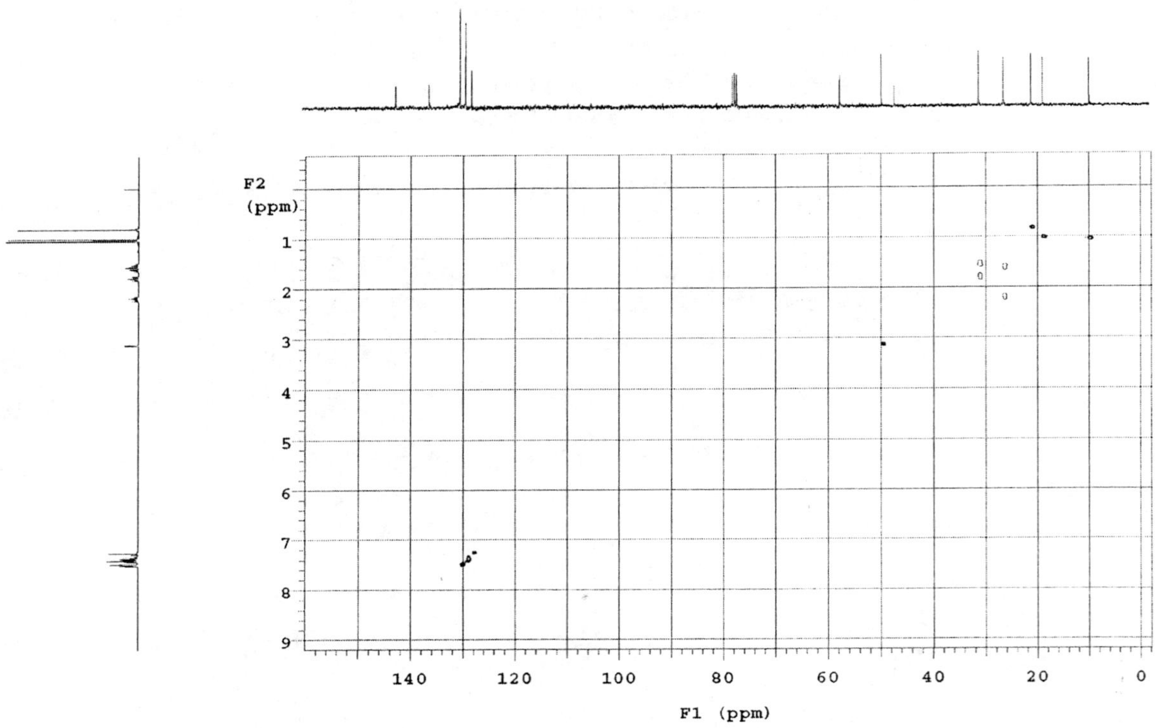

gHSQC Spectrum in $CDCl_3$

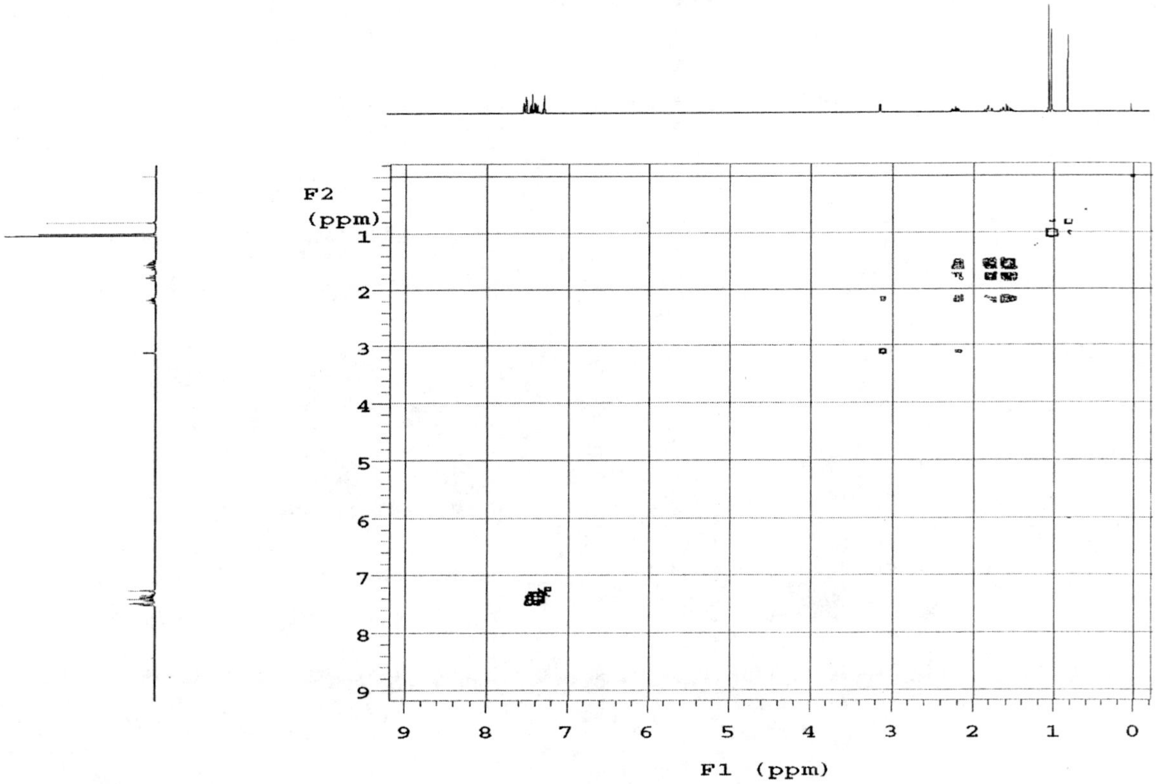

gCOSY Spectrum in $CDCl_3$

Problem 4.18 Given the spectroscopic data below (Proton, Carbon, gHSQC, gCOSY, and gHMBC), provide the structure of the molecule. (MW: 110 g/mol)

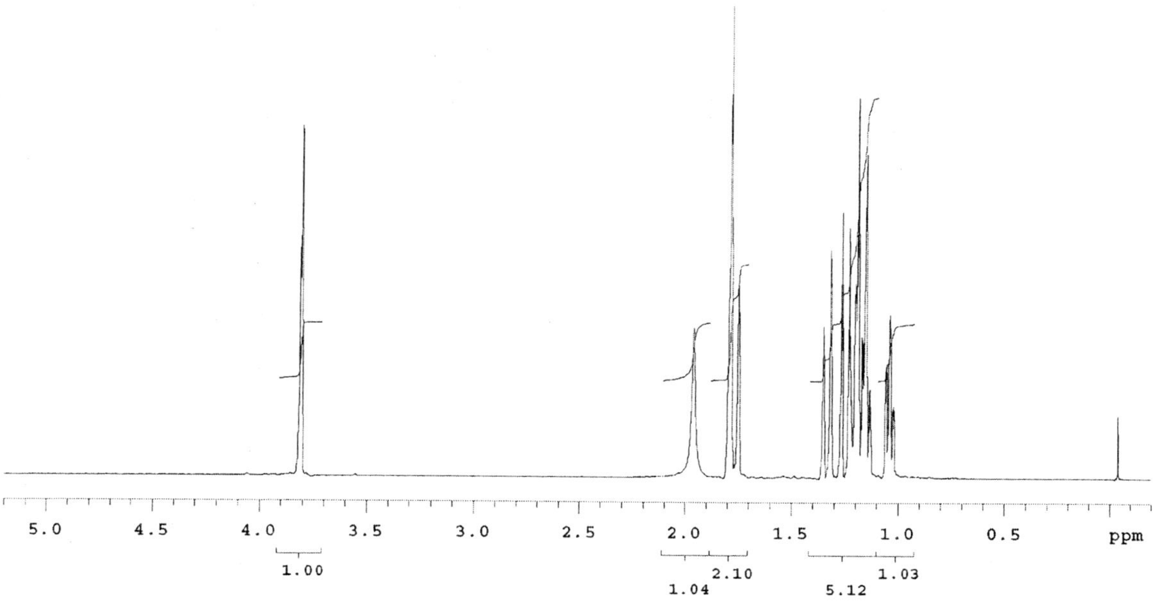

^{1}H-NMR Spectrum in $CDCl_3$

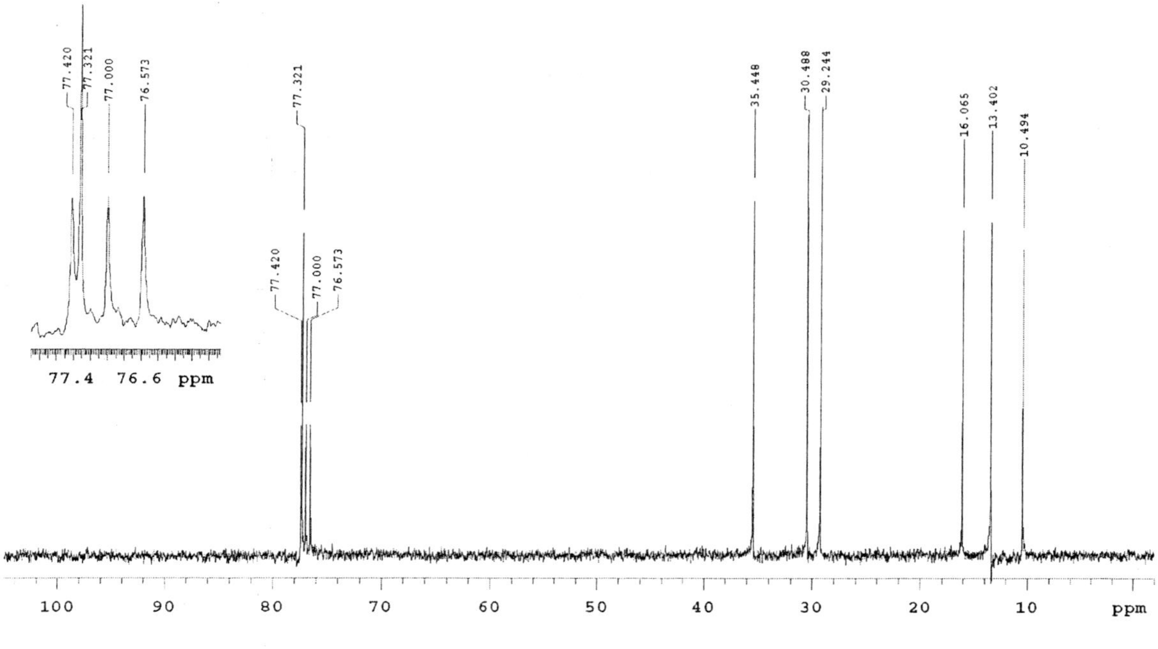

^{13}C-NMR Spectrum in $CDCl_3$

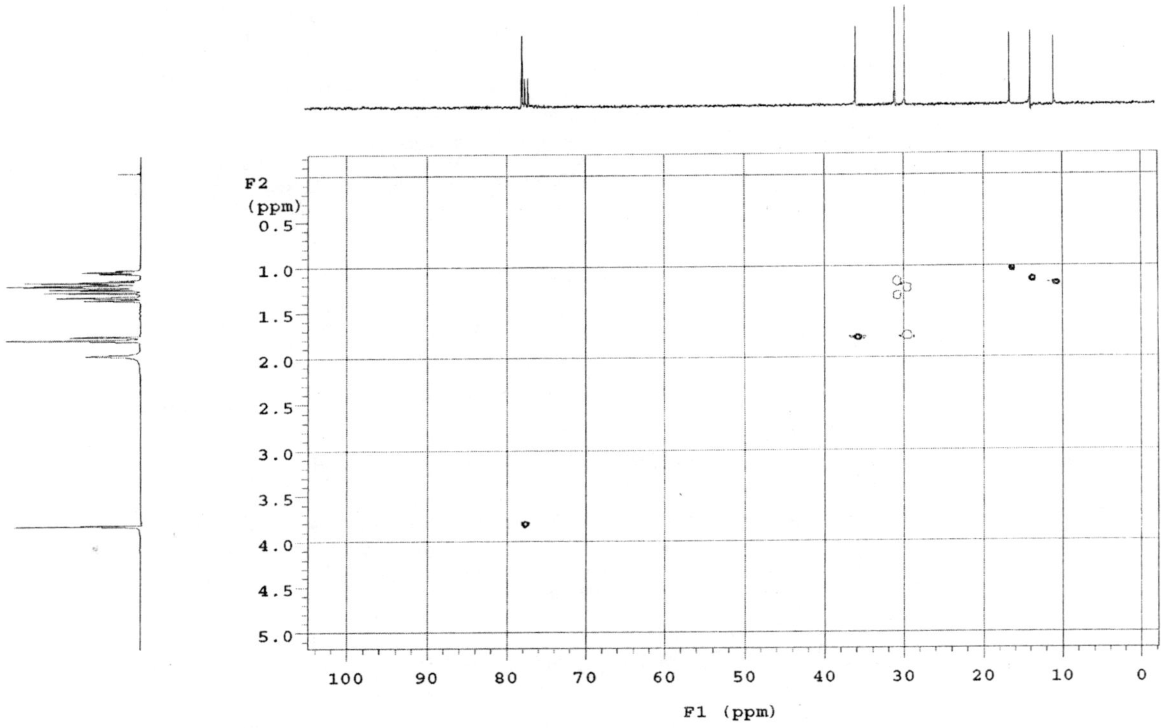

gHSQC Spectrum in $CDCl_3$

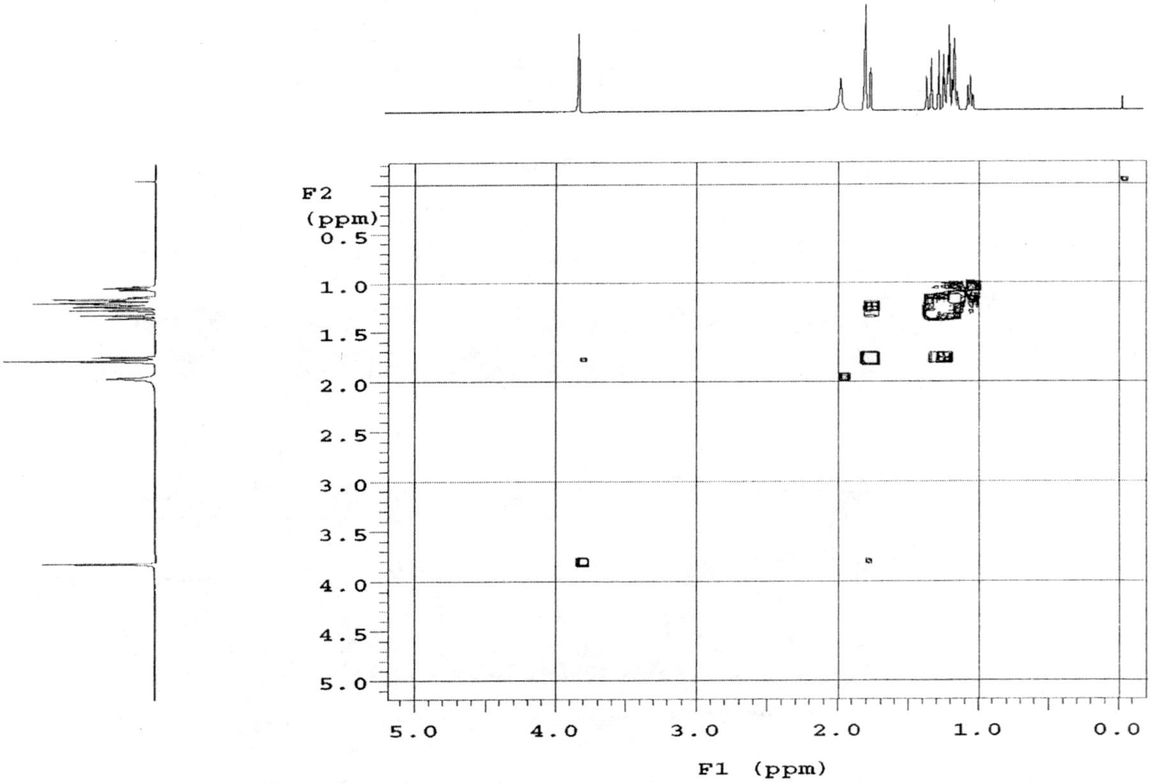

gCOSY Spectrum in $CDCl_3$

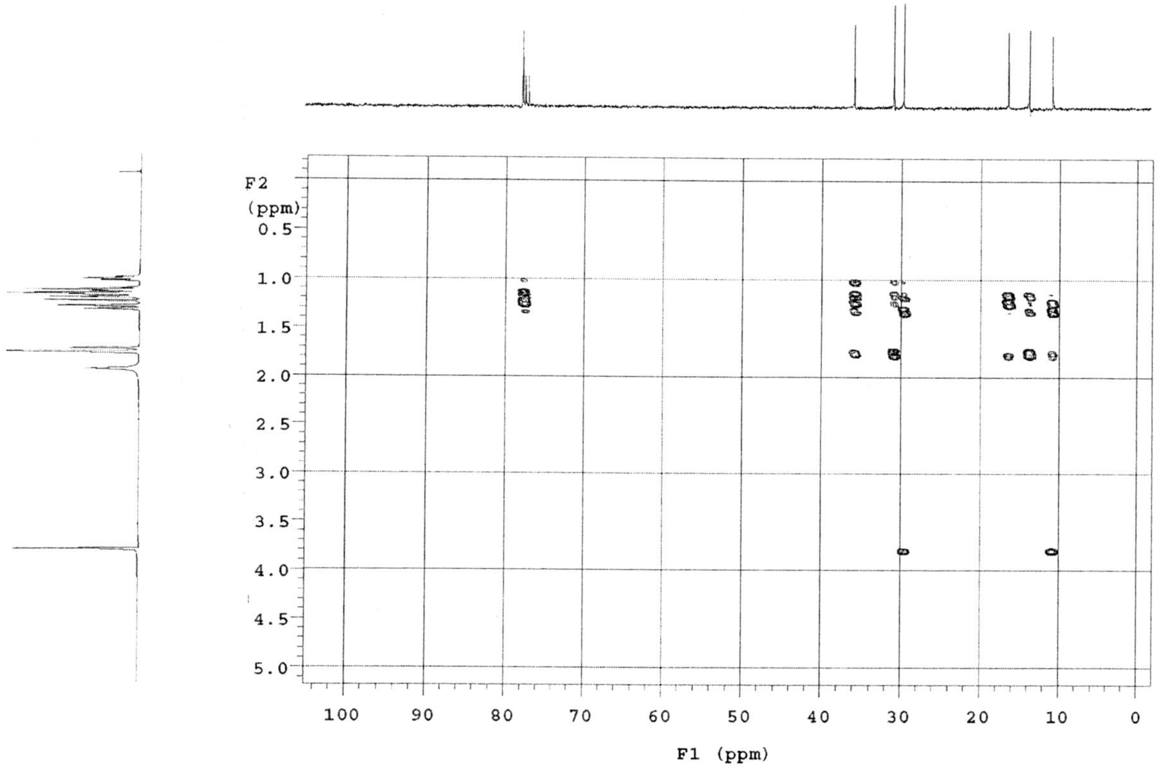

gHMBC Spectrum in $CDCl_3$

Problem 4.19 Given the spectroscopic data below (Proton, Carbon, gHSQC, gCOSY, and gHMBC), provide the structure of the molecule. ($C_{10}H_{16}OS$) HINT: See **3.18**.

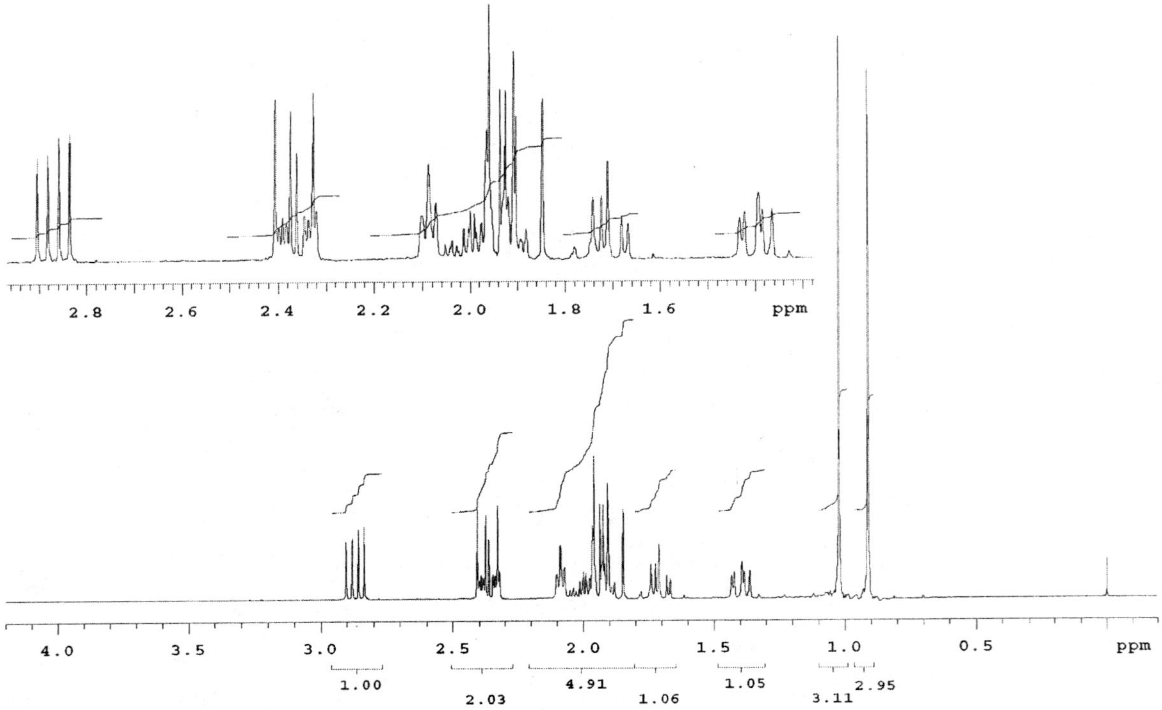

^{1}H-NMR Spectrum in $CDCl_3$

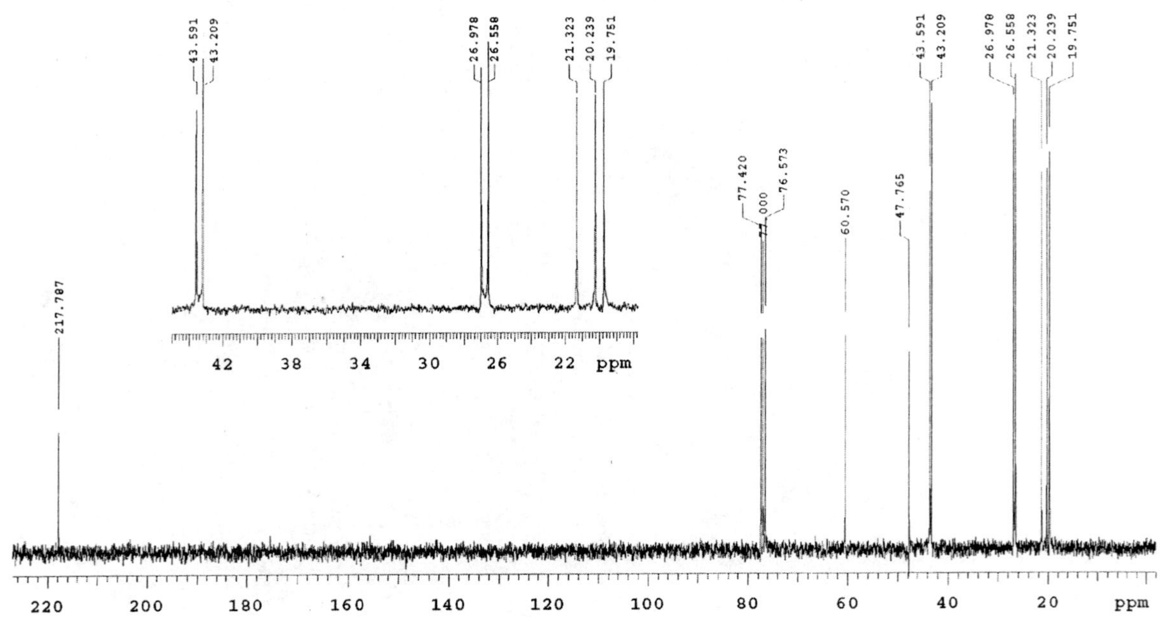

^{13}C-NMR Spectrum in $CDCl_3$

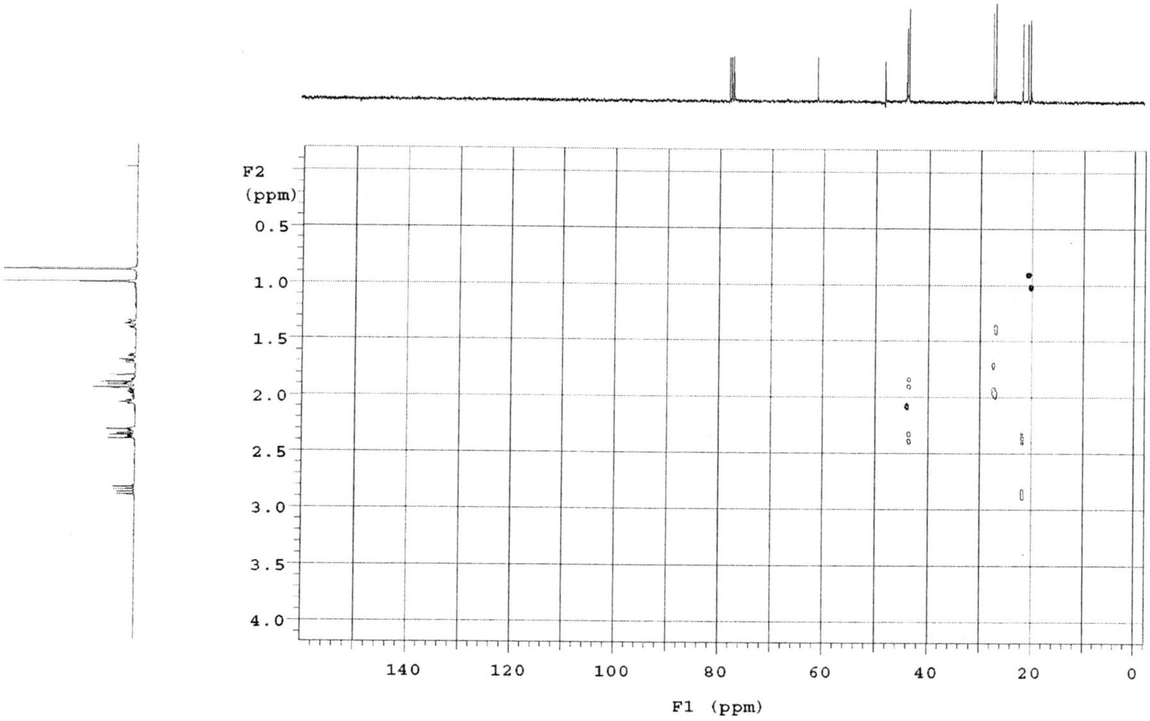

gHSQC Spectrum in $CDCl_3$

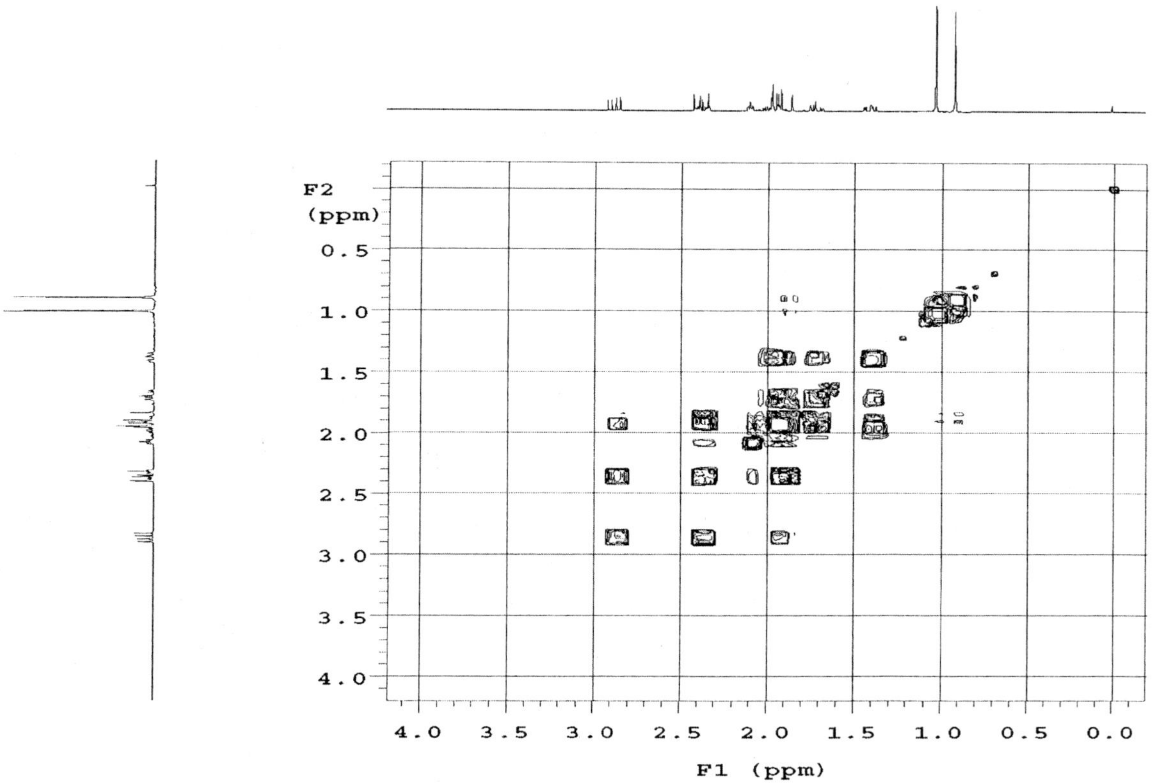

gCOSY Spectrum in $CDCl_3$

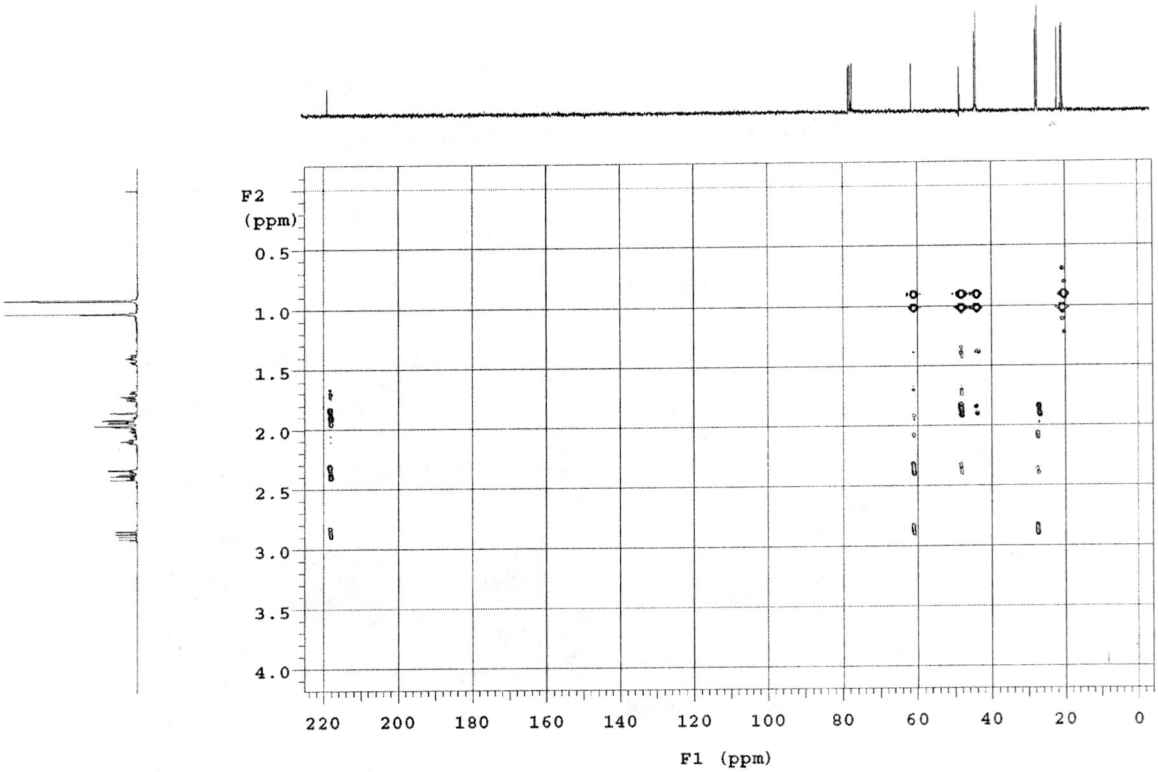

gHMBC Spectrum in $CDCl_3$

Problem 4.20 Given the spectroscopic data below (Proton, Carbon, gHSQC, and gCOSY), provide the structure of the molecule.

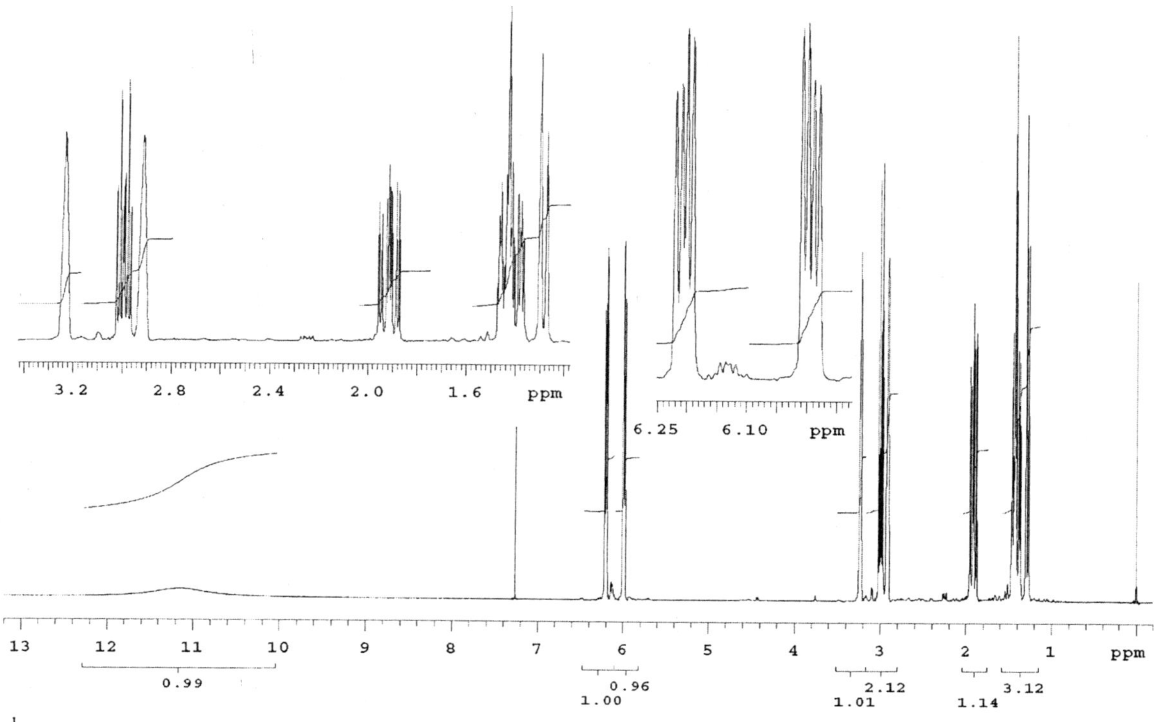

^{1}H-NMR Spectrum in $CDCl_3$

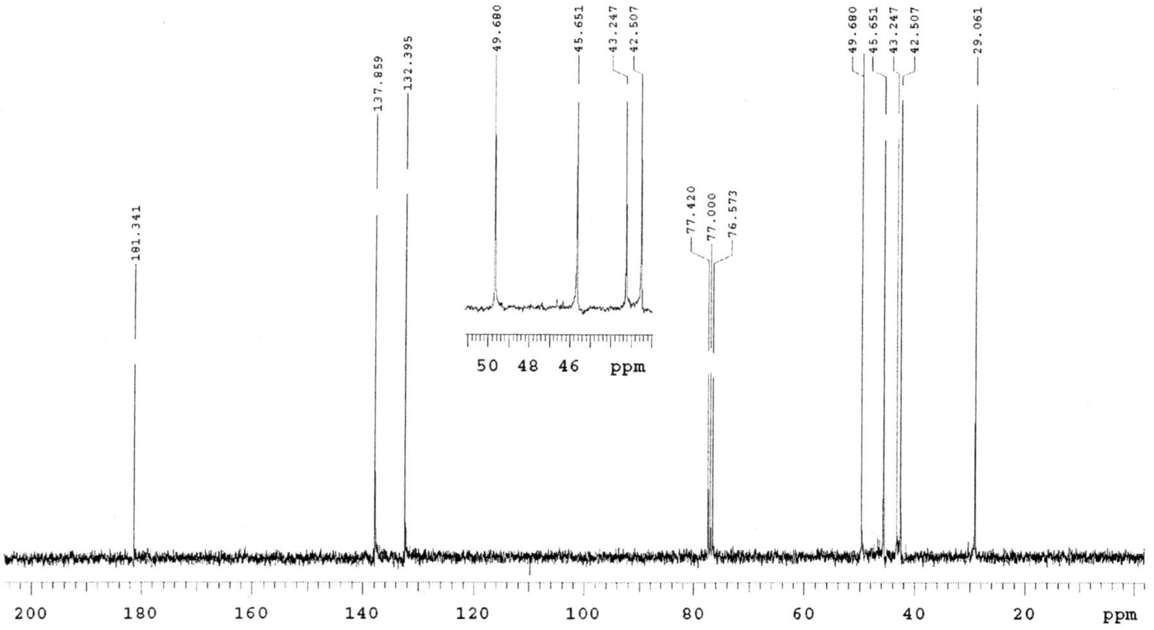

^{13}C-NMR Spectrum in $CDCl_3$

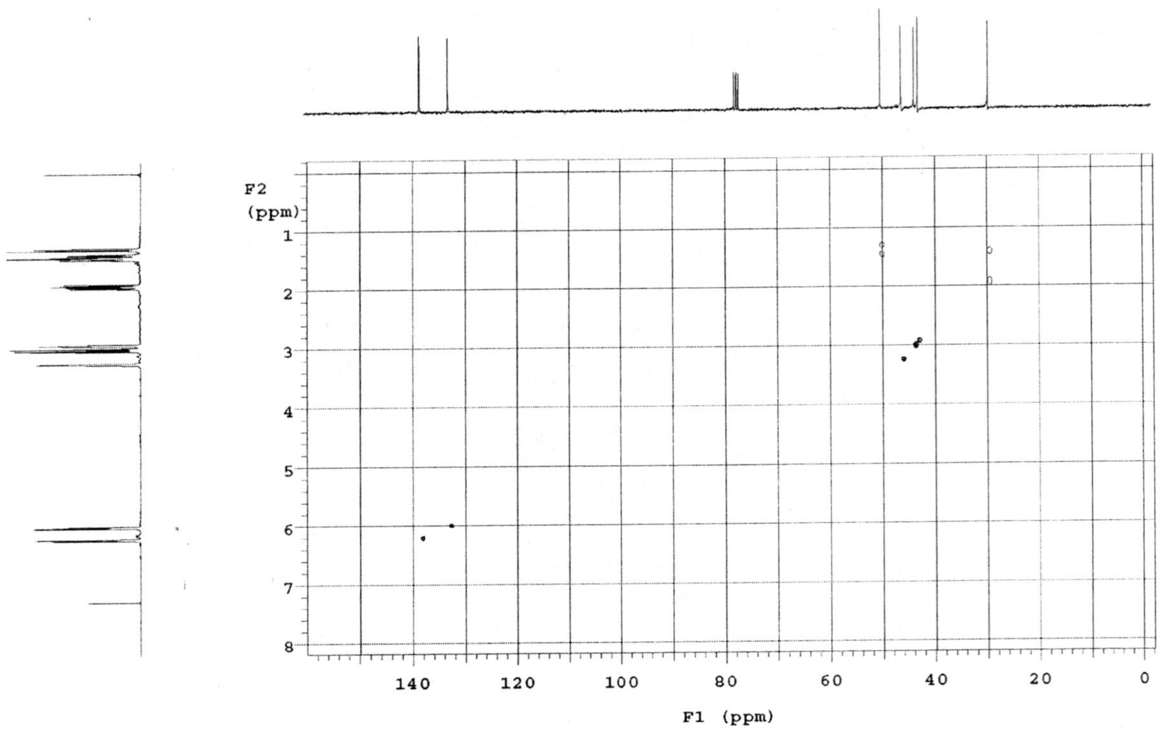

gHSQC Spectrum in $CDCl_3$

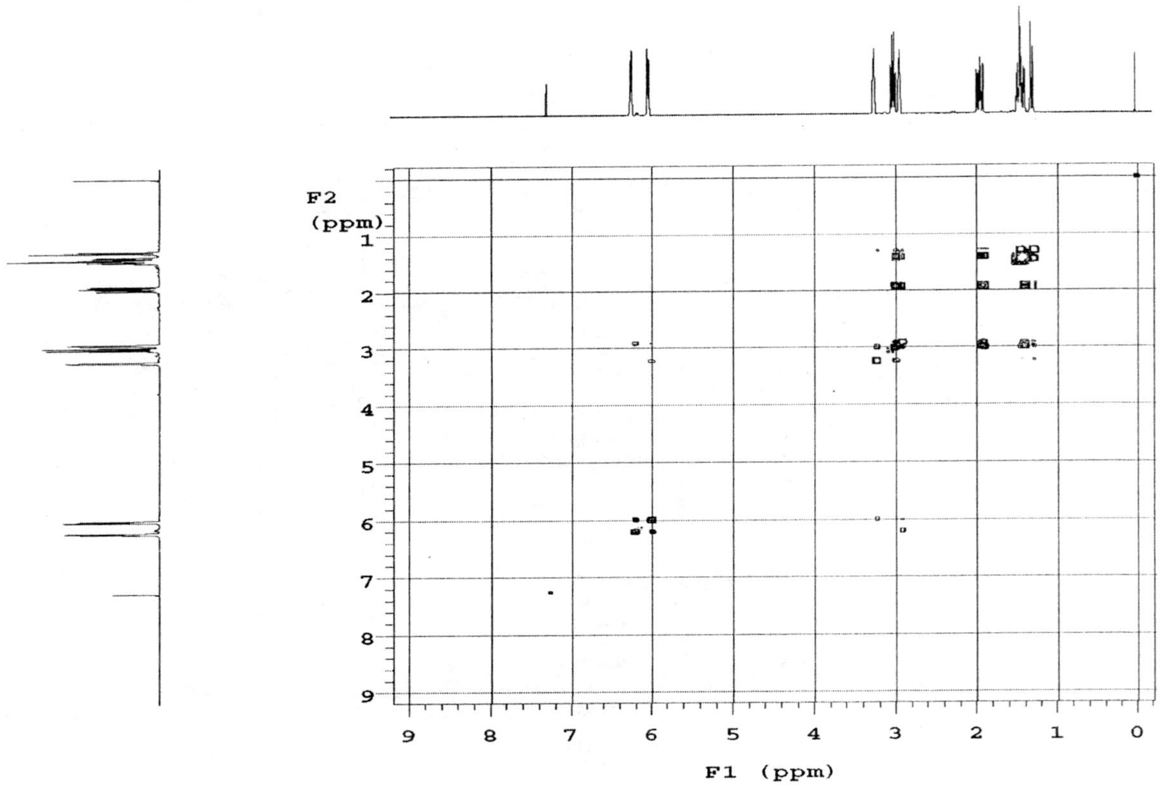

gCOSY Spectrum in $CDCl_3$

Problem 4.21 Given the spectroscopic data below (Proton, Carbon, gHSQC, and gHMBC), provide the structure of the molecule. ($C_9H_{10}O_4$)

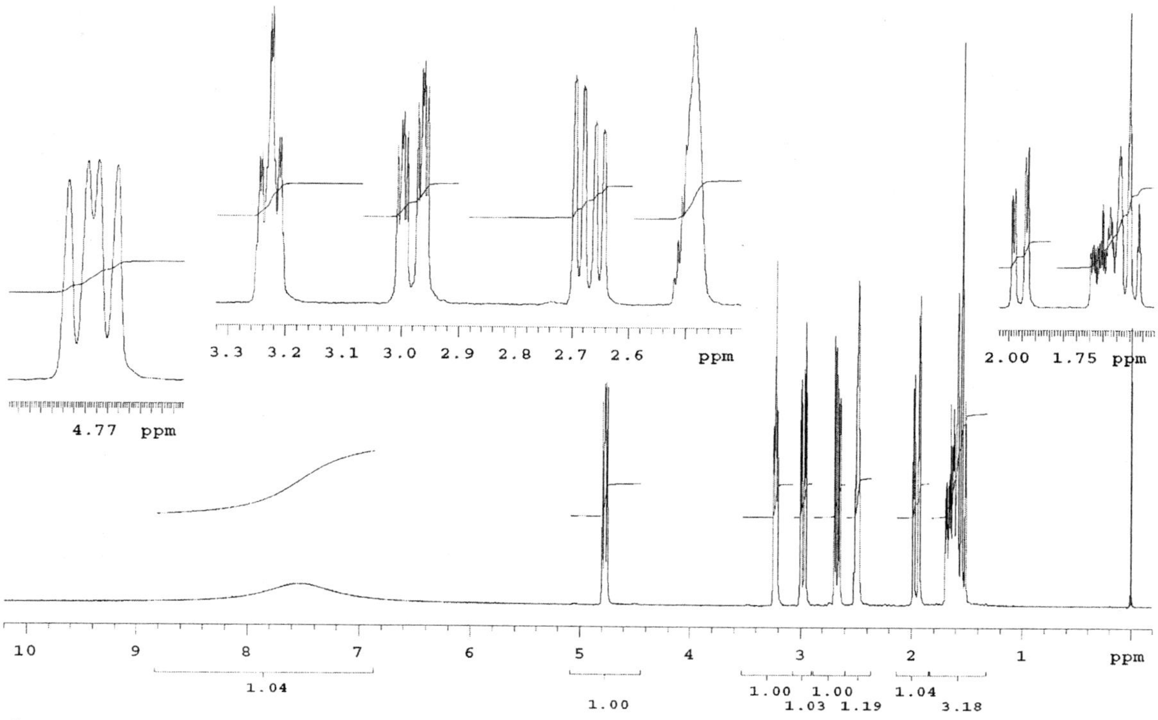

^{1}H-NMR Spectrum in DMSO-d_6

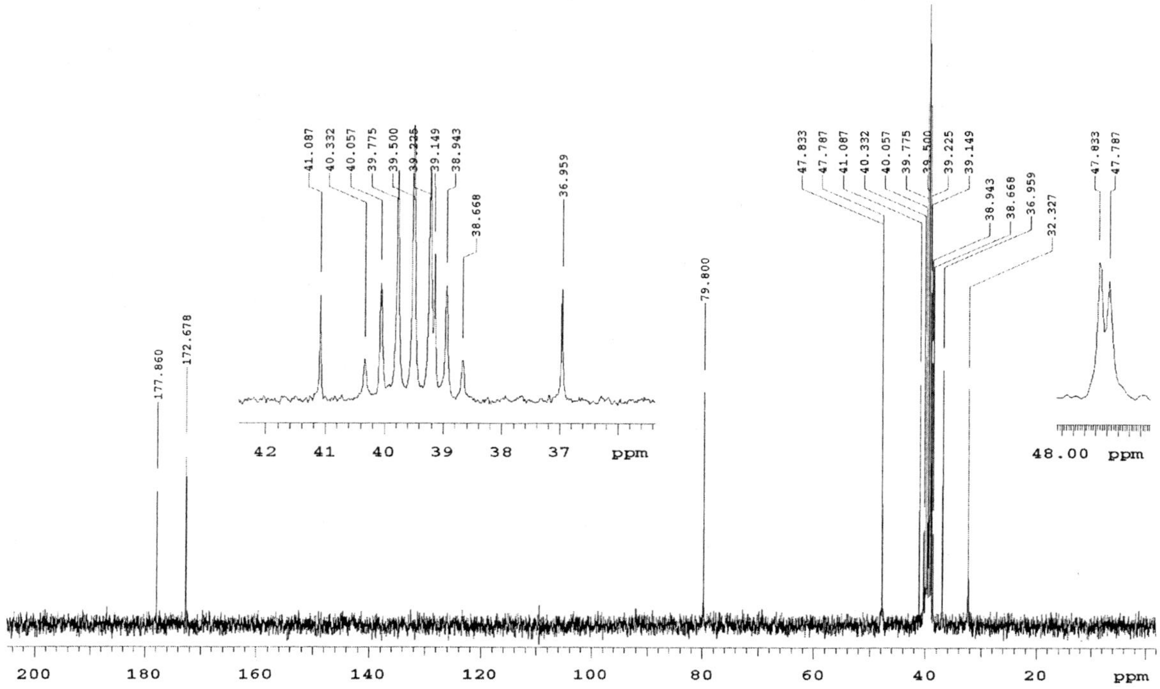

^{13}C-NMR Spectrum in DMSO-d_6

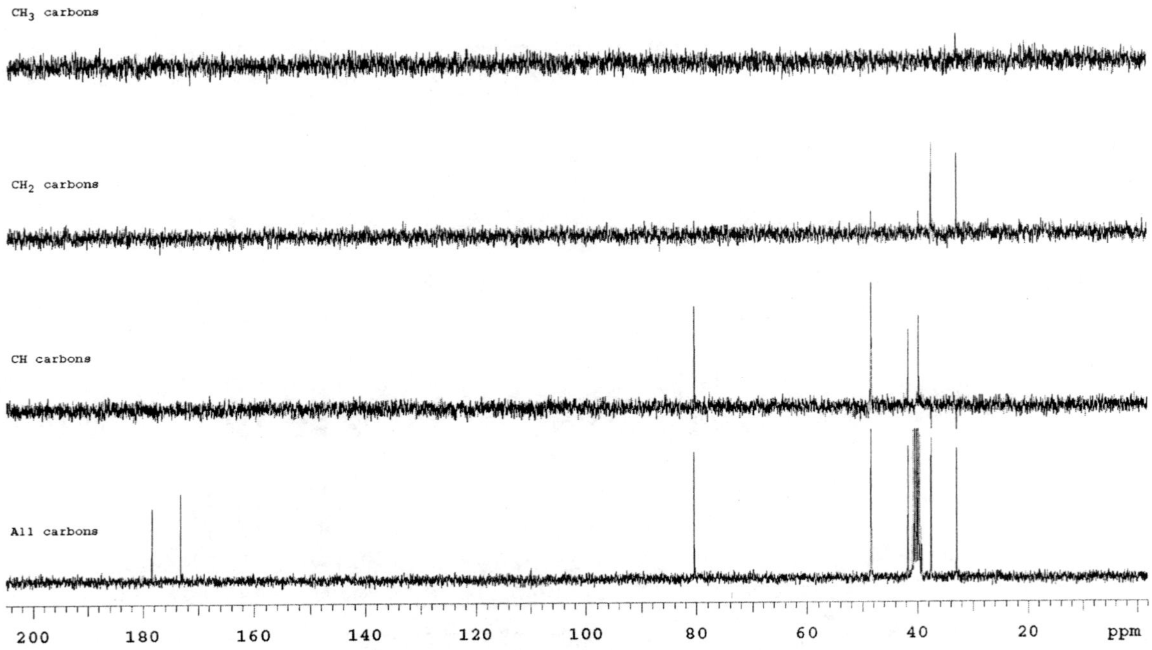

Edited DEPT ^{13}C-NMR Spectrum in DMSO-d_6

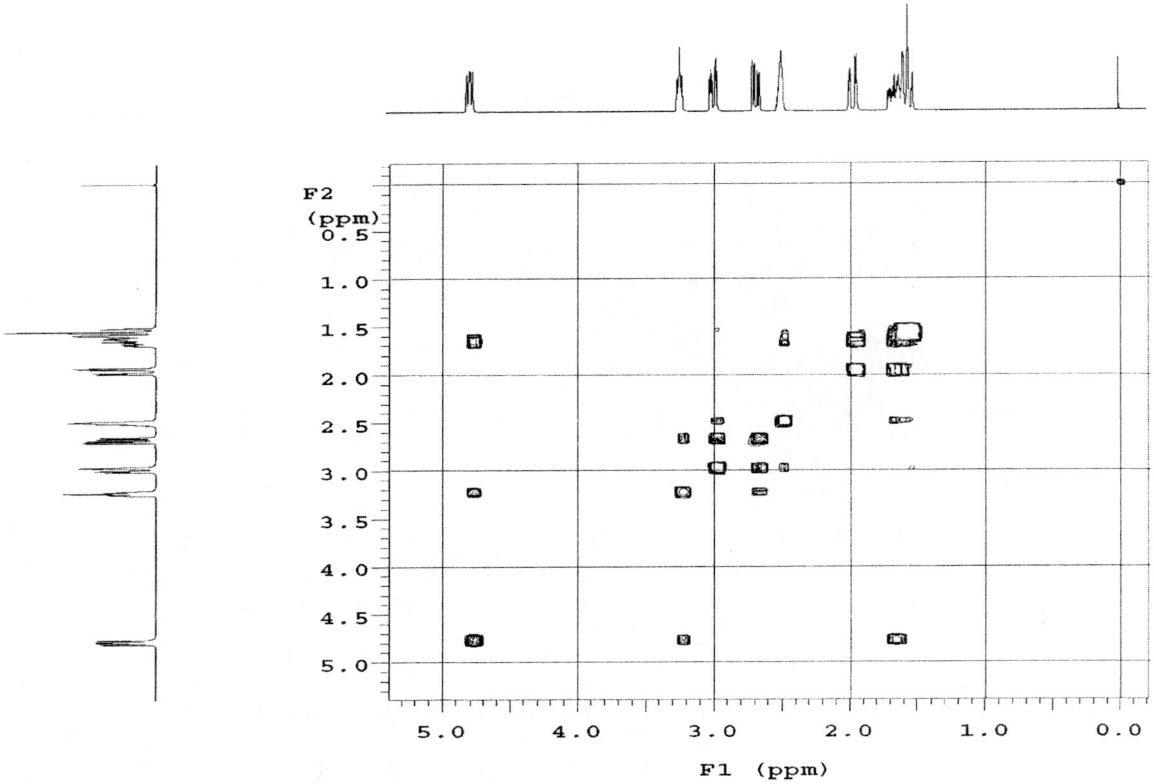

gCOSY Spectrum in DMSO-d_6

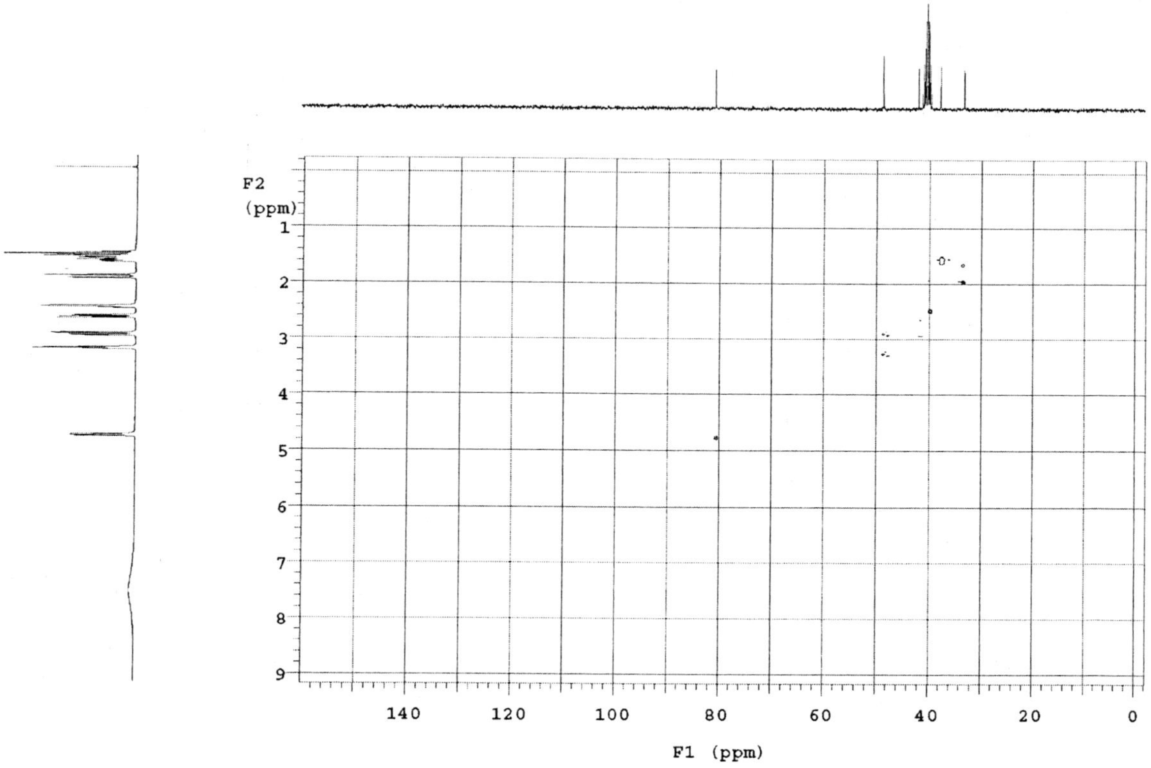

gHSQC Spectrum in DMSO-d_6

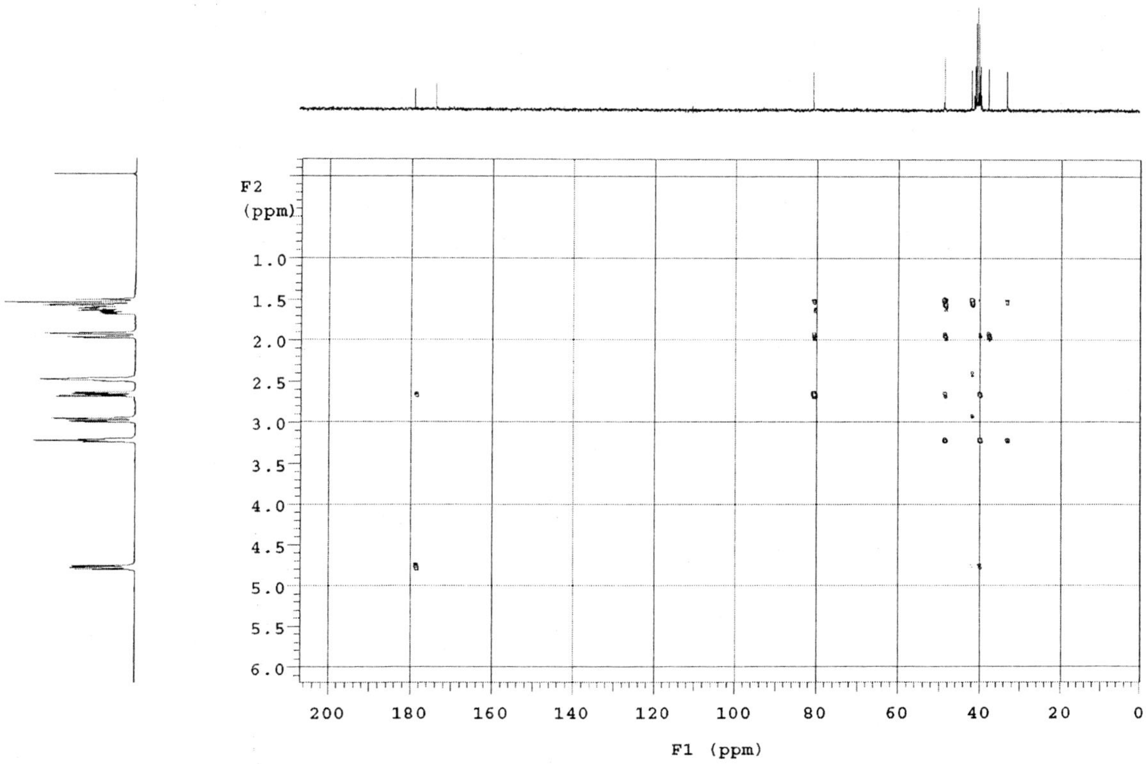

gHMBC Spectrum in DMSO-d_6

Problem 4.22 Given the spectroscopic data below (Proton, Carbon, gHSQC, gCOSY, and gHMBC), provide the structure of the molecule.

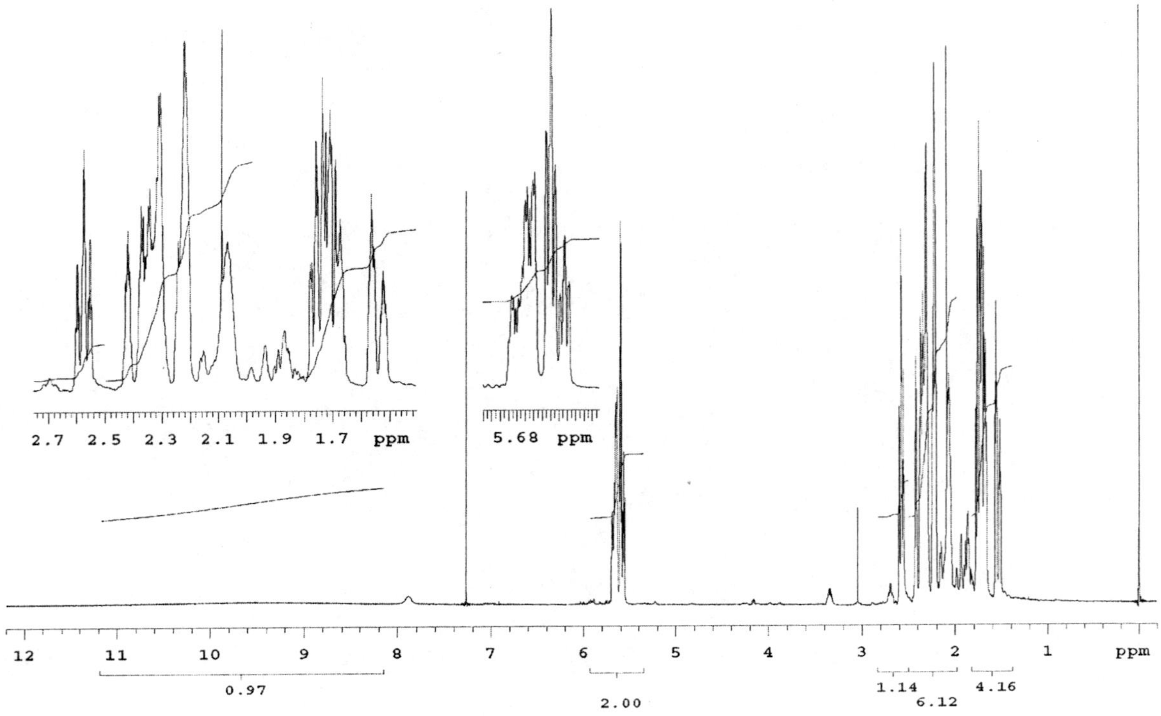

^{1}H-NMR Spectrum in $CDCl_3$

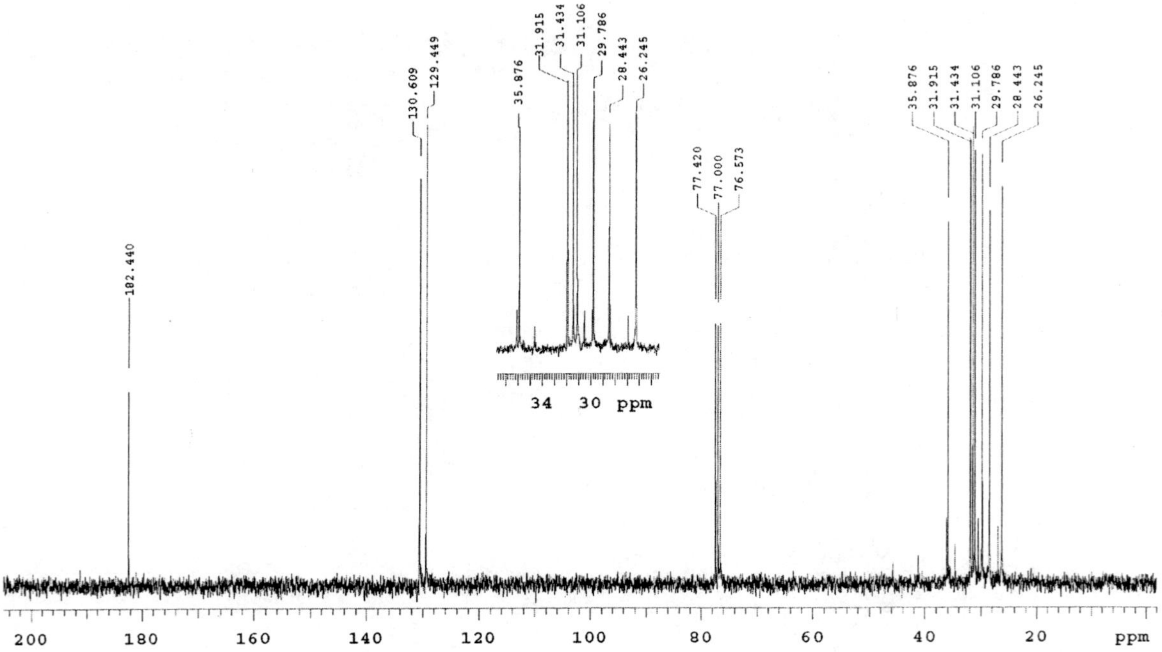

^{13}C-NMR Spectrum in $CDCl_3$

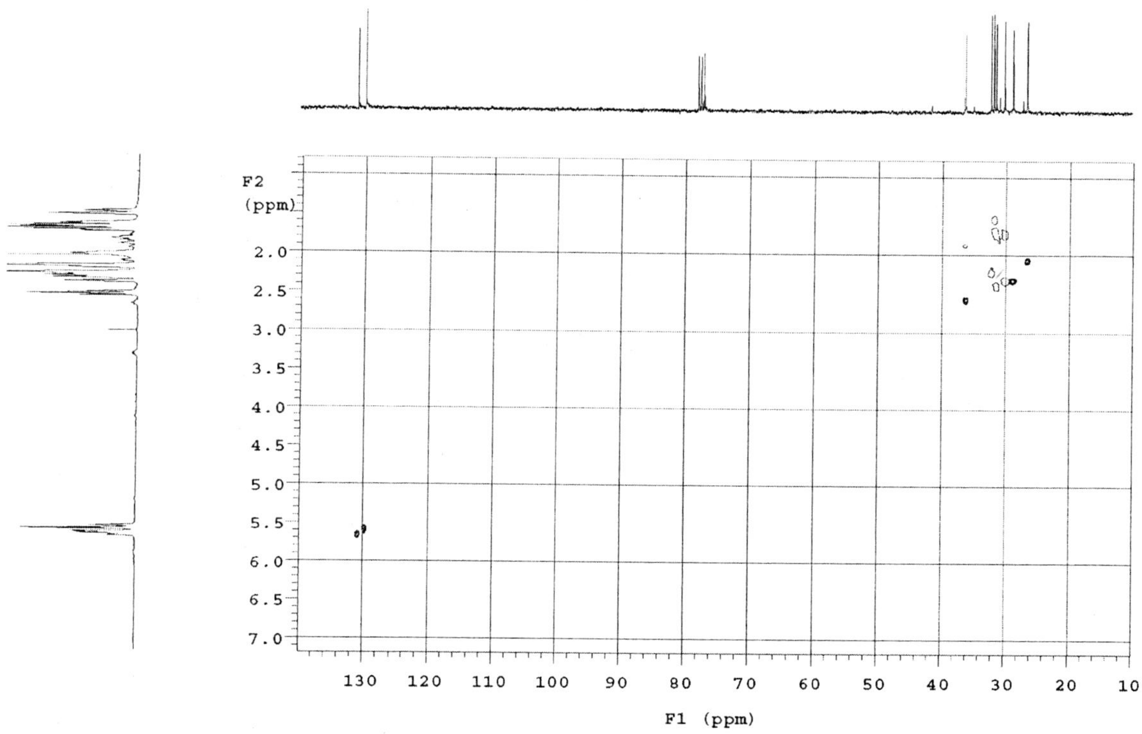

gHSQC Spectrum in $CDCl_3$

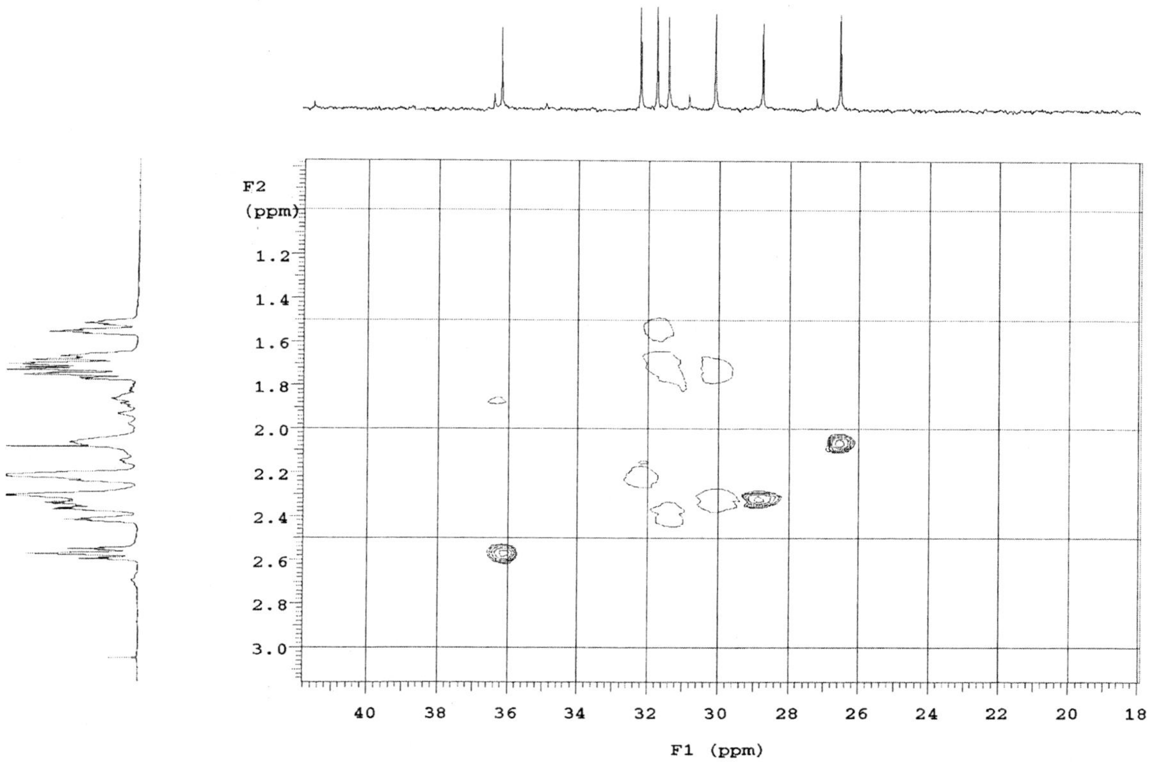

Expansion of gHSQC Spectrum in $CDCl_3$

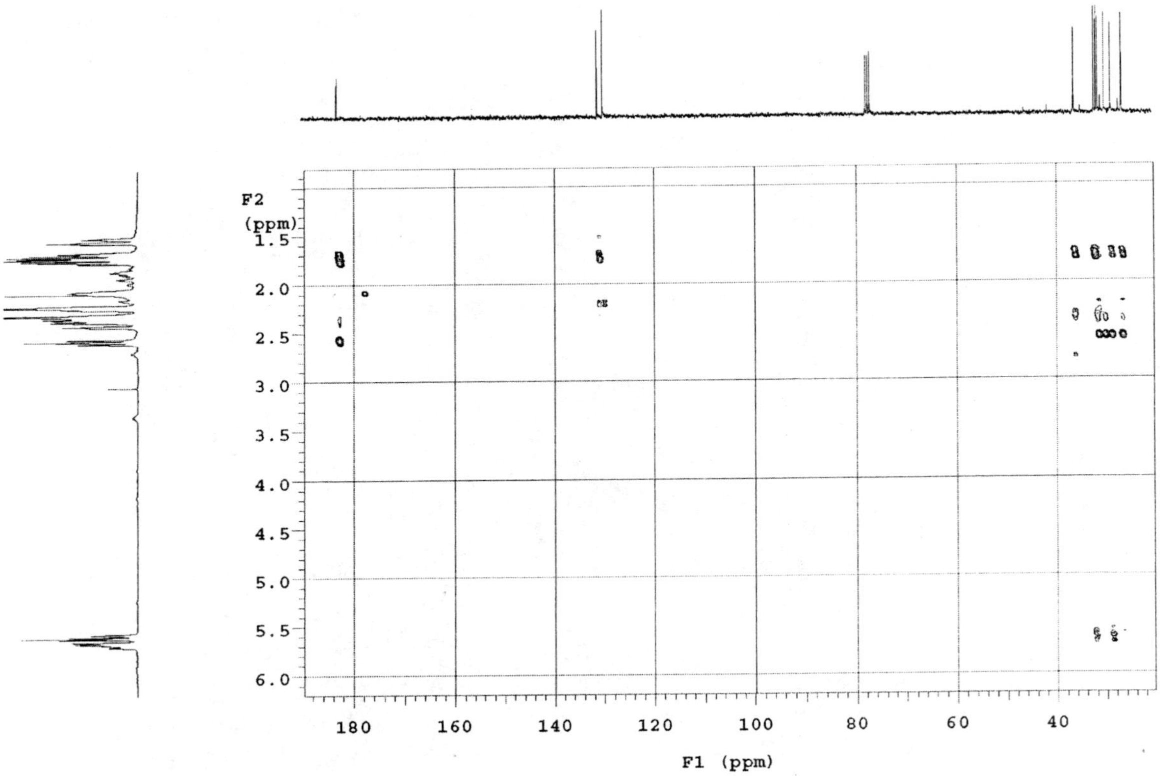

gHMBC Spectrum in $CDCl_3$

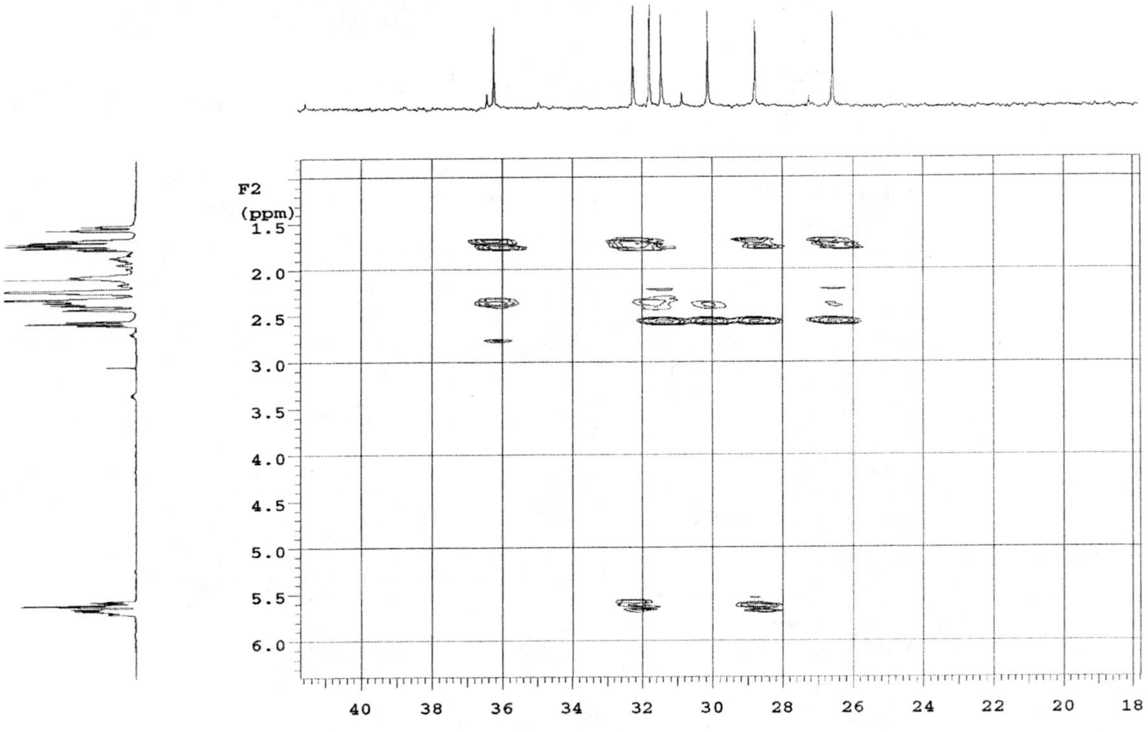

Expansion of gHMBC Spectrum in $CDCl_3$

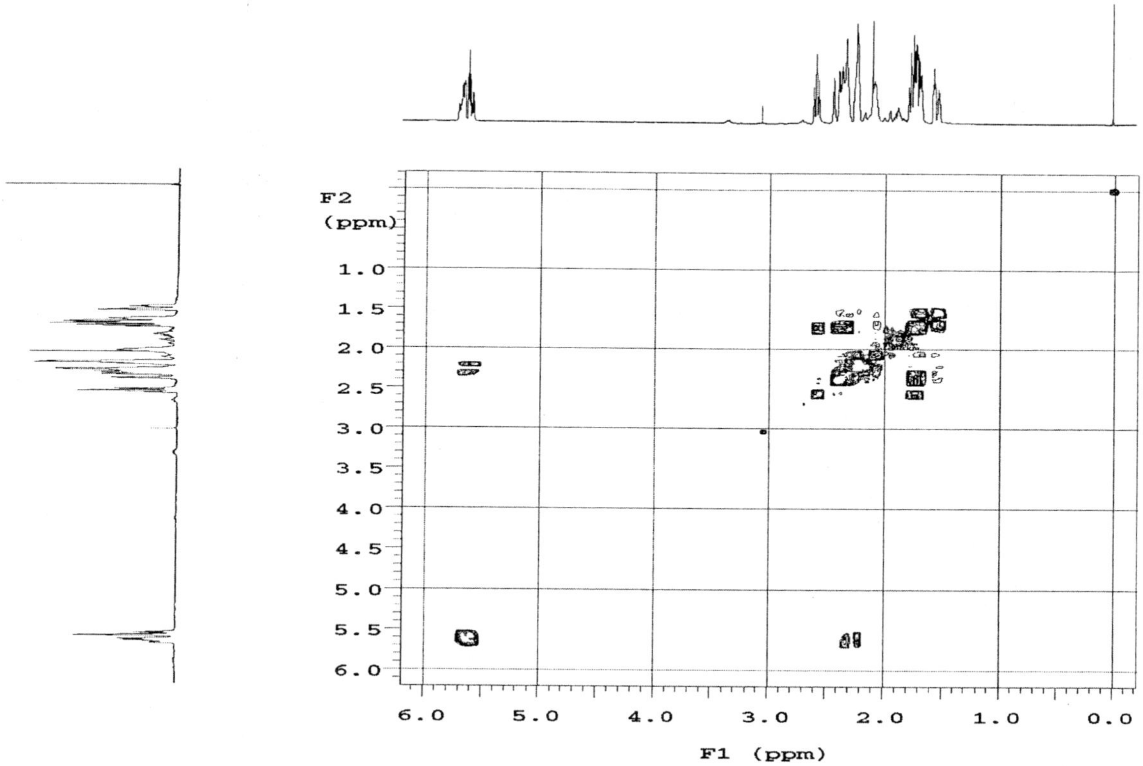

gCOSY Spectrum in $CDCl_3$

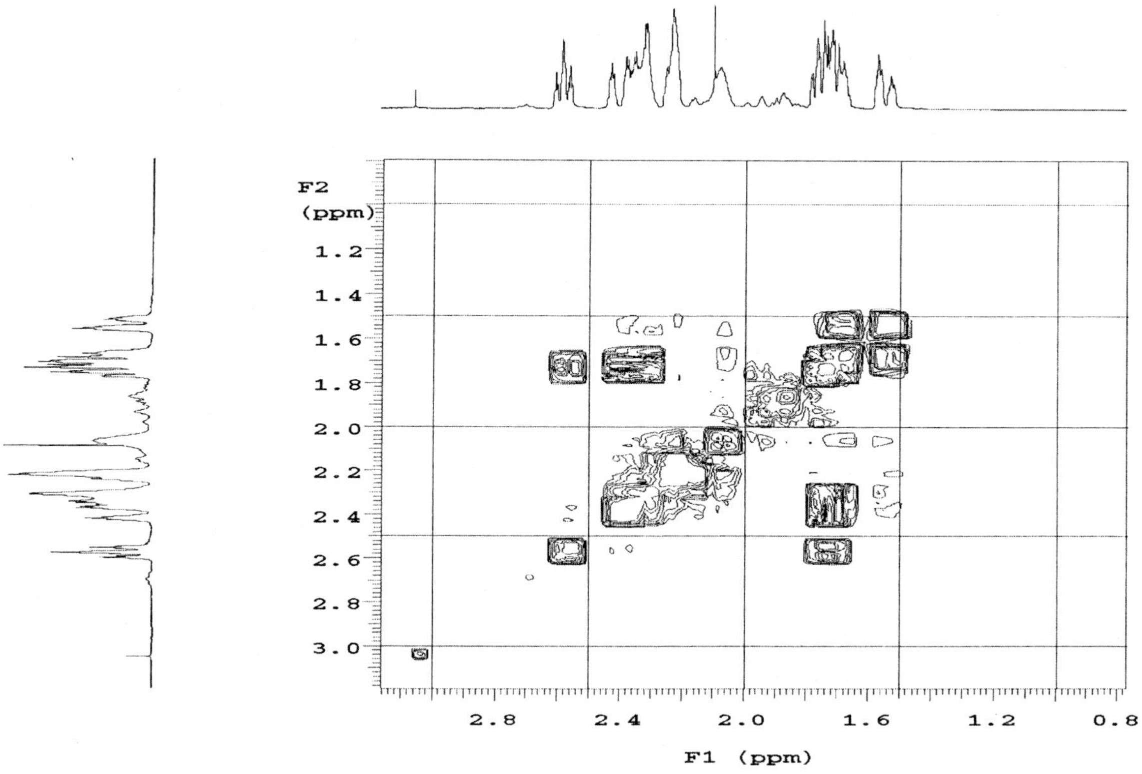

Expansion of gCOSY Spectrum in $CDCl_3$

Problem 4.23 Given the spectroscopic data below (Proton, Carbon, gHSQC, gCOSY and gHMBC), provide the structure of the molecule. (MW: 230 & 232 g/mol)

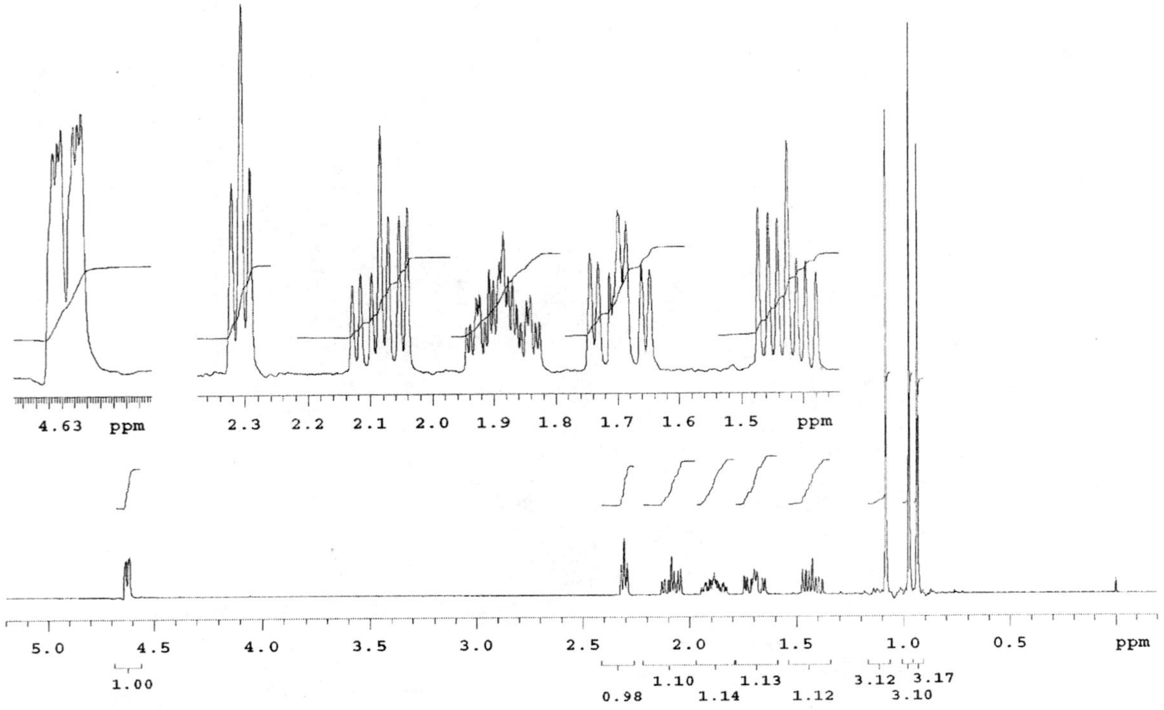

^{1}H-NMR Spectrum in $CDCl_3$

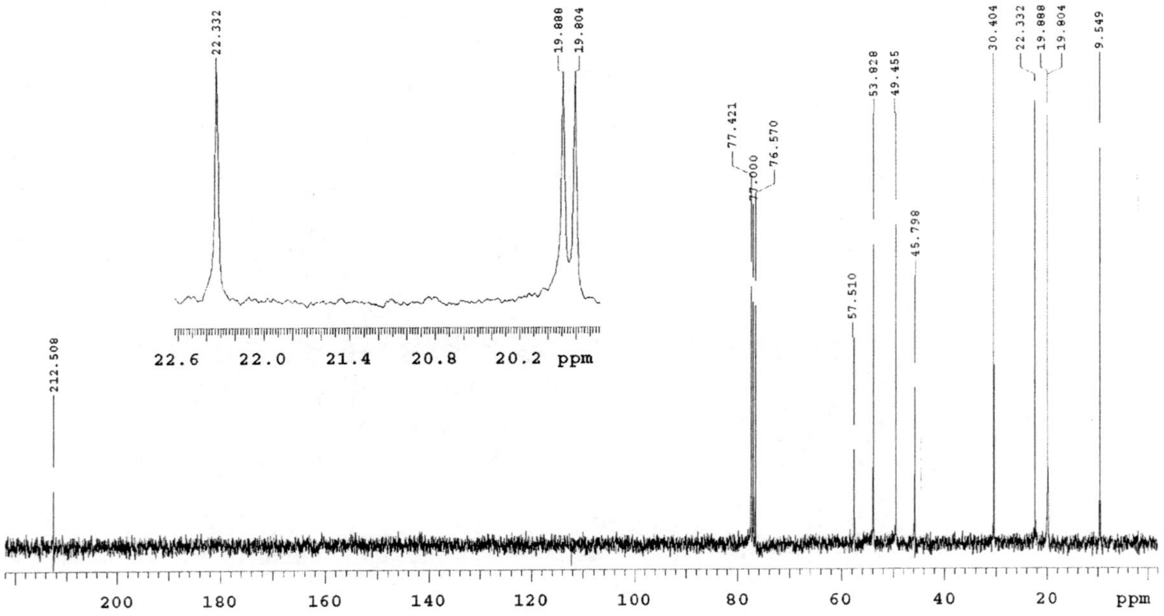

^{13}C-NMR Spectrum in $CDCl_3$

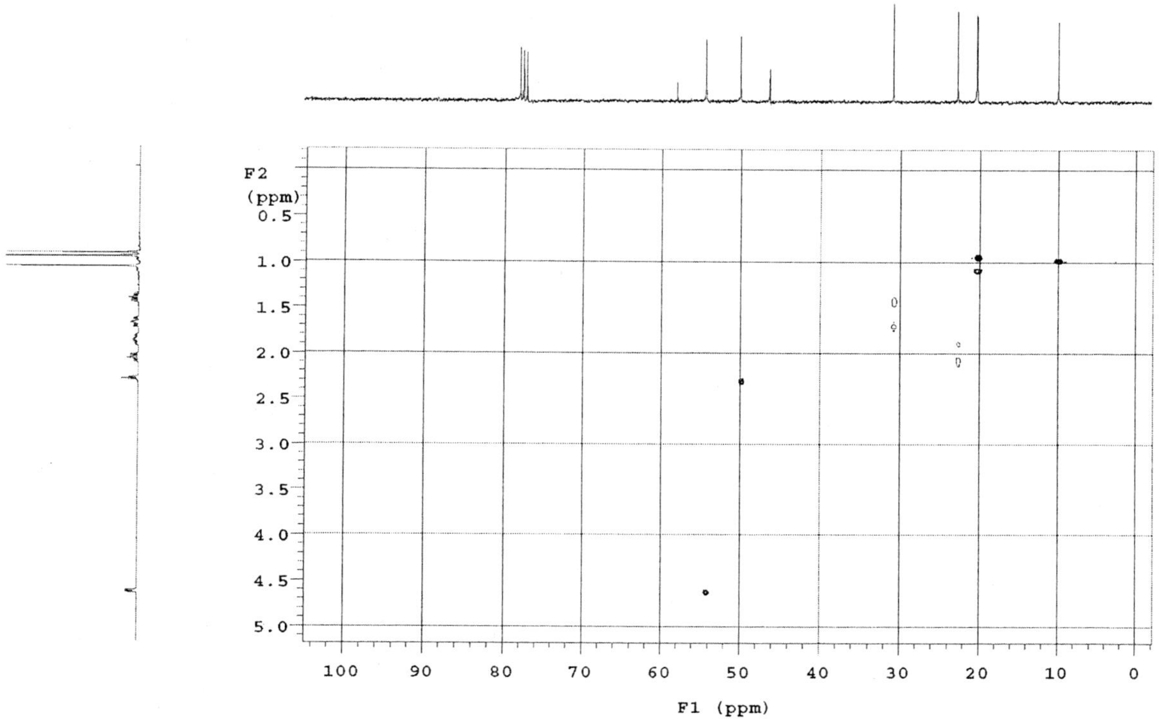

gHSQC Spectrum in $CDCl_3$

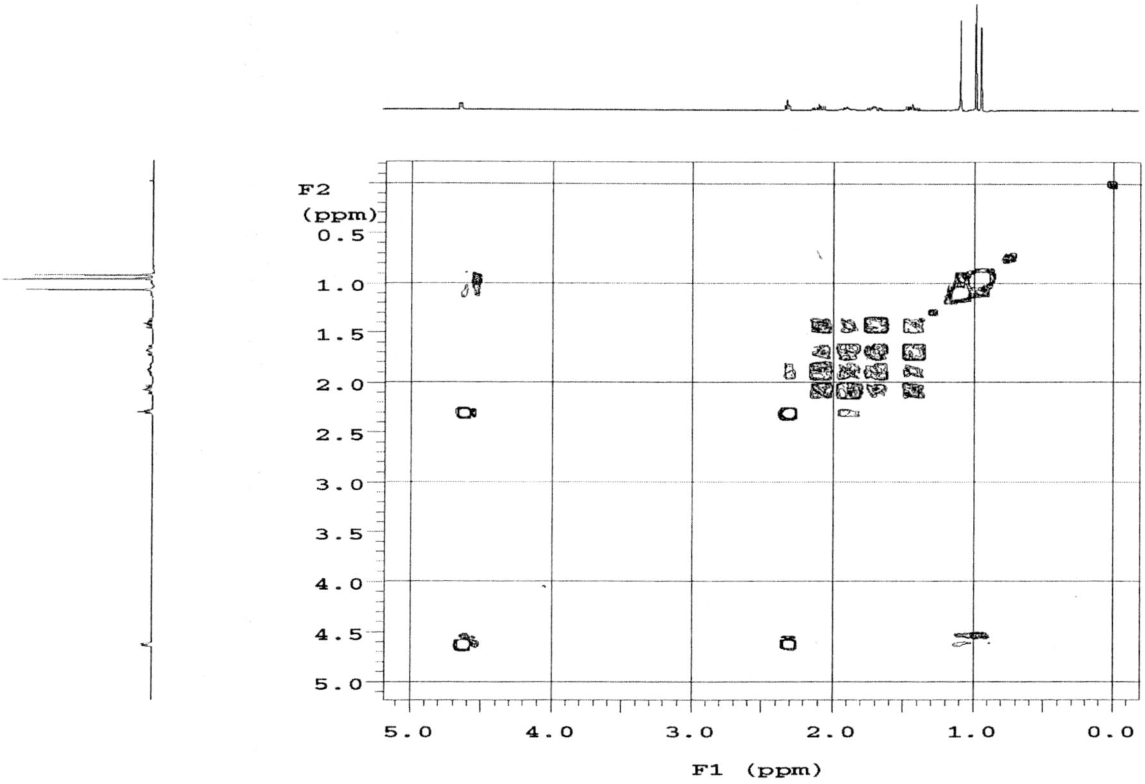

gCOSY Spectrum in $CDCl_3$

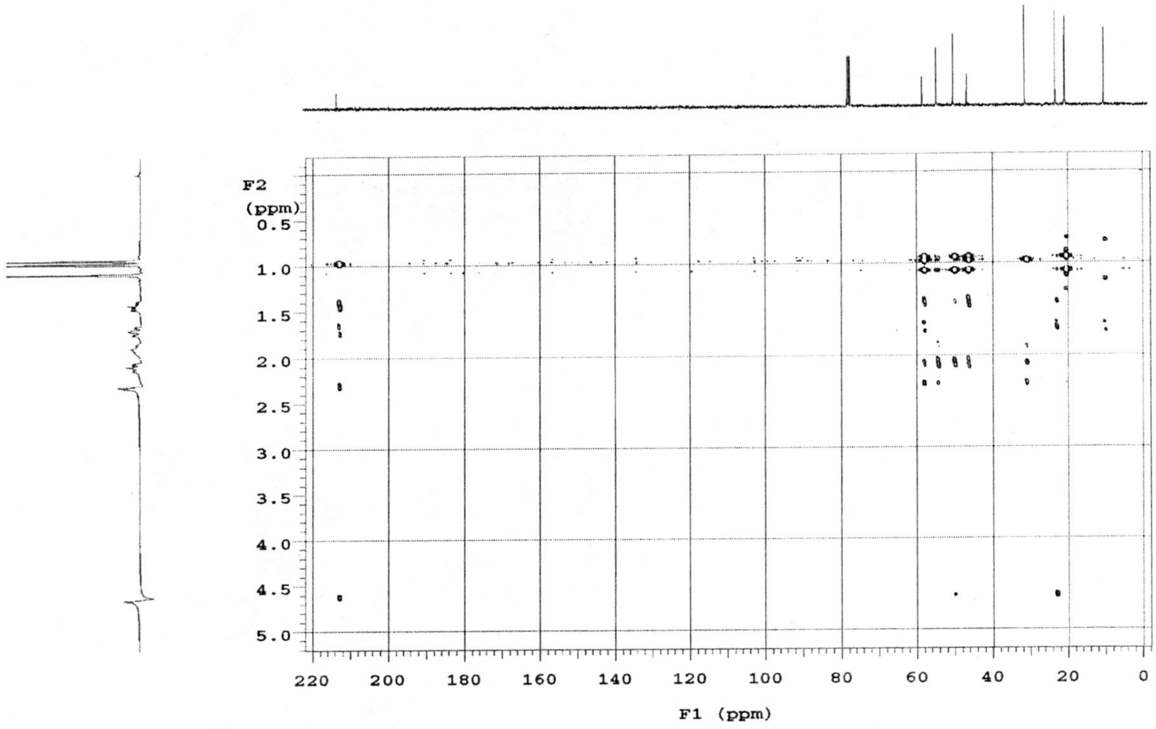

gHMBC Spectrum in $CDCl_3$

Problem 4.24 Given the spectroscopic data below (Proton, Carbon, APT, gCOSY, gHSQC, and gHMBC), provide the structure of the molecule. (HR-MS: 386.3548 amu) The compound is biosynthesized and found widely distributed in animals. It has been linked to arteriosclerosis.

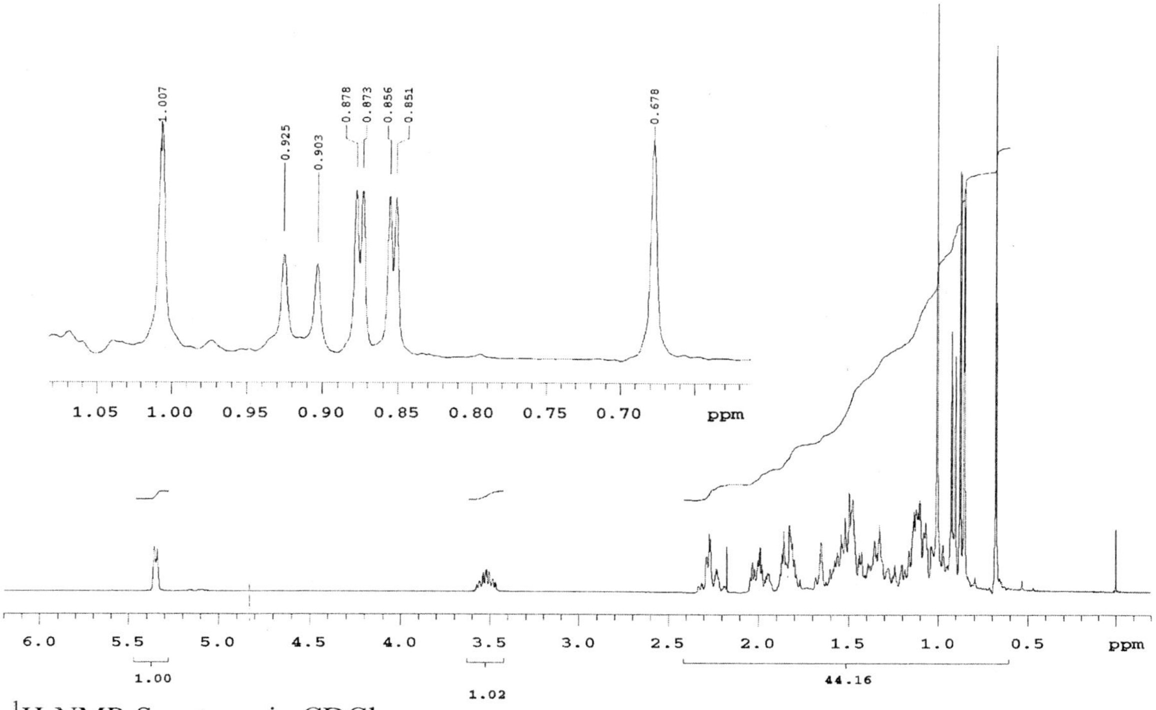

^{1}H-NMR Spectrum in $CDCl_3$

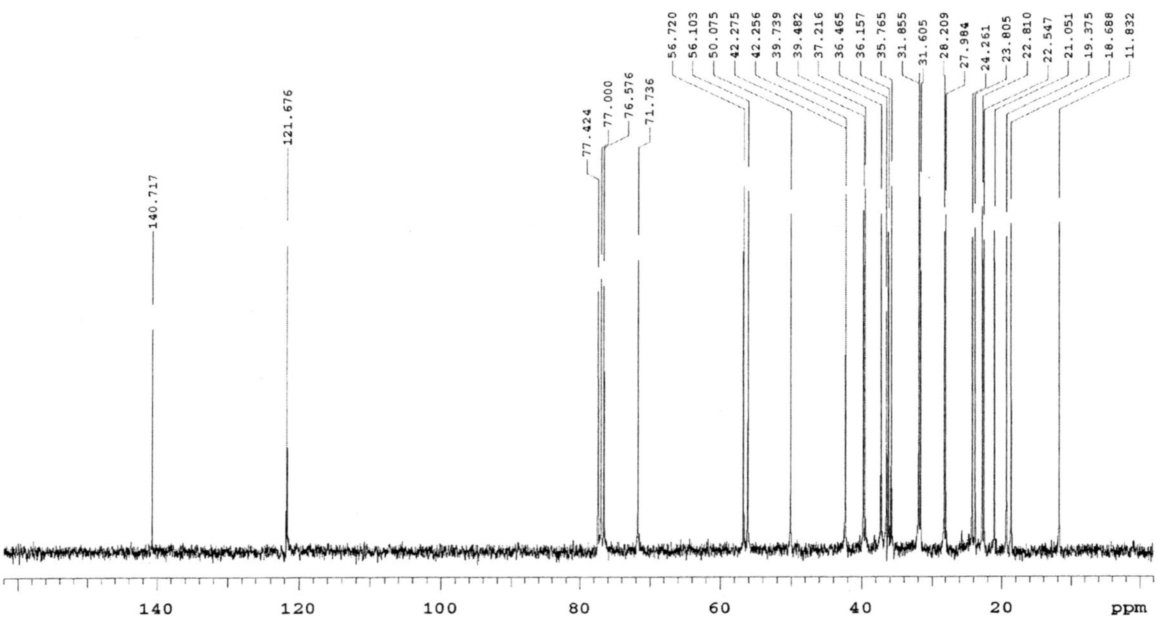

^{13}C-NMR Spectrum in $CDCl_3$

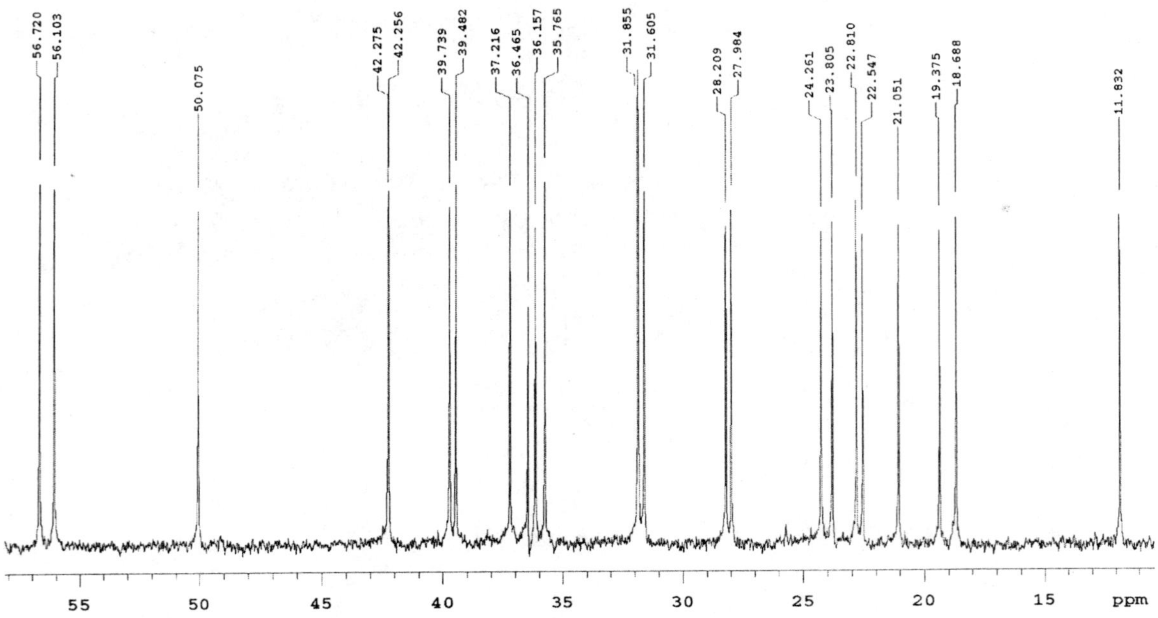

Expansion of ^{13}C-NMR Spectrum in $CDCl_3$

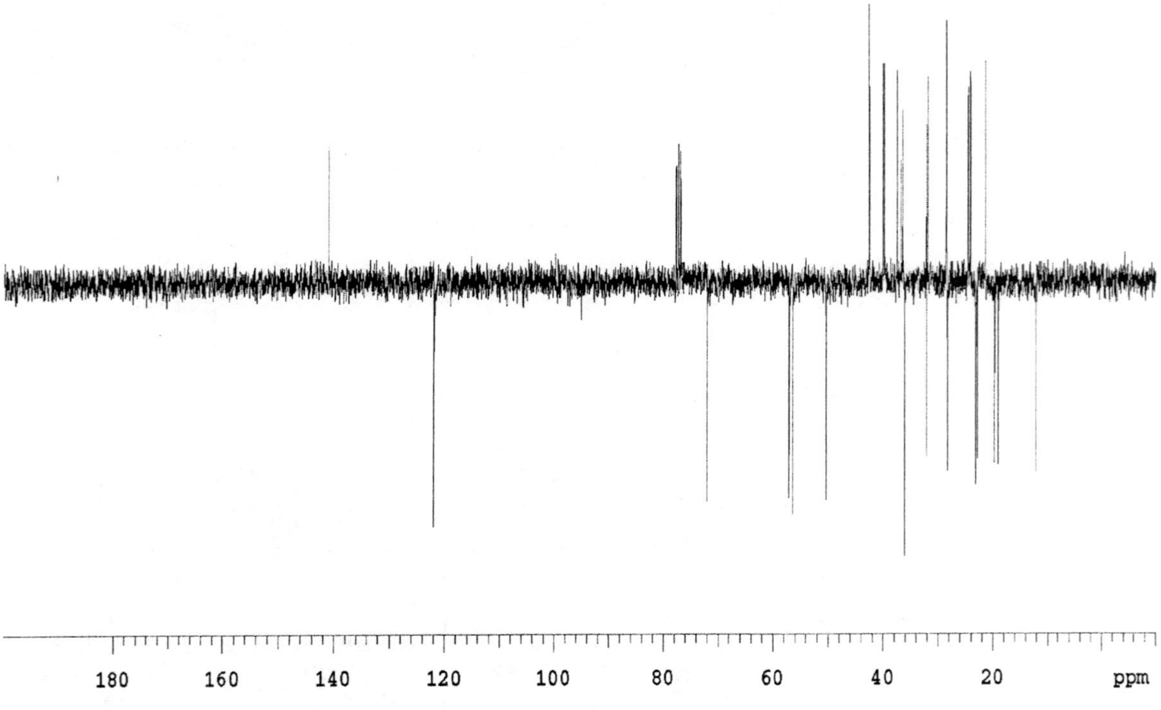

APT ^{13}C-NMR Spectrum in $CDCl_3$

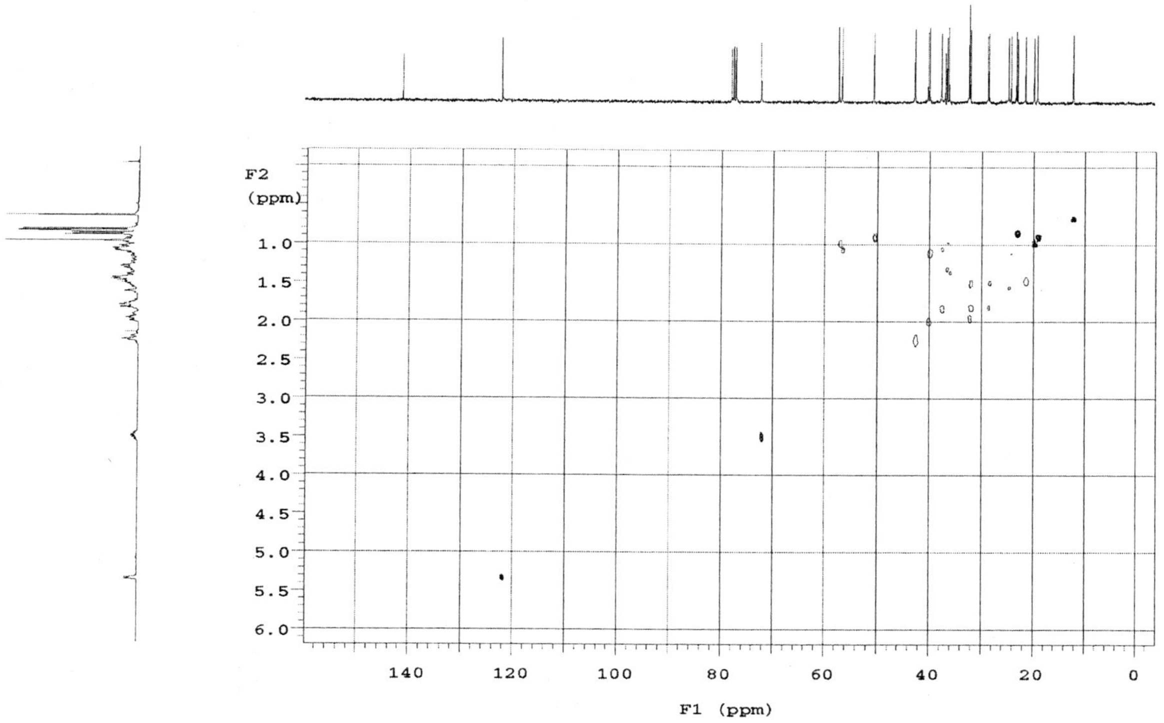

gHSQC Spectrum in $CDCl_3$

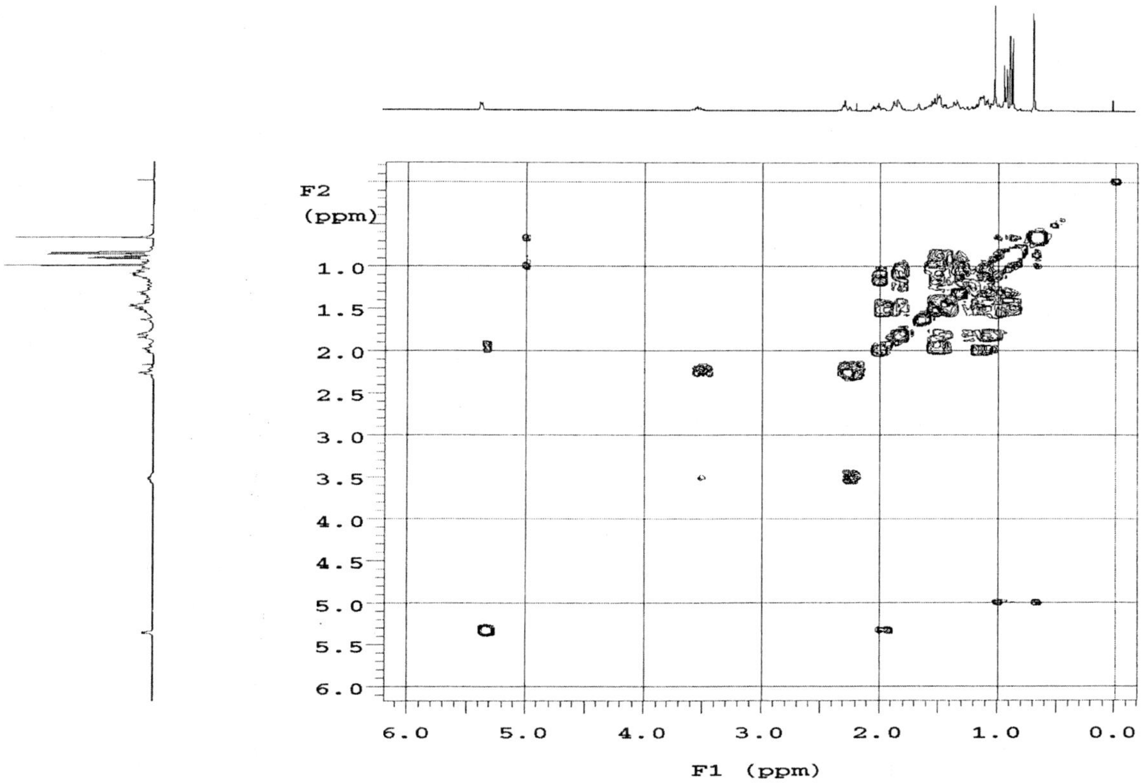

gCOSY Spectrum in $CDCl_3$

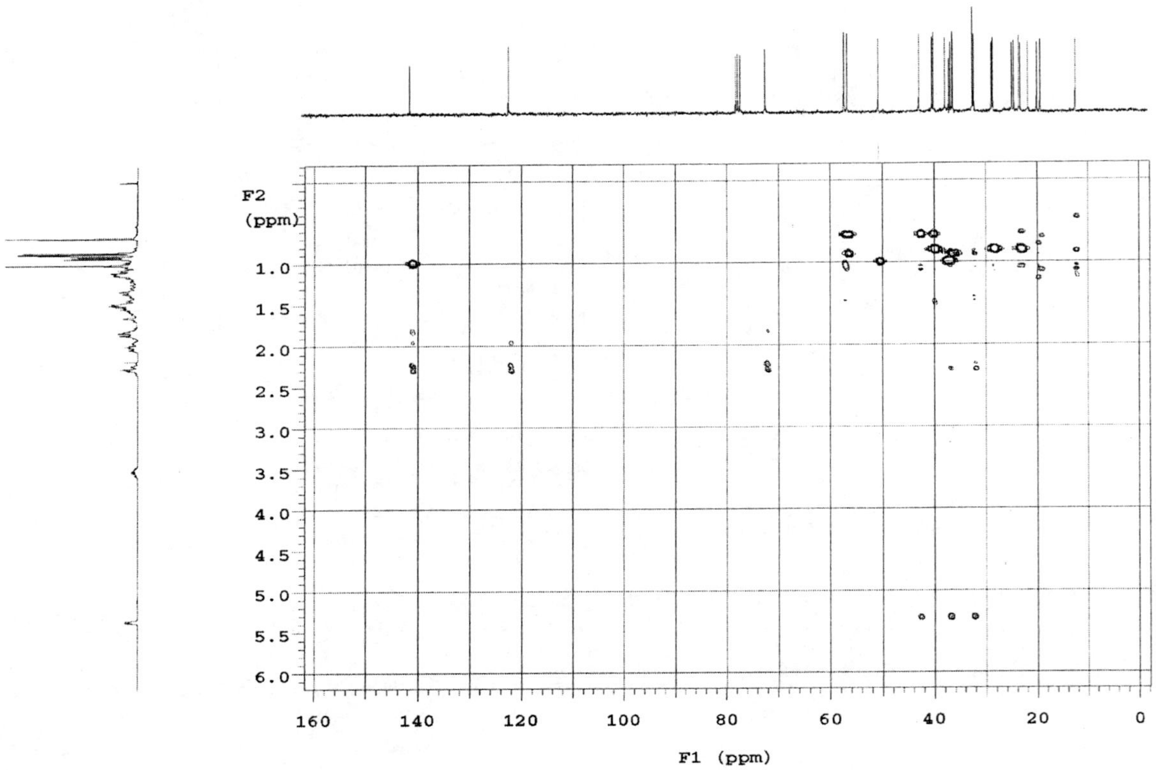

gHMBC Spectrum in $CDCl_3$

Problem 4.25 Given the spectroscopic data below (Proton, Carbon, gCOSY, gHSQC, and gHMBC), provide the structure of the molecule. (HR-MS: 428.3654 amu) This compound was prepared from **4.24** via treatment with acetic anhydride and pyridine.

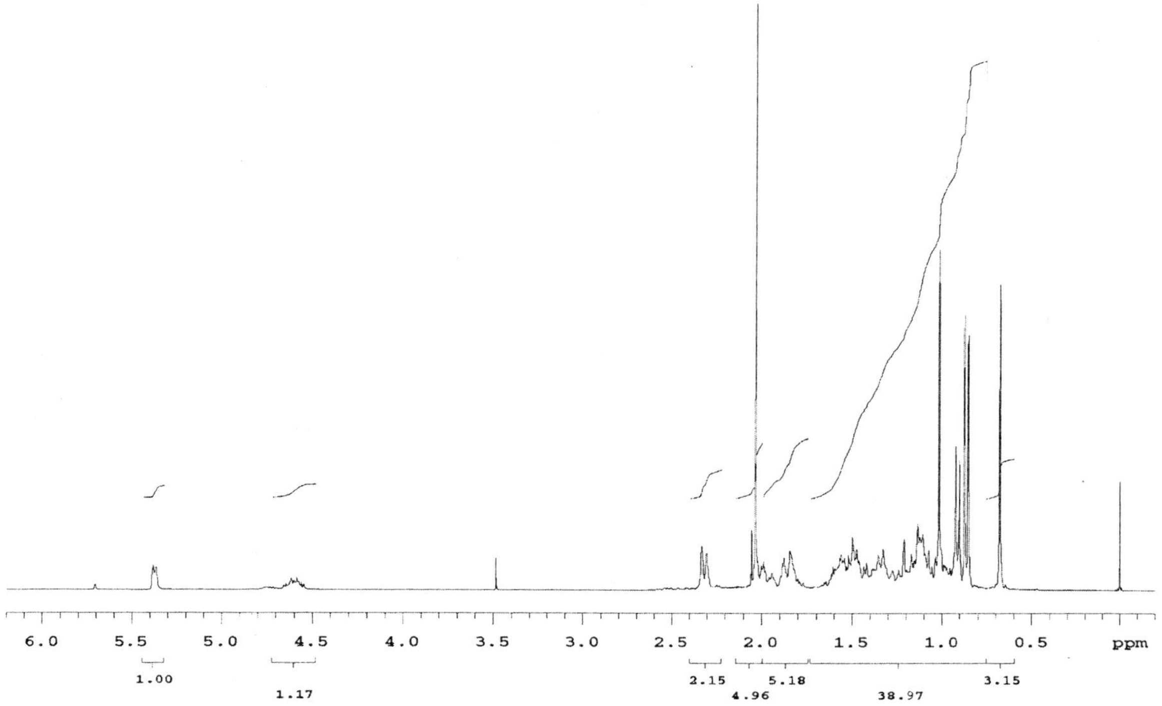

^{1}H-NMR Spectrum in $CDCl_3$

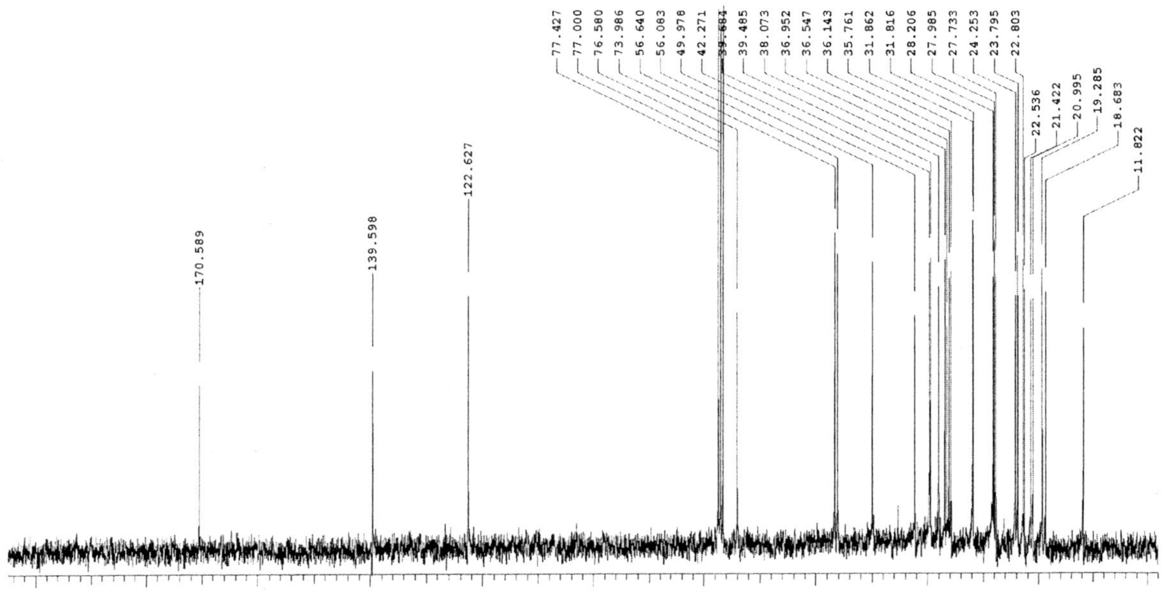

^{13}C-NMR Spectrum in $CDCl_3$

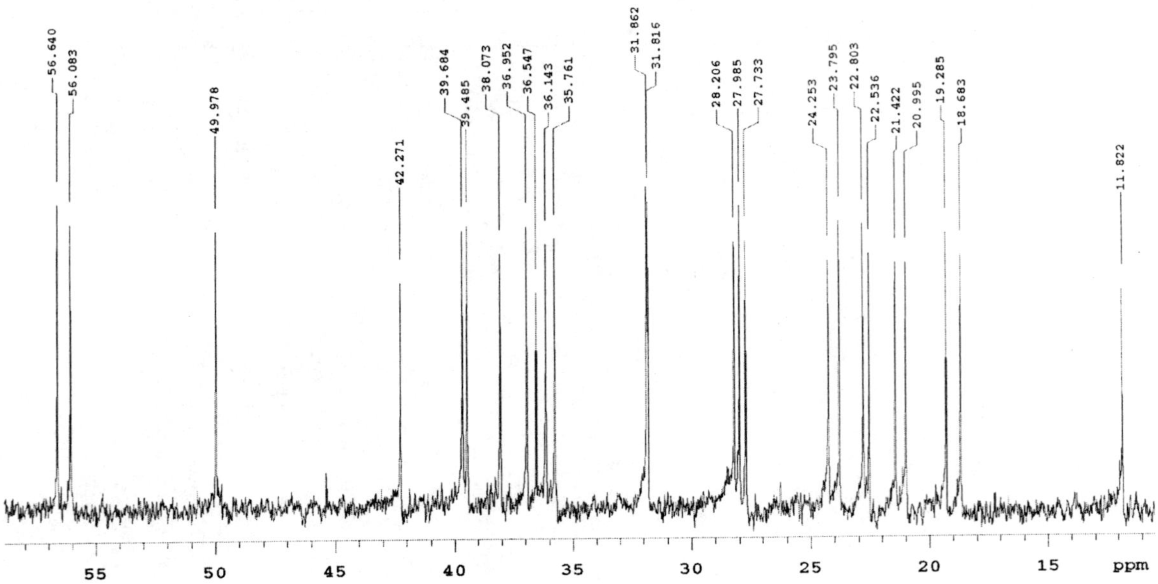

Expansion of ^{13}C-NMR Spectrum in $CDCl_3$

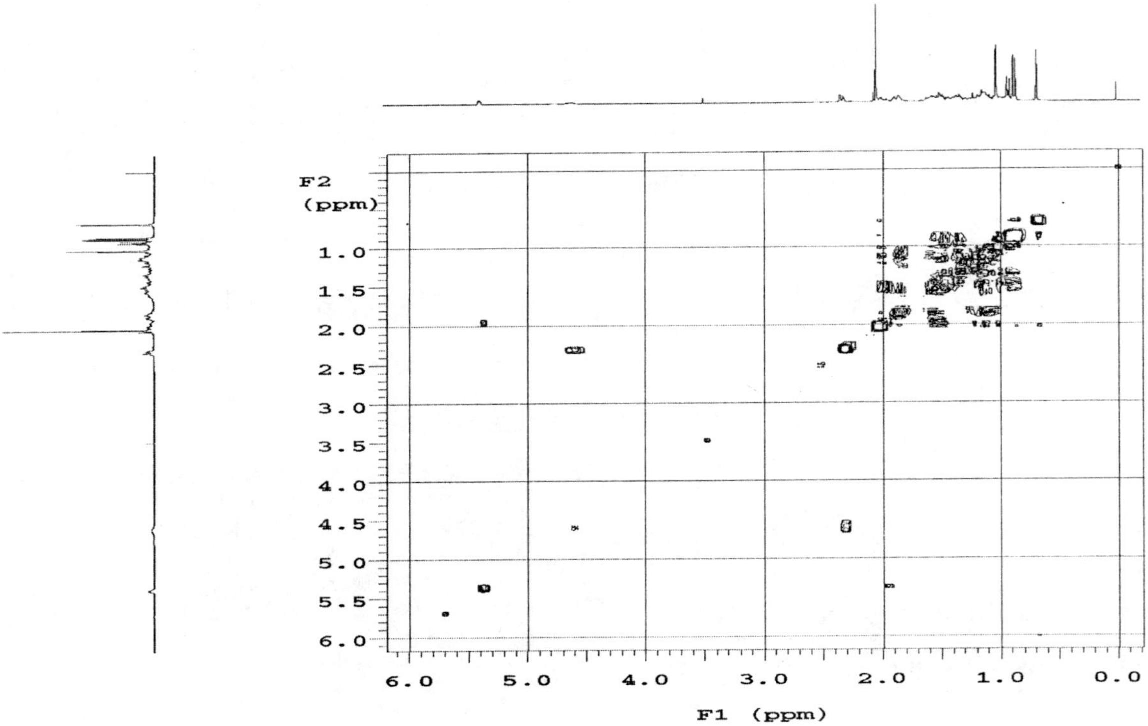

gCOSY Spectrum in $CDCl_3$

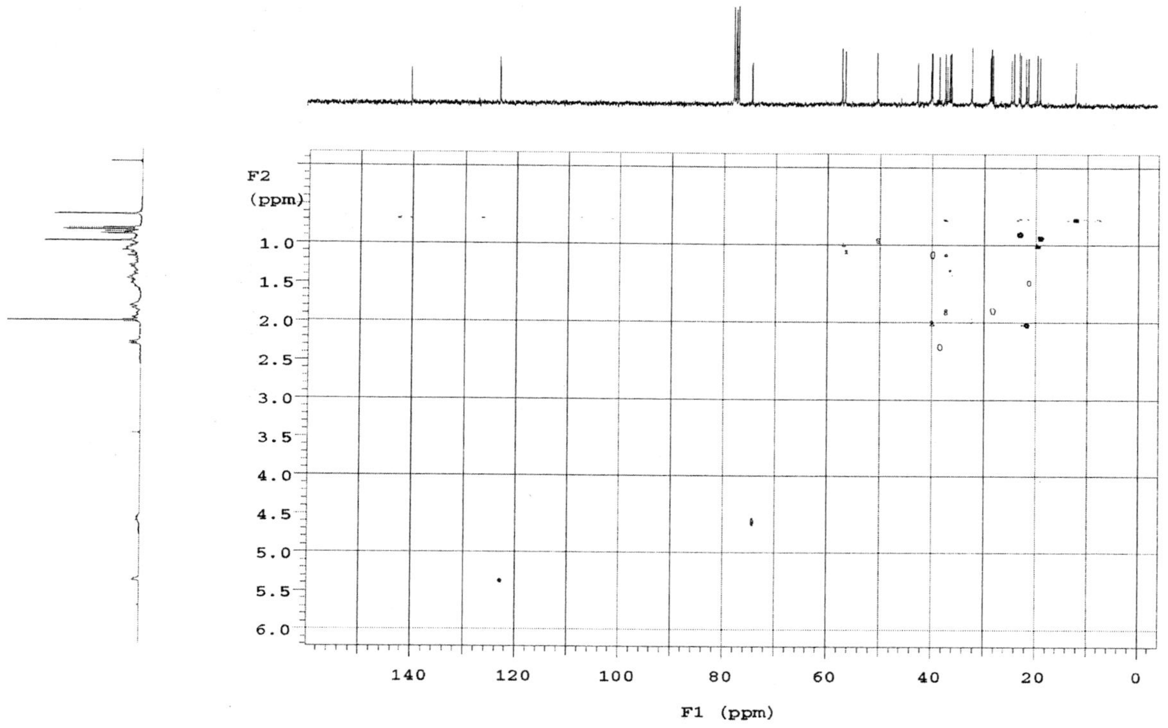

gHSQC Spectrum in $CDCl_3$

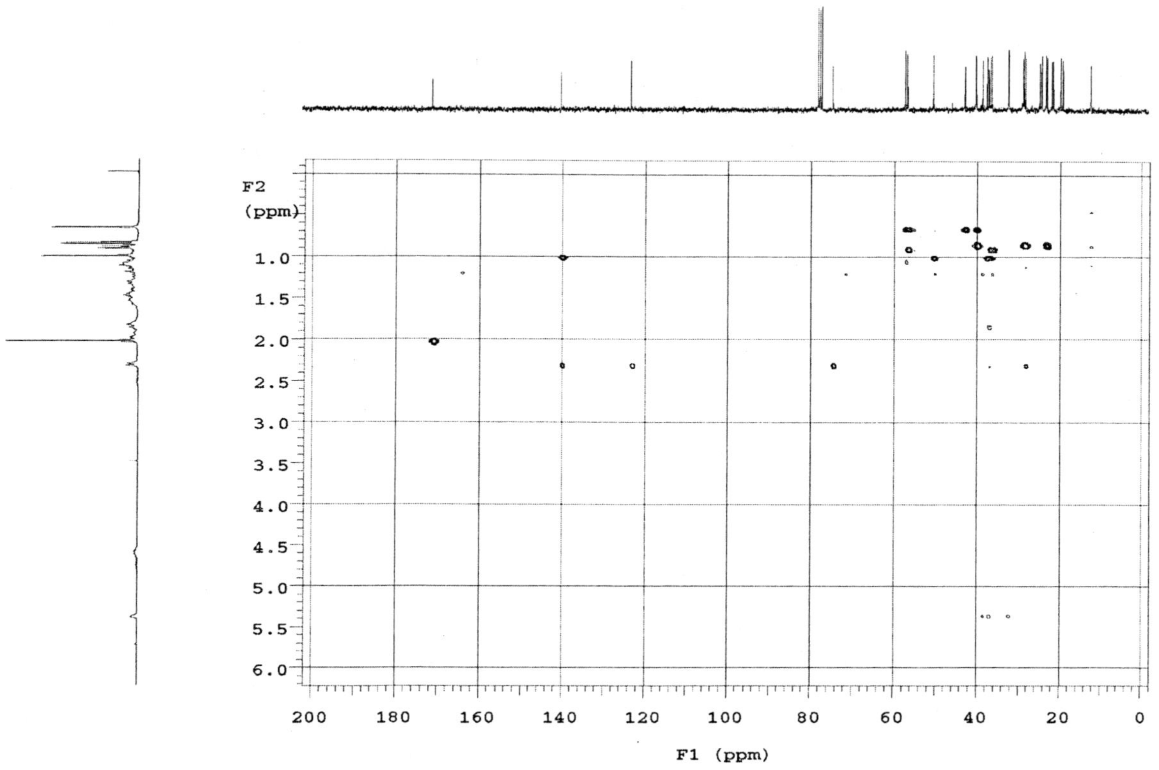

gHMBC Spectrum in $CDCl_3$

Chapter 5
Special Problems

Problem 5.01 Given the data below for 2-bromoadamantane (Proton, Carbon, gHSQC, and gCOSY), make a complete assignment of the spectroscopic signals.

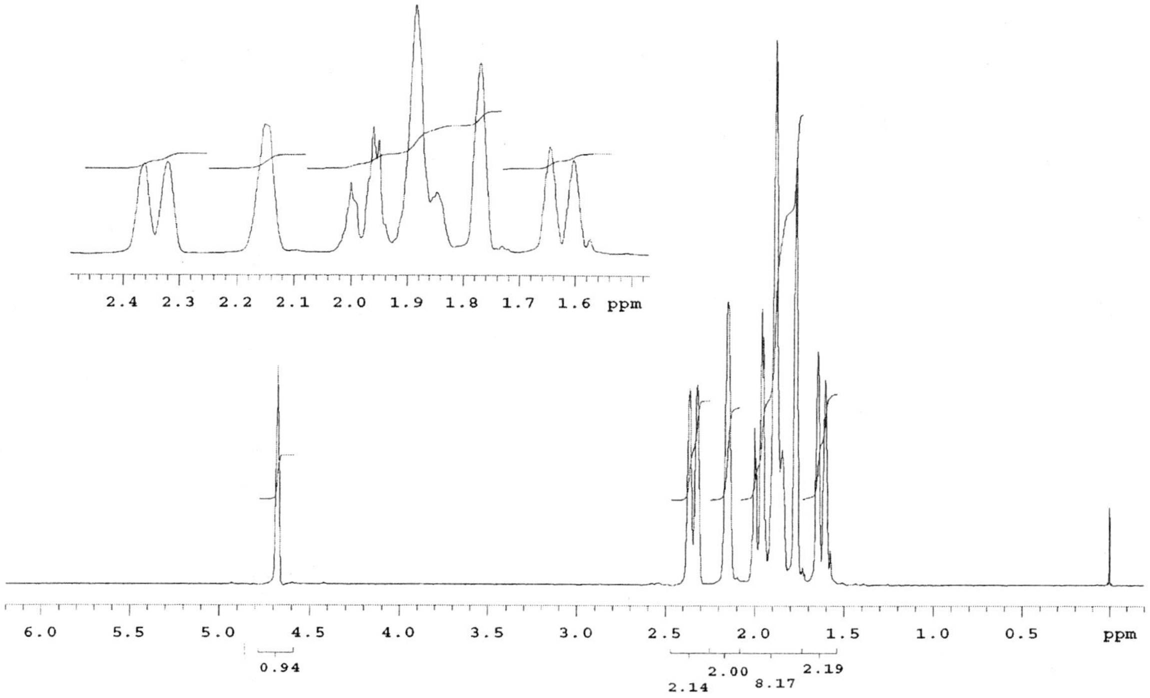

^{1}H-NMR Spectrum in $CDCl_3$

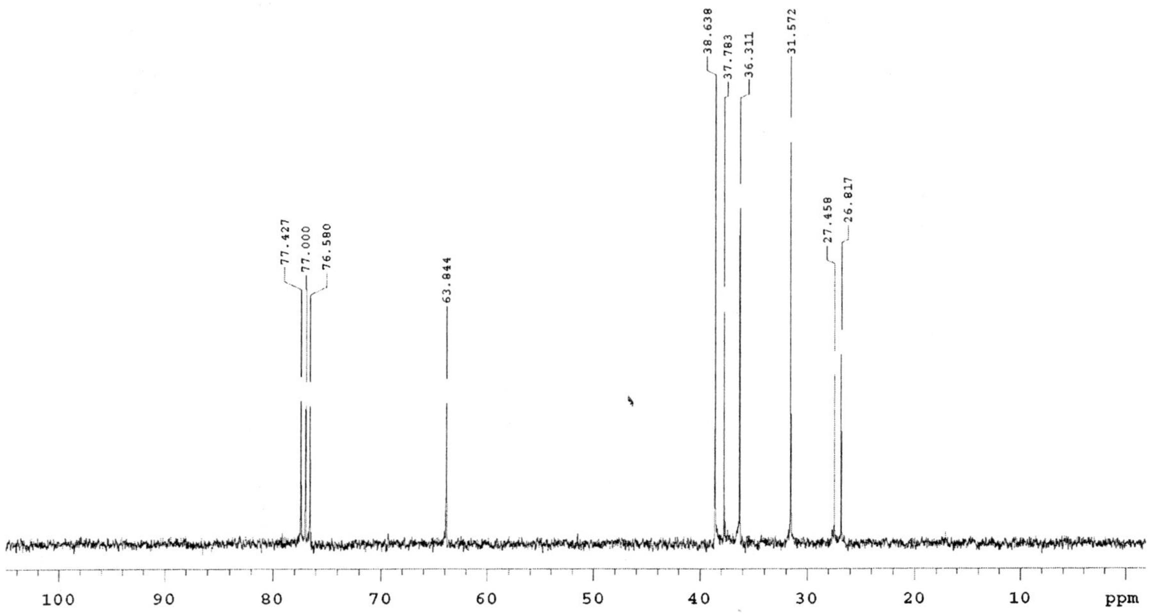

^{13}C-NMR Spectrum in $CDCl_3$

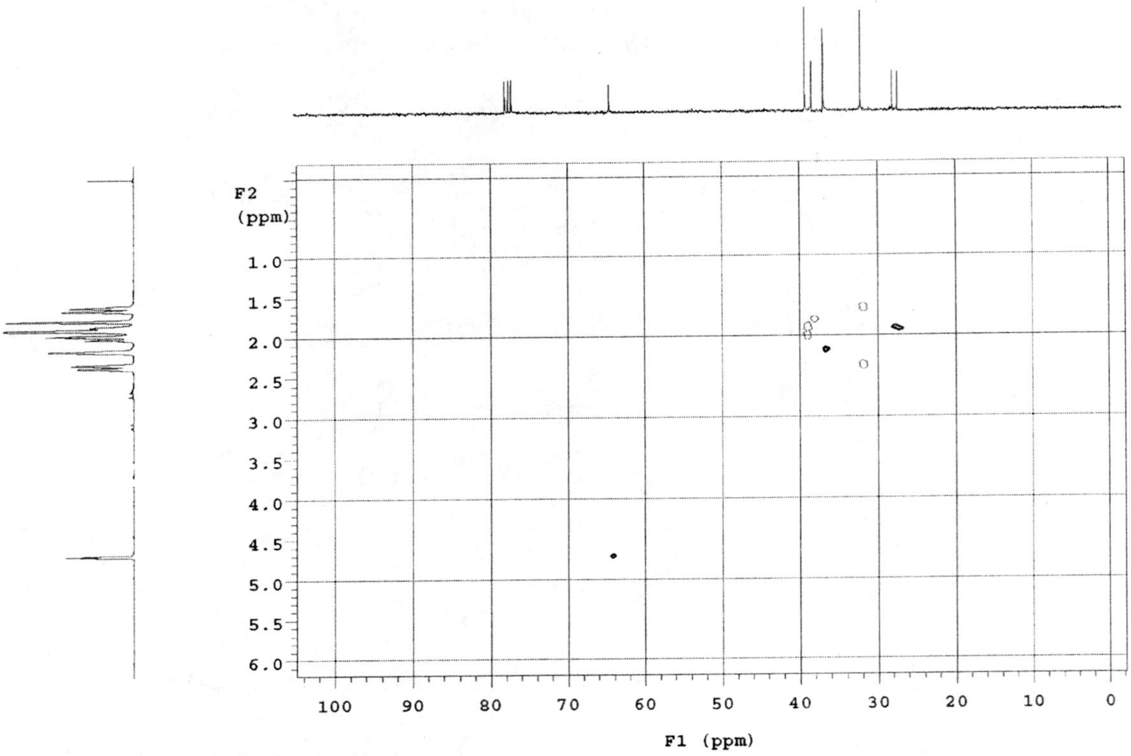

gHSQC Spectrum in $CDCl_3$

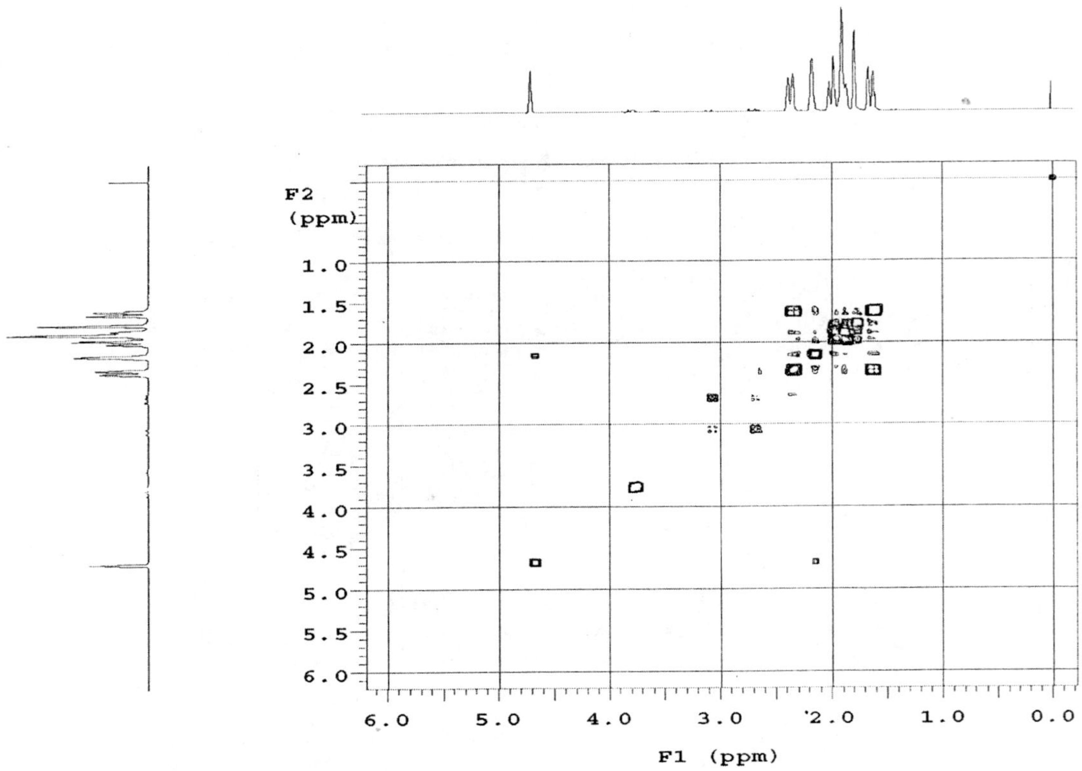

gCOSY Spectrum in $CDCl_3$

Problem 5.02 Given the data below for ethyl 4-(2-(methoxycarbonyl)ethyl)-5-formyl-3-methyl-1H-pyrrole-2-carboxylate [Structure shown on page 353.] (Proton, Carbon, gHSQC, gCOSY, and gHMBC), make a complete assignment of the spectroscopic signals. Compound **5.02** was obtained by the oxidization of **3.37** with ceric ammonium nitrate (CAN) in aqueous acetic acid.

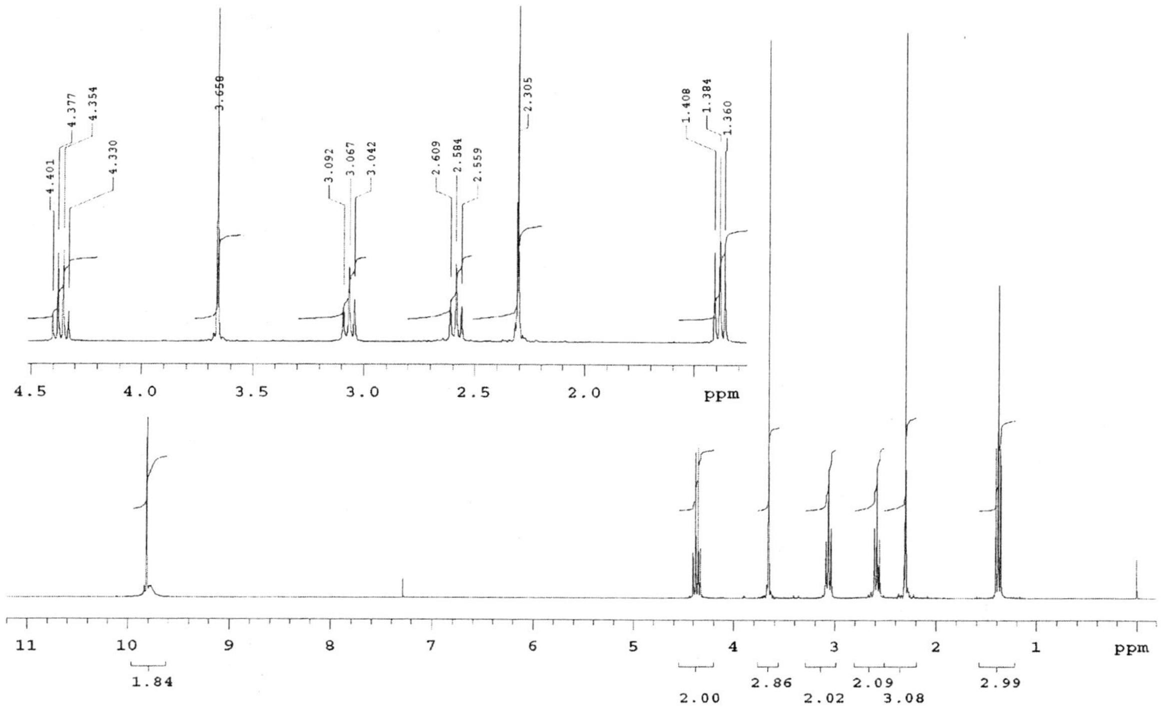

^{1}H-NMR Spectrum in $CDCl_3$

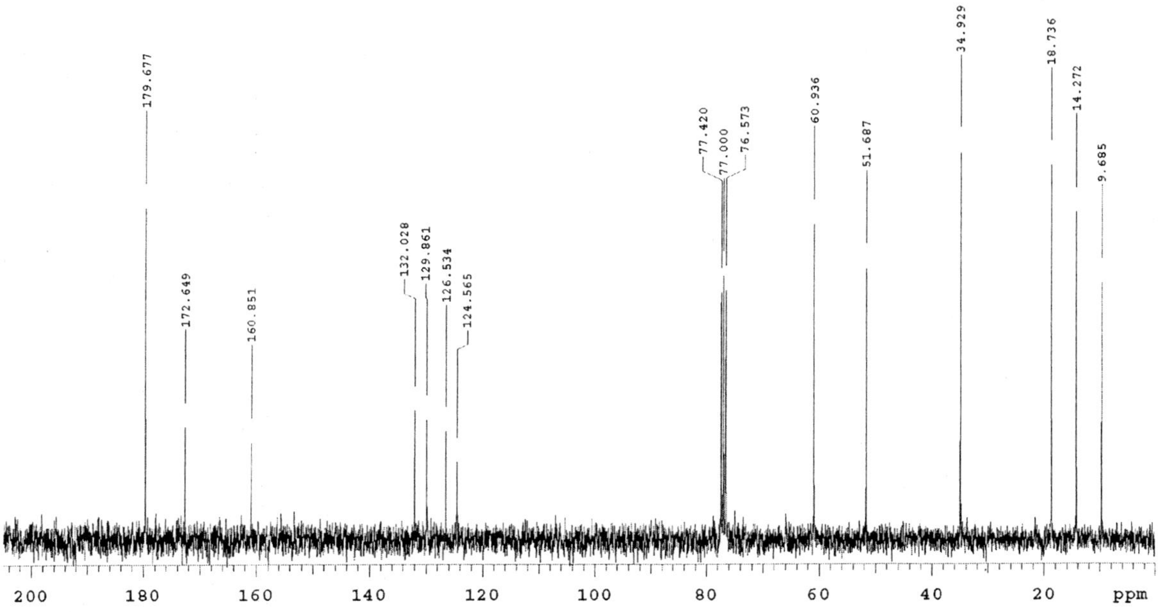

^{13}C-NMR Spectrum in $CDCl_3$

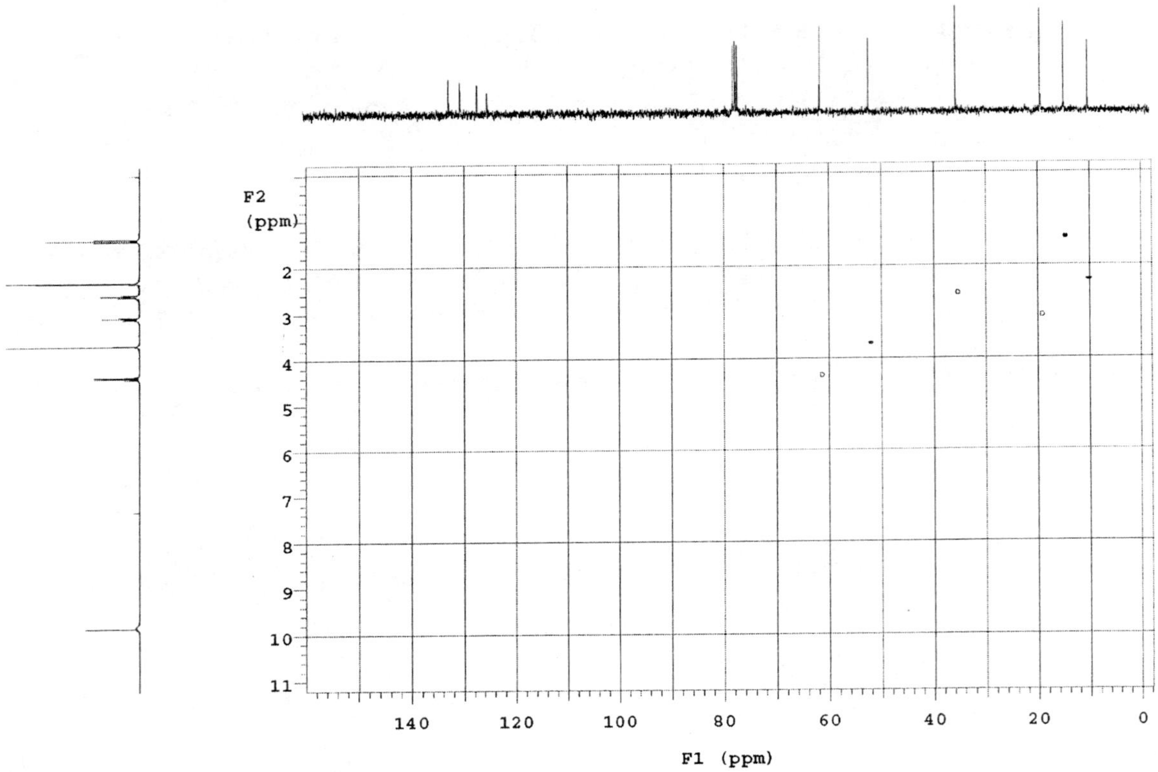

gHSQC Spectrum in $CDCl_3$

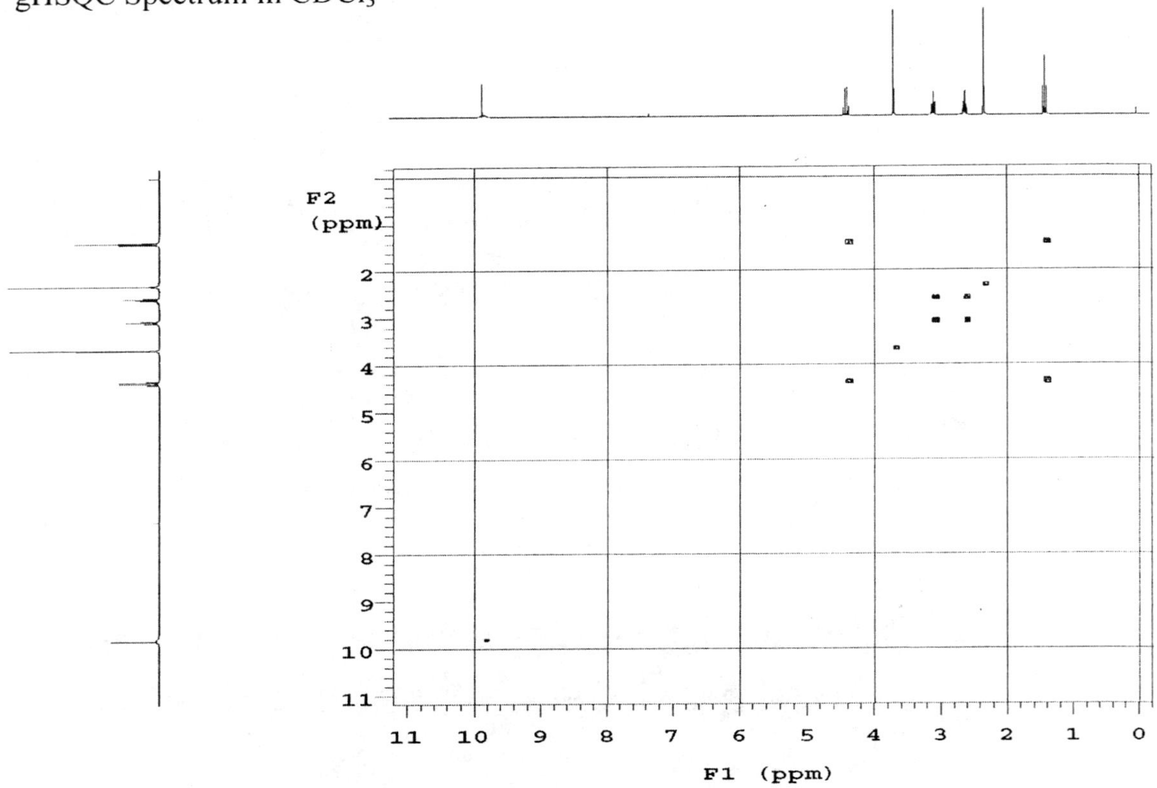

gCOSY Spectrum in $CDCl_3$

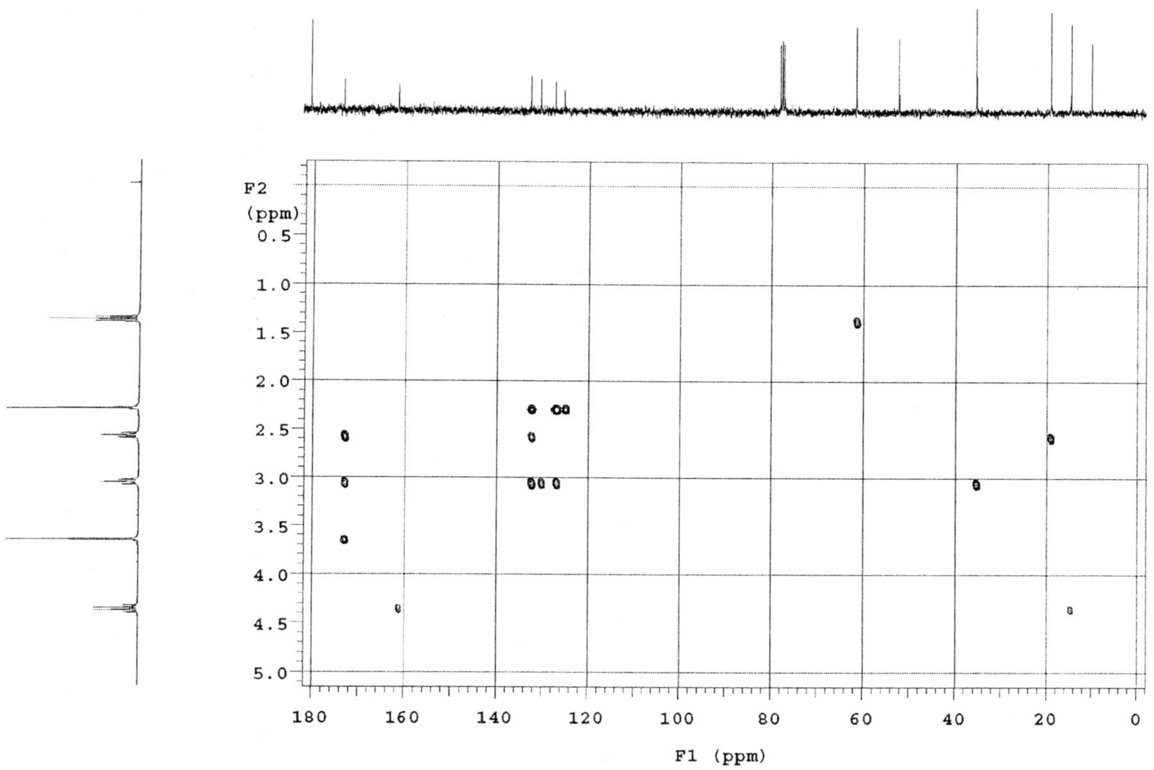

gHMBC Spectrum in $CDCl_3$

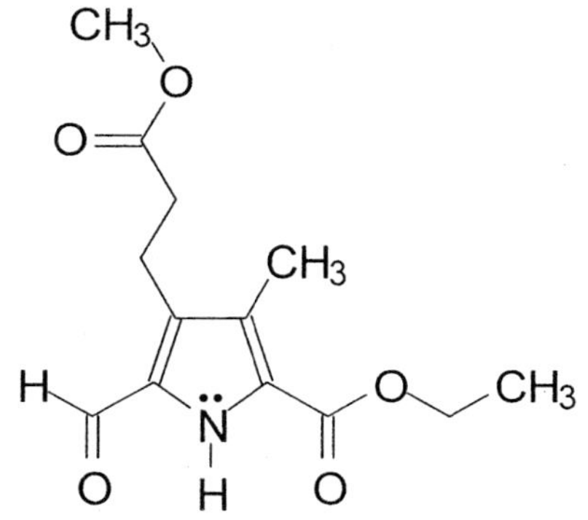

Ethyl 4-(2-(methoxycarbonyl)ethyl)-5-formyl-3-methyl-1H-pyrrole-2-carboxylate

Problem 5.03 Given the data below (Proton, Carbon, gHSQC, gCOSY, and gHMBC), provide the structure of the molecule. ($C_{10}H_{17}NO$)

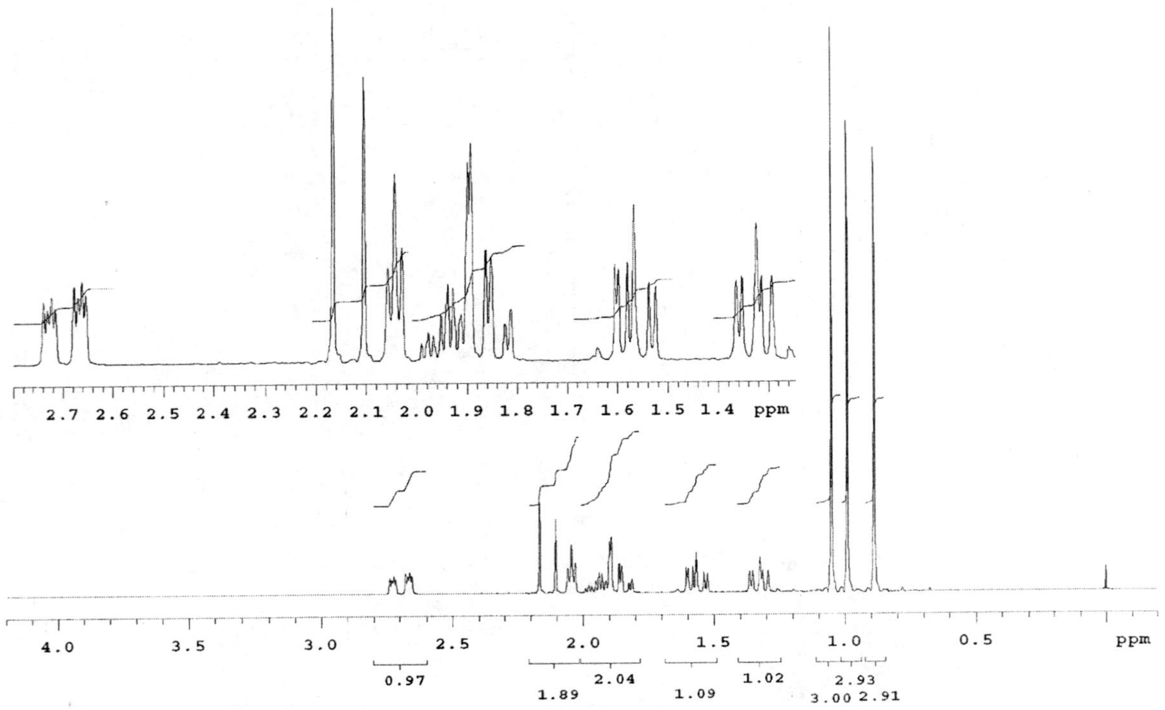

^{1}H-NMR Spectrum in $CDCl_3$

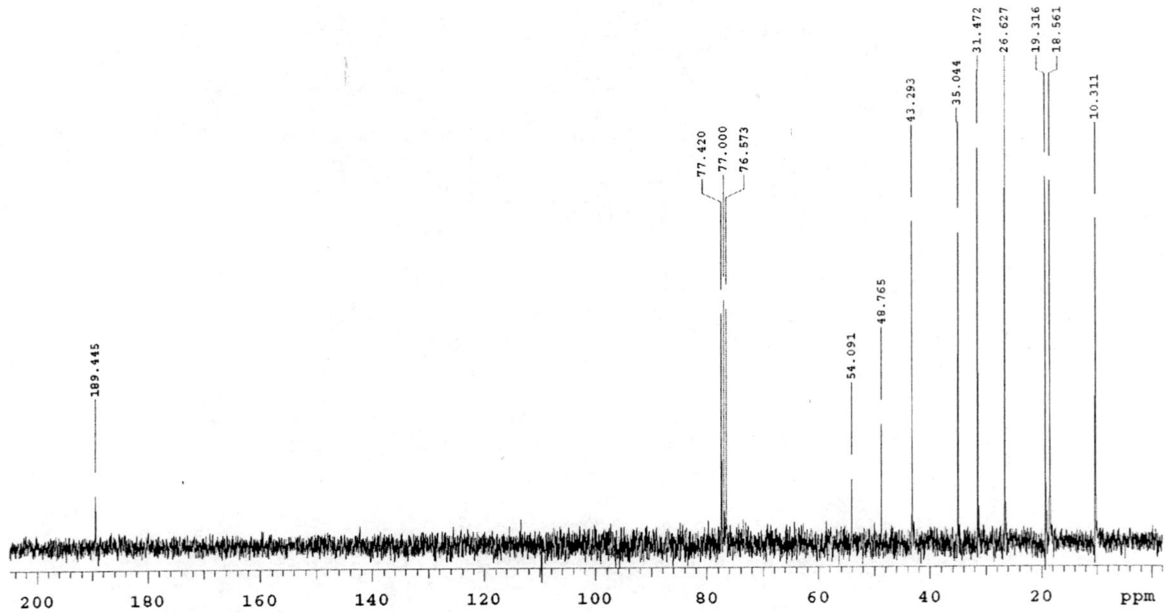

^{13}C-NMR Spectrum in $CDCl_3$

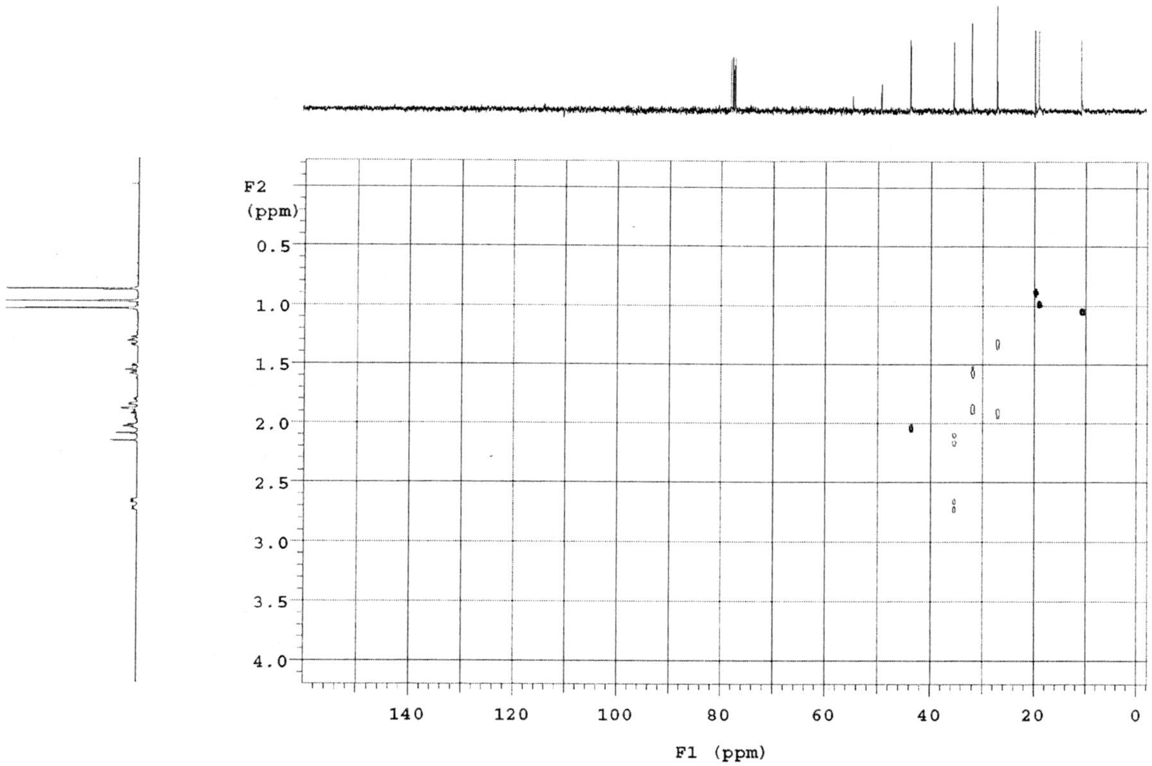

gHSQC Spectrum in $CDCl_3$

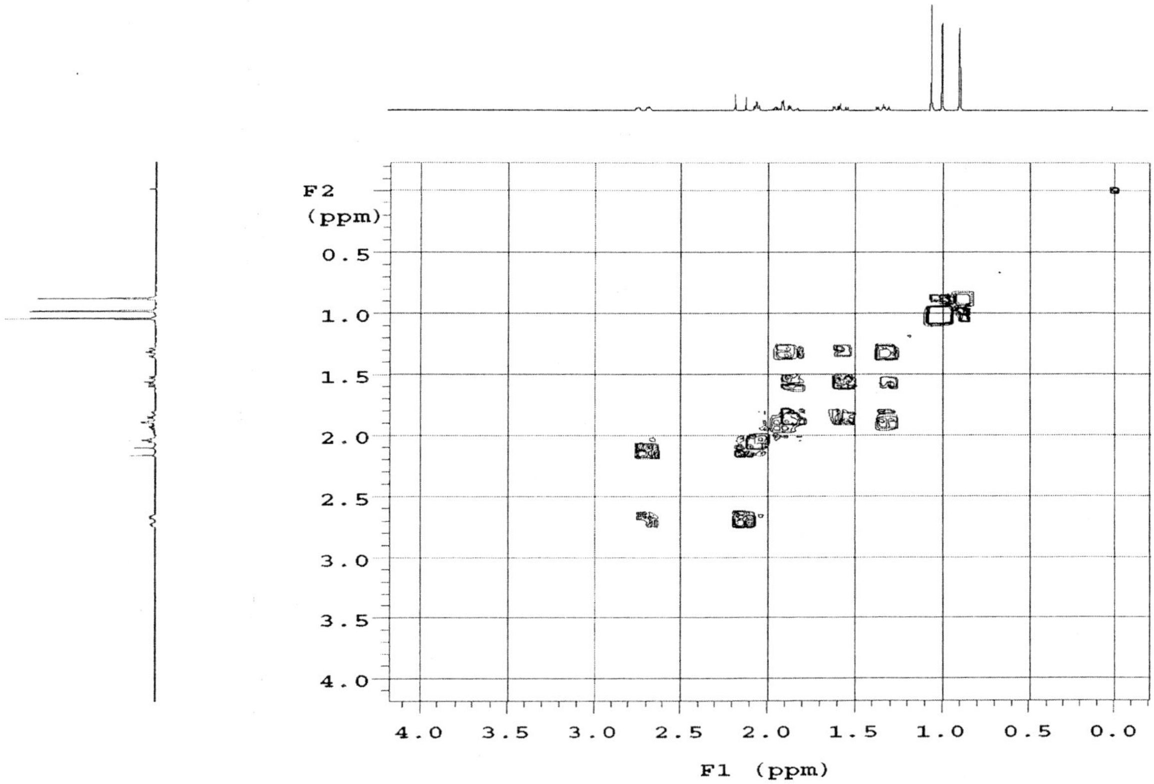

gCOSY Spectrum in $CDCl_3$

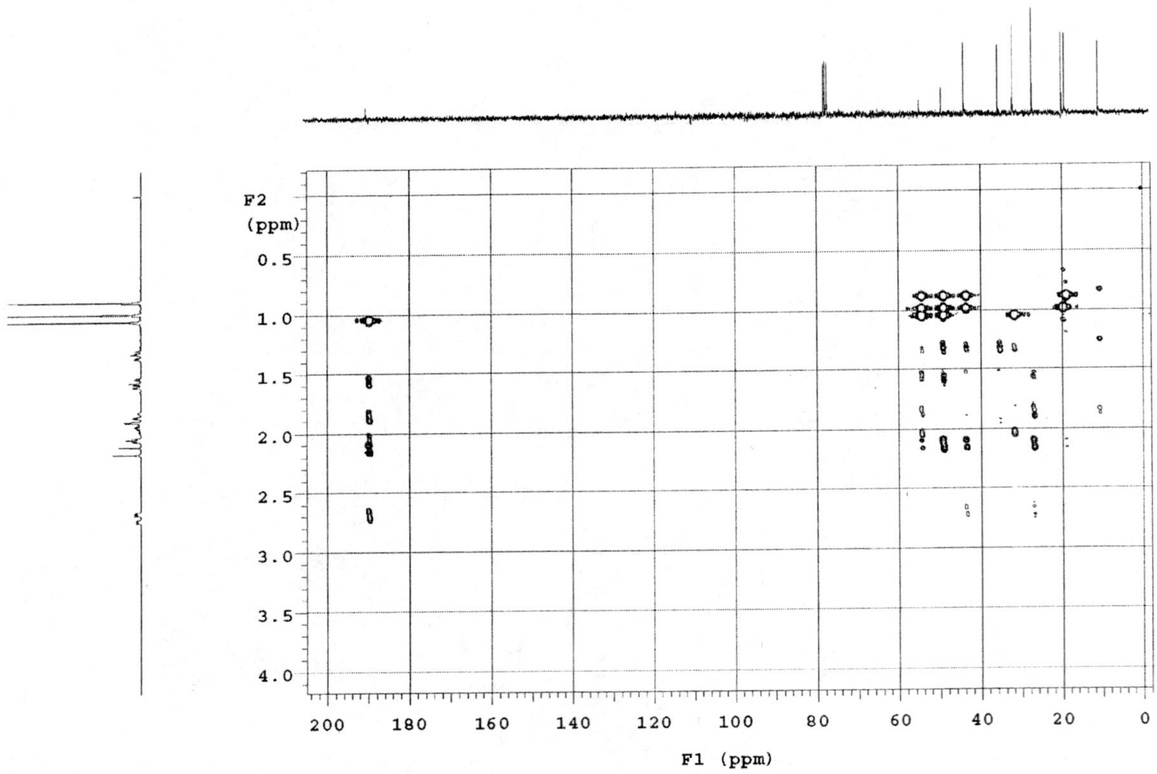

gHMBC Spectrum in $CDCl_3$

Problem 5.04 Given the data below for 3-ethyl-4-methyl-2-tosyl-1H-pyrrole [see page 359 for the structure] (Proton, Carbon, gHSQC, gCOSY, and gHMBC), make a complete assignment of the spectroscopic signals.

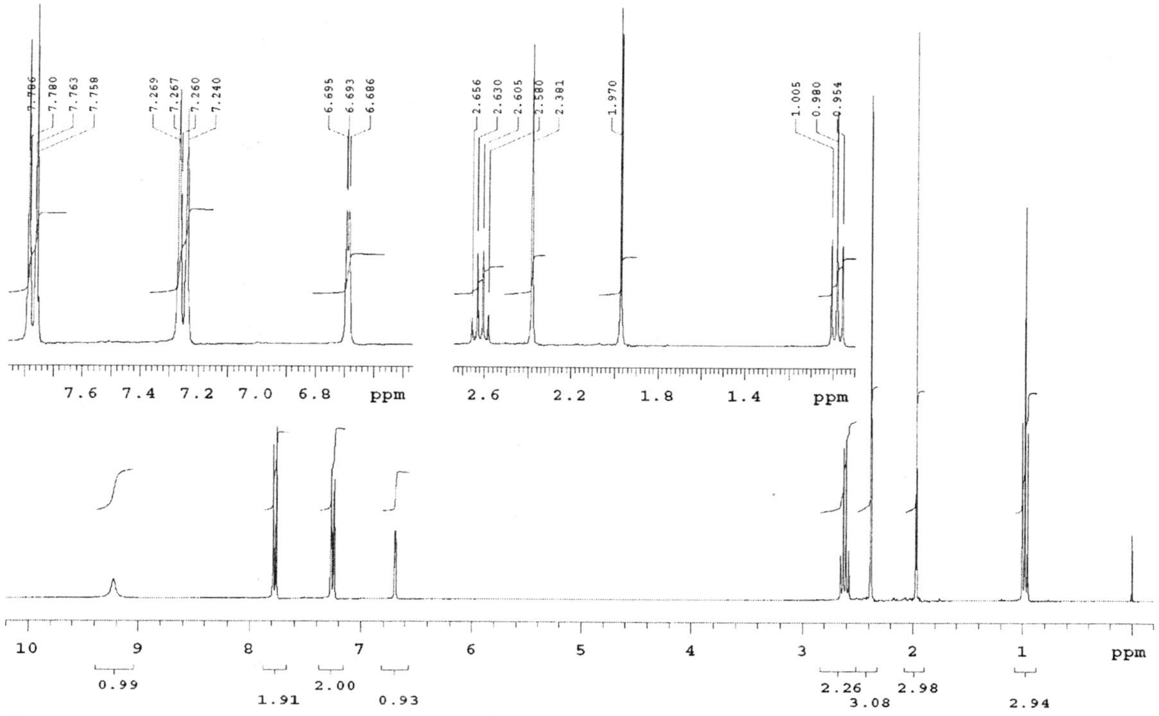

^{1}H-NMR Spectrum in $CDCl_3$

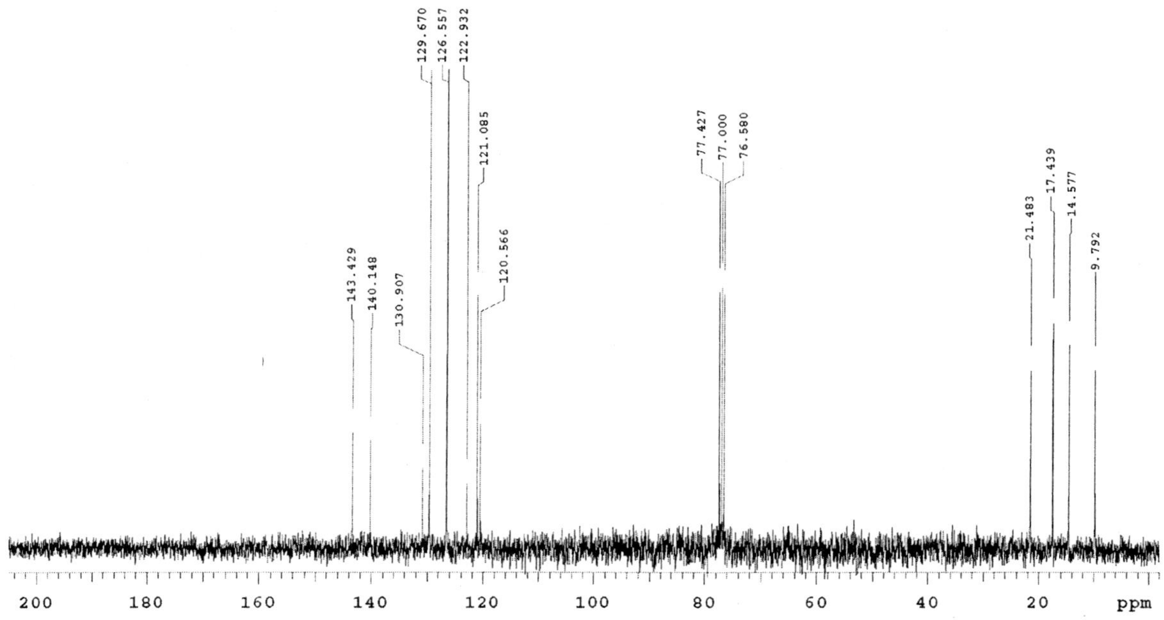

^{13}C-NMR Spectrum in $CDCl_3$

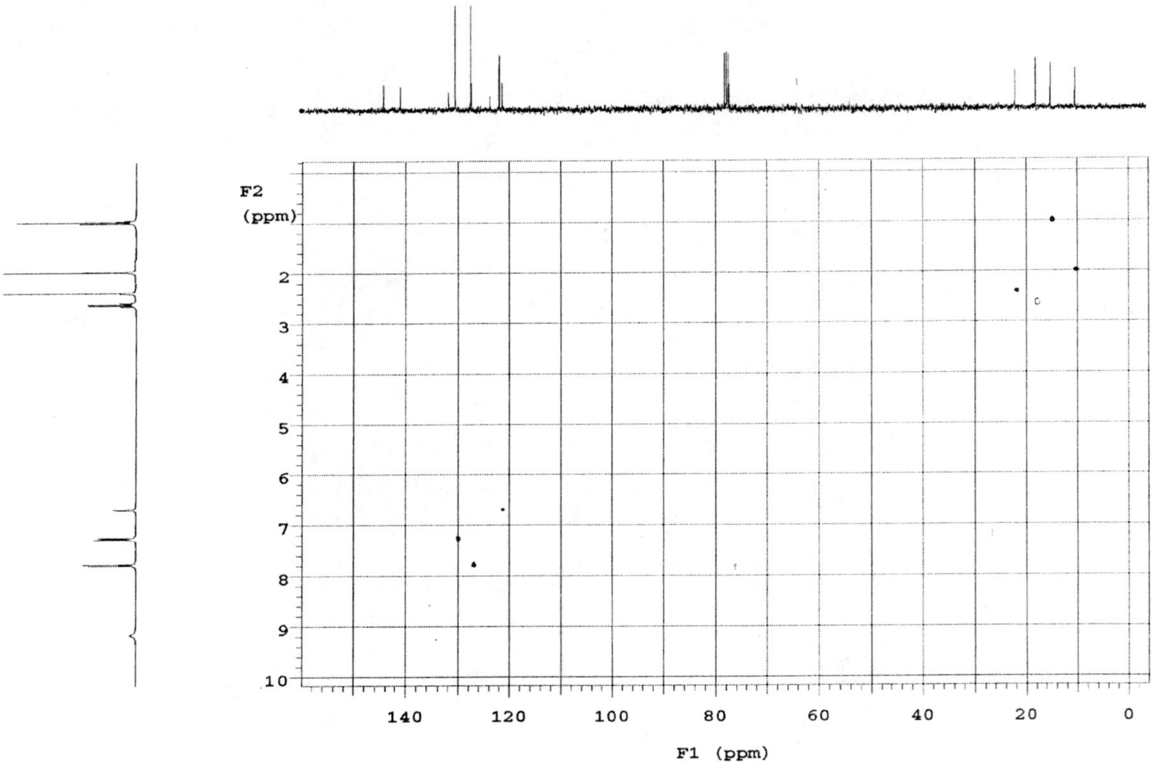

gHSQC Spectrum in $CDCl_3$

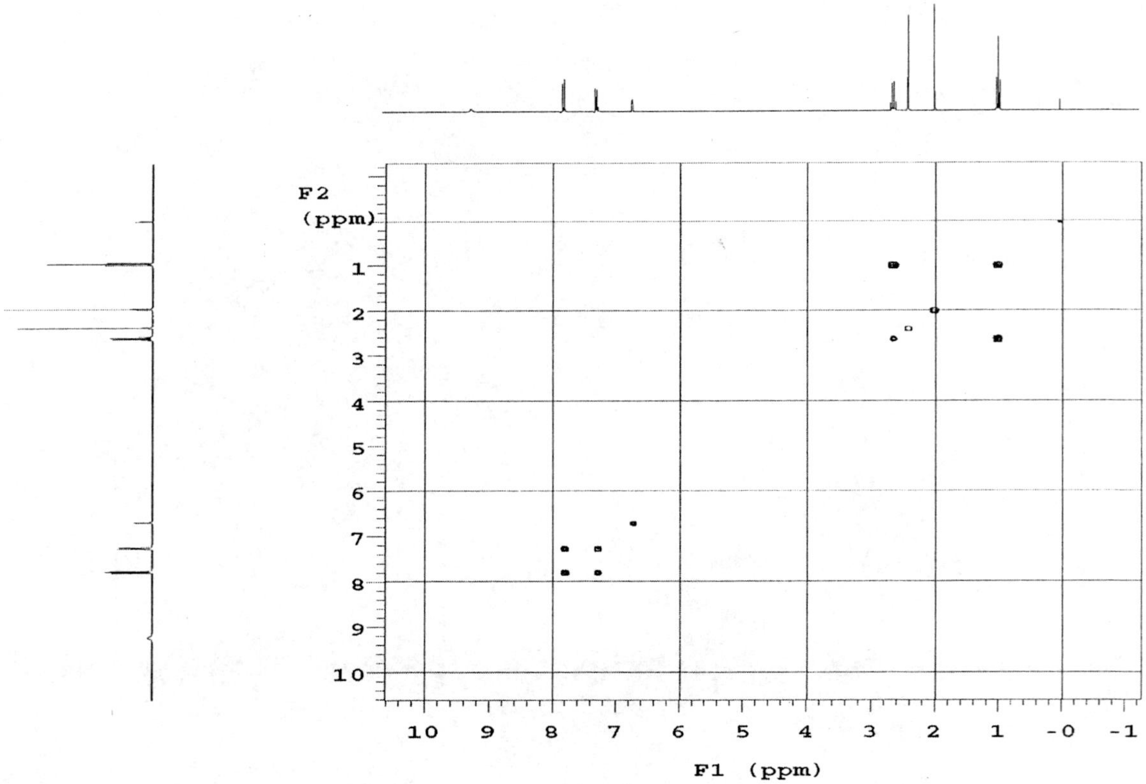

gCOSY Spectrum in $CDCl_3$

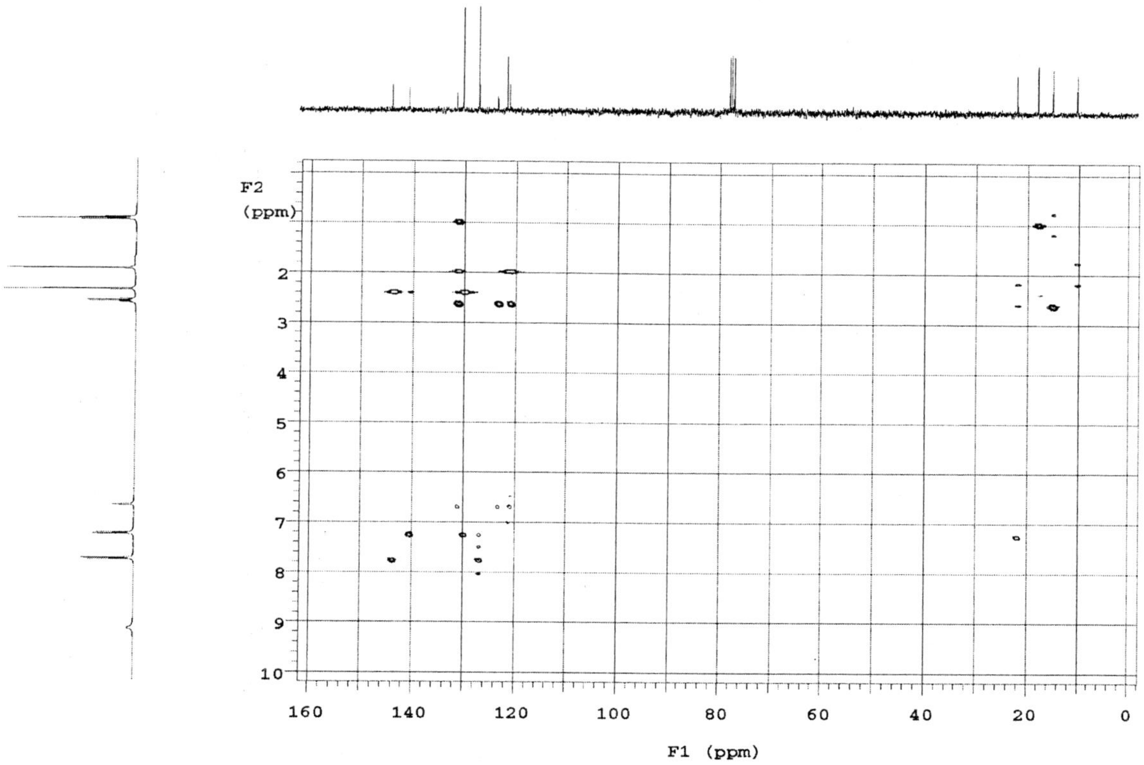

gHMBC Spectrum in $CDCl_3$

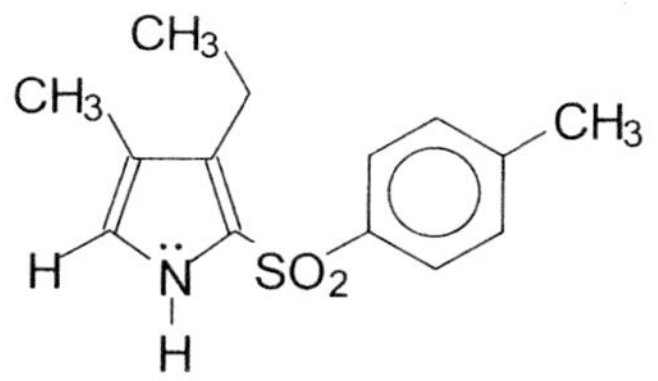

3-Ethyl-4-methyl-2-tosyl-1H-pyrrole

Problem 5.05A Given the data below (Proton, Carbon, gHSQC, gCOSY, and gHMBC), provide the structure of the molecule. Ethyl acetoacetate was α-oximated (using potassium nitrite and acetic acid) and the intermediate oxime was reacted in hot acetic acid with pentane-2,4-dione in the presence of zinc dust to afford **5.05A**.

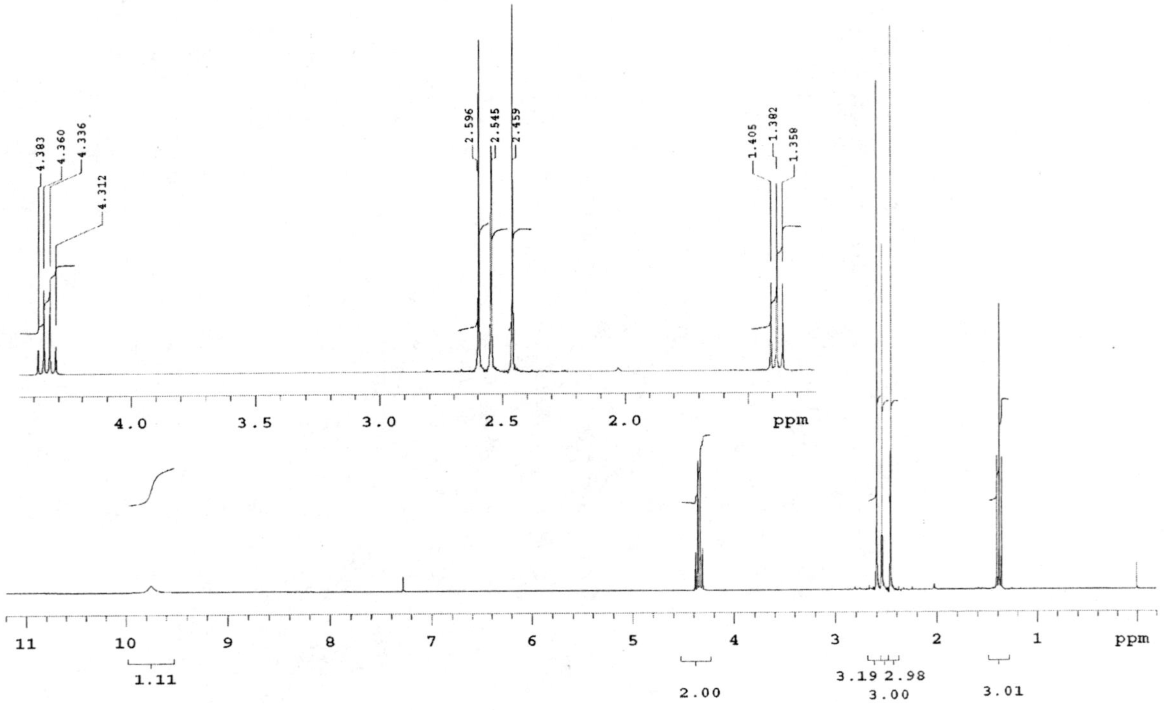

^{1}H-NMR Spectrum in $CDCl_3$

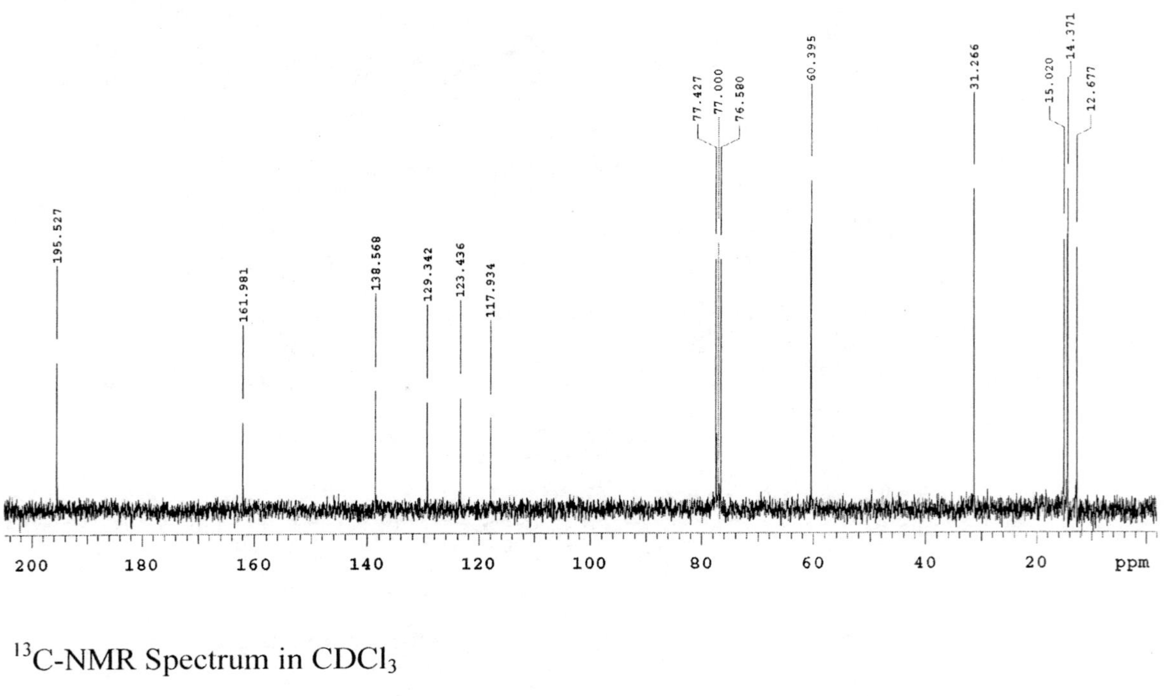

^{13}C-NMR Spectrum in $CDCl_3$

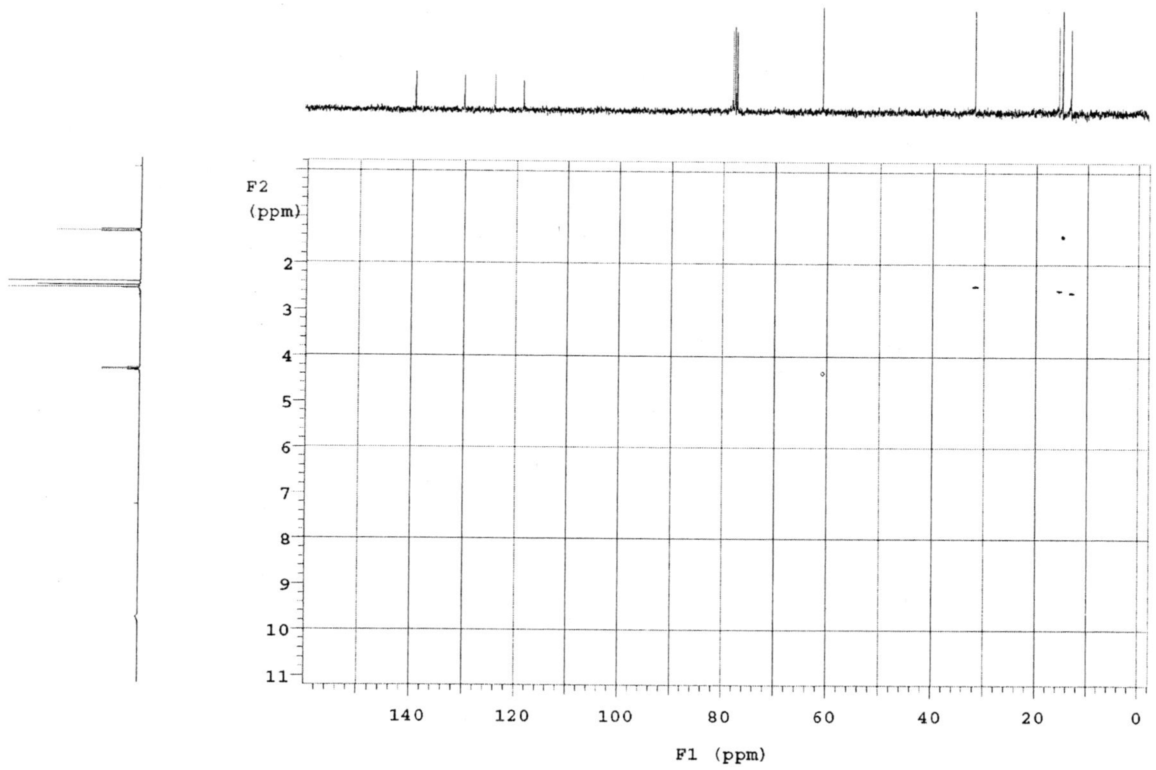

gHSQC Spectrum in $CDCl_3$

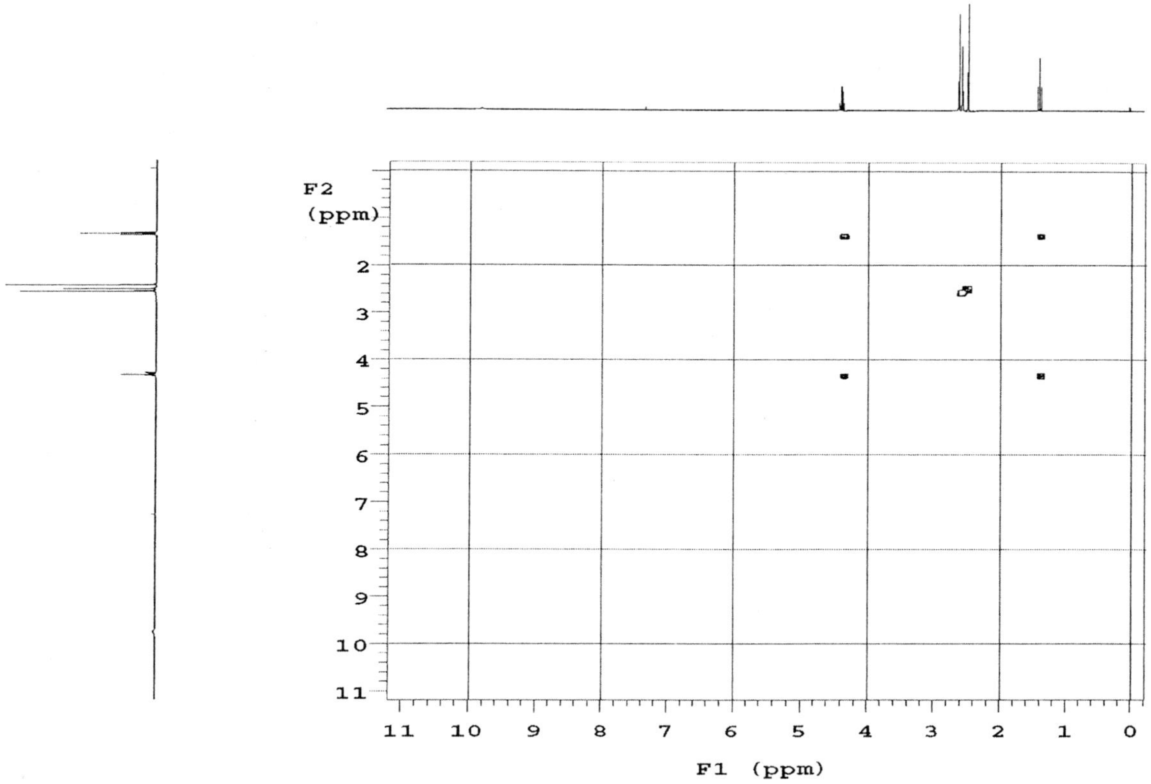

gCOSY Spectrum in $CDCl_3$

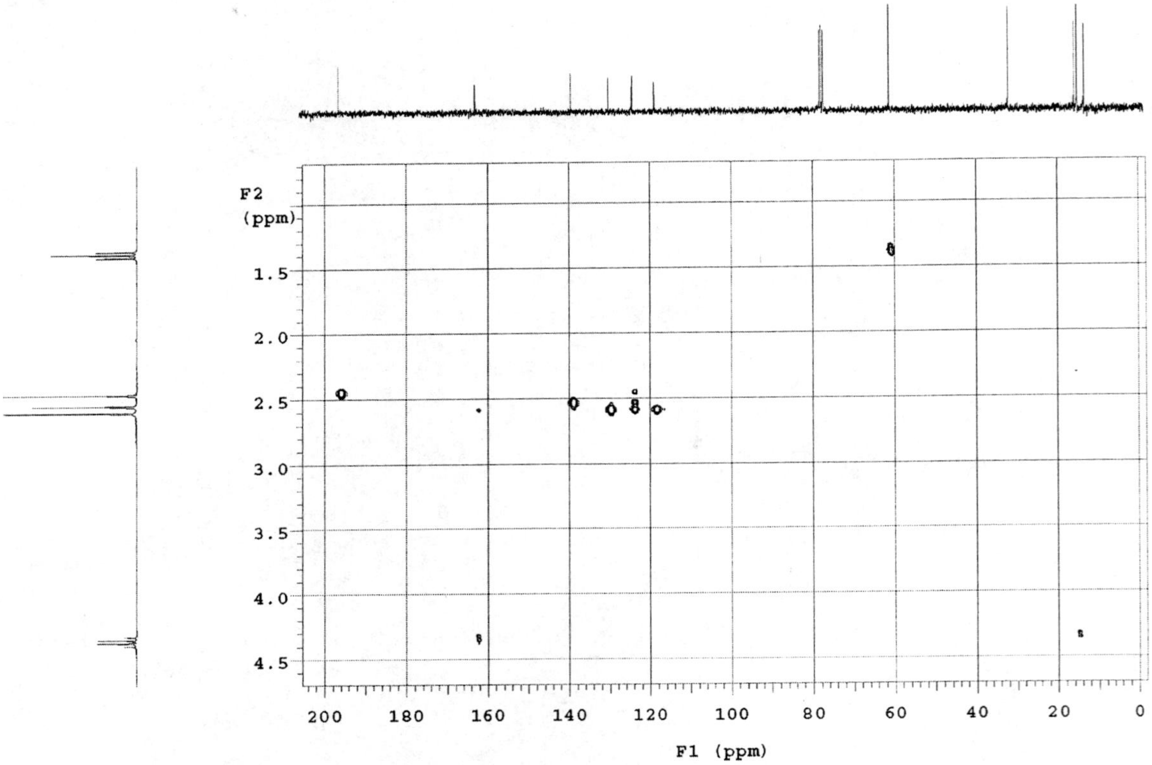

gHMBC Spectrum in $CDCl_3$

Problem 5.05B Given the data below (Proton, Carbon, gHSQC, gCOSY, and gHMBC), provide the structure of the molecule. Compound **5.05A** was heated in ethanol in the presence of granulated Zn to reduce it to **5.05B**.

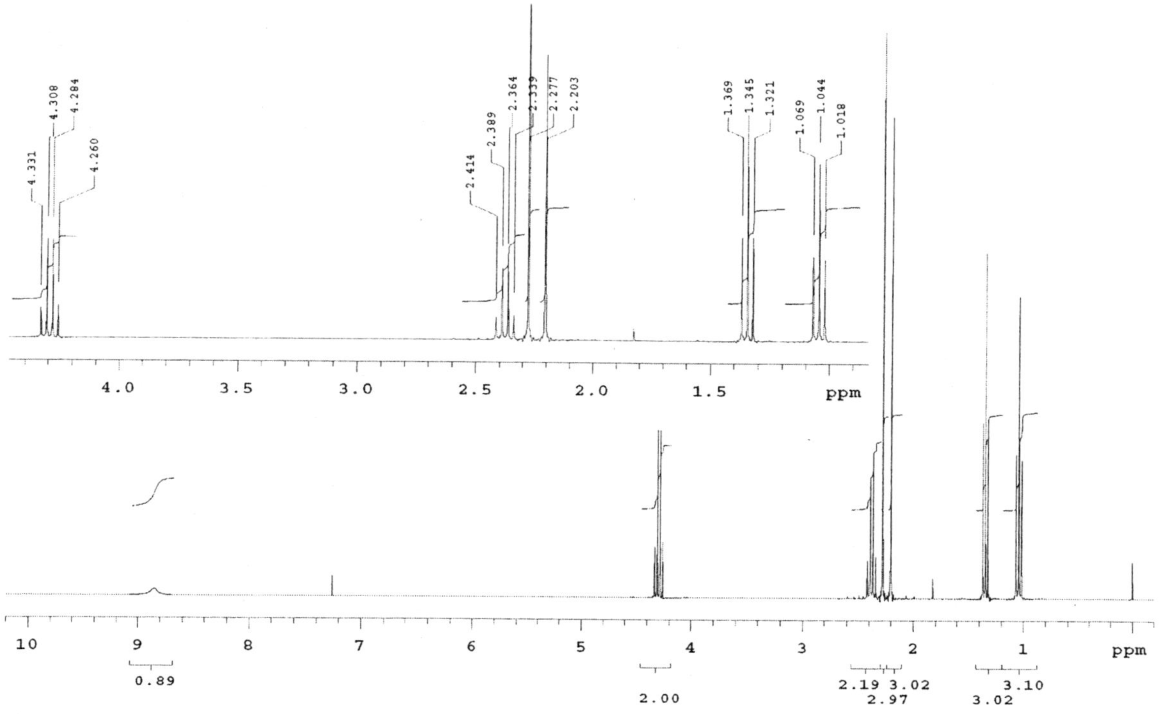

^{1}H-NMR Spectrum in $CDCl_3$

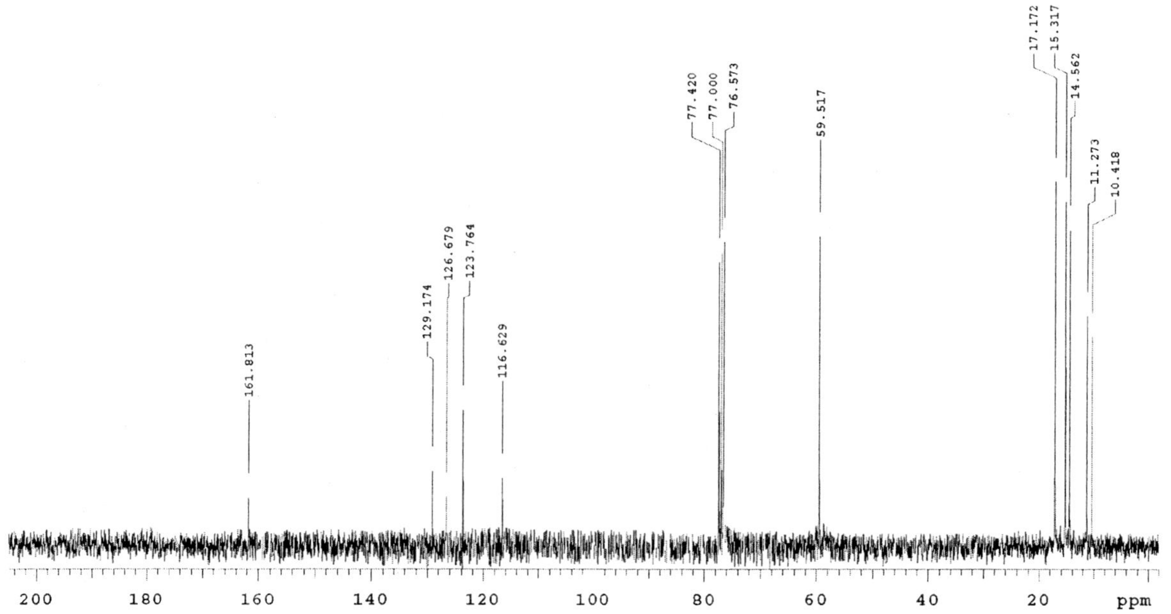

^{13}C-NMR Spectrum in $CDCl_3$

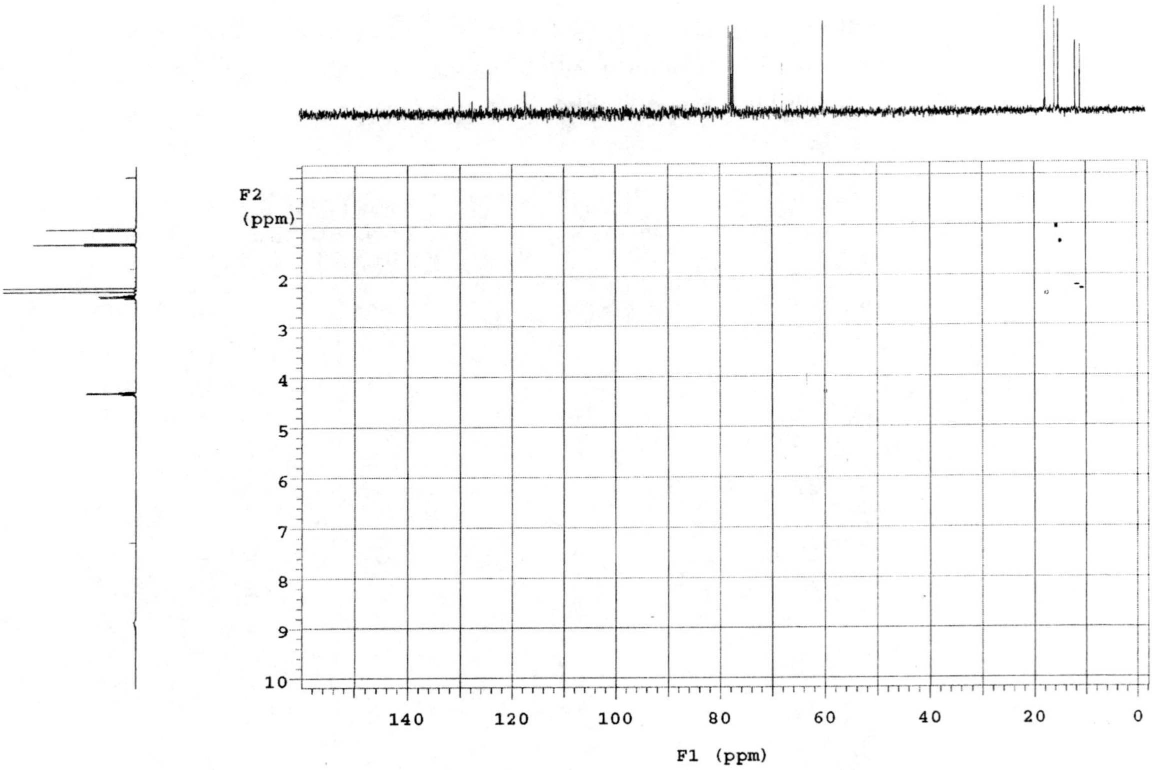

gHSQC Spectrum in $CDCl_3$

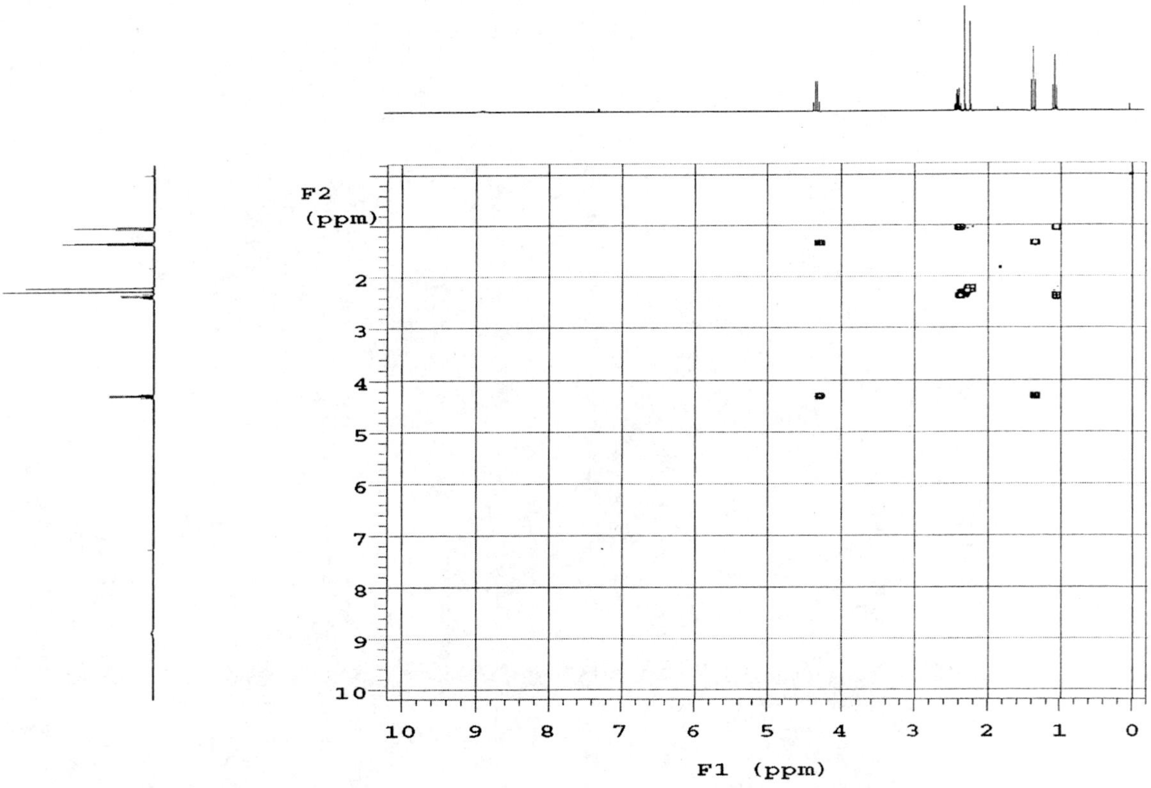

gCOSY Spectrum in $CDCl_3$

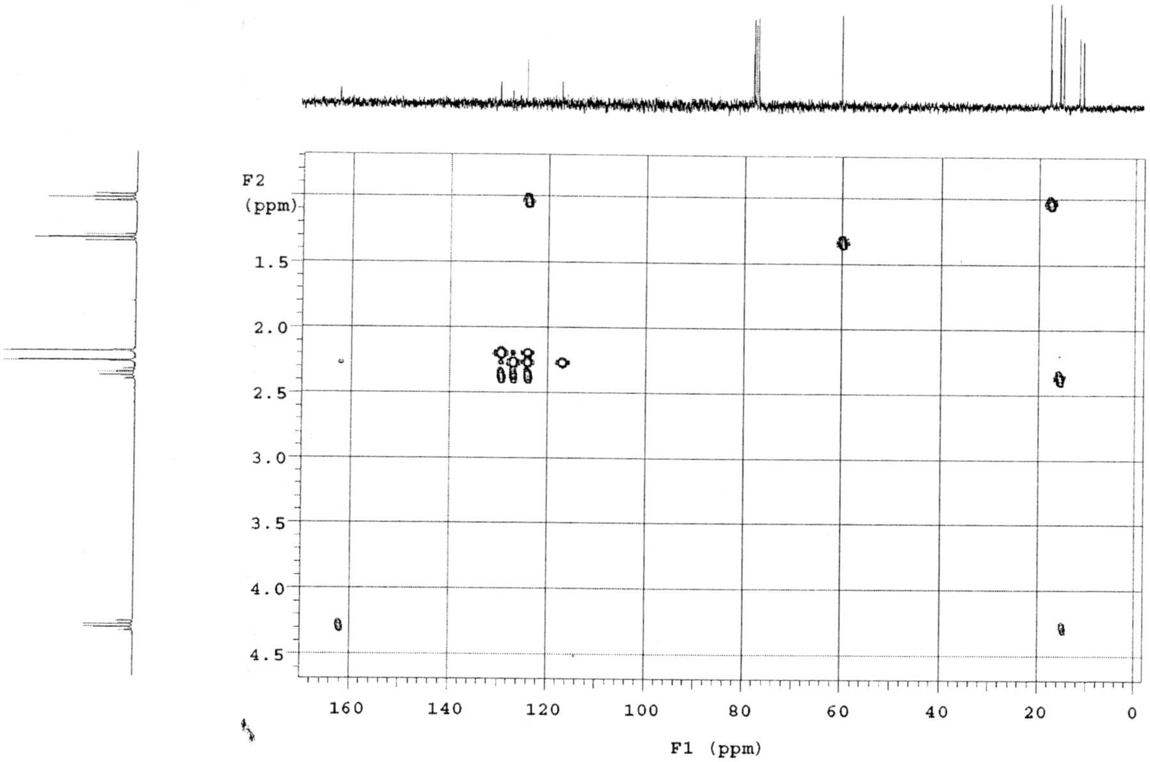

gHMBC Spectrum in $CDCl_3$

Problem 5.05C Given the data below (Proton, Carbon, gHSQC, gCOSY, and gHMBC), provide the structure of the molecule. Compound **5.05B** was oxidized with ceric ammonium nitrate (CAN) in aqueous acetic acid to yield **5.05C**.

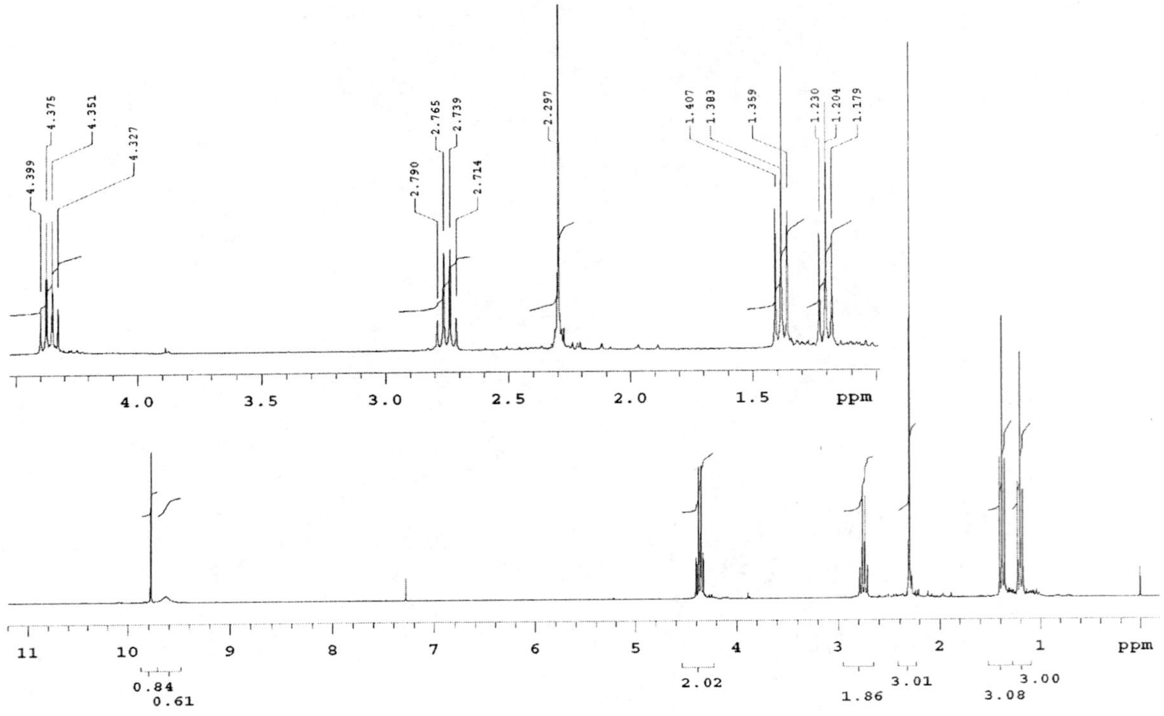

^{1}H-NMR Spectrum in $CDCl_3$

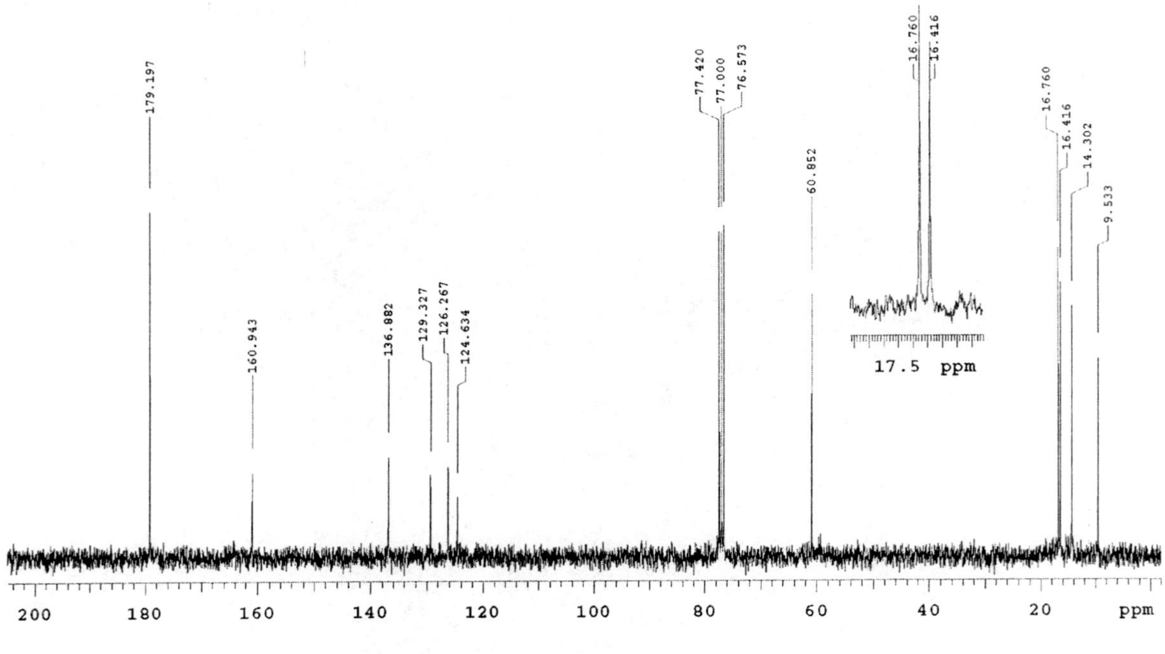

^{13}C-NMR Spectrum in $CDCl_3$

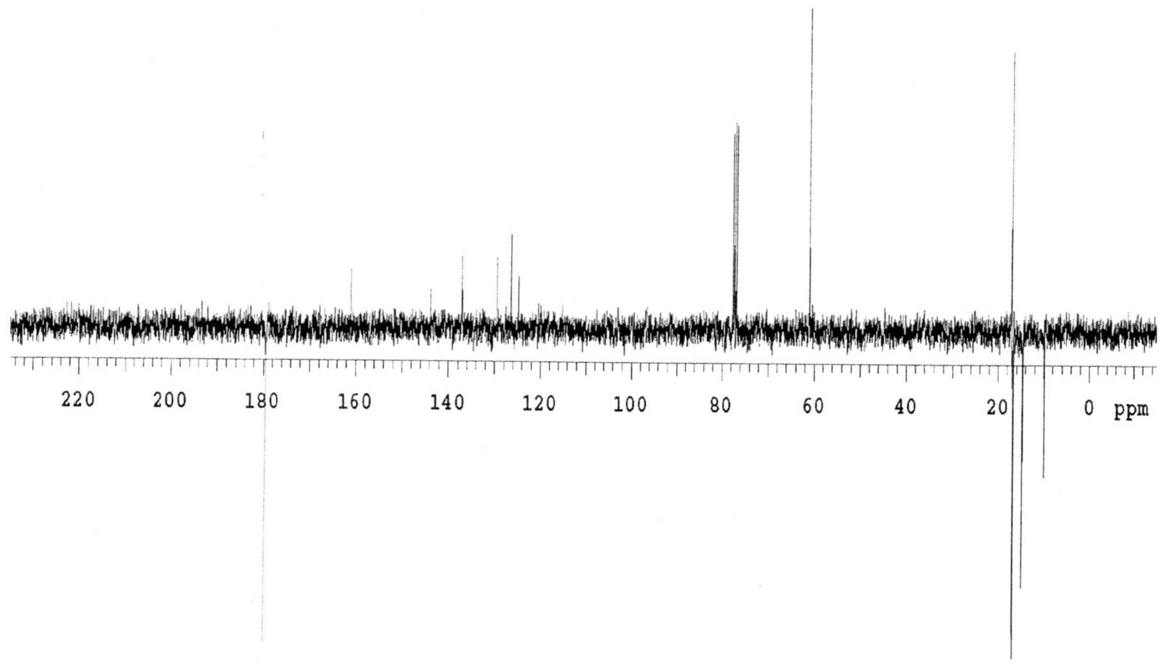

APT Spectrum in $CDCl_3$

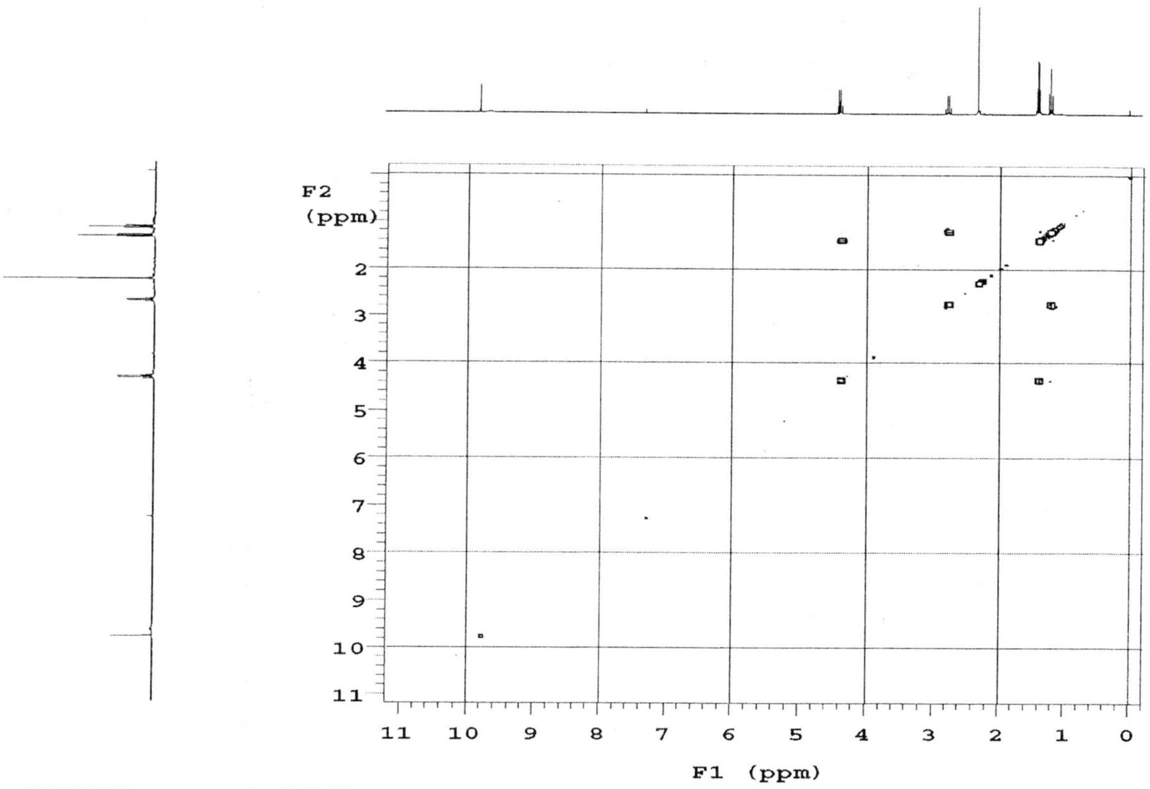

gCOSY Spectrum in $CDCl_3$

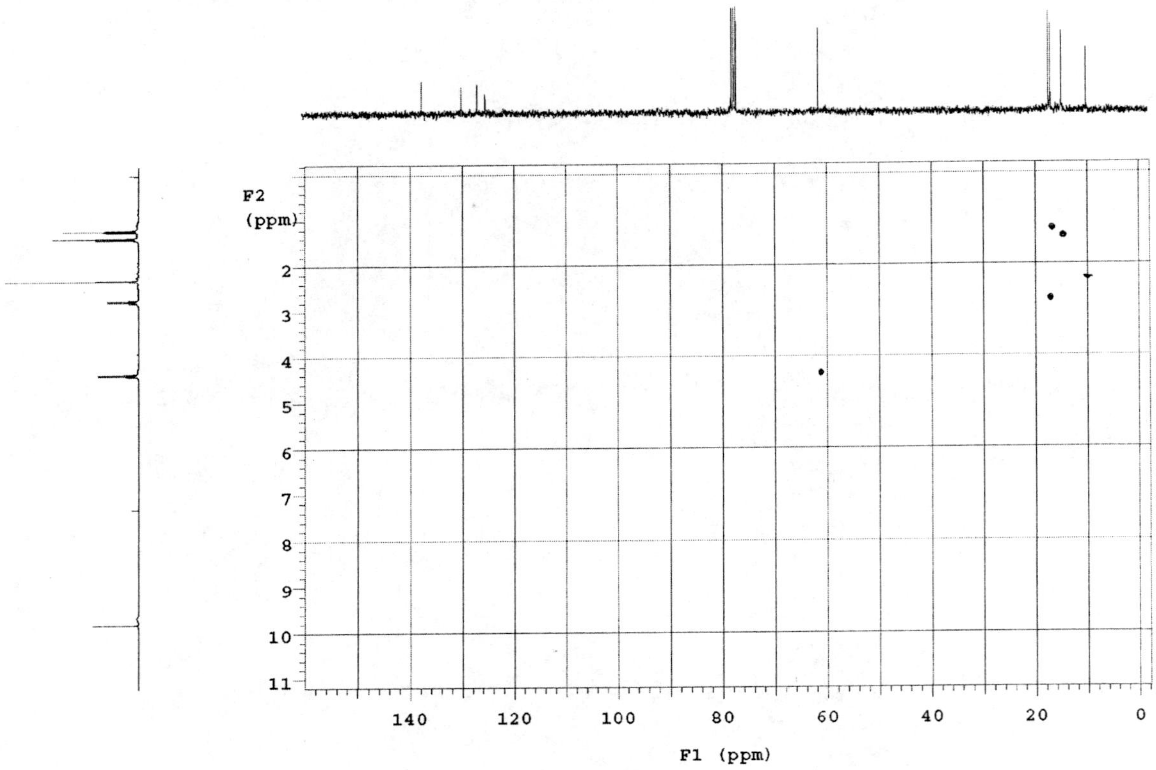

gHSQC Spectrum in $CDCl_3$

Problems 5.06-5.10 Using the spectroscopic data, determine the product of each reaction. Assign all ^{1}H and ^{13}C-NMR signals. Using the NOESY spectra, determine the configuration and conformation of each product. (See *J. Chem. Educ.*, **2007**, *84* (3), 471-474.)

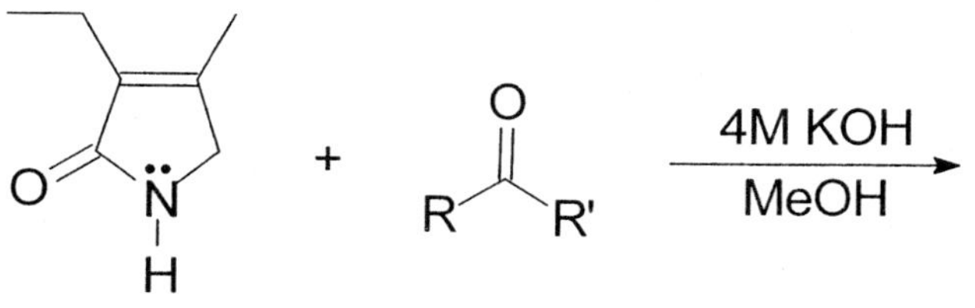

Problem	R	R'
5.06	2-Pyridyl	CH_3
5.07	2-(N-methylpyrrolyl)	H
5.08	2-Pyrrolyl	H
5.09	2,4-dimethoxyphenyl	H
5.10	Phenyl	H

Problem 5.06 See page 373 for details.

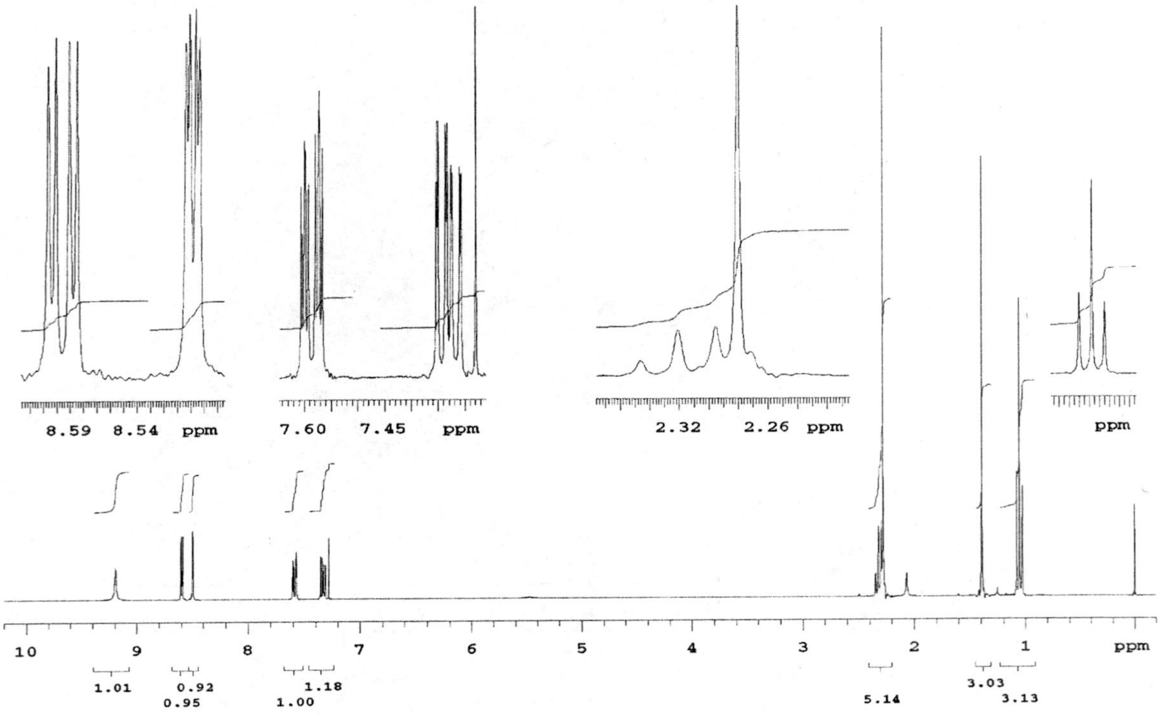

^1H-NMR Spectrum in $CDCl_3$

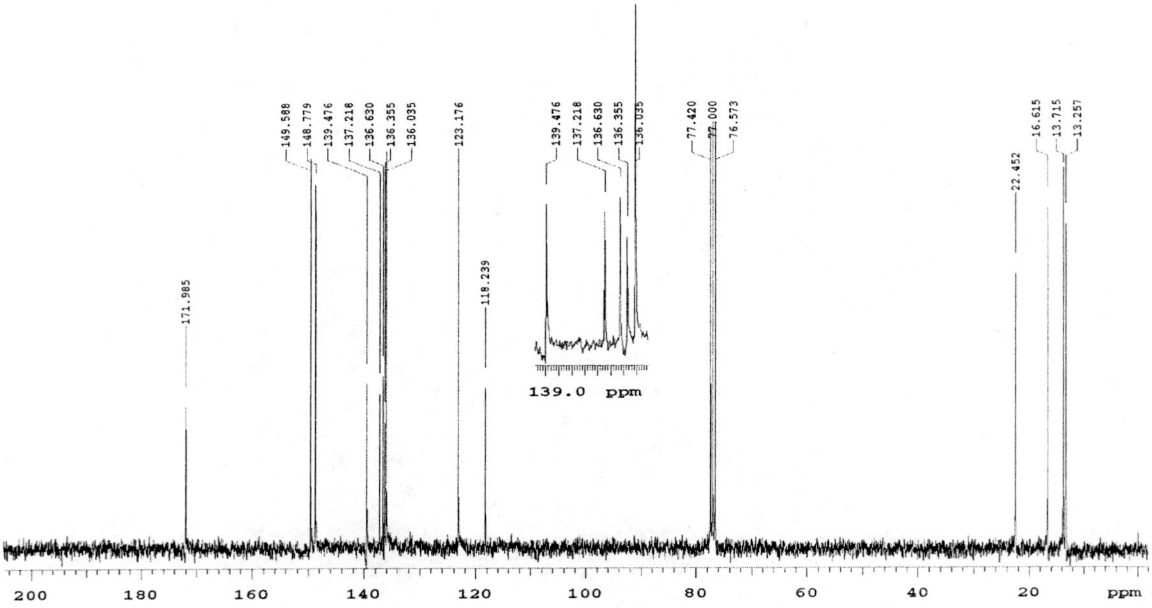

^{13}C-NMR Spectrum in $CDCl_3$

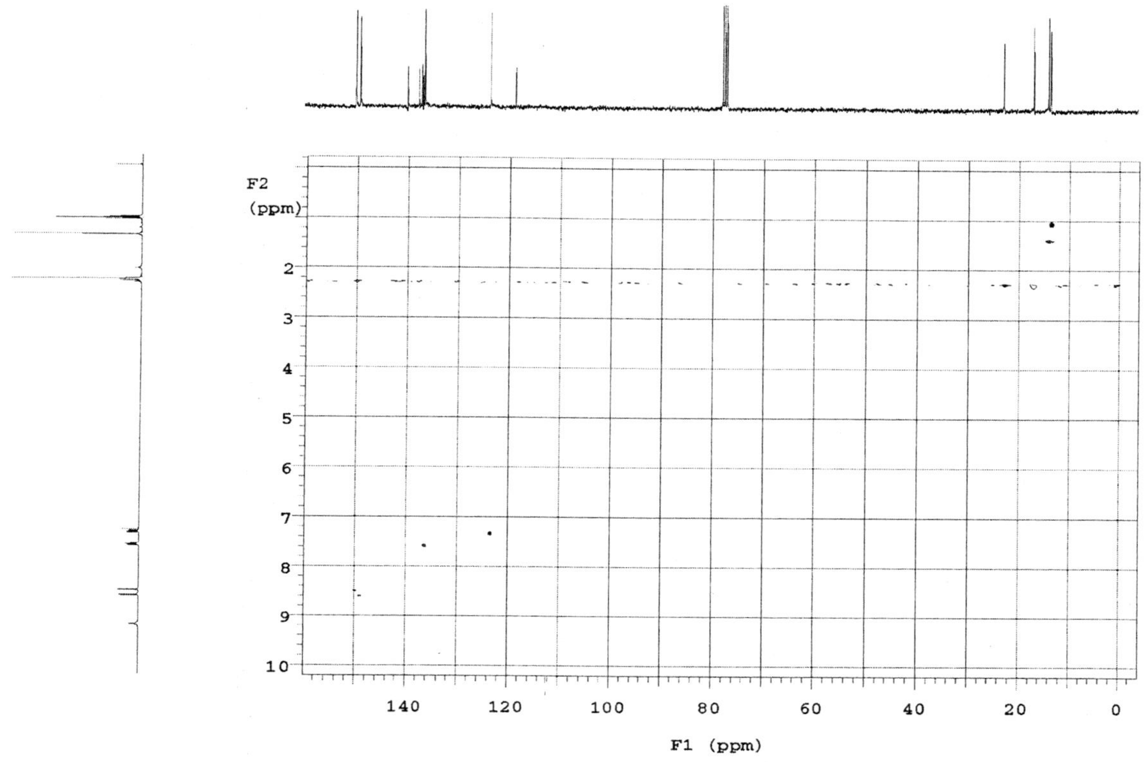

gHSQC Spectrum in $CDCl_3$

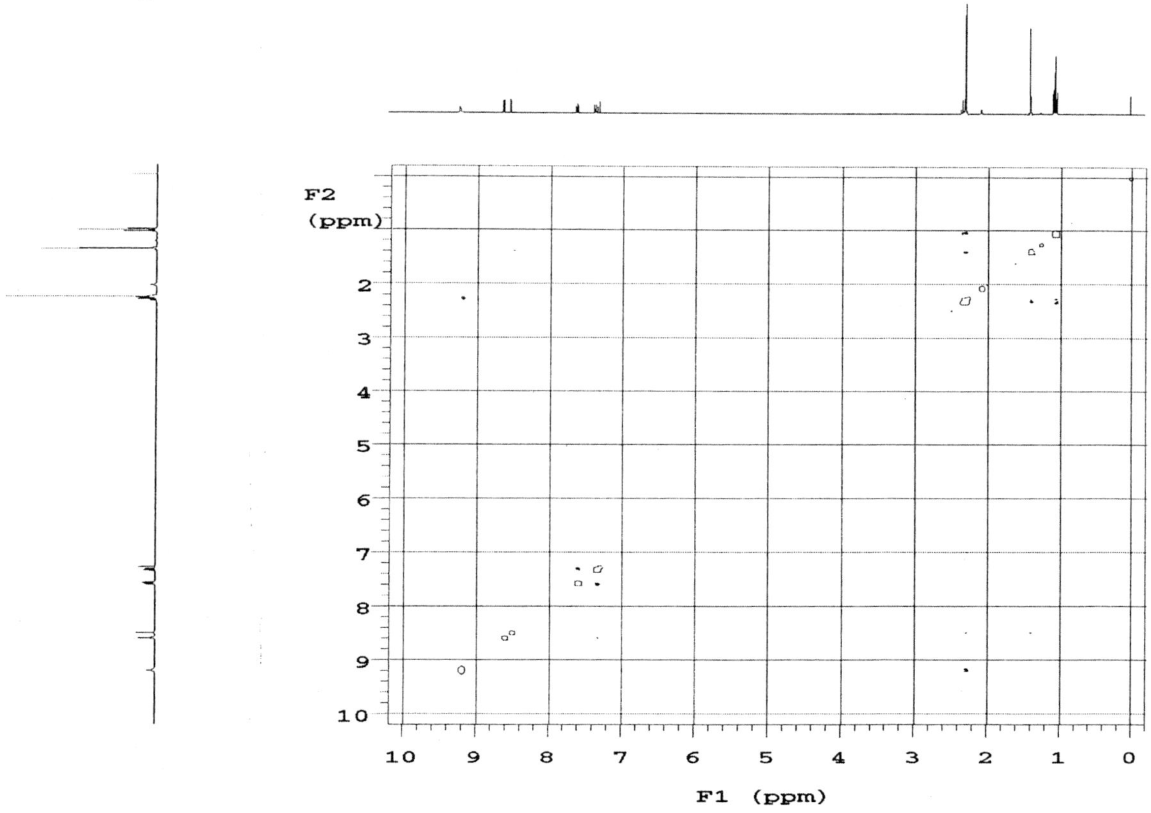

NOESY Spectrum in $CDCl_3$

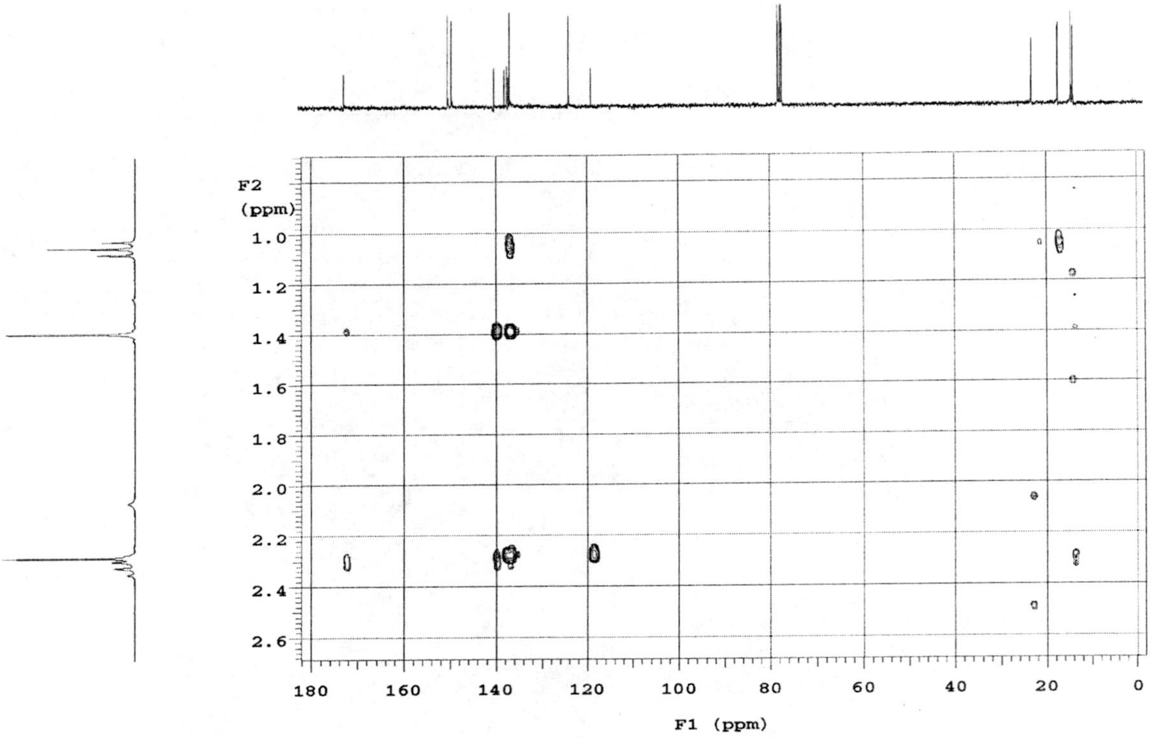

gHMBC Spectrum of the upfield region in $CDCl_3$

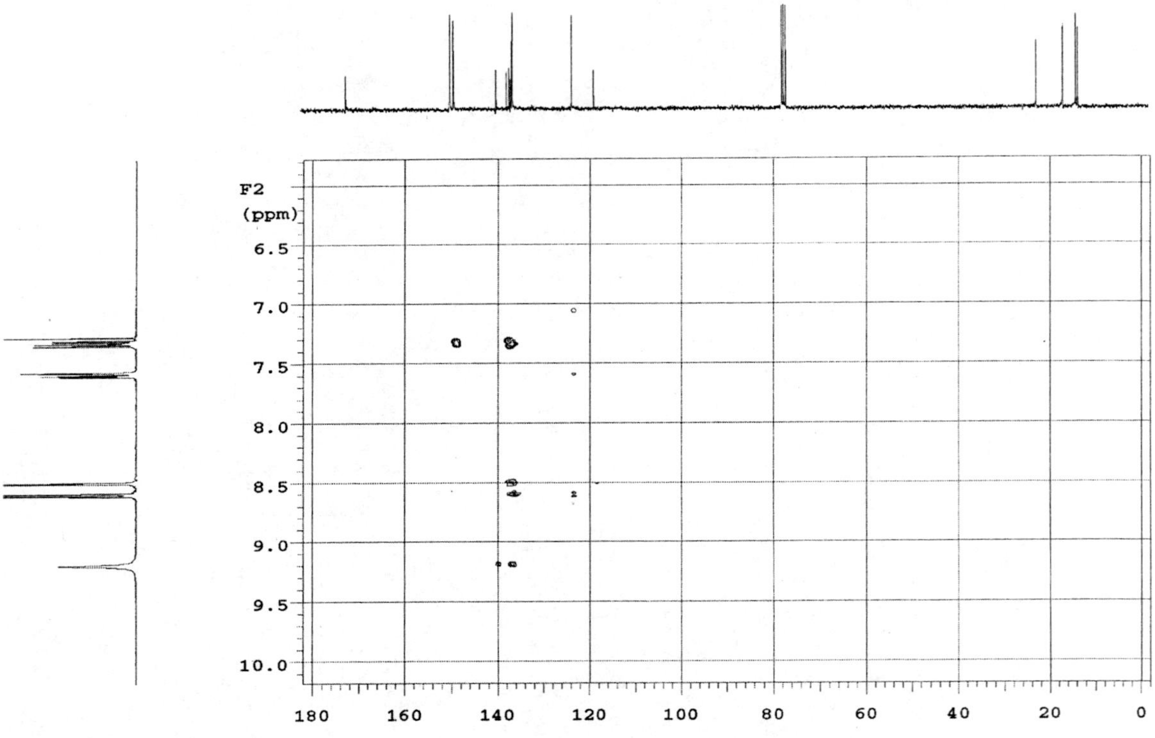

gHMBC Spectrum of the downfield region in $CDCl_3$

Problem 5.07 See page 373 for details.

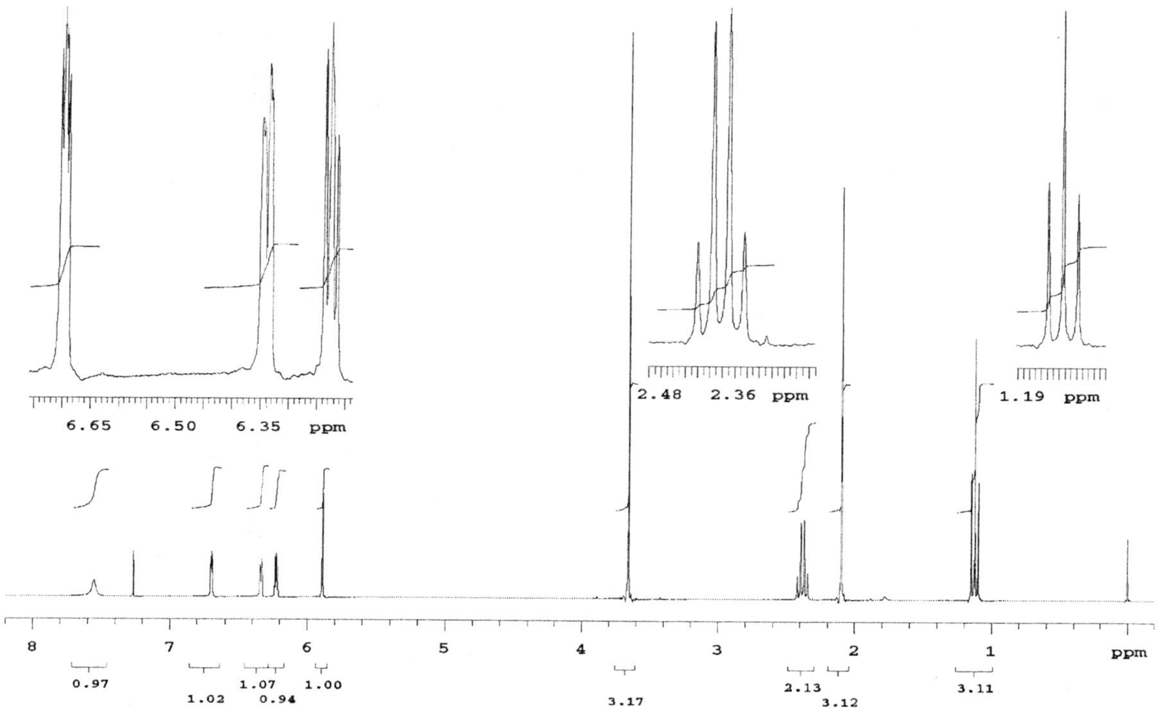

^{1}H-NMR Spectrum in $CDCl_3$

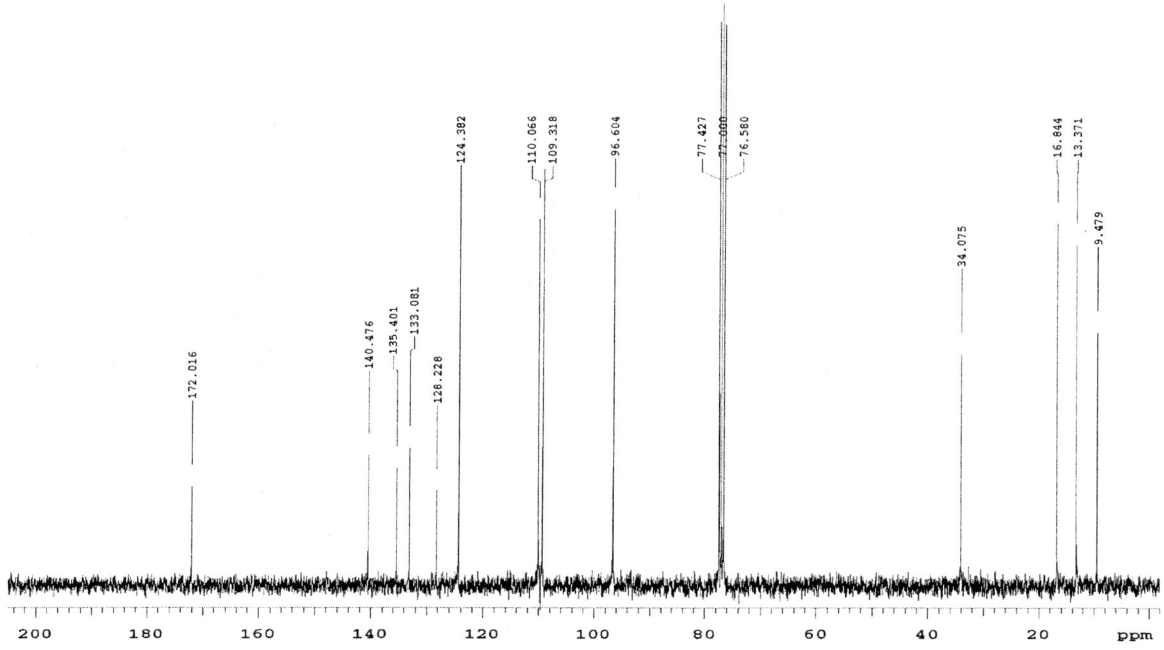

^{13}C-NMR Spectrum in $CDCl_3$

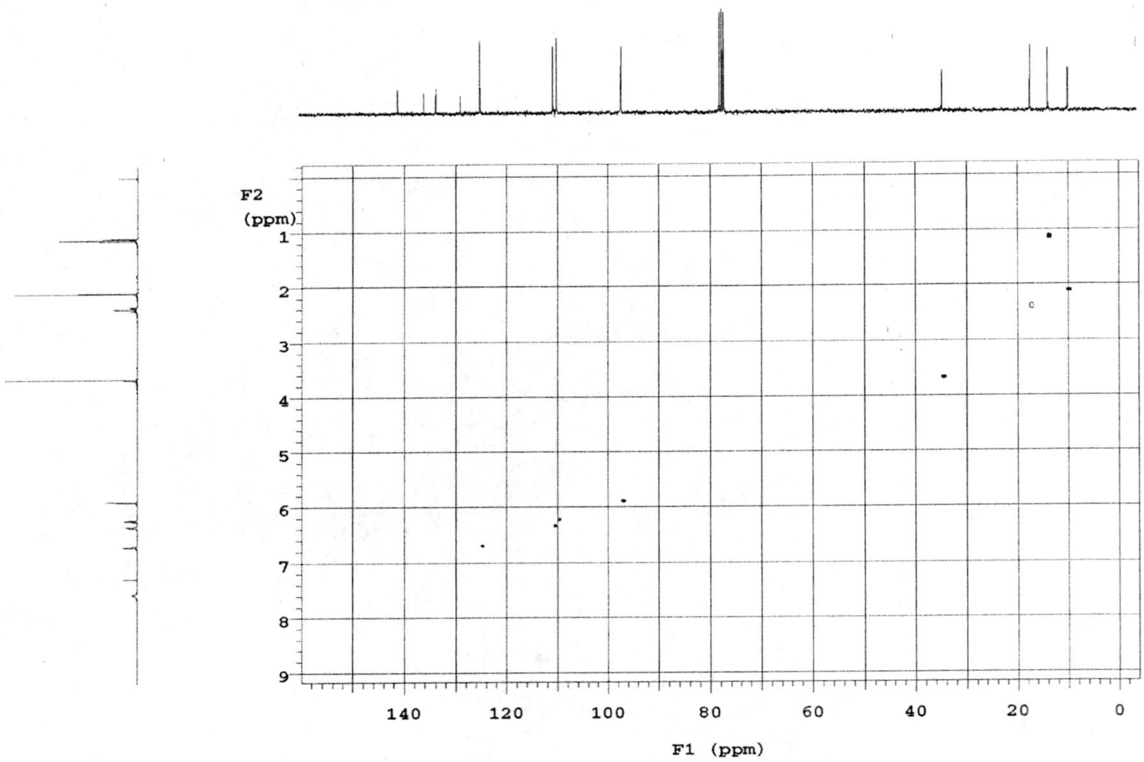

gHSQC Spectrum in $CDCl_3$

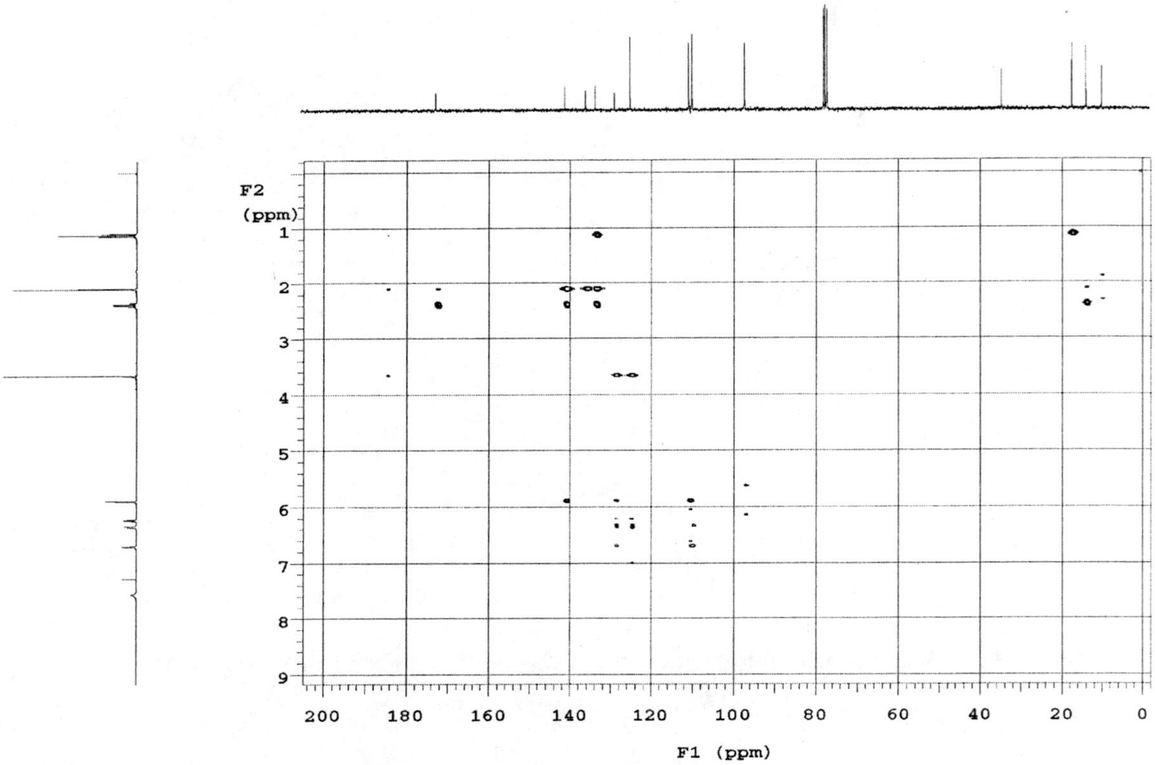

gHMBC Spectrum in $CDCl_3$

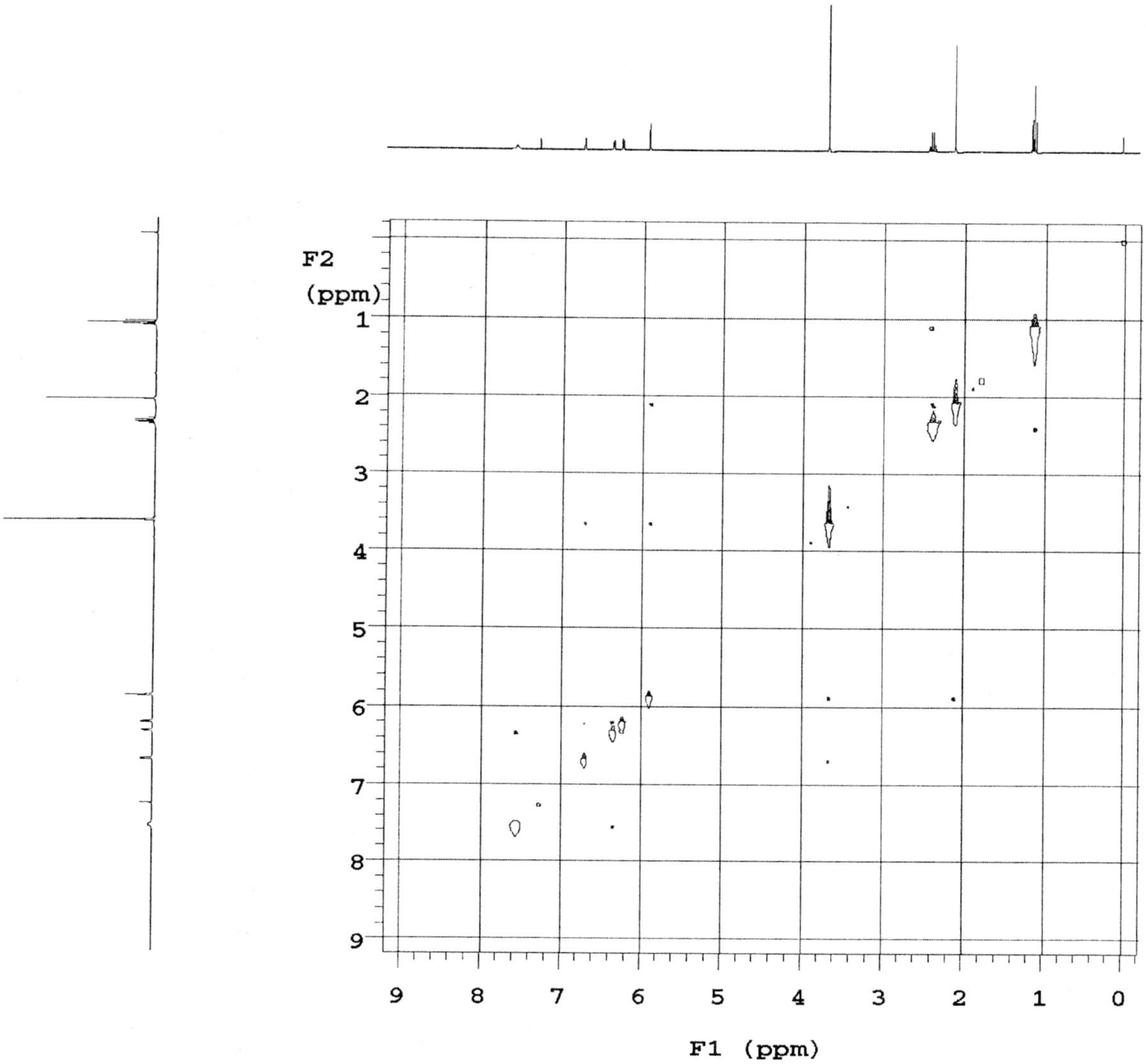

NOESY Spectrum in $CDCl_3$

Problem 5.08 See page 373 for details.

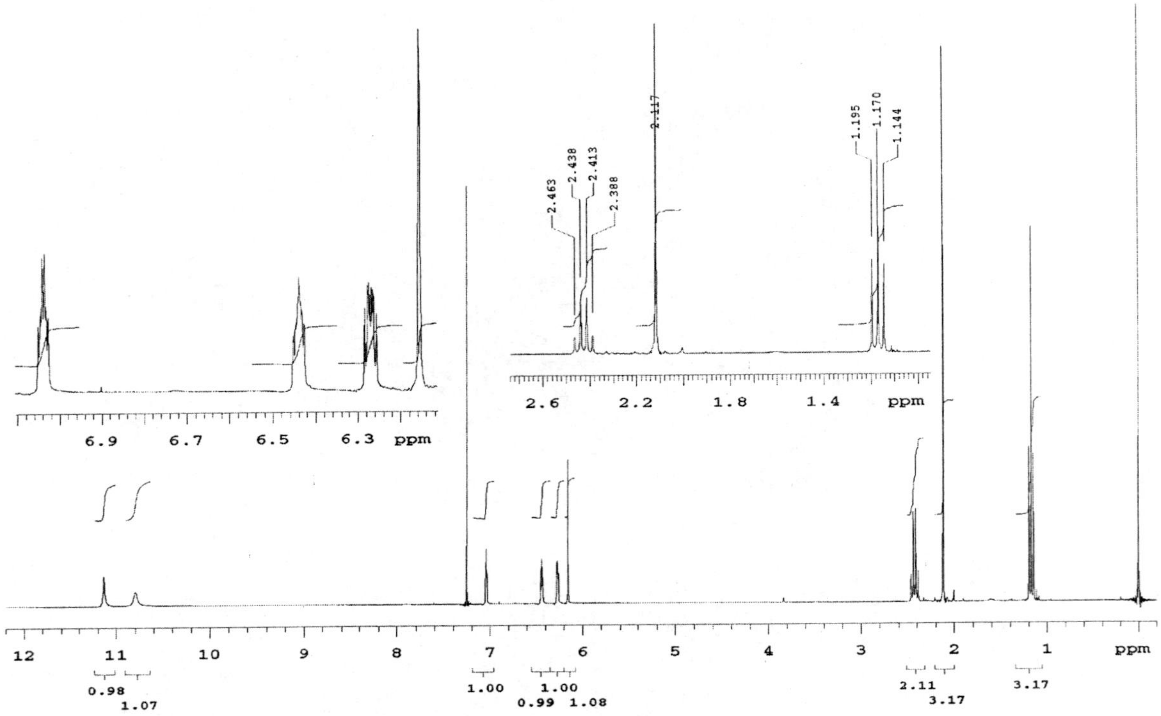

^{1}H-NMR Spectrum in $CDCl_3$

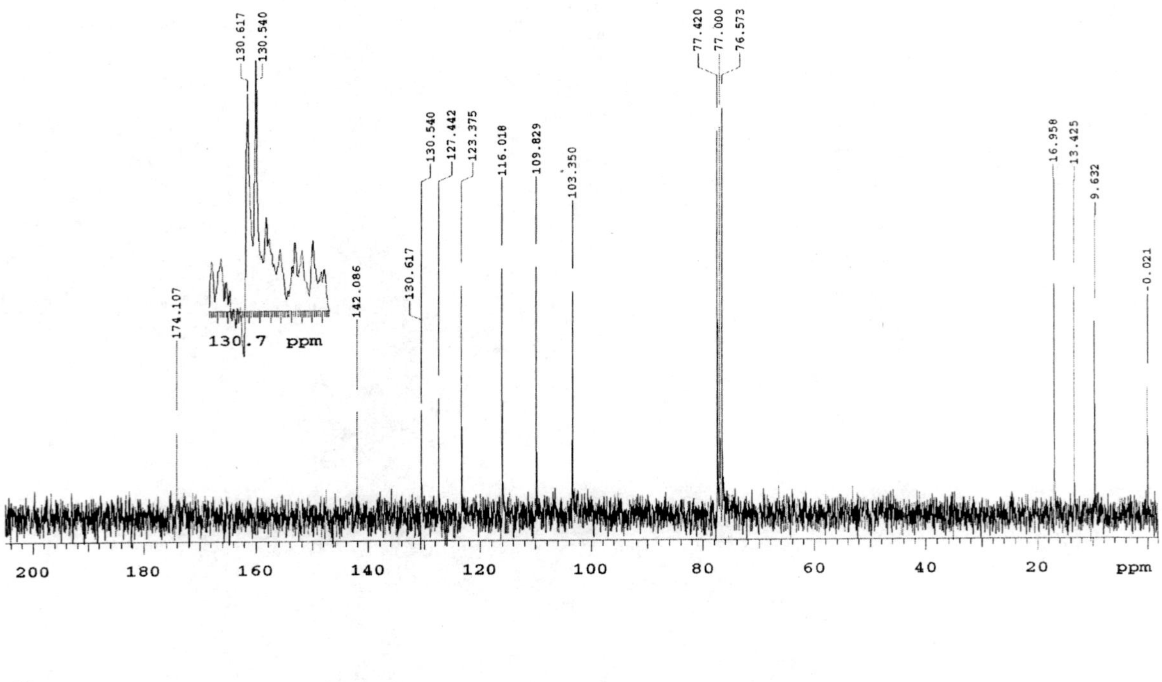

^{13}C-NMR Spectrum in $CDCl_3$

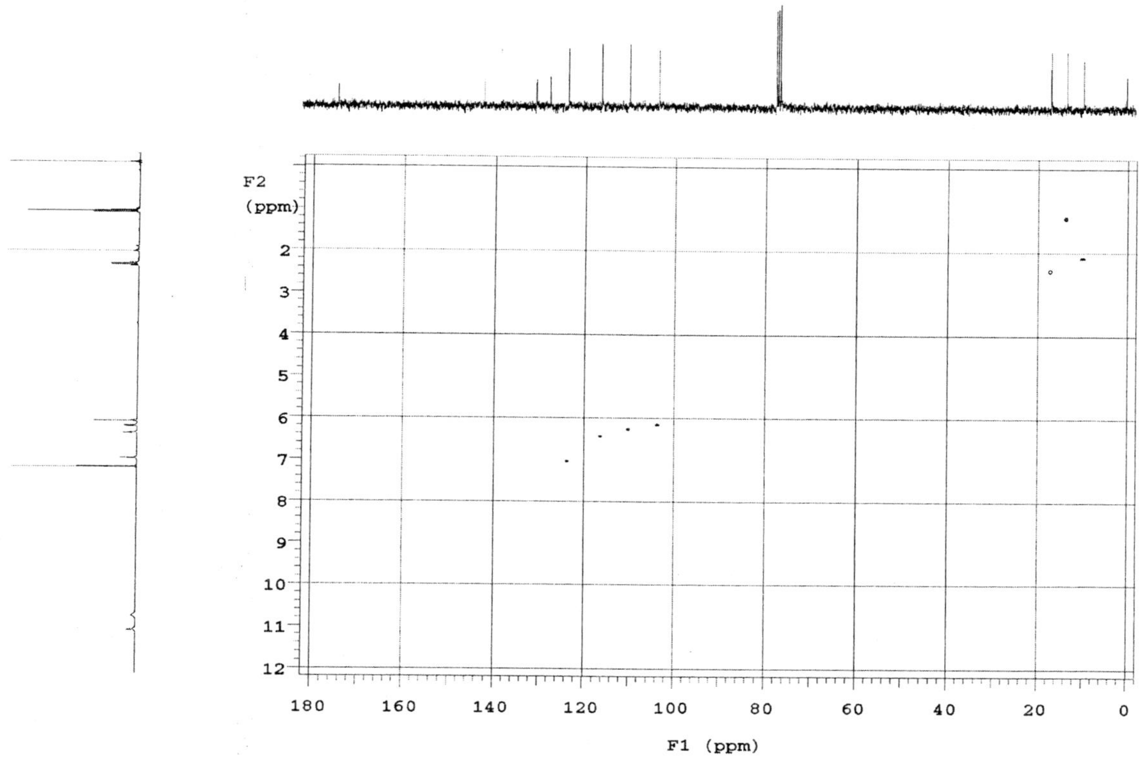

gHSQC Spectrum in $CDCl_3$

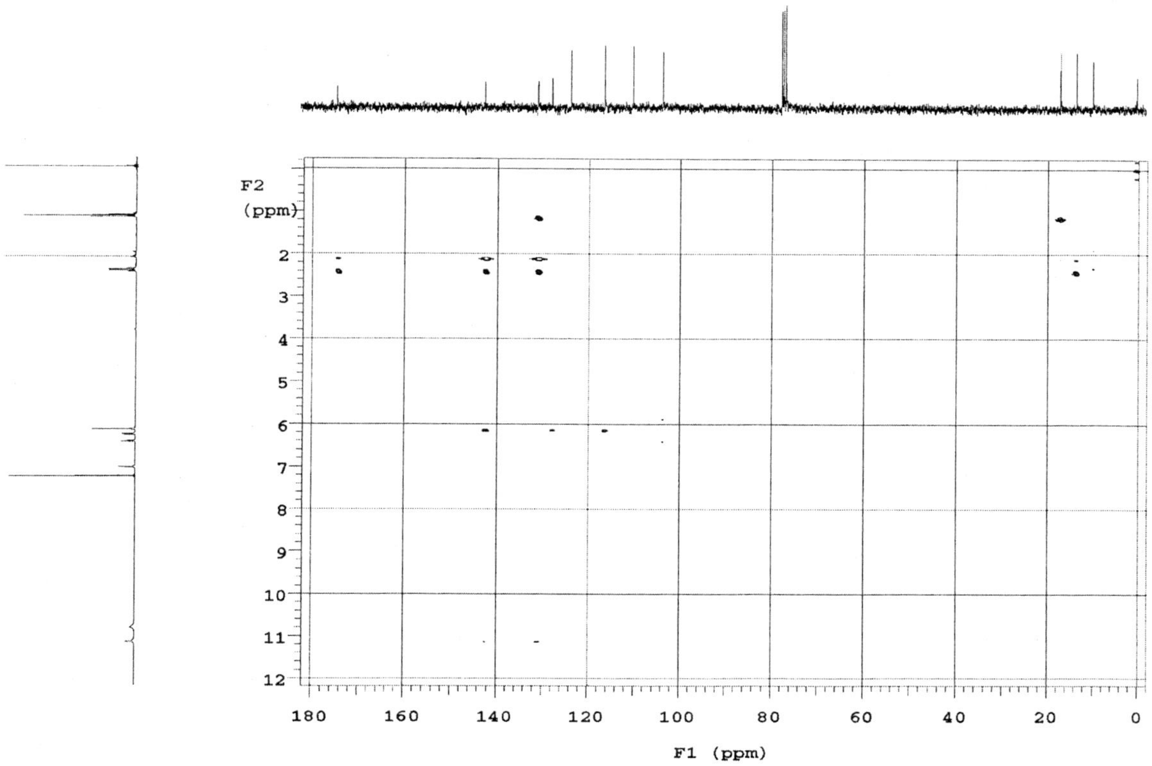

gHMBC Spectrum in $CDCl_3$

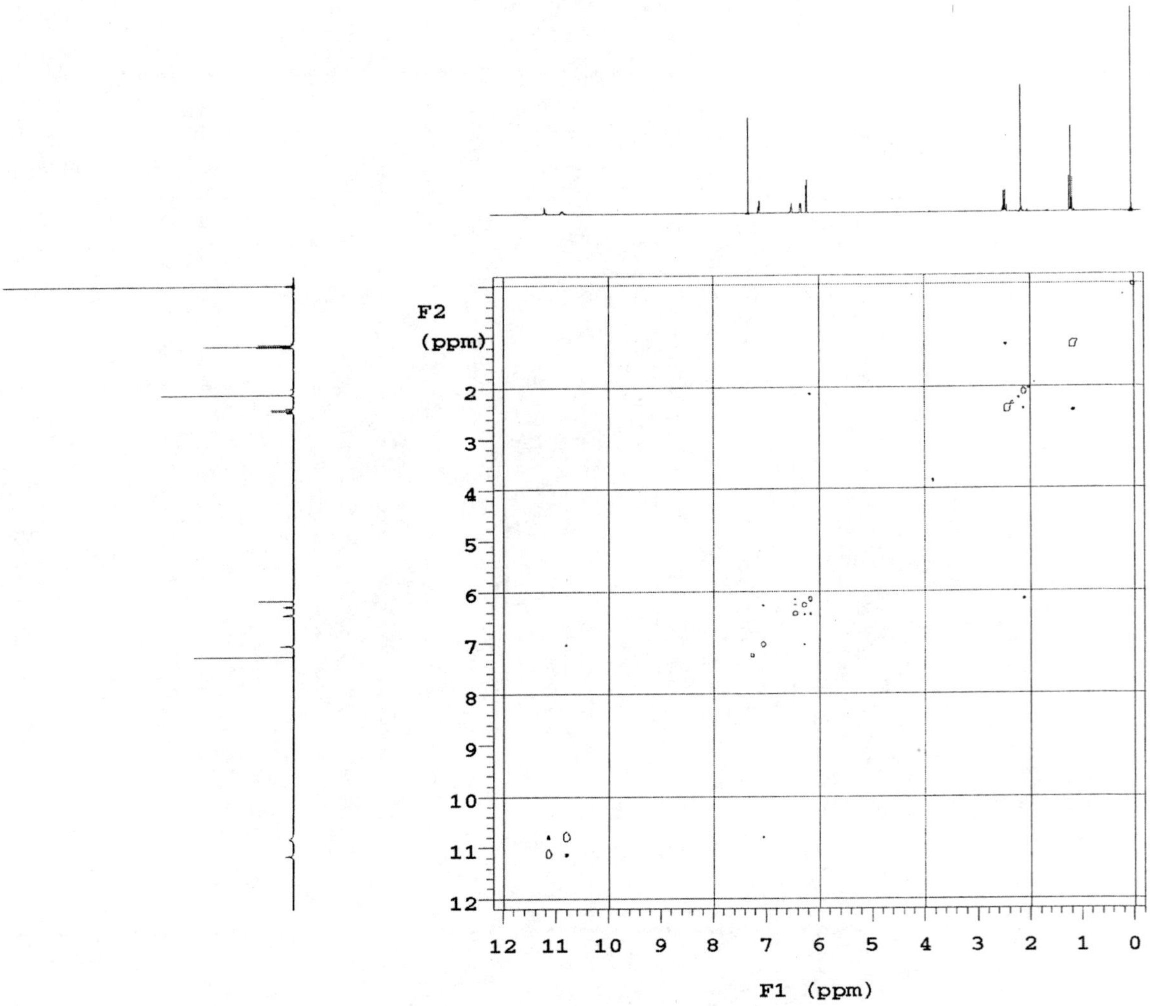

NOESY Spectrum in $CDCl_3$

Problem 5.09 See page 373 for details.

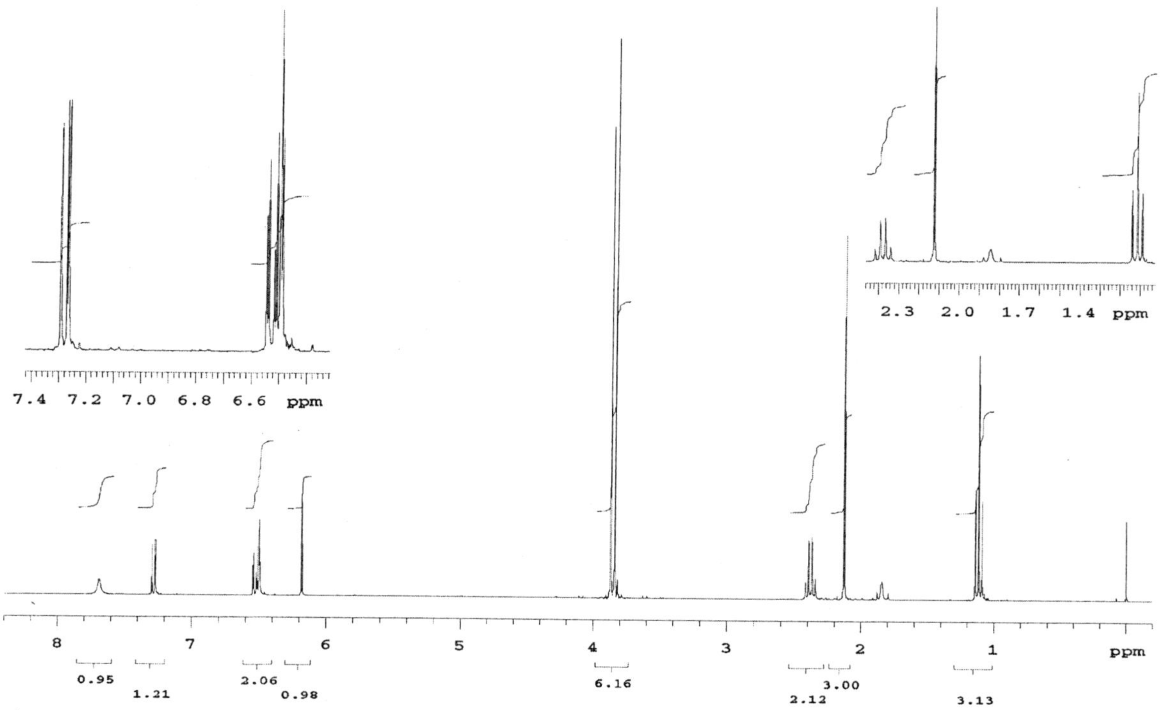

^{1}H-NMR Spectrum in $CDCl_3$

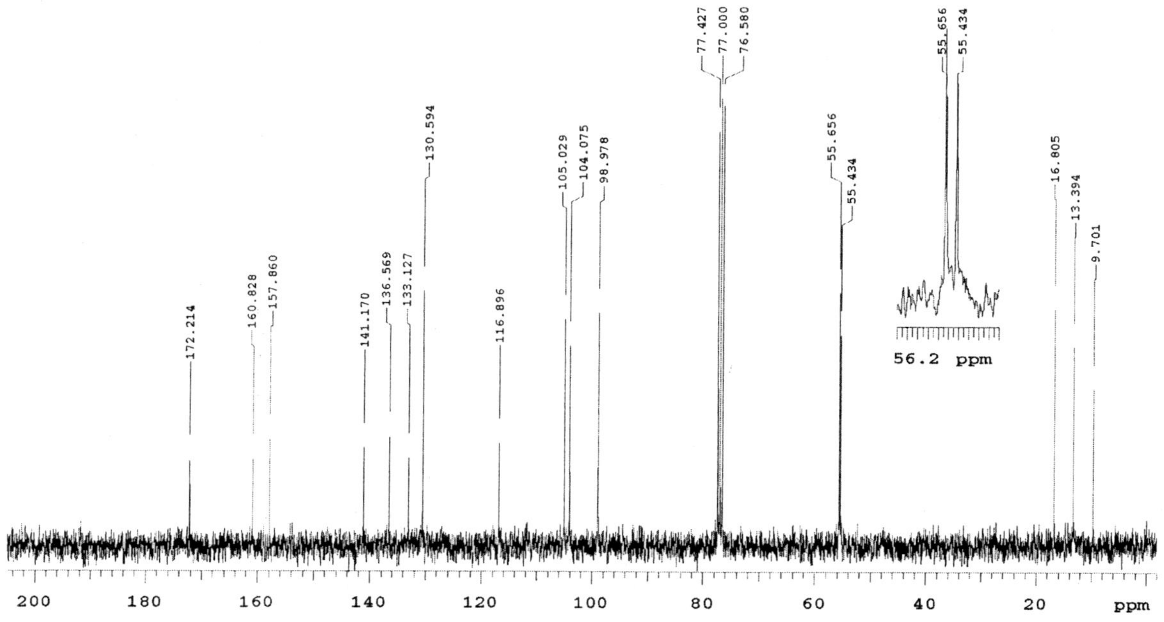

^{13}C-NMR Spectrum in $CDCl_3$

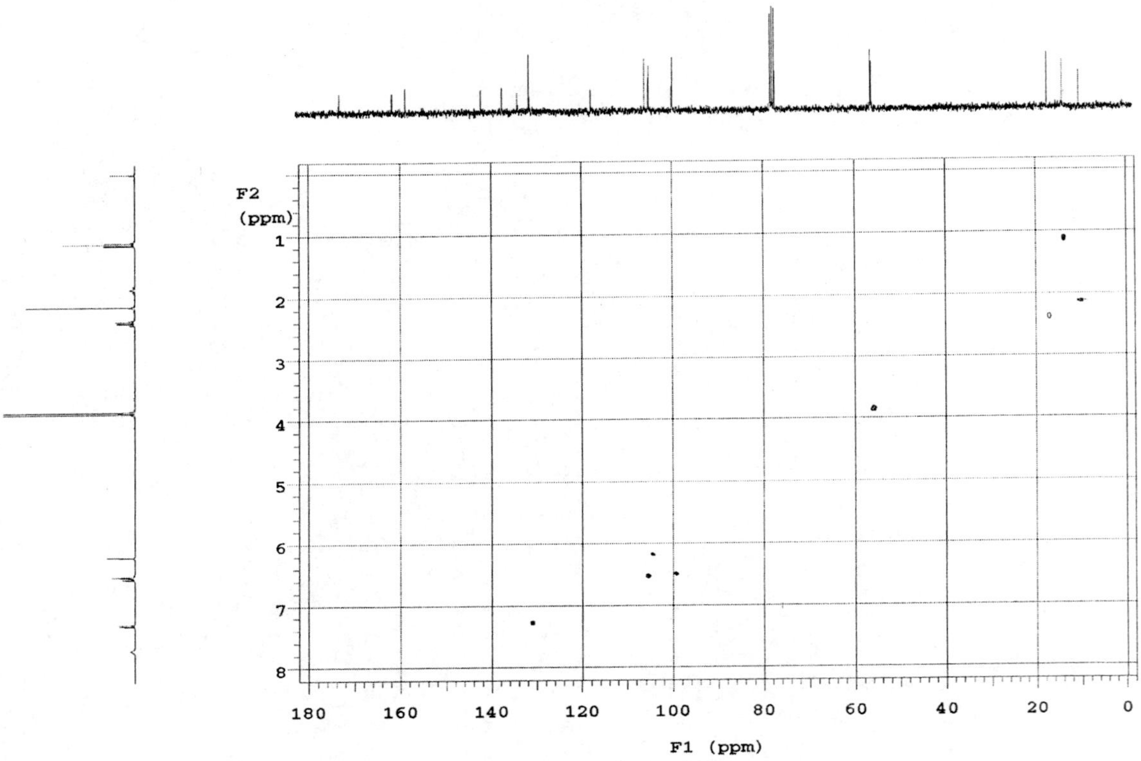

gHSQC Spectrum in $CDCl_3$

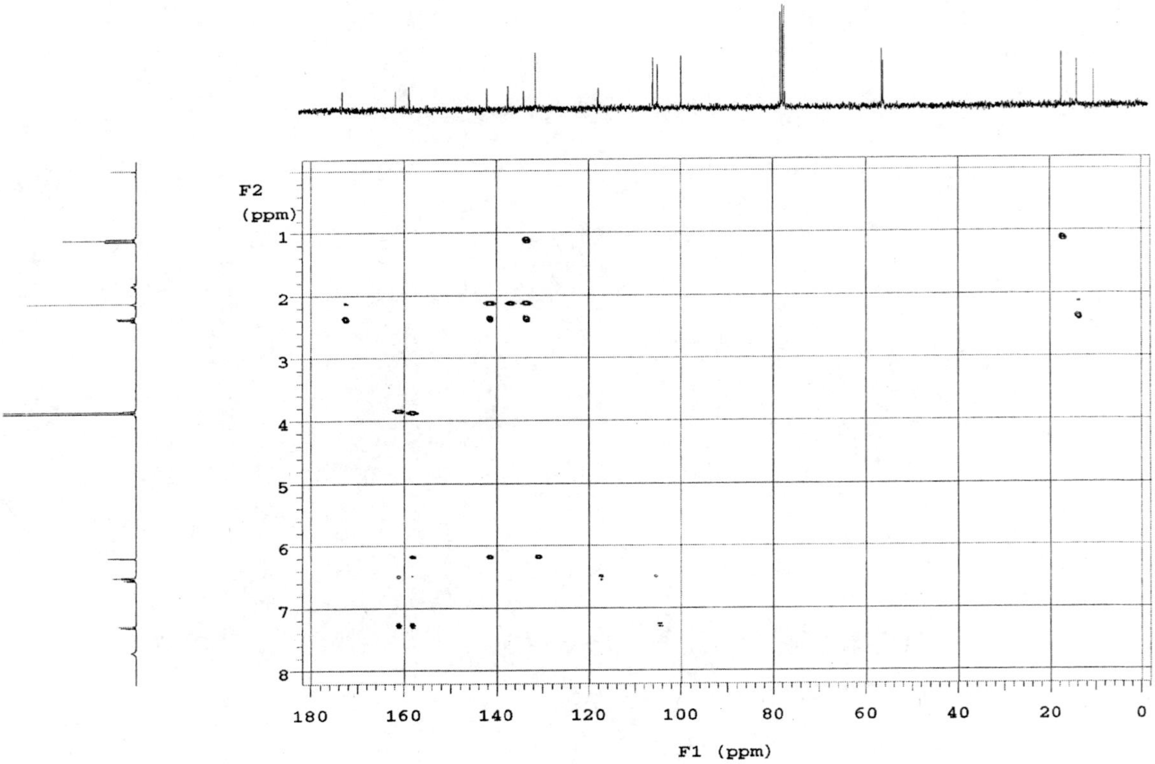

gHMBC Spectrum in $CDCl_3$

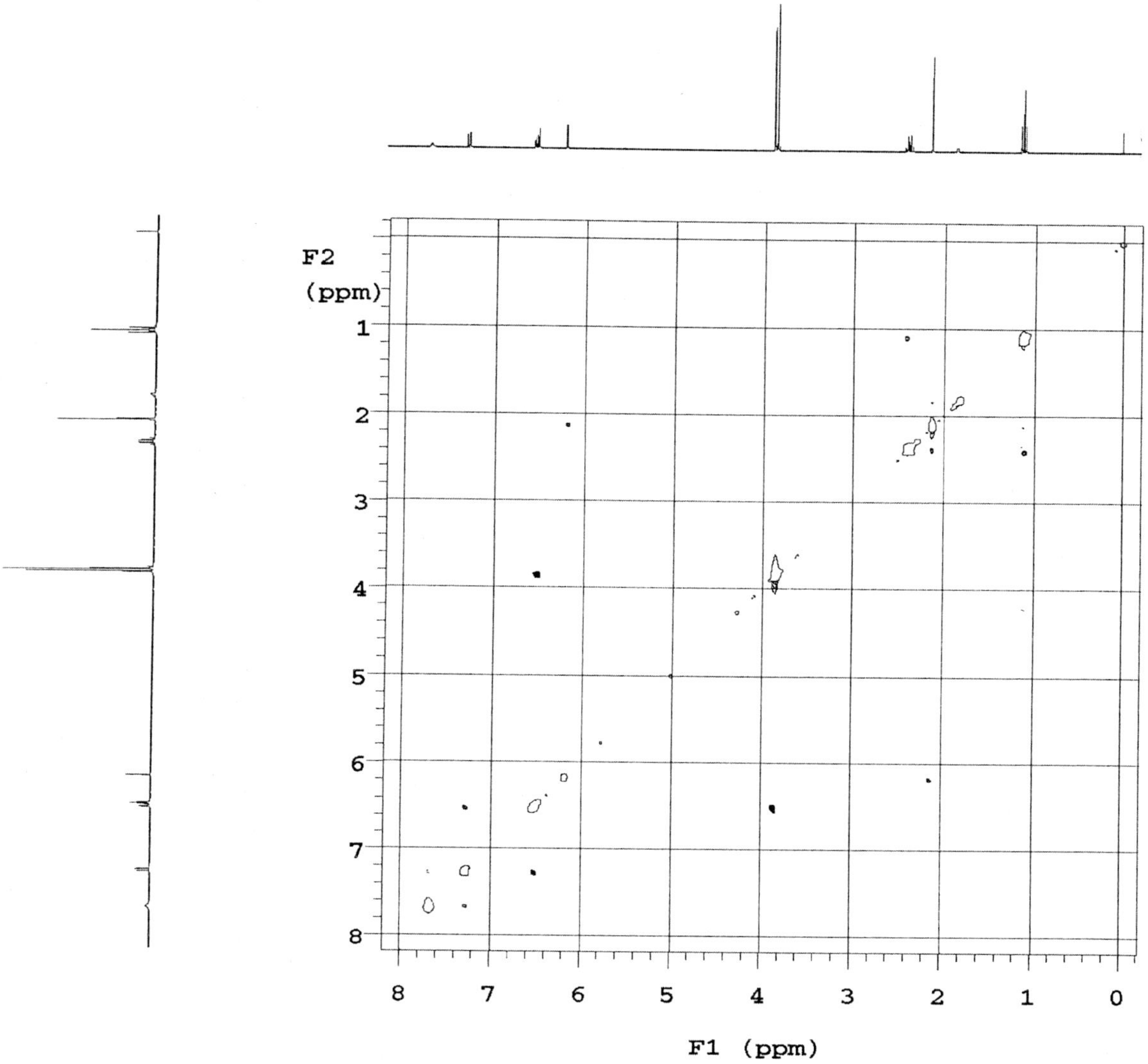

NOESY Spectrum in $CDCl_3$

Problem 5.10 See page 373 for details.

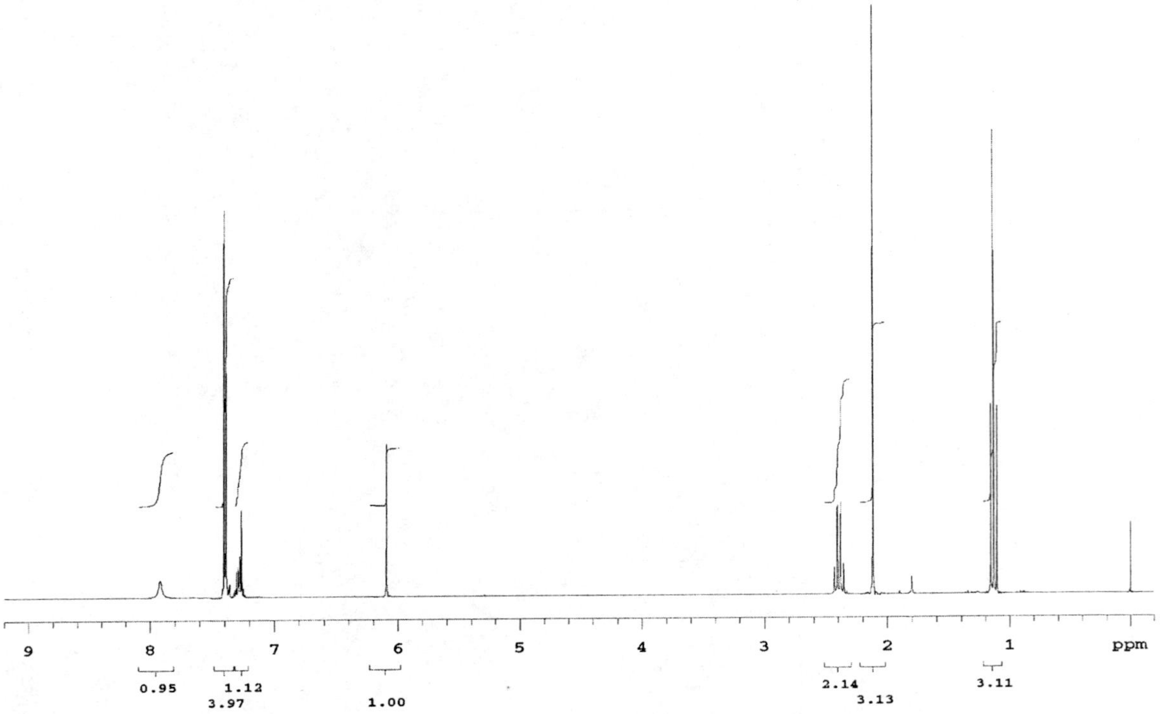

^1H-NMR Spectrum in $CDCl_3$

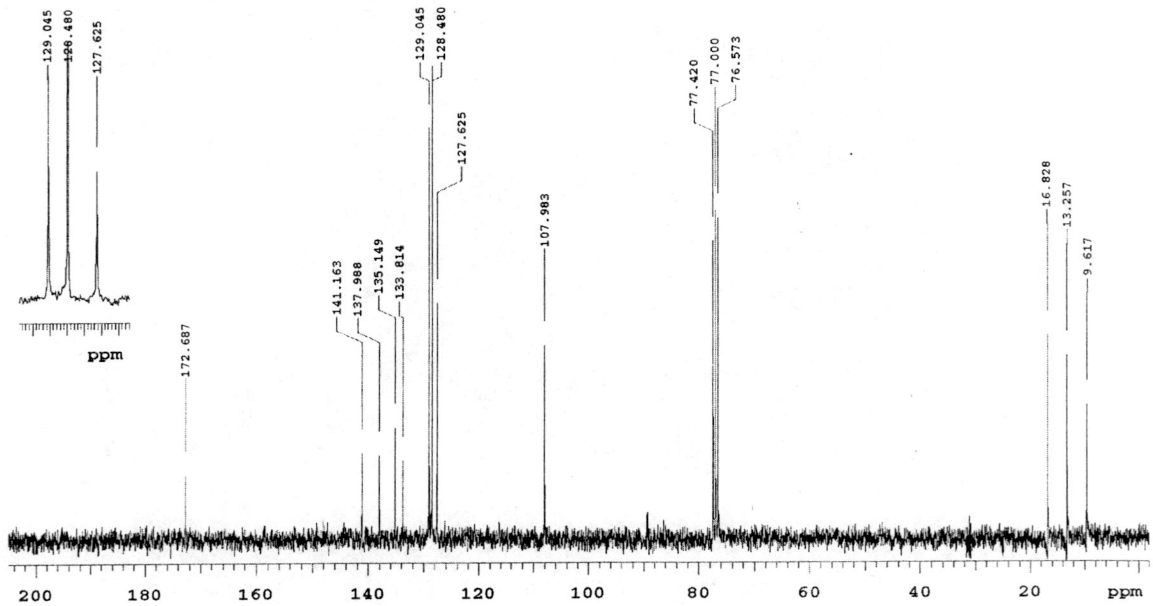

^{13}C-NMR Spectrum in $CDCl_3$

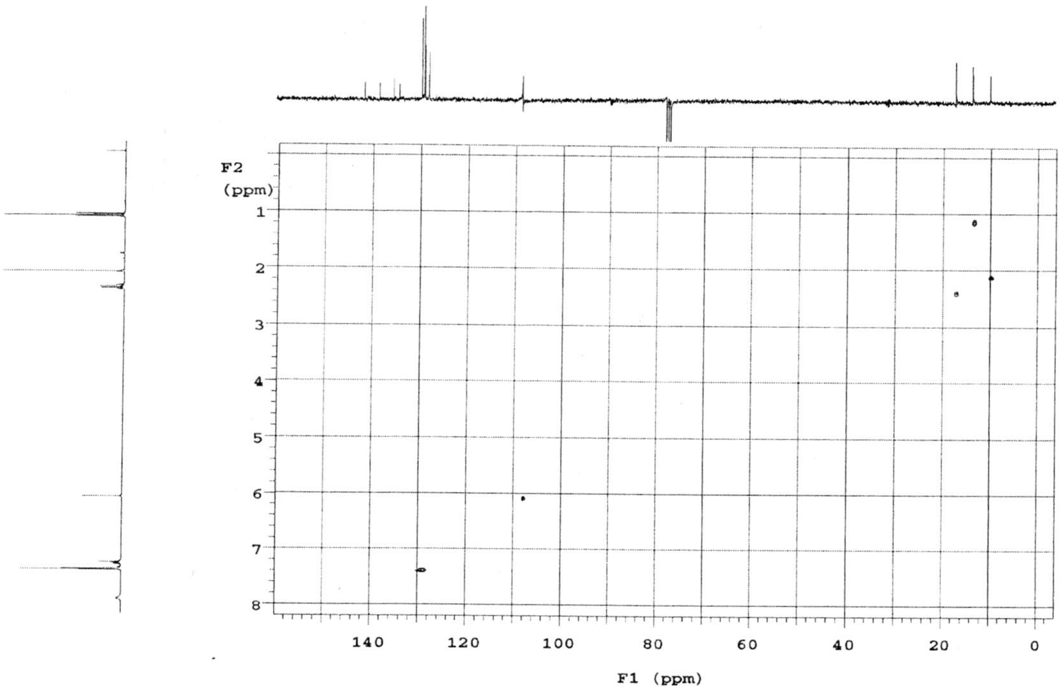

gHSQC Spectrum in $CDCl_3$

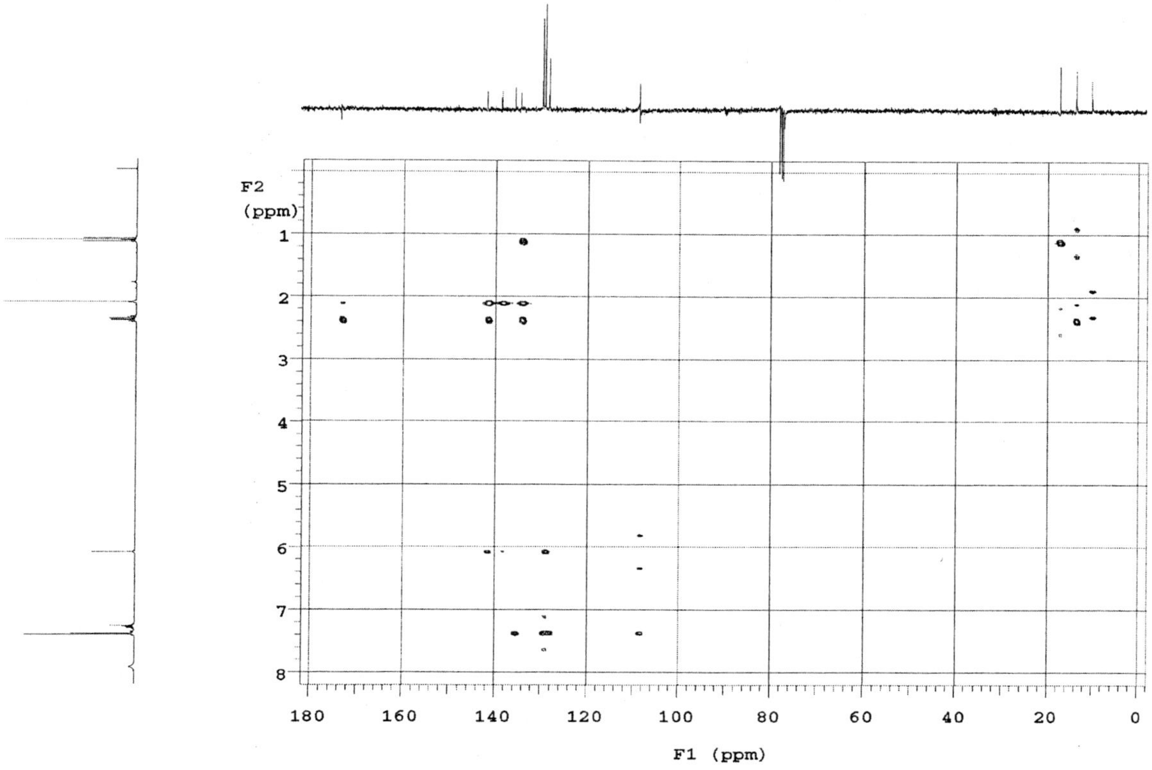

gHMBC Spectrum in $CDCl_3$

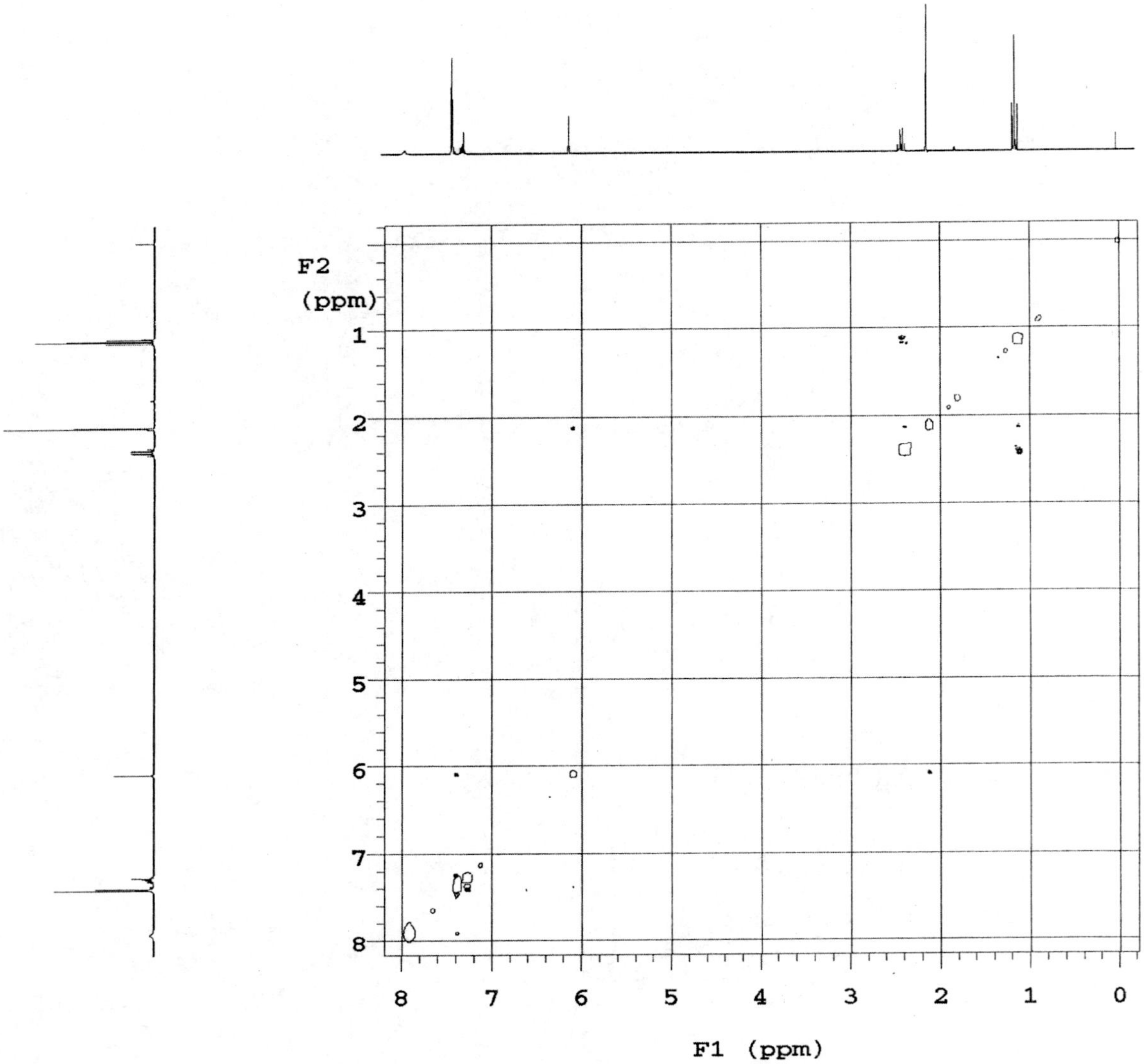

NOESY Spectrum in $CDCl_3$

Problem 5.11 Given the data below (Proton, Carbon, gHSQC, gCOSY, and gHMBC), provide the structure of the molecule. Compound **4.24** upon oxidation under Oppenauer conditions led to **5.11**.

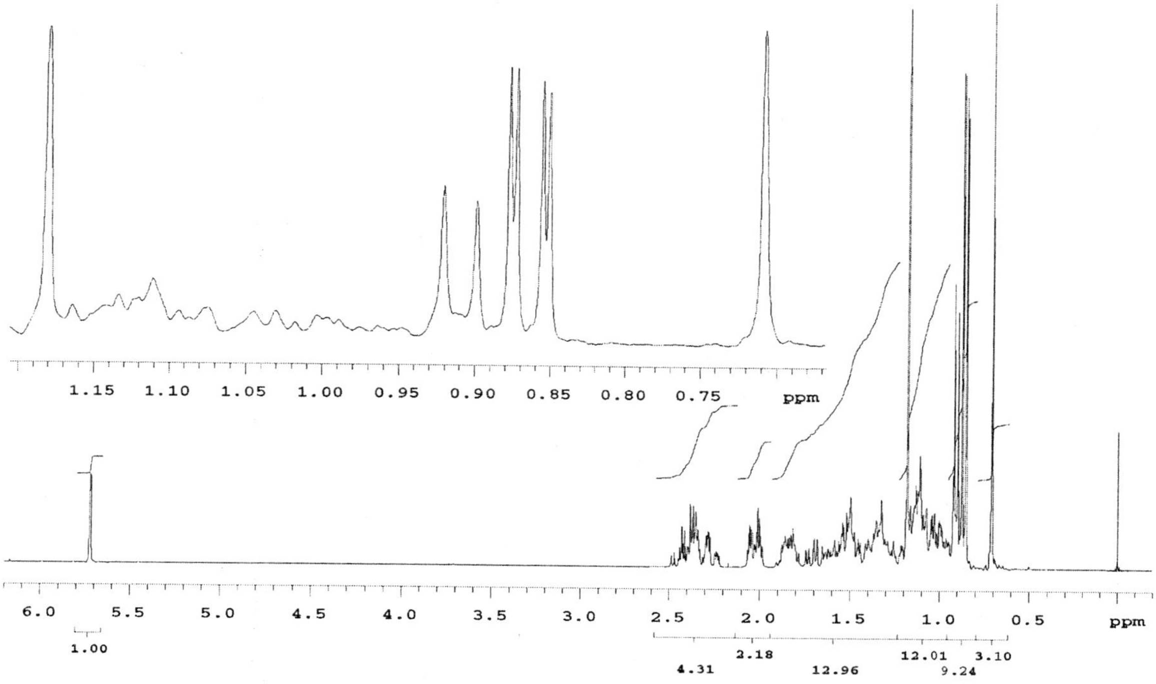

^{1}H-NMR Spectrum in $CDCl_3$

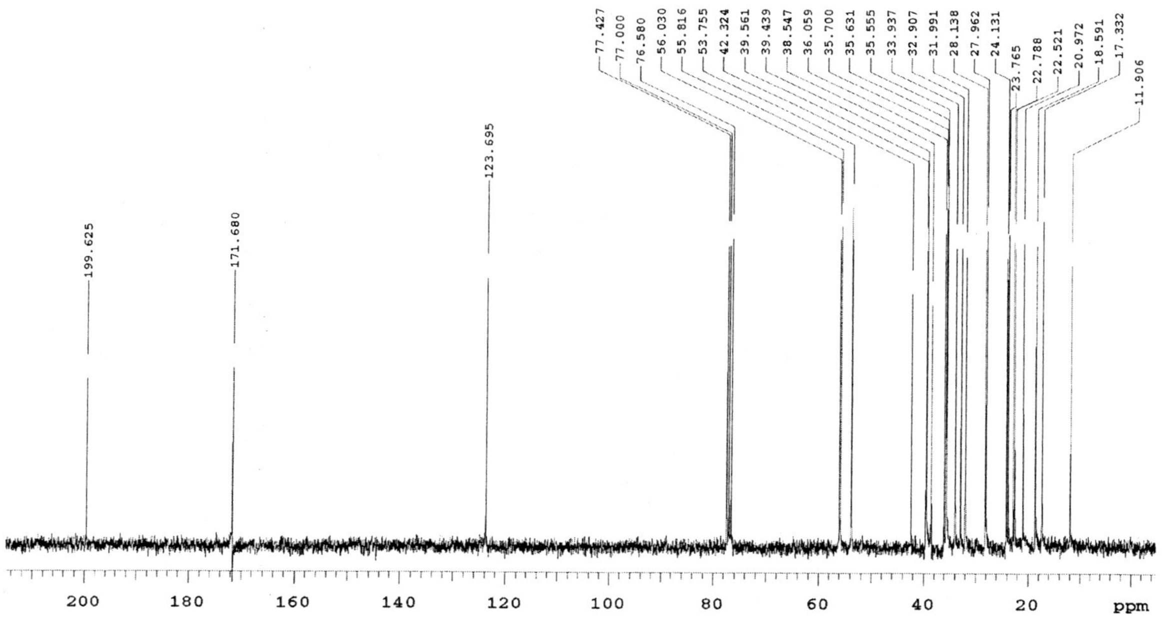

^{13}C-NMR Spectrum in $CDCl_3$

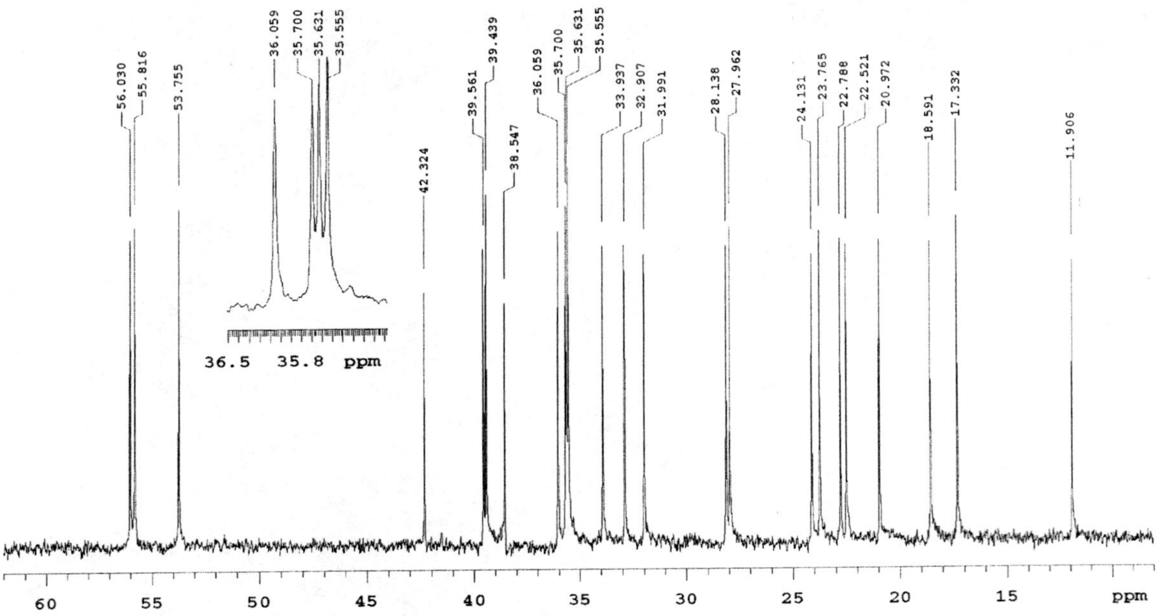

Expansion of ^{13}C-NMR Spectrum in $CDCl_3$

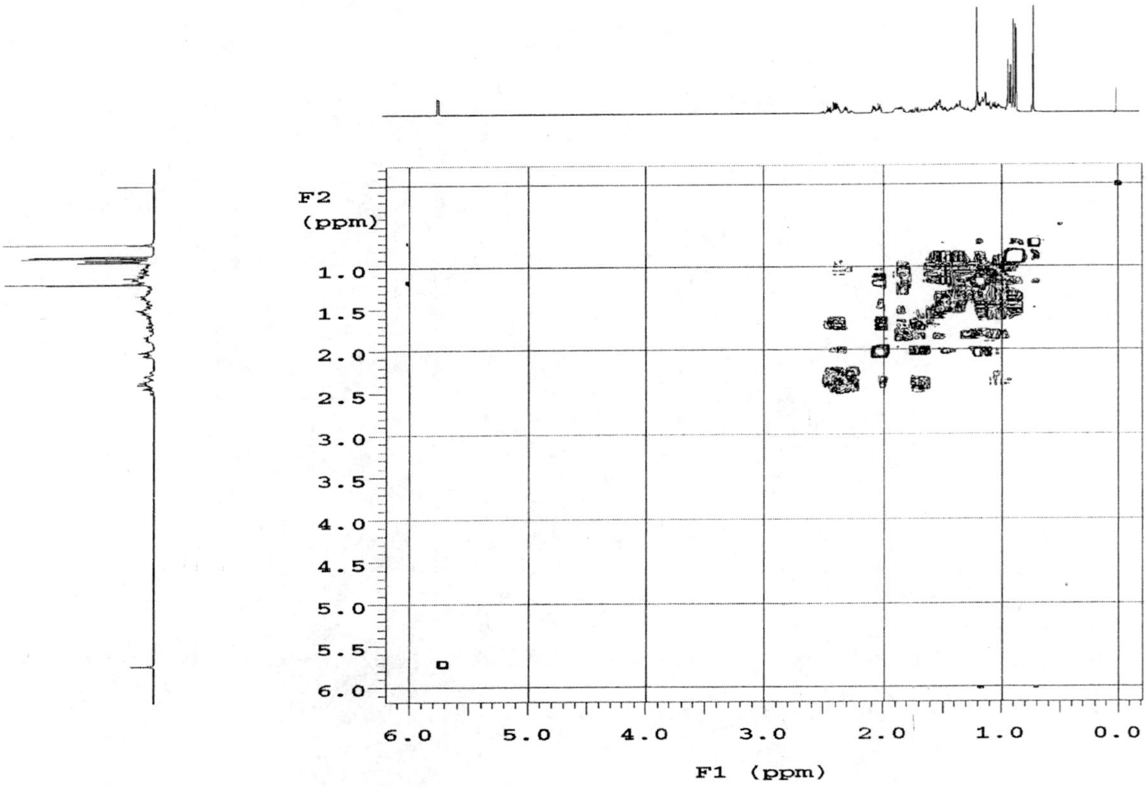

gCOSYSpectrum in $CDCl_3$

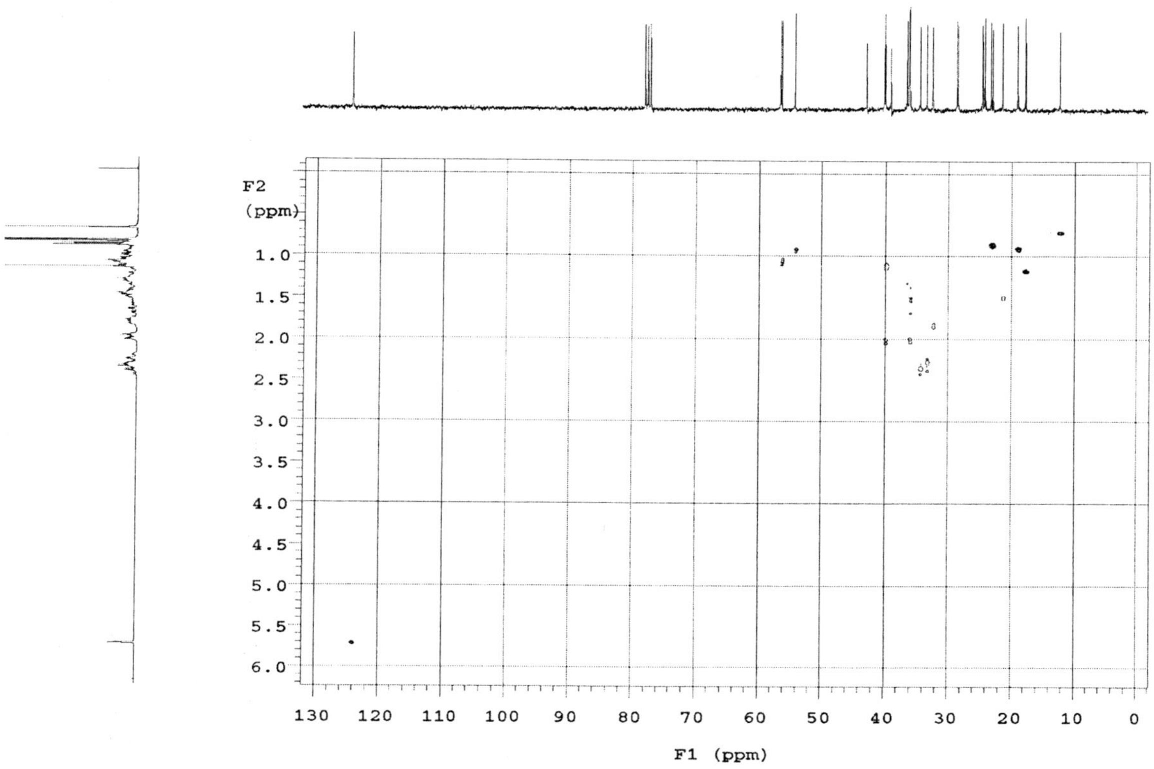

gHSQC Spectrum in $CDCl_3$

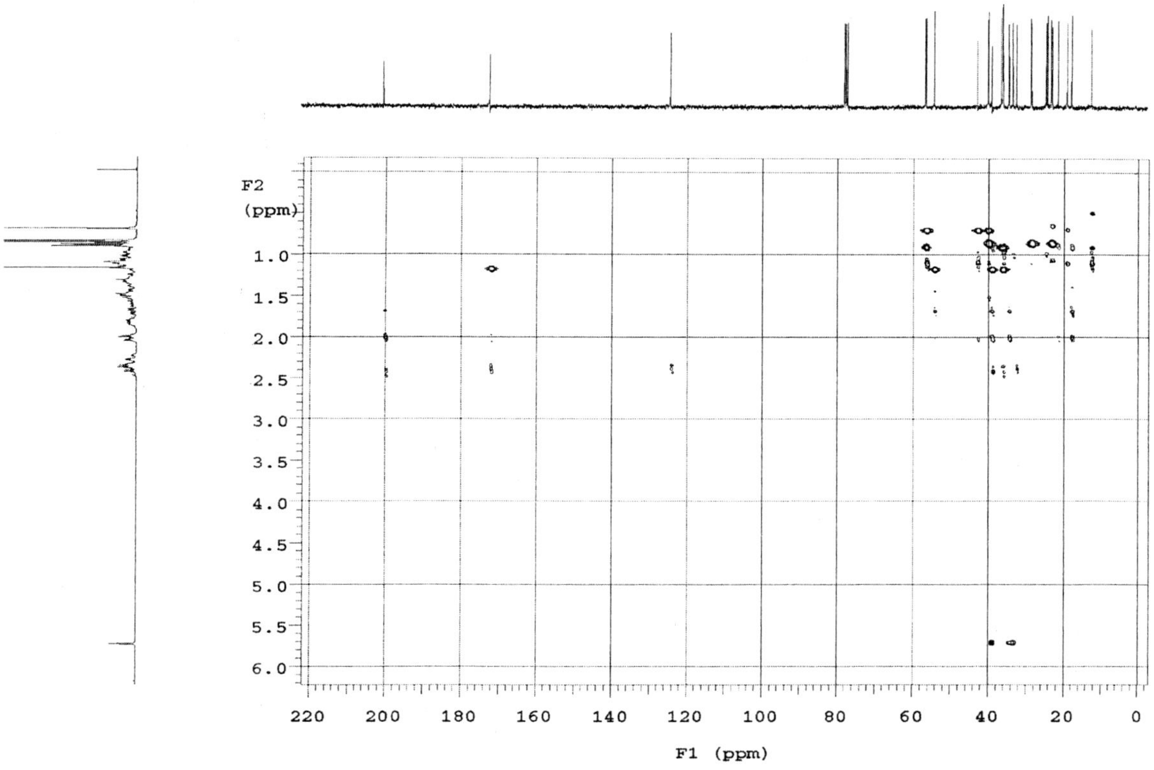

gHMBC Spectrum in $CDCl_3$

Problem 5.12 Given the data below (Proton, Carbon, gHSQC, gCOSY, and gHMBC), provide the structure of the molecule. Compound **5.11** was reacted with 1,2-ethanedithiol in the presence of boron trifluoride etherate to give **5.12**.

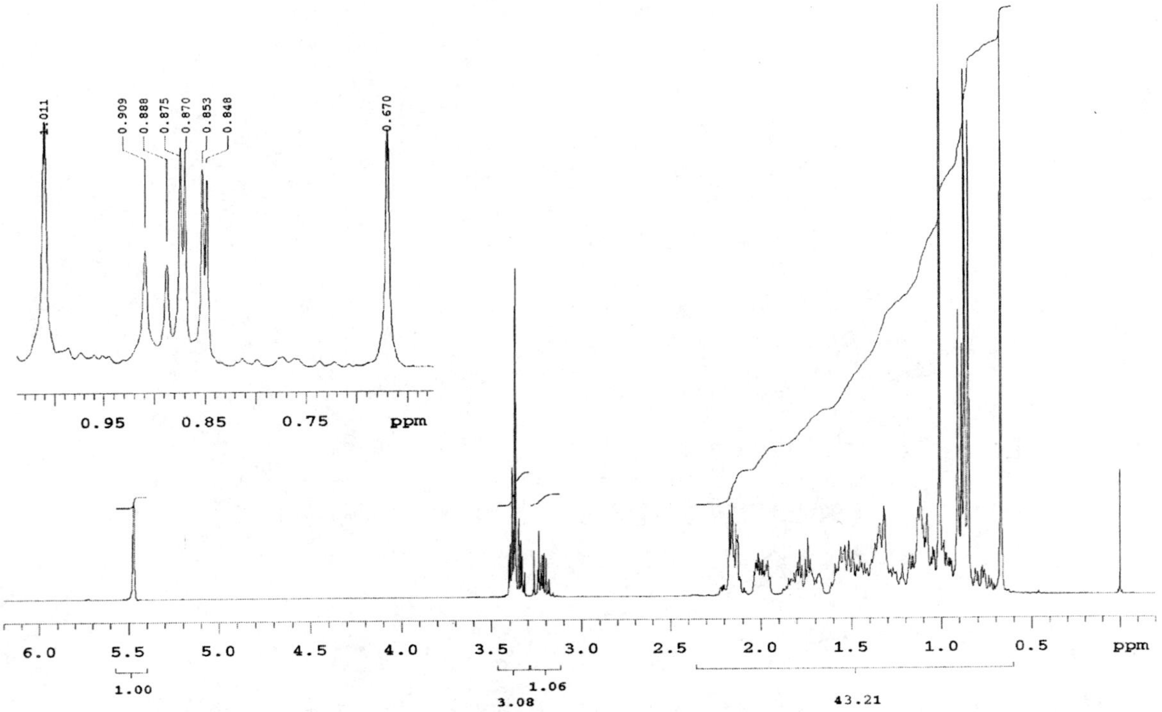

^{1}H-NMR Spectrum in $CDCl_3$

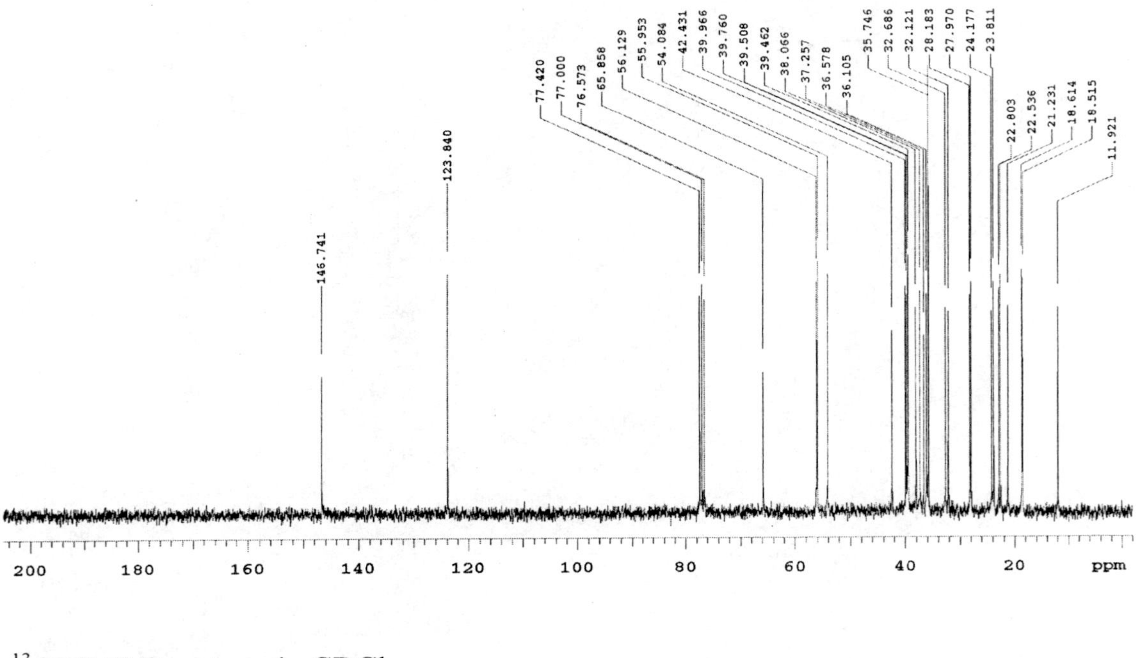

^{13}C-NMR Spectrum in $CDCl_3$

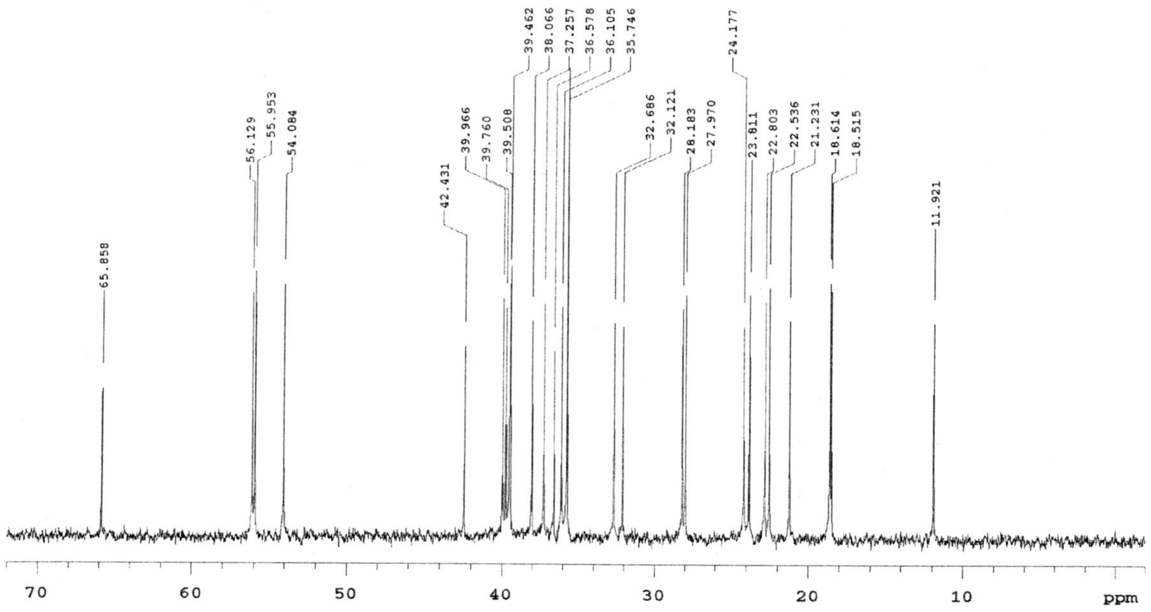

Expansion of ^{13}C-NMR Spectrum in $CDCl_3$

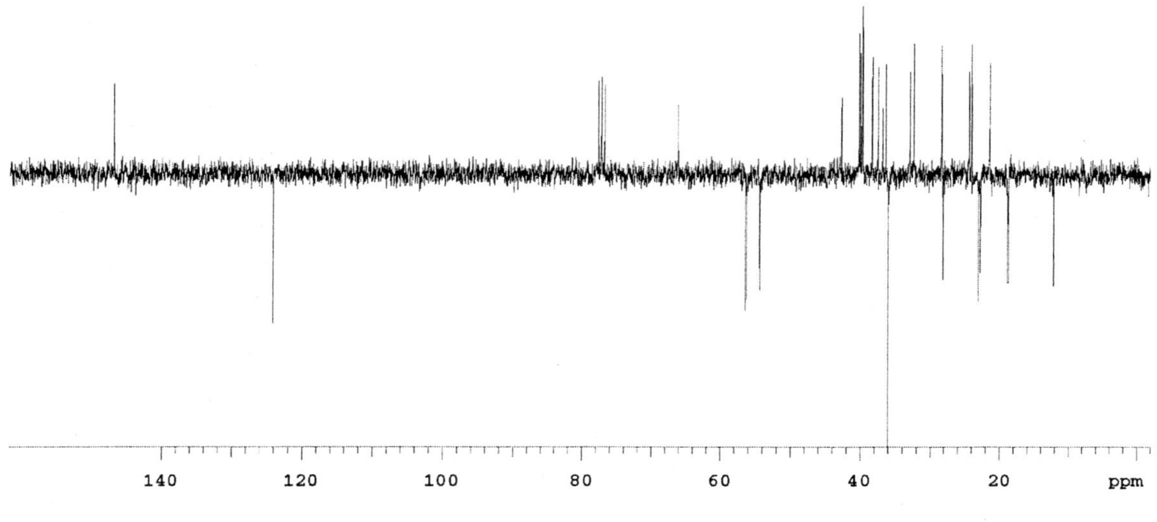

APT Spectrum in $CDCl_3$

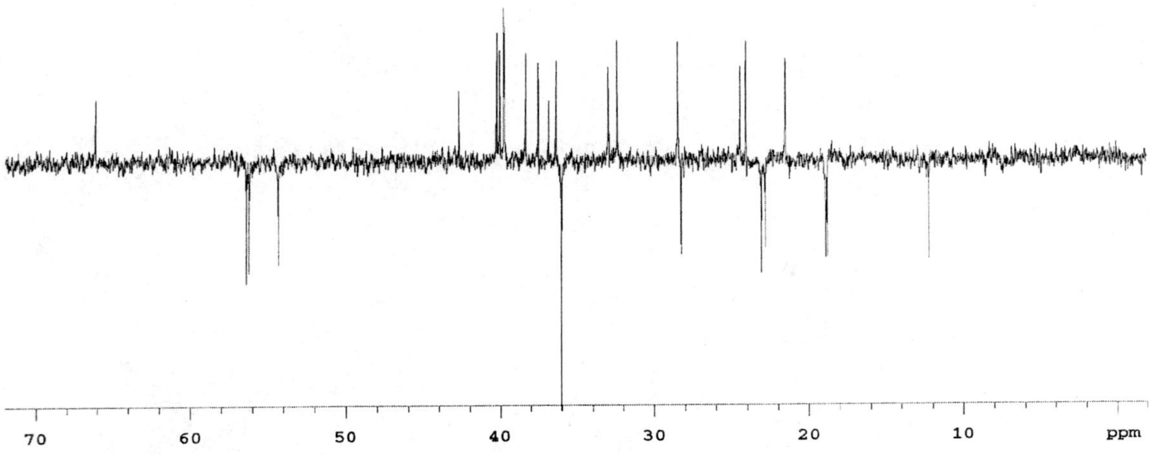

Expansion of APT Spectrum in CDCl$_3$

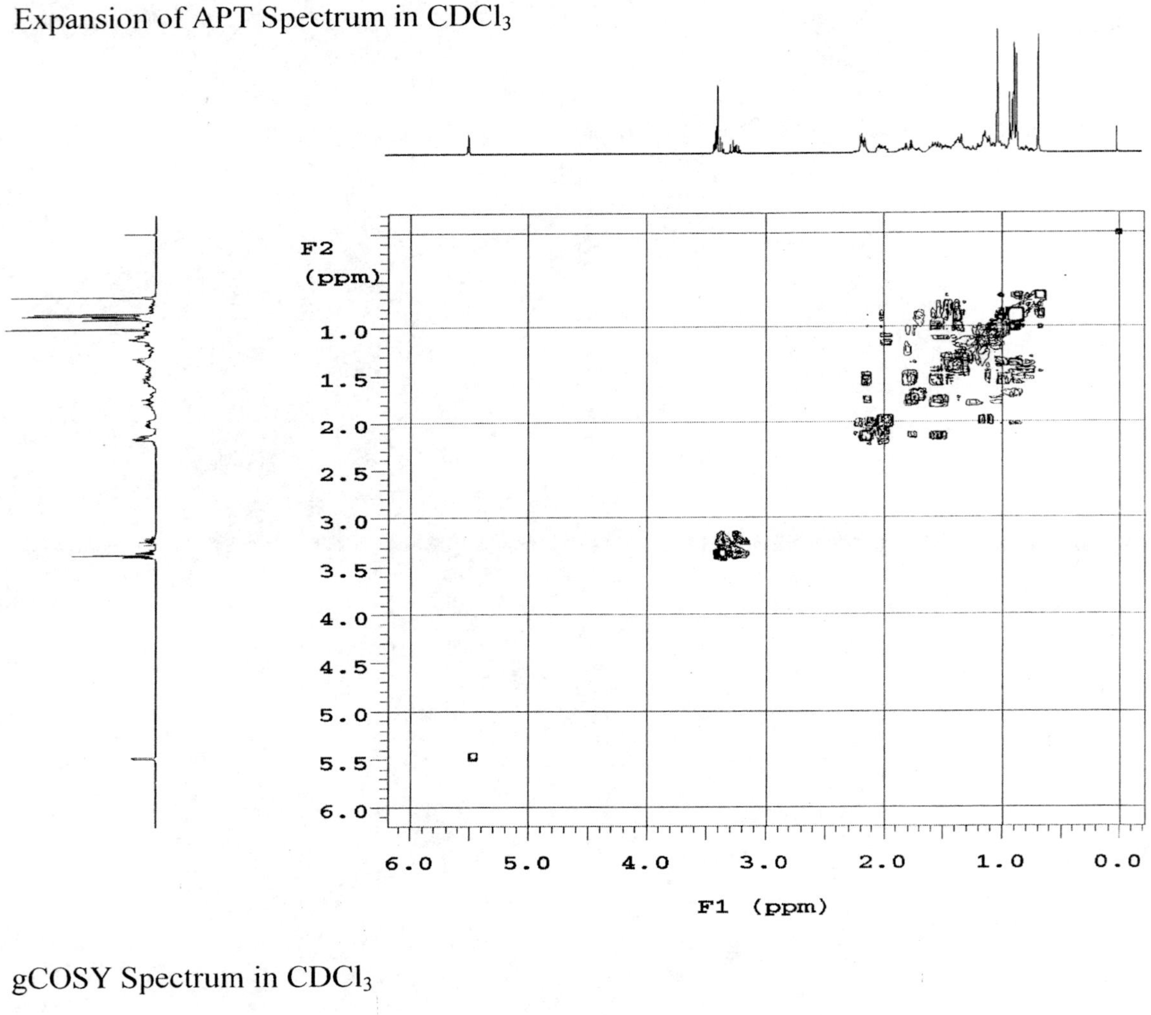

gCOSY Spectrum in CDCl$_3$

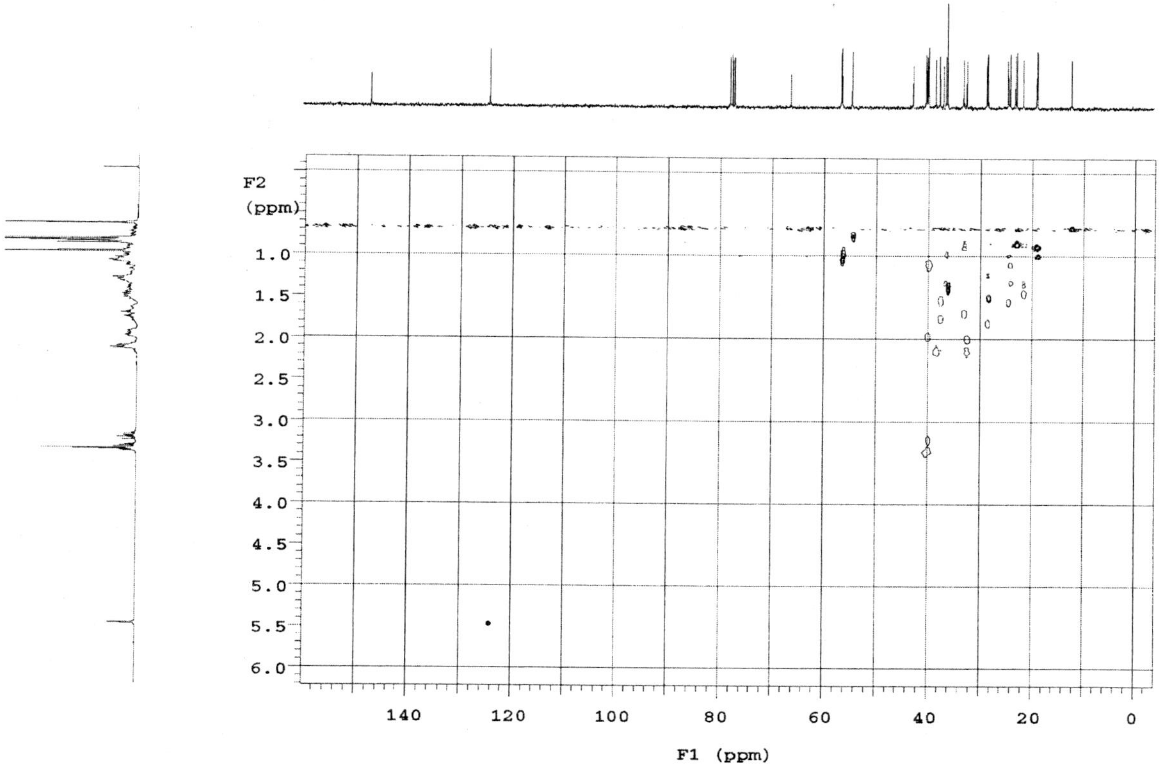

gHSQC Spectrum in $CDCl_3$

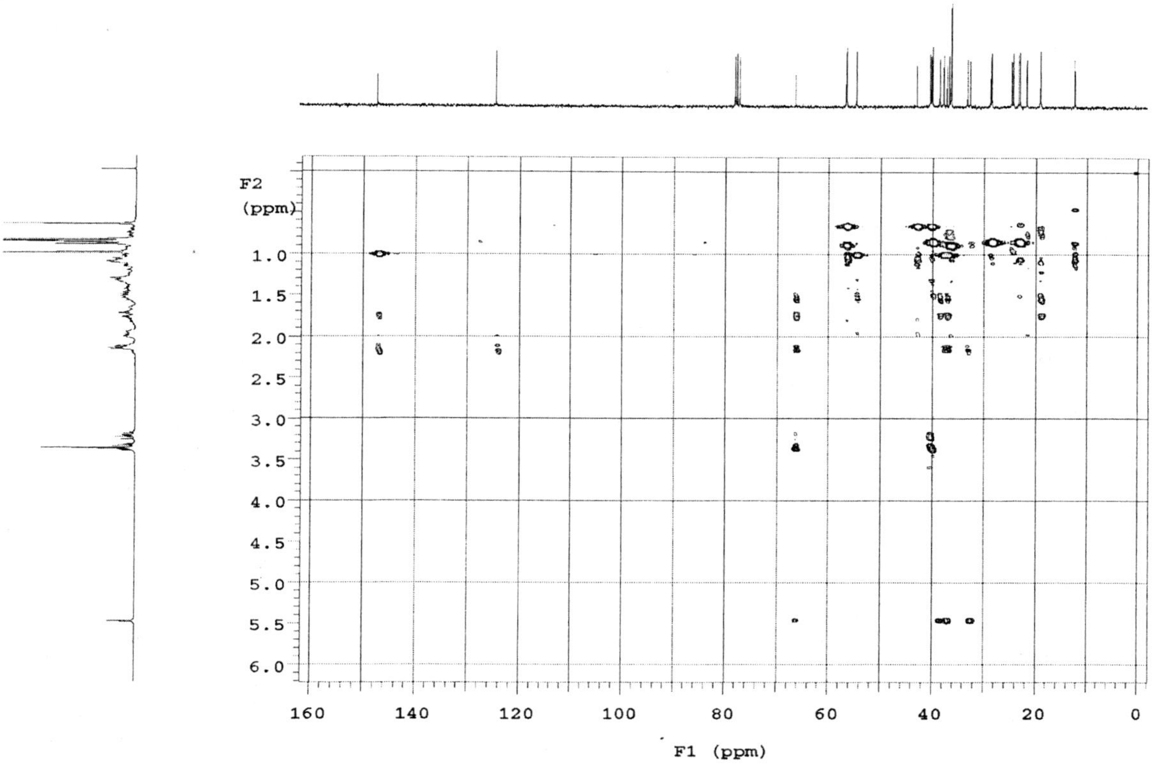

gHMBC Spectrum in $CDCl_3$

Problem 5.13 Given the data below (Proton, Carbon, gHSQC, gCOSY, and gHMBC), provide the structure of the molecule. Compound **5.12** was desulfurized using Na in liquid ammonia to afford **5.13**.

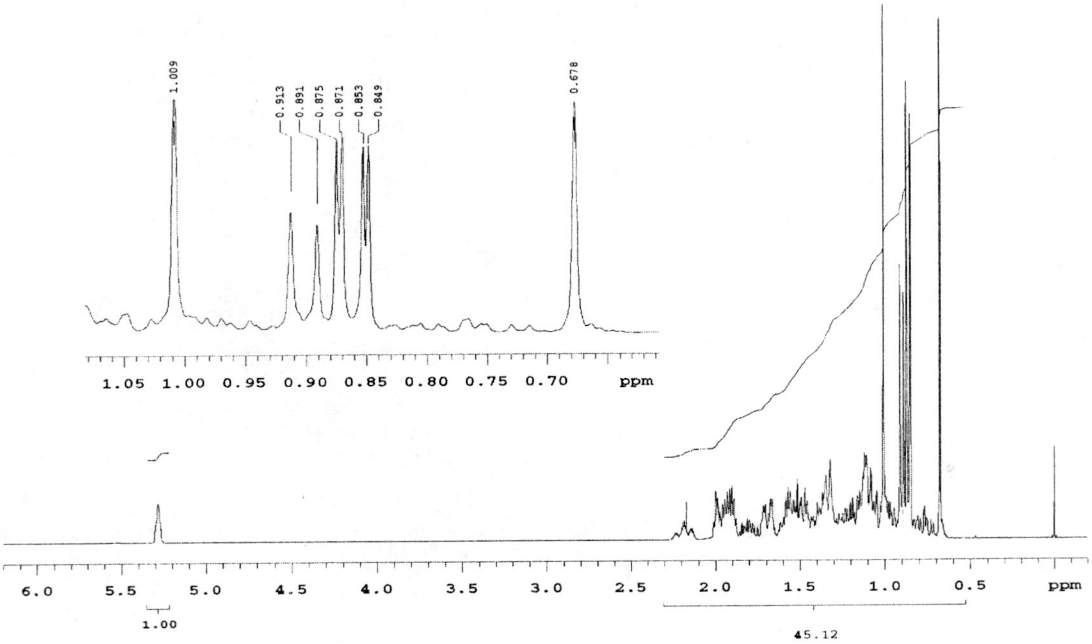

^{1}H-NMR Spectrum in $CDCl_3$

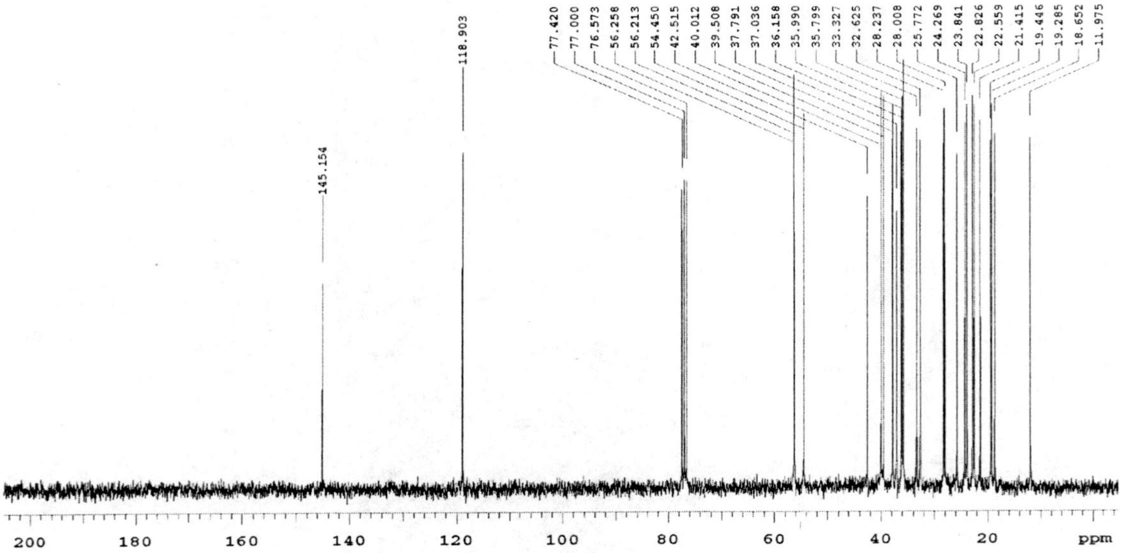

^{13}C-NMR Spectrum in $CDCl_3$

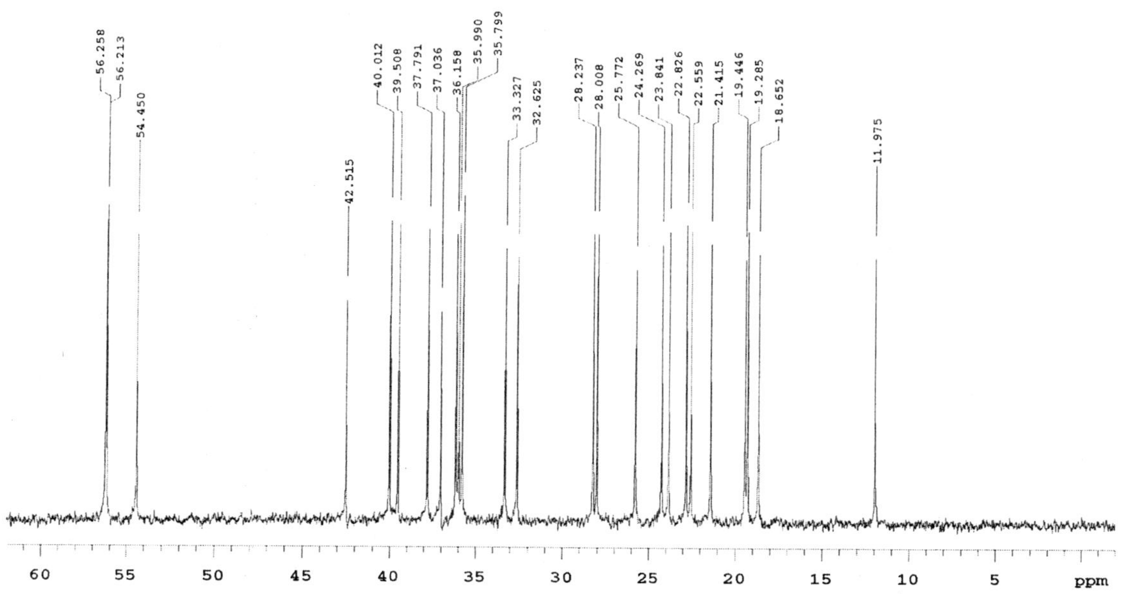

Expansion of ^{13}C-NMR Spectrum in $CDCl_3$

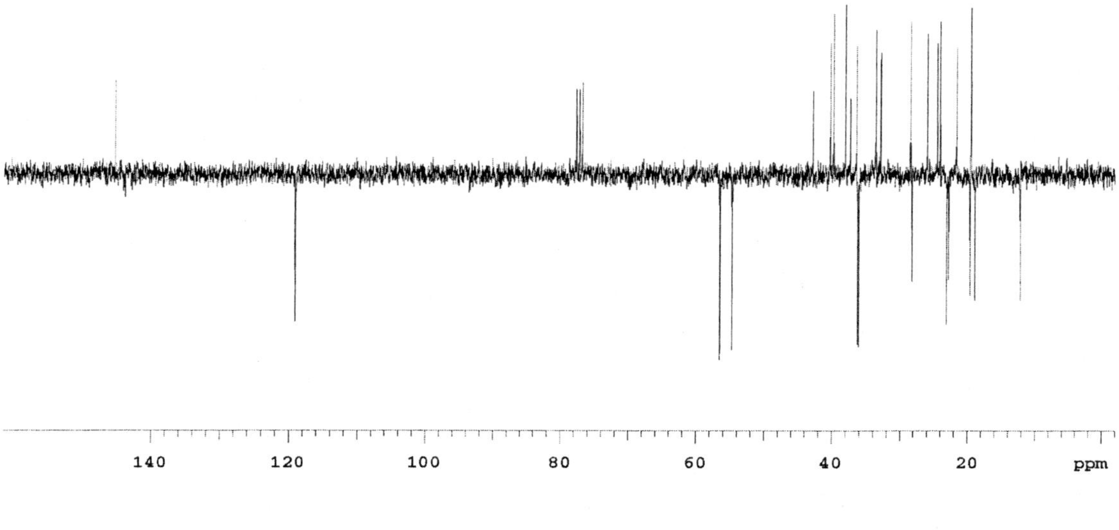

APT Spectrum in $CDCl_3$

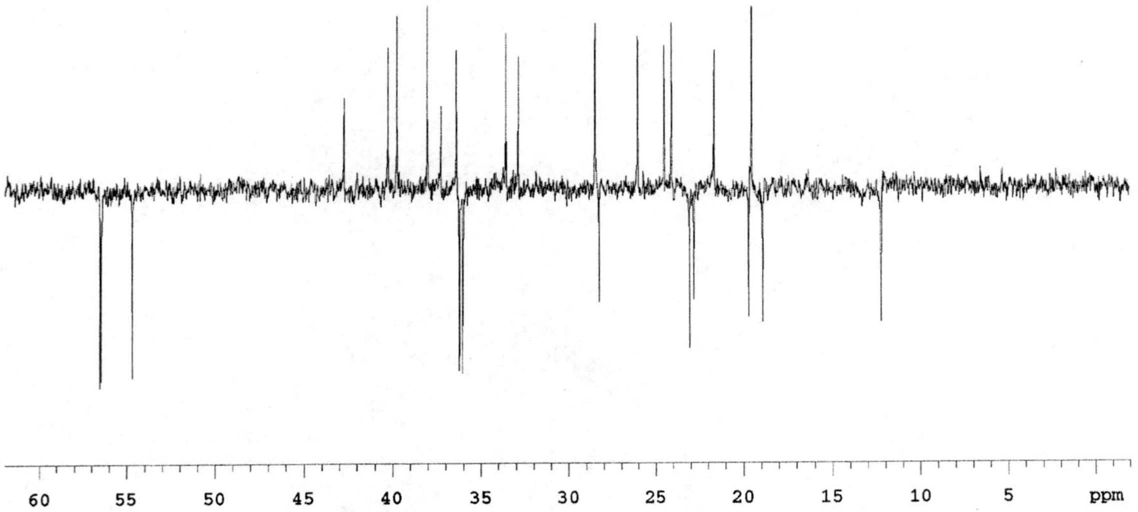

Expansion of APT Spectrum in $CDCl_3$

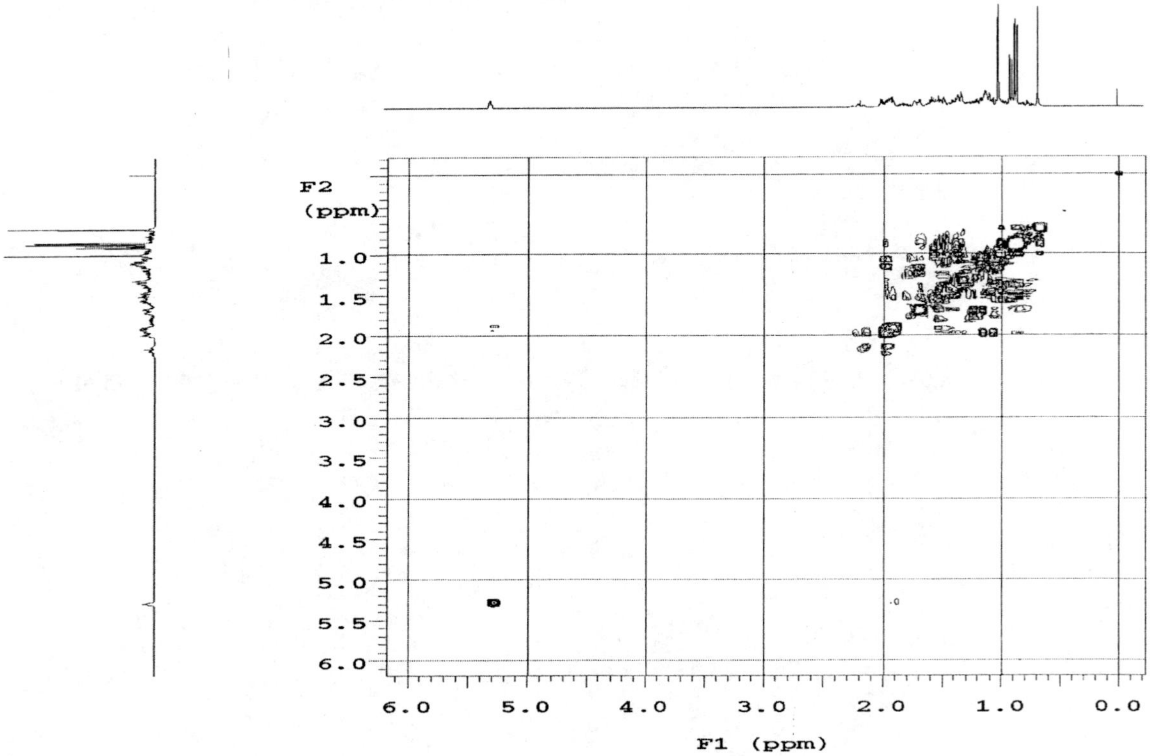

gCOSY Spectrum in $CDCl_3$

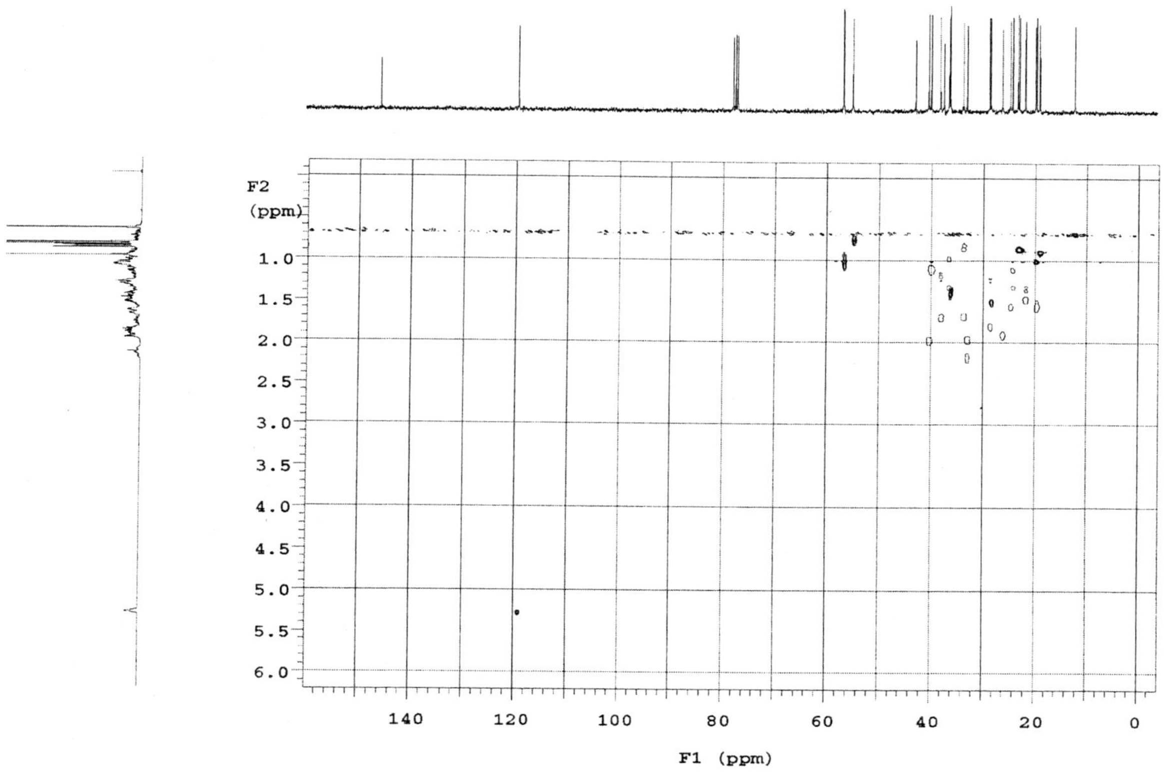

gHSQC Spectrum in $CDCl_3$

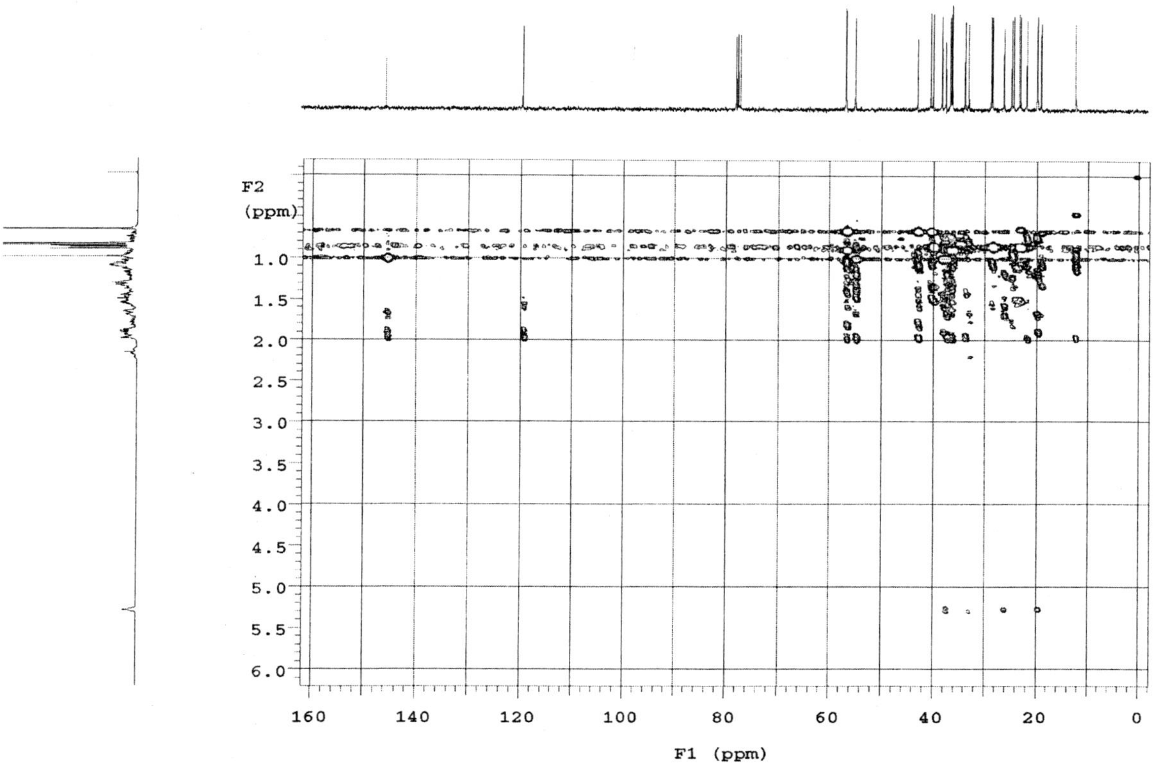

gHMBC Spectrum in $CDCl_3$

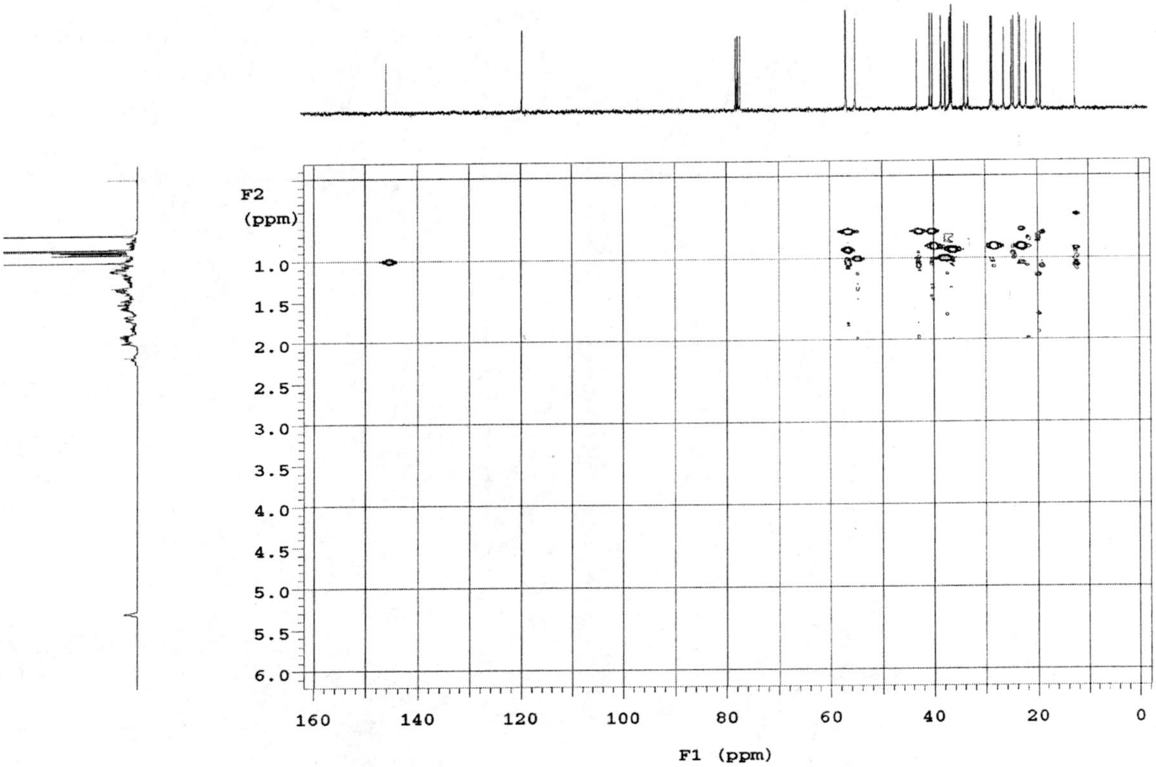

gHMBC Spectrum in $CDCl_3$ at different intensity level

Problem 5.14 Given the data below (Proton, Carbon, gHSQC, gCOSY, and gHMBC), provide the structure of the molecule. Also, compound **5.11** was reacted with methyl iodide (2 equivalents) in base to afford **5.14**.

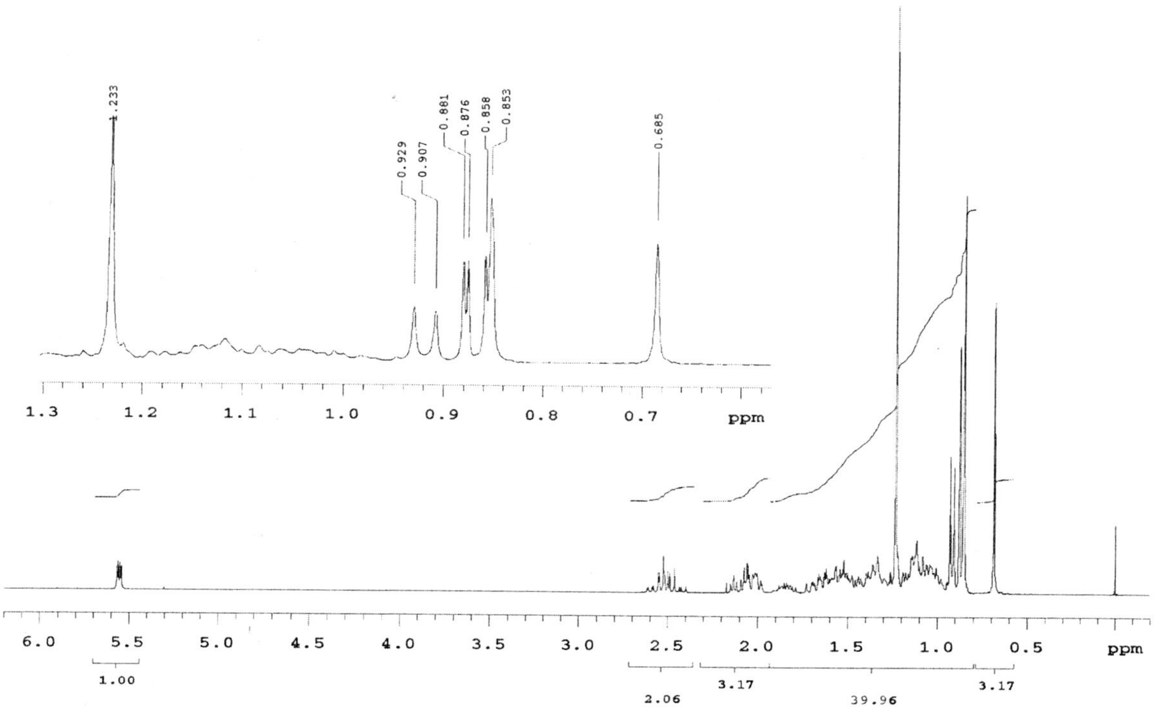

^{1}H-NMR Spectrum in $CDCl_3$

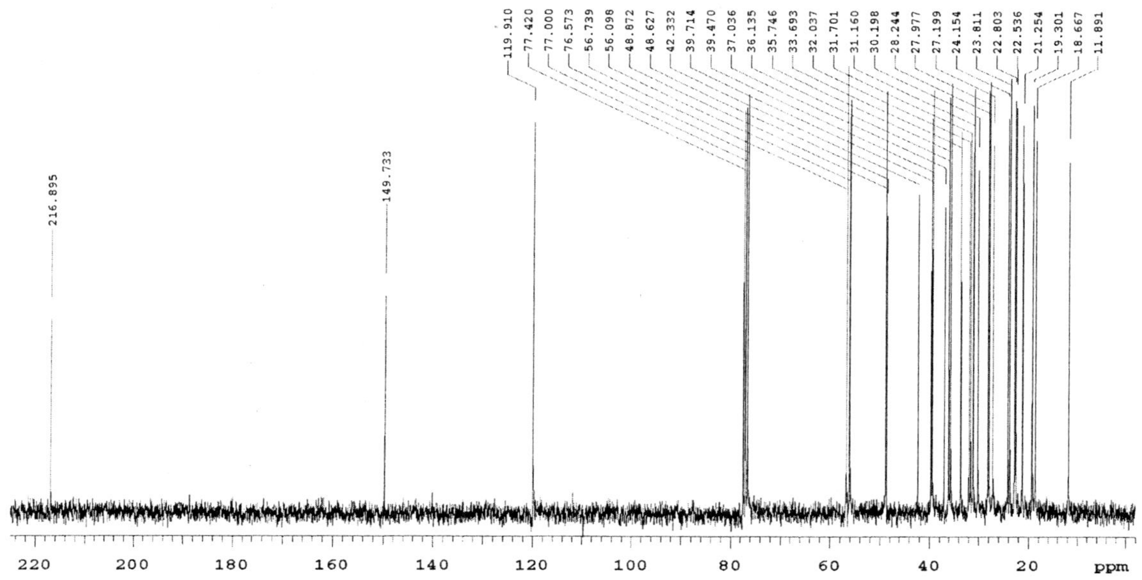

^{13}C-NMR Spectrum in $CDCl_3$

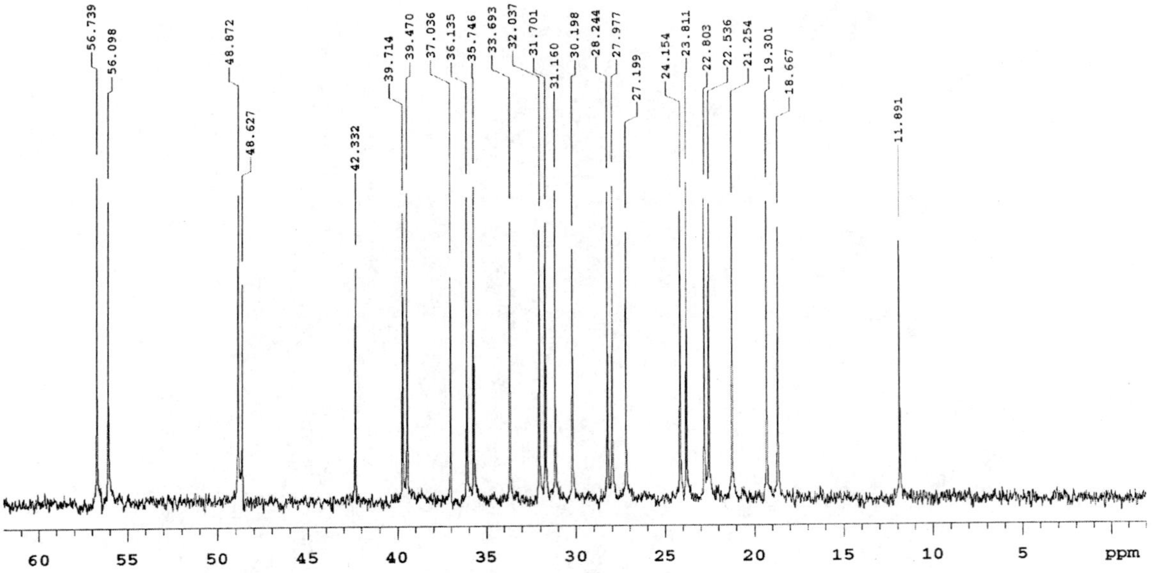

Expansion of ^{13}C-NMR Spectrum in $CDCl_3$

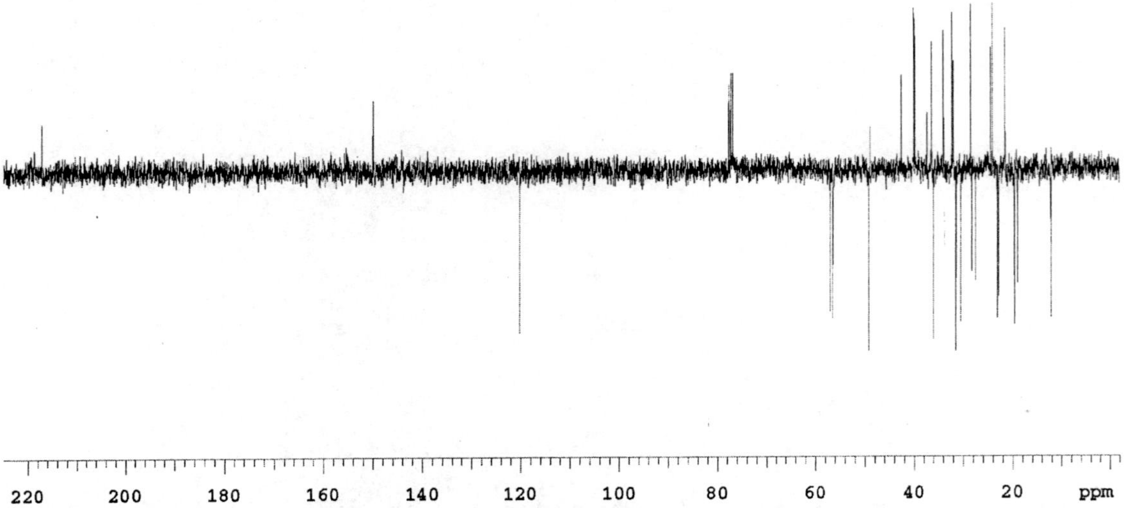

APT Spectrum in $CDCl_3$

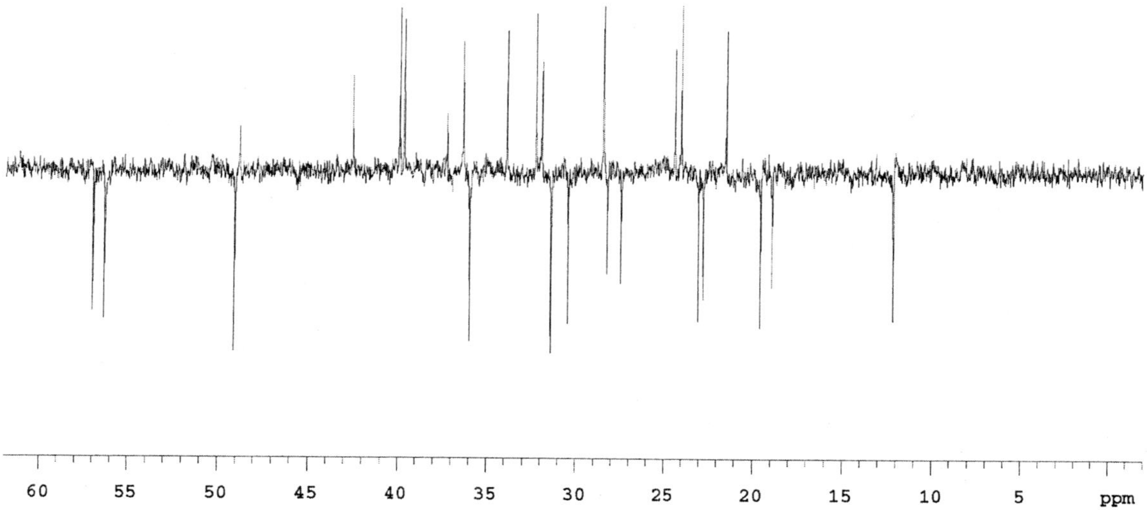

Expansion of APT Spectrum in $CDCl_3$

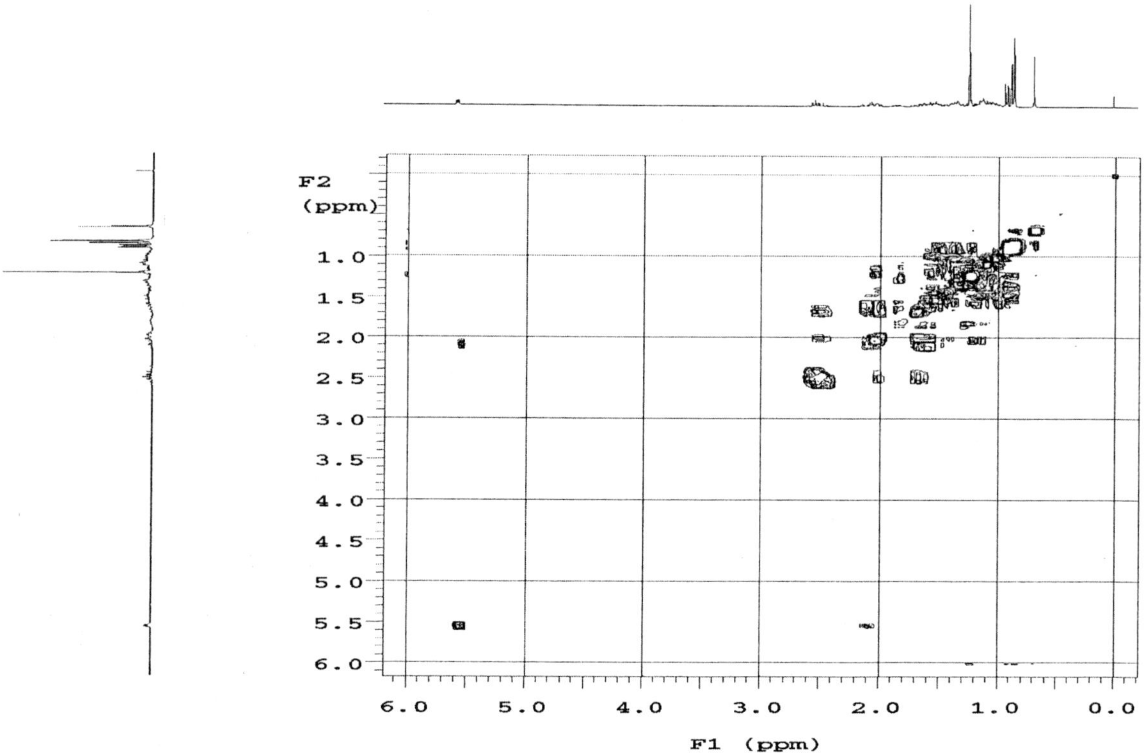

gCOSY Spectrum in $CDCl_3$

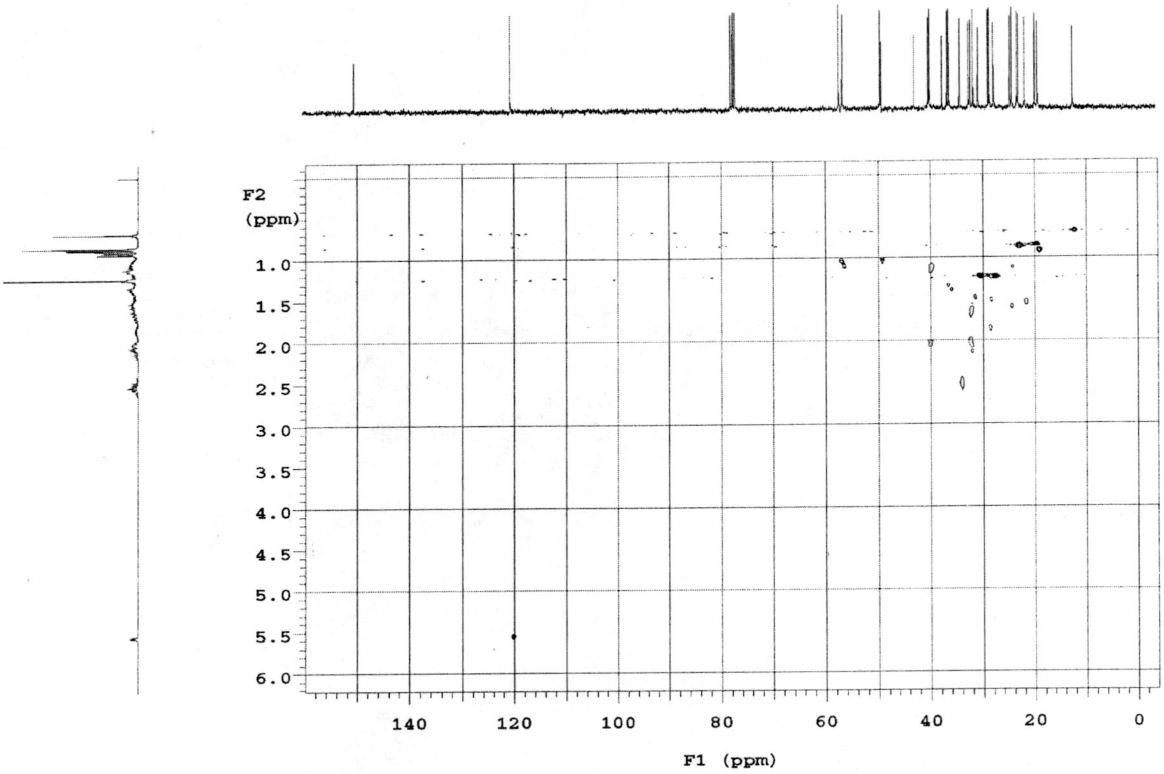

gHSQC Spectrum in $CDCl_3$

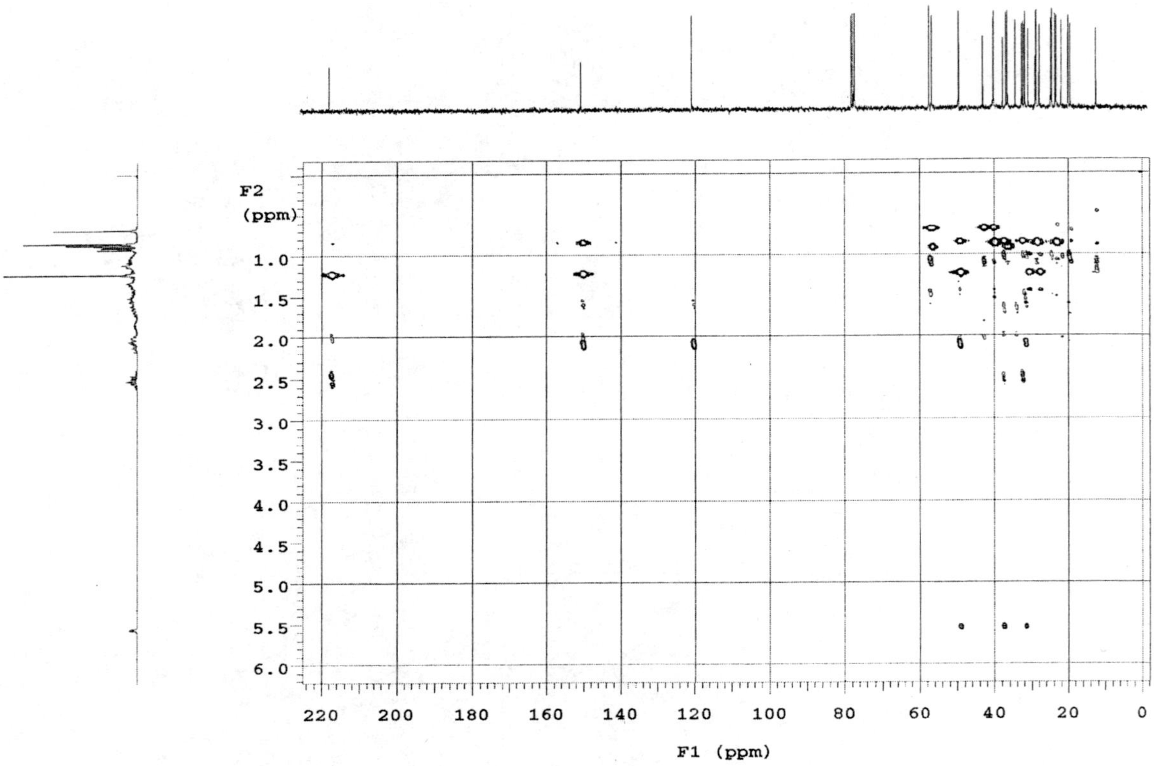

gHMBC Spectrum in $CDCl_3$

Problem 5.15 Given the data below (Proton, Carbon, gHSQC, gCOSY, and gHMBC), provide the structure of the molecule. Compound **4.24** reacted with thionyl chloride to afford compound **5.15**.

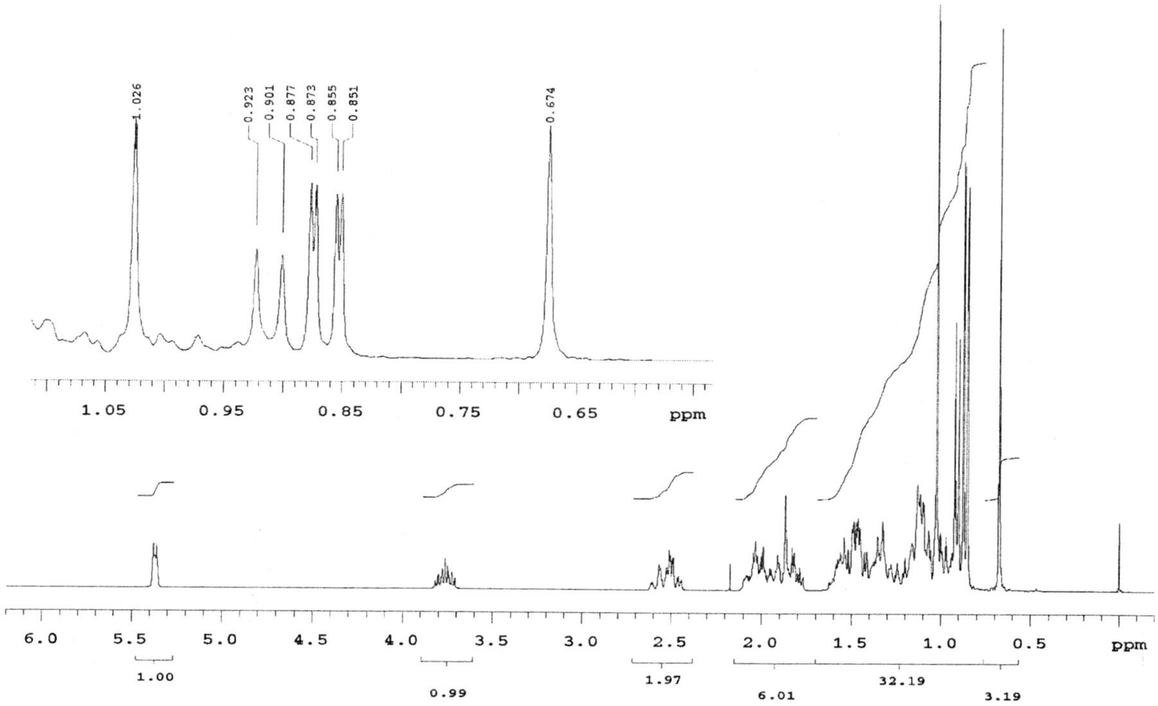

^{1}H-NMR Spectrum in $CDCl_3$

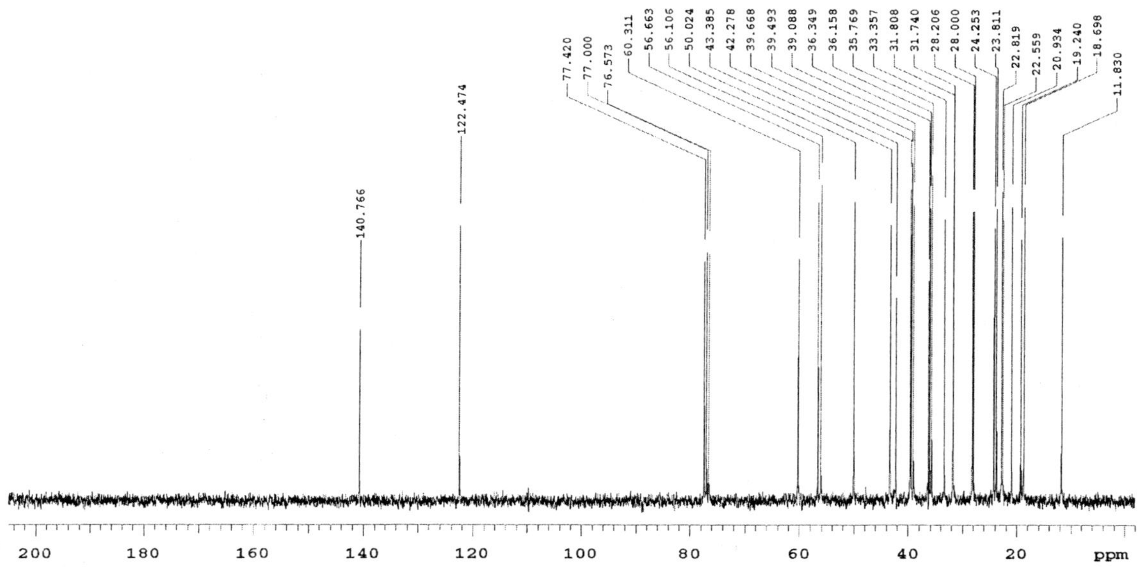

^{13}C-NMR Spectrum in $CDCl_3$

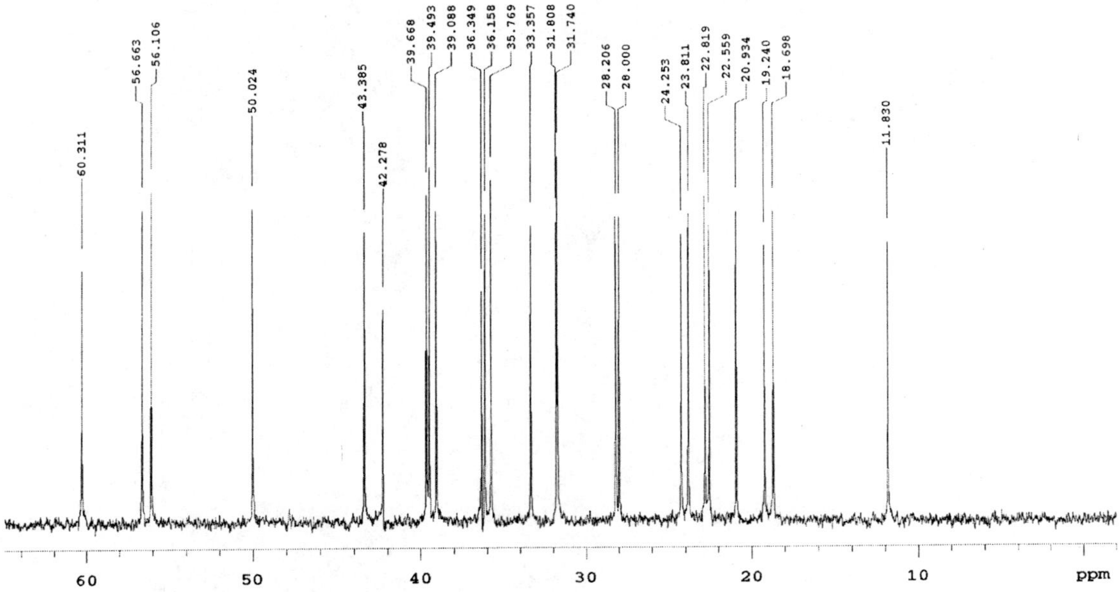

Expansion of ^{13}C-NMR Spectrum in $CDCl_3$

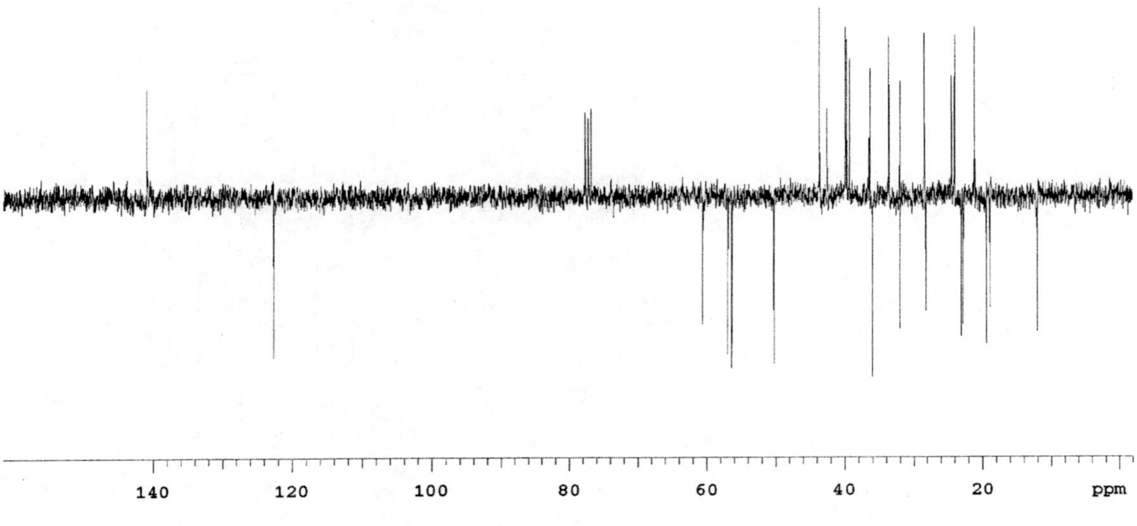

APT Spectrum in $CDCl_3$

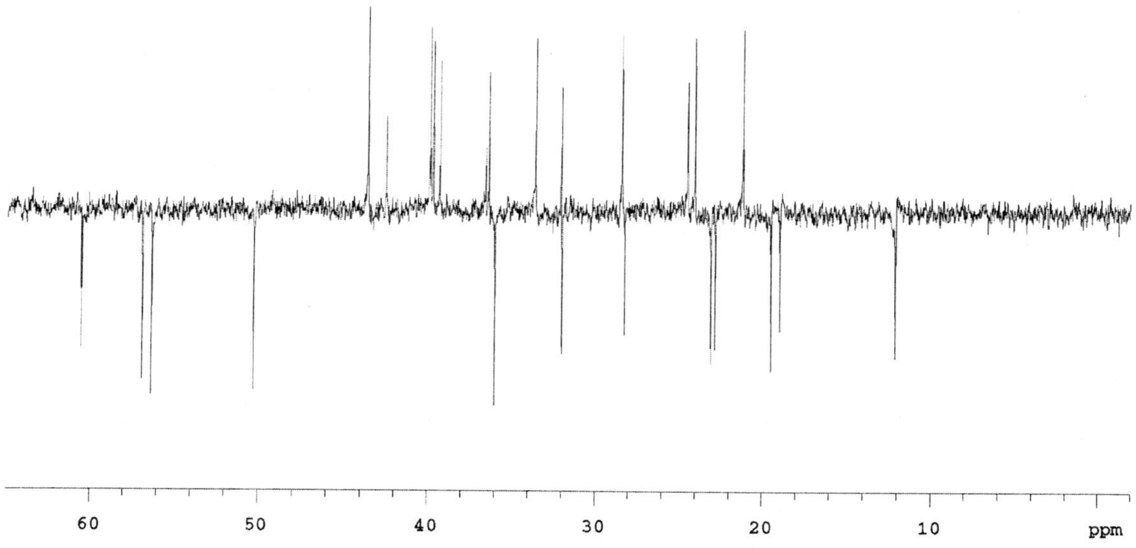

Expansion of APT Spectrum in $CDCl_3$

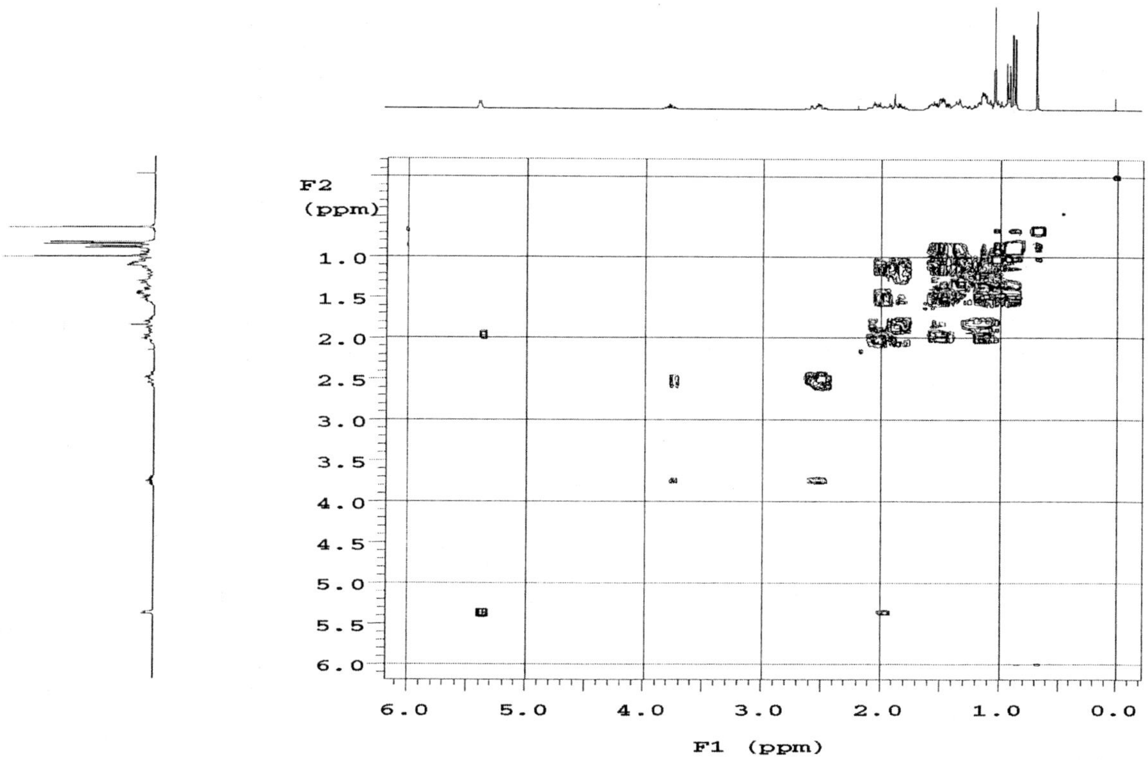

gCOSY Spectrum in $CDCl_3$

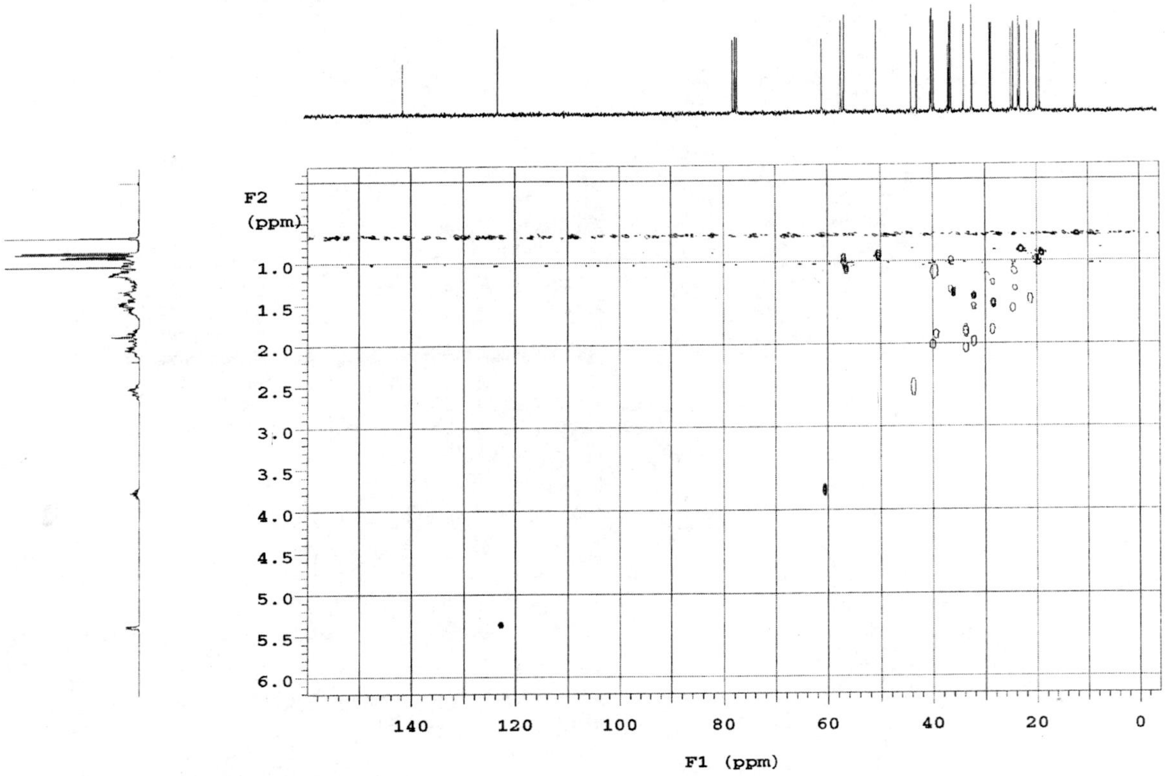

gHSQC Spectrum in $CDCl_3$

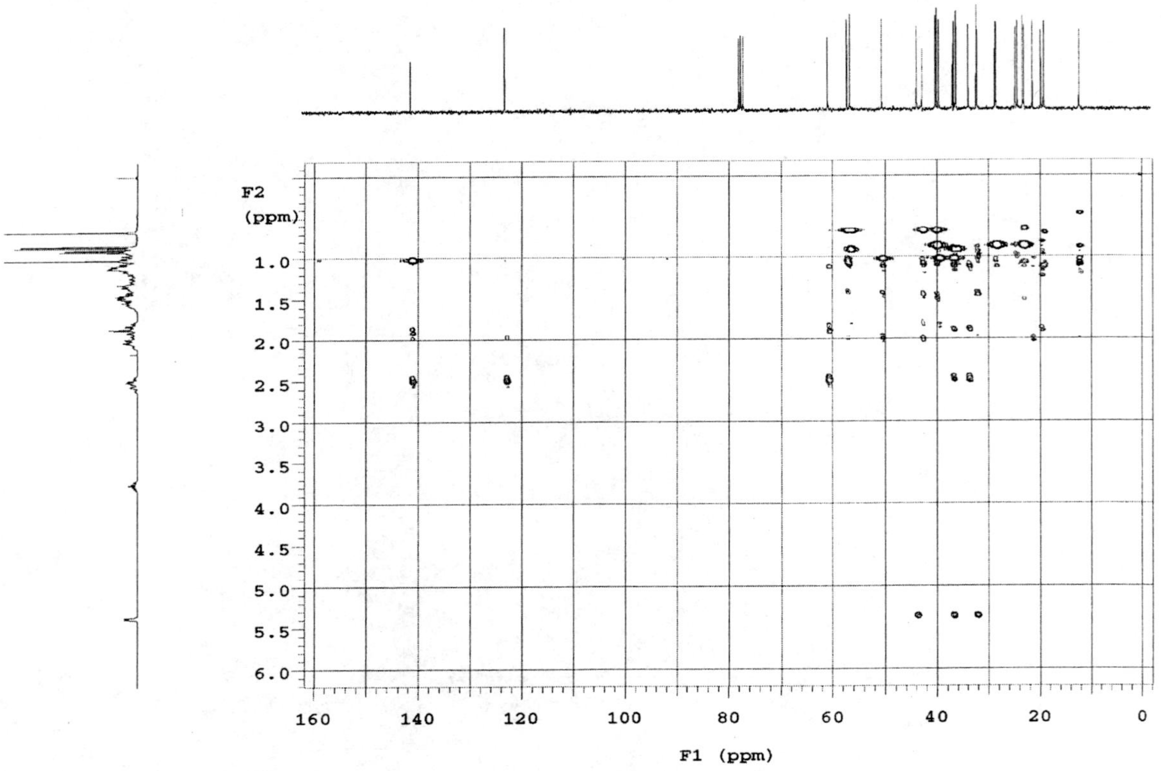

gHMBC Spectrum in $CDCl_3$

Problem 5.16 Given the data below (Proton, Carbon, gHSQC, gCOSY, and gHMBC), provide the structure of the molecule. Compound **5.15** was dechlorinated using Na in liquid ammonia to afford **5.16**.

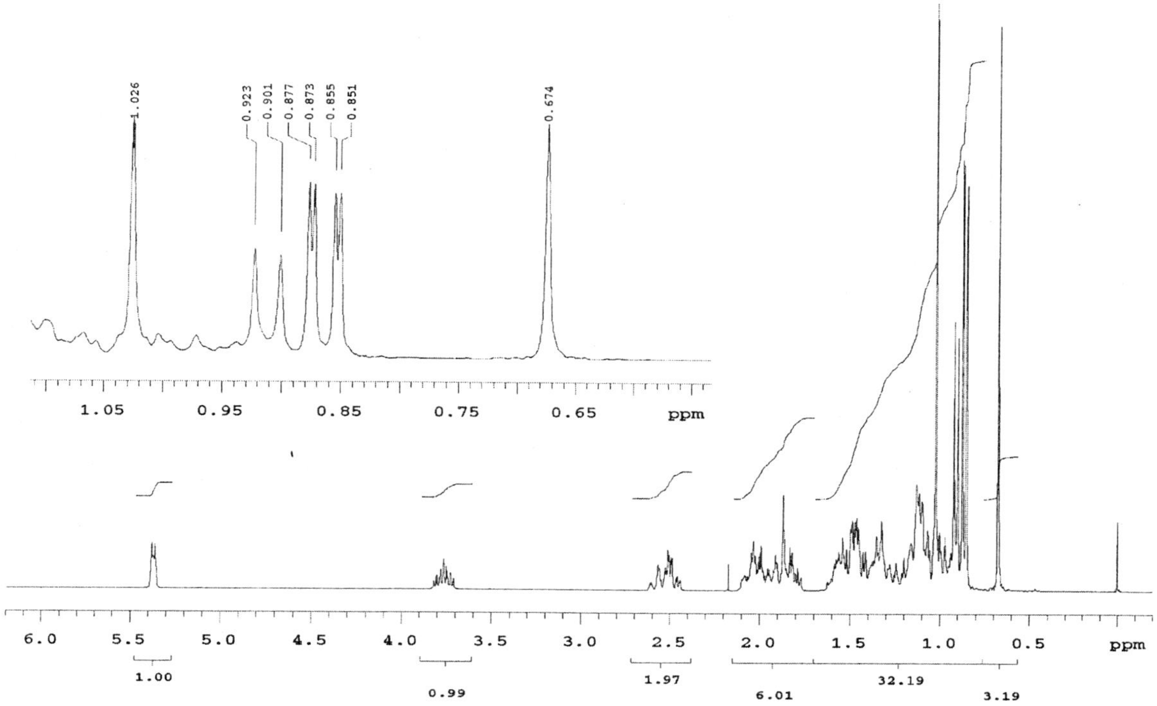

^{1}H-NMR Spectrum in $CDCl_3$

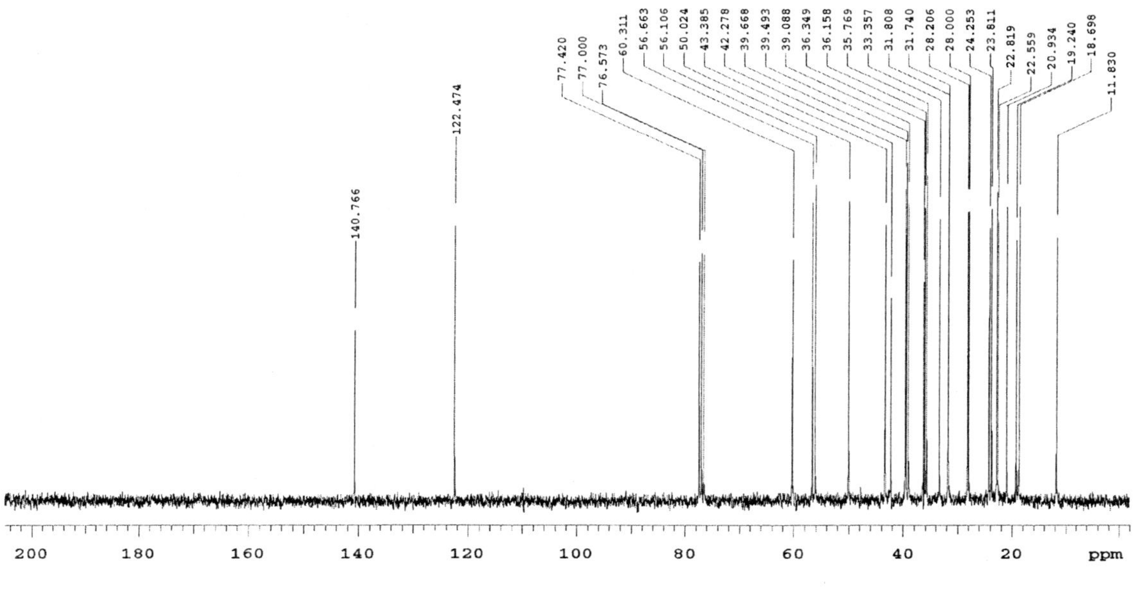

^{13}C-NMR Spectrum in $CDCl_3$

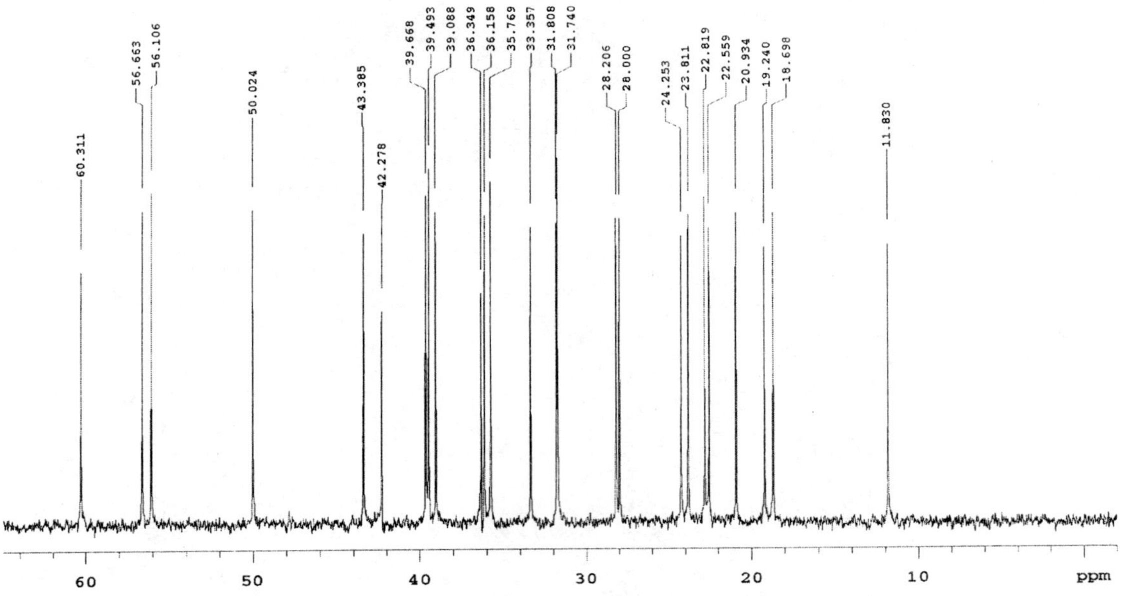

Expansion of ^{13}C-NMR Spectrum in $CDCl_3$

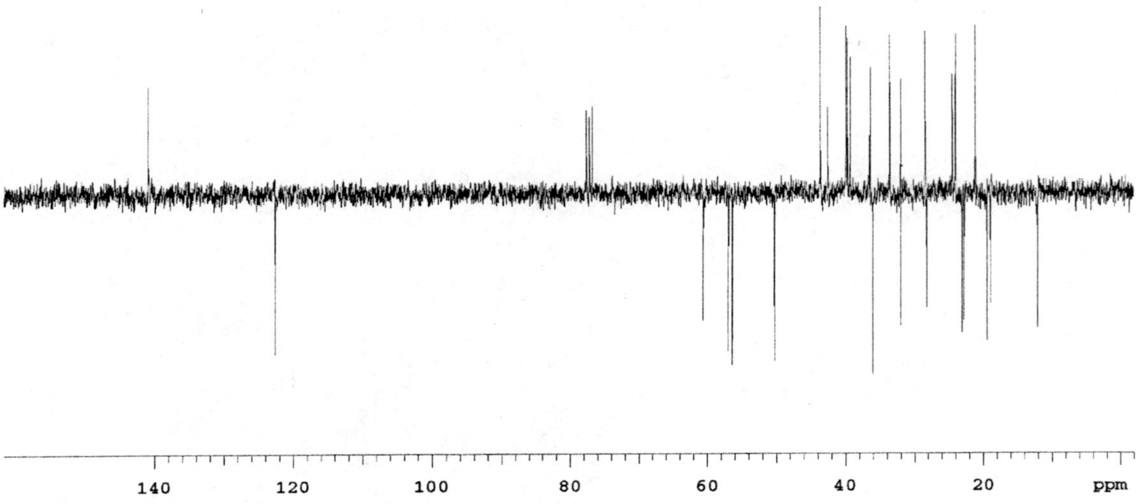

APT Spectrum in $CDCl_3$

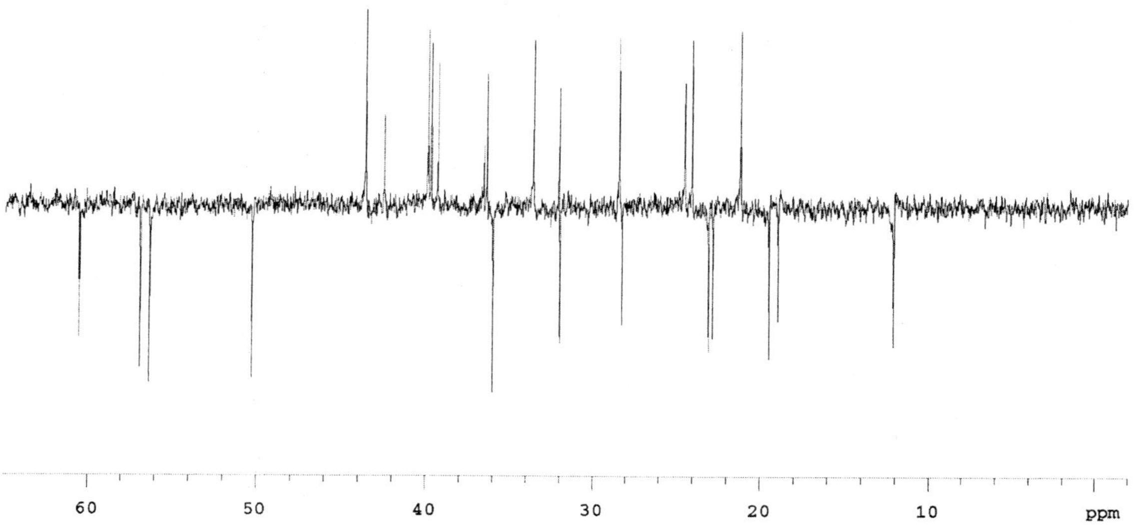

Expansion of APT Spectrum in $CDCl_3$

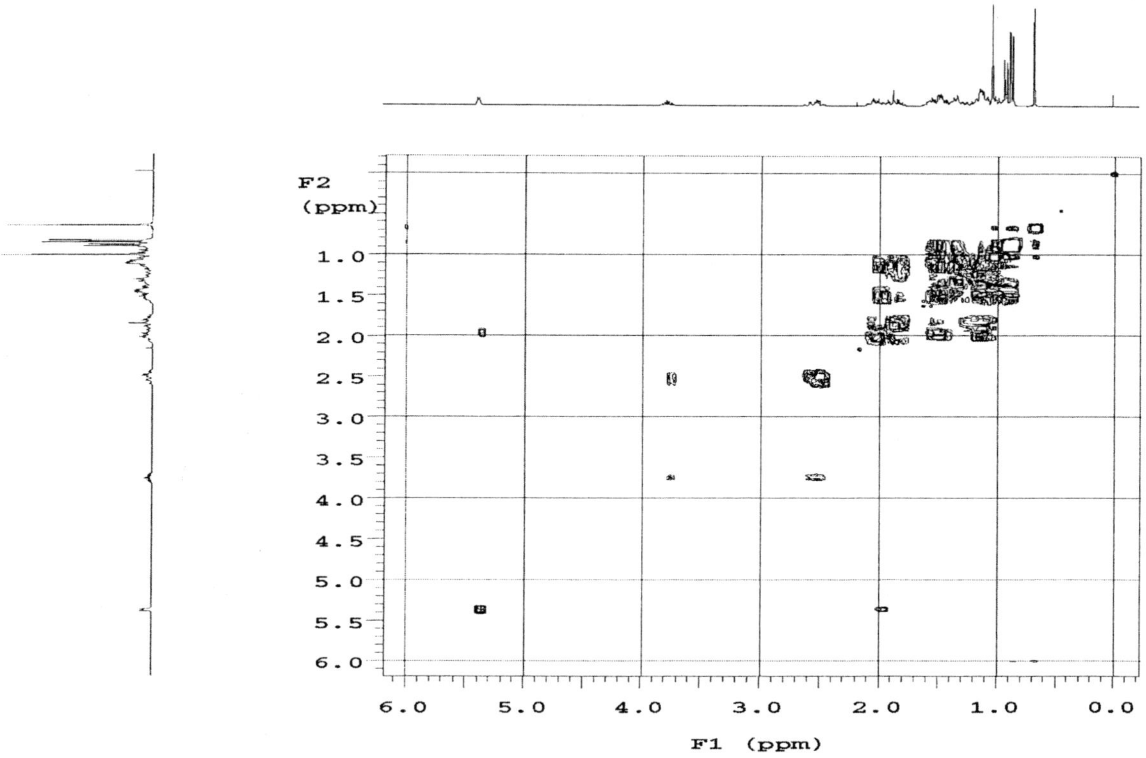

gCOSY Spectrum in $CDCl_3$

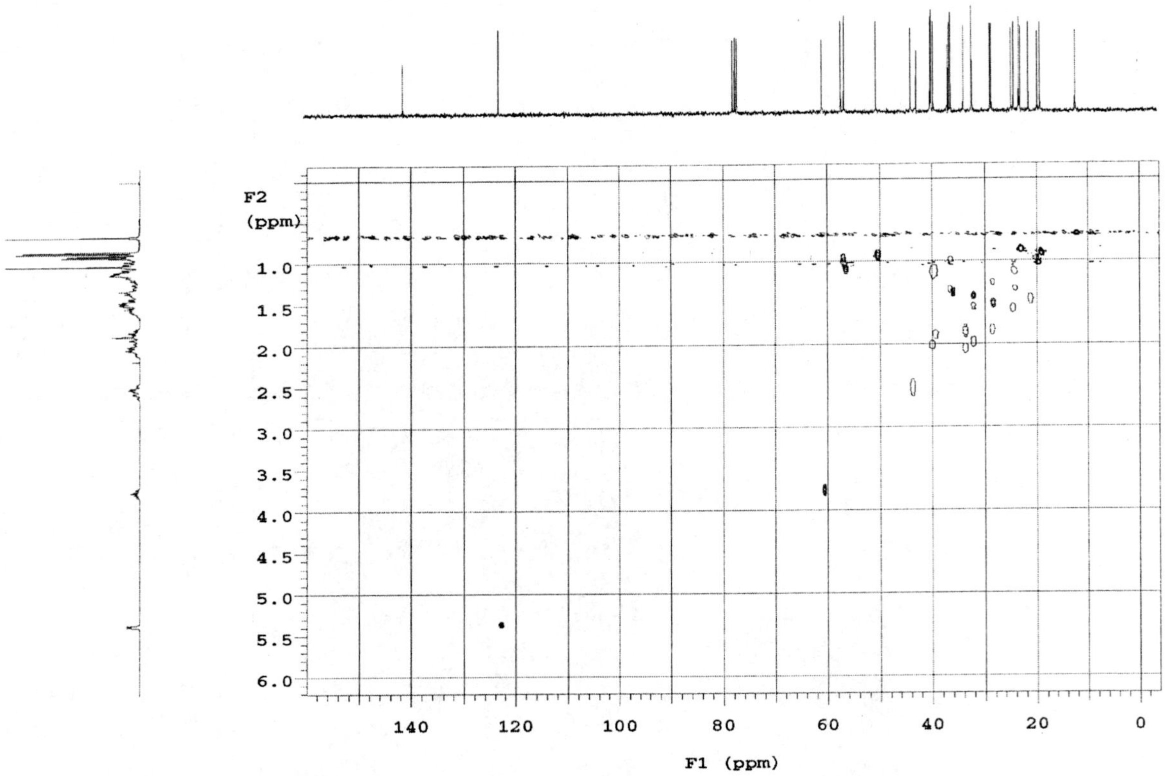

gHSQC Spectrum in $CDCl_3$

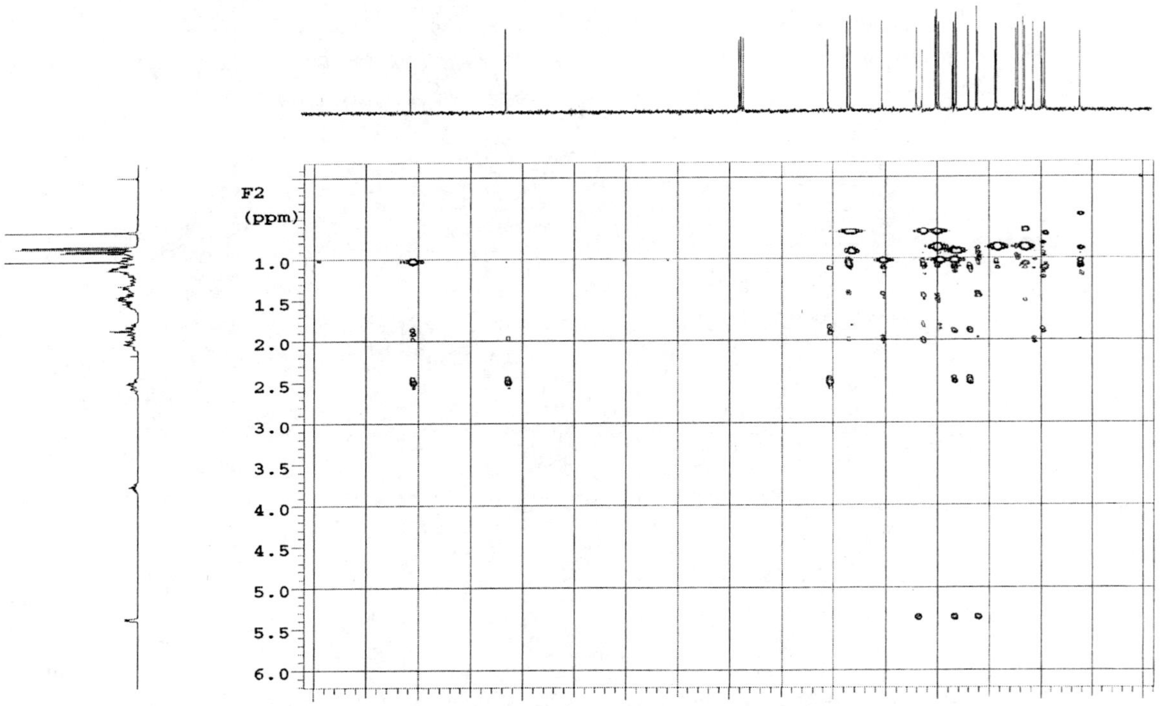

gHMBC Spectrum in $CDCl_3$

Answers to First 10 Questions of Each Chapter

Prob #	Name / Structure	Prob #	Name / Structure
2.01	1,2-Di-*ortho*-tolylethane	2.06	Diphenylethanedione [Benzil]
	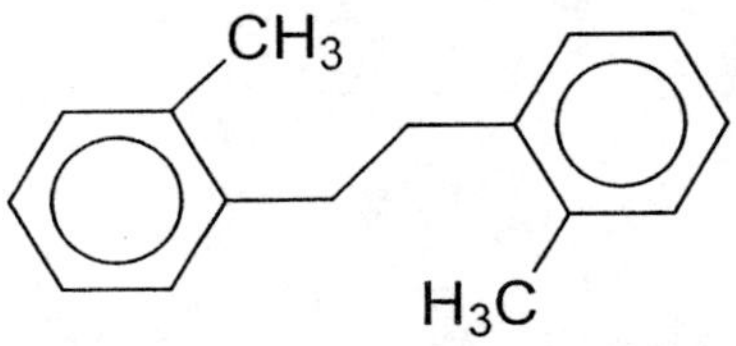		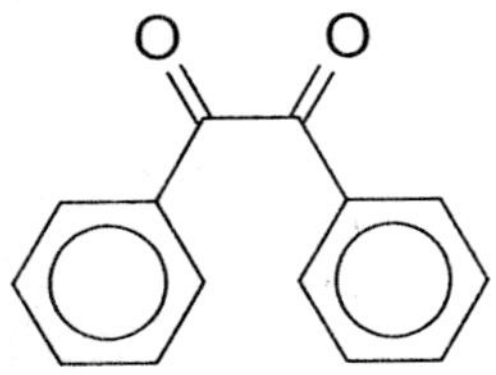
2.02	1,4-*bis*-[1-Methyl-1-hydroxymethyl]-benzene	2.07	2***H***-Chromen-2-one (Coumarin)
	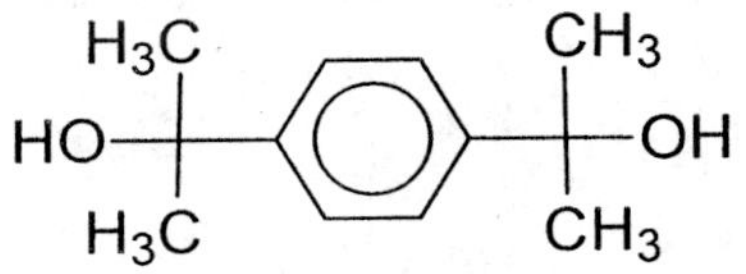		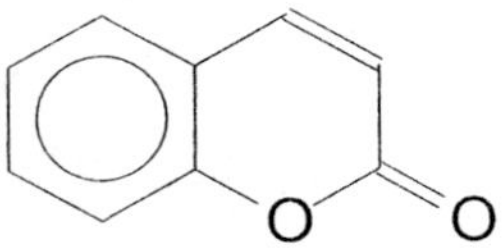
2.03	3-Ethyl-3-pentanol	2.08	2,9-Decanedione
	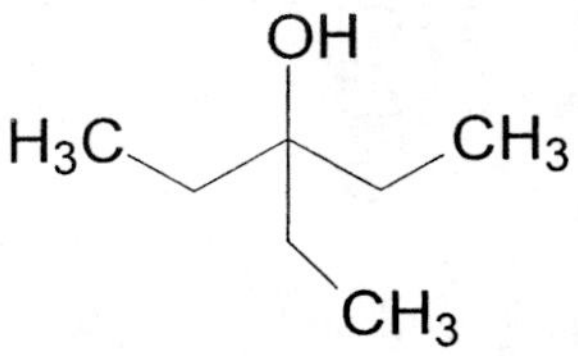		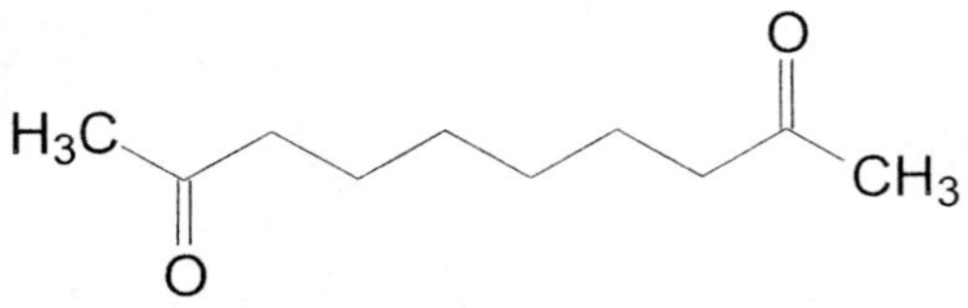
2.04	Acenaphthenedione	2.09	4-Methyl-2-pentanone
	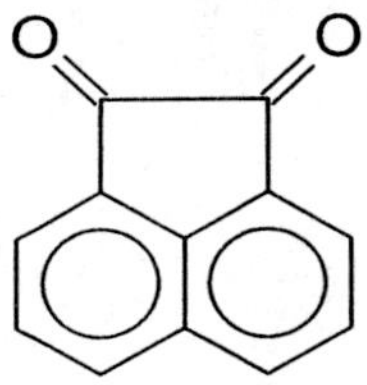		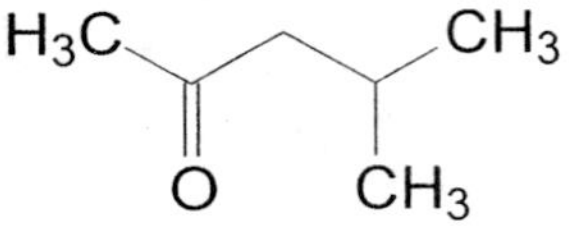
2.05	Phenanthrene-9,10-dione	2.10	*n*-Butyl propanoate
	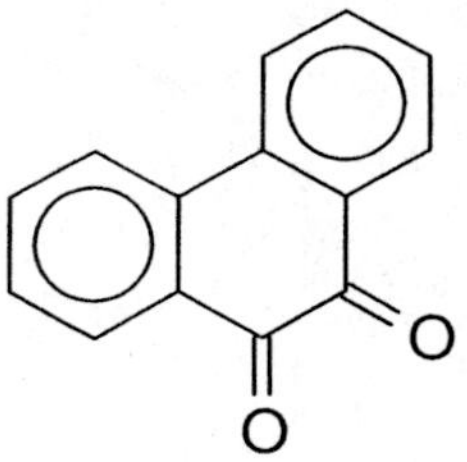		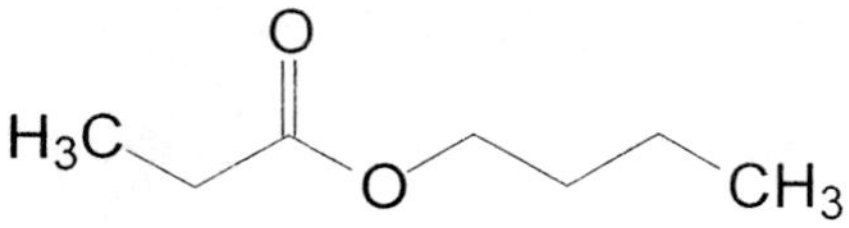

Prob #	Name	Structure	Prob #	Name	Structure
3.01	3,5-Dimethyladamantanol	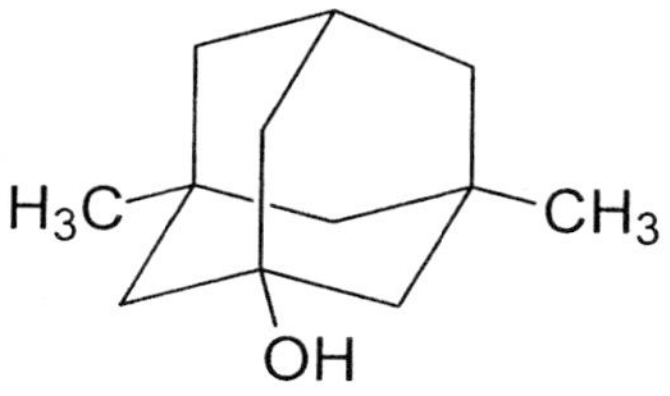	3.06	Dibenzo[a,e]-*cyclo*octanone	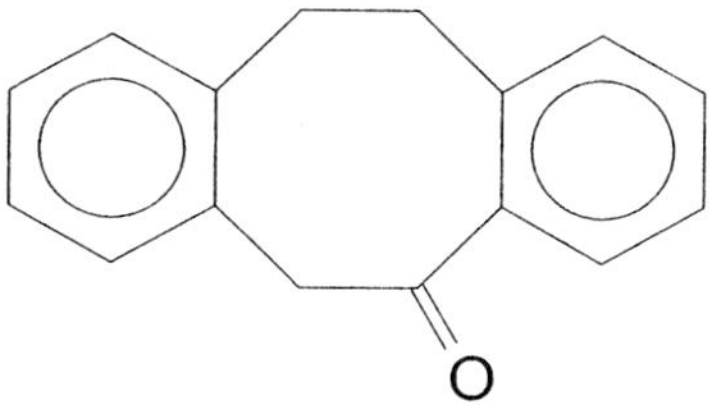
3.02	Dibenzo[a,e]-2,2-dimethyl*cyclo*octanol	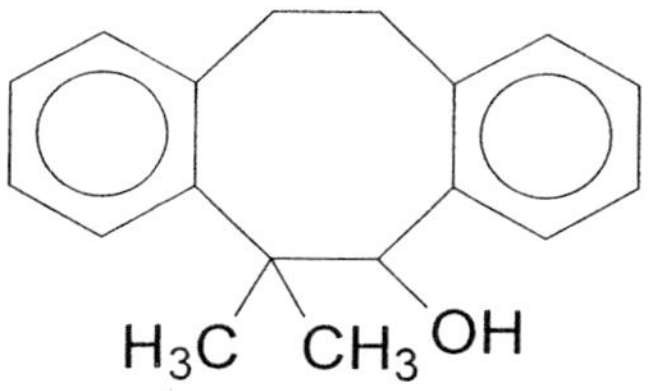	3.07	*para*-Phenylbutyrophenone	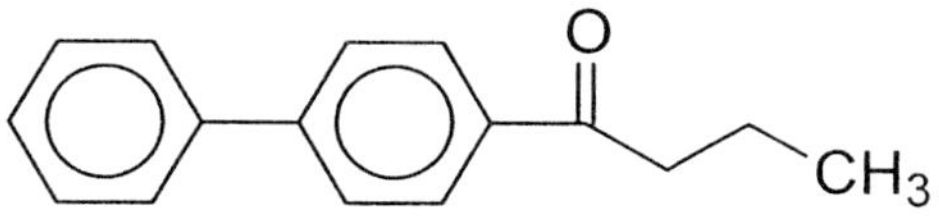
3.03	1-(2-Hydroxy-4,6-dimethylphenyl)ethanone	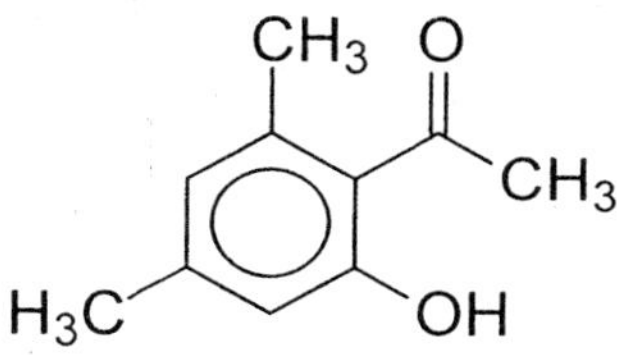	3.08	*para*-(4'-Methylphenyl)butyrophenone	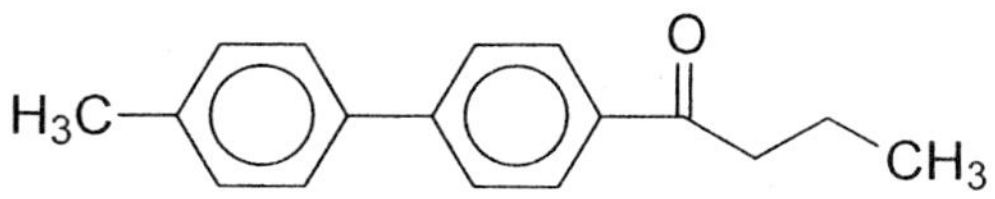
3.04	*l*-Menthol	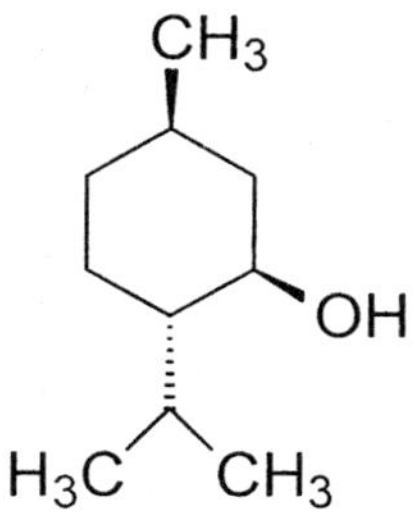	3.09	*para*-(4'-Anisyl)butyrophenone	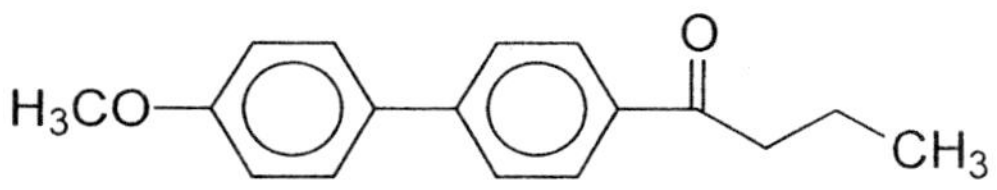
3.05	*ortho*-Chloro-α-chloroacetophenone	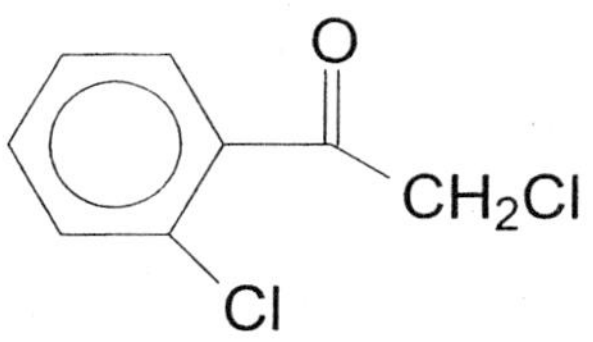	3.10	*para*-(4'-Aminophenyl)butyrophenone	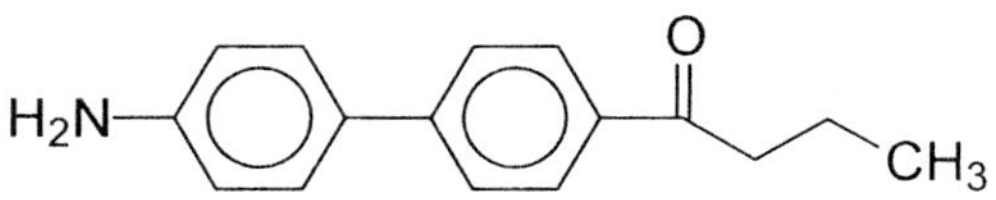

Prob #	Name	Prob #	Name
	Structure		Structure
4.01	Adamantane	4.06	Methyl α-phenyl*cyclo*pentylglycolate
	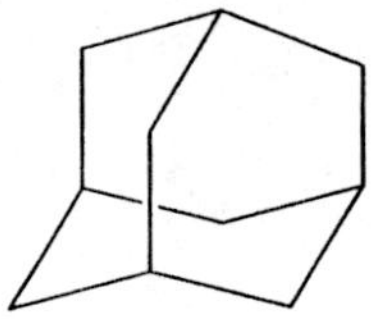		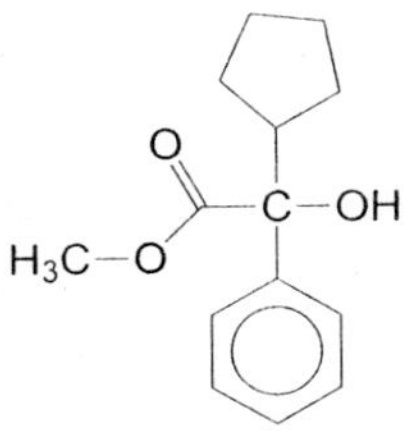
4.02	1-Bromoadamantane	4.07	Δ^3-Carene
	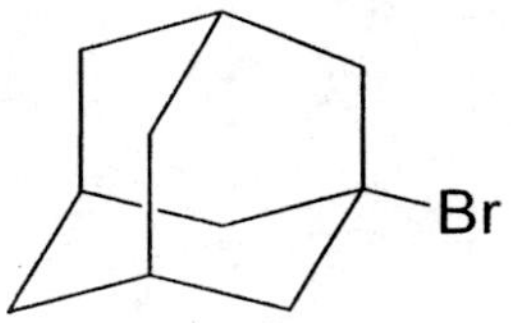		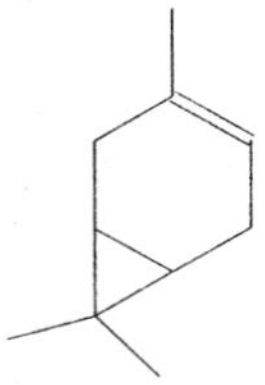
4.03	1,3-Dibromoadamantane	4.08	Δ^2-Carene
	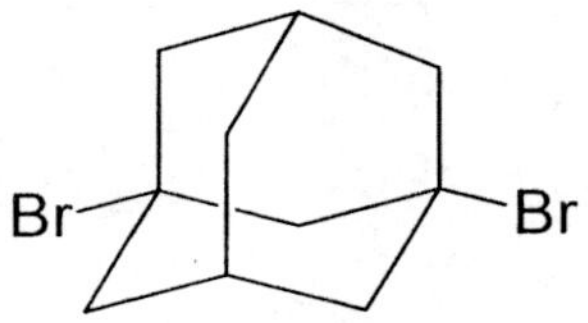		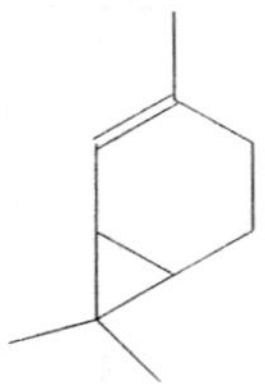
4.04	Dibenzo[a,e]-*cyclo*octanol	4.09	(+)-Fenchyl alcohol
	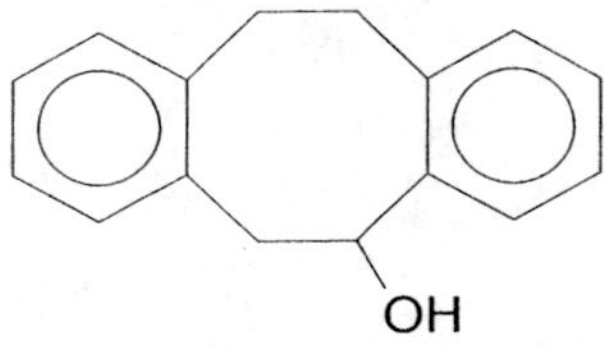		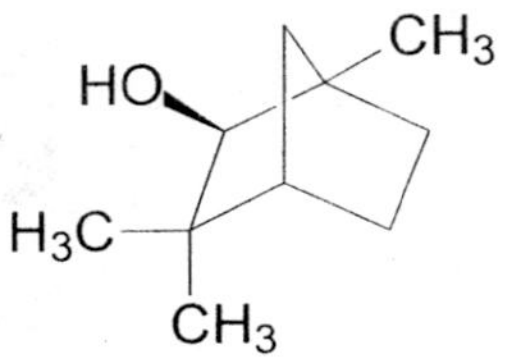
4.05	2-Methyl-4-phenylbutyrophenone	4.10	1,3,3-Trimethylbi*cyclo*[2•2•1]heptane-2-one
	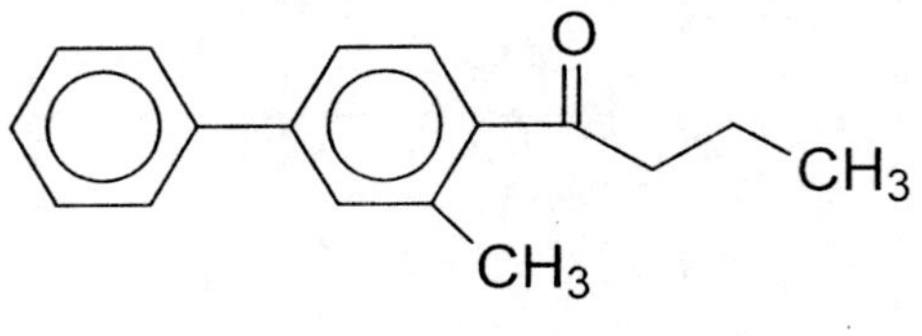		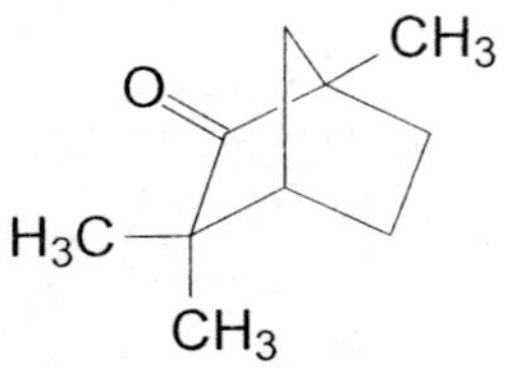

Prob #	Name / Structure	Prob #	Name / Structure
5.01	2-Bromoadamantane	5.05A	Ethyl 4-acetyl-3,5-dimethyl-1***H***-pyrrole-2-carboxylate
	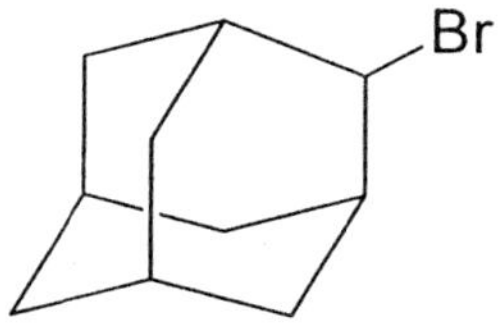 *See table following Answers for Peak Assignments*		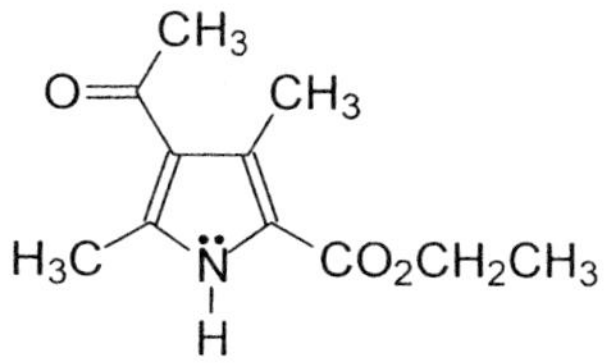
5.02	Ethyl 4-(2-(methoxycarbonyl)-5-formyl-3-methyl-1***H***-pyrrole-2-carboxylate	5.05B	Ethyl 4-ethyl-3,5-dimethyl-1***H***-pyrrole-2-carboxylate
	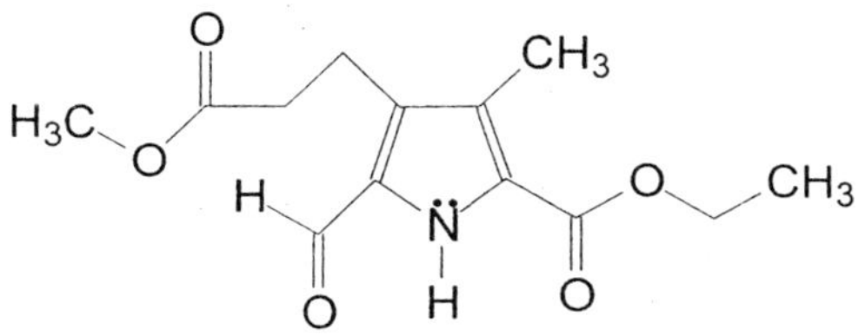 *See table following Answers for Peak Assignments*		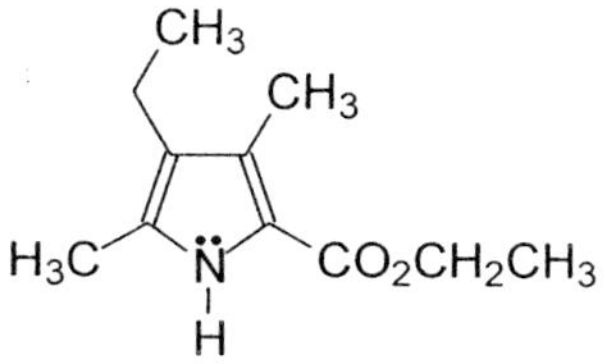
5.03	Camphor oxime	5.05C	Ethyl 4-ethyl-5-formyl-3-dimethyl-1***H***-pyrrole-2-carboxylate
	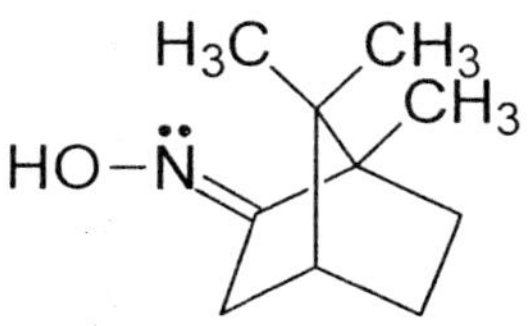		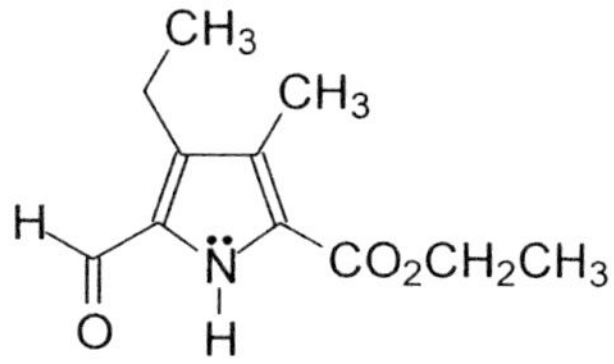
5.04	3-Ethyl-4-methyl-2-*p*-tosyl-1***H***-pyrrole	5.06	(5***E***)-3-Ethyl-4-methyl-5-(1-(pyridin-2-yl)ethylidene)-1***H***-pyrrol-2(5***H***)-one
	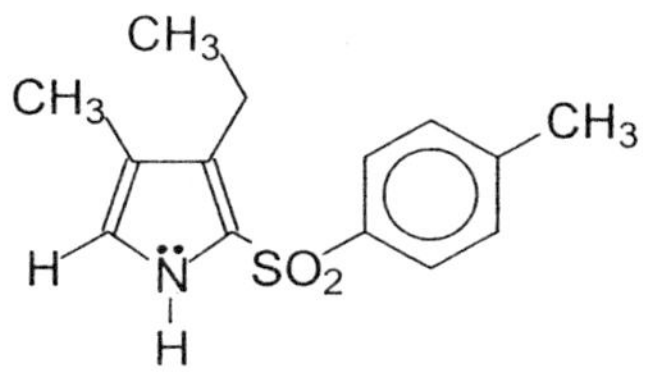 *See table following Answers for Peak Assignments*		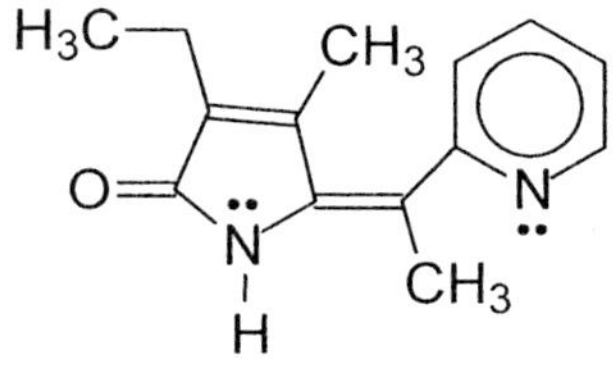 *Correct Stereochemistry Shown*

Table of Peak Assignments for 2-Bromoadamantane (Problem 5.01)

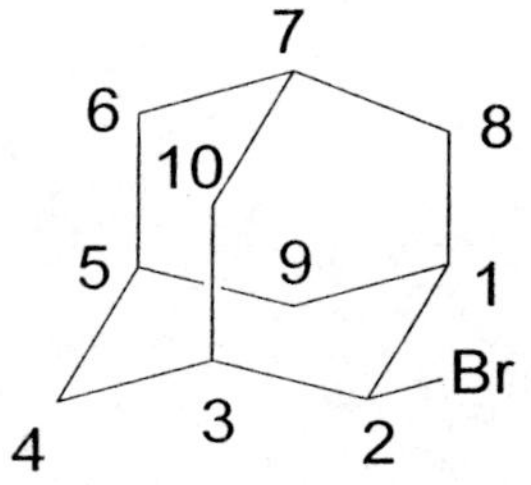

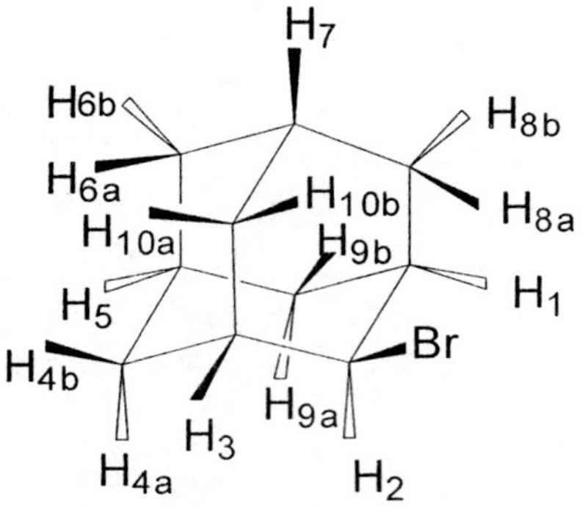

Carbon Assignments (ppm or δ)		**Proton Assignments (ppm or δ)**	
1	36.11	**1**	2.15
2	63.04	**2**	4.7
3	36.11	**3**	2.15
4	38.64	**4a**	2.34
5	26.03	**4b**	1.86
6	37.78	**5**	1.88
7	27.45	**6a**	1.76
8	31.57	**6b**	1.76
9	38.64	**7**	1.88
10	31.57	**8a**	2.34
		8b	1.62
		9a	2.34
		9b	1.86
		10a	2.34
		10b	1.62

Table of Peak Assignments for **ethyl 4-(2-(methoxycarbonyl)ethyl)-5-formyl-3-methyl-1H-pyrrole-2-carboxylate *(Problem 5.02)***

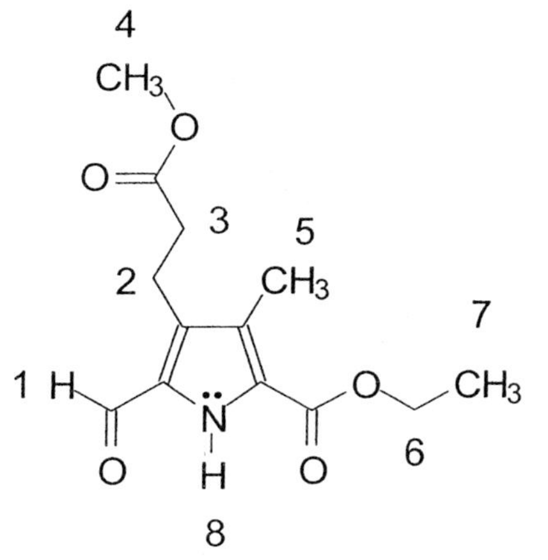

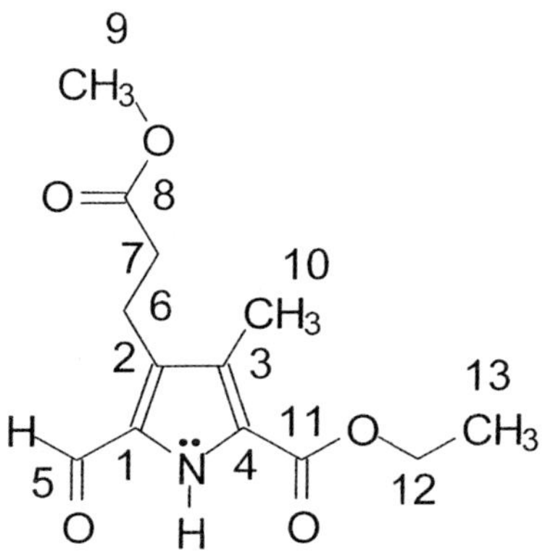

Proton Assignments (ppm or δ)		Carbon Assignments (ppm or δ)	
1	9.8 (s, 1 H)	**1**	129.86
2	3.07 (t, 7.5 Hz, 2 H)	**2**	132.03
3	2.58 (t, 7.5 Hz, 2 H)	**3**	126.53
4	3.65 (s, 3 H)	**4**	124.57
5	2.31 (s, 3 H)	**5**	179.68
6	4.36 (q, 7.0 Hz, 2 H)	**6**	18.74
7	1.39 (t, 7.0 Hz, 3 H)	**7**	34.93
8	9.7 (bs, 1 H)	**8**	172.65
		9	51.69
		10	9.69
		11	160.85
		12	60.94
		13	14.27

Table of Peak Assignments for **3-Ethyl-4-methyl-2-tosyl-1H-pyrrole (Problem 5.04)**

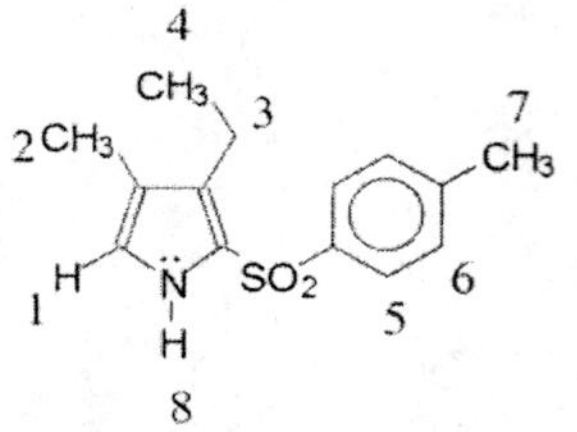

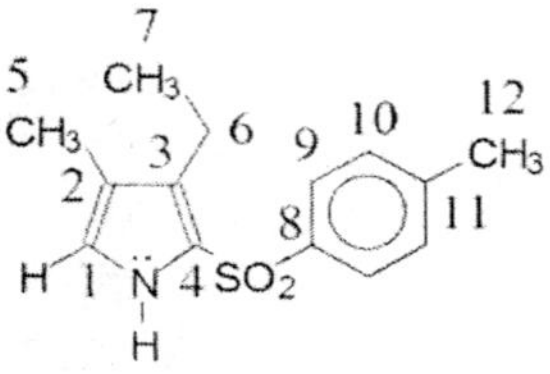

Proton Assignments (ppm or δ)		**Carbon Assignments (ppm or δ)**	
1	6.69 (d, 2.5 Hz, 1 H)	**1**	121.09
2	1.97 (s, 3 H)	**2**	120.57
3	2.62 (q, 7.5 Hz, 2 H)	**3**	130.91
4	0.98 (t, 7.5 Hz, 3 H)	**4**	122.93
5	7.78 (m, 2 H)	**5**	9.79
6	7.26 (m, 2 H)	**6**	17.44
7	2.38 (s, 3 H)	**7**	14.58
8	9.2 (bs, 1 H)	**8**	140.15
		9	126.56
		10	129.67
		11	143.43
		12	21.48